Wide Spectra of Quality Control
Edited by Isin Akyar

Published by InTech
Janeza Trdine 9, 51000 Rijeka, Croatia

Publishing Process Manager Niksa Mandic

Technical Editor Teodora Smiljanic

Cover Designer InTech Design Team

Additional hard copies can be obtained from orders@intechopen.com

Wide Spectra of Quality Control, Edited by Isin Akyar
p. cm.
ISBN 978-953-307-683-6

For EU product safety concerns:
IN TECH d.o.o., Prolaz Marije Krucifikse Kozulić 3, 51000 Rijeka, Croatia,
info@intechopen.com or visit our website at intechopen.com.

Meet the editor

Dr. Isin Akyar graduated from School of Medicine, Cukurova University, Adana, in Turkey in 1989. She finished her Medical Microbiology Specialty training in Gazi University, Ankara in Turkey in 1999. She joined the Acibadem Labmed Clinical Laboratories in Istanbul in Turkey as a Specialist of Microbiology in 2004. She became Coordinator of Microbiology in 2007. Since 2004 she had several Quality Control trainings. She works in the first accredited laboratory according to ISO 15189 for clinical laboratories in Turkey. In 2008, she joined the Department of Medical Microbiology at Acibadem University in Istanbul in Turkey. In 2011, she was promoted to serve as an Assistant Professor. Her special interests are laboratory quality control, molecular microbiology, parasitology and proteomics studies.
Currently she is both serving as Microbiology Coordinator and Assistant Professor. She has been serving as an Associate Editor for the Journal of Acibadem University Science of Health since 2009.

Contents

Preface

Quality control is a standard which certainly has become a style of living. With the improvement of technology every day, we meet new and complicated devices and methods in different fields. In order not to be back on the wrong horse we should really be on the ball, this means we should be aware of what is happening and able to react to the situation quickly and cleverly. In order to get rid of errors and mistakes there should be well organized procedures. Total quality philosophy is the acceptance that the necessities of a program or a project will be fully met based on set up quality policies and procedures. Total Quality Management (TQM) is an attitude that institutions use to ameliorate their internal procedures and increase customer satisfaction. When it is correctly put into action, this style of management can lead to lowered costs correlated to corrective or preventative maintenance, better complete performance, and an increased number of happy and faithful customers.

Nevertheless, TQM is not something that happens in a day. While there are a number of software solutions that will assist institutions immediately by putting a quality management system into action, there are some fundemental philosophies that the company must implement throughout every department of the company and at every level of management. Whatever other resources we use, we should accept the critical principles of Total Quality Management as a basis for all our activities. Quality can and must be managed. Various institutions have gotten stuck in a repetitive cycle of chaos and customer accusations. They admit that their operations are simply too large to efficiently control the level of quality. The first step in the TQM process, then, is to become aware that there is a problem and that it can be controlled.

In fact, the real problems are the processes, not the people. If our process is riddled with problems, it won't matter how many times we hire new employees or how many training sessions we put them through. We should fix the process and then train our people on these new procedures. We should not treat symptoms but rather look for the cure. If we just stuck on the basic problems in the process, we will never be able to fully reach our capacity. We should seek the source to fix the problem. Every employee is authoritative for quality. Everyone in the institution, from the workers on the line to the upper management, must appreciate that they have a particular role in guaranteeing high levels of quality in their products and services. Everyone has a customer to delight, and they must all accelerate and take responsibility for them.

Quality must be measurable. You can not manage something that you do not know and measure. A quality management system is seldom productive when we cannot quantify the results. We need to see how the process is started and if it is having the wanted effect. This will help us set our goals for the future and guarantee that every department is working towards the same result. Quality improvements must be continuous. Total Quality Management is not something that can be done once and then forgotten. It's not a management "phase" that will end after a problem has been fixed. TQM is a long-term investment, and it is designed to help us find long-term success. Quality control refers to the measures that must be included during each assay run to verify that the test is working properly. Quality Assurance is defined as the overall program that ensures that the final results reported by the laboratory are correct. The aim of quality control is simply to guarantee that the results generated by the test are correct. However, quality assurance is concerned with much more: that the right test is carried out on the right specimen, and that the right result and right interpretation is delivered to the right person at the right time. Quality control explains the directed use of testing to measure the achievement of a specified standard. Quality control is the process, procedures and authority used to accept or reject all components, drug product containers, closures, in-process materials, packaging material, labeling and drug products and the authority to review production records to assure that no errors have occurred, that they have been fully investigated. The quality and reliability of test data rely on the state and condition of the test system which is used in its production.

"Trust is Good, Control is Better" says an old proverb. The quality which is supposed to be achieved is not a concept which can be controlled by easy, numerical or other means, but it is the control over the intrinsic quality of a test facility and its studies. The aim of this book is to share useful and practical knowledge about quality control in several fields with the people who want to improve their knowledge.

Dr Isin Akyar
Acibadem University, School of Medicine,
Department of Medical Microbiology, Istanbul,
Turkey

Part 1

General Quality Control

1

Analytical Method Validation

Pedro Lopez Garcia[1], Ernesto Buffoni[1],
Fabio Pereira Gomes[1] and Jose Luis Vilchez Quero[2]
[1]*Instituto de Aperfeiçoamento Farmacêutico (IAF)*
[2]*Department of Analytical Chemistry, Faculty of Sciences, University of Granada*
[1]*Brazil*
[2]*Spain*

1. Introduction

In pharmaceutical industries, the validation of analytical methods is used to demonstrate that the method is fitted for its purpose; it must follow a plan which includes scope, performance characteristics, and acceptance limits. Analytical methods need to be validated or revalidated prior to their introduction into routine analyses (release of batch). The overarching philosophy in current good manufacturing practices (cGMPs) of the twenty - first century and robust modern quality systems is the quality that it has to be built into the product, and testing alone cannot be relied to ensure the quality of the product. From the analytical perspective, it will mean that analytical methods used to test products should have quality attributes built into them. In order to apply quality attributes into the analytical method, fundamental quality attributes have to be applied by the bench - level scientist. This is a paradigm shift that requires the bench - level scientist to have a scientific and technical understanding, product knowledge, process knowledge, and/or risk assessment ability to appropriately execute the quality functions of analytical method validation. In addition, it requires the following procedures: (a) an appropriate training of the bench - level scientist to understand the principles involved with method validation, validate an analytical method, and understand the principles involved with the method validation, (b) proper documentation and understanding and interpreting data, and (c) cross – an understanding functional of the effect of their activities on the product and to customers (the patient). Management has a responsibility of verifying that gained skills from the training are implemented in routine analyses performance.

This chapter gives a review and strategy for the validation of analytical methods in-house, recommendation in documentation and completion of method validation in the pharmaceutical environmental.

2. Regulatory agencies

In 1990, Europe, United States of America, and Japan harmonized the submission requirements for new pharmaceuticals, a forum for constructive dialogues between regulatory authorities and industry was initiated and called, "International Conference on the Harmonisation" (ICH) (ICH, 2005). One of the first topics into the quality section was analytical validation and the ICH was very helpful in harmonizing terms and definitions as well as determining the basic requirements.

Regulatory authorities in the United States of America (Food and Drug Administration (FDA) and the United States Pharmacopeia (USP)) and Brazil (Agência Nacional de Vigilância Sanitária (ANVISA)) include guidelines for analytical procedures and methods validation. In the United States of America environment, two guidelines by the FDA were proposed, one of those for applicants (Food and Drugs Administration, 1987) and another one for inspectors and reviewers (Food and Drugs Administration, 1994). The first one is also intended to ensure that the analytical procedures have to be applied in an FDA laboratory with detailed description of the procedure, for instance, reference materials, as well as a discussion of the potential impurities, and others. The second guideline focuses on reversed-phase chromatography and provides the whole details regarding critical methodological issues as well as indications of acceptability results. The USP has published a specific guideline in the chapter 1225. It focuses on the validation of compendia procedures by giving definitions and approaches to validate each analytical parameter, in addition, it also provides a table which separates the methods into four categories based on their use (United States Pharmacopeia, 2011). In Brazil, ANVISA has also proposed an industry guidance for analytical methods validation. The RE 899 of 2003 (Brasil, 2003) is an approach to validation and it is varied and opened to interpretation. Also it provides a table, the same as chapter 1225 of the USP which separates methods into four categories based on their use. For example, Category I covers quantitation of active ingredients in dosage forms and indicates that accuracy, precision, specificity, linearity, and range are required for method validation while limits of detection and quantitation are not necessary.

The USP 1225 chapter covers only the most common categories of tests which validation data should be required. Those categories can be observed on the following lines:

Category I: Analytical procedures for quantitation of major components in bulk drug substances or active ingredients (including preservatives) in finished pharmaceutical products.

Category II: Analytical procedures for determination of impurities in bulk drug substances or degradation compounds in finished pharmaceutical products. These procedures include quantitative assays and limit tests.

Category III: Analytical procedures for determination of performance characteristics (e.g., dissolution, drug release, etc.).

Category IV: Identification tests.

In Table 1, the required validation characteristics for various types of analytical procedures according to USP can be observed.

The required validation characteristics for each type of analytical procedures according to ICH are illustrated in Table 2. The discussion of the validation of analytical procedures is directed to the four most common types of analytical procedures:

- Identification tests.
- Quantitative tests for impurities' content.
- Limit tests for the control of impurities.
- Quantitative tests of the active moiety in samples of drug substance or drug product or other selected component(s) in the drug product.

Even though there are many other analytical procedures, such as dissolution testing for drug products or particle size determination for drug substance, these are no addressed in the initial text for validation of analytical procedures. The validation of these additional analytical procedures is equally important to those listed herein, and it may be addressed in subsequent documents.

Analytical performance characteristics	Category I	Category II		Category III	Category IV
		Quantitative	Limit tests		
Accuracy	Yes	Yes	*	*	No
Precision	Yes	Yes	No	Yes	No
Specificity	Yes	Yes	Yes	*	Yes
Detection Limit	No	No	Yes	*	No
Quantitation Limit	No	Yes	No	*	No
Linearity	Yes	Yes	No	*	No
Range	Yes	Yes	*	*	No

*It may be required, depending on the nature of the specific test.

Table 1. Data elements required for validation according to USP

Type of analytical procedure Characteristics	Identification	Testing for impurities		Assay
		Quantitative	Limit	
Accuracy	-	+	-	+
Precision				
Repeatability	-	+	-	+
Interm. precision	-	+ (1)	-	+ (1)
Specificity (2)	+	+	+	+
Detection Limit	-	- (3)	+	-
Quantitation Limit	-	+	-	-
Linearity	-	+	-	+
Range	-	+	-	+

(-) characteristic is not normally evaluated. (+) characteristic is normally evaluated.
(1) in some cases where reproducibility (see glossary) has been performed, intermediate precision does not need to be done. (2) lack of specificity in one analytical procedure could be compensated by other supporting analytical procedure(s). (3) it may be needed in some cases.

Table 2. Data elements required for validation according to ICH

A brief description of the types of considered tests in this document is provided below.

- Identification tests are intended to ensure that analyte is into of sample. This is normally achieved by comparing the property of the sample (e.g., spectrum, chromatographic behavior, chemical reactivity, etc) with a reference standard.
- Testing for impurities can be either a quantitative test or limit test for the impurity in a sample. Both tests are intended to accurately reflect the purity characteristics of the

sample. More different validation characteristics are required for a quantitative than for a limit test.

- Assay procedures are intended to measure the presented analyte in a given sample. In the context of this document, the assay represents a quantitative measurement of the major component(s) in the drug substance. For the drug product, similar validation characteristics are also applied when assaying an active or other selected component(s). The same validation characteristics may also be applied to associated assays with other analytical procedures (e.g., dissolution).

In Brazil, ANVISA has the same form as ICH and USP, it covers the most common categories of tests for analytical method validation: category I, category II, category III, and category IV. The Table 3 shows performance characteristics of an analytical method validation according to resolution RE 899.

Parameter	Category I	Category II		Category III	Category IV
		Quantitative	Limit tests		
Specificity	Yes	Yes	Yes	*	Yes
Linearity	Yes	Yes	No	*	No
Range	Yes	Yes	*	*	No
Precision					
Repeatability	Yes	Yes	No	Yes	No
Interm. precision	**	**	No	**	No
Detection Limit	No	No	Yes	*	No
Quantitation Limit	No	Yes	No	*	No
Accuracy	Yes	Yes	*	*	No
Robustness	Yes	Yes	Yes	No	No

* it may be required, depending on the nature of the specific test.

** If there is evidence of reproducibility is not required intermediate precision.

Table 3. Data elements required for validation according to RE 899 of ANVISA

3. Strategy for analytical method validation

The validation of a specific method must be demonstrated through laboratory experiments by routinely analyzing samples. The preparation and execution have to follow a validation protocol, preferably written in a step-by-step instruction format. Possible steps for a complete method validation are listed in Table 4. This proposed procedure assumes that the instrument has been selected, and the method has also been developed. It meets criteria such as easy of use, ability to be automated and controlled by computer systems, costs per analysis, sample throughput, turnaround time, environmental, health, and safety requirements. Successful acceptance of the validation parameters and performance criteria by all involved parties requires a cooperative effort of several departments, including analytical development, quality control, regulatory affairs, and individuals requiring the analytical data. The operating procedure or Validation Master Plan (VMP) must clearly define the rules and responsibilities of each involved department in the validation of analytical methods.

The validation experiments should be carried out by an expert analyst to avoid errors due to lack of experience. The analyst should also be very familiar with the technique and operation of the instrument. Before an instrument starting validating a method, its performance specifications must be verified by using generic chemical standards. Satisfactory results for a method can be only obtained with equipment that is performing well. A special attention must be paid to equipment characteristics which are critical for the method. For instance, if the detection limit is critical for a specific method, the specification of the instrument for baseline noise and some detectors must be verified.

1	Develop a validation protocol, an operating procedure, or a validation master plan for the validation
2	For a specific validation project, define owners and responsibilities
3	Develop a validation project plan
4	Define the application, purpose, and scope of the method
5	Define the performance parameters and acceptance criteria
6	Define validation experiments
7	Verify relevant performance characteristics of equipment
8	Qualify materials, e.g. standards and reagents for purity, accurate amounts, and sufficient stability
9	Perform pre-validation experiments
10	Adjust method parameters and/or acceptance criteria if necessary
11	Perform full internal (and external) validation experiments
12	Develop standard operation procedures for executing the method in the routine
13	Define criteria for revalidation
14	Define type and frequency of system suitability tests and/or analytical quality control checks for the routine
15	Document validation experiments and results in the validation report

Table 4. Strategies and steps in analytical method validation

There are no official guidelines on the correct sequence of validation experiments, and the optimal sequence may depend on the method itself.
Based on the experience of, the following sequence has proven to be very helpful:

1. Specificity (optimizing separation and detection of standard mixtures if selectivity is insufficient).
2. Linearity, range, limit of detection, and limit of quantitation.
3. Precision (Repeatability, short-term precision, intermediate precision, and reprodutibility).
4. Accuracy at different concentrations.
5. Robustness.

The time-consuming experiments, such as accuracy and ruggedness, are included toward the end. Some of the parameters, as listed under (2) to (6), can be measured by combining experiments. For example, when the precision of peak areas is measured over the full concentration range, the data can be used to validate the linearity. During method validation, the parameters, acceptance limits, and frequency of ongoing system suitability tests or quality control checks should be defined. Criteria must be defined to indicate when the method and system are beyond statistical control. The aim is to optimize these experiments so that, with a minimum number of control analyses, the method and the complete analytical system will provide long-term results to meet the objectives defined in the scope of the method.

4. Documentation

4.1 Validation protocol

Depending upon of the culture of the company, a method validation protocol could be simple (listed below) or exhaustive (in addition to the listed items, each parameter to be validated is described in detail): How solutions are going to be made, the experimental design, how the calculations are going to be performed, any software to be utilized (e.g., Excel). If a full-length protocol is required within a particular company, then the writing of this protocol and approval of the protocol would need to be completed prior to the commencement of the validation work. Otherwise, there may be many deviations to the protocol which will be needed to be referenced in the final method validation report. Some companies also have templates for the validation reports, thereby allowing for facile population of the results. Once populated, the file is reviewed to determine if all validation parameters and acceptance criteria were met. If they were not met, a deviation is added and the proper justification must be given. If it is deemed that the justification is not appropriate, then an action plan for the specific figure of merit in question is determined (i.e., repeat analysis, change of the analytical procedure, and revalidation). Also, if the analytical method has not been approved at the time of writing the validation protocol yet, it is recommended to attach a final draft of the method to the protocol.
Before starting the experimental work, the protocols must be written by a qualified person and approved by a quality assurance department. Some of the items that are necessary to be specified in the validation protocol are listed below:

- The analytical method for a given product or drug substance.
- The test to be validated.
- The test parameters for each test, including type and number of solutions and number of tests.
- The acceptance criteria for each parameter based on an internal standard operational procedure (product or method-specific adaptations may be necessary and are acceptable, if justified).
- List of batches of drug substance and/or drug products.
- For a drug product the grade/quality of the excipients used in the formulation.
- List of reference materials to be used in the validation experiments.
- Information of the instruments and apparatus to be used.
- Responsibilities [author, chemists, analytical research project leader, quality assurance, and etc.].

4.2 Validation report

The validation report must contain reference to the analytical methods (specific code number used as identifier within the pharmaceutical organization) and the corresponding drug substance or product name. Note that for early-phase method validation reports the results may be filled in a predefined table and compared with the acceptance criteria. However, for late-phase validation, more explicit reports are generated by explaining every experiment, with detailed steps of sample and standard preparation. The list of reference materials (reference standards with the appropriate certificate of analysis) as well as the list of calibrated and qualified instruments used in the validation experiments should be documented in the report. For drug substances the list of the batches of drug substances, notebook number/reference number for any individual impurities, or solutions or used intermediates should also be listed. For drug products, the list of the batches of drug substances, drug product, and the grade/quality of excipients should be listed. The test parameters and acceptance criteria must be listed together with the results for each test, and the passed or failed results should be indicated. The validation report should also contain whether the method validation was successful and if any changes should be applied to the analytical method, and then the final analytical method must be resubmitted for quality assurance approval.

Once the method has been developed and validated, a validation report should be prepared with the following information:

- Objective and scope of the method (applicability, type).
- Summary of methodology.
- All chemicals, reagents, reference standards, quality control samples with purity, grade, their source, or detailed instructions on their preparation.
- Procedures for quality checks of standards and chemicals used on that.
- Safety precautions.
- Method parameters.
- Critical parameters from robustness testing.
- Listing of equipment and its functional and performance requirements, e.g., cell dimensions, baseline noise, and column temperature range. For complex equipment, a picture or schematic diagram may be useful.
- Detailed conditions on how the experiments were conducted, including sample preparation. The report must be detailed enough to ensure that it can be reproduced by a competent technician with comparable equipment.
- Statistical procedures and representative calculations.
- Procedures for quality control in routine analyses, e.g., system suitability tests.
- Representative plots, e.g., chromatograms, spectra, and calibration curves.
- Method acceptance limit performance data.
- The expected uncertainty of measurement results.
- Criteria for revalidation.
- The person(s) who developed and validated the method.
- References (if any).
- Summary and conclusions.
- Approval with names, titles, date, and signature of those responsible for the review and approval of the analytical test procedure.

4.3 Revalidation

Some method parameters have to be changed or adjusted during the life of the method if the method performance criteria fall outside their acceptance criteria. The question is whether such change requires revalidation. In order to clarify this question up front, operating ranges should be defined for each method, either based on experience with similar methods or else investigated during method development. These ranges should be verified during method validation in robustness studies and also be part of the method characteristics. Availability of such operating ranges makes it easier to decide when a method should be revalidated. A revalidation is necessary whenever a method is changed and the new parameter lies outside the operating range. If, for example, the operating range of the column temperature has been specified to be between 30°C and 40°C, the method should be revalidated if, for whatever reason, the new operating parameter is 41°C. Revalidation is also required if the scope of the method has been changed or extended, for instance, if the sample matrix changes or operating conditions change. Furthermore, revalidation is necessary whether the intention is to use instruments with different characteristics and these new characteristics have not been covered by the initial validation. For example, an HPLC method may have been developed and validated on a pump with a delay volume of 5mL, but the new pump has a delay volume of only 0.5 mL. Part or full revalidation may also be considered if system suitability tests, or the results of quality control sample analysis, lie outside preset acceptance criteria where the source of the error cannot be traced back to the instruments or any other cause. Whenever there is a change that may require part or full revalidation, the change should follow a documented change control system. The change should be defined, authorized for implementation, and documented. Possible changes may include:

- New samples with new compounds or new matrices.
- New analysts with different skills.
- New instruments with different characteristics.
- New location with different environmental conditions.
- New chemicals and/or reference standards, and
- Modification of analytical parameters.

An evaluation should determine whether the change is within the scope of the method. If so, no revalidation is required. If the change lies outside the scope, the parameters for revalidation should be defined. After the validation experiments, the system suitability test parameters should be investigated and redefined, if necessary.

5. Parameters for method validation (figures of merit)

The type and degree of validation depends on the nature of the test. In particular, methods described in pharmacopeias may not have to be validated, but those should be verified. Different test methods require different validation parameters; as development of the project progresses and as analytical and product-specific information is acquired, the analytical methods evolve gradually updated. Each company has its own approach and set of acceptance criteria for different analytical assays, but these approaches must be within the confines of their line unit quality assurance department and be in accordance with any regulatory provisions. In this section, a description for each of the parameters to be validated (figures of merit) is described in details.

5.1 Specificity

An investigation of specificity should be conducted during the validation of identification test, the determination of impurities, and the assay. The procedures used to demonstrate specificity will depend on the intended objective of the analytical procedure. It is not always possible to demonstrate that an analytical procedure is specific for a particular analyte (complete discrimination). In this case, a combination of two or more analytical procedures is recommended to achieve the necessary level of discrimination (e.g., optically active substances, in addition to an achiral HPLC method, a chiral HPLC method may be added).

Identity is a general requirement for dosage forms. When determining specificity for identity, the assay and related substances or the content uniformity methods can be used. Assay and content uniformity methods are quantitated by external reference standard. This identity test confirms that the correct active ingredient (s) is presented in correct ratio if multiple variants are available. The method could also be used for post-packaging analysis. The general requirements are that the sample and standard chromatograms should correspond in retention time and normalized peak area within ±10%. The easiest way to perform specificity for any HPLC method is to perform this test in conjunction with a forced decomposition study. The utilization of mass spectrometry (MS) detector (in series) after a Photo Diode Array (PDA) detector to obtain more information is encouraged (in terms of mass-to-charge ratio of parent ions, initial fragmentation pattern, and peak purity). Specificity is confirmed when an API peak is "pure" (confirmed by PDA and/or MS) and there is no interference from placebo solution (placebo dissolved in sample preparation solvent) at the retention time of an API peak.

5.1.1 Forced degradation studies (solid state and solution)

Forced degradation studies are usually performed during the salt selection process for the drug substance. In drug product development, the forced degradation studies of drug substance in the presence of excipients are firstly performed during the pre-formulation stage to assist in the selection of the most formidable compounds and excipients. This may lead to the development of more suitable formulations, packaging, and change in storage and manufacturing conditions as the optimal formulation is defined to be used in clinical studies. Forced degradation testing is often repeated when the final drug substance route and market formulation is defined or the compound enters phase 3 clinical trials. A good overview of forced degradation testing according to the regulatory guidance documents, with emphasis on what should be considered for late clinical phases and for registration application dossiers (i.e., marketing authorization applications or new drug applications), is provided by the Impurity Profiling Group (Klick et al., 2005). Forced degradation studies (sometimes referred to as stress testing) are also performed in order to demonstrate specificity during the development and validation of stability-indicating methods. These studies are usually performed at conditions exceeding that of accelerated storage conditions. Forced degradation studies may provide information to degradation pathways and degradation products that could form during storage of the drug substance or the drug product. The main goal of forced degradation studies is to effectively produce samples containing representative and realistic degradation products. These degradation products should be assessed whether they are (a) related to the drug substance or the excipients or (b) due to drug substance–excipient interactions under certain forced degradation conditions.

A delicate balance of efficiency and severity/duration of stress conditions is needed. Overstressing can destroy relevant compounds or generate irrelevant compounds. Under stressing may fail to generate important degradation products. The extent of targeted degradation should be approximately anywhere from 5% to 10%. The other goal is that the potential degradation products that are generated should be resolved from the active component during development of a stability-indicating HPLC method. The assessment of peak purity using diode array and LC-MS detection are usually employed. These degradation products that are generated during the forced degradation studies can be identified, and the determination of degradation pathways and mechanisms for the drug substance and drug product can be elucidated. Forced degradation studies are carried out either in the solution state and/or in the solid state. Usually the forced degradation testing is carried out on one batch of drug substance and/or one formulation blend (capsules and tablets). This forced degradation testing should not be part of a formal stability program.

Stress Test	Conditions	Duration
Thermal (closed container)	50°C and 80°C (ambient RH)	1 wk and 2wks
Thermal/oxidative (open container)	50°C and 80°C (ambient RH)	1 wk and 2wks
Thermal/humidity (open container)	40°C/75% RH	1 wk and 2wks
Light (closed container)	Ambient	Maximum 1.2 million lux hours and 200 watt hours/square meter
Light/oxidative (open container)	Ambient	Maximum 1.2 million lux hours and 200 watt hours/square meter

Table 5. Solid-state forced degradation studies

Test factor	Conditions	Duration
pH	10mg in 2mL water 10mg in 2 mL of 0.1M HCl 10mg in 2 mL of 0.1M NaOH All in amber volumetric flasks and at room temperature	1 day and 3 days
Oxidation (H_2O_2)	10mg/2mL 3% H2O2 At 5°C and room temperature in amber volumetric flasks. If DS is not soluble, then pH modification may be necessary	1, 2, and 3 days
Oxidation (metal ion catalyzed)	10mg/2mL water containing 100ppm Fe3+, Ni2+, Cu2+ saturated with bubbled oxygen in amber volumetric flasks	1, 2, and 3 days
Oxidation (saturated with oxygen)	10mg/2mL saturated with bubbled oxygen in amber volumetric flasks	1, 2, and 3 days
Light	50mg/10mL wáter ambient	Maximum 1.2 million lux hours and 200 watt hours/square meter, 6hr, 1 day and 2 days
Heat	10mg in 2 mL water at 50°C	6 hours, 1 day and 2 days

Table 6. Solution forced degradation studies

The temperature/humidity used conditions may be more severe than the typical accelerated stability testing conditions in order to generate potential degradation products in a reasonable time. The typical forced degradation conditions include thermolytic, hydrolytic, oxidative, photolytic (in excess of ICH conditions), high pH (alkaline conditions), and low pH (acidic conditions). Outlined in Table 5 and Table 6 are some solid-state and solution forced degradation studies, respectively, that could be conducted.

5.2 Linearity

The linearity of an analytical procedure is its ability (within a given range) to obtain test results which are directly proportional to the concentration (amount) of analyte in the sample. It may be directly demonstrated on the analyte, or on spiked samples using at least five concentrations over the whole working range. Moreover a visual evaluation of the analyte signal as a function of the concentration, appropriate statistical calculations are recommended, such as a linear regression. The parameters slope and intercept, residual sum of squares and the coefficient of correlation should be reported. A graphical presentation of the data and residuals is recommended.
The terminology for this validation characteristic is somewhat misleading because linearity in the inner sense, i.e., a linear relationship between analyte concentration and test results is certainly preferable, but not essential. A better term would have been analytical response. Some analytical procedures have intrinsic non-linear response functions, such as quantitative TLC, fluorescence detection, etc., but they can of course be validated. The primary objective is to validate or verify the calibration model. Consequently, the requirements and the relevant parameters depend on the intended mode of calibration. The response function of a given analytical procedure is an intrinsic property of the respective analyte. That means, with respect to validation, that the answer is of a qualitative kind: Can the intended calibration be applied, yes or no? Therefore, solutions of the analyte itself are sufficient and there is no need to repeat linearity. Potential influences by the matrix, i.e., the linearity of the analytical procedure would be better addressed in accuracy. Often, the fundamental response function is known for a given type of analytical procedure, such as a linear function for LC with UV detection, according to the Lambert-Beer law. In some cases, validation of linearity can be regarded more as a verification of the assumed response function, i.e., the absence of (unacceptable) deviations. Primarily, this should be performed by means of graphical evaluation of the deviations of the experimental data from the assumed response model (residual analysis), known as residual plots. The evaluation of numerical parameters is only sensible after verification of the response function, i.e., if only random errors exist.
Frequently, the linearity is evaluated graphically, in addition to or as an alternative to mathematical evaluation. The evaluation is made by visually inspecting a plot of signal height or peak area as a function of analyte concentration. Because deviations from linearity are sometimes difficult to detect, two additional graphical procedures can be used. The first is to plot the deviations from the regression line versus the concentration or versus the logarithm of the concentration if the concentration range covers several decades. For linear ranges, the deviations should be equally distributed between positive and negative values. Another approach is to divide signal data by their respective concentrations, yielding the relative responses. A graph is plotted with the relative responses on the y-axis and the corresponding concentrations on the x-axis, on a log scale. The obtained line should be horizontal over the full linear range. At higher concentrations, there will typically be a

negative deviation from linearity. Parallel horizontal lines are drawn on the graph corresponding, for example, 95 percent and 105 percent of the horizontal line. The method is linear up to the point where the plotted relative response line intersects the 95 percent line. Figure 1 shows a comparison of the two graphical evaluations on a sample of caffeine using HPLC.

A plot of the data should be included in the report. In addition, an analysis of the deviation of the actual data points from the regression line may also be helpful for evaluating linearity. In order to establish linearity, a minimum of five concentrations is recommended. Other approaches should be justified. Plotting the sensitivity (response/amount) gives clear indication of the linear range. Plotting the amount on a logarithmic scale has a significant advantage for wide linear ranges.

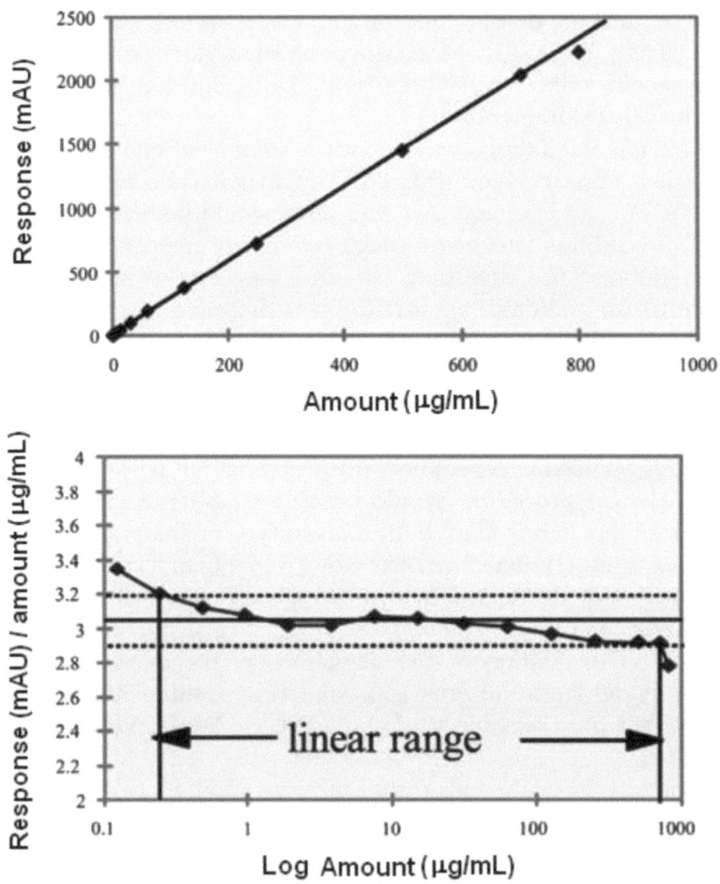

Fig. 1. Linearity plot of an analyte sample

5.2.1 Range

The range of an analytical method is the interval between the upper and lower levels (including these levels) that have been demonstrated to be determined with precision, accuracy, and linearity using the method as written. The range is normally expressed in the same units as the test results. For assay test, it requires a minimum of specified range to be

80 to 120 percent of the test concentration, and for the determination of an impurity, the range to extend from the limit of quantitation, or from 50 percent of the specification of each impurity, whichever is greater to 120 percent of the specification.

5.2.2 Limit of detection

The limit of detection is the point at which a measured value is larger than the uncertainty associated with it. It is the lowest concentration of analyte in a sample that can be detected, but not necessarily quantified. The limit of detection is frequently confused with the sensitivity of the method. The sensitivity of an analytical method is the capability of the method to discriminate small differences in concentration or mass of the test analyte. In practical terms, sensitivity is the slope of the calibration curve that is obtained by plotting the response against the analyte concentration or mass. In chromatography, the detection limit is the injected amount that results in a peak with a height at least two or three times as high as the baseline noise level. Furthermore this signal/noise method describes three more methods (ICH):

- *Visual inspection:* The detection limit is determined by the analysis of samples with known concentrations of analyte and by establishing the minimum level at which the analyte can be reliably detected.
- *Standard deviation of the response based on the standard deviation of the blank:* Measurement of the magnitude of analytical background response is performed by analyzing an appropriate number of blank samples and calculating the standard deviation of these responses.
- *Standard deviation of the response based on the slope of the calibration curve:* A specific calibration curve is studied using samples containing an analyte in the range of the limit of detection. The residual standard deviation of a regression line, or the standard deviation of y-intercepts of regression lines, may be used as the standard deviation.

5.2.3 Limit of quantitation

The limit of quantitation is the minimum injected amount that produces quantitative measurements in the target matrix with acceptable precision in chromatography, typically requiring peak heights 10 to 20 times higher than the baseline noise. If the required precision of the method at the limit of quantitation has been specified, the EURACHEM approach can be used. A number of samples with decreasing amounts of the analyte are injected six times. The calculated RSD percent of the precision is plotted against the analyte amount.

The amount that corresponds to the previously defined required precision is equal to the limit of quantitation. It is important to use not only pure standards for this test but also spiked matrices that closely represent the unknown samples. For the limit of quantitation, the ICH recommends, in addition to the procedures as described above, the visual inspection and the standard deviation of the response and the slope of the calibration curve. Any results of limits of detection and quantitation measurements must be verified by experimental tests with samples containing the analytes at levels across the two regions. It is equally important to assess other method validation parameters, such as precision, reproducibility and accuracy, close to the limits of detection and quantitation. Figure 2 illustrates both the limit of detection and the limit of quantitation.

5.3 Precision

Precision provides an indication of random errors and can be broken down into repeatability and intermediate precision. This procedure should only be performed when

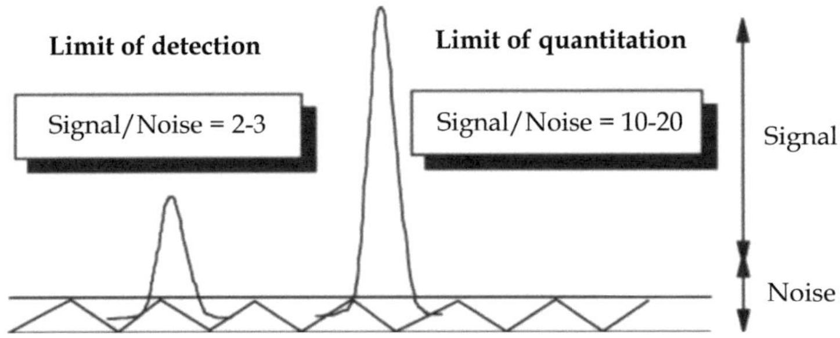

Fig. 2. Limit of detection and limit of quantitation via signal to noise

the entire analytical method procedure is finalized. Repeatability represents the simplest situation and involves analysis of replicates by the same analyst, generally one injection after the other. Repeatability tests are mandatory for all tests delivering numerical data. Repeatability is divided into two parts: injection repeatability and analysis repeatability (multiple preparations). Validation of the precision of an HPLC method occurs at three stages. The first stage is injection precision (injection repeatability) based on multiple injections of a single preparation of a sample on a particular sample on a given day. The set of criteria is given for area (% area normalization) methods (drug substance and drug product) based on %RSD of peak area. The second stage is analysis repeatability where multiple preparations and multiple injections of a sample are analyzed by the same chemist on the same day. The third stage is intermediate precision and is usually performed by different analysts, on a different system, on a different day on the same drug substance or drug product batch to determine the variability of the analytical test. The intermediate precision test may give indications to potential issues that may arise during method transfer. Relative standard deviation or coefficient of variation (Srel or %RSD) is used to assess whether the adequate precision has been obtained. If automation is utilized, then an intermediate precision test is required to compare results obtained through manual testing versus automated testing (if all solvent composition and analyte concentrations of all actives are identical in both methods).

5.3.1 Repeatability

In addition to the system precision, this short-term variability includes the contributions from the sample preparation, such as weighing, aliquoting, dilution, extraction, homogenisation, and etc. Therefore, it is essential to apply the whole analytical procedure (as described in the control test), rather merely to injecting the same sample solution six times. This is also the reason for using authentic samples because only then can the analytical procedure be performed exactly as in the routine application. There may be exceptions, but these should be demonstrated or cautiously justified. For example, analysing degradants near of the quantitation limit, where the variance contribution of the sample preparation can be neglected, injection precision and repeatability are identical. For some applications, where precision can be regarded as less critical, such as in early development or if the variability demands only a small part of specification range (less than approximately 10 %), or if the expected content of impurities is far away from the specification limit, artificially prepared (spiked) samples may be used, allowing several validation characteristics (linearity, precision and accuracy) to be addressed simultaneously. Repeatability can be calculated using Eq.(1) and Eq. (2) from a larger number of repeatedly prepared samples (at least 6).

$$s = \sqrt{\sum_{i=1}^{n}(x_i - \overline{x})^2 / (n-1)} \quad (1)$$

$$\%RSD = \frac{s}{\overline{x}} \cdot 100\% \quad (2)$$

Where s is the standard deviation, xi is the individual measurement in a set of n measurements and $\overline{x}$ is the mean of all the measurements. %RSD is the standard deviation in percentage.

5.3.2 Intermediate precision

Intermediate precision includes the influence of additional random effects according to the intended use of the procedure in the same laboratory and can be regarded as an (initial) estimate for the long-term variability. Relevant factors, such as operator, instrument, and days should be varied. Intermediate precision is obtained from several independent series of applications of the (whole) analytical procedure to (preferably) authentic, identical samples. In some cases of relative techniques, the preparation and analysis of the reference standard is an important variability contribution. Therefore, it is not appropriate to determine intermediate precision from the peak area of the sample alone (analysed on different days or even several concentrations only), as it is sometimes reported in validation literature. Apart from ignoring the contribution of the reference standard, any signal shift of the instrument will be falsely interpreted as random variability. In order to reflect the expected routine variability properly, the calibration must be performed exactly as described in the control test. Reproducibility, according to the ICH definition, is obtained by varying further factors between laboratories and is also particularly important in the assessment of "official" compendia methods or whether the method is applied at different sites. However, understood in the long-term perspective, both intermediate precision and reproducibility approach each other, at least in the same company. Reproducibility from collaborative trials can be expected to include additional contributions due to a probably larger difference of knowledge, experience, equipment, and etc. among the participating laboratories.

5.3.3 Reprodutibility

Reproducibility represents the precision obtained between different laboratories. The objective is to verify that the method will provide the same results in different laboratories. The reproducibility of an analytical method is determined by analyzing aliquots from homogeneous lots in different laboratories with different analysts, and by using operational and environmental conditions that may differ from, but they are still within, the specified parameters of the method (interlaboratory tests). Validation of reproducibility is important whether the method is to be used in different laboratories. The main factors affecting the reprodutibility are:

- Differences in room temperature and humidity.
- Operators with different experience and thoroughness.
- Equipment with different characteristics, e.g., delay volume of an HPLC system.
- Variations in material and instrument conditions, e.g., HPLC, composition of mobile phase, pH, and flow rate of mobile phase.
- Variation in experimental details not specified by the method.
- Equipment and consumables of different ages.
- Columns from different suppliers or different batches.
- Solvents, reagents, and other materials with varying quality.

5.4 Accuracy

The accuracy of an analytical method is the extent to which test results generated by the method and the true value agree. Accuracy can also be described as the closeness of agreement between the value that is adopted, either as a conventional, true, or accepted reference value, and the value found. For the quantitative approaches, at least nine determinations across the specified range should be obtained, for example, three replicates at three concentration levels each. The percentage recovery or the difference between the mean and the accepted true value together with the confidence intervals are recommended. The expected recovery (Table 7) depends on the sample matrix, the sample processing procedure, and the analyte concentration. The AOAC Guidelines for Single Laboratory Validation of Chemical Methods for Dietary Supplements and Botanicals (AOAC, 2002) includes a table with estimated recovery data as a function analyte concentration.

Active Ingredient (%)	Analyte ratio	Unit	Mean recovery (%)
100	1	100%	98 - 102
≥ 10	10^{-1}	10%	98 - 102
≥ 1	10^{-2}	1%	97 - 103
≥ 0.1	10^{-3}	0.1%	95 - 105
≥ 0.01	10^{-4}	100 ppm	90 - 107
≥ 0.001	10^{-5}	10 ppm	80 - 110
≥ 0.0001	10^{-6}	1 ppm	80 - 110
≥ 0.00001	10^{-7}	100 ppb	80 - 110
≥ 0.000001	10^{-8}	10 ppb	60 - 115
≥ 0.0000001	10^{-9}	1 ppb	40 - 120

Table 7. Analyte recovery at different concentrations

In drug substance, it can be rather difficult to demonstrate accuracy for a drug substance appropriately, especially if no (independently) characterised reference standard is available. In addition, independent analytical procedures are often not readily found. Nevertheless, every effort should be made to identify a suitable method for comparison because this is the only way to verify accuracy objectively. Instead of quantitative comparison, the results could also be supported by another method, for example, the verification of a very high purity of a drug substance by differential scanning calorimetry. Sometimes in validation literature, recovery is reported for a drug substance. However, recovery from simple solutions does not provide meaningful information (at least if all determinations are traced back to a reference standard characterized with the same analytical procedure) and it, therefore, is not appropriate to demonstrate accuracy.

In drug product, the accuracy is usually validated by analysing a synthetic mixture of the drug product components which contain known amounts of a drug substance, also termed spiking or reconstituted drug product. The experimentally obtained amount of active substance is then compared to the true, added amount (recovery). It can be calculated either at each level separately as a percentage recovery or as a linear regression of the found analyte versus the added one (recovery function). Sometimes the term "recovery" is misused when reporting in active content of the drug product batches. This is misleading because in some cases, the true amount of active is influenced by manufacturing the variability and is not exactly known. Preferably, the result should be termed "% label claim". The analyst should be aware of two important aspects regarding respect to recovery. First of all, it is based on the (validated) accuracy of the drug procedure substance otherwise the added amount will already be wrong. Secondly, by preparing the reconstituted drug product, the analyst deviates (more or less) from the routine analytical procedure. Of course, there is no other possibility of adding exactly known amounts, but consideration should be given to the possible implications. If, for example, solutions of the placebo are spiked with a stock solution of the active substance, the influence of the missed sample preparation steps, such as grinding, extracting, and etc. on the analysis should be considered. Here, the information obtained during the development of the method is very helpful (for example, homogeneity or extraction investigations). If some steps are of importance, any problems related to them will not influence the experimental recovery, and therefore, they are not identified.

5.5 Robustness

Robustness tests examine the effect that operational parameters have on the analysis results. For the determination of robustness, a number of method parameters are varied within a realistic range, and the quantitative influence of the variables is determined. If the influence of the parameter is within a previously specified tolerance, the parameter is said to be within the robustness of the method range. Obtaining data on these effects helps to assess whether a method needs to be revalidated, for instance, when one or more parameters are changed to compensate the column performance over time. In the ICH document, it is recommended to consider the evaluation of robustness of the method during the development phase, and any results that are critical for the method should be documented. In summary, an analyst must have a critical look at three different types of parameters when robustness and ruggedness are investigated:

- Internal parameters (e.g., temperature, pH, and etc., in the case of HPLC).
- External parameters (e.g., different analysts, instruments, laboratories, and etc.).
- Basic parameters (e.g., stability of test solutions and etc.).

From these lists, one point becomes very clear. A test for robustness is an individual test and depends very much on the analytical technique and equipment applied. As a rule of thumb, it is recommended to examine at least those parameters which are part of the operational qualification of the respective equipment. Then, the set of parameters investigated in a robustness study can be arbitrarily extended to those specific to the method defined in the procedure. The usual way to performing robustness testing is first to define the parameters with reasonable maximum variation. Then, each parameter is successively varied whereas the others are held constant (at nominal setting). For instance, six parameters each one at two levels would require twelve experiments when one parameter is changed and the others are always set to nominal levels. If more parameters are included, more experiments must be conducted. This classical approach is called one-factor-a-time (OFAT) approach. Certainly, this kind of robustness testing has disadvantages, such as many experiments and time. In addition, only limited information is made available from such studies, since possible interactive effects which occur when more than one parameter (factor) is varied, cannot be identified. Nowadays, an experimental design approach (DOE: design of experiment) is often preferred for robustness testing. The aim of an experimental design is to obtain as much as possible relevant information in the shortest time from a limited number of experiments. Different designs can be used in robustness testing, e.g. including full- and fractional - factorial designs as well as Plackett-Burman designs. The latter have become very popular in method robustness testing during recent years. The choice of a design depends on the purpose of the test and the number of involved factors. Experimental designs in robustness testing can be employed for all analytical techniques.

6. Conclusion

This chapter described the fundamentals and figures of merit for method validation in pharmaceutical analysis. The validation process is to confirm that the method is suited for its intended purpose and to prove the capabilities of the test method. The definitions of method validation parameters are well explained by health authorities. Although the requirements of validation have been clearly documented by regulatory authorities, the approach to validation is varied and opened to interpretation, and validation requirements differ during the development process of pharmaceuticals.

7. Acknowledgment

The authors acknowledge Instituto de Aperfeiçoamento Farmacêutico (IAF) for the scientific discussions and financial support.

8. References

AOAC International. (2002). AOAC Guidelines for Single Laboratory Validation of Chemical Methods for Dietary Supplements and Botanicals, Arlington, VA. Available from http://www.aoac.org/Official_Methods/slv_guidelines.pdf

BRASIL. (2003). Resolução RE n.899, de 29 de maio de 2003. Determina a publicação do Guia para validação de métodos analíticos e bioanalíticos. Diário Oficial da União, Brasília, 02 de junho de 2003. Available from http://www.anvisa.gov.br/legis/resol/2003/re/899_03re.htm

CDER Guideline on Validation of Chromatographic Methods. (1994). Reviewer Guidance of Chromatographic Methods, US Food and Drug Administration, Centre for Drugs and Biologics, Department of Health and Human Services

EURACHEM. (1998). *A Laboratory Guide to Method Validation and Related Topics: The Fitness for Purpose of Analytical Methods*, ISBN 0-948926-12-0, Teddington, Middlesex, United Kigdom.

Guidelines for Submitting Samples and Analytical Data for Methods Validation. (1987). US Food and Drug Administration, Centre for Drugs and Biologics, Department of Health and Human Services.

International Conference on the Harmonization of Technical Requirements for Registration of Pharmaceuticals for Human Use (ICH). Validation of Analytical Procedures: Text and Methodology Q2 (R1). (2005). Available from http://www.ich.org/fileadmin/Public_Web_Site/ICH_Products/Guidelines/Quality/Q2_R1/Step4/Q2_R1_Guideline.pdf

Klick, S.; Muijselaar, P.G.; Waterval, J.; Eichinger, T.; Korn, C.; Gerding, T.K.; Debets, A.J.; Sänger-van de Griend; van den Beld, C.; Somsen, G.W and De Jong, G.J. (2005). Toward a Generic Approach for Stress Testing of Drug Substances and Drug Product. *Pharmaceutical Technology*, Vol.29, No.2, pp. 48-66, ISSN 1543-2521

United States Pharmacopeia. (2011). Chapter 1225: Validation of Compendial Methods. United States Pharmacopeia 33, National Formulary 28. Rockville, MD.

2

General Introduction to Design of Experiments (DOE)

Ahmed Badr Eldin
Sigma Pharmaceutical Corp., Egypt

1. Introduction

Experimental design and optimization are tools that are used to systematically examine different types of problems that arise within, e.g., research, development and production. It is obvious that if experiments are performed randomly the result obtained will also be random. Therefore, it is a necessity to plan the experiments in such a way that the interesting information will be obtained.

2. Terminology

Experimental domain: the experimental 'area' that is investigated (defined by the variation of the experimental variables).
Factors: experimental variables that can be changed independently of each other
Independent Variables: same as factors
Continuous V*ariables:* independent variables that can be changed continuously
Discrete V*ariables:* independent variables that are changed step-wise, e.g., type of solvent.
Responses: the measured value of the result(s). from experiments
Residual: the difference between the calculated and the experimental result

3. Empirical models

It is reasonable to assume that the outcome of an experiment is dependent on the experimental conditions. This means that the result can be described as a function based on the experimental variables[2],
Y= (*f*) *x*. The function (*f*) *x*. is approximated by a polynomial function and represents a good description of the relationship between the experimental variables and the responses within a limited experimental domain. Three types of polynomial models will be discussed and exemplified with two variables, *x1* and *x2*.
The simplest polynomial model contains only linear terms and describes only the linear relationship between the experimental variables and the responses. In a *linear model*, the two variables *x1* and *x2* are expressed as:

$$y = b_0 + b_1 x_1 + b_2 x_2 + residual.$$

The next level of polynomial models contains additional terms that describe the interaction between different experimental variables. Thus, a *second order interaction model* contains the following terms:

$$y = b_0 + b_1x_1 + b_2x_2 + b_{12}x_1x_2 + residual.$$

The two models above are mainly used to investigate the experimental system, i.e., with screening studies, robustness tests or similar.
To be able to determine an optimum (maximum or minimum). quadratic terms have to be introduced in the model. By introducing these terms in the model, it is possible to determine non-linear relationships between the experimental variables and responses. The polynomial function below describes a *quadratic model* with two variables:

$$y = b_0 + b_1x_1 + b_2x_2 + b_{11}x_1^2 + b_{22}x_2^2 + b_{12}x_1x_2 + residual.$$

The polynomial functions described above contain a number of unknown parameters $(b_0, b_1, b_2, etc.)$ that are to be determined. For the different models different types of experimental designs are needed.

4. Screening experiments

In any experimental procedure, several experimental variables or factors may influence the result. A screening experiment is performed in order to determine the experimental variables and interactions that have significant influence on the result, measured in one or several responses.[3]

5. Factorial design[4]

In a factorial design the influences of all experimental variables, factors, and interaction effects on the response or responses are investigated. If the combinations of k factors are investigated at two levels, a factorial design will consist of 2k experiments. In Table 1, the factorial designs for 2, 3 and 4 experimental variables are
shown. To continue the example with higher numbers, six variables would give 2^6 = 64 experiments, seven variables would render 2^7 = 128 experiments, etc. The levels of the factors are given by – (minus) for low level and + (plus) for high level. A zero-level is also included, a centre, in which all variables are set at their mid
value. Three or four centre experiments should always be included in factorial designs, for the following reasons:

- The risk of missing non-linear relationships in the middle of the intervals is minimised, and
- Repetition allows for determination of confidence intervals.

What - and + should correspond to for each variable is defined from what is assumed to be a reasonable variation to investigate. In this way the size of the experimental domain has been settled. For two and three variables the experimental domain and design can be illustrated in a simple way. For two variables the experiments will describe the corners in a quadrate (Fig. 1), while in a design with three variables they are the corners in a cube (Fig. 2).

Two variables			Three variables				Four variables				
Exp. no.	Variables		Exp. no.	Variables			Exp. no.	Variables			
	x_1	x_2		x_1	x_2	x_3		x_1	x_2	x_3	x_4
1	−	−	1	−	−	−	1	−	−	−	−
2	+	−	2	+	−	−	2	+	−	−	−
3	−	+	3	−	+	−	3	−	+	−	−
4	+	+	4	+	+	−	4	+	+	−	−
			5	−	−	+	5	−	−	+	−
			6	+	−	+	6	+	−	+	−
			7	−	+	+	7	−	+	+	−
			8	+	+	+	8	+	+	+	−
							9	−	−	−	+
							10	+	−	−	+
							11	−	+	−	+
							12	+	+	−	+
							13	−	−	+	+
							14	+	−	+	+
							15	−	+	+	+
							16	+	+	+	+

Note that all variables are changed simultaneously in a controlled way, to ensure that every experiment in each design is a unique combination of variable levels.

Table 1. Factorial designs

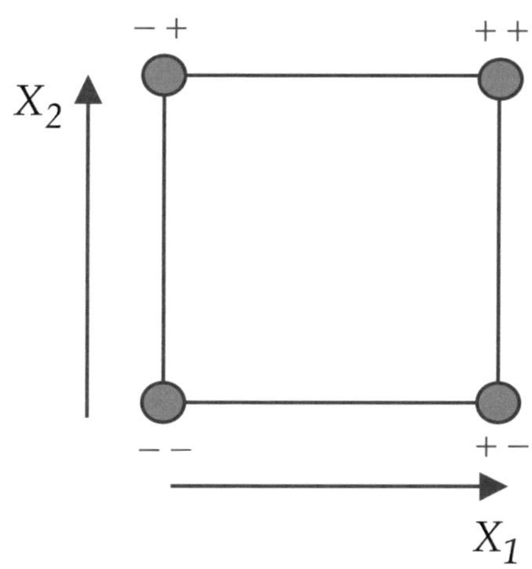

Fig. 1. The experiment in a design with two variables

6. Signs of interaction effects[5]

The sign for the interaction effect between variable 1 and variable 2 is defined as the sign for the product of variable 1 and variable 2 (Table 2). The signs are obtained according to normal multiplication rules. By using these rules it is possible to construct sign columns for all the interactions in factorial designs.

Example 1: A 'work-through' example with three variables

This example illustrates how the sign tables are used to calculate the main effects and the interaction effects from a factorial design. The example is from an investigation of the influence from three experimental variables.

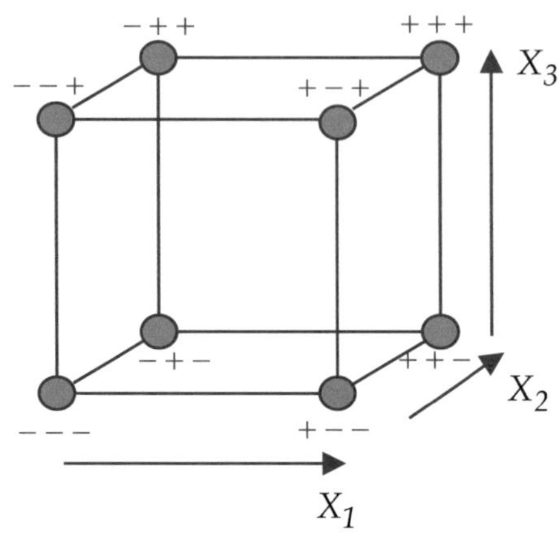

Fig. 2. The experiment in a design with three variables

7. Fractional factorial design

To investigate the effects of *k* variables in a full factorial design, 2*k* experiments are needed. Then, the main effects as well as all interaction effects can be estimated. To investigate seven experimental variables, 128 experiment will be needed; for 10 variables, 1024 experiments have to be performed; with 15 variables, 32,768

experiments will be necessary. It is obvious that the limit for the number of experiments it is possible to perform will easily be exceeded, when the number of variables increases. In most investigations it is reasonable to assume that the influence of the interactions of third order or higher are very small or negligible and can then be excluded from the polynomial model. This means that 128 experiments

are too many to estimate the mean value, seven main effects and 21 second order interaction effects, all together 29 parameters. To achieve this, exactly 29 experiments are enough. On the following pages it is shown how the fractions (1/2, 1/4, 1/8, 1/16 . . . 1/2 p) of a factorial design with 2 k-p experiments are defined, where

k is the number of variables and *p* the size of the fraction. The size of the fraction will influence the possible number of effects to estimate and, of course, the number of experiments needed. If only the main effects are to be determined it is sufficient to perform only 4 experiments to investigate 3 variables, 8 experiments for 7 variables, 16 experiments for 15 variables, etc. This corresponds to the following

response function:

$$v = \beta_n + \sum \beta_i x_i + \varepsilon$$

It is always possible to add experiments in order to separate and estimate interaction effects, if it is reasonable to assume that they influence the result. This corresponds to the following second order response function:

$$y = \beta_0 + \sum \beta_i x_i + \sum \sum \beta_{ij} x_i x_j + \varepsilon$$

In most cases, it is not necessary to investigate the interactions between all of the variables included from the beginning. In the first screening it is recommended to evaluate the result

and estimate the main effects according to a linear model (if it is possible to calculate additional effects they should of course be estimated as well.).
After this evaluation the variables that have the largest influence on the result are selected for new studies. Thus, a large number of experimental variables can be investigated without having to increase the number of experiments to the extreme.

8. Optimization

In this part, two different strategies for optimization will be introduced; simplex optimization and response surface methodology. An exact optimum can only be determined by response surface methodology, while the simplex method will encircle the optimum. simplex is a geometric figure with (k+1) corners where k is equal to the number of variables in a k-dimensional experimental domain. When the number of variables is equal to two the simplex is a triangle (Fig. 16.).

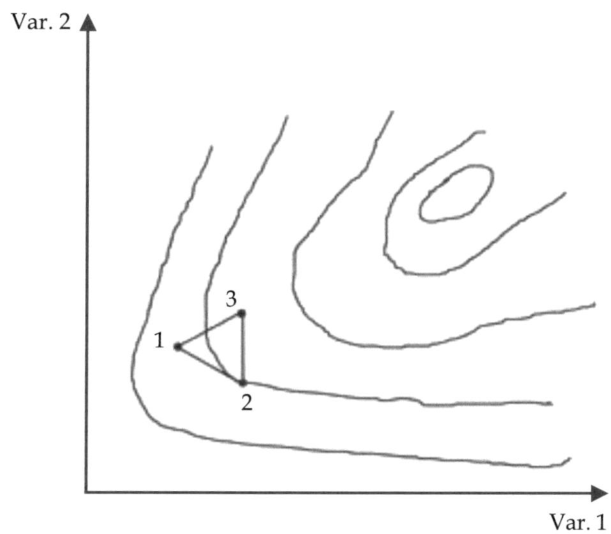

Fig. 3. A simplex in two variables

Simplex optimization is a stepwise strategy. This means that the experiments are performed one by one. The exception is the starting simplex in which all experiments can be run in parallel. The principles for a simplex optimization are illustrated in Fig. 17. To maximize the yield in a chemical synthesis, for example, the first step is to run k+1 experiments to obtain the starting simplex. The yield in each corner of the simplex is analyzed and the corner showing the least desirable result is mirrored through the geometrical midpoint of the other corners. In this way, a new simplex is obtained. The co-ordinates (i.e., the experimental settings) for the new corner are calculated and the experiment is performed. When the yield is determined,
the worst of the three corners is mirrored in the same way as earlier and another new simplex is obtained, etc. In this way, the optimization continues until the simplex has rotated and the optimum is encircled. A fully rotated simplex can be used to calculate a response surface. The type of design described by a rotated simplex is called a Doehlert design.

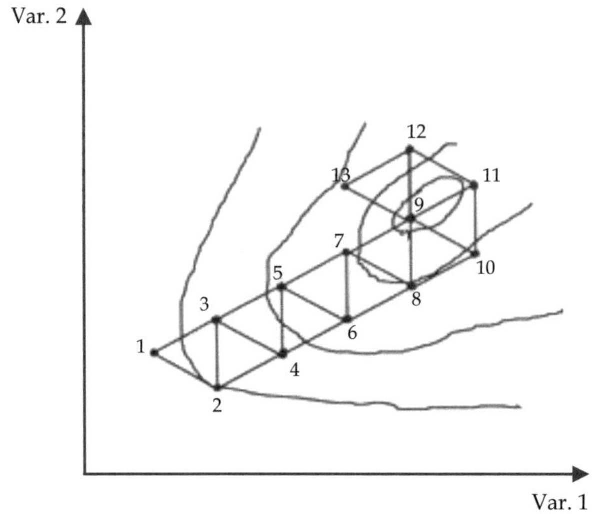

Fig. 4. Illustration of a simplex optimization with two variables

9. Rules for a simplex optimization

With *k* variables *k*+1 experiments are performed with the variable settings determined by the co-ordinates in the simplex. For two variables the simplex forms a triangle. For three variables it is recommended to use a 2^{3-1} fractional factorial design as a start simplex.

10. References

[1] Experimental design and optimization, Chemometrics and Intelligent Laboratory Systems 42 _1998. 3–40

[2] R. Sundberg, Interpretation of unreplicated two-level factorial experiments, Chemometrics and intelligent laboratory system, 24 _1994. 1–17.

[3] Atkinson, A. C. and Donev, A. N. Optimum Experimental Designs Clarendon Press, Oxford p.148.

[4] Kowalski, S.M., Cornell, J.A., and Vining, G.G. (2002) "Split Plot Designs and Estimation Methods for Mixture Experiments with Process Variables," Technometrics 44: 72-79.

[5] Goos, P. (2002) The Optimal Design of Blocked and Split-Plot Experiments, New York: Springer

Part 2

Quality Control in Laboratory

3

Good Clinical Laboratory Practice (GCLP) for Molecular Based Tests Used in Diagnostic Laboratories

Raquel V. Viana and Carole L. Wallis
Lancet Laboratories
South Africa

1. Introduction

Over the past decade there has been an expansion in molecular based technologies in the diagnostic environment. These molecular based technologies almost always involve *Polymerase Chain Reaction* of either DNA (PCR) or RNA (RT-PCR), but can also include isothermal amplification and/or sequencing. These molecular tests can be used for rapid qualitative or quantitative analysis for:

- Detection of infectious disease
- Viral load monitoring (HIV, HBV, HCV etc...)
- HIV diagnosis in paediatrics
- Translocations
- Mutations
- Gene rearrangements
- Forensic medicine

Several important steps need to be followed to ensure that a quality service is offered by a molecular laboratory. The quality of the test result is linked to a number of factors. It is reliant on activities that both directly and indirectly impact on the quality of the test ensuring that reliable and accurate results are obtained. There are several benefits to having a quality system in place, it allows for monitoring of the entire system, detects and limits errors, improves consistency among different testing sites and helps to contain costs.

Good Laboratory Practice (GLP) is defined in the Organisation for Economic Co-operation and Development (OECD) as "a quality system concerned with the organisational process and the conditions under which non-clinical health and environmental safety studies are planned, performed, monitored, recorded, archived and reported". The purpose of the Principles of Good Laboratory Practice is to promote the development of quality test data and provide a tool to ensure a sound approach to the management of laboratory studies, including conduct, reporting and archiving. Good Clinical Practice is an international ethical and scientific quality standard for designing, conducting, recording and reporting trials that involve the participation of human subjects. Compliance with this standard provides public assurance that the rights, safety and well-being of trial subjects are protected; consistent with the principles that have their origin in the Declaration of Helsinki, and that the clinical trial data is credible. The conduct of the laboratory work involving diagnostic testing requires a hybrid of GLP and GCP requirements referred to as Good Clinical Laboratory Practice (GCLP). This would revolve around the application of those GLP principles that are relevant to the analyses of samples while ensuring the purpose and objectives of the GCP principles are maintained.

General GCLP principles, which also hold for Molecular GCLP, such as: *Organisation & Personnel Responsibilities*, ensure that there are quality policies and standards in place. Organisational charts and job descriptions should give an immediate idea of the way in which the laboratory functions and the relationships between the different departments and posts. Also by describing a defined list of responsibilities it ensures that there are sufficient resources established and clearly defined roles resulting in accountability for all steps in the laboratory. Furthermore, all involved in the process should be committed to a culture of quality. *Personnel* are an integral aspect of GCLP as this ensures that there are enough well qualified people to perform the assays. To aid this, systems need to be in place to plan for the number of staff required, employment and retention of existing staff using continual development programs and training of the staff. To ensure staff retention there should be active supervision and performance management of all the staff. *Data Management* is vital for a laboratory to work efficiently and therefore needs an information flow scheme established and a data collection and management system in place which also ensures patient privacy and confidentiality. A crucial part of data management is the adequate training of staff, so they can use it effectively.

Another important component of running a quality laboratory is the establishment of *Standard Operating Procedures* (SOPs). This ensures that assay techniques and processes in the laboratory are standardised thereby contributing to reproducibility. Each SOP should detail one task in a clear and accurate fashion while also informing the operator of everything that needs to be known and how to do it. All SOPs and other documents in a laboratory need to be reviewed and approved by the laboratory manager on a regular basis to certify that all procedures used in the laboratory are up to date and accurate. To do this there needs to be a record of the number of copies (distribution list) of the SOPs and other documents in circulation within the laboratory. It therefore helps to number these documents in a consistent fashion so that there can be *Document Control* aiding in the location and removal of such documents from the laboratory when they are no longer in use. It is important that there is a *Stock Management* system in place. This allows for efficient management of reagents and consumables to ensure the continued ability to perform the assays the laboratory offers. To aid stock management there should be a procurement system in place, a mechanism of recording and managing the stock and adequate space to store the reagents and consumables correctly. There should also be appropriate *Facilities* to perform the assays (more details are described below), and to ensure quality results all the *Methods used should be Validated*, and appropriate *quality control* measures established and followed.

To ensure all of the above mentioned steps are followed it is important there be a Management *Review Process*, errors should be recorded (*Corrective Actions*), and all processes in the laboratory monitored through *Audits* (both Internal and External). This forms part of the Quality Assurance (QA) process. QA is defined as a team of persons charged with assuring management that GCLP compliance has been attained in the test facility as a whole and in each individual study. QA must be independent of the operational conduct of the studies, and functions as a witness to the entire process. Moreover, the above mentioned criteria to run a quality service, there are additional specific requirements for performing molecular based assays and supplying accurate and reliable results. These requirements are a direct result of the basis of the molecular technologies which use the ability of PCR to make millions of amplicons of the desired gene of interest (Figure 1).

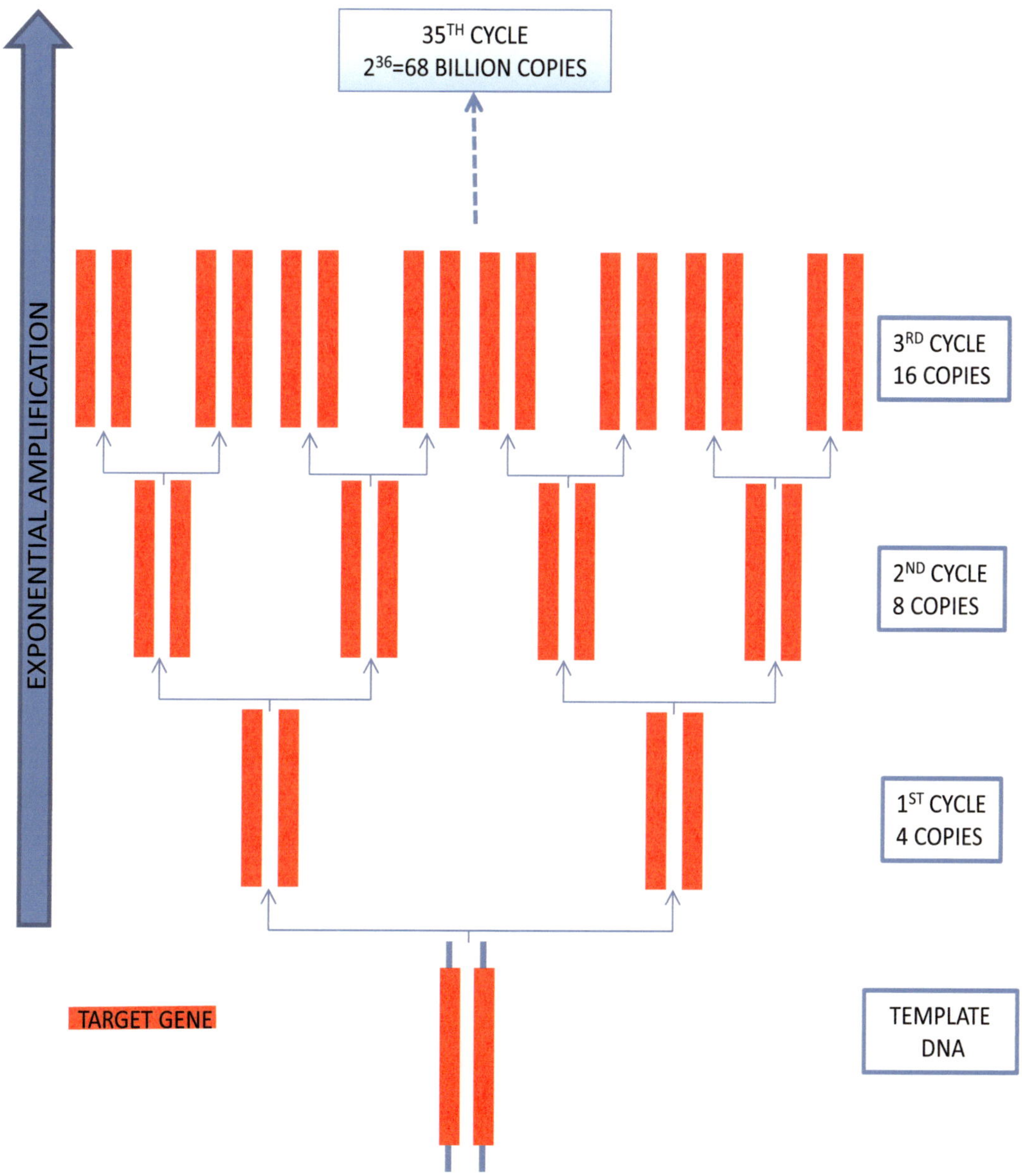

Fig. 1. The exponential amplification of a gene of interest during PCR (http://users.ugent.be/~avierstr/ principles/pcr.html)

The major limiting factor for PCR based technologies is contamination, a direct result of either the highly sensitive nature of PCR amplification and/or the large amount of amplified target obtained. The aim of this chapter is therefore to provide useful information for the appropriate set-up of a molecular laboratory and the steps that need to be taken to ensure good quality results are produced.

2. Scope

This chapter is intended to serve as a guide for diagnostic companies planning on setting up a molecular laboratory, following acceptable quality control standards. The limiting factors of contamination and technique sensitivity have resulted in several specific recommendations for the use of these molecular based technologies in diagnostics. These recommendations will be described in this chapter and include:

Section A:

Guidelines for working in a molecular diagnostic laboratory-this section will cover Sample Collection, Molecular Laboratory Layout, Staff Requirements and Competency, Quality Control around Equipment and Consumables, Laboratory Maintenance.

Section B:

Molecular Assay Development and Quality Control-this section will cover appropriate technique selection, primer design, Appropriate Reagent and Enzyme Usage, Assay Validation and Measure of Uncertainty of Molecular Assays.

Section C:

Controls to Monitor for Molecular Assay Performance-this section will ensure that contamination has not occurred and that the molecular technique is performing optimally. The following type of controls will be discussed: internal control, no template control, negative and positive control. Furthermore, corrective actions around the performance of the above mentioned controls will be discussed, including root cause analysis.

Section D:

Data Tracking and Auditing of a Molecular Sample, this section will cover the three steps of processing a sample: Pre-analytical Phase (the recording of sample receiving), Analytical Phase (sample processing and assay analysis) and the Post-Analytical Phase (result recording and interpretation) and the quality control of the results.

3. Guidelines for working in a molecular diagnostic laboratory

3.1 Sample collection

The type of collection device used for collection of specimens that will be tested using molecular diagnostic techniques is very important. The reason for this is that some collection devices are coated with a substance that can result in inhibition of the molecular assay. For example, some coagulates such as heparin result in inhibition of the molecular assay and long and cumbersome methods are required to remove the heparin prior to starting any molecular assay. Therefore the preferred method of collection is in an EDTA coated tube.

Swabs and Dry blood spots (DBS) are also appropriate collection devices, however caution needs to be taken with swabs that are collected in a formalin based collection medium as this also inhibits PCR and must be removed prior to testing.

Depending on the nucleic basis of the test, RNA versus DNA, this will also impact on the time between specimen collection and sample storage. If the sample required is plasma to be used in an RNA based assay, whole blood should be spun down and plasma removed for storage at -70°C until it can be tested. Some samples arrive in a storage medium, which allows for storage at room temperature for a certain amount of time prior to testing or long term storage. Whole blood and dried blood spots can be stored at 4°C for up to 24 hours for DNA based testing, but long term storage should be at -20°C.

3.2 Molecular laboratory layout

It is vital that the correct workflow is followed in a molecular laboratory in order to minimise contamination and ensure good laboratory practises are followed. It is the responsibility of all laboratory staff to ensure that the workflow is followed. PCR is extremely sensitive and thus poses a HUGE risk of contamination. During each step of a molecular assay multiple copies accumulate and are compounded as one progresses through the different steps of the methodology. To minimize this and thereby reduce contamination the different areas in a molecular laboratory should be physically separated. Depending on the nature of the molecular assay the ideal number of separations differs. Firstly, there should be two major separations between the work done prior to amplification (PRE-PCR) generally known as the clean area and that performed after amplification (POST-PCR) generally known as the dirty area (Figure 2). Between these two areas the work flow should be uni-directional (Figures 2, 3, and 4) and the relative air pressure and direction should differ. The equipment, consumables and laboratory coats should be dedicated to each area. If possible it is helpful to colour code racks, pipettes and laboratory coats in the different areas to be able to easily monitor movement between the different areas. Furthermore, powder-free gloves should be used throughout the process in all the different areas as the power on powered gloves results in assay inhibition.

Clean area/room

The clean area is divided into two additional areas, namely, *specimen processing laboratory* and the *no template laboratory* (Figure 3). The air pressure should be positive and blow out of the rooms. The *specimen processing laboratory* is where specimens are received, stored, total nucleic acid is extracted and the generation of complimentary DNA (cDNA) is performed. The *no template lab* is where reagents are stored and mastermix preparation for cDNA and amplification are made. The clean areas **must** be kept free of amplicon at all times, to ensure this occurs there should be no movement back from the dirty area to the clean area. If under extreme circumstances a consumable or reagent needs to be moved backwards it must be thoroughly decontaminated with bleach and ethanol. Returning racks should be soaked in 1% bleach overnight before soaking in distilled water and placing in the clean area.
In the *sample processing laboratory* the following equipment would most likely be present: -80°C and -20°C freezers and a fridge for sample storage (depending on the specimens received in the laboratory), a biohazard hood for sample extraction (especially if infectious specimens are processed in the laboratory), a centrifuge (if required for specimen extraction), automated extraction platform, a PCR workstation (a contained area that contains a UV light with or without a timer), a thermocyler (for cDNA synthesis only), dedicated pipettes, dedicated vortex, a dedicated place to hang laboratory coats and the appropriate safety materials (eye wash, medical aid box, shower). If chemicals are stored in this area appropriate facilities and storage requirements should be in place.

In the *no template laboratory* the following equipment would most likely be present: -20°C freezers and fridge for reagent storage, dedicated pipettes, dedicated vortex, dedicated microfuge, a PCR workstation (a contained area that contains a UV light with or without a timer), a dedicated place to hang laboratory coats and the appropriate safety materials (eye wash, medical aid box, shower). If chemicals are stored in this area appropriate facilities and storage requirements should be in place.

To ensure that no specimen contamination in the *no template laboratory* occurs it is vital to discard ones powder-free gloves worn in the *specimen processing laboratory* and change ones laboratory coat. This **MUST** occur before you enter the *no template laboratory*. It is therefore useful to place a biohazard bin outside the *no template laboratory* where gloves can be discarded and a hook for the laboratory coat to be hung up prior to entering the *no template laboratory*. Furthermore, nothing may enter the *no template laboratory* from the *sample processing laboratory*; this includes racks, tubes and open reagents. If possible disposable lab coats are useful in these areas.

Dirty area/room

Depending on the molecular methods performed in the laboratory the dirty area can be divided into one or two areas, namely, *post-amplification laboratory* and the *nested PCR laboratory* (Figure 3). The air pressure should be slightly positive for the *nested PCR laboratory* and neutral for the *post-amplification laboratory* and blow into both the rooms. The *post-amplification laboratory* is where the amplification reaction and detection of amplicon occurs. The detection of amplification can occur on a real-time PCR platform, gel electrophoresis, ELISA based detection and sequencing. **To note**, the *post-amplification laboratory* can be further divided into different rooms by each detection method, depending on the number of specimens and molecular assays run by a laboratory (Figure 4). In the *nested PCR laboratory* second-round amplification is set-up and a thermocycler is located there for this function. Nothing from these areas should move back into the clean area, without being completely decontaminated (as described above), under any circumstances. Gloves and laboratory lab coats must be removed when leaving this area.

In the *post-amplification laboratory* the following equipment would most likely be present: -20°C freezer and fridge for amplicon and reagent storage, a centrifuge (if required for the molecular assay performed), a PCR workstation (a contained area that contains a UV light with or without a timer), any equipment required for amplification, gel electrophoresis, sequencing or other amplicon detection methodology, dedicated pipettes, dedicated vortex, a dedicated place to hang laboratory coats and the appropriate safety materials (eye wash, medical aid box, shower). If chemicals are stored in this area appropriate facilities and storage requirements should be in place.

In the *nested PCR laboratory* the following equipment would most likely be present: -20°C freezer and fridge for reagent storage, dedicated pipettes, dedicated vortex, dedicated microfuge, a PCR workstation (a contained area that contains a UV light with or without a timer), a thermocycler, dedicated place to hang laboratory coats and the appropriate safety materials (eye wash, medical aid box, shower). If chemicals are stored in this area appropriate facilities and storage requirements should be in place.

To ensure minimal movement between areas during the running of molecular assays, it is optimal to have dedicated storage (freezer, fridge and room temperature) for each area. Furthermore, prior to starting the assay one must check that they have sufficient consumables and reagents to perform the test.

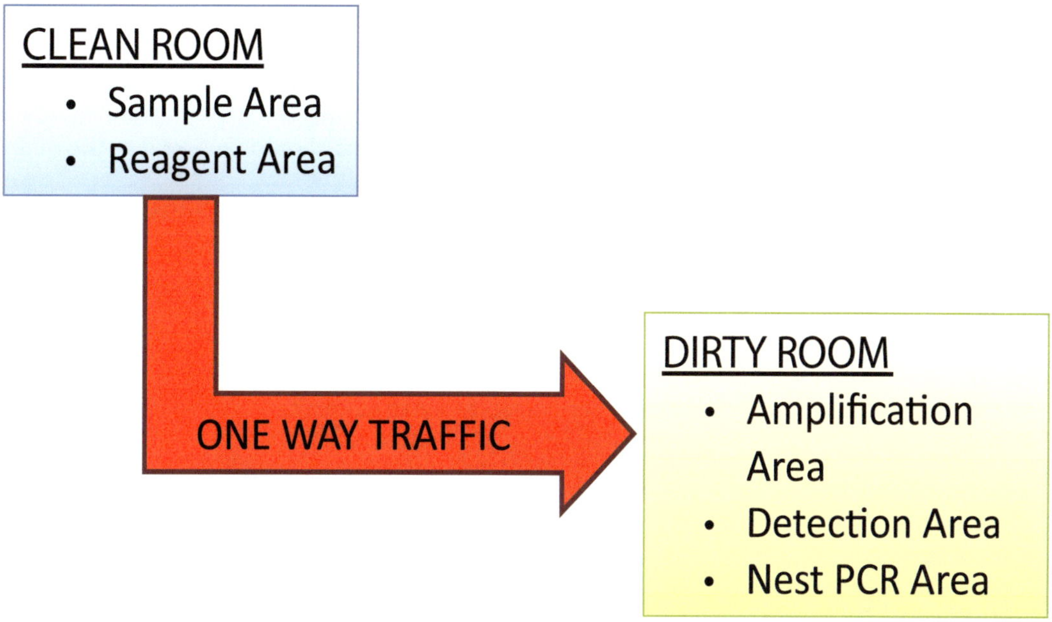

Fig. 2. Two room option for molecular lab layout. This is comprised of a clean area (for pre-analytical and sample preparation) and a dirty room (for analytical and post-analytical)

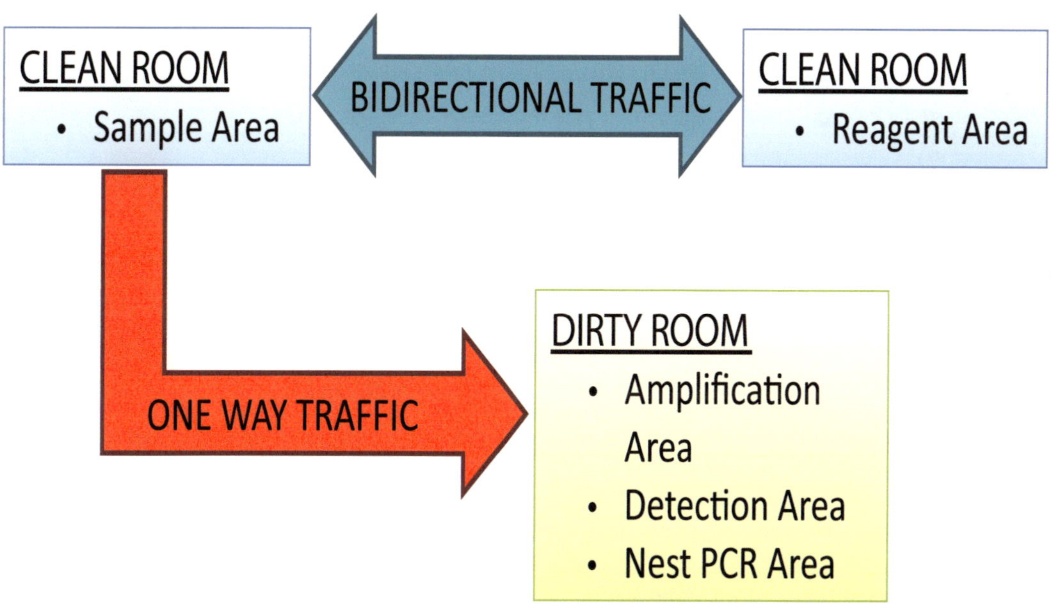

Fig. 3. Three room option for molecular lab layout. This is comprised of a clean area, which is divided into two rooms for 1) samples receiving and samples preparation and 2) room for preparation of reagents. As in the two room layout the dirty room (for analytical and post-analytical) remains the same

NOTE: To help in ensuring the above points are followed, it is important that each staff member organize their workflow as to ensure there is as little movement between clean and dirty areas during a shift and the laboratory policies should be incorporated and be well explained in a SOP that is easily accessible to all staff (including laboratory cleaners). Each work space should be kept tidy (minimal clutter) and each area should be closed to the other (with a door).

The above description of the different areas of a molecular laboratory describes the ideal laboratory layout. However, sometimes this is not always possible due to cost and space constraints, it is acceptable to divide the molecular area into just a clean and dirty area (Figure 2).

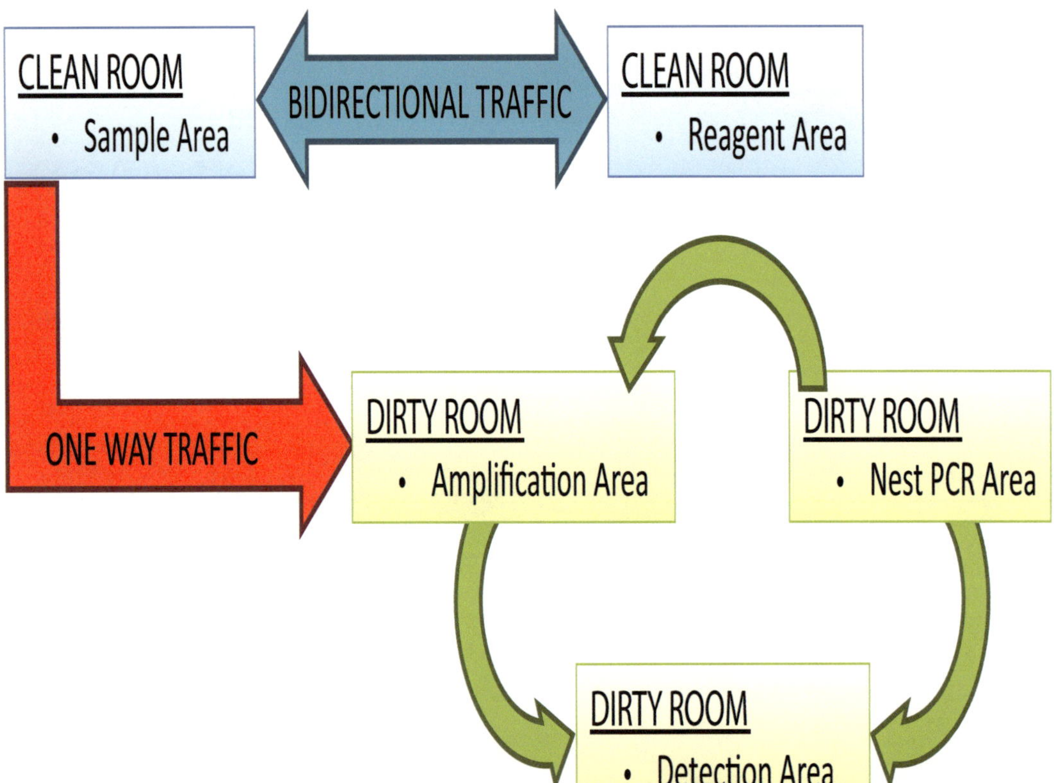

Fig. 4. Multiple room option for molecular lab layout. This is comprised of a clean area, which is divided into two rooms for 1) samples receiving and samples preparation and 2) room for preparation of reagents. The dirty area is divided into multiple rooms each with a specific function

3.3 Staff requirements and competency

Most molecular tests require highly skilled and well-trained staff. To achieve this all staff must be trained and then deemed competent prior to starting testing in the laboratory. Furthermore, it is advisable to assess the competency of the staff on an on-going basis using either external or internal quality control programs as described in Section 5. Once this has been completed the laboratory manager should formally approve the staff member competent to conduct testing.

The procedure for staff training should include the following steps: a new staff member should be given an orientation of the facility. It is vital that all new staff also be trained in laboratory specific biosafety, biohazard waste management, personal protective equipment, and general laboratory safety including the procedures that need to be followed for all chemicals used in the laboratory. Once the new staff member has passed the above training they should be given an overview of all the tests performed in the laboratory. This will ensure the staff member has an understanding of the process (including PCR) in the laboratory and give them an understanding as to why certain measures should be followed. During this initiation orientation the staff member should also be advised of the correct work flow of a molecular laboratory. The staff member should also be given an overview of the maintenance required in the molecular laboratory (Section 3.4 and Section 3.5) and read, understand and sign all SOPs used in the molecular laboratory.

New staff members should then be trained on the methodologies they are required to perform. Firstly, the new staff member should observe the procedure whilst following the SOP, during this time the new staff member is able to ask questions and is given a brief explanation of each step and the importance of it. Secondly, the new staff member then performs the methodology under supervision of the trainer. Once they are able to successfully perform the assay under supervision the new staff member should perform the methodology independently on previously tested samples and the results compared for accuracy by the laboratory manager. This training should be done for all tests that the new staff member will be performing. The records for this training are then kept in the new staff members training file.

Once a staff member is trained the competence of the staff member needs to be performed. The criterion for competence needs to be determined prior to assessing it. Competency assessments should be done on all staff members on a continual basis, but it is recommended it be carried out at least once a year on each test the staff member is performing. Competency is assessed in one of the following ways:

- Completion of an external quality assurance panel.
- Comparison of results across staff members:
  a. This can be performed in several ways, for example, staff can analyze the results of a molecular assay and these results are compared and similarity determined.
  b. Parallel testing, this is where staff members perform the entire assays on the same samples. The results obtained from each staff member are compared and the similarity determined.
- The method used to determine competency is determined by the laboratory manager.

If the staff member is deemed to be incompetent they should be retrained on the appropriate methodologies and competency reassessed.

NOTE: The qualifications of laboratory staff and the training and experience are critical in ensuring a quality service is offered in a molecular laboratory, because the training and experience of staff can positively influence the rate of human errors in the laboratory.

3.4 Quality control around equipment and consumables

Prior to setting-up a molecular assay in the laboratory it is important that one assess the equipment and reagents that are required. Each piece of equipment must meet the required specification of the laboratory and where the equipment can be sourced from. The

laboratory must ensure they have the correct space, electrical and plumbing facilities for the equipment. Consideration must be taken when determining who will supply the equipment. Are they reliable? Will they be able to support this piece of equipment and can they supply spare parts? All these factors will impact on the efficiency and reliability of the laboratory.
Once a piece of equipment is purchased, an SOP must be written defining how to use the machine, who is responsible and what the maintenance (daily, weekly, monthly and annual) procedure is. The maintenance must cover the routine checking that the machine is working correctly, if it is not, the appropriate troubleshooting is required and this must be recorded and regularly reviewed (see Table 1, an example of a maintenance chart). Furthermore, it must be determined if the piece of equipment requires a service or calibration by an external party and if so how frequently.
It is vital to train all staff on the machine (and when new ones are purchased) as correct operating of the equipment will lower the cost and regularity of repair, thereby preventing delays of tests and maintaining productivity.

Equipment	Calibrated	Maintenance			
		Daily	Weekly	Monthly	Annual
Pipettes					
Laminar Flow Hood					
Centrifuge					
Heating Block					
Waterbath					
Thermocycler					
Scale					
Plate Reader					
Sequencer					
Fridge/Freezers					

Table 1. Common equipment used in a molecular laboratory and the maintenance and calibration required

A similar process for supply of reagents needs to be followed with regard to assessing the need and establishing a reliable supplier. In the molecular laboratory one of the staff members should be put in charge of monitoring the stock levels, ordering (ensuring there is sufficient still left to perform the tests prior to running out) and ensuring regents and consumables for each test are stored appropriately. The level of consumable and reagent wastage should also be recorded so that the efficiency of the tests and laboratory can be monitored.

3.5 Laboratory maintenance

All work surfaces should be cleaned prior to use with 1% bleach solution contained in an opaque vessel which inactivates pathogenic agents and destroys nucleic acids. Residual bleach may affect stainless steel counter tops and the Perspex in hoods as well as contribute to inhibition of specific assays therefore it is advisable to then wipe down with distilled water to remove residual bleach that could form crystals. Finally 70% ethanol is used to further prevent transfer of pathogens. All cleaning solutions should be prepared daily. There are several commercially available products, such as DNA or RNA Away, that are specifically designed for removing nucleic acids or nucleases as well as pathogens, which can also be utilized for surface cleaning. Racks and trays should be soaked in the 1% bleach solution and then thoroughly rinsed with distilled water daily. Equipment such as thermocyclers and centrifuges should be cleaned with 1% bleach solution followed by 70% ethanol whenever contamination is suspected. Another means of decontaminating hoods, reagents, pipettes, tubes, and various other consumables, is exposure to UV light. Most biological safety cabinets are equipped with a UV light source. It is generally accepted that UV exposure at 254nm for a minimum of 5 minutes is sufficient for decontamination including the deactivation of nucleases and destruction of extraneous DNA on surfaces. Laboratory SOPs often include UV exposure steps as long as 30 minutes before and after use of hoods for PCR work. Wiping with bleach and/or detergents is still warranted as the penetrating power of UV light is minimal.

All equipment should be properly calibrated and maintained to ensure reliable and accurate performance. Records of repairs and routine maintenance as well as non-routine maintenance should be kept. Routine maintenance records should be documented in such a way that users of equipment can be assured that it is reliable and not outside its service interval. A good way of ensuring this is by attaching a service label to the equipment and by making provision for a clear service plan. Early warning that equipment is malfunctioning is important therefore the checking interval should be assigned to assure this. Alarms are very useful, especially if a problem occurs at a time when staff are not present. Back up for vital equipment should be available whenever possible as well as back up (generator) in the event of service failures such as power cuts. Records of equipment calibration, checking and maintenance demonstrate that the respective SOPs have been followed and that the equipment used was adequate for the task and operating within its specifications. The records should also demonstrate that the required action was taken if the equipment failed these checks and that staff were aware of this and took appropriate remedial action.

4. Molecular assay development and quality control

4.1 Appropriate technique selection

PCR has been adapted to fit several applications, including detection of target DNA, sequencing stretches of target DNA, and amplification and detection of mRNAs, ribosomal RNAs, and viral RNAs after using reverse transcriptase to make complimentary DNA. There are currently five common types of PCR used:

4.1.1 Conventional PCR

This type of PCR uses a thermostable DNA polymerase to make multiple copies of a target region of DNA defined at each end (3′ and 5′) by a specific primer.

PCR typically consists of three basic steps:

Step 1. **Denaturation**, requires that the sample DNA become a single-stranded template. To achieve this, the sample DNA is typically heated between 94°C and 97°C for 15 to 60 seconds, to separate or denature the two strands of the DNA.

Step 2. **Annealing** step, in which the reaction temperature is lowered typically between 47°C and 60°C for 30 to 60 seconds, to allow the oligonucleotide primers to bind to the single-stranded template.

Step 3. **Elongation**, during which the temperature is raised, typically to 72°C, allowing the polymerase enzyme to make a complimentary copy of the template. The length of the elongation step (30 seconds to three minutes) is determined by the speed of the enzyme, its ability to continue moving down the template DNA referred to as *processivity*, and the length of the DNA segment to be amplified.

One repetition or thermal cycle of these three abovementioned steps theoretically doubles the amount of DNA present in the reaction. The number of repetitions needed for a PCR application is determined by the amount of DNA present at the start of the reaction and the number of amplicon copies desired for post-PCR applications. Typically 25 to 40 cycles are performed.

4.1.2 Real-time PCR

Real-time PCR detects and measures the amplification of target nucleic acids as they are produced. Real-time PCR requires the use of primers similar to those used in conventional PCR, but in addition also requires an oligonucleotide probe labelled with fluorescent detection chemistry, and a thermocycler able to measure the fluorescence. Typically, the binding of a dye-labelled probe to the template sequence causes fluorescence to increase in direct proportion to the concentration of the PCR product being formed. A real-time machine monitors the fluorescence increase and calculates a cycle threshold (CT) value. This value, which represents the first cycle in which there is a detectable increase in fluorescence above the background level, is used to measure relative or absolute target quantities. In the absence of an absolute standard, the starting copy numbers of nucleic acid targets from different samples can be determined in a relative sense (e.g., sample one has 20 times more target than sample two). If an absolute standard, which contains known quantities of the target nucleic acid, is run to generate a standard curve, the starting copy number in the test samples can be estimated. Real-time PCR also differs from conventional PCR in that the target selection for real-time PCR is more restricted due to requirements of a smaller target fragment and the need to select probes with a higher melting temperature than the primers to ensure that the probe is fully hybridized during primer extension. In addition, the annealing and elongation temperatures are usually combined in a two-step PCR process that is performed at an intermediate temperature (e.g., 60°C) for one to two minutes.

There are several different fluorescent detection chemistries used for real-time PCR, including the following:

- *SYBR® Green I*, a fluorescent dye, is frequently used in real-time detection chemistry. This dye intercalates into double-stranded DNA, including PCR products and fluoresces. Therefore when used to detect amplification the level of fluorescence increases with each amplification cycle. This detection chemistry is not target sequence specific and is therefore more versatile than probe-based detection, but is susceptible to false positives due to the formation of non-specific PCR products or primer-dimers. Melting curve analyses are often used as an additional confirmation of product size for procedures using SYBR® Green.
- *Dual-labeled fluorogenic oligonucleotide probes* are most frequently used. These probes (e.g., TaqMan® probes) have a reporter fluorescent dye at the 5' end and a quencher dye at the 3' end. The probes are added to the PCR master mix along with the PCR

primers. During the PCR, if the target sequence is present, the probe anneals downstream from a primer site and is cleaved by the 5' nuclease activity of *Taq* DNA polymerase during polymerization. This cleavage releases the reporter dye from the probe and away from the quencher dye, resulting in fluorescence that is detected by the instrument. These probes can be modified with a minor groove binding (MGB) protein, allowing for shorter probes to be designed, which increases specificity in assays detecting a single nucleotide change.

- *Fluorescent resonance energy transfer (FRET) probes* involve the hybridization of two probes to adjacent sequences within the amplified product. The upstream probe has a fluorescent dye at the 3′ end and the adjacent probe has a fluorescent dye at the 5′ end. Correct hybridization of these probes brings the two dyes into close proximity. The laser excites the first fluorescent dye, which emits light at a different wavelength. This light then excites the second fluorescent dye by FRET between the adjacent probes. The real-time PCR machine detects the wavelength of light emitted by the second fluorescent dye.
- *Molecular beacon probes* use a variation of this same process, wherein reporter and quencher dyes are held together by a hairpin structure in the probes but become sufficiently separated by linearization of the probe after annealing with the template to allow the reporter dye fluorescence to be detected.

4.1.3 Multiplex PCR

Multiplex PCR involves the amplification of two or more different PCR products within the same reaction. This type of PCR is a modification of a conventional or real-time PCR with the use of multiple sets of primers in each reaction. Multiplex PCR requires less time and effort in amplifying multiple target templates or regions than individual reactions and may be utilised as an effective screening assay. While multiplex PCR provides potential time saving by allowing simultaneous detection of multiple targets, significant optimization is required to obtain all of the products with equal efficiency and sensitivity. Extra precaution must be taken to the design and concentration of the primers so that they do not interact or compete with each other.

4.1.4 Reverse transcription (RT)-PCR

RT-PCR is used to amplify RNA target sequences, such as messenger RNA and viral RNA genomes. This type of PCR involves an initial incubation of the sample RNA with a reverse transcriptase enzyme and a DNA primer. DNA primers that are used commonly include oligos dT (an oligo consisting of only thymidine residues), random hexamers (primers made of six random nucleotides), or a sequence specific primer. Oligos dT will hybridize to the poly-A tail of messenger and certain viral RNAs and prime DNA from the 3'-end of the RNA molecule as a consequence of this amplification of RNA near the 5'-end of the molecule may not occur. Random hexamers work with any RNA, but require an extra initial incubation at 25°C. Specific primers can be either the PCR primer that hybridizes to the RNA at the 3' side of the amplification region or a primer that hybridizes further downstream from the PCR primers. RNase inhibitors should be added to RT reactions to prevent the degradation of the RNA target sequence by RNase present in the sample or introduced as contamination. The reverse transcription and the PCR amplification can be performed in a one- or two-step process. In general, the two-step process is more sensitive, while the single-step reactions are less likely to be contaminated, because the tube is not opened after reverse transcription. The determination of which process should be used depends on the level of sensitivity required and the likelihood of contamination. There are many types of reverse transcriptases available for RT-PCR. The characteristics of the enzymes make some better suited for a one- or two-step reaction and other downstream applications. Some enzyme characteristics that impact the type of reverse transcriptase used for RT-PCR include: the presence or absence of RNase H activity that degrades RNA in an RNA:cDNA hybrid, processivity of the enzyme, divalent ion requirements, specificity and sensitivity, ability to incorporate dUTP for UNG carryover contamination, and optimum temperature for function.

4.1.5 Nested PCR

Nested PCR is a conventional PCR with a second round of amplification using a different set of primers annealing within the first round amplicon which helps increase the specificity and sensitivity of the target amplicon. The use of a second amplification step with the "nested" primer set results in a reduced background from products amplified during the initial PCR due to the nested primers' additional specificity to the region. The amount of amplicon produced is increased as a result of the second round of amplification. Used correctly, the multiple rounds of nested PCR should increase both the sensitivity and specificity of the PCR. However, this technique also increases the chance of carryover or cross-contamination because of the additional interaction with the first amplicon. The following precautions need to be followed to limit the chance of sample contamination and false-positives:

- Never opening more than one tube at a time.
- Adding an additional negative control for the second-round of amplification.
- Including first-round negative controls in the second-round of amplification to check for false-positives.
- Designating a fourth room or separate area for sample preparation after the first amplification (see Figure 3).

4.2 Primer design

Well-designed primers are essential for ensuring accurate and efficient detection of the desired gene of interest in a molecular assay. Primers are essential in PCR analysis and are short segments of chemically synthesized DNA (which are called oligonucleotides or, more commonly, "oligos"). A length of 18-27 base pairs, ensures adequate specificity and are short enough to ensure easy binding to the template during annealing. Primer sets are oligos with nucleotide sequences that are designed specifically to prime the amplification of a portion of a targeted nucleic acid. Hybridization probes are oligos with specific nucleotide sequences that are internal to the sequences of the primers and which are used to confirm the amplification of the target or quantitate it. Design and selection of the specific primer and probe set to be used for an experiment is based on the application, the type of PCR and hybridization that will be performed, and the segment of the target nucleic acid sequence that is known. Primers should be designed to amplify only the DNA or RNA of interest and be specific for that region. Primer melting temperature (Tm) is by definition "the temperature that the one half of the DNA duplex dissociates and becomes single stranded, thereby indicating the duplex stability". The optimal Tm range is 52-58°C, primers with melting temperatures above this (65°C) are prone to secondary binding. The Tm is directly linked to the GC content of the primer.

As a general rule, well designed primers are characterized by the following:

- Length of 18 to 27 base pairs
- No homology within or between primers, especially at the 3'end to avoid primer-dimer formation.
- No guanine-cytosine (GC) stretches greater than four base pairs
- *GC content*: (the numbers of C's and G's in the primer as a percentage of all the primer nucleotides) of 40% to 70%.
- *GC Clamp*: to promote specific binding there should be a G or C nucleotide present within five bases of the 3' end of the primer.
- Tm of the two primers should be as close as possible, however, a Tm of between 52-58°C tends to give the best result.
- Secondary primer structures:
    - A *hairpin* is formed by intramolecular interaction within the primer and reduces binding to the target, therefore, no hairpin loops with a Gibbs Free Energy of -2 kcal/mol or less.
    - *Self-Dimer*: this is formed when two primers in the same direction bind as a result of intermolecular interactions. To reduce self-dimers a primer should have a 3'end dimer of less than a Gibbs Free Energy of -5 kcal/mol or less and an internal self-dimer with a Gibbs Free Energy of -6 kcal/mol or less.
    - *Cross-Dimer*: this is formed when two primers of two different directions bind as a result of intermolecular interactions. To reduce cross-dimers a primer should have a 3'end dimer of less than a Gibbs Free Energy of -5 kcal/mol or less and an internal self-dimer with a Gibbs Free Energy of -6 kcal/mol or less. Nucleotide repeats should be avoided.

A variety of computer programs are available to aid in the creation of the best possible primers and probes, such as Primer Premier and PrimerPlex. These programs can help determine the optimum annealing temperature for newly created oligos and check for the formation of intra- and intermolecular dimers and hairpin loops. Laboratories should consider repeating the design process with more than one computer program, because these programs represent a simulated environment that may not include all the variables that affect oligo design.

For laboratories that are performing real-time PCR, the software provided with the real-time PCR instrument may be used for primer design. New primers and probes should always be tested experimentally for sensitivity and specificity before use in any method. The specificity of a chosen sequence should be evaluated using BLAST (Basic Local Alignment Search Tool) or its equivalent. Versions of BLAST are available on the WEB at a number of sites, including www.ncbi.nlm.nih.gov. BLAST compares the designed oligo sequences to known nucleic acid databases such as GenBank and EMBL. The search determines the potential of hybridization of the chosen oligo with sequences from other organisms. The results of this search should be used to define any relevant, closely matched sequences for specificity testing. The primer concentrations used in each newly developed PCR assay should be optimized to obtain maximum amplification efficiency. Optimization of primer concentrations is especially important when performing multiplex PCR.

4.3 Appropriate reagent and enzyme usage

Taq DNA polymerase, which is isolated from the thermophilic bacterium *Thermus aquaticus*, is the primary enzyme used in the amplification of DNA in nearly all procedures.

Modifications of this enzyme or other DNA polymerases with specific functions and unique properties, including different extension rates, processivity, greater proofreading ability, and different temperature tolerances, generally expressed as a half-life at the denaturing temperature, may be more appropriate for some PCR applications. Hot-start DNA polymerases are enzymes that are inactive until a specified temperature is reached. Use of hot-start enzymes reduces the production of non-specific products by preventing the elongation of primers that have non-specifically annealed to the template at lower annealing temperatures (which may happen during master mix preparation or in the first step of PCR cycling). When selecting an enzyme type for a method or study, the analyst should evaluate the different strengths and weaknesses of the DNA polymerases available to determine which individual polymerase, or combination of polymerases, will work with their template nucleic acid. Records of lot numbers of all reagents should be captured and stored. Reagents should be aliquoted to avoid excessive freeze-thawing and to protect stock reagents from contamination. These reagent tubes should be clearly labelled. All reagents containing fluorescent probes should not be exposed to excessive light to prevent degradation by photo-bleaching.

4.4 Assay validation

Before an assay can be implemented in the laboratory, several performance factors should be considered such as intended use of the test (target gene, sequences and mutations); the population the test will target; the test methodology and the type of sample that will be used. To establish the performance factors of an assay relevant information from scientific studies should be obtained, the correct testing population should be defined, the correct test method to obtain the desired result chosen, the quality control parameters decided and that the results obtained from the test can be interpreted.

Once the parameters have been chosen and prior to implementation of the molecular assay it is important to perform a validation on the assay to ensure quality results are obtained. To do this certain parameters should be accessed depending on the assay this includes both samples and analytical performance specification.

Samples used in the validation are very important. Specimens chosen for the validation should be chosen to test a specific parameter and samples used should be well characterised prior to starting the validation. The samples should represent the type of samples expected to be tested routinely. Furthermore, there should be a variety of samples to ensure that the test results can be interpreted for specific patient conditions. The results from this will allow for limitations of the methodology and results to be known. When selecting the samples for a validation the following guidelines should be followed:

- Irrespective of the prevalence of the disease/mutations/variation being tested the sample number of samples should be chosen and consistent across tests.
- All sample types expected for the assay should be included.
- Samples that include all the possible results should be included.
- Controls and calibration materials should be obtained and included.

The analytical performance parameters that can be included are: precision, accuracy, reproducibility, analytical specificity, analytical sensitivity (amplification sensitivity and sensitivity of variants), linearity, reportable range of test results, reference range and lot specific testing. Once the acceptable level for each criteria is set it cannot be modified once the validation has begun. Furthermore, the validation should be performed using the same conditions that will be used for the processing of routine specimens. This includes: the facility, equipment, reagents and personnel. The same holds true for when external quality assurance panels are performed (refer to Section 5).

Precision is the ability to obtain the same result from multiple replicates (5-10) of a sample. The closeness of each result for the same sample is then accessed. Depending how close the replicate results are will determine the precision of the assay. **NOTE:** the replicate results **might not be close** to that obtained from a different method.

Accuracy is performed to determine how close the results from two different tests are and is used for both a quantitative and qualitative assay. To access accuracy it is best to use a very well characterised sample. This parameter is easy to measure; however, it is often difficult to determine the criteria prior to performing the validation of a new assay.

Reproducibility is the ability to produce the same result even though there are changes to the conditions the assay are performed under. For example, if a different staff member performs the assay, different lot numbers or different thermocyclers are used. This assessment, unlike precision and accuracy requires a larger number of samples with fewer replicates and often over a longer time period.

Analytical Specificity is the ability of the assay to determine only the target analyte that is being detected and that there are no interfering substances. Interfering substances can be associated with specimen type and patient associated factors such as clinical condition, medication or disease stage.

Analytical sensitivity (*to note these two types of sensitivity criteria are not mutually exclusive*).

- **Amplification sensitivity** is determined by what is the minimum starting material (lower limit of detection) required to generate an accurate result. Importance should be placed on the different types of samples that can be obtained for the test.
- **Sensitivity of variants** is determined by what percentage of a specific target must be present to be detected. For example, the amount of tumour versus non-tumour cells to detect the presence of a mutation.

NOTE: A validation is also required if during the routine running of a methodology there is a change to the procedure. The change will dictate the level of validation that is required. If there is a minor change to the method - equivalency testing can be performed. For example, if the primer sequences are modified the results obtained from the previous and new primers should be compared. However, if there is a major change - a complete re-evaluation should be performed. This could include a new extraction or detection methods or chemistries.

4.5 Measure of uncertainty of molecular assays

For all assays introduced into a molecular laboratory the limitations of the assay need to be determined. For example, is there a limit of detection of the assay? Does it require a certain amount of sample for the assay to be accurately detected? And what steps have been taken to overcome these limits.

Limitations to molecular assays can include the following:

a. Collection devices and sample storage
   - The assays are normally validated on a certain collection device, such as in an EDTA tubes.
   - Specimens collected in Heparin are not suitable.
   - Once plasma has been isolated from the sample, it must be stored at -70°C.
   - Accurate and reliable results are dependent on proper sample collection and storage prior to testing.

- Samples are only processed if they are collected in the correct collection device.

b. Contamination
- The PCR based protocol has the limitation of contamination.
- To reduce contamination the following steps have been taken:
  - Three separate areas for reagent and sample preparation and post-PCR analysis and sequencing.
  - Powder-free and filter tips are essential when performing PCR.
  - Cross-contamination is further reduced by adding AmpErase UNG

c. RNA degradation
- For genotyping success it is critical that you prevent RNA degradation by RNases.
  - Sources of RNases contamination are: skin, hair, general laboratory glassware and contaminated solutions.
    - To reduce RNA degradation power-free gloves should be worn at all times.
    - RNases Inhibitor is added to the Reverse Transcriptase step.

d. Internal controls:
- Internal controls (positive and negative) should be included in every run:
  - No template Control (blank).
  - Positive control.
  - Negative Control.

e. Sequencing and interpretation of results
- Accurate and reliable results are dependent on good sequence quality. High background and noisy data can interfere with precise base calling.
- Population based sequencing only detects a population that is present in greater than about 25%.
- Some assays the interpretation system is based on a proprietary algorithm and these may need to be updated regularly and the software may at times be outdated.

f. Staff
- The level of skill and training required for staff members.

g. Limit of detection
- The assay can be dependent on the amount of initial sample material required or the amount of amplicon added to the detection step.

h. Analytical specificity
- How accurate is the method you are using? Does it detect other substances or viruses present in the assay and can this impact on the results.

5. Controls to monitor for molecular assay performance

There are several ways to monitor a molecular assay and these range from procedures put in place in the laboratory (internal) or external panels performed.

External Quality Assurance (EQA) panels assess the results of samples supplied to the laboratory to monitor the performance of the assay. The panels should be treated the same as routine samples in the laboratory, and should be rotated through all staff performing the assay. This type of proficiency testing monitors and enables a platform for improving the quality of a laboratory. EQA panels normally occur twice a year. Within ones molecular laboratory, you can run similar panels across staff members and include several controls as described below to assess the procedures in the laboratory on a more routine basis.

Internal Quality Controls need to be in place to ensure that the laboratory can control the procedures of the molecular assay. This includes the number, type and frequency of controls used for the assay. This enables the laboratory to detect errors caused by the test, environment or operator. The exact controls that can be used are described in Section 5.1.

5.1 Contamination control

5.1.1 Internal control

This is a control that is run in the same tube as the sample. Its level of amplification ensures there is nothing in the PCR that is resulting in inhibition.

5.1.2 No template control

The reaction is set-up without the presence of the starting material and DNase and RNase water is used it its place. With this sample you do not expect any amplification to occur. If there is amplification it indicates that one of the reagents or consumables are contaminated with a sample containing the desired target. In this case, you cannot use any results obtained in this run. Therefore, immediate action must be taken by the staff member by contacting the laboratory manager and all areas should be cleaned with 1% bleach and 70% ethanol and the contaminated reagents discarded. The test must then be repeated with only the controls prior to running samples to confirm that the contamination has been eliminated.

5.1.3 Negative control

The negative control is when starting material is added that does not have the target gene present. With this sample you do not expect any result if a result does appear then this indicates that either there is contamination from another sample or reagent in the run or the assay is not specific for the gene target (this should have been determined in the assay validation described Section 4.4). If this occurs immediate action must be taken by the staff member contacting the laboratory manger and the entire run must be repeated. If there is still a contamination the entire Pre-PCR area must be decontaminated using a 1% bleach solution, followed by 70% ethanol and distilled water. The PCR must then be repeated with only the positive and negative controls. If the contamination is still present the kit must be discarded and a new kit opened.

5.1.4 Positive control

The positive control is a sample that contains the target of interest and is known to work. If the positive control does not work, but the samples amplify, one can continue with the assay if it is a qualitative assay-but the reason for the control not amplifying should be investigated to ensure it does not happen regularly. However, if the positive control and the samples do not amplify further investigation is required. The assay should be repeated with only the positive and negative controls, if there is still no amplification the storage of the reagents must be checked as well as the expiry dates. It should be determined if there is anything that could be inhibiting the PCR. If everything appears in order the laboratory head will contact the appropriate reagent rep.

For qualitative assays it is **very important** that the positive control works as it identifies the amplification efficiency of the assay.

5.2 Corrective actions around the performance

The objective of a corrective action for either an internal or external quality control (QC) failure is vital as quality control is an important measure of the analytical and interpretive performance of the laboratory. Any failures are therefore an indicator of potential problems in the system and should be dealt with as a priority. QC also severs as an educational process, identifying areas of deficiency in knowledge and facilitating correction thereof through supervised feedback sessions. All corrective actions must be documented and filed in a designated place (such as a corrective action file) once it has been reviewed and signed off by the laboratory manager or head.

NOTE:

- For all corrective actions the following should be checked:
    - Are all reagents stored in the correct place and temperature?
    - Is the person performing the assay trained to perform it?
    - Check the expiry date of all reagents used in the assay.
    - Have all instruments been serviced and correct maintenance followed?

5.3 Root cause analysis

A root cause analysis is important to monitor problems that occur regularly in the laboratory and make appropriate changes to prevent them from reoccurring in the future. This is normally performed monthly by the laboratory manager. Corrective actions linked to Samples, Procedural/Technical Issues, Laboratory Equipment, Assuring Quality and Laboratory Safety should be monitored for the number of events and duplicate events. This allows for re-occurring events to be identified. Once a re-occurring problem has been identified it should be investigated and a resolution determined and actioned.

6. Data tracking and auditing of a molecular sample

It is important to have a system in place that can be used to document and record the movement of samples within the laboratory, from the pre-analytical, to analytical phase and finally to post-analytical phase. During these phases the samples processed need to have all reagent lots recorded, storage and staff member handling the samples recorded. Furthermore, all these procedures need to be written in a document controlled SOP that has been read and signed by all staff and that is accessible to staff in the laboratory on a daily basis.

6.1 Pre-analytic phase

The pre-analytical phase includes: test selection, ordering, specimen collection, processing handling and delivery to the testing site and sample receiving. Studies have shown that this step (in conjunction with the post-analytic phase) has a high level of errors and therefore needs to be closely monitored in the laboratory to ensure quality results are reported. One major contributor to this high error rate is the inaccurate selection of the molecular test to be performed. To reduce the number of incorrectly ordered molecular tests it is vital that the laboratory inform the health care professionals using this service of the tests that are offered and the diagnostic purpose (including assay limitations) of each assay. It is also vital to provide the correct collection, specimen handling, transport and submission of specimen information. Furthermore, if any patient specific information, such as treatment history, is required to interpret the test results this should be provided.

Another contributor to the error rate of the pre-analytic phase is specimen handling errors. When a sample is received in a laboratory it is given a unique number. This unique number allows for the correct test to be assigned to the sample and allows the movement of the sample through the assay steps in the laboratory to be monitored. This unique number should also be used for short or long term storage once the sample is received and/or processing is complete. During the entering of specimen information of this unique number, data entry errors can occur. Furthermore, specimens can be stored incorrectly prior to sample testing which could impact on the test. To ensure this does not occur and thereby reduce the error rate, it is important that all staff are adequately trained on sample receiving, and defined SOPs are in place to aid staff. The laboratory should have a data checking system in place to help reduce data entry errors.

During sample receipt in the laboratory the person receiving the specimen should check that the correct sample was received for the test, the correct collection device was used and there is adequate sample to perform the test. These parameters of sample acceptance or rejection should be well defined by the testing laboratory in a SOP available and understood by all staff.

6.2 Analytical phase

The analytical phase includes the sample processing and testing. Once a sample has been received, a staff member can begin processing the sample. To ensure there are no errors during the processing of samples it is important to have defined SOPs for the method being performed and that these procedures are correctly followed. Controls for the assay must be included in each run. Reagents must be prepared correctly and the appropriate safety precautions followed throughout the test.

The following should be recorded for each sample processed in the molecular lab (Figure 5):

- Test to be processed.
- Operator.
- Date for each step (if the assay occurs over multiple days).
- Lot numbers of the reagents used (each reagent used should be recorded).
- Controls used in the run (any information about the control that is important in the test).
- Specific equipment used during the assay that could impact on the test outcome.
- List of samples processed together.
- Area for review by a manager.

These sheets are commonly known as **record sheets** and can be made to suit the molecular assay being performed in the laboratory and can be test specific or generic depending on the assay requirements.

6.3 Post-analytic phase

The post-analytical phase includes assay analysis, result recording and reporting. During *assay analysis* it is important to ensure that all staff members processing samples analyse and interpret the results in a standardised manner. To control for this a detailed document controlled analysis SOP should be in place for each assay performed in a molecular laboratory. The use of a defined analysis procedure minimises the individual variances that could occur during the result analysis, thereby ensuring reproducible and accurate results are obtained and released.

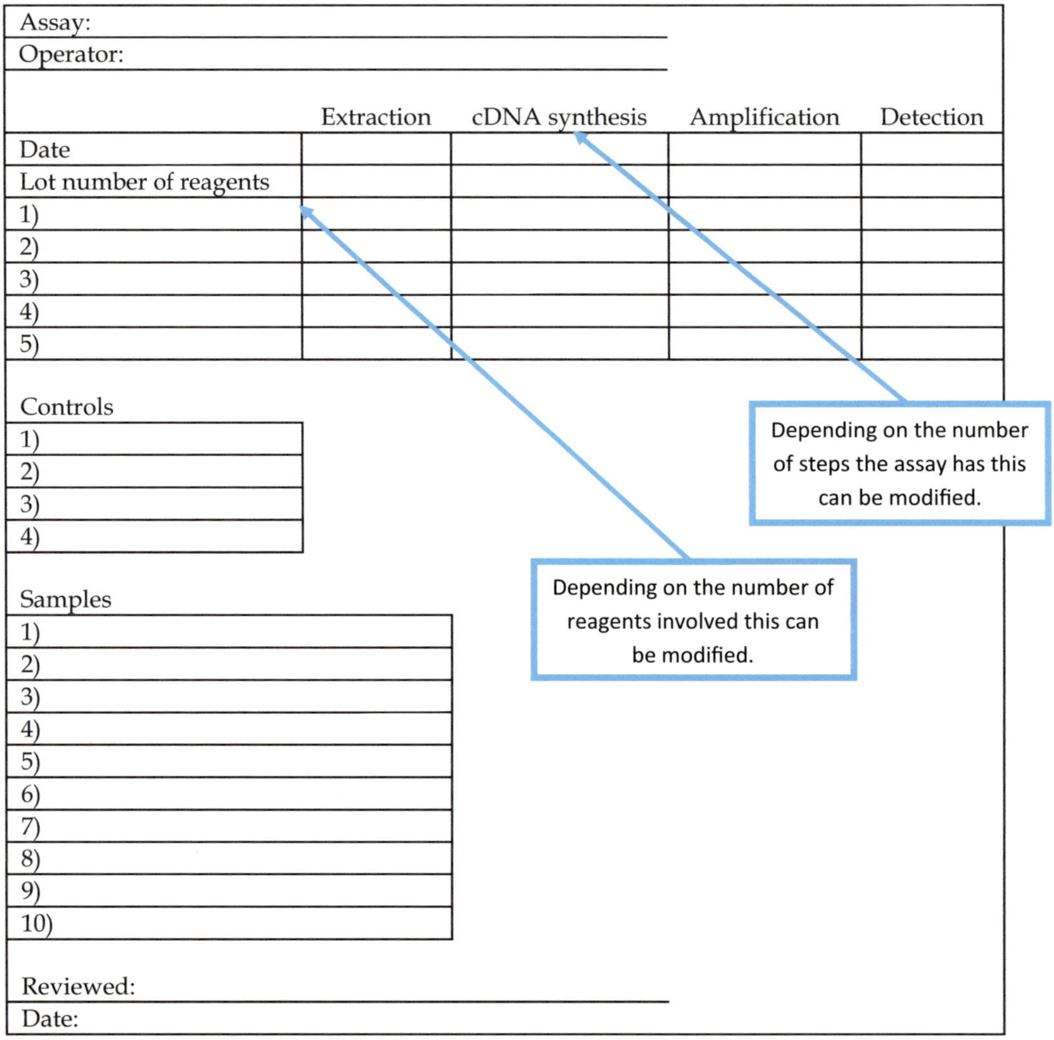

Assay:

Operator:

	Extraction	cDNA synthesis	Amplification	Detection
Date				
Lot number of reagents				
1)				
2)				
3)				
4)				
5)				

Controls

1)
2)
3)
4)

Samples

1)
2)
3)
4)
5)
6)
7)
8)
9)
10)

Reviewed:

Date:

Fig. 5. Example of record sheet

Result recording: Once the molecular assay has been completed on the samples and the results analysed. The results need to be reviewed. This should be done in the following manner:

a. The results from the controls of the run are checked to determine they are correct or in range. For a quantitative test the controls should indicate that there has been successful amplification and detection of the target region. For qualitative tests the controls need to be within the appropriate ranges.
b. Each sample identifier is checked and confirmed to ensure no data entry or clerical errors occurred during the assay.
c. The results then need to be reviewed (normally by the laboratory manager or laboratory head).
d. The specimen results should also be checked for any outliers or unusual results that do not fit the clinical picture and/or previous results obtained.

A study may have to be reconstructed many years after it has ended therefore storage of records must enable their safekeeping for long periods of time without loss or deterioration and preferably in a way which allows for quick retrieval. Access to the archive data should be restricted to a limited number of personnel. Records of the people entering and leaving the archives as well as the documents logged in and out should be kept.

6.4 Interpretation and the quality control of the results

To ensure accurate results of tests performed in a molecular laboratory are reported, additional analysis is required. For example, with sequencing to minimise the chances of sample contamination or mix-up one can align the sequences in a program such as Clustalw2 program (http://www.ebi.ac.uk/Tools/msa/clustalw2/) that is freely available on the internet. This program aligns the sequences and draws either a phenogram or cladogram which can be used for a crude analysis. *Parameters to look for are* if there are multiple sequences from the same sample do they cluster together? If you are using a positive control does it cluster with previous positive controls? (if the same sample is used as a positive control). Do samples from the same region cluster together (normally the case for infectious diseases)? Are any sequences very closely related or identical as these should be investigated further.

Once the results have been checked, the testing report should also include additional information that differs for each test but provides an accurate understanding and interpretation of the test results. All reports should contain the following information (according to CLIA guidelines):

- Patient name, Unique Laboratory Number used throughout the test and patient date of birth.
- Name and Address of the testing laboratory.
- Test performed and the date it was performed.
- Specimen information.
- Patient management recommendations (for genetic testing for heritable conditions).
- Name of referring doctor.
- Test methodology.
- Test limitations.
- Test result and interpretation of the result.

7. Conclusion

The recommendations described in this chapter should be considered in conjunction with Good Laboratory Practice and other regulatory guidelines in country. When deciding to set-up a molecular laboratory or to introduce a new test it is important to consider the requirements such as infrastructure, staff, equipment, supplier support, what are the current molecular tests that are available and will these tests complement and/or improve those that are currently in use. The clinical validity of the assay also needs to be assessed during implementation and then through the running of the assay.

The quality management approach described in this chapter allows for the monitoring and continual assessment of the assays through a defined quality control process. Furthermore, the information provided in this chapter can be used to set-up a new molecular laboratory or enhance an existing molecular laboratory. The guidelines described can be adapted for use in different settings and depending on the assay requirements.

To summarize:

a. It is important to ensure health care workers referring specimens understand the use of molecular tests.
b. To achieve Molecular GCLP the attitude of those in charge is vital.
c. To get staff to comply to the above mentioned criteria one must write brief and clear SOPs and ensure all staff read, acknowledge and observe the SOPs.
d. Be meticulous with sample labeling.
e. Ensure all quality control parameters are implemented and followed.
f. Ensure all maintenance in the laboratory is routinely performed.
g. Ensure the housekeeping guidelines are followed.
h. Everything needs to be documented (if it is not written down….it did not happen).
i. Assay design, choice and implementation must be considered carefully as this directly impacts on quality of the tests performed.

8. References

Centre for Disease Control and Prevention. Good laboratory Practices for Molecular Genetic Testing for Heritable Disease and Conditions. Morbidity and Mortality Weekly Report, June 2009, p.1-37 Vol. 58, No. RR-6 www.cdc.gov/mmwr.

PCR Primer Design Guidelines. http://www.premierbiosoft.com/tech_notes/PCR_primer_Design.html.

PPD and DAIDS. Global Solutions for HIV. DAIDS Guidelines for Good Clinical Laboratory Practice Standards. 2008. http://www3.niaid.nih.gov/research/resources/DAIDSClinRsrch/Labs/

Burd, EM. Validation of Laboratory-Developed Molecular Assays for Infectious Diseases. CLINICAL MICROBIOLOGY REVIEWS, July 2010, p. 550–576 Vol. 23, No. 3.

Principles and guidance reports for Good Laboratory Practice. Organisation for Economic Co-operation and Development (OECD). http://www.oecd.org/ehs/

GLP Handbook (2nd Edition). World Health Organisation. http://apps.who.int/tdr/publications/training-guideline-publications/good-laboratory-practice-handbook/pdf/glp-handbook.pdf.

Quality Assurance/Quality Control Guidance for Laboratories Performing PCR Analyses on Environmental Samples, EPA doc number 815-B-04-001, October 2004. http://www.epa.gov/ogwdw/ucmr/ucmr1/pdfs/guidance_ucmr1_qa-qc.pdf.

4

Quality of the Trace Element Analysis: Sample Preparation Steps

Maja Welna, Anna Szymczycha-Madeja and Pawel Pohl
Wroclaw University of Technology, Chemistry Department, Analytical Chemistry Division, Wroclaw, Poland

1. Introduction

Current status of elemental analysis performed using atomic spectroscopy techniques is to reach the best results in the shortest time and with minimal contamination and reagent consumption. Various spectroscopic methods such as flame- and graphite furnace atomic absorption spectrometry (F- and GF-AAS), inductively coupled plasma optical emission spectrometry (ICP-OES) or inductively coupled plasma mass spectrometry (ICP-MS) have been used for many years for determination of elements, since they met needs required in analytical applications. Constant progress in detector technology can still been observed, *e.g.* in terms of lowering quantification limits. Despite these advantages, quality of results does not follow the same tendency and sample preparation is recognized to be a critical point and the most important error source in modern analytical method development. This is especially true for solid samples that have to be brought into solution before measurements. It is dictated by instrumentation requirements dedicated to analysis of liquid samples. Determination of analyte concentrations in solid materials is not an easy task and several factors should be considered in order to minimize uncertainty in sample preparation and to achieve real objectives of analysis. It includes sample type and its matrix composition responsible mainly for the degree of difficulties during sample preparation and analyte determination. Therefore, the good choice of sample treatment and confidence of its application become a key ensuring to obtain reliable results.

2. Analytical sample

Samples to be analyzed can be divided generally into two main groups: **liquids** and **solids** (Hoenig, 2001).

- **Liquid samples** represent those that are already in an aqueous solution (*e.g.*, various waters, beverages, milk, blood, urine) or in other liquid form (*e.g.*, oils, fuels, organic solvents);
- **Solid samples** can be categorized due to their matrix composition as follows: those of organic nature (*e.g.*, plants, animal tissues and organs, excrements, plastics) or those with advantage of inorganic composition (*e.g.*, soils, sediments, dusts, metals).

It is well known that in most cases **sample preparation** step is needed for analysis based on atomic spectrometry techniques and leads to conversion of samples into homogenous forms like aqueous or acidic solutions. Despite aqueous solutions, which can be directly analyzed without any special pre-treatment, solid samples must be solubilised by an appropriate dissolution method, depending on the sample composition (main matrix, content of trace elements).

3. From sampling to reporting – steps of analytical process

Routine chemical element analysis involves several succeeding steps. It starts with planning a suitable strategy for a given analyte in a particular matrix, followed by representative sampling, sample pre-treatment, preparation procedure and instrumental measurement. It ends with interpretation of obtained data. A schematic diagram of the whole analytical process is drafted in Figure 1.

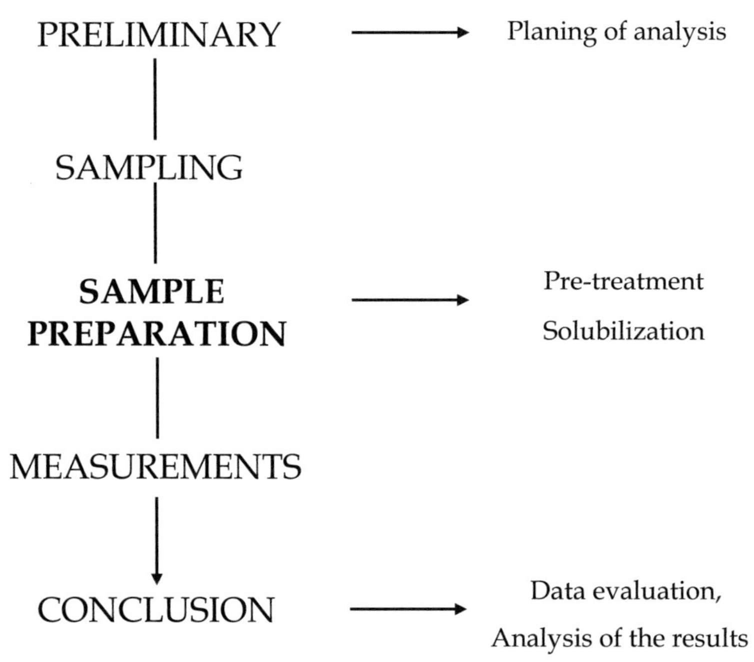

Fig. 1. Steps in analytical process (based on Hoenig, 2001)

An ideal method would allow performing all steps in one single, simple and quick process. In practice, each step in the analytical protocol contains an error, which affects reproducibility and accuracy of results. **Sample preparation** is recognized to be the largest source of errors and one of the most **critical points** of each analysis. Precisely, the **sample matrix** responds mainly for a difficulty of analysis. The sample matrix may impose a relatively pronounced effect during the preparation step or interferences during measurements, thus, eliminating or overcoming the troublesome matrix influence is necessary. Unfortunately, because of a wide number of analytes and a variety of sample types, there is no unique sample preparation technique that would maintain all requirements of analysts. Among strategies of sample preparation, dilution, acid digestion, extraction, slurry sampling or direct solid sample analyses are those that are mostly considered.

4. Quality assurance (QA) and quality control (QC)

Selection of the proper sample preparation method heavily depends on several factors. Availability of a variety of analytical techniques and instrumentation in addition to a great assortment of samples and preparation procedures make that selection of the right analytical approach is critical for method development. The incorrect sample preparation, *i.e.*, due to incomplete digestion or analyte losses, commonly can not be compensated by a versatile analytical technique and/or instrumentation. On the other hand, limitations of the instrumentation should be also taken into account since even for well-prepared sample they can lead to inadequate and untrue results. There is no doubt that the analyst should decide when his method of sample preparation used satisfies quality criteria and when results can be accepted. It is not an easy task and several different concerns can occur. However, at present, normally asked questions can lead to simple answers as follows:

Question: Which method of sample preparation should be used?
Answer: Check it.
Question: When the set of results can be accepted?
Answer: When their quality/accuracy is well demonstrated/verified.
Question: How it can be achieved?
Answers: Quality assurance and quality control concept.

Quality assurance (QA) claims to assure the existence and effectiveness of procedures that attempt to make sure that expected levels of quality will be reached (Rauf & Hanan, 2009). A particular attention should be paid to intermediate steps of an analytical protocol such sample treatment (preparation) that strongly contributes to total uncertainty of measurements. It should be improved, guaranteed and recorded by the analyst. Sample preparation is prone to errors like contamination, degradation or analyte losses and matrix interferences, which may, however, go unobserved by the analyst and affect final results.

Quality control (QC) refers to procedures that lead to control different steps in measurement process (Rauf & Hanan, 2009). It includes specific activities ensuring control of the analytical procedure. Among key points to be included during sample preparation, the most important is to demonstrate adequacy of the investigated method, *i.e.*, (1) accuracy, (2) precision, (3) efficiency and (4) contamination.

- *Accuracy* is the measurement of how close an experimental value is to the true value. It is realized by use of **control samples** with known compositions, which are treated in the same way as routine samples. Control samples allow monitoring the performance of the whole analytical procedure, including all sample preparation steps. Accuracy is based on the absence of systematic errors and the uncertainty of results corresponds to coefficients of variation. Nowadays, to demonstrate accuracy of the method, analysis of (standard, certified) **reference materials** (RMs) is the most commonly used. Another way to confirm accuracy of the method of interest is to compare results with those obtained with well established (**reference**) and independent **procedures**;
- *Precision* (reproducibility) is the degree to which further measurements or calculations show the same or similar results. It is expressed by means of relative standard deviation of measurements (RSD). The smaller RSD value, the higher precision is obtained;
- *Efficiency* in analyte determination may be demonstrated by adequate **recovery** using the **method of standard additions**. Analysis of spiked samples also allows to demonstrate accuracy of the method and recognize possible **interference effects**, which could lead to erroneous results;

- *Contamination* is a common source of error, especially in all types of environmental analysis. It can be reduced by avoiding manual sample handling and by reducing the number of discrete processing steps, however, the best way to asses and control the degree of contamination at any step of sample treatment is to use **blank samples**.

5. Sample preparation procedures

5.1 Liquid samples

In general, aqueous samples can be introduced to analysis directly and without any previous special pre-treatment, *i.e.* total or partial decomposition, as long as measured concentrations using spectrometric methods are reliable and satisfactory while possible interferences are under control.

In most cases only very little sample preparation is required and the easiest way is simple sample dilution. The dilution factor used in this case depends on concentrations of analytes and main matrix components; knowledge about the sample composition could be very helpful. Such an approach certainly reduces the analysis time and sample handling. It leads to low reagent consumption and generation of minimal residue or waste. Such simplification in sample manipulation decreases the risk of contamination and analyte losses. To minimize possible matrix interferences, standard additions and matrix-matched standards are proposed for calibration. Direct determinations from liquid samples (*e.g.*, waters, beverages) with minimal sample treatment such as dilution, degassing or matrix components evaporation provide a viable alternative to digestion as a mean of sample preparation:

El-Hadri et al. (2007) developed a highly sensitive and simple method for direct determination of the total As using HG-AFS in refreshing drink samples (colas, teas and fruit juices). Concentrations of As were directly determined in samples after pre-reduction with KI and acidification with HCl. Cola samples needed a more care, *i.e.*, degasification by magnetic stirring and sonication before analysis. Accuracy of the developed procedure was confirmed by recovery study and by comparison with a well established (reference) dry ashing digestion procedure. Quantitative recoveries (94-101%) were obtained with variation coefficients within 0.1-9%. The detection limit (DL) for As ranged from 0.01 to 0.03 ng mL^{-1}. In addition, no blank correction was required.

Matusiewicz & Mikołajczak (2001) proposed the method of direct determination of the total As, Sb, Se, Sn and Hg in untreated beer and wort samples using HG-ET-AAS. Samples were analyzed with little erased preparation: degassing by filtration for beer and sonication for wort. Calibration was made by standard additions. Accuracy and precision were ensured by using five well-established reference materials (SRMs or CRMs) and microwave (MW)-assisted digestion with HNO_3. Precision was typically better than 5% as RSD. DLs were restricted by variations in blank absorbance readings. Nevertheless, sub-ng mL^{-1} values were obtained. The problem of analytical blanks for ultrasensitive techniques was also discussed. Additionally, in terms of minimizing the risk of sample contamination, several procedures for removing CO_2 from beer were examined, including filtration, shaking, stirring, sitting overnight, storing with acid in open vessels overnight and ultrasonication.

Karadjova et al. (2005) develop a simple and fast procedure of sample preparation for the total As determination by HG-AFS directly in diluted undigested wine samples. Application of an appropriate wine dilution factor allowed minimizing ethanol interferences on HG-AFS measurements. Depressive effects by the small ethanol content (2–3% (V/V)) could be tolerated in 5–10- fold diluted samples by using solvent-matched calibration standard solutions. The method was validated through recovery studies and comparative analyses by means of HG-AFS and ET-AAS after MW digestion. Recoveries were in the range of 97–99% and precision was varied between 2 and 8% as RSD.

In the work of Tašev and co-workes (2005) simple ethanol evaporation was the only pre-treatment procedure proposed for direct wine samples analysis on the content of inorganic As species (As(III) and As(V)) by HG-AAS. Accuracy of this procedure was proved by recovery study and comparative analysis using ET-AAS. The total As content was determined after microwave digestion. Also here, preliminary evaporation of ethanol was recommended to avoid over-pressure and ensure better conditions for complete mineralization of wine organic matter. DLs of 0.1 mg L^{-1} were achieved for both species. Precision for this procedure (as RSD for ten independent determinations) varied between 8 and 15% for both As species present in the range of 1–30 mg L^{-1}. Accuracy of the aforementioned procedure (in terms of the total As content) was proved by recovery study and comparative analysis using ET-AAS.

Nevertheless, some types of liquid samples necessitate a particular caution before being introduced into detection systems. For example, blood coagulates in contact with some chemical compounds like $PdCl_2$ or $Pd(NO_3)_2$ (often used as modifiers in ET-AAS analyses) and this may partially or totally clog an autosampler capillary. Milk can not either be directly analyzed if HG is used as a sample introduction technique. The treatment with HCl (required for HG measurements) involves protein precipitation and creates a solid phase that can contain or partially retain elements under study. In this case slurry sampling (SS) is recommended.

The direct introduction of non-aqueous samples, however possible, significantly depends on their viscosity. In F-AAS analysis viscosity should be similar to that of water and organic solvents as ethanol or methyl isobutyl ketone fulfill this condition. In ET-AAS any organic solvents can be used due to similarity of analyte responses to those obtained in aqueous solutions. In ICP-OES several types of organic liquids can be introduced but an increase of the RF power is required to maintain a stability of the plasma (Hoenig & de Kersabiec, 1996).

5.2 Solid samples

Compared to liquids, preparation of solid samples is more complex. In general, unless the analytical method involves direct analysis of solid samples, they need to be in solution before analysis. Major concerns in selection of a solid sample preparation method for elemental analysis are requirements of the analytical technique used for detection, the concentration range of analytes and the type of matrix in which analytes exist. Many types of solid samples are converted into aqueous solution and therefore dissolution of sample matrices prior to determination is a vital stage of analysis aimed at releasing analytes into simple chemical forms.

The composition of sample matrices varies from purely inorganic (*e.g.*, ash, rocks, metallurgical samples) and purely organic (*e.g.*, fats) to mixed matrices (*e.g.*, soils, sediments, plant and animal tissues). Dissolution of inorganic matrices leads to clear solutions, where analytes are in their ionic forms. Both, purely organic and mixed matrices are more troublesome and dissolution does not guarantee complete matrix decomposition. Analytes may still be partially incorporated in organic molecules and masked from determination. In such case undecomposed organic matter may interfere in analysis leading, in consequence, to decrease in quality of final results. Of the methods responded for total decomposition of organic samples and normally used for sample preparation are (1) wet digestion and (2) dry ashing procedures. Alternatively, extraction of analytes from samples without total matrix destruction was proposed.

5.2.1 Dry ashing

Dry oxidation or ashing eliminates or minimizes the effect of organic materials in mineral element determination. It consists of ignition of organic compounds by air at atmospheric pressure and at relatively elevated temperatures (450-550°C) in a muffle furnace. Resulting ash residues are dissolved in an appropriate acid.

Dry ashing presents several useful features: (1) treatment of large sample amounts and dissolution of the resulting ash in a small acid volume resulted in element pre-concentration; (2) complete destruction of the organic matter, which is a prerequisite for some detection techniques (*e.g.*, ICP-OES); (3) simplification of the sample matrix and the final solution condition (clearness, colourless and odourless); (4) application to a variety of samples. Nevertheless, dry ashing presents either some limitations: (1) high temperature provokes volatilization losses of some elements; to avoid losses of volatile As, Cd, Hg, Pb and Se, and improve procedure efficiency, ashing aids (high-purity $Mg(NO_3)_2$ and MgO) are used; (2) on the other hand, the addition of ashing aids significantly increases the content of inorganic salts, which may be a problem in subsequent determinations of trace elements and contribute to contamination that necessitates careful blank control; (3) it does not ensure dissolution of silicate compounds and consequently of all elements associated with them (it can be encountered during plant analysis); after a procedure without elimination of Si (by evaporation with HF), poor recoveries for some elements can be observed, particularly traces; (4) open dry ashing exposes samples to airborne contamination (Hoenig, 2001; Sneddon et al., 2006).

Reliability of dry ashing procedures was demonstrated in some recent papers:

Vassileva et al. (2001) investigated the application of dry ashing for determination of the total As and Se in plant samples. The proposed method was a combination of dry ashing, conventional wet digestion with HNO_3 and HF and (in some cases) addition of a Mg containing solution as the ashing aid. The resulting ash was dissolved in HNO_3. It was established that plants of terrestrial origin may be mineralized using the dry ashing procedure without any As and Se losses. This was confirmed by analyses of several reference terrestrial plant and laboratory control samples in addition to direct analysis of the same plants using SS-ET-AAS. The addition of ashing aids seemed to be dispensable as errors observed were negligible. Unfortunately, more volatile As and Se species were present in plants of aquatic origin (*e.g.*, alges) and a separate wet digestion procedure remained unavoidable.

Grembecka et al. (2007) determined concentrations of 14 elements (Ca, Mg, K, Na, P, Co, Mn, Fe, Cr, Ni, Zn, Cu, Cd, Pb) in market coffee samples after dry mineralization of both dry samples and infusions evaporated to dryness prior to F-AAS measurements. Samples were ashed in electric furnace at 540°C with a gradual increase of temperature and subsequent dissolution of residues in HCl. Reliability of this procedure was checked by analysis of certified reference materials (CRMs). Recoveries of elements analyzed varied between 73.3% and 103% and precision (as RSDs) was within 0.4–19.4%.

Matos-Reyes et al. (2010) presented a method to quantify As, Sb, Se, Te and Bi in vegetables, pulses and cereals using HG-AFS. Samples were dry ashed and ashes dissolved with diluted HCl. Accuracy was assured by analysis of CRMs. A good accordance was always found between determined and certified values. For comparison the t-test (at 99% confidence level) was used but no significant difference between both sets of data was found. In addition, recovery studies on spiked samples before dry ashing was done. Recoveries determined ranged from 90 to 100% and indicated no loss of analytes and no contamination during the whole procedure.

5.2.2 Wet ashing

Wet digestion is used to oxidize the organic part of samples or to extract elements from inorganic matrices by means of concentrated acids or their mixtures. Commonly it is carried out in open vessels (in tubes, in beakers, on a hot plate, in a heating block) or in closed systems at elevated pressure (digestion bombs) using different forms of energy: thermal, ultrasonic and radiant (infrared, ultraviolet and microwave) (Hoenig, 2001; Sneddon et al., 2006).

Compared to dry ashing, wet digestion presents a wide range of varieties, concerning the choice of reagents as well as devices used. However, the sample nature and its composition as well as the composition and concentration of the reactive mixture should be considered before analysis. It includes: strength of the acid, its oxidizing power and boiling point, solubility of resulting salts, safety and purity of the reagent. In general, HNO_3, HCl, H_2SO_4, H_3PO_4, $HClO_4$, HF and H_2O_2 are used for organic samples, alloys, minerals, soils, rocks and silicates. Concentrated HNO_3 is the most favourable oxidant for destruction of the organic matter. Unfortunately, due to relatively low oxidation potential it may lead to incomplete digestion of materials with organic-rich matrices. It easily decomposes carbohydrates, however fats, proteins and amino acids require the addition of stronger H_2SO_4 or $HClO_4$. At present, the mixture of HNO_3, H_2SO_4 and H_2O_2 is a very efficient medium for different wet digestion procedures. Main disadvantages associated with the use of H_2SO_4 are its tendency to form insoluble compounds and its high boiling point. The high boiling point makes difficult to remove its excess after completion of oxidation. While $HClO_4$ is a strong oxidizing agent, it is extremely hazardous. HCl and HF ensure dissolution of inorganic compounds. Aqua regia (HCl with HNO_3 (3:1)) is widely used to dissolve soils, sediments and sludges.

The type of acid used in the sample preparation procedure may strongly affect the measurement step. In all atomic spectrometric techniques, HNO_3 is the most desirable reagent. In general, in spite of sometimes observed signal suppressions in its presence (*e.g.*, in ICP-OES), problems associated with it at concentrations up to 10% are rather occasionally observed as far as the acidity in sample and standard solutions are similar. Also, the mixture of HNO_3 and H_2O_2 used for digestion does not decrease a quality of analytical measurements. The presence of HCl is not troublesome in ICP-OES analysis, however, its use is prohibited in ET-AAS analysis because of a possible formation of volatile and difficult to dissociate analyte chlorides leading to spectral and/or vapour-phase interferences. In consequence, the latter phenomenon reduces absorbance signals of analytes. This problem may be overcome after addition of HNO_3 during the digestion procedure. For some applications, HCl should be avoided in ICP-MS analyses due to isobaric interferences, *e.g.*, during As determinations. Because of high viscosity that may provoke interferences in transport of solutions, utilization of H_2SO_4 is usually avoided despite its great efficiency in destruction of organic matrices. Its presence is particularly undesirable in analytical techniques where the sample introduction is realized by means of aspiration or pneumatic nebulisation of sample solutions (F-AAS, ICP-OES, and ICP-MS).

Main problems associated with wet digestion methods are: (1) much lower temperatures as compared to dry ashing procedures, however minimizing volatilization losses or retentions caused by reactions between analytes and vessel materials, they may lead to incomplete solubilisation of sample constituents and (2) co-precipitation of analytes with precipitates formed by main matrix elements within reactive mixtures. Both, they represent a real danger concerning reliability of analysis and hence, a good choice of a procedure and adequate reagents is critical for QA/QC of results.

5.2.2.1 Conventional wet decomposition

Wet decomposition in open vessel system (Teflon or glass beakers or glass tubes on hot plates) has been performed for many years. It may be very useful for relatively "easy" samples as food or agricultural products and materials, but generally, it is unsuitable for

Sample	Analyte	Reagents	QA/AC	Detection technique	Reference
Composts	Cd, Cr, Cu, Mn, Ni, Pb, Zn	HNO_3	- Reference material - Accuracy (recovery test) - Spiked sample	F-AAS	Hseu, 2004
Fish, mussel	Cd, Co, Cu, Cr, Fe, Mn, Ni, Pb, Zn,	HNO_3	- Reference material - Accuracy (recovery test) - Precision (RSD)	ICP-OES	Türkmen & Ciminli, 2007
Xanthan gum	Ca, K, Mg, Na	HNO_3	- Matrix matched calibration - Independent analytical procedure - Precision (RSD)	F-AAS	Abentroth Klaic et al., 2011
Dairy products	Ca, Cr, Cu, Fe, K, Mg, Mn, Na, P, Zn	$HCl+H_2O$	- Reference material - Accuracy (recovery test) - Independent analytical procedure - Precision (RSD)	ICP-OES	Kira & Maihara, 2007
Nuts	Al, Ba, Cd, Cr, Cu, Fe, Mg, Mn, Pb, Zn	$HNO_3+H_2SO_4+H_2O_2$	- Reference material - Accuracy (recovery test) - Calibration with standard additions - Precision (RSD)	ICP-OES	Momen et al., 2007
Legumes, nuts	Al, Cd, Cr, Cu, Fe, Ni, Pb, Zn	$HNO_3+V_2O_5$	- Calibration with standard additions for blank and samples - Accuracy (recovery test) - Precision (RSD)	ET-AAS	Cabrera et al., 2003
Plant, fungs	Hg	$HNO_3+H_2SO_4$	- Reference material - Precision (RSD)	CV-AAS	Lodenius & Tulisalo, 1995
Crude oil distillation products	Cu	H_2SO_4	- Reference material - Accuracy (recovery test)	ET-AAS F-AAS ICP-MS	Kowalewska et al., 2005
Herbal medicines	Al, Cr, Fe, V	HNO_3+HClO_4+HF	- Reference material - Accuracy (recovery test)	ET-AAS ICP-OES	Gomez et al., 2007

Table 1. Conventional wet digestion for diverse samples

such samples that require lengthy dissolution times (up to 24 h). Other problems to be considered are: time consumption (hours), contamination from environment, use of large amounts of reagents (especially strong oxidizing agents), pre-concentration of reagent impurities, and evaporative loss of volatile elements. Despite these drawbacks, conventional wet digestion in open vessel system allows achieving rather reliable and accurate results (according to QA/QC standards) and some recent applications are given in Table 1.

5.2.2.2 Microwave-assisted digestion

MW-assisted sample preparation with HNO_3 or its mixtures with HCl or H_2SO_4 (with or without added H_2O_2) is these days predominantly used for decomposition of a variety of inorganic and organic materials. The interaction of microwave radiation with samples and reagents results in fast heating of reaction mixtures and their efficient decomposition. Advantages of this strategy over conventional dry or wet ashing procedures are: broad application, much shorter reaction time needed (minutes), direct heating of samples and reagents, reduced need for aggressive reagents, minimal contamination and lack of loss of volatile elements. The use of small amounts of reagents decreases signals from the blank and increases accuracy of results. Usually, a mixture of HNO_3 and H_2O_2 is used for botanic, biological and food samples, while a mixture of H_2SO_4 and H_2O_2 is mainly used for oily samples. Acid mixtures are recommended for inorganic materials such as metals, alloys, minerals and for extracts from soils and sediments. Two different systems for MW-assisted digestion are used: pressurized closed vessels and open focused vessels. MW-assisted digestion in closed vessels under pressure is the most commonly applied. It offers safety radiation, versatility, energy control and possibility for addition of solutions during digestion. The only limitation is time required for cooling before vessels can be opened (even hours). In case of open focused MW system loss of volatile elements can occur. Results for low-level elements might also be affected by higher amounts of reagents used (increased risk of sample contamination). Both drawbacks can be, however, minimized by using vapour-phase acid digestion, which has been proven to be very effective in minimizing the residual carbon content (Hoenig, 2001; Sneddon et al., 2006).

In comparison to other digestion methods, accurateness and quality of MW digestion procedures for sample treatment can be found in numerous work. Some examples are presented below:

Demirel et al. (2008) compared dry ashing, wet ashing and MW digestion for Se, Fe, Cu, Mn, Zn and Al determination in various food materials (*e.g.*, rice, nuts, mushrooms, meat, milk, wine) using the F-AAS and GF-AAS detection. It was found that MW digestion procedure yielded more accurate results, required shorter time and enabled to achieve the highest recoveries for CRM analysis. Moreover, it allowed quantitative recoveries of volatile elements such as Se. For wet and dry ashings only 60 and 22% recoveries of Se were obtained. Poor recoveries (86%) were either obtained for Al when dry ashing was adopted. RSD values were below 10% and the proposed MW-assisted digestion procedure was free from matrix interferences.

Aydin (2008) tested dry, wet and MW digestion procedures for quantification of Co, Ni, Zn, Cu, Mn, Cd, Pb, Cr, Fe, Na, K, Ca and Mg in wool samples using ICP-OES. Different digestion mixtures, temperatures, dissolution times and proportions of HNO_3 and H_2O_2 were examined. The chosen MW-assisted digestion procedure maintained satisfactory recoveries, detection limits and precision for trace element determination in wool samples. For dry and wet ashings respective RSD values were considerably higher.

Du Laing et al. (2003) examined six destructive methods for determination of heavy metals (Cd, Cu, Pb, Zn, Ni, Cr, Fe and Mn) in red plants with atomic absorption detection. QC for concentration measurements was performed by analyzing adequate CRMs. MW digestion using HNO_3 yielded the best overall recoveries, whereas dry ashing was proved to be totally inappropriate for trace metal analyses of red plants (very poor recoveries). In case of Cr and Ni, the MW digestion procedure was the only one acceptable. It was concluded that red plants presented a difficult matrix and analysis of CRMs is needed for QC.

Szymczycha-Madeja & Mulak (2009) tested four digestion procedures for determination of major and trace elements (Al, Ba, Cd, Co, Cr, Cu, Fe, Mn, Mo, Ni, Pb, Sr, Ti, V and Zn) by ICP-OES in a spent catalyst. Two MW-assisted and two conventional hot-plate wet digestion procedures were applied. MW digestion with an HCl, HNO_3 and H_2O_2 mixture was the most effective. Quality of results was evaluated by analysis of CRM (CTA-FFA-1, fine fly ash). The proposed method provided a better solubilization of the matrix and much increased reproducibility. Results were sufficiently precise and accurate (RSD <5%). In contrast, MW digestion with a HNO_3 and HF mixture was found to be not suitable for proper determination of examined elements; errors in analysis of catalyst samples were encountered.

Do Socorro Vale et al. (2009) studied the effect and compared different procedures to treat the gum (deposits found in internal combustion engines) prior to determination of various elements (Al, Ca, Cd, Cr, Cu, Fe, K, Mg, Na, Ni, Pb, Si and Zn) by ICP-OES. To evaluate the best decomposition methodology, experiments were performed with one gum sample called a "reference sample". Two procedures were tested: (1) dry ashing followed by high temperature dissolution with HF and (2) MW digestion with a HNO_3 and H_2SO_4 mixture The latter procedure was found to be less time-consuming as compared to dry ashing and showed high recovery efficiencies in Cr, Cu, Fe, K, Ni, Pb, Si and Zn determinations.

5.3 Ultrasound-assisted extraction

Wet and dry digestion procedures, however excellent for sample decomposition, entail tedious, time-consuming and laborious steps, in addition to possible loss of analytes and contamination of samples. In consequence, obtained results can be far from true values. Today, ultrasound (US)-assisted procedures are considered as other alternatives for solid sample pre-treatments. They were found to be superior in facilitating and accelerating such sample preparation steps as dissolution, fusion and leaching. Chemical effects of US are attributed to acoustic cavitation, that is, bubble formation and subsequent disruptive action. It leads to generating local high temperature (*ca.* 5000 K) and pressure (*ca.* 10 GPa) gradients and to mechanical action between solid and liquid interfaces, which help in sample preparation. In US-assisted procedures diluted acid media are normally used for leaching element ions from powdered materials, thus, decreasing blank values, reagent and time consumptions and preventing analytes' losses. Smaller sample amounts can be used as well. Extractions are realized in ultrasonic baths or with sonoprobes, which are commonly employed for decomposition of organic compounds. However, a rigorous experimental control is strongly recommended to avoid losses of precision and accuracy. Uncontrolled US extraction procedures can provoke decomposition of analytes and hinder in this way extraction of organic compounds. When inorganic species are considered, ultrasonic irradiation does not present any decomposition risk; excellent results are obtained for diverse matrices (Santos Jr. et al., 2006).

Recently, ultrasonic effects have been exploited for sample preparations in agricultural, biological and environmental applications in order to improve analytical throughput. Nascentes et al. (2001) proposed a fast and accurate method for extraction of Ca, Mg, Mn and Zn from vegetables. Optimized conditions of such procedure were: 1 L of water, 25°C and 2% (v/v) detergent concentration. The best conditions for extraction were: 0.14 mol L^{-1} HNO_3, 10 minutes of sonication and a sample particle size <75 μm. Accuracy of this procedure was assessed by analyzing CRMs, as well as comparing results with those achieved with wet digestion. Recoveries determined were from 96 to 102%.

The US-assisted extraction procedure for estimation of major, minor and trace elements in lichen and mussel samples (IAEA lichen 336 and mussel tissue NIST 2976) using ICP-MS and ICP-OES was developed by Balarama Krishna & Arunachalam (2004). Parameters affecting extraction, including extractant concentration, sonication time and ultrasound amplitude, were optimized to get quantitative recoveries of elements. The procedure using a 1% (v/v) HNO_3 was fast (15 minutes) and accurate for most of elements. Solubilization of elements was achieved within 4 minutes of sonication at 40% sonication amplitude and a 100 mg sample weight. Overall precision was better than 10%.

In contrast, Maduro et al. (2006) pointed out some limits of US-assisted procedures affecting quality of analytical results. They compared three different ultrasonic-based sample treatment approaches, the automated ultrasonic SS, the ultrasonic assisted acid solid–liquid extraction (ASLE) and the enzymatic probe sonication (EPS) for determination of Cd and Pb by ET-AAS in CRMs of biological samples (spruce needles, plankton, white cabbage, oyster tissue, algae). The sample mass was 10 mg and the liquid volume was 1 mL of diluted HNO_3 (1 mol L^{-1}). Accuracy was evaluated by comparing results with those obtained using total acid digestion. The best results were obtained with the SS procedure with which accurate and precise determinations of the Cd and Pb content was possible in case of four from five analyzed CRMs. A good performance (quantiative extraction) of ASLE for Cd was only achieved in case of two from four CRMs, whereas total Pb recovery was only possible in case of three from four CRMs. Quantitative extraction with the EPS procedure was only obtained for Cd in oyster tissue. Neither ASLE nor EPS procedures were able to extract Cd or Pb from spruce needles. The Pb concentration obtained after EPS was found to be highly dependent on sample centrifugation speed and time.

5.4 Slurry sample preparation

The use of conventional wet acid digestion or dry ashing is time consuming and usually requires excessively hard sample treatment strategies. Recently, several methods for direct analysis of complex matrices by atomic spectrometric techniques have been developed and the SS approach as an alternative way of sample preparation is highly recommended (Cava-Montesinos et al., 2004; Bugallo et al., 2007). SS means preparation of a suspension of solid powdered particles of a sample in a liquid phase. Usually, after grinding the solid sample, the slurry is formed in water or in diluted acid (mainly HNO_3) in order to partially or totally extract analytes to the aqueous phase. It is possible to change the slurry concentration by simple dilution; hence, SS combines advantages of both liquid and direct solid sampling (Hoenig, 2001).

Main advantages of the SS procedure are: (1) elimination of a tedious and time-consuming step of sample dissolution; (2) avoidance of use of concentrated reagents and dilutions introducing contaminants; (3) safety and simplification of operation; (4) minimization of analytes' losses (especially volatile) and (5) possibility of use of smaller amounts of samples (1-100 mg in most common analyses). In addition, calibration performed using simple aqueous standards can be used. Nevertheless, several disadvantages affecting accuracy and precision of measurements and such variables as: (1) stabilization of the slurry; (2) its homogeneity; (3) sample particle size and (4) sedimentation must be carefully considered. Slurried samples must be stirred periodically by magnetic stirring or ultrasonic mixing before introduction to a measurement device. This helps to avoid sedimentation of sample particles, which may result in unrepresentative sample weight. Settling of solid particles in liquid-suspended samples can also be overcome by preparation of more stable slurries in a viscous medium or by using thickening agents. Concerning sample representativeness, only very fine particles in the slurry may ensure correct results; the presence of larger particles was found to be the most critical factor in analysis. For that reason, an intensive grinding of samples prior to analysis is of a great importance.

The SS procedure may be helpful in analysis of microsamples (*e.g.*, dust) or samples hardly soluble in common acid (*e.g.*, minerals). This procedure may be useful for the QC purpose of another sample preparation technique.

Recently, a lot of work has been done to maintain minimal sample manipulations with simultaneous assurance of reliability of results and at this field, SS has been proved to be quite suitable for this purpose:

Cava-Montesinos et al. (2004) developed a simple and fast procedure for determination of As, Bi, Sb, Se and Te in milk samples using HG-AFS. Samples were treated with *aqua regia* for 10 minutes in an US water bath and pre-reduced with KBr or with KI/ascorbic acid for total Se and Te or As and Sb determinations. Hydrides were generated from slurries in the presence of Antifoam A using a $NaBH_4$-HCl mixture. Calibration solutions were prepared and measured in the same way as samples. Obtained results were well comparable with those found after MW-assisted digestion. The advantage of the method was that only 1 mL of milk was required for analysis.

Matusiewicz & Ślachciński (2007) developed a SS procedure for simultaneous determination of hydride forming (As, Bi, Ge, Sb, Se, Sn), vapors (Hg) and conventional (Ca, Fe, Mg, Mn, Zn) elements in biological and environmental CRMs and real samples (coal fly ash, lake sediment, sewage) using a dual-mode sample introduction system (MSIS) coupled with MIP-OES detection. The slurry concentration up to 4% (m/v) was prepared in 10% HNO_3 containing 100 µL of decanol by ultrasonic agitation. Calibration was carried out by standard additions. An ultrasonic probe was used to homogenize the slurry. DLs below µg g^{-1} and good recoveries for all elements were obtained. Memory effects were not observed and hence, long washing times between samples were not needed. This sample pretreatment was minimal and involved only the slurry preparation procedure.

Bugallo et al. (2007) proposed a novel MW-assisted slurry tprocedure for Ca, Cu, Fe, Mg and Zn determination in fish tissues by F-AAS. The suspension was optimized for each analyte and it was established that MW irradiation in HNO_3 containing 0.3% glycerol for 15-30 s at 75-285 W permitted efficient recoveries for Ca, Fe, Mg and Zn. Only Fe recoveries were not higher than 46%, however, reduction of matrix interferences was realized by additional short MW-assisted suspension treatment. For Cu, an HCl suspension medium and homogenization with magnetic stirring (5 minutes) was found to be the most appropriate. Results obtained using SS were not significantly different from those achieved with MW-assisted digestion. Accuracy was checked using a CRM.

Da Silva et al. (2008) combined a cryogenic grinding and SS for Cu, Mn and Fe determination in seafood samples by F-AAS. Samples (80 mg) were grounded in a cryogenic mil, diluted with 1 mol L^{-1} HNO_3/HCl and sonicated for 30 min. Calibration curves had been prepared using element standards in the same suspension medium. DLs below µg g^{-1} and precision expressed as RSD lower than 4% were obtained. Accuracy of the procedure was confirmed by analysis of a CRM of oyster tissue; reliability by comparing it with ICP-OES after complete wet digestion in a HNO_3/H_2O_2 mixture. The proposed method offered the low contamination risk, simple handling and possibility of standardization using aqueous reference solutions.

5.5 Direct solid sampling

Another good alternative to wet digestion procedures used in elemental analysis is direct solid sampling (DSS). In addition, it is the most widely used technique in metallurgical laboratories. Among different techniques that can be used for DSS in combination with AAS, ICP-OES or ICP-MS there are laser ablation (LA) and electrothermal atomization or vaporization (ETV). Nowadays, direct analysis of solid samples using graphite furnace atomic absorption spectrometry (DSS-GF-AAS) has been shown to be the most attractive and convenient technique (Vale et al., 2006).

Main attributes of this method are: (1) low DLs; (2) minimal sample manipulation; (3) operational simplicity; (3) short time required to obtain results; (4) higher accuracy since errors due to analyte loss or contamination can significantly be reduced and (5) higher sensitivity due to the lack of any sample dilution. In most cases aqueous standards can be used for calibration. Drawbacks are associated with (1) quite short linear working ranges in AAS, which limits analysis to determination of low concentrations and, in consequence, of low sample weights (in many cases solid powdered samples must be diluted with graphite powder and re-homogenized before analysis); (2) natural samples inhomogeneity resulting in precision of results of order of 10% and (3) enhanced interferences as compared to analysis of dissolved samples, where matrix is simplified as a result of mineralization. Both small and large amounts of samples used for analysis can lead to overestimation or underestimation of final results.

Very recently, high-resolution continuum source atomic absorption spectrometers (HR-CS-AAS) for DSS have been proposed. By this, the entire spectral environment of analytical lines at high resolution can be observed and allows to detect, correct and avoid many spectral interferences.

Many researchers consider these exceptional facilities of DSS and according to QC/QA present very consistent results:

Sahuquillo et al. (2003) validated determination of the total and leachable As in sediments by DSS-GF-AAS. Calibration with both liquid standard solutions and CRMs of sediments was made. Under optimised instrumental conditions the DL of As of 0.44 mg kg^{-1} and long-term reproducibility within 10-15% were obtained.

Oleszczuk et al. (2007) showed DSS-ET-AAS to be a powerful tool for determination of Co, Cu and Mn in green coffee. The method was validated by analyzing several botanical CRMs and a number of pre-analyzed samples of green coffee. Measurements with ICP-OES after MW-assisted digestion were used as a reference method. Mn and Co could be determined using aqueous standard solutions for calibration, but calibration with a CRM was necessary to get accurate results for Cu. DLs for Cu and Co were more than one order of magnitude better than in case of SS-GF-AAS due to absence of sample dilution. Moreover, DSS did not require any sample preparation besides grinding of coffee beans.

Detcheva & Grobecker (2006) determined Hg, Cd, Mn, Pb and Sn in seafood by DSS-GF-AAS with Zeeman-effect background correction and an automatic solid sampler (except for Hg). A calibration range was extended using a three-field dynamic mode. Very high concentrations of elements could be determined without need for dilution of solid samples. Calibration with CRMs of organic matrices was applied. Under optimized conditions no matrix effects were observed and obtained results were in a good agreement with certified values.

Ribeiro et al. (2005) investigated determination of Co in biological samples (*e.g.*, fish) by comparison DSS-GF-AAS and tetramethylammonium hydroxide (TMAH) sample dissolution followed by conventional GF-AAS with HR-CS-GF-AAS. It was found that analysis of samples is much easier when using HR-CS-GF-AAS, however, the best DL of 5 ng g^{-1} was obtained with both DSS and HR-CS-GF-AAS.

6. Conclusion

Measurements of elements in various materials are the only way to get the knowledge about their composition. A variety of instrumental techniques including atomic, emission or mass spectrometries gives a possibility to perform reliable and accurate trace and ultra-trace determinations. It was expected that more and more sensitive detectors would guarantee and assure accuracy of analytical results. In fact, the key to the success of the whole analysis is selection of the sample preparation method. Appropriate sample preparation allows obtaining required and reliable information about element concentration of samples. There are several aspects to be considered when selecting a given sample preparation procedure like: kind and amount of samples, sample matrices, quantities of elements, need of total or partial digestion, instrumental methods for element determinations as well as traceability and uncertainty of measurements. All operations undertaken during sample preparation should be kept under control to properly represent the original status of analyzed samples. The analyst should decide when his method satisfies quality criteria and when obtained results can be accepted at expected probability. The concept of QA and QC is the best way to achieve this goal.

7. References

Abentroth Klaic, P.M.; Nunes, A.M.; da Silveira Moreira, A.; Vendruscolo, C.T. & Ribeirp, A.S. (2011). Determination of Na, K, Ca and Mg in xanthan gum: Sample treatment by acid digestion. *Carbohydrate Polymers*, Vol.83, No.4, (February 2011), pp. 1895-1900, ISSN 0144-8617

Aydin, I. (2008). Comparison of dry, wet and microwave digestion procedures for the determination of chemical elements in wool samples in Turkey using ICP-OES technique. *Microchemical Journal,* Vol.90, No.1, (October 2008), pp. 82-87, ISSN 0026-265X

Balarama Krishna, M.V. & Arunachalam, J. (2004). Ultrasound-assisted extraction procedure for the fast estimation of major, minor and trace elements in lichen and mussel samples by ICP-MS and ICP-AES. *Analytica Chimica Acta,* Vol.522, No.2, (September 2004), pp. 179-187, ISSN 0003-2670

Bugallo, R.A.; Río Segade, S. & Fernández Gómez, E. (2007). Comparison of slurry sampling and microwave-assisted digestion for calcium, magnesium, iron copper and zinc determination in fish tissue samples by flame atomic absorption spectrometry. *Talanta,* Vol.72, No.1, (April 2007), pp. 60-65, ISSN 0039-9140

Cabrera, C.; Lloris, F.; Giménez, R.; Olalla, M. & López, M.C. (2003). Mineral content in legumes and nuts: contribution on Spanish dietary intake. *The Science of the Total Environment,* Vol.308, No.1-3, (June 2003), pp. 1-14, ISSN 0048-9697

Cava-Montesinos, P.; Cervera, M.L.; Pastor, A. & de la Guardia, M. (2004). Determination of As, Sb, Se, Te and Bi in milk by slurry sampling hydride generation atomic fluorescence spectrometry. *Talanta,* Vol.62, No.1, (January 2004), pp. 175-184, ISSN 0039-9140

da Silva E.G.P.; Hatje, V.; dos Santos, W.N.L.; Costa, L.M.; Nogueira, A.R.A. & Ferreira, S.L.C. (2008). Fast method for the determination of copper, manganese and iron in seafood samples. *Journal of Food Composition and Analysis,* Vol.21, No.3, (May 2008), pp. 259-263, ISSN 0889-1575

Demirel, S.; Tuzen, M.; Saracoglu, S. & Soylak, M. (2008). Evaluation of various digestion procedures for trace element contents of some food materials. *Journal of Hazardous Materials,* Vol.152, No.3, (April 2008), pp. 1020-1026, ISSN 0304-3894

Detcheva, A. & Grobecker, K.H. (2006). Determination of Hg, Cd, Mn, Pb and Sn in seafood by solid sampling Zeeman atomic absorption spectrometry. *Spectrochimica Acta Part B: Atomic Spectroscopy,* Vol.61, No.4, (April 2006), pp. 454-459, ISSN 0584-8547

do Socorro Vale, M.; Lopes, G.S. & Gouveia, S.T. (2009). The development of a digestion procedure for the determination of metals in gum obtained from deposits in internal combustion engines by ICP-OES. *Fuel,* Vol.88, No.10, (October 2009), pp. 1955-1960, ISSN 0016-2361

Du Laing, G.; Tack, F.M.G. & Verloo, M.G. (2003). Performance of selected destruction methods for the determination of heavy metals in reed plants (*Phragmites australis*). *Analytica Chimica Acta,* Vol.497, No.1-2, (November 2003), pp. 191-198, ISSN 0003-2670

El-Hadri, F.; Morales-Rubio, A. & de la Guardia, M. (2007). Determination of total arsenic in soft drinks by hydride generation atomic fluorescence spectrometry. *Food Chemistry,* Vol.105, No.3, pp. 1195-1200, ISSN 0308-8146

Gomez, M.R.; Cerutti, S.; Sombra, L.L.; Silva, M.F. & Martinez, L.D. (2007). Determination of heavy metals for the quality control in argentinian herbal medicines by ETAAS and ICP-OES. *Food and Chemical Toxicology,* Vol.45, No.6, (June 2007), pp. 1060-1064, ISSN 0278-6915

Grembecka, M.; Malinowska, E. & Szefer, P. (2007). Differentiation of market coffee and its infusions in view of their mineral composition. *The Science of the Total Environment,* Vol.383, No.1-3, (September 2007), pp. 59-69, ISSN 0048-9697

Hoenig, M. & de Kersabiec, A-M. (1996). Sample preparation steps for analysis by atomic spectroscopy methods: present status. *Spectrochimica Acta Part B: Atomic Spectroscopy,* Vol.51, No.11, (September 1996), pp. 1297-1307, ISSN 0584-8547

Hoenig, M. (2001). Preparation steps in environmental trace element analysis - facts and traps. *Talanta,* Vol.54, No.6, (July 2001), pp. 1021-1038, ISSN 0039-9140

Hseu, Z.-Y. (2004). Evaluating heavy metal contents in nine composts using four digestion methods. *Bioresource Technology,* Vol.95, No.1, (October 2004), pp. 53-59, ISSN 0960-8524

Karadjova, I.B.; Lampugnani, L.; Onor, M.; D'Ulivo, A. & Tsalev, D.L. (2005). Continuous flow hydride generation-atomic fluorescence spectrometric determination and speciation of arsenic in wine. *Spectrochimica Acta Part B: Atomic Spectroscopy,* Vol.60, No.6, (July 2005), pp. 816-823, ISSN 0584-8547

Kira, C.S. & Maihara, V.A. (2007). Determination of major and minor elements un dairy products through inductively coupled plasma optical emission spectrometry after wet partial digestion and neutron activation analysis. *Food Chemistry,* Vol.100, No.1, pp. 390-395, ISSN 0308-8146

Kowalewska, Z.; Ruszczyńska, A. & Bulaska, E. (2005). Cu detrmination in crude oil distillation products by atomic absorption and inductively coupled plasma mass spectrometry after analyte transfer to aqueous solution. *Spectrochimica Acta Part B: Atomic Spectroscopy,* Vol.60, No.3, (March 2005), pp. 351-359, ISSN 0584-8547

Lodenius, M. & Tulisalo, E. (1995). Open digestion of some plant and fungus materials for mercury analysis using different temperatures and sample size. *The Science of the Total Environment,* Vol.176, No.1-3, (December 1995), pp. 81-84, ISSN 0048-9697

Maduro, C.; Vale, G.; Alves, S.; Galesio, M.; Gomes da Silva, M.D.R.; Fernandez, C.; Catarino, S.; Rivas, M.G.; Mota, A.M. & Capelo, J.L. (2006). Determination of Cd and Pb in biological reference materials by electrothermal atomic absorption spectrometry: A comparison of three ultrasonic-based sample treatment procedures. *Talanta,* Vol.68, No.4, (February 2006), pp. 1156-1161, ISSN 0039-914

Matos-Reyes, M.N.; Cervera, M.L.; Campos, R.C. & de la Guardia, M. (2010). Total content of As, Sb, Se, Te and Bi in Spanish vegetables, cereals and pulses and estimation of the contribution of these foods to the Mediterranean daily intake of trace elements. *Food Chemistry,* Vol.122, No.1, pp. 188-194, (September 2010), ISSN 0308-8146

Matusiewicz, H. & Mikołajczak, M. (2001). Determination of As, Sb, Se, Sn and Hg in beer and wort by direct hydride generation sample introduction-electrothermal AAS. *Journal of Analytical Atomic Spectrometry,* Vol.16, No.6, (May 2001), pp. 652-657, ISSN 1364-5544

Matusiewicz, H. & Ślachciński, M. (2007). Simultaneous determination of hydride forming (As, Bi, Ge, Sb, Se, Sn) and Hg and non-hydride forming (Ca, Fe, Mg, Mn, Zn) elements in sonicate slurries of analytical samples by microwave induced plasma optical emission spectrometry with dual-mode sample introduction system. *Microchemical Journal,* Vol.86, No.1, (June 2007), pp. 102-111, ISSN 0026-265X

Momen, A.A.; Zachariadis, G.A.; Anthemidis, A.N. & Stratis, J.A. (2007). Use of fractional factorial design for optimization of digestion procedures followed by multi-element determination of essential and non-essential elements in nuts using ICP-OES technique. *Talanta,* Vol.149, No.1, (January 2007), pp. 443-451, ISSN 0039-9140

Nascentes, C.C.; Korn, M. & Arruda, M.A.Z. (2001). A fast ultrasound-assisted extraction of Ca, Mg, Mn and Zn from vegetables. *Microchemical Journal,* Vol.69, No.1, (May 2001), pp. 37-43, ISSN 0026-265X

Oleszczuk, N.; Castro, J.T.; da Silva, M.M.; Korn, M.G.A., Welz, B. & Vale, M.G.R. (2007). Method development for the determination of manganese, cobalt and copper in green coffee comparing direct solid sampling electrothermal atomic absorption spectrometry and inductively coupled plasma optical emission spectrometry. *Talanta,* Vol.73, No.5, (October 2007), pp. 862-869, ISSN 0039-9140

Rauf, M.A. & Hanan A. (2009). Quality Assurance Considerations in Chemical Analysis. *Quality Assurance Journal,* Vol.12, No.1, (January 2009), pp. 16-21, ISSN 1099-1786

Ribeiro, A.S.; Vieira, M.A., da Silva, A.F.; Borges, D.L.G.; Welz, B.; Heitmann, U. & Curtius, A.J (2005). Determination of cobalt in biological samples by line-source and high-resolution continuum source graphite furnace atomic absorption spectrometry using solid sampling or alkaline treatment. *Spectrochimica Acta Part B: Atomic Spectroscopy,* Vol.60, No.5, (June 2005), pp. 693-698, ISSN 0584-8547

Sahuquillo, A.; Rauret, G.; Rehnert, A. & Muntau, H. (2003). Solid sample graphite furnace atomic absorption spectroscopy for supporting arsenic determination in sediments following a sequential extraction procedure. *Analytica Chimica Acta,* Vol.476, No.1, (January 2003), pp. 15-24, ISSN 0003-2670

Santos Jr., D.; Krug, F.J., de Godoi Pereira, M. & Korn, M. (2006). Currents on Ultrasound-Assisted Extraction for Sample Preparation and Spectroscopic Analytes Determination. *Applied Spectroscopy Reviews,* Vol.41, No.3, pp. 305-321, ISSN 1520-569X

Sneddon, J.; Hardaway, C.; Bobbadi, K.K. & Reddy, A.K. (2006). Sample Preparation of Solid Samples for Metal Determination by Atomic Spectroscopy-An Overview and Selected Recent Applications. *Applied Spectroscopy Reviews,* Vol.41, No.1, pp. 1-14, ISSN 1520-569X

Szymczycha-Madeja, A. & Mulak, W. (2009). Comparison of various digestion procedures in chemical analysis of spent hydrodesulfurization catalyst. *Journal of Hazardous Materials,* Vol.164, No.2-3, (May 2009), pp. 776-780, ISSN 0304-3894

Tašev, K.; Karadjova, I. & Stafilov, T. (2005). Determination of Inorganic and Total Arsenic in Wines by Hydride Generation Atomic Absorption Spectrometry. *Microchimica Acta,* Vol.149, No.1-2, (February 2005), pp. 55-60, ISSN 0026-3672

Türkmen, M. & Ciminli, C. (2007). Determination of metals in fish and mussel species by inductively coupled plasma-atomic emission spectrometry. *Food Chemistry,* Vol.103, No.2, pp. 670-675, ISSN 0308-8146

Vale, M.G.R.; Oleszczuk, N. & dos Santos, W.N.L. (2006). Current Status of Direct Solid Sampling for Electrothermal Atomic Absorption Spectrometry-A Critical Review of the Development between 1995 and 2005. *Applied Spectroscopy Reviews,* Vol.41, No.4, pp. 377-400, ISSN 1520-569X

Vassileva, E.; Dočekalová, H.; Baeten, H.; Vanhentenrijk, S. & Hoenig, M. (2001). Revisitation of mineralization modes for arsenic and selenium determinations in environmental samples. *Talanta,* Vol.54, No.1, (March 2001), pp. 187-196, ISSN 0039-9140

5

Aspects of Quality and Project Management in Analyses of Large Scale Sequencing Data

Björn M. von Reumont, Sandra Meid and Bernhard Misof
Zoologisches Forschungsmuseum Alexander Koenig,
Adenauerallee 160, 53113 Bonn,
Germany

1. Introduction

We describe step-by-step the outline of a project, in which the evolutionary history of pancrustaceans (crustaceans and hexpods) was revisited using molecular methods. It was part of a larger program, the 'Deep Metazoan Phylogeny' priority program of the Deutsche Forschungsgemeinschaft (DFG), wich aimed to reconstruct the metazoan tree of life involving more than 30 subprojects. This chapter should be understood as a backbone, that clarifies important points to plan and to conduct projects in molecular biology, also using next generation sequencing data. The text is divided in four parts: 1) theoretical aspects to projects in molecular biology, 2) the process from the collection of material in the field to the final sequencing, 3) the process from the sequence to the reconstructed topology with a special emphasis on data quality, and 4) the conclusions to prevent pitfalls.

1.1 Fascination and complexity of molecular evolutionary biology

Working in molecular evolution to reconstruct the evolutionary history of organisms is a very fascinating, but also very complex issue. Per definition evolutionary biology, and respectively molecular evolutionary biology, is the division in science, which overlaps and intersects mostly with other areas of natural sciences, like chemistry, physics, informatics, mathematics, bioinformatics, geography but also philosophy and history. Exactly that complexity and intersection creates the fascination and addiction of many scientists to work in that area.

Being on field excursions and collecting specimens in their natural habitats is like travelling back in time into the century and time of classic field biology, geography and history. If once the laboratory part has started, technicial and laboratory skills are demanded, while in parallel the amount of characterized sequences starts to force one to become a sophisticated software user, partly applying bioinformatics knowledge or (the often much faster alternative) cooperating with bioinformaticians. The analyses, interpretation and discussion of the results represent the climax of the project by some (at least) publications in highly respected journals.

1.2 General management strategies applicable for scientific projects in molecular evolution

In general, scientists are highly educated in their specific disciplines, but are often 'freshmen' in managing projects with all involved aspects.

These eventually less developed soft skills can cause an underestimation of possible volume of work and subsequently lead to a massive lack of time, which finally degrades the results and the quality of the scientific project. A rigorous project management as conducted in economics featuring a global, yet detailed intersected time schedule with 'milestones' as anchor points and deadlines (including buffer-time in reserve) as general frame in a project roadmap is mandatory for a solid project. The 'golden triangle' of project management (e.g. Kerzner, 2009; Litke et al., 2010) illustrates interrelations that affect projects and their quality management: A) goals and qualitative results, B) planned time schedule and C) calculated costs. If one edge of that triangle becomes delicate, all could be at risk, and the quality of the project is affected (see figure 1).

Fig. 1. The golden triangle of project management adapted to molecular projects. The red arrows indicate where the points written outside the second (red) triangle have most impact. However, some points have an impact on more than just one edge. Laboratory difficulties for example cost primarily time, but also stress the budget. If things go wrong (and mostly they unfortunately follow the law of Murphy in the scientific business) goals might also be affected by laboratory difficulties. The core triangle pictures the three main components, which are interwoven. If one edge is affected, the other ones are affected either. A major specification is probably, that A and B generally are more connected with each other in most aspects, while the budget is constant or not directly affected (golden arrows). If e.g. computational analyses of phylogenetic trees do not work or cause difficulties, a delay in the time schedule is created, that primarily affects the results, but not directly the budget

If a larger project is conducted, in which more persons are directly involved or third parties included (e.g. by outsourcing of sequencing to companies, etc.), additional aspects play a veritable role. Who is directly or indirectly involved in and linked to the project? Which interests and influence (negative and positive) have the different persons or parties in the project? All of these involved persons (with different expectations and interests) are stakeholders of the project. In general, a stakeholder analysis in the planning phase is extremely crucial and a standard approach in economics (Weaver, 2007; Freeman, 2010; Litke et al., 2010). Which risks might rise by involved persons? In science, competion between work groups must be considered. Is cooperation possible, which is always to prefer. If no cooperation is feasible, which risks exist subsequently for the project? If third parties are involved by outsourcing of e.g. sequencing, an exact analysis of possible candidate companies and their interests and capability are important (see also additionally paragraph 2.3). Last but not least, if you are a PhD student or postdoc do not forget one very important or even the key stakeholder (Bourne, 2010), the PI or supervisor. What are his interests, which are yours? Is there a risk or conflict you might have to deal with or to solve? What are his expectations? Perhaps an agreement on objectives is necessary. One major factor is an open discussion, regular (scheduled) communication and time for additional, intermediate meetings; also a clearly communicated agreement on objectives avoids difficulties or even disappointment of one or both parties in the project.

The communication strategy is a further key factor (Bourne, 2010), it is important to prevent typical pitfalls like 'just reporting', 'flood of detailed information' and that 'no feedback' is given. See also general principles of communication to transport information (Chapter 1.3.5/1.3.6 in: Wägele, 2005; Bourne, 2010). Communication is quite clearly time consuming, but it pays off. All points of the golden triangle are linked to communication, including budget and quality of results. Communication skills improve the general quality of the project, can save costs and time, and eventually most importantly: control and enhance the motivation of the involved persons.

Several software packages to coordinate communication, interaction and project work exist to provide an effective platform and frame to conduct and coordinate projects. Examples are Teamwork, OpenLab, Italy; Teamlab, Ascensio System (open source); Clarizen (web based); Endeavour software project management, Ezequiel Cuellar (open source). If you are a bioinformatician, the last package might be respectively interesting.

A characteristic of scientific projects is that new open questions and potentially new fields of methodologies are explored. Respectively, if additionally laboratory work is included, the risk to end without any or absolutely unexpected results (latter one might result in the desired nature paper) is part of the scientific business and in general hard to evaluate. That has to be calculated in advance and should be reflected in the time and risk management.

However, there is also a clear difference between projects in economics and science: scientific projects aim in most cases for fundamental and theoretical insights instead for a direct financial benefit of involved parties. Changing and evaluating laboratory methods for example, might be unexpected time consuming, but necessary and can at the same time establish a new state of the art method. Time and space to walk open minded on paths that seem to be ineffective, not suitable or even out of topic at first glance might bring the breakthrough and must be possible. Louis Pasteur (1822-1895) quoted on his accidentally discovery of penicillin, "chance favours the prepared mind", but one condition for this famous quote is, that the scientist needs the (mentally) freedom to meet chance. A too rigid framework and control might hinder that. Contrariwise many scientists focus often too much on details (as being trained for) and loose their track on the overall relations of the project, which provokes a rather high inefficiency. Consequently a compromise between efficiency and creativity/innovation has to be made. This is easy to write, but hard to transfer and to realize, as personally experienced.

2. Project phases from species collection in the field to sequencing

2.1 Collection and fixation of samples in the field – RNAlater or sooner?

Normally, the planned molecular project starts with the extraction of molecules (DNA or RNA) from specimens (see figure 2) and every true biologist will do his very best to collect and preserve these specimens by himself in the field.
If the specimens or the tissue is preserved in Ethanol for DNA based work, 94% (or higher), ethanol p.a. should be used. This is true for every tissue collected in the field. Despite the rumour, that crustaceans are tricky to sequence in the laboratory, because the aggressive enzymes of the exocrine glands rapidly degrade the DNA, this specific experience was never made working with 94% ethanol p.a.. Working with material collected and sent by colleagues, difficulties appeared and could be linked to the quality (not p.a.) or low concentration of ethanol. Especially material of larger, vessel based expeditions, is obviously often stored in ethanol, which has been diluted due to ethanol shortage during the cruise. If you expect to join an expedition, plan enough quantities of 94% ethanol (and you better hide some of the ethanol in case colleagues did not properly calculate their ethanol contingents, they seem to tend to desperate actions in these situations). Storing the samples in -20 °C probably keeps degradation processes at a low level, but fieldtrip cooling is not obligatory to preserve high quality DNA.
However, cooling plays a veritable role, if you have to collect samples in the field for RNA based analyses. RNA as a single stranded molecule can be degraded very fast (and unfortunately very efficiently) by a group of enzymes, called RNAses. These enzymes are nearly omnipresent in our body including e.g. perspiration liquid. They have to be inhibited by cold temperatures or chemicals (or both) to stop RNA degradation. The best procedure to ensure good quality of RNA samples is consequently to collect the specimen and to extract the RNA immediately. Unfortunately this is in most cases not possible in the field. For example, many groups of crustaceans live in remote habitats.
For example, remipedes live in anchialine cave systems (see figure 2, top right picture) and require cave diving expeditions. They were collected by BMvR on the Yucatan peninsula in Mexico. Even the organization of the cooling chain to freeze the samples directly in the field and to ship them to the laboratory for RNA extraction was not possible: logistic companies that could have shipped the samples in time did not ship dry ice due to regulations of the International Air Transport Association (IATA), In general, the dry ice transportation by airplane is not officially authorized and problematic in some countries. Awareness and integration of such eventual logistic problems are eminent for a realistic project plan and time schedule.
Using RNAlater for RNA isolation is one solution to collect specimens. It is a non toxic, non flammable liquid that can be transported everywhere without any problems (even in airplanes) and it preserves RNA at room temperature at least for 5-7 days (Grotzer et al., 2000; product descriptions of e.g. Qiagen, Applied Biosystems) without loss of quality compared to frozen samples (Grotzer et al., 2000; Mutter et al., 2004; Gorokova, 2005). A closed cooling chain is not mandatory. For preservation of microcrustaceans of zooplankton like copepods, up to a month of storage time is possible without any losses of RNA quality if RNAlater is used (Gorokhova, 2005). Own experiences corroborate this study with samples

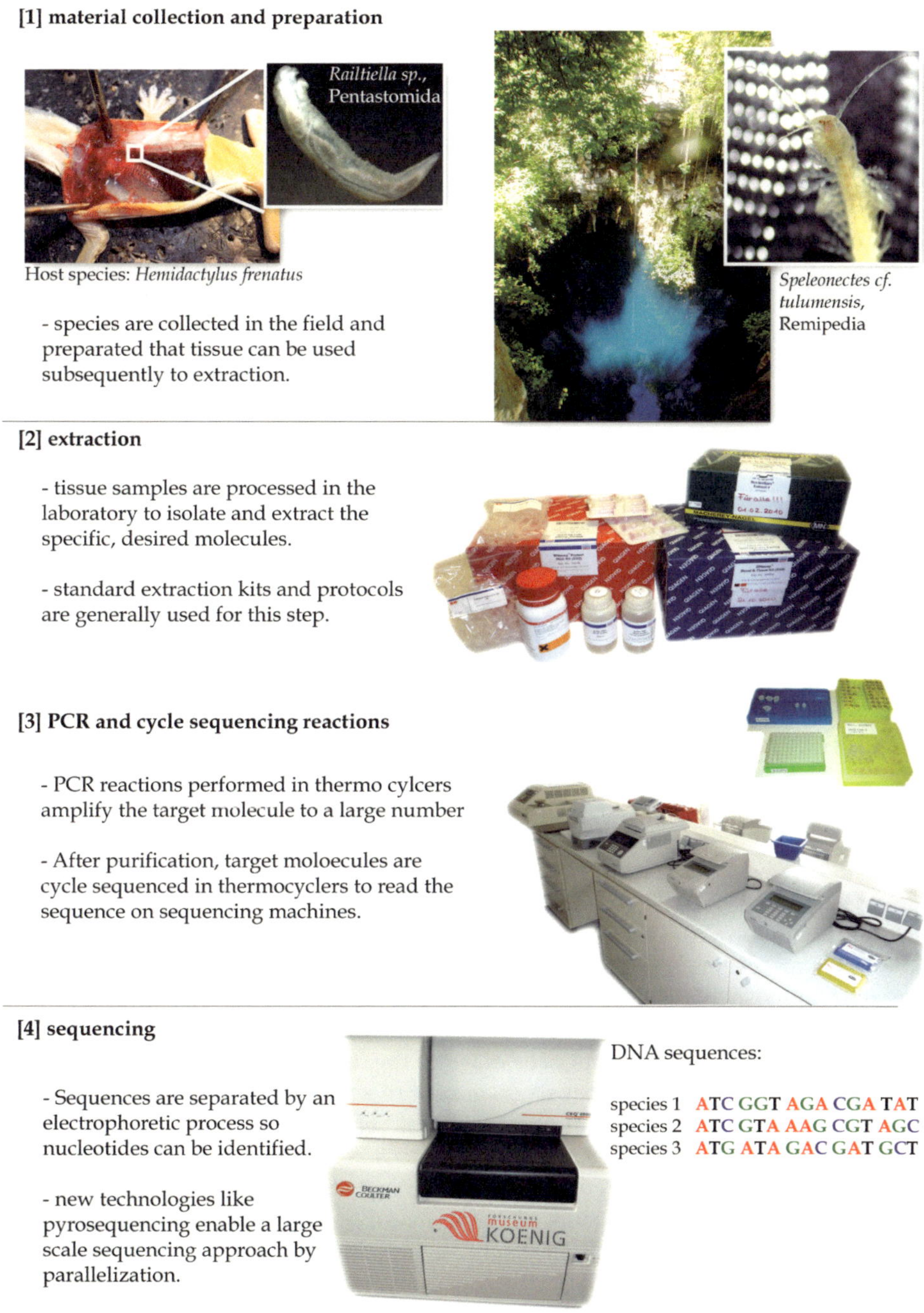

Fig. 2. Overview of the typical phases within a molecular project that start with material collection in the field and end with the final sequences. The two pictures in the left on top [1] show a dissected house gecko (*Hemidactylus frenatus*), which was parasitized by tongue worms (Pentastomida, small picture) in his lunge tract. On the right, a remote anchialine cave system in Mexico is shown. Within these caves live the enigmatic Remipedia (*Speleonectes cf. tulumensis*) that were collected by cave diving

of different sizes like copepods, ostracods, remipedes, and leptostracans, which were stored at room temperature for up to 14 days after collection (including transportation and shipping time). High temperatures may harm the sample quality despite RNAlater preservation, depending on the general temperature conditions of the expedition area. Good experiences were made with standard fridges (about 4°C), they are easy to organize and the sample is cooled, but not frozen.

RNAlater should have room temperature for preservation of tissue samples to enable a thorough penetration, and the liquid should not be cooled before and directly after preservation of material. Before preservation, tissue has to be cut into little fragments, additionally use a pestle (even some smaller crustaceans have a carapace that has to be cracked) to ensure a fast diffusion of the liquid into the tissue. After a few hours or a day, RNAlater can be moderately cooled. If frozen away after one day, a cooling chain must be guaranteed.

For marine organisms a careful sorting or sample preparation is crucial before the preservation of tissue to prevent larger amounts of salt water to dilute and affect the preservation liquid. In general, RNAlater should be sufficiently added to the sample, about 1:5-10 between sample and RNAlater (according to manufacturer protocols) turned out to be insufficient. Even for smaller specimens 15-25 ml tubes were at least used, depending on the collected numbers.

However, contrary to own good experience with RNAlater, other projects using RNAlater to preserve representatives of evolutionary early hexapod lineages report frustrating results, gaining degraded RNA or only very few EST sequences. As stated, the best method has to be tested for each species group. In that special case the best choice was liquid nitrogen, with all subsequent difficulties in the field. An interesting effect is, that RNAlater perfectly preserves DNA (Gorokova, 2005; Vink et al., 2005), which makes it an ideal alternative to ethanol preservation.

The main goal of many projects in molecular biology is the reconstruction of the evolutionary history of species. In this context so called large-scale next generation sequencing approaches have recently been used applying RNA based sequencing (see paragraph 2.3). The approach aims to randomly sequence expressed genes of a specimen when the tissue or specimen was collected and preserved ('transcriptome shot'). One quality criterion to achieve a good coverage of different genes is, how fast the specimen was preserved. If the stress level of the specimen was high, a relatively high level of stress response proteins are the consequence, biasing the quantity but also quality of finally sequenced genes. Always ensure that stress is kept to a minimum level for organisms before preservation to guarantee a maximum number of represented genes. Another important method to achieve a maximum intersection of expressed genes is the collection of different larval and/or development stages of an organism to cover possibly different gene expression patterns. If parasitic forms are sampled, like tongue worms, that parasitize the respiratory tract of vertebrates (Pentastomida, see figure 2 top left picture), a careful preparation of the tissue is necessary to prevent contamination by the host tissue.

Collected specimens should carefully be determined before preservation. Additionally, collected and stored voucher species might enable a second identification after sequencing, if unexpected results or difficulties occur. This specific point is often forgotten. An approach to centralize the storage of voucher specimens and DNA including the linked collection and laboratory data is the DNA bank network (Gemeinholzer et al., 2011). This platform provides an efficient and practical solution to access and exchange data and tissue in an extended form, compared to classical accession sheets like in GenBank. This storage allows a general traceability of DNA sequences, and their quality concerning specimen identification and the DNA itself, like concentration, signal strength, electropherogram etc. In most cases this information is missing in published NCBI data (see figure 4).

2.2 Extracting DNA, RNA and subsequent amplification of the molecules

The extraction of DNA or RNA from tissue follows standard protocols and available kits (e.g. Mülhardt, 2008; Sambrook & Russel, 2000). Eventually it is reasonable, to test different kits and protocols to be time efficient.

A fast and specifically tested method is needed to isolate RNA from tissue. Only few studies mainly from the medical/clinical field are published, which show that quality and quantity of RNA yields are dependent on used preservation/isolation method and extraction kits; additionally both parameters can improve using RNAlater (Forster et al., 2008; Hemmrich et al., 2010, see also Gorokhova, 2005). One serious consideration should be outsourcing of RNA extraction and subsequent sequencing. Time is saved if one party or company provides service from extraction to the final sequences, also in cases of difficulties with the samples.

The PCR method is an established method and several specific adaptations exist to ensure the maximum sensitivity to amplify the desired fragments (e.g. Mülhardt, 2008; Palumbi in: Hillis et al., 1996).

Everyone who works in a molecular lab performing PCR knows that this step is the most sensitive and delicate one for possible contamination. Consequently, a rigorous management should be conducted to maintain high standards in working procedures (Mülhardt, 2008; Sambrook & Russel, 2000). The awareness that contamination can happen despite all efforts is important. If that is considered and influences a general risk management, in consequence all sequences, which are finally included in analyses are blasted in a standard procedure. Exactly this step is the last bastion to guarantee as first step the quality of phylogenetic analyses. If a contamination occurred, the contaminated sequences must be identified and excluded (see figure 4 and paragraph 3.1).

2.3 The sequencing process – A typical case for outsourcing

The term phylogenomics was coined by Eisen (1998) and is recently used for analyses including large scale sequencing data and large numbers of genes derived from cDNA libraries (see also Philippe et al., 2005). A new strategy is the sequencing of the 'transcriptome', which represents the set of expressed genes in an organism, that are encoded by mRNA molecules. Most mRNA molecules are tagged by Poly-A tails and thus easily to fish by specific adaptors if total RNA was isolated. These fished mRNA molecules are reverse transcribed in cDNA and finally libraries are reconstructed that represent ideally all expressed genes in an organism. These mRNA fragments are called expressed sequence tags (ESTs) because of their poly-A tail 'tag' (excellent reviews on that topic: Jongeneel, 2000; Bouck & Vision, 2007). With the new technology of pyrosequencing the possibility arose to directly sequence cDNA molecules in a large scale sequencing approach. Pyrosequencing is not based on the principles of the Sanger sequencing with chain termination reactions, but instead on an enzyme cascade, which generates light if deoxynucleotides are added and pyrophosphate is separated. This difference enables a highly miniaturized and parallelized procedure and technique (see figure 3). For more details see Ronaghi, 2001; Shendure et al., 2004; Ellegren, 2008; Hudson, 2008; Petterson et al., 2009; Voelkerding et al., 2009.

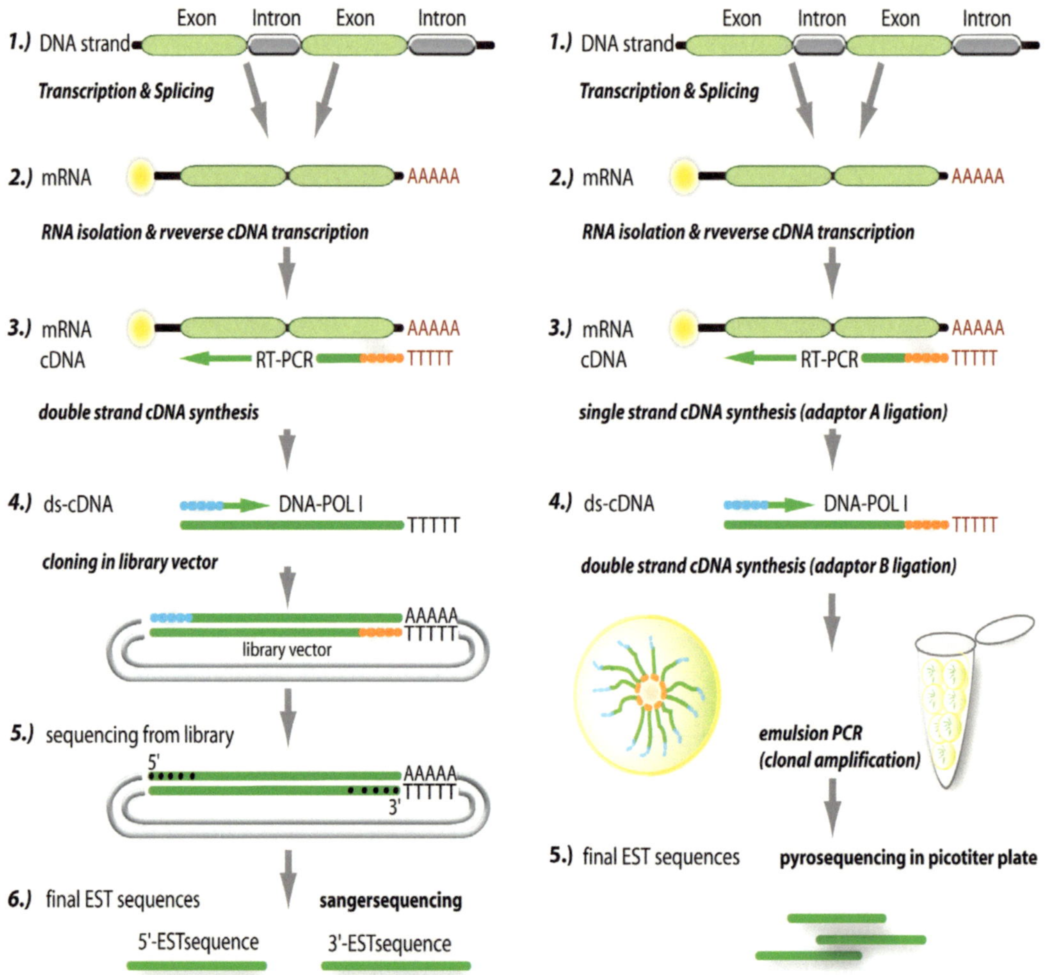

Fig. 3. Differences between standard sanger-sequencing (on the left) and the new pyrosequencing technology (on the right) of next generation sequencing (NGS). Both technologies use mRNA specific target sequences to extract mRNA form the total RNA, which is isolated from tissue. The main difference is that the time and cost intensive step of fragment cloning and sequencing from a subsequently picked library is skipped for pyrosequencing. Depending on the precise technology, double stranded cDNA is generated by an emulsion PCR, in which fragments are amplified in micro compartiments. The sequence fragments are finally transferred on picotiter plates for a massive parallel sequencing

Sequencing is frequently outsourced, which offers a price level that is hard to beat by do-it-yourself sequencing at universities or other research institutions. Focused on large scale or next generation sequencing, some points should be considered. In most companies laboratory procedures and steps are ISO certified ensuring a guaranteed high level of quality and reproducability.

It is a specific quality of molecular biological studies that often unique samples of species with rather unknown evolutionary history are analysed. The collection of these specimens is

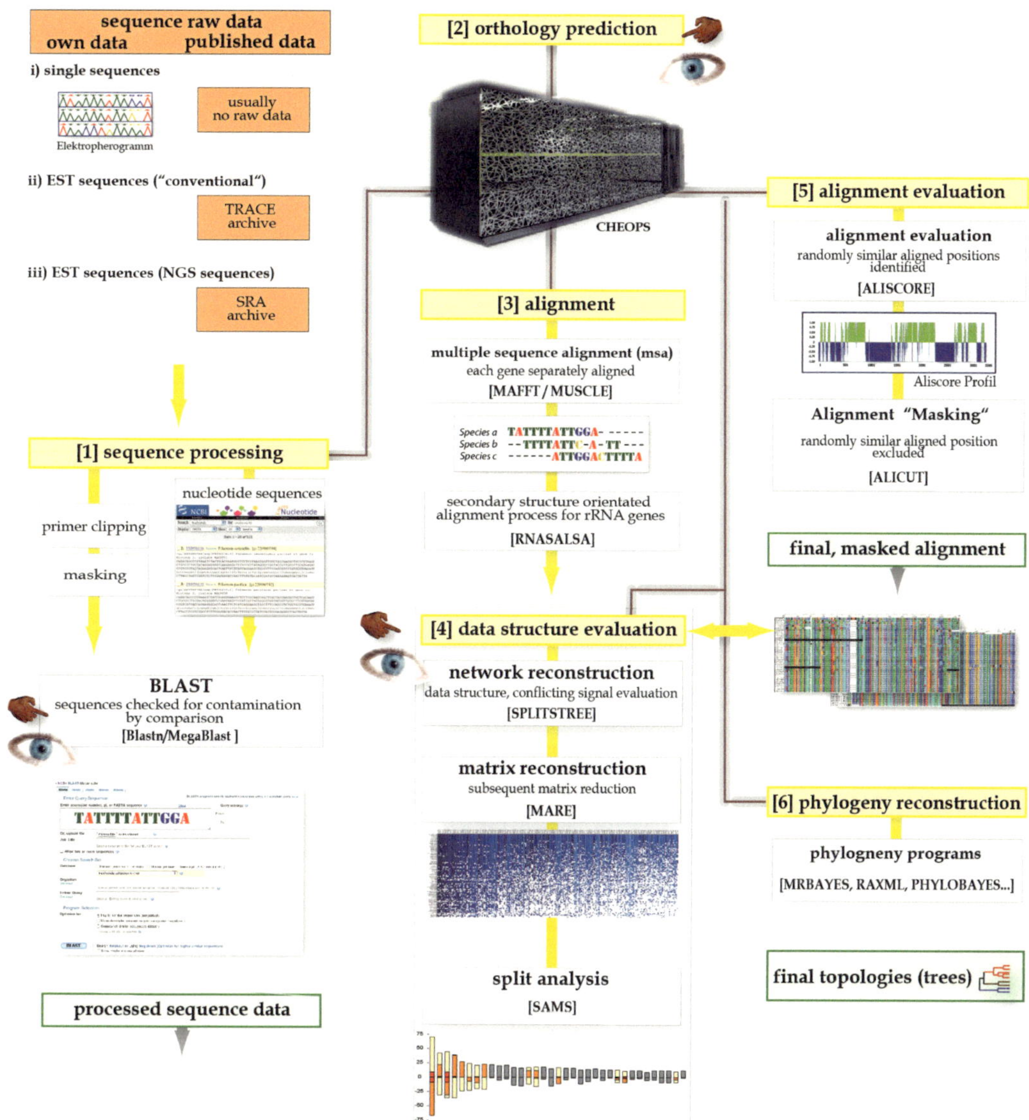

Fig. 4. Working flow of a typical phylogenetic analysis, which starts from scratch with the raw data (gained sequences) and ends with the final topology. Finger and eye symbols pinpoint crucial points to control not only the quality of the process, but also the data quality in the meaning of potential information or conflicts within gene sequences (data structure). A major aspect is, that large scale sequencing and phylogenomic data requires enormous computational power. Supercomputers (in this case CHEOPS: Cologne High Efficiency Operating Platform for Science, RRZK University of Cologne) or large cluster systems (ZFMK Bonn) are an essential requisite in the conducted analyses. Bold bars shaded in grey with internal brown lines symbolize circuit paths and represent steps that are constraint by computational limitations. Own sequence raw data and published data (orange) are processed and quality controlled

often difficult and dependent on single favourable unpredictable conditions. Thus, if anything goes wrong during sequencing, the loss may be irreversible. The second aspect is that samples must not be contaminated by other samples before and after sequencing. If contamination happens, it might not be detectable at all with desastrous consequences. This aspect must be integrated in process flows of sequencing facilities, for example by using tagging techniques applied on each library prior to sequencing to identify immediately eventual contamination. BLAST procedures against other processed project samples or libraries must be a second manadatory strategy.

3. Quality management during molecular analyses

For phylogenomic data the presented figure 4 illustrates only a rough scheme or framework of analysis. Depending on applied techniques and the choice of different software packages an adaptation is needed. Detailed descriptions of the working process to analyse rRNA and phylogenomic data with an emphasis on data quality are given in: von Reumont et al., (2009), von Reumont, (2010) and Meusemann et al., (2010).
[1] Sequences from different sources are processed in software pipelines, quality checked and controlled. It is problematic, that normally electropherograms are not available for published single sequences selected from public databases i). Therefore sequence errors cannot be discovered in these data. ii) EST sequences are normally stored in the TRACE archive in NCBI including the trace files. These represent the raw data and are in general not quality checked. iii) NGS raw data is stored in the Short Read Archive (SRA), which accounts for the difference of sequences from next generation sequencing to the 'conventional' EST sequences. [2] Respectively for the phylogenomic data the prediction of putative ortholog genes is eminent important. This step is computationally intensive and different approaches can be used, see paragraph 3.2. [3] Processed sequence data is aligned applying multiple sequence alignment programs. In case of rRNA genes a secondary structure-based alignment optimization is suggested. [4] A first impression of the data structure is gained by phylogenetic network reconstructions. That point becomes problematic with phylogenomic datasets comprising hundreds of genes and alignment sizes larger than 100 MB! Consequently, a method to evaluate the structure for these datasets could be the software MARE that reconstructs graphics of the data matrix based on the tree-likeness of single genes for each taxon (Misof & Meyer, 2011). Subsequently, a matrix reduction is possible after the alignment evaluation. [5] The final alignment evaluation and processing is applied for each gene with ALISCORE (Misof & Misof, 2009) to identify randomly similar aligned positions and those positions are subsequently excluded (=masking) by ALICUT (www.utilities.zfmk.de). Single, masked alignments are concatenated to the final alignment or supermatrix. A matrix reduction for phylogenomic datasets is performed applying MARE to enlarge the relative informativeness and to exclude genes that are uninformative (Misof & Meyer, 2001; www.mare.zfmk.de). For most analyses it could be useful to compare data structure before and after the alignment process in a network reconstruction or unreduced matrix [4]. Information content in respect of signal that supports different splits in the alignment can be visualized by SAMS (Wägele & Mayer, 2007). [6] After this the phylogenetic tree reconstruction is performed with several software packages.

3.1 The processed sequences and their quality

Most phylogenetic studies use own and published sequences in their analyses. However, in both cases a rigorous control of the quality of the sequence is crucial. This is conducted in the steps of sequence processing (see figure 4, [1]). Different software tools guarantee quality by threshold value settings. A completely different aspect of quality is that the finally included sequence is indeed linked to the supposed species. Either misidentification of the specimen or the sequence can evoke serious bias in a subsequent analysis. If reaction in the laboratory were contaminated, the sequence is linked to the wrong species depending on the source of contamination. Both kinds of misidentification can be identified in general by careful BLAST procedures (Altschul et al., 1997, Kuiken & Corber, 1998). Yet, they are time intensive and in some cases difficult to interpret. For example, if you work with closely related species. In this case, the misidentification or contamination is rather impossible to detect, in particular if one species is unknown or only few or no sequences have been published. Other sources of data (like morphology) can also help to identify contamination (Wiens, 2004).

Several studies report that possible contaminations of taxa played a veritable role in studies, which proposed new evolutionary scenarios, but were actually based on contaminated sequences (von Reumont, 2010; Waegele et al., 2009; Koenemann et al., 2010). A careful control of sequence quality or a more critical interpretation of the reconstructed topologies could have prevented the (eventually repeated) inclusion of the contaminated sequences and subsequent publication of such suspicious phylogenetic trees. If contaminated sequences of older studies from rarely sequenced species are tacitly included into new analyses, this indeed can obscure phylogenetic implications. That is probably the case with the Mystacocarida, a crustacaean group with an still unclear phylogenetic position. They are rarely sequenced and the first and only published 18S rRNA sequence by Spears and Abele (1998) is very likely a contamination (von Reumont, 2010; Koenemann et al., 2010), which was impossible to identify for the authors in that study of 1998, which constituted the first larger analysis of crustaceans at all. A new study with completely sequenced 18S rRNA genes (von Reumont et al., 2009) including a new 18S rRNA gene sequence of the Mystacocarida revealed the contamination of the published sequence (von Reumont, 2010).

The search for contamination reaches a new dimension in phylogenomic data. A recent study (Longo et al., 2011) describes, that some non-primate genome databases, like the NCBI trace archive, provide sequences with human DNA contaminations, which can be traced back to pre-sequencing errors and/or low quality standards. Consequently, cross checking with published data might not help to be 100 percent sure about your own sequences. If you read the last sentence think about your own laboratory routines. Are they sufficient? If you outsource EST sequencing to an external company, which quality standard do they have and which risk management to handle possible contaminations?

This is respectively worrisome in cases of cross species analyses and genome analyses and indicates, that a better screening is generally needed (Phillips, 2011). The response of NCBI was, that trace archive data represents the raw data, which is not quality checked (http://www.ncbi.nlm.nih.gov/About/news/18feb2011.html). A careful processing of these sequences is obligate before analyses, including the control for possible contamination. An important conclusion is that every sequence from public databases should be treated suspiciously and a careful processing procedure is necessary to prevent errors by contamination. Do not trust your own data, but also do not trust public data.

3.2 Orthology prediction

Only homologous genes can be used in molecular phylogenetic studies. Homologous genes are further distinguished in two different classes: i) ortholog genes which originate in a single speciation event, and ii) paralog genes that originated from gene duplications independently of speciation events (Fitch, 1970; Sonnhammer & Koonin, 2002; see review: Koonin, 2005). The prediction of ortholog genes in the era of large scale and next generation sequencing is a very delicate and computationally intensive process. An overview of commonly used methods for prediction of putative ortholog genes and their efficiency assessment is given in Roth et al. (2008) and Altenhoff and Dessimoz (2009).

A difficulty for phylogenetic reconstructions within arthropods is that only few data bases include sufficient numbers of complete arthropod genomes (Altenhoof & Dessimoz, 2009). INPARANOID and OMA are the two leading projects concerning the number of included arthropods. For that reason the orthology prediction for an arthropod dataset (Meusemann et al., 2010; von Reumont, 2010) and a further pancrustacean dataset (von Reumont et al., 2011) were based on INPARANOID 6 and 7 (Ostlund et al., 2010). Identified ortholog gene sets were extended using the HaMStR approach (Ebersberger et al., 2009) relying on the INPARANOID project. A set of orthologous genes was constructed using the InParanoid transitive closure (TC) approach in HaMStR described by Ebersberger et al. (2009). This set based on proteome data of so called 'primer taxa', which are completely sequenced genome species. Sequences of primer taxa were aligned within the set of orthologs and used to infer profile hidden Markov models (pHMMs). Subsequently, the pHMMs were used to search for putative orthologs among the translated ESTs of all taxa in the data set.

For the pancrustacean dataset pre-analyses were performed to compare the influence of using the OMA or INPARANOID projects with the same settings in HaMStR and the previous processing pipeline. For both analyses the same five primer taxa (*Aedes aegypti, Apis mellifera, Daphnia pulex, Ixodes scapulatis, Capitella* sp.) were used in HaMStR to train hidden markov models to extent the putative orthologs for all included taxa. Relying on OMA, 344 putative ortholog genes were identified in contrast to 1886 genes using INPARANOID. The resulting, reduced topologies (RAXML, -f, a, PROTCATWAG, 1000 BS) differ clearly in their resolution: the OMA based topology shows less resolution.

However, these results demonstrate the importance of further, more detailed studies on the impact of ortholog gene prediction. The quality of the trees might be severely influenced in this step of the analysis. A problem is the enormous computational power needed for comparative analysis of phylogenomic datasets.

3.3 Evaluation of data structure and data quality

All steps described so far are important to obtain in a standardized, rigorous processing high quality of the data and finally gene sequences, which are subsequently aligned and used for phylogenetic analyses.

The term *data quality*, however, addresses a different level of quality. A given multiple sequence alignment (MSA, synonymously often named data matrix) can include processed genes that are finally (after the processing procedure) of high quality, but for the phylogenetic goal to reconstruct a specific evolutionary history maybe not usable, if not informative. *Data quality* indeed refers to the scale of information or signal within the alignment. The term *data structure* is sometimes used synonymously to the term *data quality*. Multiple substitution processes generally change sequences with time caused by random substitution processes, however, the extent of substitutions differs for parts of the DNA. In some parts of the DNA this substitution process erodes the former phylogenetic signal by multiple exchanges of nucleotides. After a long time nucleotides that represented synapomorphic characters to a sister taxon are by chance multiple substituted in the process of signal erosion (Wägele & Mayer, 2007). By this process a different, random signal (noise) can arise, that in most cases is in conflict (and obscures) the historical, phylogenetic signal. In contrast, other genes are extremely conservative and nucleotides barely change with time. In this case a phylogenetic signal is hardly to detect either, caused by too few substitutions or synapomorphic characters. The mathematical substitution models, which are applied to reconstruct phylogenetic trees from multiple sequence alignments, try to implement several aspects of the briefly described processes. However, they are always an approximation and respectively are unable to differ between phylogenetic signal and noise. For further details see (Felsenstein, 1988; Wägele, 2005; Wägele & Mayer, 2007).

A first and fast evaluation of the structure in a dataset is feasible with network reconstructions, in which conflicts are visualized that are not illustrated by the (forced) bifurcations in phylogenetic trees (Holland et al., 2004; Huson & Bryant, 2006). It was the first time proposed by Bandelt and Dress (1992) to combine every phylogenetic analysis with a non-approximative method, which allows not compatible, alternative groupings contrary to bifurcting phylogenetic trees. One approach, the method of split decompositon, was developed by Bandelt and Dress (Bandelt & Dress, 1992). Hendy, Penny and Steel published a second method, the split analysis (Hendy & Penny, 1993; Hendy et al., 1994). Both methods work with so called bifurcations or splits.

A split is a couple of two groups of taxa, which are distinct subsets of the whole taxaset. Within the molecular phylogenetic context splits are distinguished by the occurence of nucleotide bases within sites. For a set of n taxa, exist 2^{n-1} possible bipartitions, in real datasets occur normally fewer splits. If there is only split signal for one unique dichotomous tree within a dataset, the number of splits is of the same value as the edges of a possible phylogeny. Given a taxon quartet (A, B), (C, D) few synapomophies between B and C can cause a split for second, alternatively supported topology (A, D) (B, C). This split migth not be visualized in a reconstructed tree-topology. Software packages offering non-approximate methods are SplitsTree (Huson & Bryant, 2006), Spectrum (Charleston, 1998), Spectronet (Huber et al., 2002) and SAMS (Wägele & Mayer, 2007).

SAMS is a software approach that was developed by Wägele and Mayer (2007) to perform a split analysis on the alignment. It accounts for all states of bases but analyses the columns of an alignment for occurring splits in a efficient way. Hence you can generate a split spectrum showing conflicting signal simultaneously obtaining a good overview on the data quality. Real splits are additionally differentiated from the conflicting ones. The method is currently under development, at the moment large datasets are difficult to analyze. Additionally, only nucleotide data is possible as input format. Further development is necessary and in progress to establish a new system, which evaluates all sites of an alignment and weights them according to contrast and homogeneity aspects to address these aspects.

Yet, network reconstruction and split analysis is limited by the size of a dataset and with larger or phylogenomic datasets still beyond abilities of available programs. Additionally, networks give only a rough overview and illustrate the present data structure, answering the question if a conflict or noise exists. More details are often not to analyze, for example which single genes or partitions create a conflict within an alignment. This part becomes additionally delicate handling 'supermatrices' that are composed of phylogenomic data.

Several strategies exist to handle 'supermatrices', which mostly are data sets with a large number of taxa and genes, but also missing information or gaps. Often, concatenated 'supermatrices' are filtered and reduced using predefined thresholds of data availability

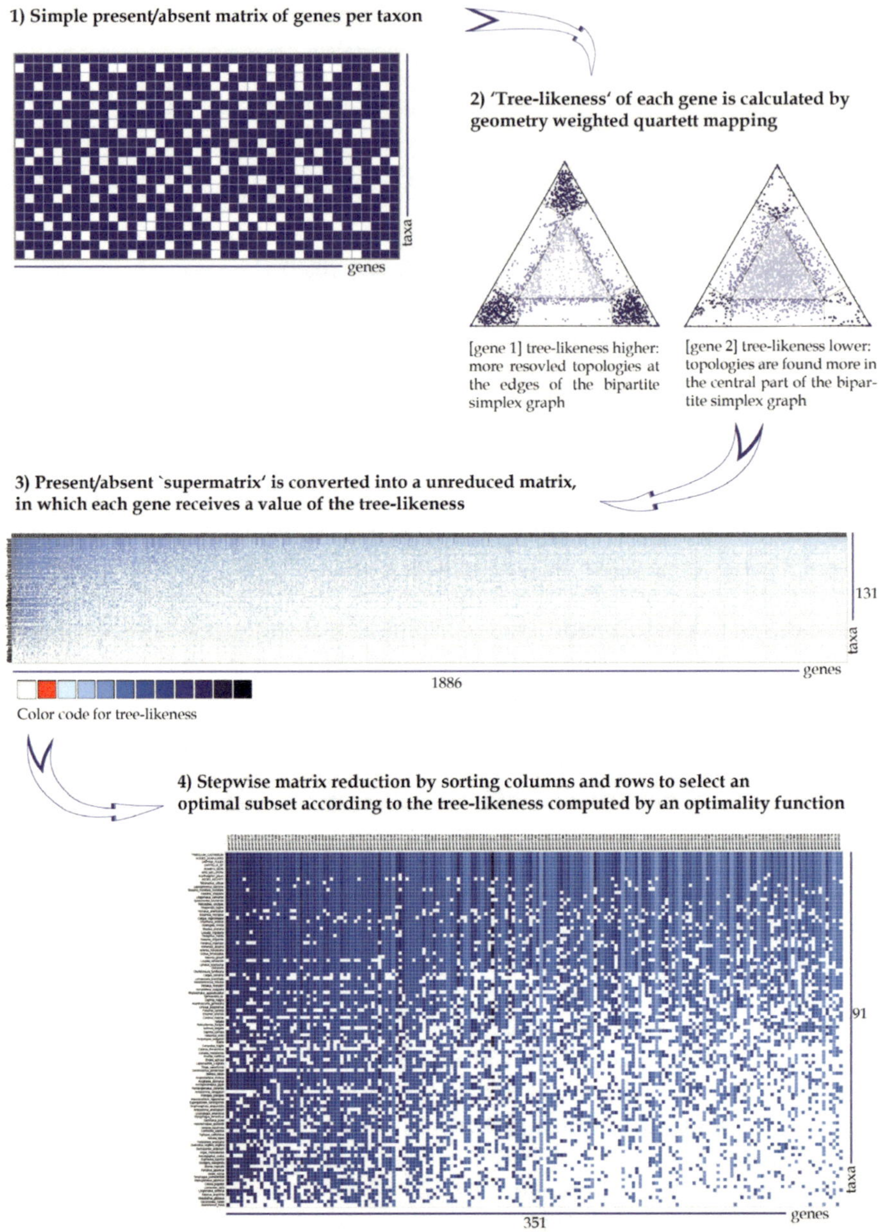

Fig. 5. Work flow of the MARE software. All genes are concatenated to a supermatrix, which is transformed into a `supermatrix' composed of all genes that are represented by tree-likness value. A tree-likeness is calculated in the step before via geometry weighteed quartet mapping. This supermatrix` is reduced by selecting an optimal subset of genes and taxa relying on the calculated value of the tree-likeness. The reduction is stepwise performed using an optimality function. The matrices composed of the tree-likeness values for each gene are colour coded. White symbolizes an absent gene, red a value of 0. From light to dark blue the value increases, dark blue represents a value of 0.9 -1.0

(Dunn et al., 2008; Philippe et al., 2009) depending on the relational number of present genes for a taxon. Taxa are excluded, if they are represented by less genes than accepted with the defined threshold value. Software tools like MARE are a first step to evaluate the data more detailed and enable an objective reduction of 'supermatrices' (large MSA´s of phylogenomic data), by selecting subsets of genes. MARE utilizes an alternative approach to data reduction selecting a subset of genes and taxa from a supermatrix based on information content and data availability (Meyer & Misof, 2010; http://mare.zfmk.de; Meusemann et al., 2010; von Reumont et al., 2011). The approach yields a condensed data set of larger information content by maximizing the ratio of signal to noise, and reducing uninformative genes or poorly sampled taxa.

MARE evaluates in a first step the 'tree-likeness' of each single gene. Tree-likeness reflects the relative number of resolved quartets for all possible (but not more than 20,000) quartets of a given sequence alignment or alignment partitions. The process is based on geometry-weighted quartet mapping (Nieselt-Struwe & von Haeseler, 2001), extended to amino acid data. For each gene a value for the tree-likeness is calculated by summarizing the support values for each of the three possible topologies during the quartet mapping procedure. After this step the previous present/absent matrix is changed to a matrix that contains values of tree-likeness for each gene per taxon. In the second step the matrix reduction is performed. The connectivity of the matrix (the gene and taxa overlap) is monitored during this step: two genes must have connection with at least three taxa. The matrix is reduced stepwise, with each reduction a new matrix is generated. Within each reduction step the column or row with the lowest information content (sum of values for tree-likeness) is excluded. The procedure is guided by an optimality function, which represents a trade off between matrix density and retained taxa and genes. For further details on the procedure and the algorithm, see: (Meyer & Misof, 2011; http://mare.zfmk.de).

4. Conclusions

When conducting or managing a project in molecular evolution use the available elements of project managing to prevent mistakes at this basic level. Important are the time schedule and milestones with sufficient backup time. A careful stakeholder analysis provides a detailed risk analysis, which is important in general, respectively if many persons or working groups are involved. Fieldtrips and appropriate preservation methods of the collected species must be carefully planned either, to start the molecular analysis with qualitative successful isolated material.

A process flow with a rigorous concept of quality control contributes to the quality of the gained sequences or data. The final sequences should have been checked for contamination. If techniques of next generation sequencing or expressed sequence tags are used, pay sufficient attention to select the best strategy for the prediction of ortholog genes. The aligned sequences should always be processed in the multiple sequence alignment for each gene or partition. Software like ALISCORE identifies randomly aligned alignment positions. Before the reconstruction of phylogenetic trees the *data quality* should be evaluated applying software to visualize the data structure and potential conflicts. Software for a more specific split analysis capable of larger data is e.g. SAMS, which is still under development. Assessing the data structure and quality is an essential strategy to identify conflict in phylogenetic trees or their eventual inability to reflect the 'real' evolutionary history of a species group.

Large data matrices or MSAs should be reduced to subsets, which were selected by the tree-likeness of each gene applying the software MARE. The software MARE is a first step to utilize objective criteria to select informative subsets of genes from a partially 'supermatrix'. However, several aspects are still to address further in future. Procedures of orthology prediction and matrix reduction need for example further investigation.

5. Acknowledgement

BMvR and SAM thank J-W Wägele for the chance and support to conduct the projects within the DMP framework. We would like to thank all colleagues who have been involved in the priority program SPP 1174 'Deep Metazoan Phylogeny' of the Deutsche Forschungsgemeinschaft (DFG) and the members of the molecular lab and Zentrum für molekulare Biodiversität (zmb) at the Zooloogischen Forschungsmuseum Alexander Koenig (ZFMK), Bonn. Respectively cooperation with Karen Meusemann was prosperous. Open discussions and exchange of experiences was extremely fruitful in all fields, not only the molecular area. Michael Kube from the Max Planck Institute of Molecular Biology and Genetics, Berlin, Germany gave eminent help and tips for the work with RNA. For detailed explanations and answers regarding the NGS projects we would like to thank colleagues from following companies: GATC, Konstanz, Germany and LGC, Berlin, Germany. The work for this manuscript is granted by the DFG proposals WA530/34, WA530/33.

6. References

Altschul, S. F.; Schäffer, A. A.; Zhang, J.; Zhang, Z.; Miller, W. & Lipman, D. J. (1997). Gapped BLAST and PSI-BLAST: a new generation of protein database search programs, *Nucleic Acids and Research,* 25, 3389-3402

Altenhoff, A. M. & Dessimoz, C. (2009). Phylogenetic and functional assessment of orthologs inference projects and methods, *PLoS Computational Biology,* 5, 1

Bandelt, H. J. & Dress, A. W. (1992). Split decomposition: a new and useful approach to phylogenetic analysis of distance data. *Molecular Phylogenetics and Evolution* 1:242-252.

Bouck, A. & Vision, T. (2007). The molecular ecologist's guide to expressed sequence tags. *Molecular Ecology,* 16, 907-924

Bourne, L. (2010). Beyond reporting. The communication strategy, *PMI Global Congress Proceedings,* Melbourne, Australia

Budd, G.E & Telford, M.J. (2009). The origin and evolution of arthropods, *Nature,* 457, pp. 812-817

Charleston M. (1998). Spectrum: spectral analysis of phylogenetic data, *Bioinformatics (Oxford, England)* 14, 1, 98-9

Forster, J.L.; Harkin, V.B.; Graham, D.A. & McCullough, S.J. (2008). The effect of sample type, temperature and RNAlater (TM) on the stability of avian influenza virus RNA, *Journal of Virological Methods,* 149, pp. 190-194

Ebersberger, I.; Strauss, S. & Von Haeseler, A. (2009). HaMStR: profile hidden markov model based search for orthologs in ESTs, *BMC Evolutionary Biology,* 9, 157

Edgecombe, G.D. (2010). Arthropod phylogeny: An overview from the perspectives of morphology, molecular data and the fossil record, *Arthropod Structure and Development,* 39, pp. 74-87

Eisen, J. A. (1998). Phylogenomics: improving functional predictions for uncharacterized genes by evolutionary analysis, *Genome Research,* 8, 163-7

Ellegren, H. (2008). Sequencing goes 454 and takes large-scale genomics into the wild, *Molecular Ecology,* 17, 1629-1631.

Felsenstein, J. (1988). Phylogenies from molecular sequences: inference and reliability. *Annu. Rev. Genet.* 22:521-565.

Fitch, W. M. (1970). Further improvements in the method of testing for evolutionary homology among proteins, *Journal of Molecular Biology,* 49, 1-14.

Freeman, E.R. (2010). *Strategic management: a stakeholder approach.* ISBN 978-0521151740, Cambridge University Press (first published by Pitman Publishing, 1984)

Gemeinholzer, B.; Droege, G.; Zetzsche, H.; Knebelsberger, T.; Raupach, M.; Borsch, T.; Klenk, H.-P.; Haszprunar, G. & Waegele; J.-W. (2011). The DNA Bank Network: the start from a German initiative. Biopreservation and Biobanking. April 2011, 9 (1):51-55, available at http://www.dnabank-network.org

Gorokhova, E. (2005). Effects on preservation and storage of microcrustacenas in RNAlater™ on RNA and DNA degradation, *Limnology and Oceanography: Methods,* 3, 143-148

Grotzer, M.A.; Pati, R.; Georger, B.; Eggert, A.; Chou, T.T. & Philips, P.C. (2000), Biological stability of RNA isolated from RNAlater™-treated brain tumor and neuroblastoma xenografts, *Medical Pediatric Oncology,* 34:438-442

Hemmrich, K.; Denecke, B.; Paul, N.E.; Hoffmeister, D. & Pallua, N., (2010). RNA Isolation from Adipose Tissue: An Optimized Procedure for High RNA Yield and Integrity, *Labmedicine,* 41 (2), pp 104-106

Hendy, M. & Penny, D., (1993). Spectral analysis of phylogenetic data. Journal of Classification, 10, 1, 5-24

Hendy, M., Penny, D. & Steel, M., (1994). A discrete Fourier analysis for evolutionary trees. Proceedings of the National Academy of Sciences of the United States of America, 91, 8, 3339-43

Holland, B. R.; Huber, K. T.; Moulton, V. & Lockhart, P. J. (2004). Using Consensus Networks to Visualize Contradictory Evidence for Species Phylogeny, *Molecular Biology and Evolution,* 21, 1459-1461

Huber, K, Langton M, Penny D, Moulton V, & Hendy M., (2002). Spectronet: a package for computing spectra and median networks., Applied bioinformatics 1, 3, 159-61

Hudson, M. E., (2008). Sequencing breakthroughs for genomic ecology and evolutionary biology. *Molecular Ecology Resources,* 8, 3-17

Huson, D. H. & Bryant, D. (2006). Application of phylogenetic networks in evolutionary studies, *Molecular Biology and Evolution,* 23, 254-267

Jongeneel, C. V. (2000). Searching the expressed sequence tag (EST) databases: panning for genes. *Briefings in Bioinformatics* 1, 76-92.

Kerzner, H. (2009). *Project management: a systems approach to planning, scheduling and controlling,* ISBN 978-0470278703, John Wiley & Sons, 10th edition

Koenemann, S.; Jenner, R. A.; Hoenemann, M.; Stemme, T. & Von Reumont, B. M. (2010). Arthropod phylogeny revisited, with a focus on crustacean relationships, *Arthropod Structure and Development*, 39, 88-110

Koonin, E. (2005). Orthologs, paralogs and evolutionary genomics, *Annual Reviews of Genetics*, 39, 1, 209-338

Kuiken, C. & Korber, B. (1998). Sequence quality control, Los Alamos National Laboratory *HIV Compendium*, III, pp. 80-90

Litke, H.-D.; Kunow, I. & Schulz-Wimmer, H. (2010). *Projektmanagment*, ISBN 978-3-448-09949-2, Haufe-Lexware GmbH & Co. KG, Freiburg

Longo, M. S.; Longo, M. J.; O'Neill, R. J. & O'Neill (2011). Abundant Human DNA Contamination Identified in Non-Primate Genome Databases, *PLoS ONE*, 6, 2, e16410. doi:10.1371/journal.pone.0016410

Meusemann, K.; Von Reumont, B. M.; Simon, S.; Roeding, F.; Strauss, S.; Kuck, P.; Ebersberger, I.; Walzl, M.; Pass, G.; Breuers, S.; Achter, V.; Von Haeseler, A.; Burmester, T.; Hadrys, H.; Wagele, J. W. & Misof, B. (2010). A phylogenomic approach to resolve the arthropod tree of life. *Molecular Biology and Evolution* 27, 2451-64.

Meyer B. & Misof, B. (2011). MARE: Matrix Reduction - A tool to select optimized data subsets from supermatrices for phylogenetic inference. Zentrum für molekulare Biodiversitätsforschung (zmb) am ZFMK, Adenauerallee 160, 53113 Bonn, Germany, http://mare.zfmk.de

Misof, B. & Misof, K. (2009). A Monte Carlo approach successfully identifies randomness in multiple sequence alignments: a more objective means of data exclusion, *Systematic Biology*, 58, 1

Mülhardt, C. (2008). *Der Experimentator: Molekularbiologie/Genomics*, Spektrum Akademischer Verlag, 6. Auflage. ISBN-10: 9783827420367

Mutter, G.L.; Zahrieh; D., Liu; C.M.; Neuberg, D.; Finkelstein, D.; Baker, H.E. & Warrington, J.A. (2004). Comparison of frozen and RNAlater™ solid tissue storage methods for use in RNA expression microarrays, *BMC Genomics*, 5:88

Nieselt-Struwe K. & Von Haeseler A. (2001). Quartet-mapping, a generalization of the likelihood-mapping procedure. *Molecular Biology and Evolution* 18:1204-1219

Ostlund, G.; Schmitt, T.; Forslund, K.; Köstler, T.; Messina, D. N.; Roopra, S.; Frings, O. & Sonnhammer, E. L. L. (2010). InParanoid 7: new algorithms and tools for eukaryotic orthology analysis, Nucleid Acid Research, 38

Palumbi, S. R. (1996). Nucleic acids II: The Polymerase Chain Reaction, in: Molecular Systematics, Hillis, D. M., Moritz, C., Mable, B. K. 2nd edition, Sinauer Associates, ISBN 978-0878932825

Petterson, E.; Ludneber, J. & Ahmadian, A. (2009). Generations of sequencing technologies, *Genomics*, 93, pp. 105-111

Philippe, H.; Delsuc, F.; Brinkmann, H. & Lartillot, N. (2005). Phylogenomics, *Annual Review of Ecology and Evolutionary Systematics*, 36, 541-562

Philippe H; Derelle R; Lopez P; Pick, K.; Borchiellini, C.; Boury-Esnault, N.; Vacelet, J.; Renard, E.; Houliston, E.; Quéinnec, E.; Da Silva, C.; Wincker, P.; Le Guyader, H.; Leys, S.; Jackson, D. J.; Schreiber, F.; Erpenbeck, D.; Morgenstern, B.; Wörheide, G.

& Manuel, M. (2009). Phylogenomics revives traditional views on deep animal relationships. *Curr Biol.* 19:706-712.

Phillips, M.L. (2011). Contamination of non-primate DNA archives with human sequences indicates that better screening is needed, *nature news*, doi:10.1038/news.2011.99

Ronaghi, M. (2001). Pyrosequencing Sheds Light on DNA Sequencing, *Genome Research,* 11, pp. 3-11

Sambrook, J. & Russel, D. W. (2000). Molecular Cloning: A laboratory manual, 3rd reprint, ISBN 978-0879695774

Shendure, J.; Mitra, R.; Varma, C. & Church, G. (2004). Advanced sequencing technologies: methods and goals, *Nature Reviews in Genetics,* 5, pp. 335-344.

Sonnhammer, E. L. L. & Koonin, E. V. (2002). Orthology, paralogy and proposed classification for paralog subtypes, Trends in Genetics, 18, 12, 619-620

Spears, T. & Abele, L. G. (1998). Crustacean phylogeny inferred from 18S rDNA, In *Arthropod Relationships,* editors: R. A. Fortey and R. H. Thomas, ISBN 978-0412754203, Chapman and Hall, pp. 169-187, London

Thornton, J. W. & Desalle, R. (2000). Gene family evolution and homology: genomics meets phylogenetics, *Annual Reviews of Genomics and Human Genetics,* 1, 41-73

Vink, C.J.; Thomas, S.M.; Paquin, P.; Hayashi, C.Y. & Hedin, M. (2005). The effects of preservatives and temperatures on arachnid DNA, *Invertebrate Systematics,* 19, pp. 99-104

Voelkerding, K. V.; Dames, S. A. & Durtschi, J. D. (2009). Next-Generation Sequencing: From Basic Research to Diagnostics, *Clinical Chemestry,* 55, pp. 641-658

Von Reumont, B. M.; Meusemann, K.; Szucsich, N.; Dell'ampio, E.; Gowri-Shankar, V.; Bartel, D.; Simon, S.; Letsch, H. O.; Stocsits, R. R.; Luan, Y. X.; Wägele, J. W.; Pass, G.; Hadrys, H. & Misof, B. (2009). Can comprehensive background knowledge be incorporated into substitution models to improve phylogenetic analyses? A case study on major arthropod relationships, *BMC Evolutionary Biology* 9, 119.

Von Reumont, B. M. (2010). *Molecular insights to crustaecan phylogeny. A status quo of past, present and perspective prospects also covering phylogenomics,* ISBN 978-3-8381-1770-6, Südwestdeutscher Verlag für Hochschulschriften, Saarbrücken, Germany.

Von Reumont, B. M.; Jenner, R. A.; Wills, M. A.; Dell´Ampio, E.; Pass, G.; Ebersberger, I.; Meusemann, K.; Meyer, B.; Koenemann, S.; Iliffe, T. I.; Stamatakis, A.; Niehuis, O. & Misof, B. (2011). Pancrustacean phylogeny in the light of new phylogenomic data: support for Remipedia as a sister group to Hexapoda, accepted with minor revisions, in re-prep for MBE

Weaver, P. (2007). A Simple View of Complexity in Project Management, *Proceedings of the 4th World Project Management Week, Singapore*

Wiens, J. (2004). The Role of Morphological Data in Phylogeny Reconstruction, *Systematic Biology,* 53, 653-661

Wägele, J.-W. (2005). *Foundations of phylogenetic systematics,* ISBN-13: 9783899370560, Friedrich Pfeil Verlag, München

Wägele. J.-W. & Mayer, C. (2007). Visualizing differences in phylogenetic information content of alignments and distinction of three classes of long-branch effects, *BMC Evolutionary Biology*, 7, 147

Wägele, J. W.; Letsch, H.; Klussmann-Kolb, A.; Mayer, C.; Misof, B. & Wagele, H. (2009). Phylogenetic support values are not necessarily informative: the case of the Serialia hypothesis (a mollusk phylogeny), *Frontiers in Zoology*, 6, 12

6

Gene Markers Representing Stem Cells and Cancer Cells for Quality Control

Shihori Tanabe
National Institute of Health Sciences
Japan

1. Introduction

Populations of cells have unique characteristics and gene markers representative of each cell type, and these features are useful for identifying cell characteristics. For example, the gene expression profile of cells differs at each stage of development and differentiation. This review focuses on gene expression in stem and cancer cells to investigate the possibility of identifying cancer stem cells by such markers.

Cancer stem cells show similarities to normal stem cells in terms of self-renewal and differentiation into multiple lineages. However, cancer stem cells have an indefinite potential for self-renewal that leads to malignant tumorigenesis. The origins of cancer stem cells are not completely clear but accumulation of gene mutations and cell niches are involved in their development. This article describes the gene expression patterns of stem and cancer cells with the aim of determining gene markers for diverse cell types and culture stages for quality control in cellular therapeutics.

2. The microarray quality control (MAQC) projects

Stem cells have varied gene and protein expression profiles and it is important to identify these profiles for quality control in disease treatment, as illnesses such as cancer may cause cell feature changes. The differentiation capacity of stem cells might be altered upon malignancy and there is the possibility that cancer comes from so-called cancer stem cells. Several methods are available to detect cell marker expression, such as surface protein marker detection, intracellular protein marker detection, and gene expression detection. The MAQC project, which is a collaborative effort conducted as part of the US Food and Drug Administration's Clinical Path Initiative for medical product development is useful to detect gene markers in cells (MAQC Consortium, 2006, 2010; Fan et al., 2010; Oberthuer et al., 2010; Huan et al., 2010; Luo et al., 2008; Parry et al., 2010; Shi et al., 2010; Miclaus et al., 2010; Hong et al., 2010; Tillinghast, 2010). It began in February 2005 and aims to describe the reliability and evaluate the performance of microarrays on several platforms.

MAQC-I mainly focuses on the technical aspects of gene expression analysis, whereas MAQC-II focuses on developing accurate and reproducible multivariate gene expression-based prediction models. Possible uses for gene expression data are vast, including diagnosis, early detection (screening), monitoring of disease progression, risk assessment, prognosis, complex medical product characterisation and prediction of responses to treatment (with regard to safety or efficacy) with a drug or device labelling intent.

The MAQC-II data model prediction is dependent upon endpoints, including preclinical toxicity, breast cancer, multiple myeloma and neuroblastoma. Some endpoints are highly predictive based on the nature of the data, and other endpoints are difficult to predict regardless of the model development protocol. Clear differences in proficiency exist between data analysis teams, and such differences are correlated with the level of team experience. The internal validation performance from well-implemented, unbiased cross-validation analyses shows a high degree of concordance with the external validation performance in a strictly blinded process, and many models with similar performance can be developed from a given data set (Table 1).

	MAQC-I	MAQC-II
Aim	To address the concerns about the reliability of microarray techniques	To develop and evaluate accurate and reproducible multivariate gene expression-based predictive model
Summary	The technical performance of microarrays as assessed in the project supports their continued use for gene expression profiling in basic and applied research and may lead to their use as a clinical diagnostic tool as well.	1) Model prediction performance was endpoint dependent. 2) There are clear differences in proficiency between data analysis teams (organisations). 3) The internal validation performance from well-implemented, unbiased cross-validation shows a high degree of concordance with the external validation performance in a strict blinding process. 4) Many models with similar performance can be developed from a given data set. 5) Application of good modelling practices appeared to be more important than the actual choice of a particular algorithm over the others within the same step in the modelling process.
Reference	MAQC Consortium (2006). The MicroArray Quality Control (MAQC) project shows inter- and intraplatform reproducibility of gene expression measurements, Nature Biotechnology, Vol.24, No.9, (September 2006), pp.1151-1161	MAQC Consortium (2010). The MicroArray Quality Control (MAQC)-II study of common practices for the development and validation of microarray-based predictive models, Nature Biotechnology, Vol.28, No.8, (August 2010), pp.827-838

Table 1. The Microarray Quality Control (MAQC) projects

Applying good modelling practice seems to be more important than the actual choice of a particular algorithm over the others within the same step in the modelling process. The order of the analysis process was as follows: design, pilot study or internal validation, and pivotal study or external validation. Observations based on an analysis of the MAQC-II dataset may be applicable to other diseases. (MAQC Consortium, 2010)

3. Gene markers for stem cells

3.1 Cell surface marker genes

The stem cell expression profile varies in differentiated cells. The expression pattern may change depending on differentiation or malignancy of the disease. Endothelial cells in glioblastomas have unique gene expression profiles, and the differences between glioblastomas and lower grade gliomas suggest a more complex ontogeny of the glioblastoma endothelium (Wang et al., 2010). Quantitative *in situ* hybridisation analyses have revealed that fluorescence-activated cell-sorted CD105+ (one of the human endothelial markers) cells with more than 3 copies of the epidermal growth factor receptor (EGFR) amplicon or the centromeric portion of chromosome 7 are similar to the proportion of tumour cells with similar aberrations. CD133 is a cell surface glycoprotein, which has been used as a possible cancer stem cell marker. CD133 is also expressed in haematopoietic stem cells.

3.2 Genes for mesenchymal stem cells

3.2.1 Genes expressed in mesenchymal stem cells

CD29, CD44, CD49a–f, CD51, CD54, CD71, CD73, CD90, CD105, CD106, CD166, Stro-1 and MHC class I molecules are positively expressed in human bone marrow derived mesenchymal stem cells (MSCs), whereas CD11b, CD14, CD18, CD19, CD31, CD34, CD40, CD45, CD56, CD79α, CD80, CD86 and HLA-DR are not (Chamberlain et al., 2007; Kuroda et al., 2010; Pittenger et al., 1999; Kumar et al., 2008; Tsai et al., 2007) (Table 2). Specific markers for MSCs have not been identified. A combination of gene markers may be important to characterise the features of MSCs.

3.2.2 Genes representing the mesenchymal stem cell culture stage

MSCs are often used for treating graft-versus-host disease (GVHD) (Weng et al., 2010; Le Blanc et al., 2008), suggesting that an infusion of MSCs may be an effective therapy for patients with steroid-resistant acute GVHD. The necdin homologue (mouse) (*NDN*), EPH receptor A5 (*EPHA5*), nephroblastoma overexpressed gene (*NOV*) and runt-related transcription factor 2 (*RUNX2*) are possible markers to describe culture status, including growth capacity and differentiation (Tanabe et al., 2008). *EPHA5* and *NOV* are upregulated in late culture stage of human MSCs, whereas *NDN* and *RUNX2* are downregulated (GEO series, Tanabe et al., 2008, accession GSE7637 and GSE7888).

NOV expression in prostate cancer tends to be involved in cancer conditions, based on human prostate cancer gene expression data (Best et al., 2005). This expression is upregulated in androgen-independent primary human prostate cancer compared to untreated human prostate cancer (GEO series, Best, 2005, accession GSE2443). *NOV* might be a candidate marker for identifying the cancer state.

Human MSCs have been reported to promote growth of osteosarcomas, a common primary malignant bone tumour (Bian et al., 2010). In addition, interleukin-6 plays an important role in maintaining the 'stemness' of human MSCs and the proliferation of Saos-2. It is possible that the secretion of interleukin-6 and interaction of human MSCs and Saos-2 through interleukin-6 are essential for their proliferation; this suggests that humoural factors participate in stem cell development.

	positive	negative
Human MSC	CD29, CD44, CD49a–f, CD51, CD54, CD71, CD73, CD90, CD105, CD106, CD166, Stro-1 and MHC class I molecules	CD11b, CD14, CD18, CD19, CD31, CD34, CD40, CD45, CD56, CD79α, CD80, CD86, HLA-DR

Table 2. Positive and negative human mesenchymal stem cell genes

3.3 Marker genes in neural stem cells

Glial fibrillary acidic protein (GFAP), Musashi, nestin, excitatory amino acid transporter 1 (GLAST), PDGFR-α and CD133 are known to be expressed in neural stem cells and are generally used as their markers (Yadirgi & Marino, 2009; Jackson & Alvarez-Buylla, 2008; Gage, F.H., 2000). The expression of nestin, Dlx2, DVC, PSA-NCAM and βIII tubulin has been reported to be altered during cell development. Another report has shown that Sox2, which is believed to be a marker of the nervous system, is expressed in embryonic neural stem cells and other multipotent cells, and that it plays an essential role in mouse brain neurogenesis (Ferri et al., 2004).

Recently, it had been revealed that glia have the ability to act as stem cells (Robel et al., 2011). In this study, GFAP, vimentin, nestin, tenascin-C (TNC) and brain lipid-binding protein (BLBP) are described as immature markers; whereas, glycogen granules, glutamine synthetase, S100β, GLAST and excitatory amino acid transporter 2 (GLT1), which is involved in glutamate uptake and conversion to glutamine, are indicated to be common glial markers. GLAST and GLT1 are shared markers in astrocytes, radial glial in the developing central nervous system, and neural stem cells in the adult mammalian brain. GFAP, S100β and aldehyde dehydrogenase 1 family, member L1 (ALDL1H1) are expressed in both astrocytes and neural stem cells.

4. Genes for reprogramming stem cells

Recent findings have indicated that a set of 4 genes such as POU class 5 homeobox 1 (*POU5F1*, *OCT3/4*), sex-determining region Y-box 2 (*SOX2*), Kruppel-like factor 4 (gut) (*KLF4*) and v-myc myelocytomatosis viral oncogene homologue (avian) (*MYC*, *c-Myc*) induce fibroblast reprogramming into pluripotent stem cells (Takahashi et al., 2007; Park et al., 2008; Lowry et al., 2008). After determining that reprogramming genes actually exist to manipulate and modify human cells, these 4 genes, or some other set of genes such as *POU5F1* (*OCT4*), *SOX2* and *KLF4* have been used globally to produce induced pluripotent stem cells.

Recently, somatic cells have been suggested to be directly reprogrammed without an induced pluripotent stem cell-mediated pathway but with culture condition modifications (Han et al, 2011). In that study, fibroblasts were infected with retrovirus expressing *Oct4*, *Sox2*, *Klf4* and *c-Myc*, and directly induced into epiblast stem cells by adding basic fibroblast growth factor. The combination of gene expression and factors from outside the cells may play important roles in reprogramming cells.

4.1 Genes for generation of induced pluripotent stem (iPS) cells

Recently, it had been reported that *OCT4* is sufficient to induce alterations in the human keratinocyte differentiation pathway (Racila et al., 2011). Transfection of *OCT4*, using a plasmid, into human skin keratinocytes resulted in exhibited expression of endogenous embryonic genes and reduced genomic methylation. These *OCT4*-transfected cells could become neuronal and mesenchymal cell types. The cells have been shown to have characteristics of cultured smooth muscle or myofibroblast cells from a mesenchymal stem cell lineage. It is probable that partial reprogramming using several genes can induce transitions in cell phenotypes and features; hence, complete reprogramming of somatic cells into iPS cells would not always be required for the application of these cells in clinical therapy.

The characterization of human iPS cells, with respect to pluripotency and the ability for terminal differentiation, has been performed with 16 iPS cell lines (Boulting et al., 2011). This study revealed that all iPS cell lines examined, reprogrammed with OCT4, SOX2 and KLF4, or OCT4, SOX2, KLF4 and c-MYC showed the capacity to function as functional motor neurons after differentiation, although there was some variation in the expression of early pluripotency markers and the transgenes. iPS cell lines have been shown to express pluripotency markers, such as NANOG, OCT4, SSEA3, SSEA4, TRA-1-60 and TRA-1-81.

4.2 Involvement of genome structure in reprogramming to iPS cells

Copy number variation has been reported to be involved in the reprogramming to pluripotency (Hussein et al., 2011). The comparison of copy number variations of different passages of human iPS cells with their fibroblast cell origins and with human embryonic stem (ES) cells revealed high copy number variation levels in early-passage human iPS cells. The number of copy number variations in human iPS cell lines decreases with an increase in the number of passage. This decrease during culture passages could be due to DNA repair mechanisms or mosaicism followed by selection. The authors proposed that *de novo* generated copy number variations create mosaicism that is followed by selection of less damaged cells during culturing, because DNA repair is not considered as a sufficient explanation of the rapid decrease in copy number variation.

4.3 Involvement of epigenetic modification and methylation in iPS cells

EMT has been shown to be associated with a stem cell phenotype (Mani et al., 2008; Battula et al., 2010; Polyak & Weinberg, 2009). The tumour suppressor p53 has been suggested to regulate EMT and EMT-associated stem cell properties through transcriptional activation of miRNA (Chang et al., 2011). EMT and the reverse process, the mesenchymal-epithelial transition, are believed to be key elements in the regulation of embryogenesis. It has also been suggested that EMT activation is related to cancer progression and metastasis.

Recently, EMT has been shown to play a role in the acquisition of stem cell properties in normal and neoplastic cell populations. miRNAs are small non-coding RNA molecules and suppress gene expression by interacting with the 3′-untranslated regions (3′ UTRs) of target mRNAs. miRNAs are known to be related to EMT and cancer. The study revealed that p53 activates miR-200c, which is down-regulated in normal stem cell and neoplastic stem cell populations, and suppresses the EMT phenotype and stem cell properties represented in

CD24−CD44+ cell populations. The expression of mesenchymal stem cell markers, such as N-cadherin and ZEB1, has been shown to be suppressed by p53. The mRNA levels of KLF4 and BMI1, which are known as stemness-associated genes and RNA targets of miR-200c and miR-183, have been shown to be regulated by p53.

It has also been reported that the p53R175H mutant up-regulates Twist1 expression and promotes EMT in immortalized prostate cells (Kogan-Sakin et al., 2011). Inactivated or mutated p53 may result in the up-regulation of cell cycle progression genes, such as Twist1, which is a regulator of metastasis and EMT.

4.4 Epithelial-mesenchymal transition (EMT) and microRNAs (miRNAs)

iPS cells have been known to show reprogramming variability such as aberrant reprogramming of DNA methylation (Lister et al., 2010). From whole-genome, single-base-resolution DNA methylomic analyses of iPS cells and ES cells, the authors obtained new evidence showing that iPS cells are methylated during reprogramming, and the methylome of iPS cells generally resembles that of ES cells. In the study, a detailed interpretation of the data indicated that there were many differences in DNA methylation between ES cells and iPS cells. For example, many differentially methylated regions that were differentially methylated in either the iPS cell line or the ES cell line existed in several iPS cell lines.

5. Gene markers for cancer cells

5.1 Regulated genes in renal cancer

NCBI's Gene Expression Omnibus (GEO) database is a useful tool to profile gene expression and search for markers representing cell features (Edgar et al., 2002; Barrett, 2011). Renal tumour samples have been analysed using microarray (Yusenko et al., 2009). It was observed that loss of chromosomes 2, 10, 13, 17 and 21 discriminate chromophobe renal cell carcinomas from renal oncocytomas. These authors suggested that detecting chromosomal changes can be used for an accurate diagnosis in routine histology.

The gene expression profiles of the microarray data deposited in the GEO database (GEO series, Szponar et al., 2009, accession GSE11151) were analysed and caveolin 2 (CAV2), proteasome (prosome, macropain) subunit, beta type, 8 (large multifunctional peptidase 7) (PSMB8) major histocompatibility complexes, class I, F (HLA-F), major histocompatibility complex, class I, B (HLA-B), apoptosis enhancing nuclease (AEN), major histocompatibility complex, class I, G (HLA-G), and tumour necrosis factor receptor superfamily member 10b (TNFRSF10B) are upregulated by more than three-fold in renal tumours (n = 62) compared to normal kidney (n = 5) (Table 3).

The collaborative genome-wide study for renal cell carcinoma using SNP detection techniques has revealed that genome loci on 2p21 and 11q13.3 are genomic regions associated with renal cell carcinoma (Purdue et al., 2011). From this study, EPAS1, encoding hypoxia-inducible-factor-2 alpha at 2p21 and SCARB1, the scavenger receptor class B, member 1 at 12q24.31, were identified as feature genes that have single nucleotide polymorphism mutations in renal cell carcinoma.

5.2 Genes expressed in leukaemia

A model in which human cancers are believed to be generated hierarchically from self-renewing cancer stem cells has been reported. Human acute myeloid leukaemia (AML) is a disease that relates to the model, and AML stem cell-targeting therapy has been developed

(Majeti, 2011; Jin et al., 2006). CD25, CD32, CD44, CD47, CD96, CD123 and CLL-1 are expressed on the surface of AML stem cells. Of these genes, CD44 is suggested to be a cancer stem cell marker.

Gene Symbol	Gene Title	Chromosomal location	Entrez Gene	Function
CAV2	Caveolin-2	Chr7q31.1	858	Protein homodimerisati on activity
PSMB8	Proteasome (prosome, macropain) subunit, beta type, 8 (large multifunctional peptidase 7)	Chr6p21.3	5696	ATP binding, MHC class I protein binding
HLA-F	Major histocompatibility complex, class I, F	Chr6p21.3	3134	MHC class I receptor activity
HLA-B	Major histocompatibility complex, class I, B	Chr6p21.3	3106	MHC class I receptor activity
AEN	Apoptosis enhancing nuclease	Chr15q26.1	64782	Exonuclease activity
HLA-G	Major histocompatibility complex, class I, G	Chr6p21.3	3135	MHC class I receptor activity
TNFRSF10B	Tumour necrosis factor receptor superfamily, member 10b	Chr8p22-p21	8795	TRAIL binding, caspase activator activity

Table 3. Upregulated genes in renal tumours (Ref. GSE11151)

The concept of cancer stem cell is important in explaining cancer development from the viewpoint of stem cells (Clevers, 2011; Wang & Shen, 2011). The cancer stem cells for leukaemia were identified from a study showing that CD34+CD38− fractions of cells derived from acute myeloid leukaemia had the capacity to initiate engraftment in immunodeficient mice (Lapidot et al., 1994; Bonnet & Dick, 1997). It is known that deletion or mutation of *IKZF1* (*IKAROS*), *PAX5*, *EBF1* and *CDKN2A/B* are involved in BCR-ABL1 lymphoblastic leukaemia (Mullighan et al., 2008; Mullighan et al., 2009).

The function of human BCR-ABL1 lymphoblastic leukaemia-initiating cells in human lymphoblastic leukaemia has been studied from the point of view of genome diversity (Notta et al., 2011). Functional and genetic analysis of Philadelphia chromosome acute lymphoblastic leukaemia (Philadelphia-positive [Ph+] ALL) revealed that the frequencies of

genetic alterations in *IKZF1* (84%), *CDKN2A/B* (50%) and *PAX5* (50%) were consistent with those reported in previous studies. Complete deletion of *IKZF1* was observed in both aggressive and non-aggressive groups; whereas, there were differences in the frequencies of deletion of the *CDKN2A/B* and *PAX5* genes, which may provide markers for malignancy.
On the other hand, CD44 has been identified as a key regulator of leukaemic stem cells in AML (Jin et al., 2006). It was suggested that elimination of leukaemic stem cells, cells capable of initiating and maintaining the leukaemic clonal hierarchy, was required for a permanent cure of AML. Hence, stimulation with a CD44-specific antibody has been reported to result in the elimination of leukaemic stem cells.

5.3 Genes expressed in glioblastomas

Wang *et al.* hypothesised that the CD133+ fraction is related to the endothelial differentiation potential and analysed cells from a series of glioblastomas fractionated as follows: (1) CD144+ (endothelial cadherin)/CD133−, (2) CD144+/CD133+, (3) CD133+/CD144−, (4) CD133−/CD144−. The results of quantitative PCR with a reverse transcription analysis demonstrated that VEGFR2 and endothelial progenitor marker CD34 are enriched in the CD144+/CD133− and in the CD144+/CD133+ populations. CD105 was negative in the CD133+ and CD144+ fractions. It was also shown that mice who were administered an injection of CD133+/CD144− or CD133+/CD144+ cells from a primary glioblastoma revealed tumours.
A new mechanism was suggested in which tumour vascularisation occurs through endothelial differentiation of glioblastoma stem-like cells (Ricci-Vitiani et al., 2010). The differentiation of cancer stem-like cells may be involved in cancer malignancy, and it is possible to predict or diagnose the malignant stage of cancer using stem cell markers for quality control.
In a genome-wide association study (GWAS) of four case series on 2,251 test patients and 6,097 control patients of European ancestry, LIM domain only 1 (*LMO1*) at 11p15.4 was found to be associated with neuroblastoma and malignancy (Wang et al., 2011). An integrative genomics study to demonstrate that common genetic polymorphisms associated with cancer tendencies are also related to genomic regions that have possibility of somatic alterations which in turn influence tumour progression, revealed that mutation in *LMO1* may also be a candidate indicator of a malignant phenotype.

5.4 Surface markers for cancer stem cells

Several markers have been reported for identification of cancer stem cells (Clevers, 2011). CD19 as a surface marker for B cell malignancies, CD20 and ATP-binding cassette transporter B5 (ABCB5) for melanoma, and the following molecules for cancer stem cells in the respective cancer type have been reported: CD24 for pancreas/lung cancer, CD34 for hematopoietic malignancies, CD44 for breast/liver/head and neck/pancreas cancer, CD90 for liver cancer, CD133 for brain/colorectal/lung/liver cancer and epithelial cell adhesion molecule (EpCAM)/epithelial-specific antigen (ESA) for colorectal/pancreatic cancer (Ebben et al., 2010).

5.5 Cancer stem cell hypothesis

Cancer stem cells have capacity for self-renewal, which is also the feature to normal stem cells. Cancer stem cells are also capable of generating malignant tumours, and this property may differentiate them from normal stem cells. The origin of the cancer stem cells has not been fully revealed, however, there is a model in which cancer stem cells occur by normal stem cells or normal cells by the accumulation of gene mutations. The process of cancer stem cell derivation is considered to be involved with niche which is microenvironment around normal stem cells.

There are two models to explain tumourigenesis. The first model is stochastic model in which all cells have capacity of tumourigenesis, but the probability to enter into tumourigenesis cell cycle is relatively low. The second model is hierarchy theory in which only small population of cells in cancer has capacity of tumourigenesis and generate tumour with high probability, which lead to cancer stem cell hypothesis.

It is also notable that cancer stem cells are not necessarily related to the cell of origin in a cancer (Visvader, 2011). Although the cell of origin for a particular tumour may have the capacity to differentiate into a mature cell, cancer stem cells have the ability to maintain tumourigenesis according to the cell-of-origin model.

6. Conclusion

The recent development in molecular biology and bioinformatics technology has revealed stem cell features and their candidate marker genes. Gene expression profiles change widely and dramatically with cell development, various culture conditions and disease status. The each cell type has different gene expression profile after being differentiated, and it is known that the expression pattern alters in each disease status. Even though it seems that the stemness has distinct feature in gene expression, the cell population show various gene expression patterns in each cell lineage or even in each subset of the cell.

Until recently, targeting cancer stem cells in cancer therapy was rare because the proportion of these cells in cancer was considered very low and retaining the feature of cancer stem cells in vitro was difficult. The stem cell-targeted therapy including cancer treatment will be expected to progress further in the near future, and the role of markers would become much greater. It is important to know the precise feature and gene expression pattern for quality control in the cell-targeted therapy.

7. Acknowledgments

The author acknowledges Dr. Yoji Sato, Dr. Takayoshi Suzuki, Dr. Kazuhiro Suzuki, Dr. Taku Nagao, Dr. Teruhide Yamaguchi, Dr. Yasuo Ohno, Dr. Eriko Uchida, Dr. Tadashi Oshizawa and Dr. Masahiro Nishijima.

8. References

Barrett, T.; Troup, D.B.; Wilhite, S.E.; Ledoux, P.; Evangelista, C.; Kim, I.F.; Tomashevsky, M.; Marshall, K.A.; Phillippy, K.H.; Sherman, P.M.; Muertter, R.N.; Holko, M.; Ayanbule, O.; Yefanov, A. & Soboleva, A. (2011). NCBI GEO: archive for functional genomics data sets—10 years on, *Nucleic Acids Research*, Vol.39, Database issue, pp.D1005-D1010 (January 2011)

Battula, V.L.; Evans, K.W.; Hollier, B.G.; Shi, Y.; Marini, F.C.; Ayyanan, A.; Wang, R-Y.; Brisken, C.; Guerra, R.; Andreeff, M. & Mani, S.A. (2010). Epithelial-mesenchymal transition-derived cells exhibit multilineage differentiation potential similar to mesenchymal stem cells, *Stem Cells*, Vol.28, No.8, (August 2010), pp.1435-1445

Best, C.J.; Gillespie, J.W.; Yi, Y; Chandramouli, G.V.; Perlmutter, M.A.; Gathright, Y.; Erickson, H.S.; Georgevich, L.; Tangrea, M.A.; Duray, P.H.; González, S.; Velasco, A.; Linehan, W.M.; Matusik, R.J.; Price, D.K.; Figg, W.D.; Emmert-Buck, M.R. & Chuaqui, R.F. (2005). Molecular Alterations in Primary Prostate Cancer after Androgen Ablation Therapy, *Clinical Cancer Research*, Vol.11, No.19, pp.6823-6834 (October 2005)

Bian, Z-Y.; Fan, Q-M.; Li, G.; Xu, W-T. & Tang, T-T. (2010). Human mesenchymal stem cells promote growth of osteosarcoma: Involvement of interleukin-6 in the interaction between human mesenchymal stem cells and Saos-2, *Cancer Science*, Vol.101, No.12, (December 2010), pp.2554-2560

Bonnet, D. & Dick, J.E. (1997). Human acute myeloid leukemia is organized as a hierarchy that originates from a primitive hematopoietic cell, *Nature Medicine*, Vol.7, (July 1997), pp.730-737

Boulting, G.L.; Kiskinins, E.; Croft, G.F.; Amoroso, M.W.; Oakley, D.H.; Wainger, B.J.; Williams, D.J.; Kahler, D.J.; Yamaki, M.; Davidow, L.; Rodolfa, C.T.; Dimos, J.T.; Mikkilineni, S.; MacDermott, A.B.; Woolf, C.J.; Henderson, C.E.; Wichterle, H. & Eggan, K. (2011). A functionally characterized test set of human induced pluripotent stem cells, *Nature Biotechnology*, Vol.29, (March 2011), pp.279-286

Chamberlain, G.; Fox, J.; Ashton, B. & Middleton J. (2007). Differentiation Capacity, Immunological Features, and Potential for Homing, *Stem Cells*, Vol.25, pp.2739-2749 (November 2007)

Chang, C-J.; Chao, C-H.; Xia, W.; Yang, J-Y.; Xiong, Y.; Li, C-W.; Yu, W-H.; Rehman, S.K.; Hsu, J.L.; Lee, H-H.; Liu, M.; Chen, C-T.; Yu, D. & Hung, M-C. (2011). P53 regulates epithelial-mesenchymal transition and stem cell properties through modulating miRNAs, *Nature Cell Biology*, Vol.13, No.3, (March 2011), pp.317-323

Clevers, H. (2011). The cancer stem cell: premises, promises and challenges, *Nature Medicine*, Vol.17, No.3, (March 2011), pp.313-319

Ebben, J.D.; Treisman, D.M.; Zomiak, M.; Kutty, R.G; Clark, P.A. & Kuo, J.S. (2010). The cancer stem cell paradigm: a new understanding of tumor development and treatment, *Expert Opinion on Therapeutic Targets*, Vol.14, No.6, (June 2010), pp.621-632

Edgar, R.; Domrachev, M. & Lash, A.E. (2002). Gene Expression Omnibus: NCBI gene expression and dhybridization array data repository, *Nucleic Acids Research*, Vol.30, pp.207-210 (January 2002)

Fan, X.; Lobenhofer, E.K.; Chen, M.; Shi, W.; Huang J.; Luo, J.; Zhang, J.; Walker, S.J.; Chu, T.M.; Li, L.; Wolfinger, R.; Bao, W.; Paules, R.S; Bushel, P.R.; Li, J.; Shi, T.; Nikolskaya, T.; Nikolsky, Y.; Hong, H.; Deng, Y.; Cheng, Y.; Fang, H.; Shi, L. & Tong, W. (2010). Consistency of predictive signature genes and classifiers generated using different microarray platforms, *Pharmacogenomics Journal*, Vol.10, No.4, pp.247-257 (August 2010)

Ferri, A.L.; Cavallaro, M.; Braida, D.; Di Cristofano, A.; Canta, A.; Vezzani, A.; Ottolenghi, S.; Pandolfi, P.P.; Sala, M.; DeBiasi, S. & Nicolis, S.K. (2004). Sox2 deficiency causes neurodegeneration and impaired neurogenesis in the adult mouse brain, *Development*, Vol.131, No.15, (August 2004), pp.3805-3819

Gage, F.H. (2000). Mammalian neural stem cells, *Science*, Vol.287, (February 2000), pp.1433-1438

Han, D.W.; Greber, B.; Wu, G.; Tapia, N.; Araúzo-Bravo, M.J.; Ko, K.; Bernemann, C.; Stehling, M. & Schöler, H.R (2011). Direct reprogramming of fibroblasts into epiblast stem cells, *Nature Cell Biology*, Vol.13, No.1, pp. 66-71 (January 2011)

Hong, H.; Shi, L.; Su, Z.; Ge, W.; Jones, W.D.; Czika, W.; Miclaus, K.; Lambert, C.G.; Vega, S.C.; Zhang, J.; Ning, B.; Liu, J.; Green, B.; Xu, L.; Fang, H.; Perkins, R.; Lin, S.M.; Jafari, N.; Parl, K.; Ahn, T.; Chierici, M.; Furlanello, C.; Zhang, L.; Wolfinger, R.D.; Goodsaid, F. & Tong, W. (2010). Assessing sources of inconsistencies in genotypes and their effects on genome-wide association studies with HapMap samples, *Pharmacogenomics Journal,* Vol.10, No.4, pp.364-374 (August 2010)

Huan, J.; Shi, W.; Zhang, J.; Chou, J.W.; Paules, R.S.; Gerrush, K.; Li, J.; Luo, J.; Wolfinger, R.D.; Bao, W.; Chu, T.M.; Nikolsky, Y.; Nikolskaya, T.; Dosymbekov, D.; Tsyganova, M.O.; Shi, L.; Fan, X.; Corton, J.C.; Chen, M.; Cheng, Y.; Tong, W.; Fnag, H. & Bushel, P.R. (2010). Genomic indicators in the blood predict drug-induced liver injury, *Pharmacogenomics Journal,* Vol.10, No.4, pp.267-277 (August 2010)

Hussein, S.M.; Batada, N.N.; Vuoristo, S.; Ching, R.W.; Autio, R.; Närvä, E.; Ng, S.; Sourour, M.; Hämäläinen, R.; Olsson, C.; Lundin, K.; Mikkola, M.; Trokovic, R.; Peitz, M.; Brüstle, O.; Bazett-Jones, D.P.; Alitalo, K.; Lahesmaa, R.; Nagy, A. & Otonkoski, R. (2011). Copy number variation and selection during reprogramming to pluripotency, *Nature*, Vol.471, (March 2011), pp.58-62

Jin, L.; Hope, K.J.; Zhani, Q.; Smadja-Joffe, F. & Dicke, J.E. (2006). Targeting of CD44 eradicates human acute myeloid leukemic stem cells, *Nature Medicine*, Vol.12, No.10, pp.1167-1174 (October 2006)

Kogan-Sakin, I.; Tabach, Y.; Buganim, Y.; Molchadsky, A.; Solomon, H.; Madar, S.; Kamer, I.; Stambolsky, P.; Shelly, A.; Goldfinger, N.; Valsesia-Wittman, S.; Puisieux, A.; Zundelevich, A.; Gal-Yam, E.N.; Avivi, C.; Barshack, I.; Brait, M.; Sidransky, D.; Domany, E. & Rotter, V. Mutant p53R175H upregulates Twist1 expression and promotes epithelial-mesenchymal transition in immortalized prostate cells, *Cell Death & Differentiation*, Vol.18, No.2, (February 2011), pp.271-281

Kumar, S.; Chanda, D. & Ponnazhagan, S. (2008). Therapeutic potential of genetically modified mesenchymal stem cells, *Gene Therapy*, Vol.15, pp.711-715 (March 2008)

Kuroda, Y.; Kitada, M.; Wakako, S.; Nishikawa, K.; Tanimura, Y.; Makinoshima, H.; Goda, M.; Akashi, H.; Inutsuka, A.; Niwa, A.; Shigemoto, T.; Nabeshima, Y.; Nakahata, T.; Nabeshima, Y.; Fujiyoshi, Y. & Dezawa, M. (2010). Unique multipotent cells in adult human mesenchymal cell populations, *Proceedings of the National Academy of Sciences of the United States of America*, Vol.107, No.19, pp.8639-8643 (May 2010)

Lapidot, T.; Sirard, C.; Vormoor, J.; Murdoch, B.; Hoang, T.; Caceres-Cortes, J.; Minden, M.; Paterson, B.; Caligiuri, M.A.; Dick, J.E. (1994). A cell initiating human acute myeloid leukaemia after transplantation into SCID mice, *Nature*, Vol.367, (February 1994), pp.645-658

Le Blanc, K.; Frassoni, R.; Ball, L.; Locatelli, F.; Roelofs, H.; Lewis, I.; Lanino, E.; Sundberg, B.; Bernardo, M.E.; Remberger, M.; Dini, G.; Egeler, R.M.; Bacigalupo, A.; Fibbe, W. & Ringdén, O. (2008). Mesenchymal stem cells for treatment of steroid-resistant, severe, acute graft-versus-host disease: a phase II study. *The Lancet*, Vol.371, pp.1579-1586 (May 2008)

Lister, R.; Pelizzola, M.; Kida, Y.S.; Hawkins, R.D.; Nery, J.R.; Hon, G.; Antosiewicz-Bourgets, J. ; O'Malley, R. ; Castanon, R. ; Klugman, S. ; Downes, M. ; Yu, R. ; Stewart, R. ; Ren, B. ; Thomson, J.A. ; Evans, R.M. & Ecker, J.R. (2010). Hotspots of aberrant epigenomic reprogramming in human induced pluripotent stem cells, *Nature*, Vol. 471, No.7336, (March 2011), pp.46-47

Lowry, W.E.; Richter, L.; Yachechko, R.; Pyle, A.D.; Tchieu, J.; Sridharan, R.; Clark, A.T. & Plath, K. (2008). Generation of human induced pluripotent stem cells from dermal fibroblasts, *Proceedings of the National Academy of Sciences of the United States of America*, Vol.105, No.8, pp.2883-2888 (February 2008)

Luo, J.; Schumacher, M.; Scherer, A.; Sanoudou, D.; Megherbi, D.; Davison, T.; Shi, T.; Tong, W.; Shi, L.; Hong, H.; Zhao, C.; Elloumi, F.; Shi, W.; Thomas, R.; Lin, S.; Tillinghast, G.; Liu, G.; Zhou, Y.; Herman, D.; Li, Y.; Deng, Y.; Fnag, H.; Bushel, P.; Woods, M. & Zhang, J. (2010). A comparison of batch effect removal metehods for enhancement of prediction performance using MAQC-II microarray gene expression data, *Pharmacogenomics Journal,* Vol.10, No.4, pp.278-291 (August 2010)

Majeti, R. (2011). Monoclonal antibody therapy directed against human acute myeloid leukemia stem cells, *Oncogene*, Vol.30, No.3, pp.1009-1019 (March 2011)

Mani, S.A.; Guo, W.; Liao, M-J.; Eaton, E.N.; Ayyanan, A.; Zhou, A.Y.; Brooks, M.; Reinhard, F.; Zhang, C.C.; Shipitsin, M.; Campbell, L.L.; Polyak, K.; Brisken, C.; Yang, J. & Weinberg, R.A. (2009). The epithelial-mesenchymal transition generates cells with properties of stem cells, *Cell*, Vol.133, No.4, (May 2008), pp.704-715

MAQC Consortium (2006). The MicroArray Quality Control (MAQC) project shows inter- and intraplatform reproducibility of gene expression measurements *Nature Biotechnology*, Vol.24, No.9, (September 2006), pp.1151-1161

MAQC Consortium (2010). The MicroArray Quality Control (MAQC)-II study of common practices for the development and validation of microarray-based predictive models, *Nature Biotechnology*, Vol.28, No.8, (August 2010), pp.827-838

Miclaus, K.; Wolfinger, R.; Vega, S.; Chierici, M.; Furlanello, C.; Lambert, C.; Hong, H.; Zhang, L.; Yin, S. & Goodsaid, F. (2010). Batch effects in the BRLMM genotype calling algorithm influence GWAS results for the Affymetrix 500K array, *Pharmacogenomics Journal,* Vol.10, No.4, pp.336-346 (August 2010)

Mullighan, C.G.; Miller, C.B.; Radtke, I.; Phillips, L.A.; Dalton, J.; Ma, J.; White, D.; Hughes, T.P.; Le Beau, M.M.; Pui, C-H.; Relling, M.V.; Shurtleff, S.A. & Downing, J.R. (2008). BCR-ABL1 lymphoblastic leukaemia is characterized by the deletion of Ikaros, *Nature*, Vol.453, (May 2008), pp.110-114

Mullighan, C.G.; Su, X.; Zhang, J.; Radtke, I.; Phillips, L.A.A.; Miller, C.B.; Ma, J.; Liu, W.; Cheng, C.; Schulman, B.A.; Harvey, R.C.; Chen, I-M.; Clifford, R.J.; Carroll, W.L.; Reaman, G.; Bowman, W.P.; Devidas, M.; Gerhard, D.S.; Yang, W.; Relling, M.V.; Pharm, D.; Shurtleff, S.A.; Campana, D.; Borowitz, M.J.; Pui, C-H.; Smith, M.; Hunger, S.P.; Willman, C.L.; Downing, J.R. (2009). Deletion of IKZF1 and prognosis in acute lymphoblastic leukemia, *New England Journal of Medicine*, Vol.360, No.5, (January 2009), pp.470-480

Notta, F.; Mullighan, C.G.; Wang, J.C.; Poeppl, A.; Doulatov, S.; Phillips, L.A.; Ma, J.; Minden, M.D.; Downing, J.R. & Dick, J.E. (2011). Evolution of human BCR-ABL1 lymphoblastic leukemia-initiating cells, *Nature*, Vol.469, (January 2011), pp.362-367

Oberthuer, A.; Juraeva, D.; Li, L.; Kahlert, Y.; Westermann, F.; Eils, R.; Berthold, F.; Shi, L.; Wolfinger, R.D.; Fischer, M. & Brors, B. (2010). Comparison of performance of one-color and two-color gene-expression analyses in predicting clinical endpoints of neuroblastoma patients, *Pharmacogenomics Journal,* Vol.10, No.4, pp.258-266 (August 2010)

Park, I.-H.; Zhao, R.; West, J.A.; Yabuuchi, A.; Huo, H.; Ince, T.A.; Lerou, P.H.; Lensch, M.W. & Daley, G.Q. (2008). Reprogramming of human somatic cells to pluripotency with defined factors, Nature, Vol.451, pp.141-146 (January 2008)

Parry, R.M.; Jones, W.; Stokes, T.H.; Phan, J.H.; Moffitt, R.A.; Fnag, H.; Shi, L.; Oberthuer, A.; Fischer, M.; Tong, W. & Wang, M.D. (2010). K-Nearest neighbor models for microarray gene expression analysis and clinical outcome prediction, *Pharmacogenomics Journal,* Vol.10, No.4, pp.292-309 (August 2010)

Pittenger, M.F.; Mackay, A.M.; Beck, S.C.; Jaiswal, R.K.; Dounglas, R.; Mosca, J.D.; Moorman, M.A.; Simonetti, D.W.; Craig, S. & Marshak, D.R. (1999). Multilineage Potential of Adult Human Mesenchymal Stem Cells, *Science,* Vol.184, pp.143-147 (Apr 1999)

Polyak, K. & Weinberg R.A. (2009). Transitions between epithelial and mesenchymal states: acquisition of malignant and stem cell traits, *Nature Reviews Cancer,* Vol.9, No.4, (April 2009), pp.265-273

Purdue, M.P.; Johansson, M.; Zelenika, D.; Toro, J.R.; Scelo, G.; Moore, L.E.; Prokhortchouk, E.; Wu, X.; Kiemeney, L.A.; Gaborieau, V.; Jacobs, K.B.; Chou, W-H.; Zaridze, D.; Matveev, V.; Lubinski, J.; Trubicka, J.; Szeszenia-Dabrowska, N.; Lissowska, J.; Rudnai, P.; Fabianova, E.; Bucur, A.; Bencko, V.; Foretova, L.; Janout, V.; Boffetta, P.; Colt, J.S.; Davis, F.G.; Schwartz, K.L.; Banks, R.E.; Selby, P.J.; Harnden, P.; Berg, C.D.; Hsing, A.W.; Grubb, R.L.III; Boeing, H.; Vineis, P.; Clavel-Chapelon, F.; Palli, D.; Tumino, R.; Krogh, V.; Panico, S.; Duell, E.J.; Quirós, J.R.; Sanchez, M-J.; Navarro, C.; Ardanaz, E.; Dorronsoro, M.; Khaw, K-T.; Allen, N.E.; Bueno-de-Mesquita, H.B.; Peeters, P.H.M.; Trichopoulos, D.; Linseisen, J.; Ljungberg, B.; Overvad, K.; Tjønneland, A.; Romieu, I.; Riboli, E.; Mukeria, A.; Shangina, O.; Stevens, V.L.; Thun, M.J.; Diver, W.R.; Gapstur, S.M.; Pharoah, P.D.; Easton, D.F.; Albanes, D.; Weinstein, S.J.; Virtamo, J.; Vatten, L.; Hveem, K.; Njølstad, I.; Tell, G.S.; Stoltenberg, C.; Kumar, R.; Koppova, K.; Cussenot, O.; Benhamou, S.; Oosterwijk, E.; Vermeulen, S.H.; Aben, K.K.H.; van der Marel, S.L.; Ye, Y.; Wood, C.G.; Pu, X.; Mazur, A.M.; Boulygina, E.S.; Chekanov, N.N.; Foglio, M.; Lechner, D.; Gut, I.; Heath, S.; Blanche, H.; Hutchinson, A.; Thomas, G.; Wang, Z.; Yeager, M.; Fraumeni, J.F.Jr; Skryabin, K.G.; McKay, J.D.; Rothman, N.; Chanock, S.J.; Lathrop, M. & Brennan, P. (2011). Genome-wide association study of renal cell carcinoma identifies two susceptibility loci on 2p21 and 11q13.3, *Nature Genetics,* Vol.43, No.1, (January 2011), pp.60-65

Racila, D.; Winter, M.; Said, M.; Tomanek-Chalkley, A.; Wiechert, S.; Eckert, R.L. & Bickenbach, J.R. (2011). Transient expression of OCT4 is sufficient to allow human keratinocytes to change their differentiation pathway, *Gene Therapy,* Vol.18, No.3, (March 2011), pp.294-303

Ricci-Vitiani, L.; Pallini, R.; Biffoni, M.; Todaro, M.; Invernici, G.; Cenci, T.; Maira, G.; Parati, E.A; Stassi, G.; Larocca, L.M. & Maria, R.D. (2010). Tumour vascularization via endothelial differentiation of glioblastoma stem-like cells, *Nature,* Vol.468, pp.824-828 (December 2010)

Robel, S.; Berninger, B. & Götz, M. (2011). The stem cell potential of glia: lessons from reactive gliosis, *Nature Reviews Neuroscience,* Vol.12, (February 2011), pp.88-104

Shi, W.; Bessarabova, M.; Dosymnekov, D.; Dezso, Z.; Nikolskaya, T.; Dudoladova, M.; Serebryiskaya, T.; Bugrim, A.; Guryanov, A.; Brennan, R.J.; Shah, R.; Dopazo, J.;

Cheng, M.; Deng, Y.; Shi, T.; Jurman, G.; Furlanello, C.; Thomas, R.S.; Corton, J.C.; Tong, W.; Shi, L. & Nikolsky, Y. (2010). Functional analysis of multiple genomic signatures demonstrates that classification algorithms choose phenotype-related genes, *Pharmacogenomics Journal,* Vol.10, No.4, pp.310-323 (August 2010)

Suh, H.; Consiglio, A.; Ray, J.; Sawai, T. & D'Amour, K.A. (2007). In vivo fate analysis reveals the multipotent and self-renewal capacities of Sox2+ neural stem cells in the adult hippocampus, *Cell Stem Cell,* Vol.1, (November 2007), pp.515-528

Takahashi, K.; Tanabe, K.; Ohnuki, M.; Narita, M.; Ichisaka, T.; Tomoda, K. & Yamanaka, S. (2007). Induction of pluripotent stem cells from adult human fibroblasts by defined factors, *Cell,* Vol.131, No.5, pp.861-872 (November 2007)

Tanabe, S.; Sato, Y.; Suzuki, T.; Suzuki, K.; Nagao, T. & Yamaguchi, T. (2008). Gene expression profiling of human mesenchymal stem cells for identification of novel markers in early- and late-stage cell culture, *Journal of Biochemistry,* Vol.144, No.3, pp.399-408 (September 2008)

Tillinghast, G.W. (2010). Microarrays in the clinic, *Nature Biotechnology,* Vol.28, No.8, pp.810-812 (August 2010)

Tsai, M.S.; Hwang, S.M.; Chen, K.D.; Lee, Y.S.; Hsu, L.W.; Chang, Y.J.; Wang, C.N.; Peng, H.H., Chang, Y.L.; Chao, A.S.; Chang, S.D.; Lee, K.D.; Wang, T.H.; Wang, H.S. & Soong, Y.K. (2007). Functional Network Analysis of the Transcriptomes of Mesenchymal Stem Cells Derived from Amniotic Fluid, Amniotic Membrane, Cord Blood, and Bone Marrow, *Stem Cells,* Vol.25, No.10, pp.2511-2523 (October 2007)

Visvader, J.E. (2011). Cells of origin in cancer, *Nature,* Vol.469, (January 2011), pp.314-322

Wang, K. ; Diskin, S.J. ; Zhang, H. ; Attiyeh, E.F. ; Winter, C. ; Hou, C. ; Schnepp, R.W. ; Diamond, M. ; Bosse, K. ; Mayes, P.A. ; Glessner, J. ; Kim, C. ; Frackelton, E. ; Garris, M. ; Wang, Q. ; Glaberson, W. ; Chiavacci, R. ; Nguyen, L. ; Jagannathan, J. ; Saeki, N. ; Sasaki, H. ; Grant, S.F.A., ; Iolascon, A. ; Mosse, Y.P. ; Cole, K.A. ; Li, H. ; Devoto, M. ; McGrady, P.W. ; London, W.B. ; Capasso, M. ; Rahman, N. ; Hakonarson, H. & Maris, J.M. (2011). Integrative genomics identifies LMO1 as a neuroblastoma oncogene, *Nature,* Vol.469, (January 2011), pp.216-220

Wang, R.; Chadalavada, K.; Wilshire, J.; Kowalik, U.; Hovinga, K.E.; Geber, A.; Fligelman, B.; Leversha, M.; Brennan, C. & Tabar, V. (2010). Glioblastoma stem-like cells give rise to tumour endothelium, *Nature,* Vol.468, pp.829-833 (December 2010)

Wang, Z.A. & Shen, M.M. (2011). Revisiting the concept of cancer stem cells in prostate cancer, *Oncogene,* Vol.30, No.11, (March 2011), pp.1261-1271

Weng, J.Y.; Du, X.; Geng, S.X.; Peng, Y.W.; Wang, Z.; Lu, Z.S.; Wu, S.J.; Luo, C.W.; Guo, R.; Ling, W.; Deng, C.X.; Liao, P.J. & Xiang, A.P. (2010). Mesenchymal stem cell as salvage treatment for refractory chronic GVHD, *Bone Marrow Transplantation,* Vol.45, pp.1732-1740 (December 2010)

Yadirgi, G. & Marino, S. (2009). Adult neural stem cells and their role in brain pathology, *Journal of Pathology,* Vol.217, (January 2009), pp.242-253

Yusenko, M.V.; Kuiper, R.P.; Boethe, T.; Ljungberg, B.; van Kessel, A.G. & Kovacs, G. (2009). High-resolution DNA copy number and gene expression analyses distinguish chromophobe renal cell carcinomas and renal oncocytomas, *BMC Cancer,* Vol.9, No.152 (May 2009)

7

Study of Degradation Products and Degradation Pathways of ECD and Its Drug Product, ECD Kit

Kung-Tien Liu, Yu-Yung Lin, Yi-Chih Hsia, Jian-Hua Zhao, Chang-Yung Su, Shang-Yu Shen, Lee-Chung Men and Lie-Hang Shen
Institute of Nuclear Energy Research
Taiwan

1. Introduction

Recently, researchers have shown an increased interest in the investigations of drug stability and degradation pathway. Baertschi (2006) and Alsante et al., (2007) described an overall strategy for the prediction, identification and control of stability-related issues in active pharmaceutical ingredients (APIs) and drug products using stress testing. Forced degradation studies are part of the drug development strategy being undertaken to elucidate the intrinsic stability of the drug. Such studies are conducted under more severe and exaggerated conditions than those usually used for long-term stability tests. These studies are particularly useful to establish the drug degradation pathway and validate the suitable analytical procedures (Jocić et al., 2009).

Technetium-99m ethyl cysteinate dimer (Tc-99m-ECD or bicisate) is one of the most essential single-photon emission computed tomography (SPECT) imaging agents. In according to the practice guidelines of the American College of Radiology (ACR) (Abdel-Dayem, 2003) and the European Association of Nuclear Medicine Neuroimaging Committee (ENC) (Kapucu et al., 2009), clinical indications of Tc-99m-ECD include in evaluating the regional cerebral blood flow (rCBF) in patients with cerebrovascular diseases, transient ischemic attack, various forms of dementia, symptomatic traumatic brain injury, encephalitis, vascular spasm following subarachnoid hemorrhage, inflammation, epileptic foci and lacunar infarctions.

More indications of Tc-99m-ECD in SPECT brain perfusion imaging of neuropsychiatric disorders and chronic fatigue syndrome have not been fully characterized (Abdel-Dayem, 2003; Kapucu et al., 2009). However, investigations of the conversion in patients of mild cognitive impairment to Alzheimer's disease (AD) (Borroni et al., 2006), the functional compensation mechanism in incipient AD (Caroli et al., 2010), the mechanism for suppression of parkinsonian tremor by thalamic stimulation (Wielepp et al., 2001), the mechanism by which thyroid hormone availability affects cerebral activity (Schraml et al., 2006), brain glucose metabolism in hypothyroidism (Bauer et al., 2009), reduction in the bifrontal regions and diffusion-weighted imaging of Creutzfeldt-Jacob disease (Sunada et al., 2004; Ukisu et al., 2006), quantitation and differentiation in patients with Tourette's syndrome (Diler et al., 2002; Chen et al., 2003; Sun et al., 2001) and abnormal rCBF in patients with Sjögren's syndrome (Chang et al., 2002) were reported.

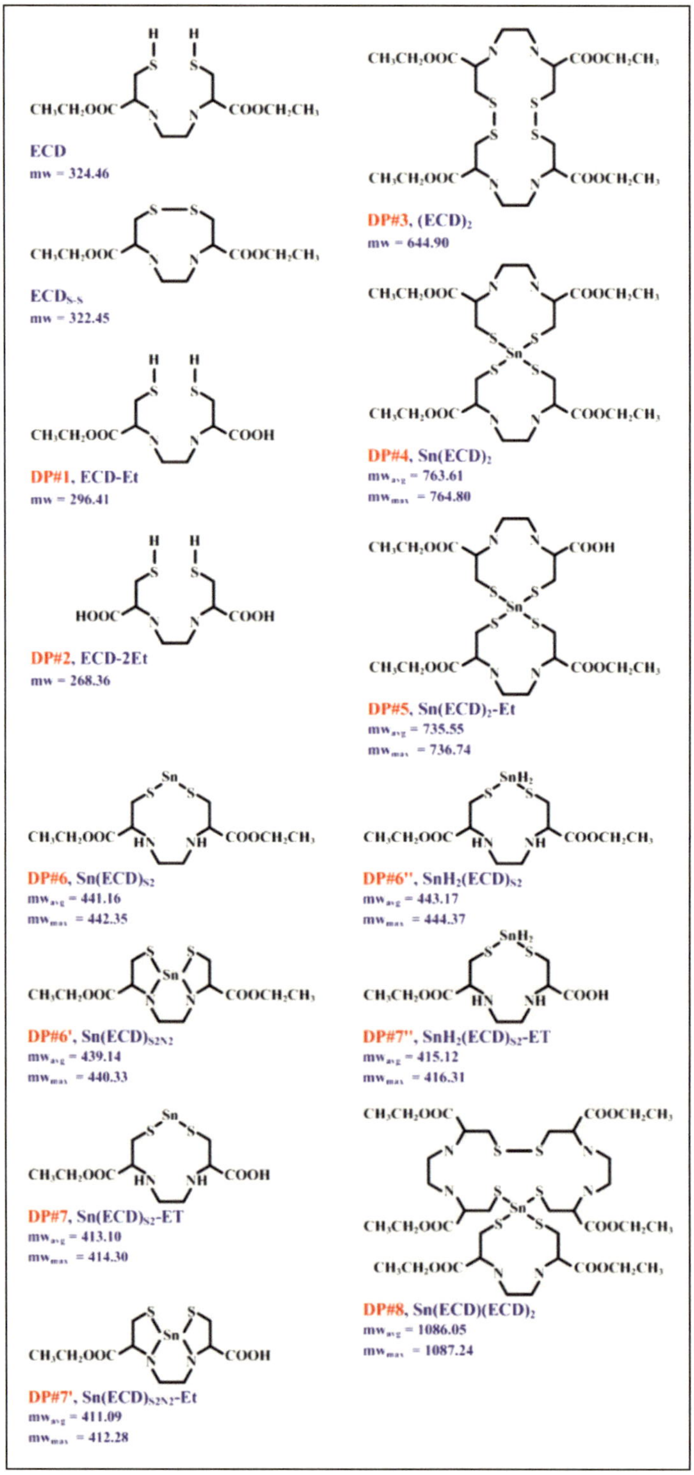

Fig. 1. Chemical structures of ECD and its DPs

For clinical implements, Tc-99m-ECD is obtained by radiolabeling of Tc-99m to its API, i.e. *L*-Cysteine, *N,N'*-1,2-ethanediylbis-, diethyl ester, dihydrochloride (ECD) (Fig. 1). Despite its safety and efficacy, a major drawback for the application of Tc-99m-ECD is the stability of ECD Kit in aqueous solution. For example, no more delay for patient injection that longer than 30 minutes is recommended by ACR and EANM (Abdel-Dayem, 2003; Kapucu et al., 2009).

So far, several studies investigating stability and degradation of ECD and Tc-99m-ECD have been reported, but there is still insufficient data for the elucidation of degradation products (DPs) structure and pathway. Mikiciuk-Olasik et al. (2000) demonstrated that ECD decomposed as soon as it was dissolved in phosphate buffer solutions. However, they offered no explanation for the structures of three decomposition products. Verduyckt et al. (2003) investigated the identity of Tc-99m-ECD using radio-LC-MS and reported five DPs for ECD including disulfide, monoacid monoester (ECM), diacid (EC) and Sn(IV) complexes with ECD (Sn-ECD) and EC (Sn-EC) derivatives. However, no systematic degradation study was reported to investigate the degradation mechanism or pathway on ECD Kit. The study of Tsopelas et al. (2005) was mainly focused on the behavior of Tc-99m and its reactions with components of ECD Kit.

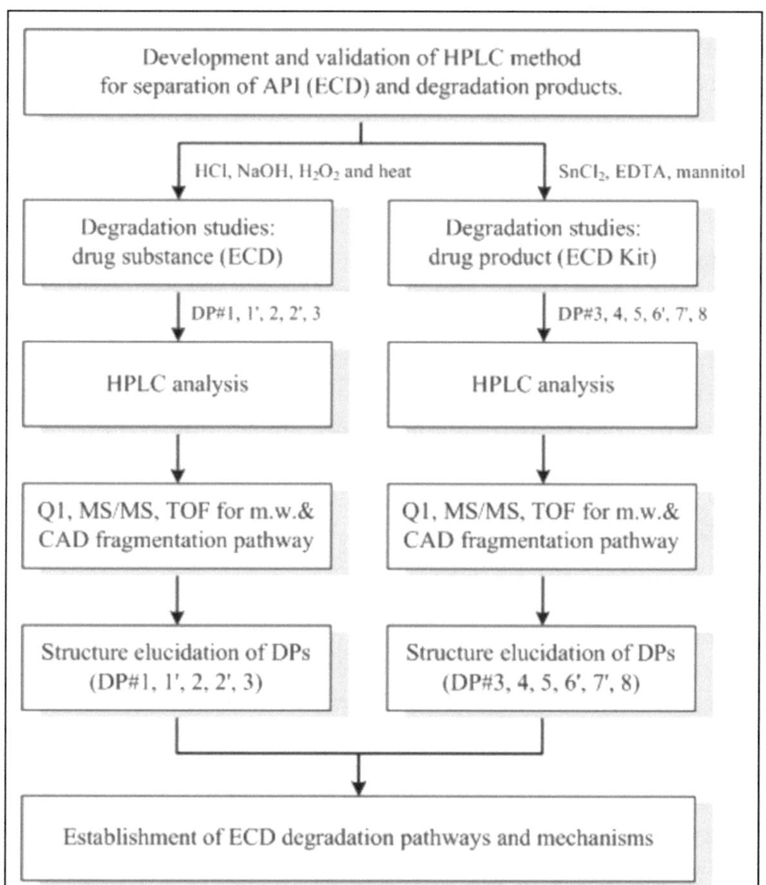

Fig. 2. Overall scheme for the ECD and ECD Kit degradation studies

Our preliminary observations showing that ECD Kit is highly unstable in (non)aqueous solution and the composition of ECD Kit is the major obstacle to determine stability of ECD (Chao et al., 2011). Therefore, the aim of this paper is to evaluate degradation kinetics and mechanism of ECD Kit in aqueous solution. No previous reports were found in the literature regarding the degradation behavior of ECD. The main issues addressed in this study were modified according to the procedures outlined by Shah et al. (Shah & Singh, 2010; Shah, et al., 2010; Raijada et al., 2010). The overall scheme in this degradation investigation (Fig. 2) includes: (i) development and validation of an HPLC method for determination of ECD and DPs, (ii) implement of the degradation and stress studies on ECD and ECD Kit, (iii) analysis of the DPs by HPLC, (iv) characterization of the molecular weights and collision activated dissociation (CAD) fragmentation pathways of ECD and DPs by Q1, MS/MS and TOF, (v) elucidation of the DPs' structures and (vi) establishment of the ECD degradation pathways and decomposition mechanisms.

2. Experimental

2.1 Materials and reagents

ECD (purity: 97.53%) was obtained from ABX (Radeberg, Germany). All chemicals and reagents were of analytical grade and used as received without further purification. Methanol (MeOH) (HPLC grade) was obtained from Merck (Darmstadt, Germany). Deionized water was purified using a Smart DQ3 reverse osmosis reagent water system (Millipore, MA, U.S.A.) with a 0.22-μm filter, TOC < 5 ppb, resistivity ≥ 18.2 MΩ-cm and endotoxin < 0.001 EU/mL.

2.2 HPLC instrumentation

An Agilent 1100 series high performance liquid chromatography (HPLC) (Agilent, Palo Alto, CA, U.S.A.) was employed, consisting of an on-line degasser, binary pump, autosampler, thermostated column oven and photodiodearray detector (PDA). Data were acquired and processed with ChemStation (Agilent, Palo Alto, CA, U.S.A.). A C-18 reversed-phase column (Zorbox Eclipse XDB-C18, 4.6 × 50 mm, 1.8 μm, Agilent) was used for the separation of ECD and DPs. An isocratic elution was achieved using a mobile phase which consisted of methanol and sodium acetate (pH 7.0, 50 mM; 60 : 40, v/v). The flow-rate was 0.5 mL/min and the injection volume was 2-5 μL. The absorbance detection wavelength was 210 nm. The column temperature was set at 25°C in all experiments performed.

2.3 LC-MS/MS and MS/TOF instrumentation

MS analysis was carried out on a 4000 QTrap LC-MS/MS system with API Analyst software of version 1.4.1 or a QSTAR Elite Hybrid QTOF with API Analyst QS software of version 2.0 (MDS Sciex, Ontario, Canada). Samples were introduced by an HPLC system (Agilent 1100 series HPLC system, Agilent, CA, U.S.A.) or a syringe pump (Harvard, Harvard Apparatus Inc., Holliston, MA, U.S.A.). The samples were ionized by a turbo spray ion source (electrospray ionization) in the positive ion mode at 5500 V. Mass spectra were obtained over the range of 50 or 100 to 2000 amu with unit resolution in Q1 and Q3. Other parameters are shown in Table 1. In all cases of 4000 QTrap LC-MS/MS studies, nitrogen was used as the nebulization, curtain and collision gas.

2.4 HPLC method development and validation

The method was validated according to the International Conference on Harmonization (ICH) guidelines for the validation of analytical methods, which includes specificity, linearity, precision, accuracy, LOD/LOQ, solution stability, robustness and system suitability and was achieved as the procedures described earlier (Liu et al., 2008; Yang et al., 2010).

2.4.1 Specificity (selectivity)

Forced degradation studies are used to evaluate the development of analytical methodology (the specificity or selectivity of the purity assay method), to gain better understanding of the stability of APIs and drug products and to provide information about degradation pathways and DPs.

Parameter	Q1 scan	MS2 scan	TOF MS
Source Type	Turbo Spray	Turbo Spray	Turbo Spray
Source Temperature (°C)	-	-	-
Scan Type	Q1 MS	Product Ion (MS2)	Positive TOF
Scan Mode	Profile	Profile	None
Polarity	Positive	Positive	Positive
Resolution (Q1 & Q3)	Unit	Unit	Unit
Nebulizer Gas (NEB)	-	-	-
Curtain Gas (CUR)	10	10	20
IonSpray Voltage (IS, V)	5500	5500	5500
Collision Gas (CAD)	-	Medium	-
Ion Source Gas 1 (GS1)	20	20	20
Ion Source Gas 2 (GS2)	0	0	0
Ion Energy 1 (IE1, V)	0.30	0.30	1.00
Ion Energy 3 (IE3, V)	-	-0.50	-
Detector Parameters	Positive	Positive	-
Deflector (DF)	-	-	-
Channel Electron Multiplier (CEM, V)	1950	1950	-

Table 1. Mass spectrometry working parameters for ECD and DPs analysis

Here, forced degradation studies of ECD were carried out under the conditions of acidic and alkaline hydrolysis, oxidation and dry heat. Samples of ECD (2 mg) were dissolved in 0.34 mL of methanol and subjected to 0.33 mL of 1 M HCl and 0.33 mL of 1 M NaOH at ambient temperature for 4 hrs and 1 hr, respectively. Acidic and alkaline hydrolysis samples were neutralized using 1 M NaOH or 1 M HCl and diluted to 2 mg/mL with methanol before HPLC analysis. Equivalent amounts (2 mg) of ECD that one portion was dissolved in 0.50 mL of methanol and subjected to 0.50 mL of 3% H_2O_2 and the other portion of solid drug was heated at 50°C (in oven over a period of 4 hrs) and were injected into the HPLC for analysis.

2.4.2 Linearity

The calibration curves of five concentrations (1.6 to 2.4 mg/mL) were obtained by plotting the respective peak areas against concentrations. The linearity was evaluated by the linear least square regression method with three determinations at each concentration.

2.4.3 Precision

In relation to the precision of the method, repeatability (intra-day), intermediate (inter-day) precision and reproducibility were investigated by performing assays of retention times, peak widths at half height, number of theoretical plates, linear least squares regression equations and correlation coefficients for the ECD standard at five concentrations and purities for one quality control (QC) sample. The repeatability and intermediate precision were evaluated by one analyst within one and two days, respectively, while the reproducibility was achieved by two analysts (Kulikov & Zinchenko, 2007).

2.4.4 Accuracy (recovery)

The accuracy of the method was determined by the recovery test. QC samples of ECD of concentration at 2.0 mg/mL ($C_{nominal}$) were analyzed by the proposed method. Experimental values (C_{exp}) were obtained by interpolation to the linear least square regression equation of a fresh newly prepared calibration curve (1.6 to 2.4 mg/mL) and comparing with the theoretical values ($C_{nominal}$).

$$\text{Recovery yield (\%)} = \frac{C_{exp}\,(\text{mg/mL})}{C_{nominal}\,(\text{mg/mL})} \times 100\%$$

2.4.5 Limit of detection (LOD) and limit of quantification (LOQ)

The LOD and LOQ of the method for impurities in ECD were determined at signal to noise ratios of 3 and 10, respectively.

2.4.6 Stability of drug (API) solution

The stability of the API solution was examined using the QC sample (2.0 mg/mL) for bench-top stability study. The QC samples were kept in the autosampler at ambient temperature for HPLC analysis over three consecutive days. Experimental data were obtained by interpolation to the linear least square regression equation of a calibration curve (1.6 to 2.4 mg/mL) newly prepared each day. Retention time, recovery yield and purity of ECD over three consecutive days were analyzed.

2.4.7 Robustness

The robustness of an analytical method is a basic measurement of its capacity to remain unaffected by small variations in method parameters. In this investigation, method robustness was evaluated through the effects of different columns (same type and manufacturer), column temperatures (± 2°C), pH values (± 0.1) and flow rates (± 0.05 mL/min) of mobile phase.

2.4.8 System suitability

The system suitability was assessed by five triplicate analyses of the drug in a concentration range of 1.6 to 2.4 mg/mL. The efficiency of the column was expressed in terms of the theoretical plates number (N), column capacity (k'), column selectivity (α) and tailing factor (t). The acceptance criteria for the N, k', α, t and percentage relative standard deviation (% R.S.D.) for the retention time of ECD were > 3000, 2-8, 1.05-2.00, 0.9-2.5 and ± 2%, respectively.

2.5 Forced degradation studies of ECD

Forced degradation studies of ECD were carried out according to the procedures described above in Section 2.4.1 Specificity (selectivity). Moreover, samples of ECD (2 mg) were dissolved in 0.50 mL of methanol and subjected to 0.25 mL of 1 M NaOH and 0.50 mL of 3% H_2O_2 at ambient temperature for kinetic studies. The structures and degradation of DPs were further characterized by HPLC and LC-MS/MS for the molecular weights and the CAD fragmentation pathways.

2.6 Degradation studies of ECD Kit

First, degradation studies of ECD Kit were carried out by subjecting samples of ECD to various components of ECD Kit for determining the effect of $SnCl_2$, mannitol and EDTA. Second, ECD (1 mg/mL, 500 μL) and $SnCl_2$ (1 mg/mL) were mixed in ratio of 12.5 : 1, 8 : 1, 4 : 1, 2 : 1 and 1 : 1 (v/v) and diluted to total volume of 1000 μL with deionized water. The mixtures were kept at ambient temperature in HPLC autosampler and in bench-top for HPLC and MS analysis, respectively. All samples were diluted to 1 ppm with methanol for MS analysis. Positive ESI-MS/MS scanning types, i.e. precursor ion scan, product ion scan and neutral loss scan were performed. The structures of DPs were proposed based on the molecular weights and the CAD fragmentation pathways.

3. Results and discussion

3.1 HPLC method development

A reversed-phase high performance liquid chromatography (RP-HPLC) method for the determination of ECD and forced degradation DPs was developed and validated. A Zorbox Eclipse XDB-C18 (4.6 × 50 mm, 1.8 μm, Agilent) reversed-phase column was selected for the separation of ECD and DPs. ECD samples at concentrations of around 2 mg/mL and 100 ppb were used to optimize conditions for HPLC and LC-ESI-MS/MS, respectively. Absorption spectra of ECD were recorded over the range of 200 to 300 nm by a post-column photodiode-array detector (PDA). A wavelength of 210 nm was found to be optimal for the detection and quantification of ECD.

Chromatographic separation of ECD was achieved using a mobile phase which consisted of methanol and sodium acetate (pH 7.0, 50 mM; 60 : 40, v/v). The typical HPLC chromatograms of ECD are shown in Fig. 3(a) and 4(a). The difference of retention time (t_R) of ECD chromatograms between degradation studies of API and drug product was due to the gradual damage of column packing materials. However, no significant efficiency of the column, such as the number of theoretical plates (N) and tailing factor (t) was found.

3.2 Mass spectrometric analysis of ECD

The proposed high-salt contained mobile phase of HPLC was not suitable for ESI-MS studies. Therefore, a syringe pump was chosen for the sample introduction for Q1 and MS/MS scan. Q1 full scans were achieved in a positive ion mode to optimize the

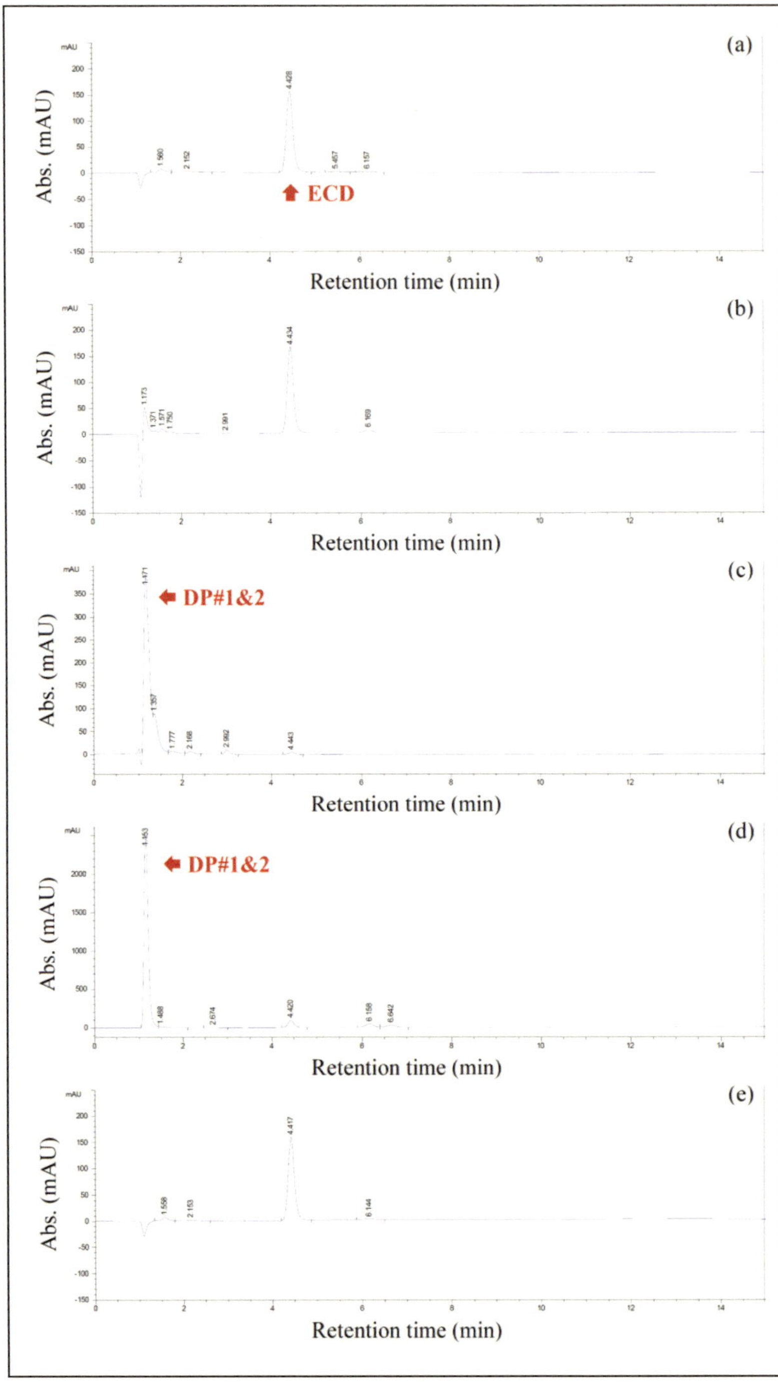

Fig. 3. Typical HPLC chromatograms of degradation studies of ECD. Samples (2 mg of ECD) were carried out under the conditions of (a) methanol (no degradation), (b) acidic hydrolysis (0.5 M HCl at ambient temperature for 4 hrs), (c) alkaline hydrolysis (0.5 M NaOH at ambient temperature for 1 hr), (d) oxidation (1.5% H_2O_2) and (e) dry heat (50°C for 4 hrs)

electrospray ionization (ESI) conditions of ECD and $(ECD)_2$ (Fig. 5(a)). The peaks at retention time (t_R) of 4.43 and 3.82 min were identified as a protonated ECD ion ($[M+H]^+$) at m/z 323.4 by ESI-MS (Fig. 5(b)). Moreover, a protonated molecular ion with m/z 645.4 at t_R of 6.17 and 5.27 min were identified as ECD dimer (DP#3), i.e. $(ECD)_2$ (Fig. 5(g)).

Both product ion and precursor ion scans were then carried out at different collision-activated dissociation (CAD) conditions to optimize the declustering potential (DP), entrance potential (EP), collision energy (CE) and collision cell exit potential (CXP). The MS/MS fragments of ECD, ECD and $ECD_{S\text{-}S}$ are summarized in Table 2.

The linearities of multiple reaction monitoring (MRM) transitions of ECD ($ECD_{S\text{-}S}$) were studied. The linear least-square regression equations and correlation coefficients of MRM transitions showed a good linearity over the calibration range. The correlation coefficients (r) were all above 0.9980, indicating the stability of these fragmentations (data not shown). Tandem mass spectrometry (MS/MS) experiments performed in a QTrap MS were used to investigate the CAD fragmentation behavior of ECD ($ECD_{S\text{-}S}$) (Fig. 6(a)).

Although precursor scan of m/z 323.50 can show its precursor ion at m/z 325.40 and 646.36, we found that intra-molecular disulfide product ($ECD_{S\text{-}S}$) is the prominent form in aqueous solution than ECD. This is consistent with previous experiment by Verduyckt et al. (2003), in which they pointed out the existence of disulfide and incomplete esterification of ethylene dicysteine derivatives.

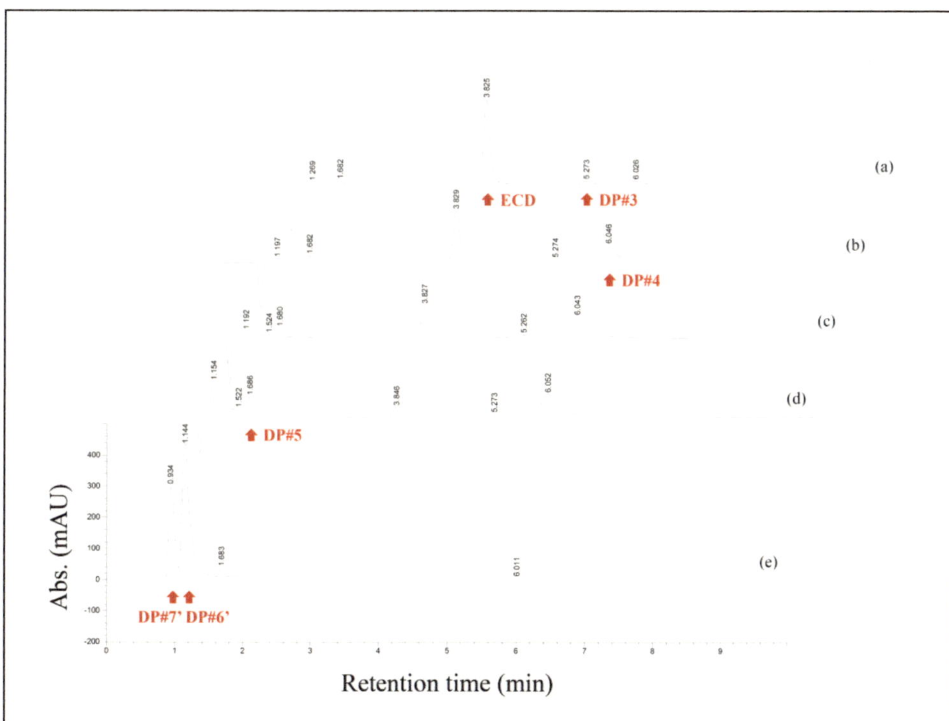

Fig. 4. Typical HPLC chromatograms of degradation studies of ECD Kit. Samples were carried out by subjecting ECD to $SnCl_2$ in ratio (v/v) of (a) 1 : 0, (b) 12.5 : 1, (c) 8 : 1, (d) 4 : 1 and (e) 2 : 1. Duration time is 7-8 hrs

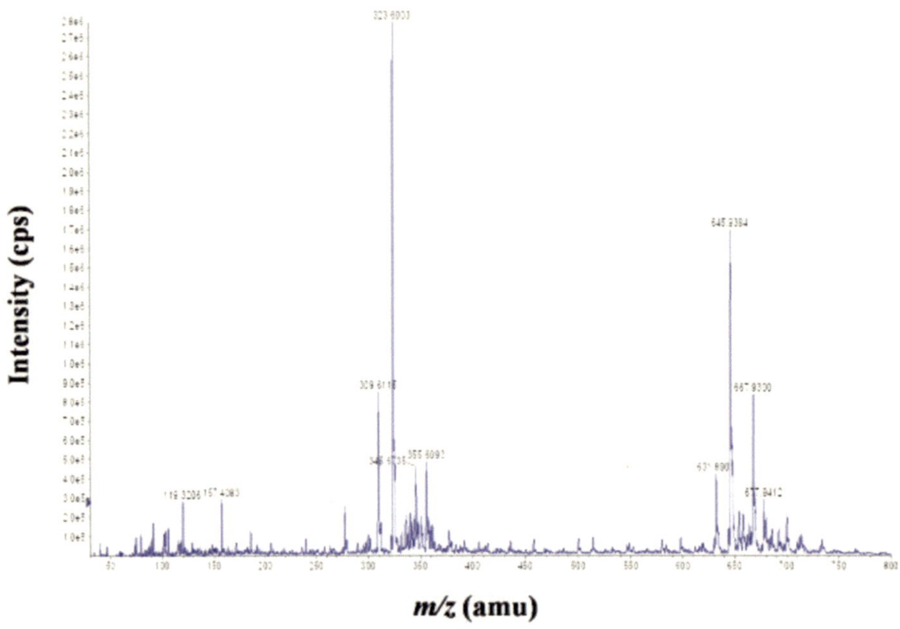

(a)

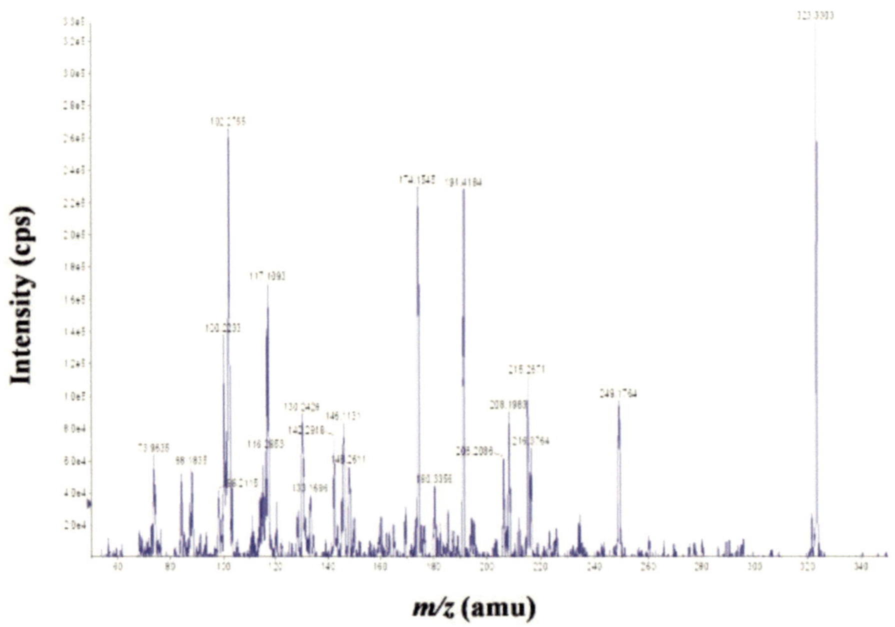

(b)

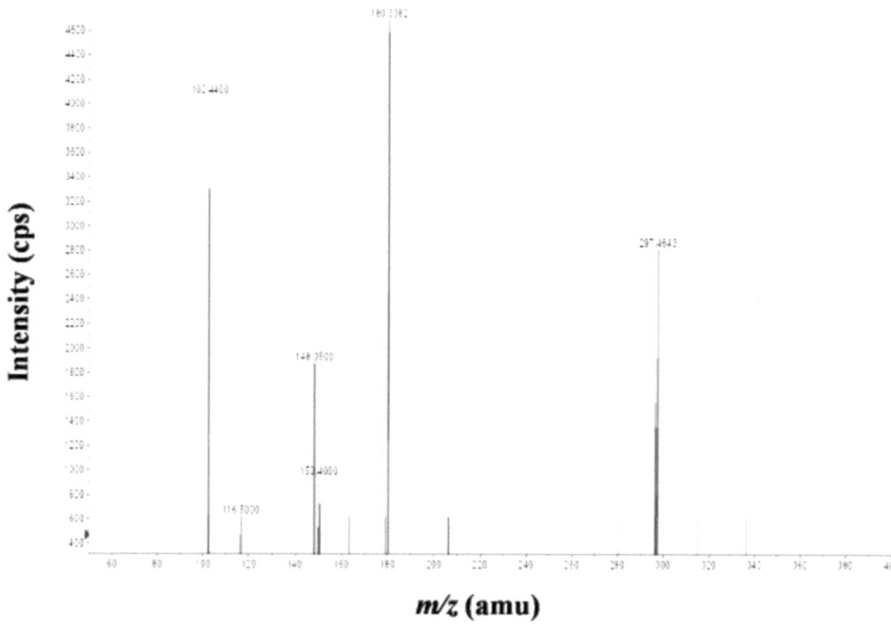

(c)

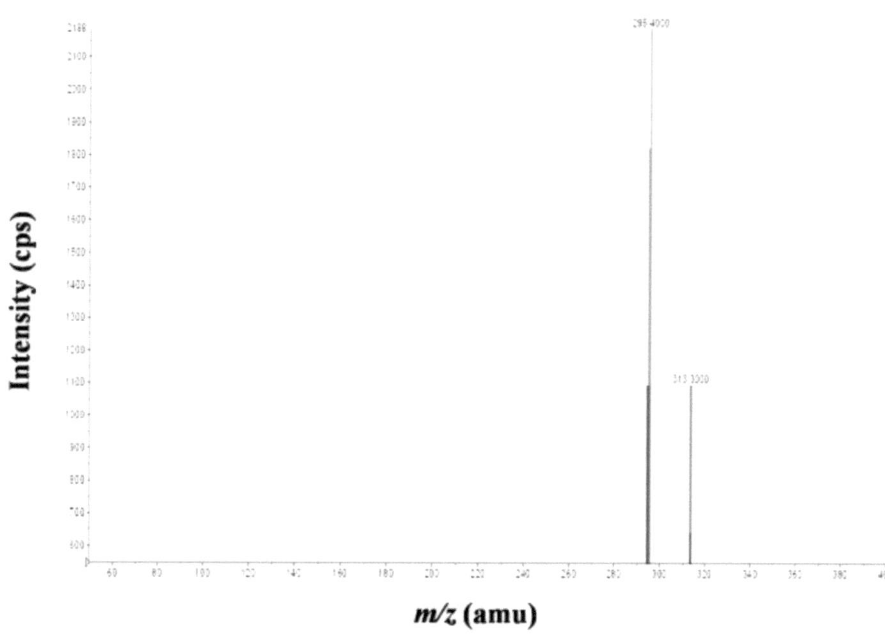

(d)

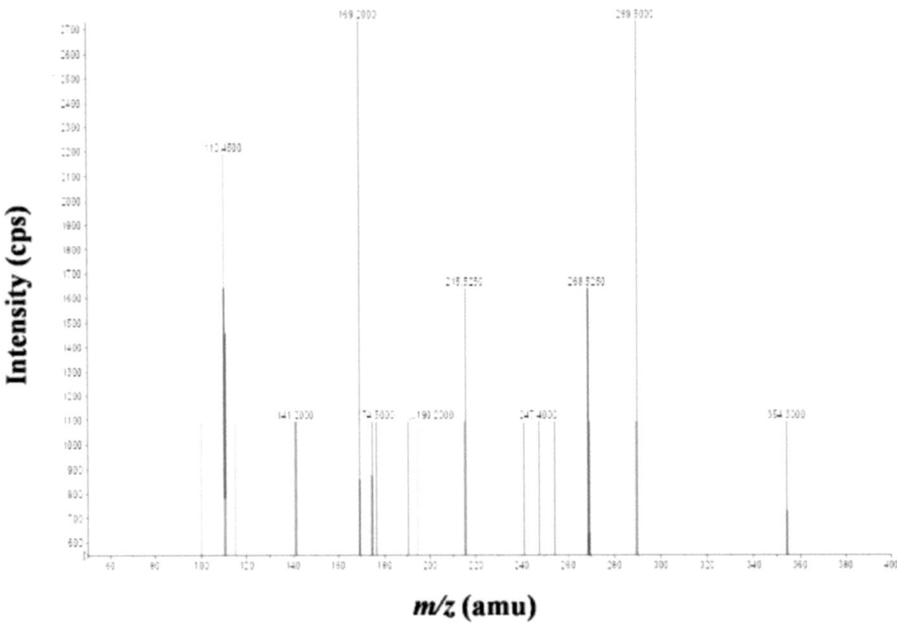

(e)

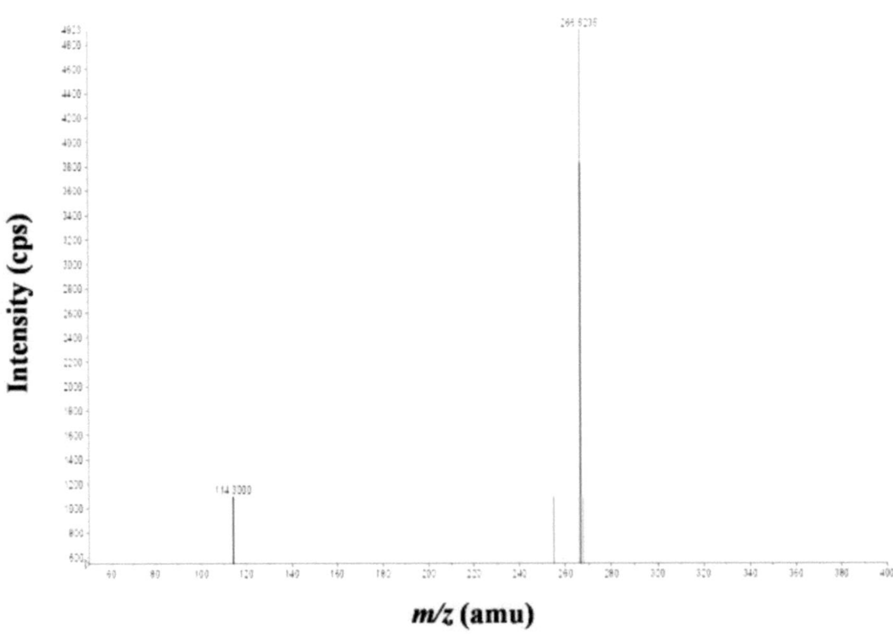

(f)

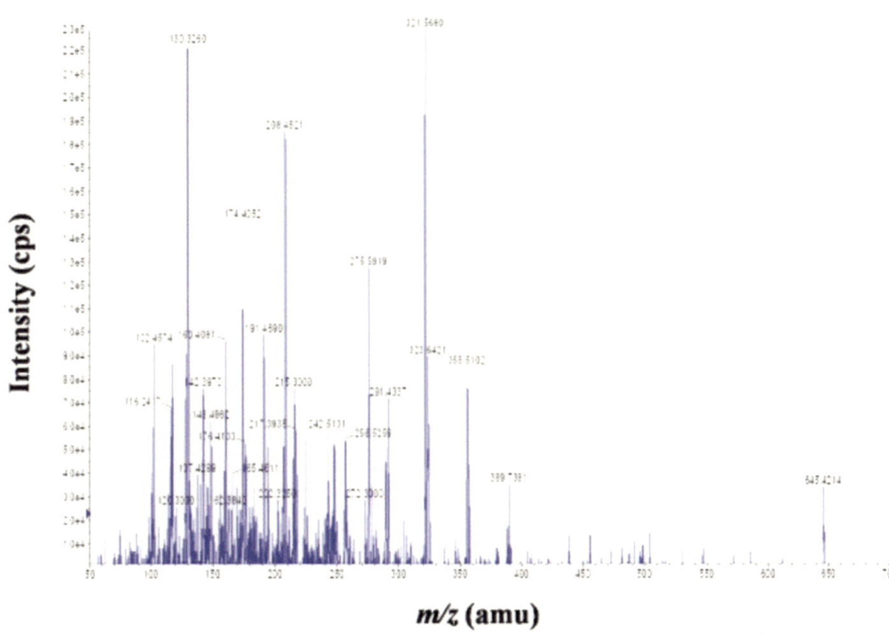

(g)

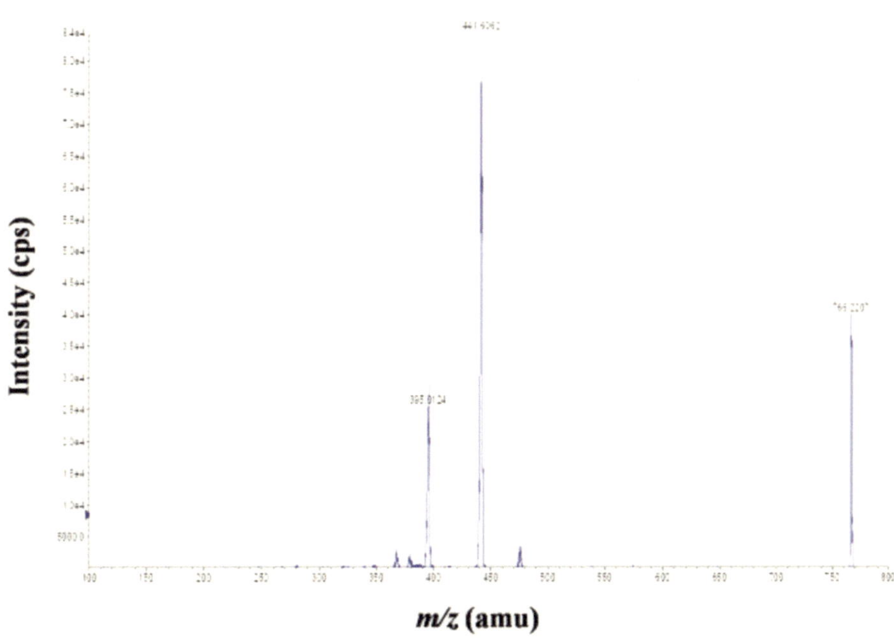

(h)

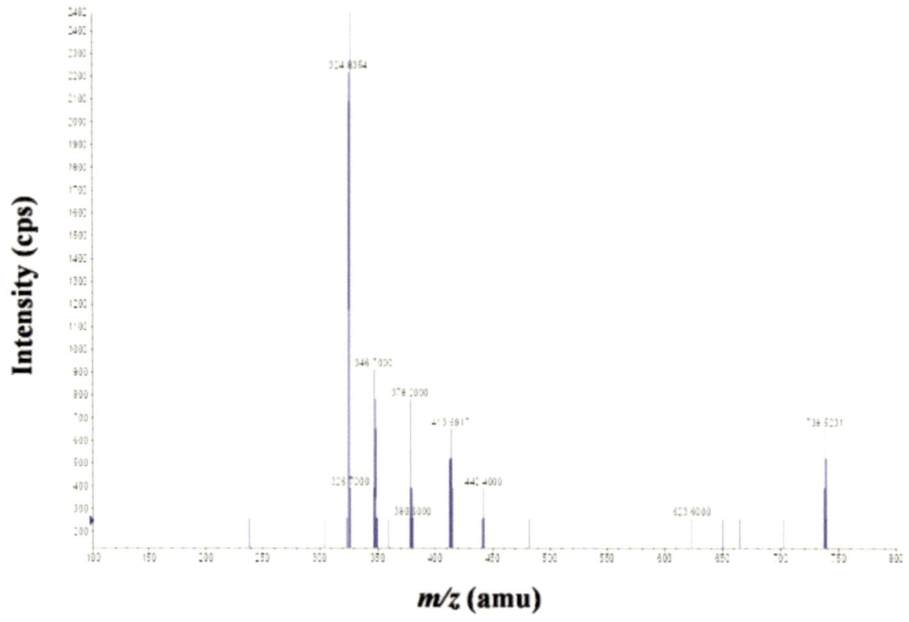

(i)

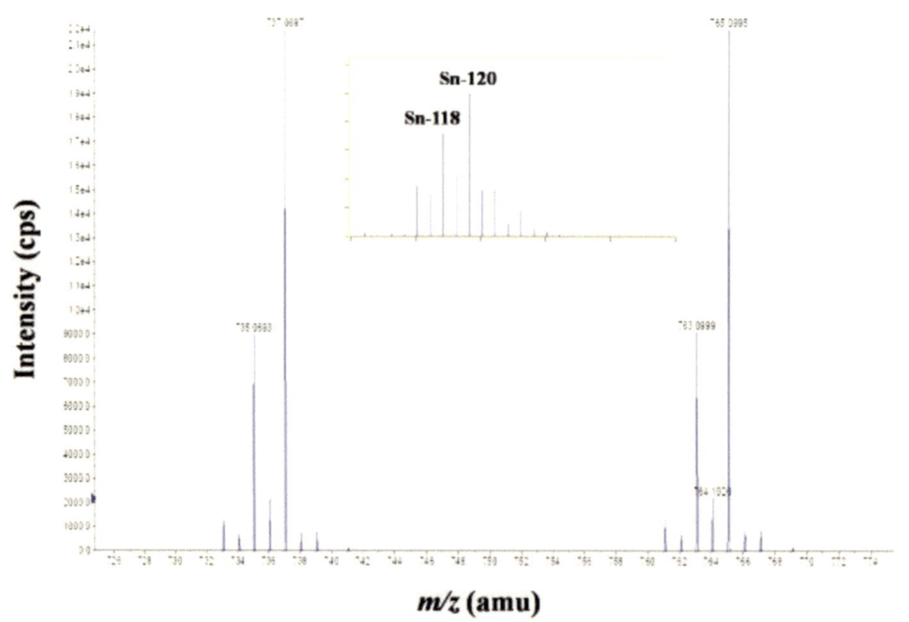

(j)

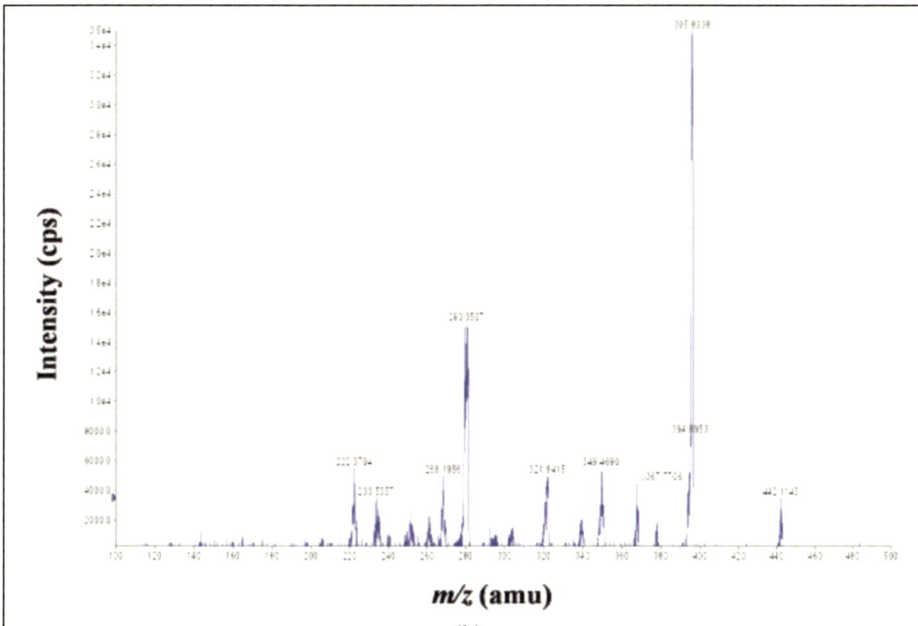

(k)

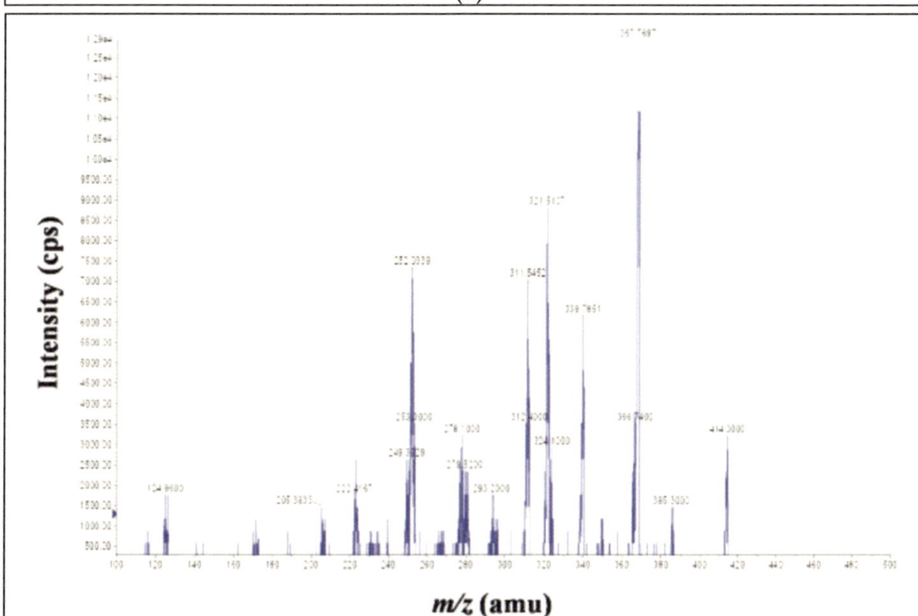

(l)

Fig. 5. (a) Typical ESI-MS Q1 spectra of ECD, typical ESI-MS/MS product ion spectra of (b) ECD_{S-S} (m/z 323.4), (c) DP#1 (ECD-Et, m/z 297.5), (d) DP#1′ ($(ECD)_{S2N2}$-Et, m/z 295.4), (e) DP#2 (ECD-2Et, m/z 268.5), (f) DP#2′ ($(ECD)_{S2N2}$-2Et, m/z 266.5), (g) DP#3 ($(ECD)_2$, m/z 645.4), (h) DP#4 ($Sn(ECD)_2$, m/z 766.4), (i) DP#5 ($Sn(ECD)_2$-Et, m/z 738.0), (j) isotopic ESI-TOF spectra of DP#4 ($Sn(ECD)_2$) and DP#5 ($Sn(ECD)_2$-Et), (k) DP#6′ ($Sn(ECD)_{S2N2}$, m/z 442.0) and (l) DP#7′ ($Sn(ECD)_{S2N2}$-Et, m/z 414.0)

ECD and DPs		Molecular Formula	mw_{avg} or mw_{max}†	Major fragments (m/z)
	ECD	$C_{12}H_{24}N_2O_4S_2$	324.46	175.72, 147.79, 132.53, 129.30, 119.47, 101.52, 86.53
	ECD_{S-S}	$C_{12}H_{22}N_2O_4S_2$	322.45	323.33, 249.18, 215.27, 208.20, 191.42, 174.15, 146.11, 130.24, 117.11, 102.28, 88.18, 73.96
DP#1	ECD-Et	$C_{10}H_{20}N_2O_4S_2$	296.41	297.46, 180.34, 148.35, 102.44, 74.30
DP#1′	ECD_{S-S}-Et	$C_{10}H_{18}N_2O_4S_2$	294.39	295.40, 313.30, 248.40, 219.20, 139.50, 117.40
DP#2	ECD-2Et	$C_8H_{16}N_2O_4S_2$	268.36	268.53, 289.50, 354.30, (322.40, 304.53), 247.40, 215.52, 190.20, 169.20, 110.45
DP#2′	ECD_{S-S}-2Et	$C_8H_{14}N_2O_4S_2$	266.34	266.52, 114.30
DP#3	$(ECD)_2$	$C_{24}H_{44}N_4O_8S_4$	644.90	389.74, 355.51, 321.57, 275.59, 215.3, 208.45, 191.47, 174.41, 130.33, 116.24, 102.46
DP#4	$Sn(ECD)_2$	$C_{24}H_{44}N_4O_8S_4Sn$	764.80†	441.61, 396.01, 367.20, 321.40, 280.40
DP#5	$Sn(ECD)_2$-Et	$C_{22}H_{40}N_4O_8S_4Sn$	736.74†	442.40, 413.69, 378.20, 346.70, 324.84
DP#6′	$Sn(ECD)_{S2N2}$	$C_{12}H_{20}N_2O_4S_2Sn$	440.33†	395.83, 367.77, 349.47, 321.84, 280.35, 268.20, 222.37
DP#7′	$Sn(ECD)_{S2N2}$-Et	$C_{10}H_{16}N_2O_4S_2Sn$	412.28†	385.30, 367.77, 339.79, 321.51, 311.55, 293.20, 279.52, 278.10, 252.03, 222.42, 205.38, 124.96

Table 2. Major MS/MS fragments of ECD and DPs. †mw_{max}: Theoretic molecular weight of maximum isotopic composition

3.3 HPLC method validation

3.3.1 Specificity (selectivity)

ECD was firstly subjected to forced degradation under the conditions of hydrolysis (acid, alkali and neutral), oxidation and thermal stress as requirements of ICH. Significant degradations of 0.5 M NaOH and 1.5% hydrogen peroxide were noticed under stress conditions. Several DPs in the chromatograms at the t_R of 6.64, 2.99, 2.17 and 1.00-1.50 min were detected as shown in Fig. 3(c) and 3(d). Fig. 3(b) and 3(e) represent the chromatograms of a sample degraded at 0.5 M HCl and 50°C for 4 hrs, respectively. No significant degradation was found in these cases. The resolutions between ECD and its degradation peaks were greater than 4.4, indicating that the proposed method was sufficiently selective for its intended purpose.

3.3.2 Linearity

Standard curves were constructed by plotting peak area against concentration of ECD and were linear over the concentration range of 1.6 to 2.4 mg/mL. The linear least squares regression equation of the standard curve correlating the peak areas (PAs) to the drug concentration (X in mg/mL) in this range was Y = 832.03X - 148.88. The correlation coefficient (r) was 0.9991.

3.3.3 Precision

The results of repeatability, intermediate precision and reproducibility were demonstrated by analysing ECD at five concentrations and one QC sample (Table 3). Although the number of theoretical plates were decreased for ~20%, no significant difference in the retention times, peak widths at half height, linear least squares regression equations and correlation coefficients were found. The difference of purities (P (%)) could be due to the stability (equilibrium) and uniformity of QC samples, but also might indicate the sufficient resolution of the proposed method.

Parameters	t_R (min)	W_{half} (min)†	N†	L eq.†	r	P (%)‡
Analyst 1, Day 1	4.42 ± 0.00 (0.05%)	0.15 ± 0.00 (1.27%)	5007 ± 129 (2.58%)	Y = 859.35X - 204.71	0.9998	100.30 ± 0.01
Analyst 1, Day 2	4.42 ± 0.00 (0.06%)	0.15 ± 0.00 (0.69%)	4933 ± 66 (1.34%)	Y = 910.18X - 244.25	0.9992	99.20 ± 0.02
Analyst 2, Day 3	4.41 ± 0.00 (0.05%)	0.16 ± 0.00 (0.81%)	4174 ± 67 (1.61%)	Y = 834.46X - 127.08	0.9990	97.42 ± 0.00

Table 3. Repeatability, intermediate precision and reproducibility of ECD analysis. †Linear range: 1.6 to 2.4 mg/mL; W_{half}: Peak width at half height; N: Number of theoretical plates; n = 15. ‡P (%): The purity of QC sample (n = 3)

3.3.4 Accuracy (recovery)

Recovery tests were achieved by comparing the concentration (C_{exp}) obtained from injection of QC samples to the nominal values ($C_{nominal}$). The intra-day recovery of ECD at concentration of 1.95 mg/mL was 99.68 ± 0.48%. The recoveries, 99.14, 99.89 and 100.03% were between 97 and 103%, indicating that there was sufficient accuracy in the proposed method. The % R.S.D. for measurement of accuracy was 0.48%.

3.3.5 Limit of detection (LOD) and limit of quantification (LOQ)

The limits of detection (LOD, S/N = 3/1) and quantification (LOQ, S/N = 10/1) for the major impurity (DP#3, average abundance in percentage of peak area = 1.32 ± 0.07%) in ECD were found to be 0.004 and 0.014 mg/mL (n = 3), respectively.

3.3.6 Stability of drug (API) solution

The stability of ECD solutions was examined by analyzing solutions over 3 days. The results of these studies are shown in Table 4, where the t_R of ECD and the recovery and purity of QC samples were within the range of 97-103%. No significant degradation or reduction in the absolute peak area was observed within three days, indicating that ECD standard solution would be stable for at least three days when kept on a bench top.

3.3.7 Robustness

The robustness of an analytical procedure is a measurement of its capacity to remain unaffected by small, but deliberate, variations in method parameters and provides an indication of its reliability during normal usage. In this case, robustness of the method was investigated by making small changes of column parameters, column temperature, mobile phase pH and flow rate. The results of the robustness studies were within acceptable range, except that one theoretical plates number (N) was less than 3000, as indicated in Table 5. However, no critical change in performance was found.

Day	Calibration range (mg/mL)	L eq.	r	t_R (min)†	P (%)†
1	1.51-2.33	Y = 859.35X - 204.71	0.9998	4.42 ± 0.00 (0.05%)	100.30 ± 0.97 (0.97%)
2	1.58-2.44	Y = 910.18X - 244.25	0.9992	4.42 ± 0.00 (0.06%)	99.66 ± 1.07 (1.08%)
3	1.54-2.36	Y = 834.02X - 114.17	0.9948	4.40 ± 0.00 (0.04%)	100.13 ± 1.14 (1.14%)

Table 4. Bench-top stability studies of ECD. †The retention time and purity results of QC samples (n = 3)

Parameters		t_R (min)	W_{half} (min)	N	L eq.	R	P (%)*
Column†	#1	4.49 ± 0.00 (0.05%)	0.19 ± 0.00 (1.46%)	n. r.#	Y = 842.24X - 138.39	0.9984	98.99 ± 0.12
	#2	4.42 ± 0.00 (0.05%)	0.15 ± 0.00 (1.27%)	5007 ± 129 (2.58%)	Y = 859.35X - 204.71	0.9998	100.30 ± 0.97
Temperature (°C)	25	4.41 ± 0.00 (0.05%)	0.16 ± 0.00 (0.81%)	4174 ± 67 (1.61%)	Y = 834.46X - 127.08	0.9990	97.42 ± 0.28
	27	4.35 ± 0.00 (0.07%)	0.17 ± 0.00 (1.83%)	3698 ± 138 (3.74%)	Y = 849.90X - 154.71	0.9996	97.13 ± 0.22
pH‡	6.9	4.41 ± 0.00 (0.08%)	0.14 ± 0.00 (0.75%)	5418 ± 84 (1.55%)	Y = 860.08X - 227.29	0.9983	99.21 ± 0.87
	7.0	4.42 ± 0.00 (0.05%)	0.15 ± 0.00 (1.27%)	5007 ± 129 (2.58%)	Y = 859.35X - 204.71	0.9998	100.30 ± 0.97
	7.1	4.40 ± 0.00 (0.09%)	0.15 ± 0.01 (5.11%)	4777 ± 465 (9.73%)	Y = 900.62X - 270.33	0.9968	99.90 ± 0.06
Flow rate (mL/min)	0.45	5.00 ± 0.00 (0.06%)	0.25 ± 0.00 (0.92%)	2249 ± 43 (1.90%)	Y = 986.87X - 253.74	0.9967	100.69 ± 0.43
	0.50	4.49 ± 0.00 (0.05%)	0.19 ± 0.00 (1.46%)	n. r.#	Y = 842.24X - 138.39	0.9984	98.99 ± 0.12
	0.55	4.08 ± 0.00 (0.10%)	0.18 ± 0.00 (1.77%)	n. r.#	Y = 808.35X - 177.83	0.9981	98.75 ± 0.18

Table 5. Robustness study of ECD calibration standard and QC samples analysis. †Column #1 and #2 refer to columns of same type, same manufacturer, but different batch. ‡The pH value of the original aqueous component. *P (%): The purity of QC sample. #n. r.: No record

3.3.8 System suitability

The theoretical plates number (N), column capacity (k'), column selectivity (α) and tailing factor (t) were 5007 ± 129 (2.58%), 2.85 ± 0.01 (0.18%), 1.31 ± 0.00 (0.00%) and 1.19 ± 0.01 (1.07%), respectively. The repeatabilities (% R.S.D.) of t_R for triplicate analysis were within the acceptance criterion range (± 2%). These results were within acceptable range.

3.4 Forced degradation studies of ECD

ECD was subjected to forced degradation under the conditions of hydrolysis (acid, alkali and neutral), oxidation and thermal stress as requirements of ICH. No significant degradation product under the stress conditions of neutral solvents, acidic hydrolysis and dry heat was found (Fig. 3(a), 3(b) and 3(e)). On the contrary, the drug was demonstrated to be liable to degradation under the alkaline hydrolysis and oxidation stress conditions. The reaction in 0.5 M NaOH and 1.5% H_2O_2 at ambient temperature was so fast that almost 100% of ECD was degraded within 1 hr and even immediately, respectively (Fig. 3(c) and 3(d)).

Several high polarity degradants of alkaline hydrolysis of esters in ECD, i.e. DP#1, DP#1′, DP#2 and DP#2′ were formed. The MS/MS spectra are presented in Fig. 5(c)-5(f) and the major fragments are summarized in Table 2. DP#1 and DP#1′ were shown to be monoacid monoester degradants of ECD and $ECD_{S\text{-}S}$, whereas DP#2 and DP#2′ were diacid degradants of ECD and $ECD_{S\text{-}S}$. These results are consistent with previous study (Verduyckt et al., 2003). The proposed structures of DP#1, DP#1′, DP#2 and DP#2′ are shown in Fig. 1

Under oxidation condition of 1.5% H_2O_2, our results also demonstrated that: (i) MS/MS fragments of DP#1, DP#1′, DP#2 and DP#2′ can be detected within duration time less than 0.5 hr, (ii) peak at t_R of 0.97 min was a mixture of DP#1, DP#1′, DP#2 and DP#2′ and (iii) MS/MS intensities of DP#2 and DP#2′ were significantly weaker than those of DP#1 and DP#1′.

Fragmentation ions at m/z 354.50, 322.40 and 304.53 (Table 2) can be detected in the precursor scan of DP#2 (mw_{avg} = 268.36) when the duration time was increased to 1.0 hr, indicating that further oxidation might result in dimer formation.

No protonated molecular ions of DP#1, DP#1′, DP#2 and DP#2′ were detected when $SnCl_2$ was added to the ECD aqueous solution, suggesting that concentrations of DP#1, DP#1′, DP#2 and DP#2′ were negligible in ECD Kit.

Comparing to the degradation rate under oxidation condition, alkaline hydrolysis was much more complicate, and several degradation intermediates were found before they were degraded to DP#1, DP#1′, DP#2 and DP#2′ (Fig. 3(c)).

3.5 Degradation studies of ECD Kit

ECD was very stable in deionized water, methanol and DMSO. The purity of ECD was kept in 95% for 45 hours, whereas ECD Kit was very unstable for quick deceasing to purity of 74.80% within 11 minutes.

ECD was subjected to various components of ECD Kit, such as $SnCl_2$, mannitol and EDTA, to investigate its degradation behavior. Bi-component mixtures of ECD and mannitol, EDTA and $SnCl_2$ in variant of ratio and duration time were analyzed by HPLC, MS and MS/MS.

Our preliminary results showed that mannitol and EDTA had no significant degradation effect in ECD and thus did not affect the purity of ECD. In contract to mannitol and EDTA, a positive correlation between ECD degradation and stannous chloride ($SnCl_2$) was found, suggesting that ECD degradation is significantly correlative to the ratio of ECD to $SnCl_2$ and duration time. These results demonstrated that $SnCl_2$ was the leading cause (key factor) for ECD degradation in ECD Kit. Therefore we prepared mixtures of ECD (1 mg/mL, 500 μL) and $SnCl_2$ (1 mg/mL) in ratio of 12.5 : 1 (the ratio of ECD to $SnCl_2$ in ECD Kit), 8 : 1, 4 : 1, 2 : 1 and 1 : 1 (v/v) and diluted to total volume of 1000 μL with deionized water. The mixtures were kept at ambient temperature in HPLC autosampler and in bench-top for HPLC and MS analysis, respectively.

Six major DPs of ECD, i.e. DP#3 - DP#7′ were numbered in sequence of the coordination number of ECD with Sn and hydrolysis of ester group in ECD. Their MS/MS spectra are shown in Fig. 5(g)-5(l). These results did not quantify the effects on $SnCl_2$ on ECD degradation in detail due to the fact that the liability of $SnCl_2$ for oxidation in aqueous from Sn(II) to Sn(IV) make it difficult to exactly control the concentration of $SnCl_2$.

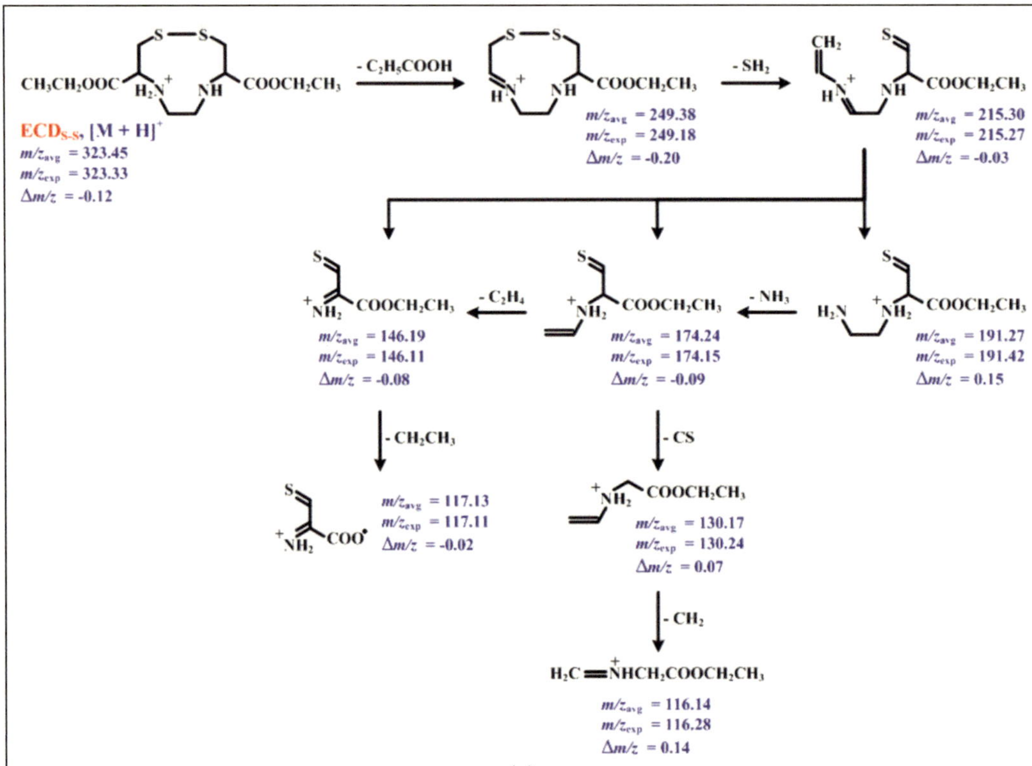

(a)

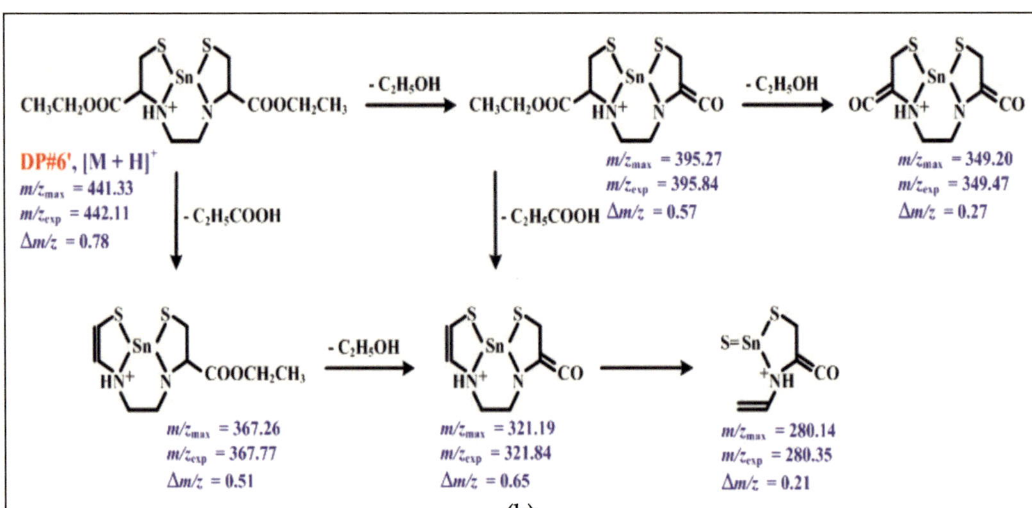

(b)

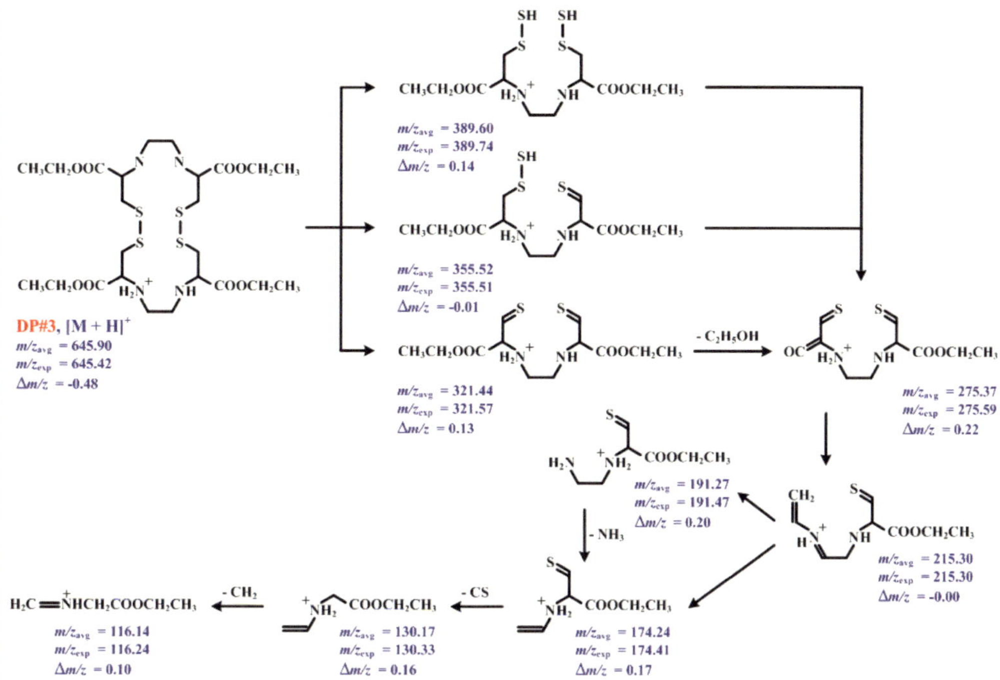
DP#3, [M + H]+
m/z_avg = 645.90
m/z_exp = 645.42
Δm/z = -0.48
m/z_avg = 389.60
m/z_exp = 389.74
Δm/z = 0.14
m/z_avg = 355.52
m/z_exp = 355.51
Δm/z = -0.01
m/z_avg = 321.44
m/z_exp = 321.57
Δm/z = 0.13
- C2H5OH
m/z_avg = 275.37
m/z_exp = 275.59
Δm/z = 0.22
m/z_avg = 191.27
m/z_exp = 191.47
Δm/z = 0.20
- NH3
m/z_avg = 215.30
m/z_exp = 215.30
Δm/z = -0.00
m/z_avg = 174.24
m/z_exp = 174.41
Δm/z = 0.17
- CS
m/z_avg = 130.17
m/z_exp = 130.33
Δm/z = 0.16
- CH2
m/z_avg = 116.14
m/z_exp = 116.24
Δm/z = 0.10

(c)

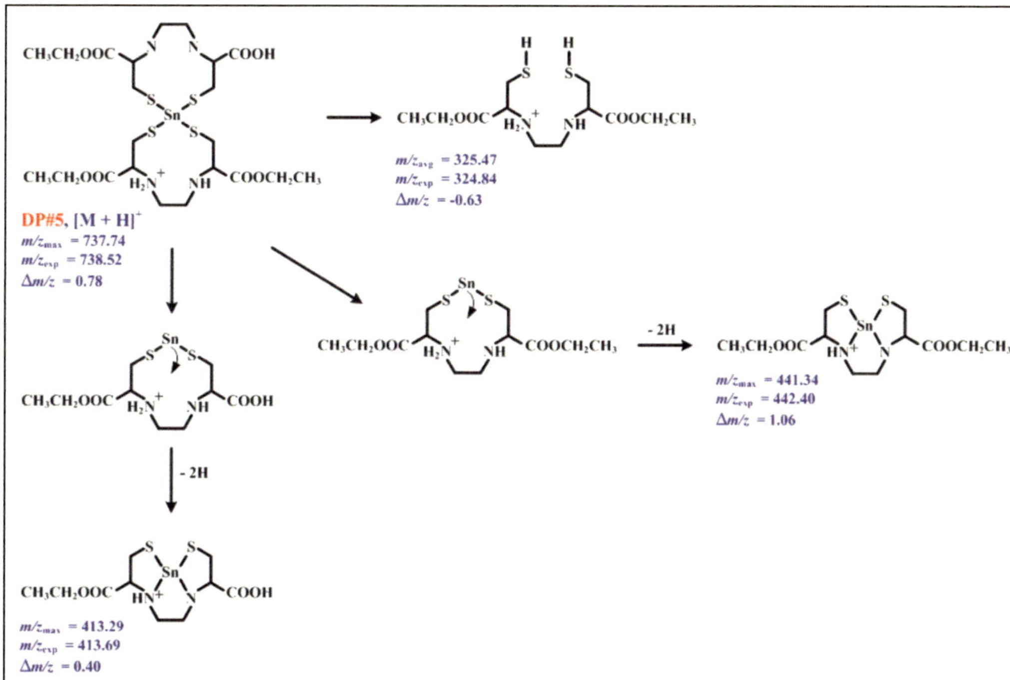
DP#5, [M + H]+
m/z_max = 737.74
m/z_exp = 738.52
Δm/z = 0.78
m/z_avg = 325.47
m/z_exp = 324.84
Δm/z = -0.63
- 2H
m/z_max = 441.34
m/z_exp = 442.40
Δm/z = 1.06
- 2H
m/z_max = 413.29
m/z_exp = 413.69
Δm/z = 0.40

(d)

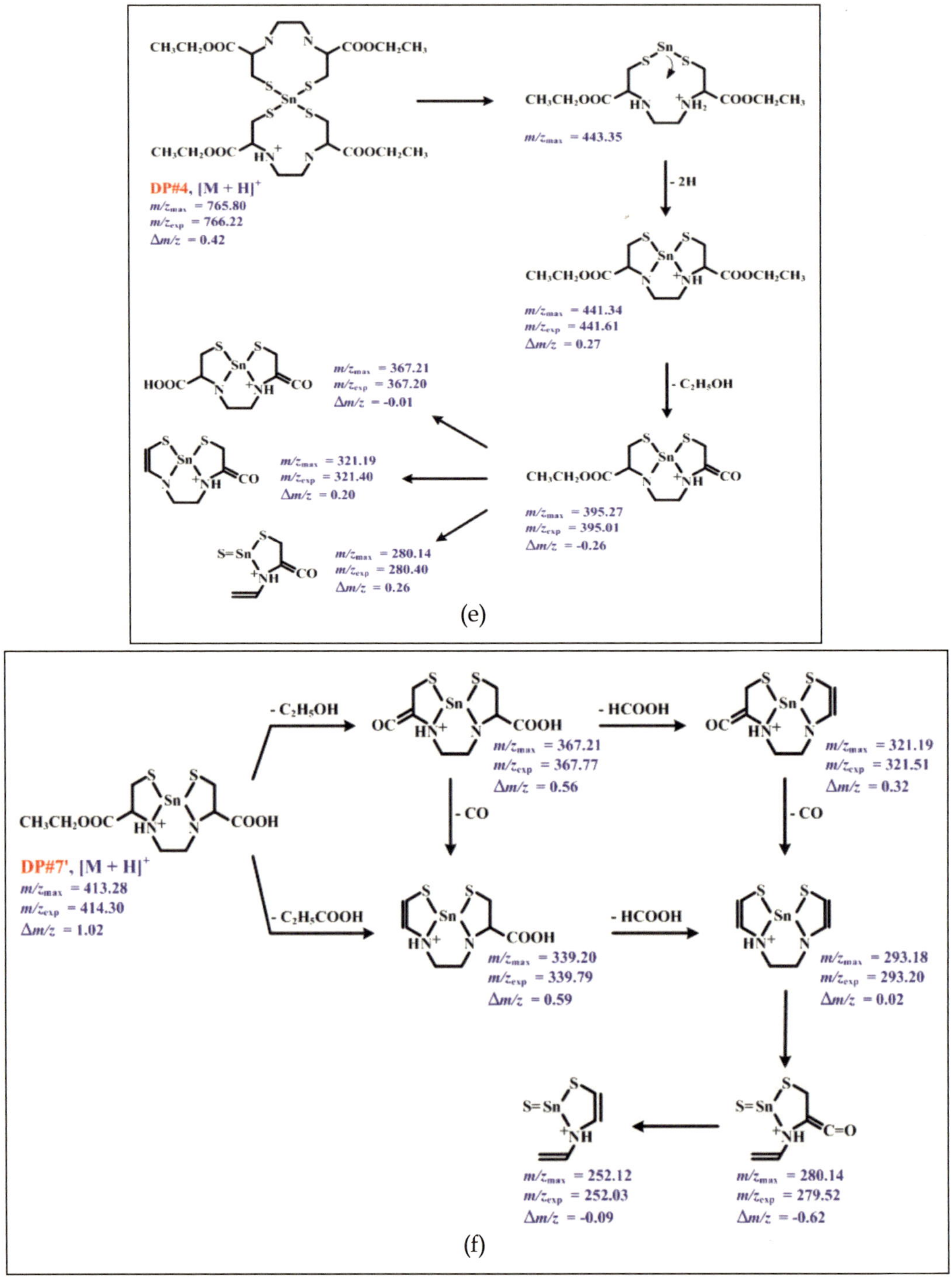

Fig. 6. Proposed CAD fragmentation pathways of the protonated molecules of (a) $ECD_{S\text{-}S}$ (m/z = 323.4), (b) $Sn(ECD)_{S2N2}$ (m/z = 442.0), (c) $(ECD)_2$ (m/z = 645.4), (d) $Sn(ECD)_2$-Et (m/z = 738.0), (e) $Sn(ECD)_2$ (m/z = 766.4) and (f) $Sn(ECD)_{S2N2}$-Et (m/z = 414.0)

3.5.1 Degradation product, DP#3

Here, we have identified the degradation production with intermolecular disulfide bond as (ECD_2), i.e. DP#3. The structure of DP#3 is shown in Fig. 1. In the HPLC chromatograms, DP#3 was found in the neutral solvents, acidic hydrolysis, oxidation, thermal degradation (Fig. 3 and 4(a)) and solutions with low concentration of $SnCl_2$ (Fig. 4(b) and (c)).

The typical product ion (MS/MS) scan spectra of protonated molecular ion with *m/z* 645.4 were identified as DP#3 (Fig. 5(g)). The MS/MS fragments of DP#3 are summarized in Table 2 and the linearities of MRM transitions were studied. The linear least-square regression equations and correlation coefficients ($r > 0.9990$) of MRM transitions showed a good linearity over the calibration range, indicating the stability of these fragmentations (data not shown).

Proposed CAD fragmentation pathways of the protonated molecules of DP#3 at m/z = 645.4 is presented in Fig. 6(c).

No significant hydrolysis product of DP#3, i.e. $(ECD)_2$-Et, $(ECD)_2$-2Et, $(ECD)_2$-3Et or $(ECD)_2$-4Et was detected in the MS scanning. Because species exchange reaction among ECD, ECD_{S-S} and $(ECD)_2$ was found in the HPLC chromatograms, we suggested that DP#3, $(ECD)_2$ can decompose reversibly into ECD or ECD_{S-S} and degrade further.

3.5.2 Degradation products, DP#4 and DP#5

In the ECD to $SnCl_2$ ratio of 12.5 : 1, 8 : 1 and 4 : 1 (v/v), one more nonpolar product (DP#4, $t_R = 6.04$ min) when compared to ECD and its polar hydrolysis product (DP#5, $t_R = 1.68$ min) were formed as indicated in Fig. 4(a)-(d). For higher concentration of $SnCl_2$ (ratio = 2 : 1 and 1 : 1), DP#4 was fast degraded and disappeared. The structures of DP#4 and DP#5 are shown in Fig. 1. The typical product ion (MS/MS) scan spectra of protonated molecular ions of DP#4 and DP#5 are shown in Fig. 5(h)-5(j). The MS/MS fragments of DP#4 and DP#5 are summarized in Table 2. Proposed CAD fragmentation pathways of the protonated molecules of DP#4 and DP#5 are shown in Fig. 6(e) and 6(d), respectively.

The peaks that appeared in the protonated molecular ions with the m/z range of 732 to 770 was further studied by TOF (Fig. 5(j)), the pattern was mainly due to the contribution of stable isotopes of tin and sulfur. Simulation spectra of DP#4 ($Sn(ECD)_2$) and DP#5 ($Sn(ECD)_2$-Et) are shown in the inset of Fig. 5(j). The isotopic distribution pattern and isotopic abundances were similar and coincident with the simulation results calculated by the software of API 'Isotopic Distribution Calculation' (Analyst, version 1.4.1, MDS Sciex, Ontario, Canada). This finding is in agreement with our earlier report which showed that highest intensity peak was mainly contributed from the stable isotope Sn-120 (Yang et al., 2010). However, in this case, isotopic composition of sulfur and tin were significantly complicated the MS spectra of $Sn(ECD)_2$-Et and $Sn(ECD)_2$ for determining of a fragmentation ion's molecular weight and m/z.

No significant DPs of $Sn(ECD)_2$-2Et was found in the MS spectra.

3.5.3 Degradation products, DP#6' and DP#7'

In the ECD to SnCl2 ratio of 2 : 1 (v/v), only two high polarity products at t_R of 0.93 and 1.14 min were left (Fig. 4(e)). It indicated that they might be partial degradation products of DP#4 and DP#5 when compared to the spectra of lower $SnCl_2$ solution (Fig. 4(d)). The typical product ion spectra and fragments of protonated molecular ions are shown in Fig. 5(k)-5(l) and summarized in Table 2. Three possible structures of Sn(ECD) (DP#6, DP#6′ and DP#6″) and Sn(ECD)-Et (DP#7, DP#7′ and DP#7″) are proposed in Fig. 1, of which

$Sn^{4+}(ECD)_{S2N2}$ (DP#6′) and $Sn^{4+}(ECD)_{S2N2}$-Et (DP#7′) were considered to be the prominent ones.
The experimental values of protonated molecular ions at m/z_{exp} = 442.11 and 414.30 supported this consideration. Moreover, there might be two possible explanations for this result.
First, the proposed net reactions of Sn(II) to Sn(IV) in the existence of dissolved oxygen or H_2O_2 are spontaneous in the forward direction. The proposed net reactions are as follows:

$$\frac{1}{2}O_2 + 2H^+ + Sn^{2+} \rightarrow H_2O + Sn^{4+} \quad E_{net} = 1.090 \text{ volt}$$

or

$$H_2O_2 + 2H^+ + Sn^{2+} \rightarrow 2H_2O + Sn^{4+} \quad E_{net} = 1.625 \text{ volt}$$

Second, both sulfur and nitrogen have lone pair electron can donate to the electrophile, Sn(IV). Sulfur is more nucleophilic than nitrogen, therefore sulfur can bond to the electrophile and react with it faster than the nitrogen does. For an irreversible reaction, the molecules do not have a chance to find the most energetically stable formation, and so they stay in whatever shape they form first and nucleophiles determine what the products are (A Crystal Clear Chemistry Concepts Tutorial). Highest amounts of DP#6′was existed in the ratio of ECD to $SnCl_2$ = 2 : 1 (v/v) and duration time of 4-7 hrs. Additionally, DP#7′ was existed only when the ratio of ECD to $SnCl_2$ (w/w) was greater than 2:1 and duration time was longer than 2 hrs. These results indicated that DP#6′ and DP#7′ were reversible thermodynamic products. Proposed CAD fragmentation pathways of the protonated molecules of DP#6′ and DP#7′ are shown in Fig. 6(b) and 6(f), respectively.
No significant DPs of $Sn(ECD)_{S2N2}$-2Et was found.

3.5.4 Degradation product, DP#8

Surprisingly, m/z 872.1, 901.0 and 975.5 can be found in the precursor scan of m/z 441.0, indicating that ECD trimer might be existed. Although no significant $Sn(ECD)(ECD)_2$ (mw_{avg} = 1086.05) can be detected in the MS spectra, it is reasonable to suggest a feasible structure and formation of DP#8 (trimer), i. e. $Sn(ECD)(ECD)_2$ shown as in Fig. 7. It seems that these results are due to labile and further decomposition of $Sn(ECD)(ECD)_2$.

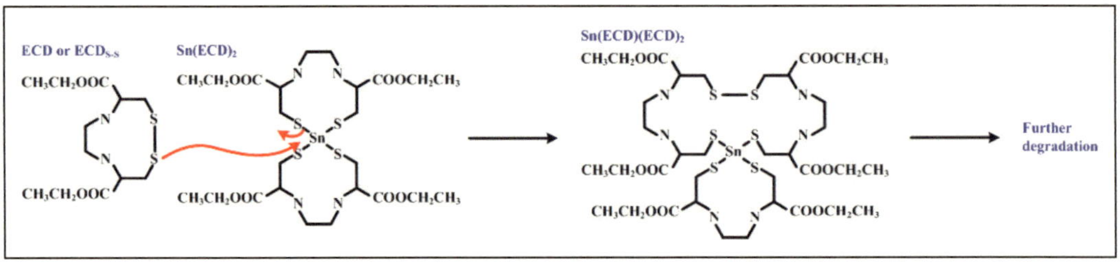

Fig. 7. Proposed formation mechanism of DP#8, $Sn(ECD)(ECD)_2$

3.6 Postulated degradation pathway of ECD and ECD Kit

The degradation pathway of ECD (API) and ECD Kit is shown in Fig. 8. Under alkaline and oxidation conditions, the drug can form DP#1, DP#1′, DP#2 and DP#2′ through ester hydrolysis and intra-molecular disulfidation. Under oxidation conditions, inter-molecular

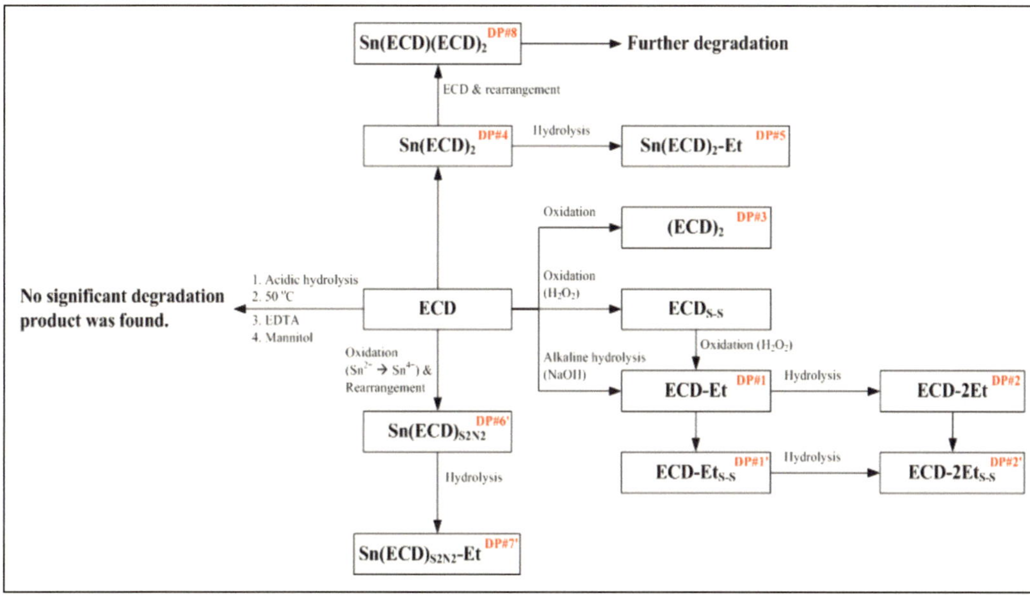

Fig. 8. Degradation pathway of ECD (API) and ECD Kit (drug product)

disulfidation of ECD also resulted in the formation of DP#3, i.e. $(ECD)_2$. For ECD Kit, the existence of $SnCl_2$ inhibited the formation of DP#1, DP#1′, DP#2 and DP#2′. In the meantime, oxidation of Sn(II) to Sn(IV) promoted the Sn(IV)-ECD complexation with coordination number of 1 and 2 to DP#6′ and DP#4, respectively. DP#6′ and DP#4 was further hydrolyzed to monoacid monoester derivatives, i.e. DP#7′ and DP#5. Moreover, the detection of Sn-trimer demonstrated the existence of DP#8, i.e. $Sn(ECD)(ECD)_2$.

4. Conclusion

The present study was designed to determine the factors affecting on the stability of ECD and ECD Kit and was given an account and the reasons for the use of Tc-99m-ECD which are suggested in practice guideline of ACR and EANM. The most interesting results emerging from the data are the degradation mechanisms and profiles of ECD. These findings enhance our understanding of ECD Kit about its stability, degradation pathways and structures of DPs. ECD is one of the diaminodithiol (DADT) derivatives to form stable complexes with radiorhenium or radiotechnetium. Therefore, the present study makes important implications for developing formulation of radiorhenium or radiotechnetium labeling pharmaceuticals. Further study for designing a more stable ECD Kit, such as a new reducing agent, reduction methodology or procedure is strongly recommended.

5. References

Abdel-Dayem, H. M. (Principal Drafter). (2002). ACR Practice Guideline for the Performance of Single-Photon Emission Computed Tomography (SPECT) Brain Perfusion Imaging, The American College of Radiology, (2002), Res. 19, pp. 487-491

Alsante, K. M., Ando, A., Brown, R., Ensing, J., Hatajik, T. D., Kong, W. & Tsuda, Y. (2007). The Role of Degradant Profiling in Active Pharmaceutical Ingredients and Drug Products. *Advanced Drug Delivery Reviews,* Vol.59, (2007), pp. 29-37, ISSN 0169-409X

Baertschi, S. W. (2006). Analytical Methodologies for Discovering and Profiling Degradation-Related Impurities. *Trends in Analytical Chemistry,* Vol.25, No.8, (2006), pp. 758-767, ISSN 0165-9936

Bauer, M., Silverman, D. H. S., Schlagenhauf, F., London, E. D., Geist, C. L., van Herle, K., Rasgon, N., Martinez, D., Miller, K., van Herle, A., Berman, S. M., Phelps, M. E. & Whybrow, P. C. (2009). Brain Glucose Metabolism in Hypothyroidism: A Positron Emission Tomography Study before and after Thyroid Hormone Replacement Therapy. *Journal of Clinical Endocrinology & Metabolism,* Vol.94, No.8, (2009), pp.2922-2929, ISSN 0021-972X

Borroni, B., Anchisi, D., Paghera, B., Vicini, B., Kerrouche, N., Garibotto, V., Terzi, A., Vignolo, L. A., Di Luca, M., Giubbini, R., Padovani, A. & Perani, D. (2006). Combined 99mTc-ECD SPECT and Neuropsychological Studies in MCI for the Assessment of Conversion to AD. *Neurobiology of Aging,* Vol.27, (2006), pp. 24-31, ISSN 0197-4580

Caroli, A., Geroldi, C., Nobili, F., Barnden, L. R., Guerra, U. P., Bonetti, M. & Frisoni, G. B. (2010). Functional Compensation in Incipient Alzheimer's Disease. *Neurobiology of Aging,* Vol.31, (2010), pp. 387-397, ISSN 0197-4580

Chang, C. -P., Shiau, Y. -C., Wang, J. -J., Ho, S. -T. & Kao, A. (2002). Abnormal Regional Cerebral Blood Flow on 99mTc ECD Brain SPECT in Patients with Primary Sjögren's Syndrome and Normal Findings on Brain Magnetic Resonance Imaging. *Annals of the Rheumatic Diseases,* Vol.61, (2002), pp. 774-778, Online ISSN 1468-2060

Chao, M. -Y., Liu, K. -T., Hsia, Y. -C., Liao, M. -H. & Shen, L. -H. (2011). Direct Determination of ECD in ECD Kit: a Solid Sample Quantitation Method for Active Pharmaceutical Ingredient in Drug Product. *Journal of Biomedicine and Biotechnology, in press,* ISSN 1110-7243

Chen, W. -Y., Lin, C. -Y., Chou, I -C., Tsai, F. -J. & Sun, S. -S. (2003). The Role of 99mTc-ECD Brain SPECT in Differentiating Tourrette's Syndrome from Chronic Tic Disorder. *Annals of Nuclear Medicine and Sciences,* Vol.16, (2003), pp. 59-63, ISSN 1022-923X

Diler, R. S., Reyhanli, M., Toros, F., Kibar, M. & Avci, A. (2002). Tc-99m-ECD SPECT Brain Imaging in Children with Tourett's Syndrome. *Yonsei Medical Journal,* Vol.43, No.4, (2002), pp. 403-410, ISSN 0513-5796

Jocić, B., Zečević, M., Živanović, L., Protić, A., Jadranin, M. & Vajs, V. (2009). Study of Forced Degradation Behavior of Eletriptanhydrobromide by LC and LC-MS and Development of Stability-Indicating Method. *Journal of Pharmaceutical and Biomedical Analysis,* Vol.50, (2009), pp. 622-629, ISSN 0731-7085

Kapucu, Ö. L., Nobili, F., Varrone, A., Booij, J., Vander Borght T., Någren, K., Darcourt, J., Tatsch, K. & Van Laere, K. J. (2009). EANM Procedure Guideline for Brain Perfusion SPECT Using 99mTc-Labelled Radiopharmaceuticals, Version 2, *European Journal of Nuclear Medicine and Molecular Imaging,* Springer, Published online: 17 October 2009.

Kulikov, A. U. & Zinchenko, A. A. (2007). Development and Validation of Reversed Phase High Performance Liquid Chromatography Method for Determination of

Dexpanthenol in Pharmaceutical Formulations. *Journal of Pharmaceutical and Biomedical Analysis*, Vol.43, (2007), pp. 983-988, ISSN 0731-7085

Liu, K. -T., Yang, H. -H., Hsia, Y. -C.; Yang, A. -S., Su, C. -Y., Lin, T. -S. & Shen, L. -H. (2008). Development and Validation of an HPLC Method for the Purity Assay of BZM, the Precursor of Striatal Dopaminergic D2/D3 Receptor SPECT Imaging Agent [123I]IBZM (Iodobenzamide). Journal of Food and Drug Analysis, Vol.16, No.5, (2008), pp. 28-38, ISSN 1021-9498

Mikiciuk-Olasik, E. & Bilichowski I. (2000). Determination of L,L-ethylene Dicysteine Di-Ethylester Stability by RP HPLC. *Chemia Analityczna (Warsaw)*, Vol.45, (2000), pp. 809-813, ISSN 0009-2223

Raijada, D. K., Prasad, B., Paudel, A., Shah, R. P. & Singh, S. (2010). Characterization of Degradation Products of Amorphous and Polymorphic Forms of Clopidogrelbisulphate under Solid State Stress Conditions. *Journal of Pharmaceutical and Biomedical Analysis*, Vol.52, (2010), pp. 332-344, ISSN 0731-7085

Schraml, F. V., Beason-Held, L. L., Fletcher, D. W. & Brown, B. P. (2006). Cerebral Accumulation of Tc-99m Ethyl Cysteinate Dimer (ECD) in Severe, Transient Hypothyroidism. *Journal of Cerebral Blood Flow & Metabolism*, Vol.26, (2006), pp. 321-329, ISSN 0271-678X

Shah, R. P. & Singh, S. (2010). Identification and Characterization of a Photolytic Degradation Product of Telmisartan Using LC–MS/TOF, LC–MSn, LC–NMR and on-Line H/D Exchange Mass Studies. *Journal of Pharmaceutical and Biomedical Analysis*, Vol.53, (2010), pp. 755-761, ISSN 0731-7085

Shah, R. P., Sahu, A. & Singh, S. (2010). Identification and Characterization of Degradation Products of Irbesartan Using LC-MS/TOF, MSn, on-Line H/D Exchange and LC-NMR. *Journal of Pharmaceutical and Biomedical Analysis*, Vol.51, (2010), pp. 1037-1046, ISSN 0731-7085

Sun, S. -S., Chou, I -C., Lai, Y. -H. & Kao, C. -H. (2001). 99mTc-ECD SPECT Image in Children with Gilles de la Tourette's Syndrome: a Preliminary Report. *Annals of Nuclear Medicine and Sciences*, Vol.14, (2001), pp. 149-153, ISSN 1022-923X

Sunada, I., Ishida, T., Sakamoto, S. & Tsuyuguchi, N. (2004). A Discrepancy between Tc-99m HMPAO SPECT and Tc-99m ECD SPECT in Creutzfeldt-Jacob Disease. *Journal of Clinical Neuroscience*, Vol.11, No.6, (2004), pp. 648-650, ISSN 0967-5868

The Difference between Nucleophiles and Bases, *A Crystal Clear Chemistry Concepts Tutorial*, Date of access: 2011/4/7, Available from: http://www.crystalclearchemistry.com/organic/concepts/nucbase.php

Tsopelas, C. & Smyth, D. (2005). Characterization and Quality Control Analysis of 99mTc-Bicisate. *Journal of Nuclear Medicine Technology*, Vol.33, No.2, (2005), pp. 89-93, ISSN 0091-4916

Ukisu, R., Kushihashi, T., Tanaka, E., Baba, M., Usui, N., Fujisawa, H. & Takenaka, H. (2006). Diffusion-Weighted MR Imaging of Early-Stage Creutzfeldt-Jakob Disease: Typical and Atypical Manifestations. *RadioGraphics*, Vol.26, (2006), pp. S191-S204, Published online: www.rsna.org/education/rg_cme.html., Online ISSN 1527-1323

Verduyckt, T., Kieffer, D., Huyghe, D., Cleynhens, B., Verbeke, K., Verbruggen, A. & Bormans, G. (2003). Identity Confirmation of 99mTc-MAG3, 99mTc-Sestamibi and

99mTc-ECD Using Radio-LC-MS. *Journal of Pharmaceutical and Biomedical Analysis,* Vol.32, (2003), pp. 669-678, ISSN 0731-7085

Wielepp, J. P., Burgunder, J. -M., Pohle, T., Ritter, E. P., Kinser, J. A. & Krauss, J. K. (2001). Deactivation of Thalamocortical Activity is Responsible for Suppression of Parkinsonian Tremor by Thalamic Stimulation: A 99mTc-ECD SPECT Study. *Clinical Neurology and Neurosurgery,* Vol.103, (2001), pp. 228-231, ISSN 0303-8467

Yang, H. -H., Liu K. -T., Hsia Y. -C., Chen, W. -H., Chen, C. -C., Men, L. -C. & Shen, L. -H. (2010). Development and Validation of an HPLC Method for Determination of Purity of Sn-ADAM, a Novel Precursor of Serotonin Transporter SPECT Imaging Agent I-123-ADAM. *Journal of Food and Drug Analysis,* Vol.18, No.5, (2010), pp. 307-318, ISSN 1021-9498

8

Analog and Digital Systems of Imaging in Roentgenodiagnostics

Dominika Oborska-Kumaszyńska
[1]*General Radiology, Interventional Radiology and Neuroradiology, University Hospital, Wroclaw,*
[2]*Wolverhampton Royal Hospitals, New Cross Hospital, Medical Physics and Clinical Ennginеering Department, Wolverhampton,*
[1]*Poland*
[2]*United Kingdom*

1. Introduction

In contemporary radiology, the carrier of the diagnostic information is the image, obtained as a result of an X-ray beam transmitted through the patient's body, with modulation of intensity of X-ry beam and processing of data collected by the image detectors. Depending on the diagnostic method used for image acquisition, signals can be detected with analog (x-ray film) or digital systems (CR, DR and DDR). The imaging systems based on digital presentation of diagnostic image have a dominating advantage in contemporary roentgenodiagnostics. Each of these methods of image acquisition due to its own technological solutions, determines a different quality of imaging (diagnostic data). Owing to that fact, quality control procedures, their scope (range), studied (evaluated) parameters as well as the evaluation of detection efficiency in these diagnostic systems are so much different.

2. Systems of imaging in roentgenodiagnostics

Imaging in roentgenodiagnostics in based on three technological solutions of diagnostic data acquisition:

- Analog systems: cassette with an intensifying screen, x-ray film, viewing box;
- CR systems (Computed Radiography) – cassettes with phosphor imaging plates, CD reader, control station with monitors for description of diagnosed images;
- Direct and indirect digital systems (DR and DDR):
  - imaging panel coated with a layer of scintillation material (e.g. cesium iodide-CsI-famous for being used in image intensifiers), with photodiodes (Si) and an active matrix TFT (Thin Film Transistor- which is an electronically controlled switch) – indirect digital detector,
  - camera CCD – optical system coated with scintillation layer placed in front of CCD camera (charge-coupled devices) – indirect digital detector
  - imaging panel with a photoconductors – a layer of amorphous selenium (Se) with an active matrix TFT – indirect digital detector.

Each of above mentioned systems and technological solution is characterized by properties influencing image parameters, both improving their quality, as well as causing the loss or deterioration of the diagnostic information.

A. Analog systems

Analog image detector of an x-ray unit consists of three components: x-ray film, intensifying screen and light-proof housing (cassette). An ancillary part of imaging system is a developing device along with reagents for photochemical processing of x-ray films (developer, fixing agent, water).

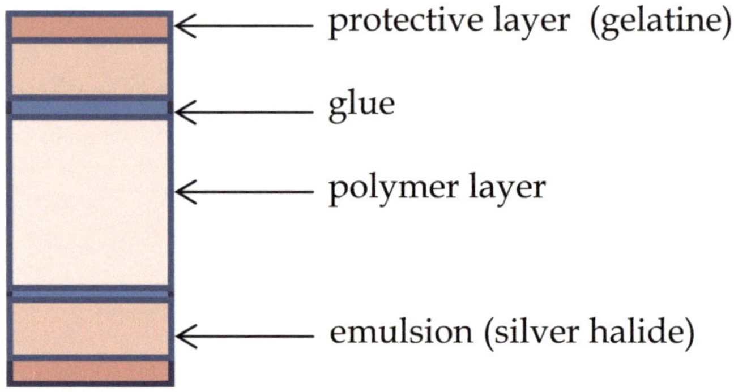

(according to: http://astrophysics.fic.uni.lodz.pl/medtech/)

Fig. 1. Scheme of cross-section through x-ray film

X-ray film is made of a thin, transparent, flexible polyester film (base material of x-ray) and a thin layer of photographic emulsion (Fig. 1). The emulsion is coated with protective polymer layers and with a dulling agent. The photosensitive elements are silver halide microcrystals. If quanta of x-rays or visible light radiation energy emitted by the intensifying screens transfer their energy, as soon as they reach the silver halide crystals. If the energy is sufficient then crystal electron can be transferred from valence band to the conduction band, where it can move freely throughout the crystal. The electron moved until it reaches the place of crystal structure distortion, where it is "trapped". The presence of the "trapped" electron with a negative electric charge causes the attraction of positively charged silver ions which further leads to origin of a metallic silver atom in that particular place. This place becomes a part of a latent image. As a result of described process, latent image sites may occur on the crystal, large enough (a few up to several metalic silver atoms), large enough for the developer to initiate the process of reduction of the whole crystal to metalic silver. The amount of the developed metallic silver determines the level of optical density, creating a real image on the film. „Revealing" of the image on x-ray film proceeds in the course of photochemical processing.

In order to increase the probability of x -ray photos detection the intensifying screens are applied (Fig. 2). The intensifying screens are polymer, on one side coated with scintillation powders (containing phosphor) which contain elements with high atomic number (e.g. calcium tungstate $CaWO_4$, gadolinium oxide sulphide Gd_2O_2S). In radiographic cassette we may usually find two screens, (an exception make mammographic cassettes with a single intensifying screen), whereas the scintillation layer is directed to inner side of the cassette, where the x-ray film is placed. Applying scintillation materials enables the reduction of radiation doses from 20 to 100 times, depending on type and screen sensitivity.

The light, generated due to influencing of X radiation with a phosphor layer, is emitted in all directions. Approximately 50% of this radiation does not reach x-ray film owing to its absorption in intensifying screens. That is why in construction of intensifying screens two solutions have been asummed (accepted)- the reflecting layer or absorbing transmitted light in the structure of intensifying screen. The reflecting layer has a function of directing the light radiation to x-ray film. It increases the sensitivity of the configuration: x-ray film/intensifying screen, which results in reducing the radiation dose delivered to the patient's body. The drawback of this solution is deterioration of clarity of obtained images. In case of applying absorbing layer the sensitivity of the configuration x-ray film/intensifying screen decreases but the clarity of the image improves.

The polymer layer (base) of intensifying screen is responsible for mechanical parameters (qualities) of the screen and other layers are applied on it. Scintillation material layer is made of phosphor crystals and bonding material. Equal density of phosphor crystals is a fundamental parameter determining homogeneity for the screen which should ensure the equal exposure of z-ray film. Each of phosphor crystal is bounded with setting (bonding material) and the rate of volume of phosphor crystals to total volume of the layer is determined as coating weight. The layer with a higher value of this factor means higher content of phosphor crystals per volume unit. This allows for applying thinner scintillation layer, therefore obtaining images with higher clarity than in case of set x-ray film/intensifying screen with the same sensitivity but lower value of coating weight factor. The other factor determining the quality of the screen is specific gravity of phosphor - theoretically higher value of specific gravity should provide thinner thickness of phosphoric layer. The thicker phosphor layer improves the sensitivity of the set screen/ film resulting from increased x radiation absorption, however, on the other hand the increased thickness of the layer leads to deterioration of image clarity, mainly due to light diffusion in this layer. Of no importance is also the phenomenon of light scattering in phosphor layer. And truly saying, from the technological point of view and qualities of the imaging set, more significant as determining the thickness of the screen is the relation between phosphor and a setting (bonding) material. In case of universal screens this relations approximately means nine phosphor units per one unit of bonding material.

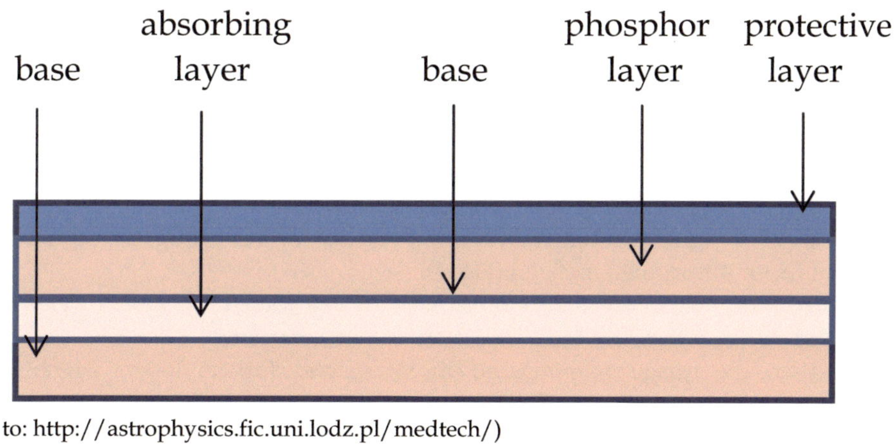

(according to: http://astrophysics.fic.uni.lodz.pl/medtech/)

Fig. 2. Detailed structure of the screen

Polymer protective layer, protects the scintillation layer against mechanical damages (e.g. during cleaning the screen) and at the same time prevents from collecting air between the phosphor layer and x-ray film. In order to provide equal clamp of the x-ray film to the intensifying screens, the magnetic plate in construction of cassettes is also applied (made on base of rubber, placed at the bottom side of the cassette) and a steel film. (on the lamp side).
The materials used for construction of intensifying screens increase their efficiency due to higher value of absorption coefficient (Fig. 3).

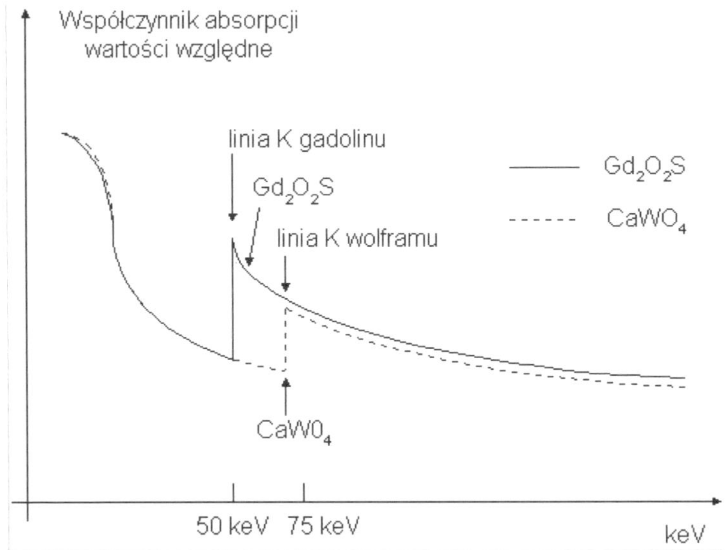

(source: http://astrophysics.fic.uni.lodz.pl/medtech/)

Fig. 3. Absorption coefficient of the screens made on base of CaWO4 and Gd2O2S

The conversion efficiency determines the applying of intensifying screens in radiography in particular kinds of research (examination). The efficiency of converting x radiation into light emission is determined as conversion efficiency coefficient. It is a measurement of light emission efficiency by phosphor layer in result of x radiation influence and efficiency of x-ray film exposure. The conversion efficiency of calcium tungstate (CaWO4) is about 5 %, while applying a pair of screens may account for, for system with low sensitivity up to 20 %, and with high sensitivity up to 40 %. The screens are made on base of rare earth metals characterize with sensitivity up to 60 %.
Summarizing, the factors determining the sensitivity of the set x-ray film/ intensifying screen belong among others, type of phosphor (conversion efficiency), the thickness of its layer, value of weight coating coefficient, presence of reflecting/absorption layer, the presence of factor absorbing (crystals) (Fig. 4).
Use of x-ray films with both side emulsion allows to increase the sensitivity of imaging systems. However, when the light emission is not completely absorbed in one layer of emulsion then the image is produced in its second (other) layer, which may cause deterioration of the image (Fig. 5, Fig. 7).
X-ray film has a certain performance characteristics, the parameters of which should be taken into account when setting the exposure conditions and the proper degree of film

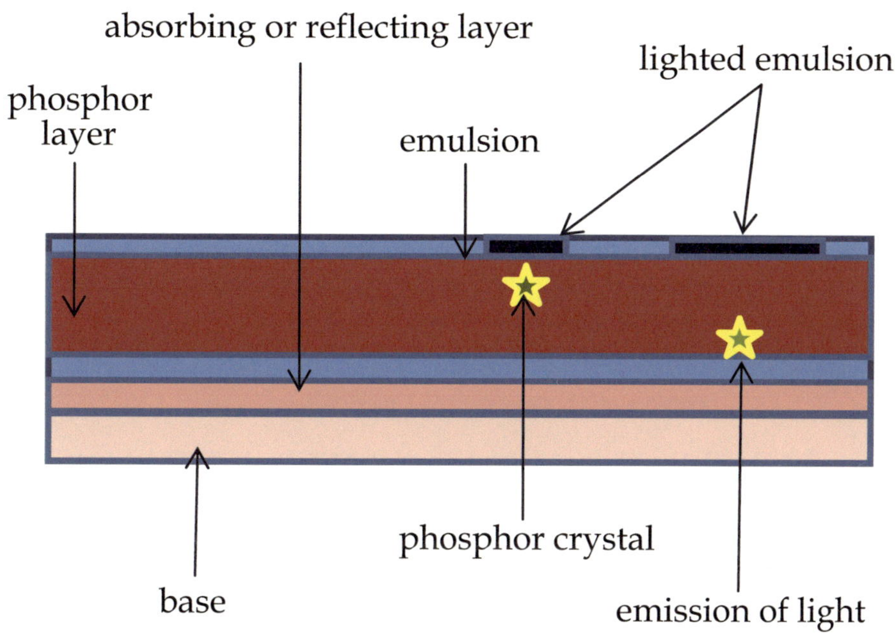

(according to: http://astrophysics.fic.uni.lodz.pl/medtech/)

Fig. 4. Influence of thickness of phosphor layer on the lighted emulsion area

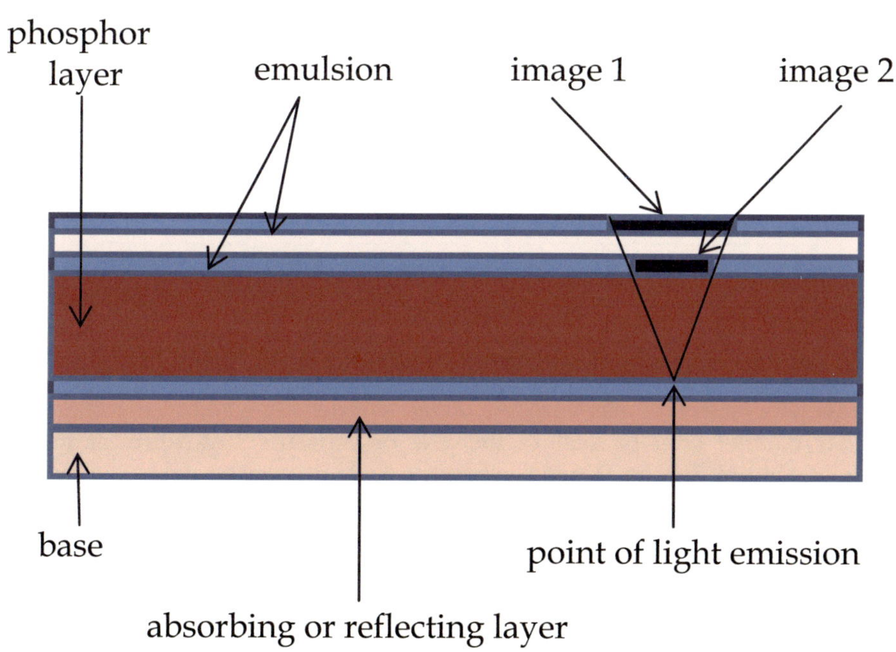

(according to: http://astrophysics.fic.uni.lodz.pl/medtech/)

Fig. 5. Influence of phosphor layer thickness on the sharpness of image in the case of both side emulsion film

utilization. Sensitometric properties of light -sensitive materials are determined by the characteristic curve (Fig. 6), which is the graph of function of the optical density over the logarithm of exposure. The exposure (E or H) is the product of illuminance (I) and exposure time (t), and the optical density is the logarithm of the quotient of incident light intensity to intensity transmitted through the exposed x-ray film. Photosensitive material does not react in areas of low exposure, and obtained optical density results from optical density of a base material coated with photosensitive layers (Dp) and from optical density of fog (D_0), which added all together, constitutes the minimum optical density (Dmin). With a certain value of exposure, the value of optical density start increasing slowly and converts into the linear dependency with a specific angle of inclination.

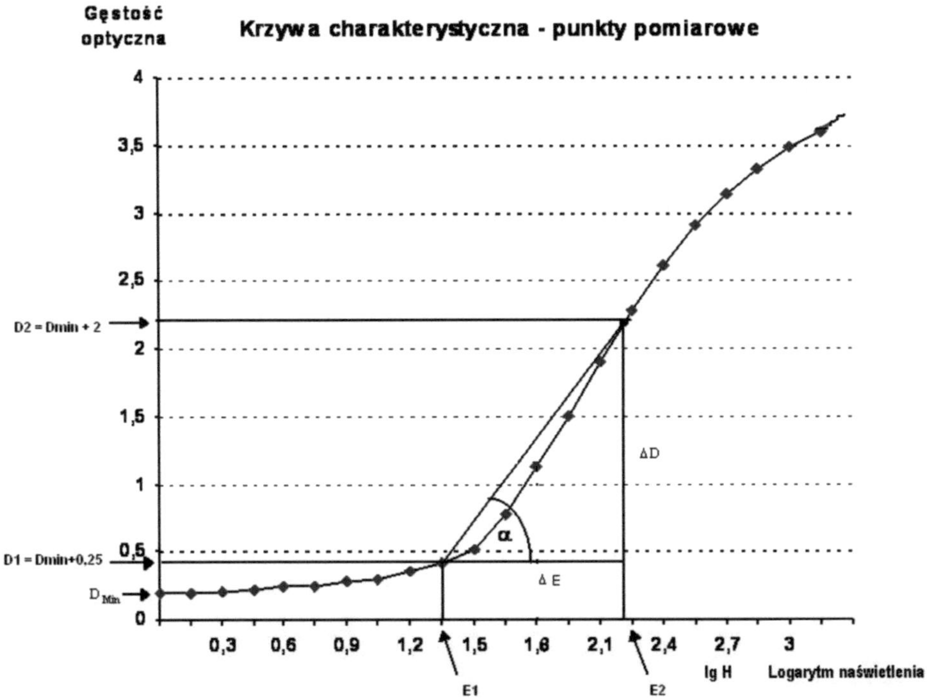

Fig. 6. X-ray film optical density curve (response curve)

Then along with increasing of exposure, the increase of density is reduced and characteristic curve reaches parallel line, parallel to the axis of exposure logarithm, at the constant density- the maximum density (Dmax), when the level of reached exposure results in the development of all crystals contained in the layer.

The basic sensitometric values determined by characteristic curve are the sensitivity coefficient as well as contrast coefficient.

Sensitivity coefficient

The sensitivity of photographic material is defined as the smallest amount of light required to produce a certain optical density. Its measure is the value of exposure logarithm essential to obtain optical density, which is defined as Dkr = Dmin + 1,0 and for this value the relevant to it, the value of exposure logarithm is read.

Contrast coefficient

The contrast is defined as the measure of differences in optical density in the image and it be calculated from the inclination of rectilinear part of characteristic curve. It is defined as slope in the point (e.g. contrast coefficient α, as trigometric function of inclination angle of tangent in the point of inflection of characteristic curve in closeness of the middle of rectilinear part) or as the average gradient which is determined as trigometric function of inclination angle of the part joining 2 critical points of optical density D1 = Dmin + 0,25 and D2 = Dmin + 2,00 (Fig. 6).

The basic values allowing for determining imaging parameters are optical density, contrast and resolution, where:

1. Optical density is the opacity in image and is defined as the value of common logarithm from converse of transmission coefficient. This coefficient can be recorded as the ratio of light intensity transmitted through certain point to light intensity reaching this point.

$$D = \log\left(\frac{1}{T}\right) = \log\left(\frac{I_{padaj.}}{I_{przep.}}\right)$$

2. Contrast is a measure of difference in optical density of particular image areas, relevant to differences in density an thickness of tissues visible in the image. The image contrast depends on: energy of radiation, structure of studies tissue or organ, sensitivity of the film and the type of intensifying screen as well as the dose of scattering radiation and optical density fog.
3. Image resolution is determined by the number of pairs of lines per 1 millimetre (no/mm), which may be imaged and possible to recognize as separated structure. Resolution determines the smallest object possible to imaging, at the same time determines the smallest, possible to be recognized, distance between two objects.

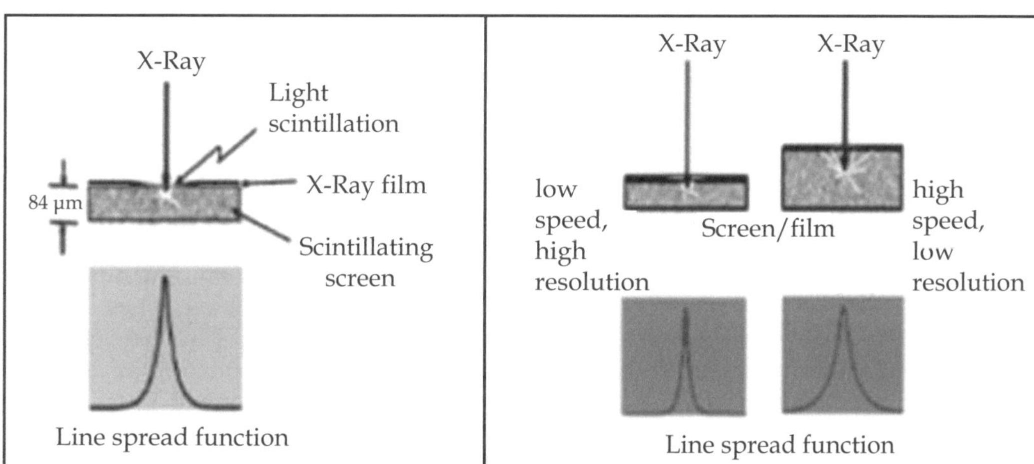

(according to: Andrew P. Smith, Fundamental Digital Mammography, Physics, Technology and Practical Considerations)

Fig. 7. Intensifying screen performance – the influence of sensitivity and scattering of imaging system

x-ray film is the detector with limited capacity of data collection, for which significantly important is the proper optimization of process of image development, starting with proper device setting (exposure management) through the process of photographic proceeding (system sensitivity, artefacts in image, level of noises), illumination conditions of dark room to proper choice of parameters of the whole imaging system (intensifying screens in range of length of emitted light, relevant to parameters of applied x-ray films). Properly setting of elements of diagnostic data development reflects creating the most beneficial conditions for proper image quality (optimization).
In analog systems quality and diagnostic evaluation takes place in descriptive rooms with use of viewing box which should absolutely meet parameters values determining respectively the illumination conditions (no more than 50 lx) as well as lumination of emitted light (cd/m2).

B. Systems CR

An imaging detector in digitized computed radiography (CR) is phosphor imaging plate. An essential detection component of its structure is a layer of luminophore (PSP-photostimulable phosphor imaging system) (Fig. 8).

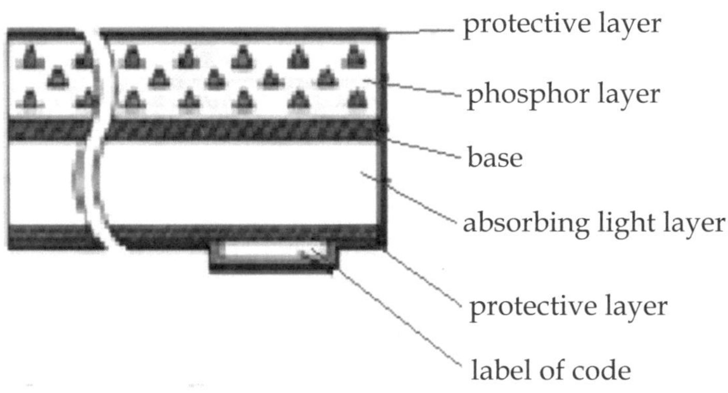

(source: http://www.fujifilm.pl/page,168.html)

Fig. 8. Construction of CR imaging plate

The imaging plate is placed in the cassette similar to one used for analog radiography. Geometry and imaging technique are similar as well.
In the system basing on phosphorous imaging plates, x-ray radiation quanta are absorbed by a phosphor layer of the imaging plate (IP). Deposited energy of x-ray radiation in the material of the imaging plates is stored in a portion of energy, located in metastable regions called F-centres. During x-ray beam exposure, the latent image is formed in phosphor layer by accumulation of energy in these centres. Reading of imaging information from CR plates bases on the phenomenon of transmitting energy to the electrons located in metastable states (F centres) and on moving them to energetic levels, causing introduction atoms of phosphor plate material in the rough state. It results from returning of the atoms to the ground state and generate photons emission from the spectrum the visible light range, which is recorded by a photomultiplier. The photomultiplier converts the light image into analog electric signal, which on the output is converted into a digital signal by an analog-digital converter. Then the signal values are intensified and with a use of mathematical algorithms are processed in segmentation, rescaling and filtering procedures.

Scanning of the image and converting into diagnostic form is performed with reader scanning imaging plates and the control computer at description unit. In case of point-scan readout in scanner (Fig. 9), the imaging plate is moved in one direction while the concentrated laser beam (diameter of the beam 50um-100um) moves perpendicularly to that direction, from one side of the imaging plate to the opposite one.

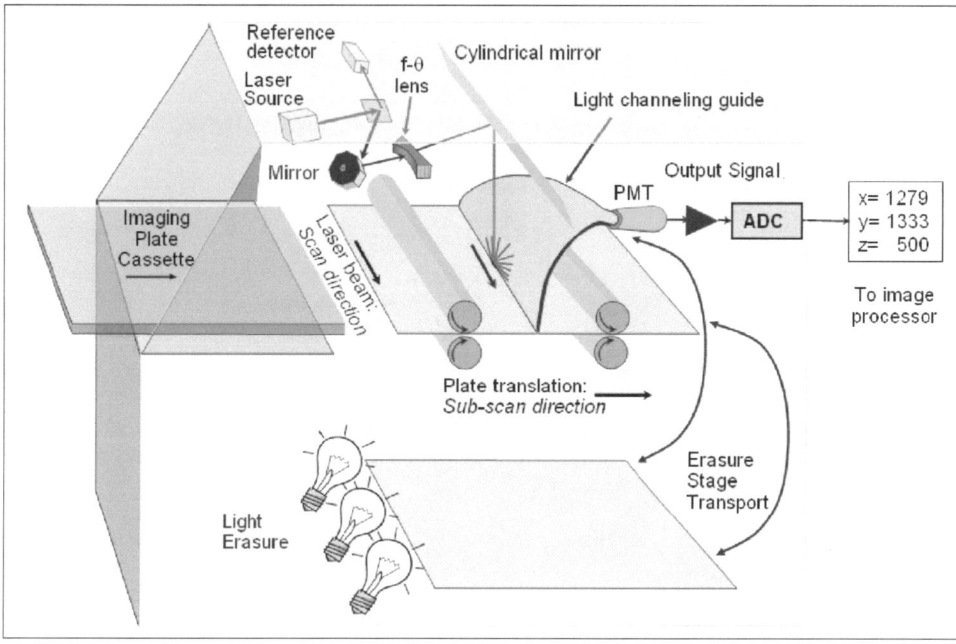

(source: AAPM Report No 93)

Fig. 9. The process of image scanning from imaging plate - point scan system

The entire surface of the plate is scanned by the laser beam and the light generated in the process of photostimulation and emitted by each point of the imaging plate, is collected by the optical fibre. The time of scanning plates depends on the size of the detector and the scanning capacity (speed) of the reader (the average time of scanning is about 60-70s). In recent technology readers, the linear laser beam is used, which increases the speed of scanning data (average scanning time is about 5-10s). In such scanners, reading imaging plate is still and the source of linear laser beam moves above its surface (Fig. 10).

Reading of imaging information from CR plates bases on the phenomenon of transmitting energy to the electrons located in metastable states (F centres) and on moving them to energetic levels, causing introduction atoms of phosphor plate material in the rough state. It result of returning of the atoms to the ground state, it leads to generating photons emission from the spectrum the visible light range, which is recorded by a photomultiplier. The amount of the recorded light from photostimulation stays in adequate proportion to the number of F-centres and thus also to the amount of x-ray radiation absorber in that point. Photomultiplier converts the light image into analog signal, which, on the output is converted into a digital signal by an analog-digital converter. Before digitization, the PMT signal is intensified, usually in non-linear manner. As the next step, „raw" signal values are processed in segmentation, rescaling and filtering procedures, using.

In order to optimize the effectiveness of imaging plate utilization within range of exposure, the digitized systems provide the pre-reading procedure, which allows for testing the sensitivity of the signal reading. Initially, a weak beam laser is used for reading a „raw" image data, appropriate reading, sensitivity and exposure conditions are determined basing on analyses of the data obtained, afterwards the proper reading proceeding takes place. The method enables normalization of the luminescence, in which the x-ray mage appears, in order to allow the conversion of digital signals, irrespectively of the object being tested and the x-ray radiation dose.

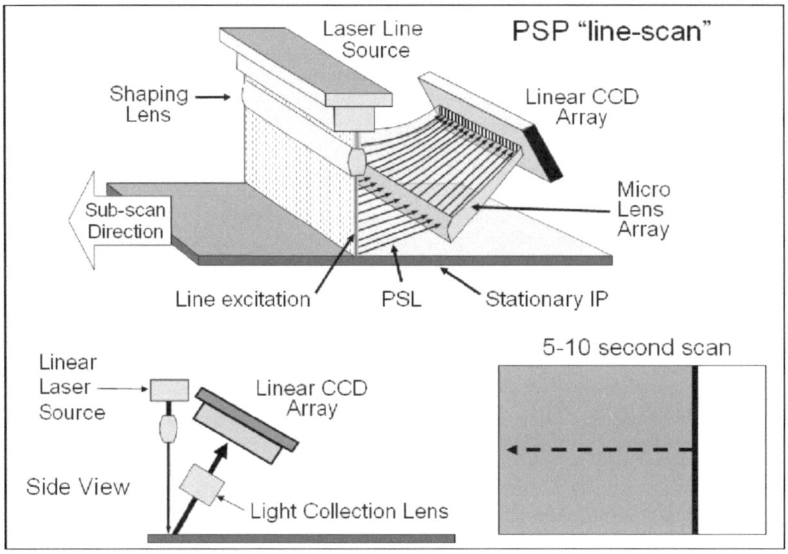

(source: AAPM Report No 93)

Fig. 10. The process of image scanning from imaging plate - line scan system

After scanning (reading) of imaging plate is completed, the imaging plate is exposed to a visible light emitted, with a high insensitivity beam, by the erasing lamp that „deletes" the x-ray image and makes the imaging plate ready for reuse.

In digital radiography in CR systems, the disadvantageous for image acquisition, phenomenon of fading is present i.e. fading of recorded signal, thus the time between exposure of imaging plate and its reading with the reader (scanner) is significant. Typical image recorder loses approximately 25% z of deposited signal in the period of time from 10 minutes to 8 hours after exposure.

C. Digital system: DR and DDR

Imaging system CCD

Detectors in CCD technology are the devices used for image recording, their performance in based on recording the lights emitted from luminophor. Matrix CCD (Charge Coupled Device - the device with coupling load) is made of series of electrodes (light-sensitive components) based on semiconductors base and constituting matrix of capacitors (Fig. 6). the number of components determines the resolution of obtained digital images.

The voltage is delivered separately (solely) to each of the electrodes, which enables generating the image detector with particular positioning system. When the surface of CCD matrix is illuminated with light emitted from luminophor, the carriers are revealed. These carriers are moved in regular electric impulses and are „recalculate" by the circuit which „traps" carriers from each light-sensitive element. Then transfers them to condensers, measures, intensifies the voltage and erases condensers again. The number of carriers gathered in this manner, within specific time depends on light intensification which is adequate to the amount of ionizing radiation reacting with luminophor layer. In the result, information on value of the voltage of light reaches each of light-sensitive components.

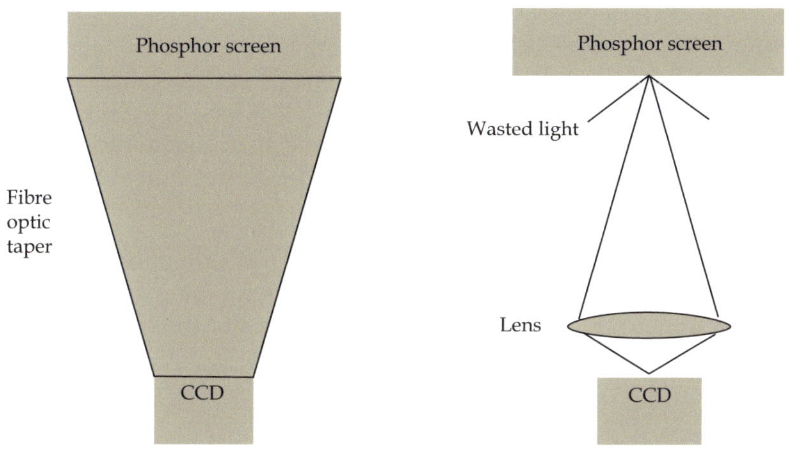

(source: IPEM, report no 32 part 7)

Fig. 11. Image detector based on CCD technology

Each element of CCD (connector MIS) has layered structure (Fig. 12). component layers are M - Metal, I - Insulator, S - Semiconductor. Electrode (M) constitutes upper layer of the MIS connector and is made of non-transparent metal with doped silicon (Me+Si). The electrode covers part of surface of the photo element reducing efficient apparatus, which determines the percentage of participation of photo element active surface in relation to its total surface.

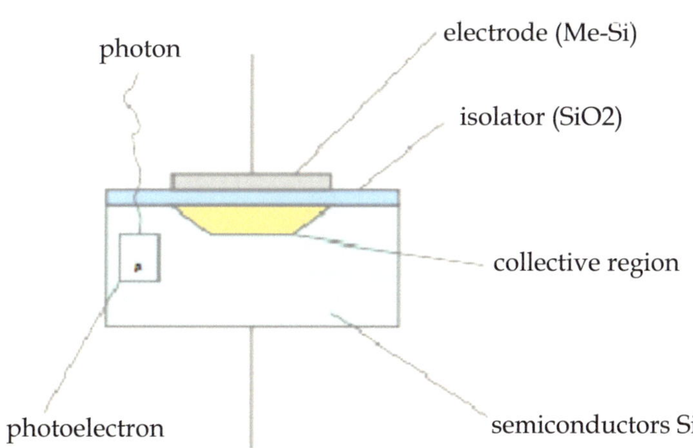

(source: http://www.fotospokojna.com/linki/www_cyfra/matryce.pdf)

Fig. 12. Scheme of single element CCD construction

The function of positive electrode is maintaining of generated during exposure electrons in the region of the photo element (Fig. 12 - collective region). It prevents from arising of phenomenon of blooming, which is blurring of the voltage on the adjoining elements. The effect regards saturation state of detector cell which overload causes effluent of collected voltage to the adjoining cells. Below the positive electrode, there is semitransparent layer of isolator (I) made of silicon dioxide. (SiO2). The function of isolator is to prevent from uncontrollable effluent of the voltage to the electrode. The light- sensitive element of MIS connector is bottom layer of silicon semiconductor (Si). The number of current carrier, released due to reacting of the light with semiconductor layer, is directly proportional to the amount (voltage and time of duration) of falling light. Reading of collected in photo elements of the matrix charges has a sequential character. Along each of matrix columns, the canal CCD is placed, in which charges move in direction to reading recorders. The electrons from the first row of sensors are transmitted to reading recorders and then signal intensifier and analog-digital convertor, where the current signal is digitalized and saved on memory carrier.

Systems DR and DDR (image panels)

In case of radiography with digital image detectors, the most common solution iare panels made of amorphous silicon or selenium (indirect digital systems) and panels based on a matrix made of electrodes separated by a layer of insulator and the active components, such as thin-film transistors. (Fig. 13, Fig. 14).

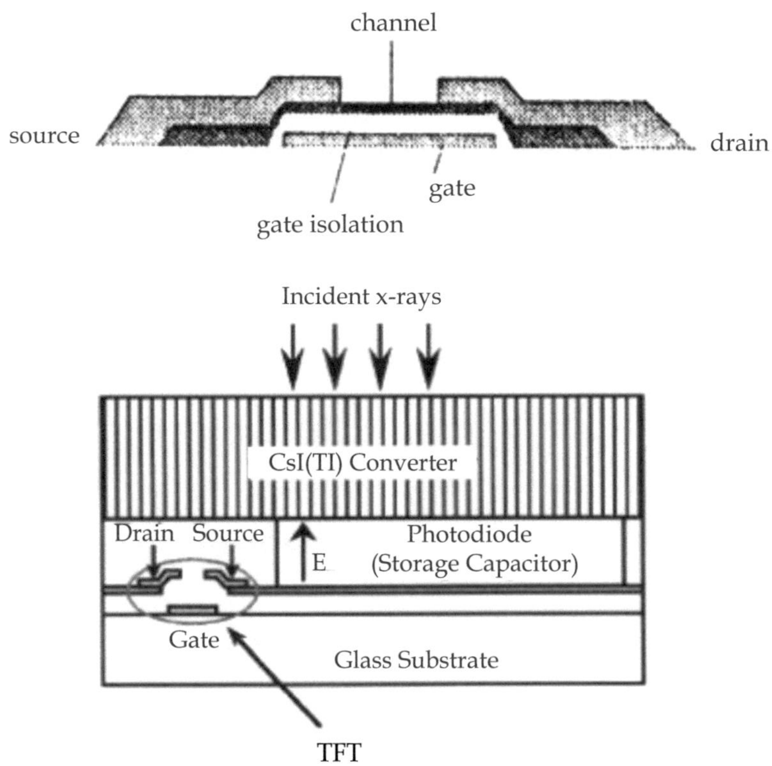

(source: http://astrophysics.fic.uni.lodz.pl/medtech/)

Fig. 13. Structure of thin-film transistor

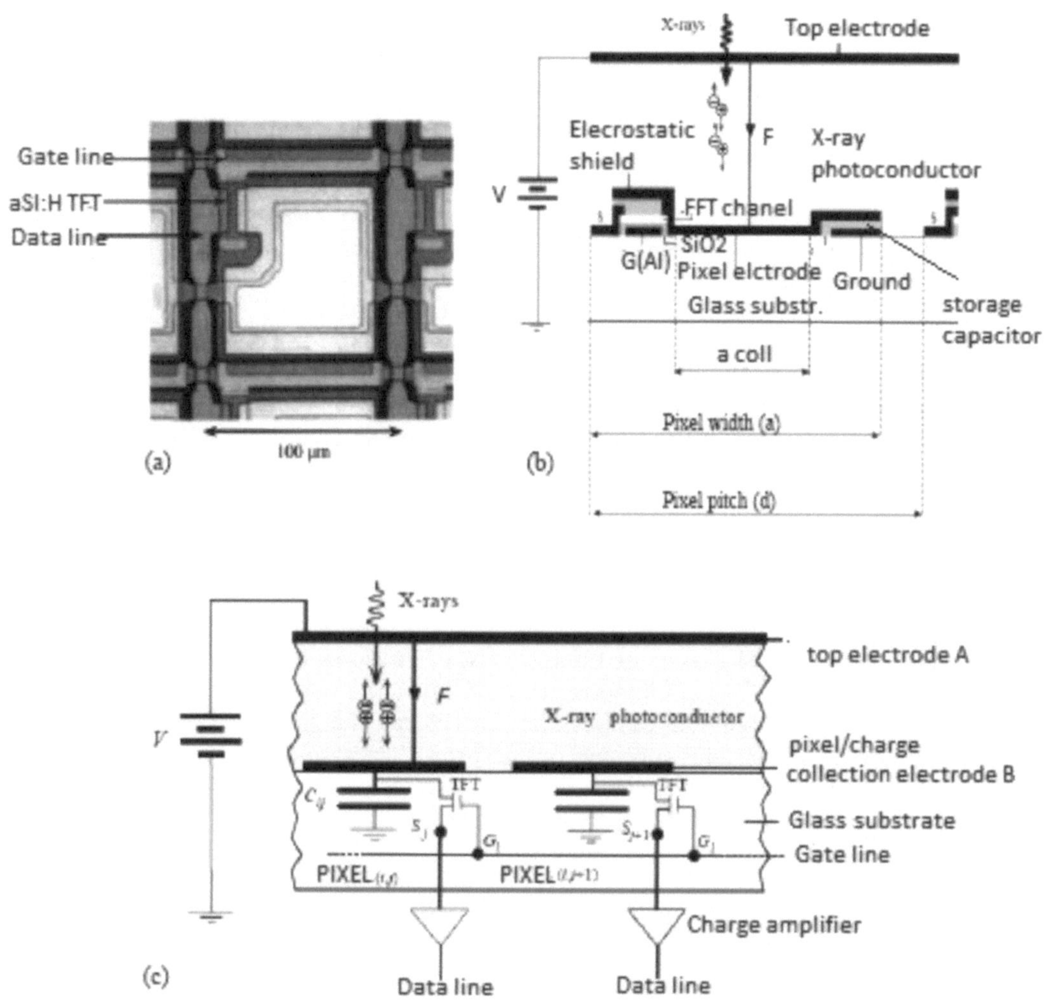

(source: Mammographic detectors, G. PANAGIOTAKIS UNIV OF PATRAS)

Fig. 14. Detector of the direct digital system: (a) microphotograph, (b) structure of the single pixel of the TFT matrix (c) schematic diagram of the structure of two pixels

The base for indirect digital systems with imaging panels are the detectors which consist of photoconductors, such as amorphous silicon or selenium. Layer of silicon detector contains a matrix of receptors, each equipped with its own control components (transistor or diode)and corresponding to one pixel of the image. Regulating (control) systems are responsible for the process of data reading: line after line, electrical signals are intensified and converted into a digital form. Silicon itself is not sufficiently sensitive to energy of x-rays radiation used in diagnostic imaging. Therefore, silicon layer is covered with a layer of scintillation material such as cesium iodide (CsI), which is characterized by a needle-like structure of a crystal, causing less side scattering of light and higher resolution of the imaging system. The thickness of the CSI crystal with its needle-like structure can be adjusted to the desired sensitivity of the system (ensuring proper level of absorbance of x-ray radiation) with the maintenance of high spatial resolution at the same time.

Photodiodes (Si:H), located under a layer of scintillation material, convert the optical signal into an electrical signal (charge), which is accumulated in a capacitive element of a pixel.
In the direct digital imaging system, the detector is made of photoconductors characterized with a high atomic number (e.g., Se or PbI_2), which cover an active area of the matrix. That kind of the structure forms a layer of photoconductor which directly converts x-ray radiation into charge carriers, drifting to collecting electrodes. The main advantage of direct digital systems, comparing to CR systems and indirect DR systems, in terms of image quality, is the lack of effects from the light photons scattering at the detector material. Electric charge, generated as the effect of x-rays radiation, is collected by a single electrode, which makes the side-scatter (diffusion) effect not significant for the process of image creation. Additionally, detector absorption efficiency can be maximized by an appropriate selection of the material of photoconductors, calibration, and a proper thickness of the layer of capacitive elements. An active matrix consists of M x N number of pixels. Each pixel has three basic elements: the TFT switch, pixel electrode and capacitor. Active matrix is determined by the pixel width, width of pixel collection and the distance between pixels (pitch) (d) (Fig. 14).
TFT elements function as switches, for each pixel individually, and control the charge. Each line of pixels is simultaneously electronically activated during the reading process. Normally, all TFT elements are deactivated, allowing the accumulation of the charges on pixels electrodes. Data can be obtained by external electronics and controlling of the TFT status by software. Each TFT contains three electrical components: Gate controlling "on" or "off" TFT status, Drain (D) connecting the pixel electrode and the pixel capacitor and Source (S) connected to a collective data transmission line. When the gate line is activated, all the elements of TFT in a particular row are 'on' and the charge collected on the electrodes is read from the data line. Parallel data are multiplexed into serial data, discretized and transferred to a computer to create the image (Fig. 15).

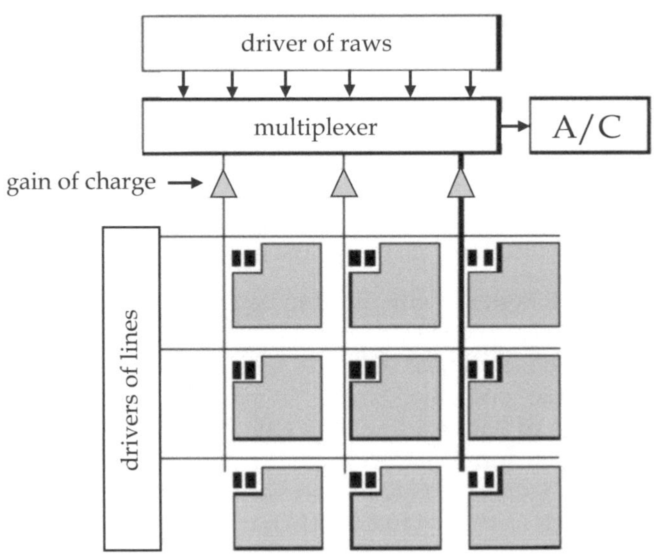

(source: http://astrophysics.fic.uni.lodz.pl/medtech/)

Fig. 15. The structure of the matrix of sensors of displays and the way of controlling the reading structure of the matrix of sensors

The undoubtful advantage of the image acquisition in the digital form is the possibility of post processing of this image.

3. Image processing

Initial image processing (pre-processing)

Raw image generated by the digital system is the image that does not have any diagnostic value. It is caused due to wide range of dynamic as well as presence of inhomogeneity of particular detective components of the image recorder. That is why, initial processing of the raw material in connected with compensation of artefacts coming from image detector.

In digital systems (DR) detectors are not homogenous regarding performance of the particular components, due to differences in intensification of recorded image (offset), the presence of defected pixels. Inhomogeneity of the detector constitutes the source of the noise in the image and is some cases geometrical uniformity.

Inhomogeneity in the image may be eliminated by applying proper correction processes:

- offset - „dark current" - generated by electronical components as the additional charge which without applying map of offset correction would add to the value of the charge, formed as the result of reacting of x -ray radiation with the detector. Correction of offset map is produced by signal recording for the image created without participation (involvement) of x-ray radiation. (black/ dark image).
- intensification - the differences in intensification for particular components of the detector result from the differences in thickness of phosphorous components, sensitivity of these elements and the difference of the intensifiers. This effect should not be reflecting the diagnostic image, therefore the gain map of intensification is applied. The map of intensification corrections is obtained as the result of averaging of a few flat images achieved in the result of detector exposure to homogenous beam of x-ray radiation. In order to obtain homogenous signal from the surface of the whole detector, recorded values of the signals for its particular components are divided into values present on gain map of intensification.
- bad pixels - digital detector of the image may have damaged or faulty (broken) detector components, both as a single as well as the whole lines of these components. The effect of presence of irregularly working components requires correction and the gain map is produced („bad pixel map"). Then the dead regions of imaging may be deleted from the diagnostic image and compensated by the assigning the pixel value as the average or median of signal from adjoining pixels.
- geometrical uniformity - for the majority of digital systems, imaging systems are not spatial uniformity in diagnostic images. However, in case of detectors based on CCD technology, using during forming image, one or more lenses, the clinical image will be distorted. During calibration of the device, the value of distortion caused by the lenses, should be measured and should be implemented fixed correlation for each image.

Diagnostic image processing (post-processing)

The process of initial image processing is used for correction of detectors characteristics. Further image processing is applied for generating the image for presentation and with parameters allowing for conducting its clinical evaluation. It is connected with identification of collimation as well as with process of processing special frequency and grey scale. The process of processing in range of frequency (e.g., accumulation of noises, edges enforcement and attaching the imaging net) is a common tool used for improving quality of the image. During the process of processing of the diagnostic image also the transformation of pixel values to new digital values is also performed– LUT („a look-up table"). LUT is mainly applied in two cases:

- digital detector usually has much wider dynamic range than the range obtained intensifications in clinical image, therefore LUT is used for extraction of the range of detector work to clinical signal and its adjustment to displayed grey scale,
- LUT is used for reinforcing the contrast of pixel values applied in clinical conditions. In clinical application non-linear LUT function may be more useful- the most common is correlation curve in shape of letter S (similar to response curve for imaging with radiographic film - OD characteristic curve).

LUT also rescales the pixels vales to the values close to the referencing values, which may sometimes cause data loss between obtained dose by the detector and the vales of grey scale (therefore, the evaluation of this relations is conducted on the image after pre-processing).

4. Factors determining image quality

Detection efficiency (DQE)

Quantitative detection efficiency (DQE) i the parameter describing image receptor regarding its radiation detection efficiency, spatial resolution and the noise. DQE describes relative efficiency of maintaining of SNR level (the ratio of the signal scale to the noise), possibly obtained in imaging process and is defined as SNR^2_{out}/SNR^2_{in}, where SNR^2_{in} is SNR of exposure reaction on the receptor and quantitative equal to the input stream. In this manner, DQE may be expressed as efficiency of transferring SNR through the system and its efficiency reflects the detection quality and image acquisition. For imaging system SF (screen film), CR (phosphor imaging plates) and DR (digital systems), quantum efficiency is determined by the thickness, density and structure (content) of absorber (image detector).

Signal transfer property (STP – signal transfer property)

Signal transfer property (STP), which determines the relations between initial parameters of the detector(usually optical density or pixel value), which is non-changeable parameter) and an air kerma, measured at the entrance of this detector, is a parameter allowing for objective evaluation of image quality. Imaging system must retain linear response or at least possibly linear in order to form proper results for quantitive analysis of the measurements, or it regards simple measurement such as homogeneity or more complex as MTF measurements. In the system is not linear (e.g. logarithmic, quadratic) the relevant inversion of STP function should be applied, corresponding the type of relation of detectors response to obtained radiation dose.

Dose indicator (DDI – Detector dose indicator)

DDI is the parameter characterizing digital form of imaging. The essential benefit of the digital imaging is separation of acquisition from the image presentation. Most of the digital detectors have a wide dynamic range and wide exposure range, which ensure good image quality. However, different exposures values may change in ambiguous way the sensitivity of the system or cause the increase of the number of situations, in which the dose received by the patient is not an optimal one. DI indicator is the parameter allowing for determining the changes in sensitivity of imaging system as well as calibration and system testing AEC (Automatic Exposure Control). Usually, there i s no linear relationship due to the dose and for needs of quantitative evaluation requires its transmission to the linear function. DDI is also the parameter depending on the radiation energy.

Dynamic range

In order to obtain the proper imaging quality in digital radiography, the image detector must have good contrast resolution in wide range of exposure intensity to X radiation. Dynamic range of the imaging system is the ratio of the largest and the smallest input intensities, which can be visualized. The smallest useful value of intensity is limited by the noise level of the system, while the highest value of intensity depends on detector saturation level.

Spatial sampling

All digital detectors sample the permanently fluctuating stream of X-rays at the input, at discrete locations, separated by gaps (pitch). In CR systems, the spacing between samples is the distance between adjacent positions of the laser beam during reading process from the imaging plate. In DR systems, pitch is the distance between centers of the spaces separating each of detecting elements. The spatial frequency in sampling, determines the digital system's ability to display high-frequency fluctuations in X-ray stream. If the influence of radiation stream with the receptor contains data of higher frequency than Nyquista frequency and the modulation transfer function (MTF) before sampling is not evanescent for these frequencies, then for low frequency, false noise may appear in the image.

MTF -Modulation Transfer Function

Modulation Transfer Function (MTF) is the response of the imaging system expressed depending on spatial frequency- i.e. it is the relationship of contrast and spatial frequency to the contrast for low frequencies (it means where the signal is not clear). Spatial frequency is expressed in cycles per pixel or pair of lines per millimeter. High spatial frequencies correspond to recognition of great number of details. MTF is determined with the pixel value as well as the distance between the centers of adjoining pixels („pixel pitch")

$$\underline{MTF(u)} - \text{sinc}(2\pi\Delta x u)$$

where:
Δx - pixel pitch,
u - spatial frequency.

MTF allows to compare in an objective way the qualities of different imaging systems. In order to perform the comparison, definition of signal transmission from communication theory is quoted (Fig. 16). if on the input, the proper signal is provided, in case of imaging the pattern object then on the output its image will be obtained. Comparing of the image, in the proper manner, with object allows to determine the imaging system characteristics. Therefore the object should be chosen in the way that the information about the system was as complete as possible. These object include among others: point image, linear image and edge image. (these are analogical terms to Dirac's delta function - unit impulse signal used in signal theory).In response to the object, the image is formed which is determined as point spread function *PSF*. analogically, in case of the object in the line form, the image is determined as the Line spread Function *LSF* (Line Spread Function). There is the relationship between *PSF* and *LSF* as well as imaging system characteristics and function *MTF* (Modulation Transfer Function). This function is defined on base of knowledge of input and output signal in area of spatial frequencies.

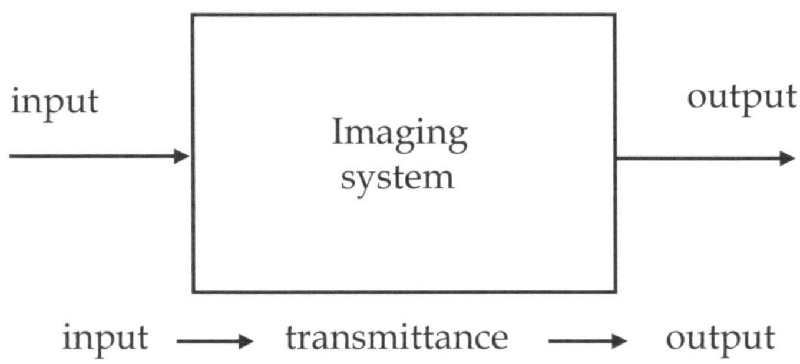

(source: http://astrophysics.fic.uni.lodz.pl/medtech/)

Fig. 16. Method of characteristic of imaging system

Spatial resolution

Spatial resolution is the ability of imaging system to visualize two adjacent structures as separate image elements, or a clear edge marking in the image (sharpness). This parameter describes the capacity of the system to imaging small objects. However, this parameter does not define how various frequencies are transmitted through detector system but this evaluation is proceeded with MTF measurement. In order to obtain the initial shape of MTF function for the system, the quadratic wavefunction transfer of contrast -SWCTF(f) may be applied. In this method, the resolution test object is used as a measurement object (lp/mm), and SWCTF(f) calculates according to the formula :

$$\mathrm{SWCTF(f)} = \frac{M_D(f)}{|M_D - M_S|}$$

where: f - spatial frequency, $M_o(f)$ - standard deviation of a region covering several line pairs, M_B – is the signal level of a region within a bar, and M_s is the signal level of a region in the spacing between bar (Fig. 17).

Losses in the spatial resolution occur due to blurring caused by geometric factors (e.g., size of a focus, scattering of light in the receptor), the effective area of the detector determined by the size of aperture, patient's movements in relation to the source of X-radiation, image detector, the thickness of the detector elements, screen, CSI crystal thickness and density of data reading.

In order to evaluate this parameter the resolution phantom is used (Fig. 17) not only the size of the detector influences the resolution in case of digital system but also the algorithm of processing of high contrast. Resolution for CR systems is also determined by the size of section of laser beam, as well as, hesitation and focusing the laser.

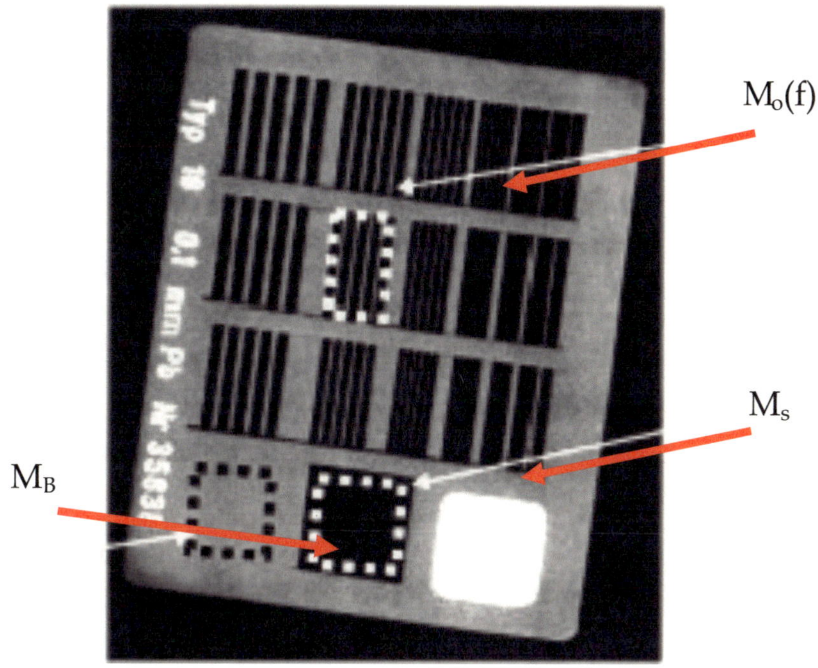

(source: IPEM report no 32 part 7)

Fig. 17. High contrast and spatial resolution test object

High contrast spatial resolution

High contrast resolution is determined in CR systems mainly by pixel distribution and value of sampling of photomultipliers in the reader (the direction of the scanning). Standard frequency of sampling in case of classic radiography is 5 - 12 pixels/mm, giving in the result the distribution of pixels in range of 200-80 um and leading to obtaining theoretical resolution limit 2.5-6 lines/mm. in case of mammographic systems the value of pixels system is 40 um. Resolution limit should be close to the Nyquist frequency. For smaller values of pixels distribution, the frequency is often below Nyquist frequency which implies that there are also other factors determining this parameter, e.g.. screen parameters and diagnostic workstation, processing process, section of laser beam, light scattering in phosphor layer etc.. Finally obtained in measurement, value of resolution, should be compared with Nyquist frequency limit, defined for 45 degrees by expression $\sqrt{2}/2^*\Delta p$, where Δp is pixels distribution.

Noise

Noise can be defined as fluctuations in the image, which do not correspond to differences in X-radiation absorption by objects. A measure of noise may be determined by estimating the noise power spectrum (NPS), which describes the correlation of spatial frequency and noise. The noise in the image is dominated by quantum (shot) noise resulting from quantum fluctuations in the X-ray and data digitization (in case of digital systems). However, all image receptors contain internal sources of noise, such as noise coming from the film grain and electronic noise in the CR and DR systems.

Internal noise of the detector, which has agreed correlation depending on the place on the receptor, is caused spatial difference in thickness of the intensifying screen in systems in SF, the efficiency of light detection depending on position in cd readers and the differences in intensification preintesifier in DR systems.

Deterioration of the image in radiography is also conditioned by the scattering of radiation, which is another source of noise and contributes to decreasing of image contrast. The solution to this problem is the use of anti-scatter grids placed in front of the image detector.

Utilization of the grip is particularly important in CR systems due to increase of sensitivity to scattered radiation of barium halide (edge K approximately is 35 keV), in ratio to system screens SF and contained in them gadolinium oxide sulphide (edge K approximately is 50 keV). However, in case of scanning systems (scanning with gap field), DR detectors have the capacity of „deleting" from registration scattered radiation and therefore they do not require the use of anti-scatter grid.

In most of detectors, the noise of the image is coherent with Poisson distribution (coefficient b should be 1.0 for Poisson noise in the image):

$$\nu = \alpha^* K^b,$$

where: K=DAK (detector air kerma); ν - variation, α i b - stable.

One of the essential parameters allowing to determine noise component in the image is defining signal to noise ratio (SNR – signal to noise ratio).

Dark noise (noise characterizing only digital systems, because is connected with electronical elements) may have a significant participation in image for regions with low level of useful signal,in particular, that similar to usage signal in registration process is intensified. Image correction for this parameter threshold contrast happens while adjusting look-up table.

One of advantages of digital imaging is the possibility of digital elimination of internal noises of image detector in post-processing stage, (obtaining the image with diagnostic values).

Contrast resolution

Contrast resolution refers to the value of the signal difference between the examined structure and the surrounding. It is the result of differences in X-ray absorption in the examined tissues. It is expressed as a relative difference in brightness between the relevant areas in the digital image shown on the monitor. Radiographic contrast is determined by the contrast of the object and receptor sensitivity. It is strongly depending on spectrum of x-ray radiation energy and presence of scattered radiation. However, in digital imaging, contrast in the image can be changed by setting the visualization parameters, independent of the acquisition conditions.

Evaluation of the system in range of its capacity of imaging regions with small values of the signal (small contrast) may be conducted on base of phantom image containing testing components with different thicknesses and diameters. During tests the visuality of these parameters in the image is determined and the diagram of detection coefficient id fixed. High value of coefficients of threshold contrast ($H_T(A) = 1 / (C_t * \sqrt{A})$, where: C_T –threshold contrast; A – region of visible element) is the measurement of high visuality of low contrast elements, depends on dose therefore imaging of testing object should be conducted for exposure values from the range of clinically applied doses.

Artefacts

Artefacts are all kinds of disorders appearing in diagnostic region. In case of SF systems they may be such as all sorts of no homogeneities resulting from developing process or lack of homogeneity of the film or intensifying screens, additional components in the image resulting from pollution of the cassettes or defects of intensifying screen. In case of digital systems apart from, origin of the artefacts in the image results from mainly from defective work of detector (in case of CR systems additionally the reader), defective processing of the signal or functioning of reconstruction algorithm.

For the imaging systems with imaging plate, the typical artefacts are „Moire patterns" ones - coming from anti-scatter grid; ghost image - resulting from unsuccessful delete of previous image, uniformity of the image; artefacts resulting from faulty cd CR. In case of DR systems, irregularity in the image may appear due to presence of faulty lines/pixels (generally they are eliminated in diagnostic image) in the process of pre- processing). They may also result from „checker board" effect - digital detectors are made of isolate panels, from which image date is connected in one entire part through electronic way. Each of panels also has a few intensifiers coating separated regions of detectors. If the response of any of these intensifiers or panels drifts then it may cause the change in the signal level and creating darker and lighter regions in diagnostic or testing image. Whereas, from combining image data from various detectors regions may result artefacts connected with accumulating of the signals or too big their separation- „stitching artefacts"- between plates of the detector may be potential gaps which size should not be significant from the point of forming diagnostic image (accepted for the general diagnostics is 100um). Artefacts appearing owing to the process of image processing is delay of the image- if the detector was exposed to high radiation exposure then initial image may be temporarily „ burnt" in the detector. Repeated calibration of the detector may cover it. However, after calibration process covered by this process" burnt" region may be revealed in next image. In this situation the detector requires performing another calibration Naturally, the artefacts in diagnostic image may also appear in result of defects of detector components, e.g., damage of phosphor layer - if phosphor or photoconductor disconnect from the TFT matrix or coupling of the light occurs then may appear region with weak signal or blurring region. The only solution in this case id the exchange of the detector.

5. References

[1] AAPM REPORT NO. 93, Acceptance testing and quality control of photostimulable storage phosphor imaging systems, 2006.

[2] AAPM REPORT NO. 96, The measurement, reporting and management of radiation dose in CT, 2008.

[3] AAPM REPORT NO. 116,,An exposure indicator for digital radiography, 2009.

[4] AAPM REPORT NO.74, Quality control in diagnostic radiology, 2002. 6) IPEM report no 32 part 7, Measurement of the Performance Characteristics of Diagnostic X -Ray Systems, Digital Imaging Systems, 2010

[5] B. Pruszyński:„Diagnostyka obrazowa. Podstawy teoretyczne i metodyka badań", PZWL, Warszawa 2001

[6] R. Kowski, M. Kubasiewicz: „Mammografia - podręcznik zachowania standardów jakości", Wydawnictwo Lekarskie, ACR, Warszawa 2001

[7] G. Panagitakis: „Mammographic detectors",
http://www.hep.upatras.gr/class/download/bio_sim_eik/mammographic_detectors.pdf
[8] Practice guideline for digital radiography, ACR practice guideline, 2007 (Res. 42)*
[9] Seibert JA, Ph., D Performance Assessment of DR Systems, UC DavisMedicalCenter Sacramento, CA, http://www.aapm.org/meetings/03AM/pdf/9787-48370.pdf
[10] http://astrophysics.fic.uni.lodz.pl/medtech/

9

Quality Assessment of Solid Pharmaceuticals and Intravenous Fluid Manufacturing in Sub-Saharan Africa

Adedibu C. Tella[1], Musa O. Salawu[2], Iyabo M. Phillips[3], Ojeyemi M. Olabemiwo[4] and George O. Adediran[5]
[1]Department of Chemistry, University of Ilorin,
[2]Department of Biochemistry, University of Ilorin,
[3]Department of Climate Change, School Advocacy Unit, Lagos State Ministry of the Environment,
[4]Department of Pure and Applied Chemistry, Ladoke Akintola University of Technology
[5]Department of Chemical Sciences, Ajayi Crowther University, Oyo, Nigeria

1. Introduction

The quality of pharmaceuticals cannot be compromised as these constitute a group of products ingested into the human and animal systems by routes such as oral, parenteral, topical etc. These groups of products therefore have direct bearings on our well being and there is therefore an absolute need to guarantee their quality, safety and efficacy. Drugs therefore have to be designed and produced such that when patients receive them for management of their ailments, they do not produce any adverse side reactions on such patients or their unborn babies.

The sub-Saharan Africa countries market are flooded with fake and adulterated drugs to such an extent that only 30 % of drugs available in these countries can be said to be genuine in terms of contents and efficacy. The side effect of fake and adulterated drugs is so serious that therapeutically, if administered, can give rise to treatment failure which at times may be serious enough to result to death. Assurance of the quality, safety and efficacy of pharmaceutical products is a continuing concern of World Health Organization. It is now recognized that stability of active components of preparations poses serious problems for many manufactured products, especially those entering international commerce and/or distributed in territories with harsh climatic conditions. These problems may arise as a consequence of

a. Improper storage (in heat, moisture, sunlight). This might lead to degradation or loss in potency. The manufacturer will always indicate the best possible storage conditions on the product, but it has been found that retailers and wholesalers do not have required facilities to achieve these conditions or some do not give regard to these warning consequently this leads to product quality deterioration before expiry dates.

In most sub-saharan Africa countries, manufacturers, retailers, wholesalers and general public persistently flout most storage instructions and thereby jeopardize the quality of the product.

b. Poor quality assessment. Due to local sourcing of raw material in a developing nation, lack of current high-tech analytical instrument or even unavailability of certain reagents used in official procedure may force the quality control analyst to develop alternative methods.

Despite efforts made around the world to ensure a supply of right quality and effective drugs, substandard, spurious and counterfeit products still compromise health care delivery in some countries especially in Sub-Saharan Africa.

Every government allocates a substantial proportion of its total health budget to drugs. This proportion tends, to be greatest in developing countries, where it may exceed 40%.

Without assurance that these drugs are relevant to priority health needs and that they meet acceptable standards of quality, safety and efficacy, any health service is evidently compromised. In Sub-Saharan Africa, drug manufacturing faces various challenges in assessment of quality of solid pharmaceuticals and intravenous fluid.

The first challenge is the deterioration of solid pharmaceuticals during distribution and storage as a result of sunny and humid climate..

The second challenge is non-availability of equipments specified in official references books (British Pharmacopeaia and United States Pharmacopeaia) in monographs for analysis of drugs. The third challenge is insufficiency of personnel with adequate technical know-how to man the quality control unit of the pharmaceutical company.

Intravenuos fluids, otherwise called infusions are fluids used in medical delivery by intravenous administration.

The most challenging quality control aspect of infusion manufacturing are sterilty and pyrogen level determination of the final product. Most intravenuous fluid product failures in Nigeria involve sterility failures and high pyrogen contents. The challenges of quality control of infusion manufacturing in Nigeria is compounded by lack of infrastructures (epileptic electric power supply) and high cost of useful test kits for sterilty and pyrogen. There is challenge of finding a more affordable and reliable test materials (kit) for pyrogen test. Most companies use the rabbit test method for pyrogen tests which has limitations in false results, delayed decision making. Since rabbit test for pyrogen is done after the terminal sterilization of products, failed product cannot be re-processed. The Limulus Amebocyte Lysate (LAL) test kits are expensive and not affordable though reliable. Nigerian infusion manufacturers require a cheaper and locally sourced test kit for in-vitro determination of pyrogen in addition to good infrastuctures for smooth operation.

It is therefore reasonable to assure that the analytical procedures involving the use of simple instruments will find greater application in Sub-Saharan Africa. Taking into consideration the aforementioned challenges, the main objectives of this paper is to carry out a review of degradation studies of common antibiotics in Sub-Saharan Africa by investigating the effect of heat, sunlight, moisture and U.V radiation on the potency of the drugs. The paper will also review some of the alternative analytical methods developed for assessment of quality of selected solid pharmaceuticals. A cheaper and locally sourced test kit for in-vitro determinations of pyrogen in intravenous fluids will be described. The chapter will also review some of the previous work done on this subject.

2. Degradation of drugs

Some of the drugs that are marketed in tropical countries are vulnerable subjected to degradation processes that can result into loss in the active component of these drugs. These problems may arise as a consequence of improper or inadequate storage and distribution of the products which can lead to physical deterioration and chemical decomposition resulting in reduced bioactivity, formation of toxic degradation products or the formation of an unstable product especially under tropical conditions of high ambient temperature and humidity. Many pharmaceutical substances are known to deteriorate during distribution and storage particularly in hot, sunny and humid climate. Tropical and subtropical climatic conditions are therefore expected to pose serious problems with respect to stability of drug. Many drugs are thermosensitive, when they are exposed to high temperature, degradation tends to occur.Many workers have investigated the effect of heat on the degradation rate profile of many pharmaceuticals.

In 1982, Kabela studied the influence of temperature on the stability of solid tetracycline HC1 measured by High Performance Liquid Chromatography in pure drug and capsules. The tetracycline hydrochloride were stored for about two years at temperature of 37°C, 50°C, and 70°C. It was found that at 37°C and 50°C, no decomposition was observed for tetracycline nor for its related substance. At 70°C, a distinct decrease in tetracycline HC1 was observed as well as a small increase in degradation products (Anhydrotetracyclinavaie HC1, 4-epitetracycline HC1 and 4-epianhydrotetracycline HC1). The degradation products are shown in figure 1.

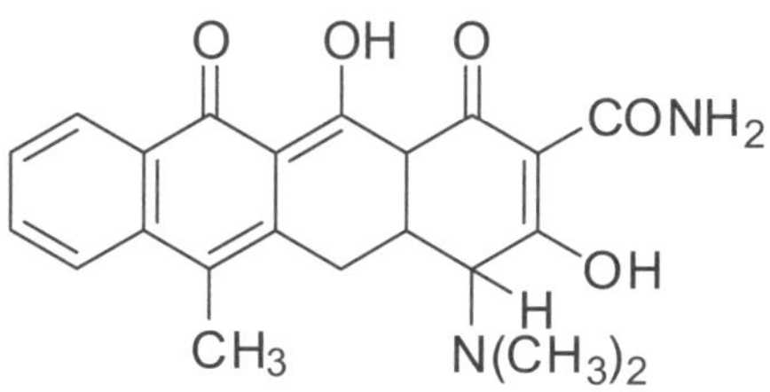

4-epianhydrotetracycline

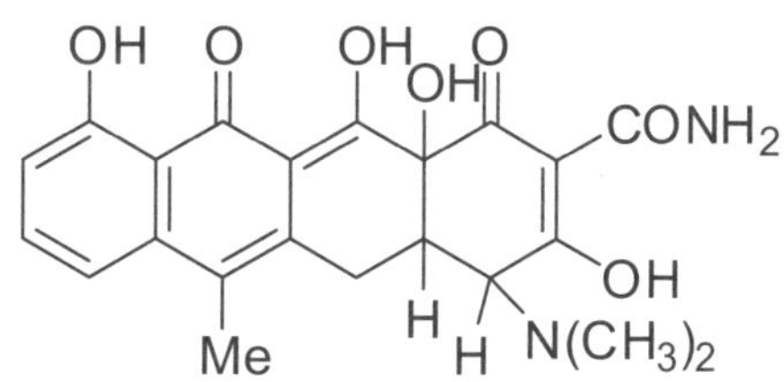

Anhydrotetracycline

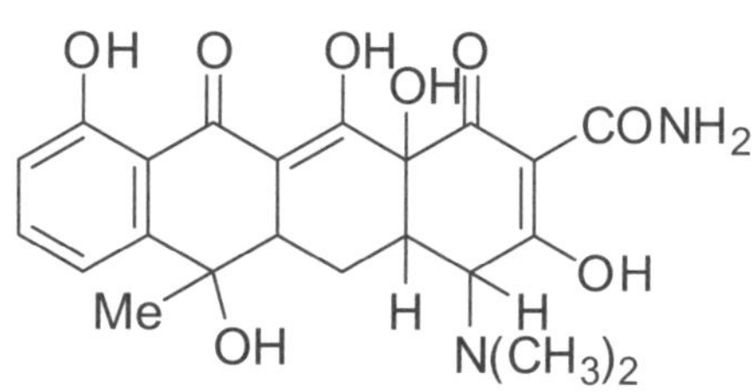

4-epitetracycline

Fig. 1. Degradation products of tetracycline

Another work on influence of temperature on drugs was reported by Matsui *et al.*, (1978) . It was observed that phenylbutazone (figure 3) formulations showed no evidence of chemical instability when stored at ambient temperature of 37°C. At temperature above 37°C measurable chemical degradation occurred with several formulations showing more than 50% degradation.

Even at temperature below 37°C , degradation can take place as shown by work carried out by Kaplan *et al*, 1976. They reported that Amikacin exposed to 25°C for 24 months showed an average of 3.9% degradation. But when the drugs were subjected to 56°C for 4 month an average of 7.2% degradation was observed. Owoyale and Elmarkby, (1989) found out that proguanil (Figure 4) which appeared not to undergo photochemical reaction was thermally degraded when subjected to heat at 45°C. The same drug when stored at room temperature of 25°C for six years was found not to undergo any decomposition.

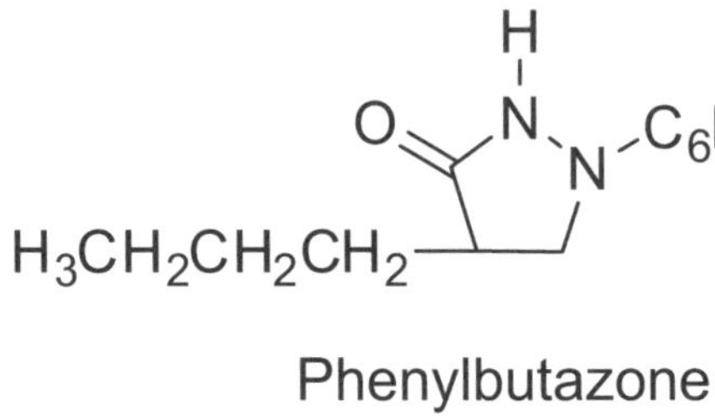

Fig. 2. Structure of Phenylbutazone

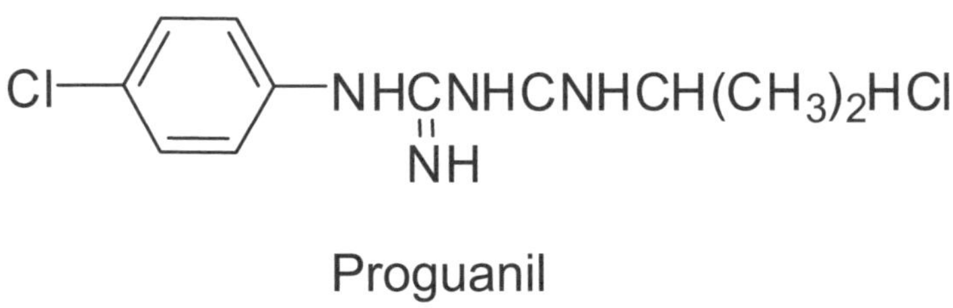

Fig. 3. Structure of Proguanil

Low temperature can sometimes have a negative effect on the stability of some drugs, for instance sulfacetamide sodium (Figure 3) in aqueous medium may be recystallized if stored at low temperature.

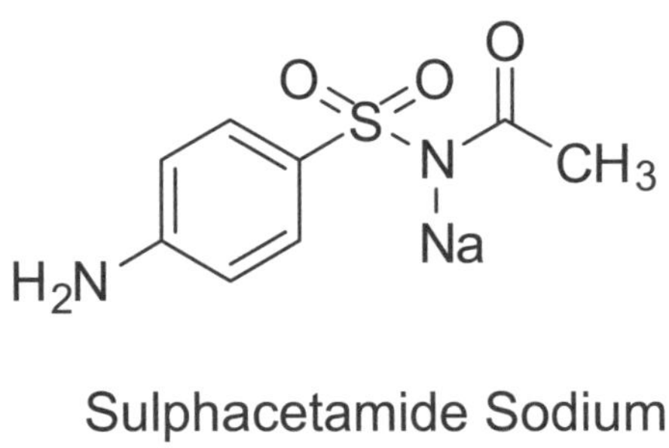

Fig. 4. Structure of Sulfacetamide Sodium

Many drugs have been discovered to be photosensitive, hence they undergo photochemical reactions when exposed to sunlight, due to absorption of U.V light (wavelength of 190 - 320mn). It is therefore not surprising to find many pharmaceutical preparation being destroyed or degraded when exposed to sunlight.

Fadiran and Grudzinski, (1987) studied photostability of chloramphenicol using TLC to detect the number of degradation products. It was shown that chloramphenicol in solid crystalline state (pure drug) and capsule on exposure to U.V light and sunlight developed a yellow colour, the intensity of which increased with increasing exposure time. During photolysis of chloramphenicol, the Beta-bond to the aromatic ring undergoes cleavage to yield one aromatic and one alkyl radical. Irradiation of drugs in solution produces a reaction that is faster than in solid state. Earlier worker preferred to study degradation of drugs in solution.

Chloramphenicol was degraded by light in 0.25% w/v aqueous solution and the solution became yellow and acid to form 2-amino-1 -(4-nitrophenyl) propane-1, 3-diol. and dichloroacetic Acid. (*Shih,1971*). Similarly, Hvalka, (1989) reported that the potency of tetracycline HC1 reduced to 50% when the solution was irradiated with U.V. light for 3 hours. The Degradation products are 4-epitetracycline, Anhydrotetracycline and 4-epianhydrotetracycline.

Drugs which contain multiple unsaturation are particularly prone to photolysis. Drugs with more double bonds are more susceptible to degradation. This assumption was proved by Hamlin *et al.* (1960), they investigated the photolytic degradation of alcoholic solution of hydrocortisone, prednisolone, and methylprednisolone exposed to Ordinary Fluorescent light. It was discovered that the degradation follows 1st order kinetics and that prednisolone and methylprednisolone showed the same rate of degradation, whereas hydrocortisone degrades 1/7th the rate of the two steroids. Hence the two double bonds present in prednisolone and methylprednisolone make these steroids more susceptible to light catalyzed degradation than the one double bond in the ring of hydrocortisone.

Solid pure drugs with ester, amide linkages deteriorate with moisture via hydrolysis pathways. The effect of moisture on degradation of drugs, are many, when deposited on drugs, especially the solid dosage forms, it provides a suitable medium for micro-organism to thrive which may eventually lead to biological degradation of the drugs. Moisture may also cause some physical changes such as swelling, dissolution, cracking and adhesion of coated tablets. Ordinarily, one expects hydrolysis to occur frequently in drugs in aqueous solution and suspension.

Leeson and Mattocks (1958) reported that a thin layer of moisture deposited on aspirin was all it needed for hydrolytic degradation to commence.

There is no restriction to the use of additives and excipient but they should be chosen in a way so as not to affect the stability of the drugs. Incompatibilities of active ingredient with additives can lead to degradation. Kornblum and Zoglio, (1967) studied the potency degradation of Aspirin suspension with lubricant-namely, Aluminum stearate, magnesium stearate, calcium stearate. It was found out that the extent of degradation was more with magnesium stearate.

From the review of the previous works done on degradation of drugs, it can be observed that few works have been reported in degradation of antibiotics, especially in solid state. The few reports that are available are not comprehensive enough especially exposure of the drugs to environmental conditions. Hence there is need to investigate and carry out extensive studies on the degradation of drugs.

Antibiotics like any other drugs show loss in potency when subjected to some environmental conditions.

In continuation of an effort on stability studies of drug, effects of moisture, sunlight, heat and UV radiation on the potency of some antibiotics (Ampicillin, Tetracycline and Chloramphenicol) were investigated by our research group (Adediran and Tella, 2000; Adediran *et al*, 2003; Tella *et al*, 2008). The pure drug of antibiotics and capsules were exposed to moisture, sunlight, temperature (37°C), (70°C) and UV (254nm) for 60 days. Percentage potency or Percentage residual amount of active ingredient were determined before and after exposure.

The three drugs showed evidence of stability with no loss of potency at 37°C, but exhibited loss in potency when exposed to moisture and heat at 70°C.

Exposure of the three drugs to sunlight and UV resulted in loss of potency except Ampicilin which showed loss in potency only at UV radiation.

Peak (cm^{-1})	Assignments
3789	Free OH
3475	N-H (str)
2920	C-H (Str)
1895	C=O (str)
1684	C=C aromatic System
1352, 1527	NO_2 vibration
1569	C-N (str)
1069	C-O (str)
978	O-H (def)
701	Presence of free adjacent protons in aromatic

Table 1. Infrared spectrum of unirradiated Chloramphenicol pure drug and its assignment

Peak (cm^{-1})	Assignment
3475	N-H (str)
1647	C=O (str) or C=C (str)
1521	presence of NO_2 vibration
1418	C-H (def) in methyl)
1069, 1105	C-O (str)
972	OH (def)
701, 815	Presence of hydrogen or Proton in aromatic

Table 2. Infrared spectrum of sunlight irradiated Chloramphenicol pure drug

The infrared spectral assignments of samples of the Chloramphenicol exposed to sunlight and unexposed chloramphenicol are shown in tables 1 and 2

Peaks such as 3789 cm^{-1} due to free OH, 2920cm^{-1} for C-H (str) in unexposed pure drug disappeared in the drug exposed to sunlight. This is in agreement with the finding of Fadiran and Grudzinki(1987) who reported that β-bond to aromatic ring present in Chloramphenicol molecule in solid state undergoes cleavage to form one aromatic and one alkyl radical when the drug was exposed to sunlight.

Also 1894cm^{-1} due to C=O (str), 978cm^{-1} due to O-H (def) in pure drug shifted to 1647cm^{-1} and 972cm^{-1} respectively in exposed drug. There is one C-O (str) peak at 1069cm^{-1} in pure drug whereas there are two in the exposed drug, one at 1069cm^{-1} and another 1105cm^{-1}. All these changes arise from peak destruction and spectra shift are indications of drug degradation.

Infra red spectroscopic analyses were carried out on ampicillin and Tetracycline before and after exposure to heat and moisture (Tables 3, 4 , 5 and 6). Ampicillin drug was exposed to heat at 70°C and moisture for 60 days. Ampicillin pure drug exposed to temperature 70°C exhibited loss in potency and degradation as evidenced by disappearance of absorption band at 1785cm^{-1} of the C=O (Str) in the Beta-lactam ring. This led to the appearance of new band at 2932cm^{-1} due to C-H (str) and 2510cm^{-1} due to S-H (str) and C=N (str) at 1563cm^{-1} in Ampicillin exposed to 70°C to form pencillenic acid as shown in Figure 5.

Peak (cm^{-1})	**Assignment**
3700	OH in carboxylic acid
3442	Free N-H (str)
1782	C=O (str) in β–lactam ring
1697	C=O (str) in the amide
1266	C-N (def)
1168	C-O (str)
651, 700,931	Presence of free adjacent protons in aromatics or C-S (str).

Table 3. Infrared spectrum of ampicillin pure drug (unexposed) and its assignment

Peak (cm^{-1})	**Assignment**
3700	OH in carboxylic acid
3451	Free N-H (str)
2931	C-H (str)
2510	S-H (str)
1660	C=O or C=C (str)
1576	C=N (str)
1508	Presence of aromatic system
1400	C-H deformation in CH_3 or CH_2
1243	C-O (str)
701	Presence of free adjacent protons in aromatics.

Table 4. Infra-red spectrum of ampicillin pure drug exposed to 70°C (assignments)

Peak (cm$_{-}$^{1})	**Assignments**
3500	N-H (str)
2360	H-X (str) in salt of hydrohalides)
1715	C=O (str)
1636	C=O/CO-NH_2 (str)
1500, 1526	C=C Stretching in Aromatic system.
1236, 1183	Presence of C-O/C-N (str)

Table 5. Infrared spectrum of tetracycline pure drug unexposed (assignments)

It can be observed from Figure 5 that the peaks due to 1715cm^{-1} and one (2360) due to H-X (str), in unexposed drug disappeared in spectrum of exposed drug. Also, new peaks at 2926cm^{-1} due C-H (str) and 3700cm^{-1} due to free OH appeared in the exposed drug, which is an indication that hydrolysis of tetracycline may have taken place.

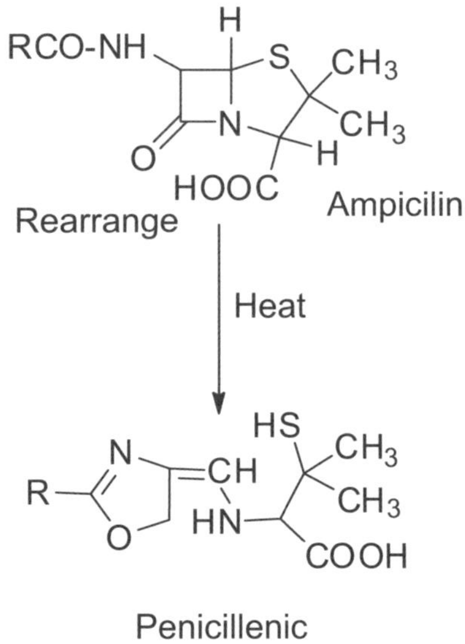

Fig. 5. Rearrangement of Ampicillin after exposure to heat

Peak (cm^{-1})	**Assignments**
3700	Free O-H
3500	N-H (str)/ OH in carboxylic acid.
2926	C-H (str)in Aromatic system
1623	C=O (str)
1521	C=O in aromatic system
1038, 1128, 1261	C-O/C-N (str)

Table 6. Infra-red spectrum of tetracycline capsule exposed to moisture

The infrared spectra of all the three drugs showed evidence of degradation when they are exposed to different environmental conditions.

3. Development of alternative analytical procedure

The awareness of populace as regards drug toxicity and effectiveness in relation to drug quality, requires stricter control of qualitative and quantitative nature of governmental agencies. It is however not possible to enforce a quality standard when there is no analytical method to determine the level of compliance with such standard.

New analytical procedure development is required due to advancement of pharmaceutical practice but problems peculiar to an environment may bring about adaptation of even old methods.

In any case, such new method has to be checked to be at least of equal performance, if not superior, to an already accepted official compendia method.

The first scientist to develop analytical method for the assay of penicillin was Alicino[24] in 1946. He reported the first general iodometric method for the assay of most penicillin. He discovered that most chemical methods of assay for the benzyl penicillin salts depend upon hydrolytic cleavage of the Beta-lactam ring to give penicilloic Acid. The cleavage can be brought about either by alkali or by the enzyme penicillinase. If the cleavage is brought by penicillase in a previously neutral and unbuffered solution the resulting acid may be titrated with alkali to give a measure of the penicillin present. Alternatively, most commonly, the liberated penicilloic acid is determined through the ability to take up iodine, a property not possessed by the parent molecule. This method has undergone various modifications and revisions from time to time.

The modification of Alicino was done by Beckett and Stenlake (1976) using benzyl penicillin for the modification. After primary hydrolysis with sodium hydroxide solution to convert the antibiotics to the corresponding penicilloic acid, treatment with acid yield D-penicillamine (and benzylpenillic Acid) which is oxidized almost quantitatively by iodine to the corresponding disulphide, excess iodine is back-titrated with 0.02M sodium thiosulfate solution. The equation of reaction is shown figure 6.

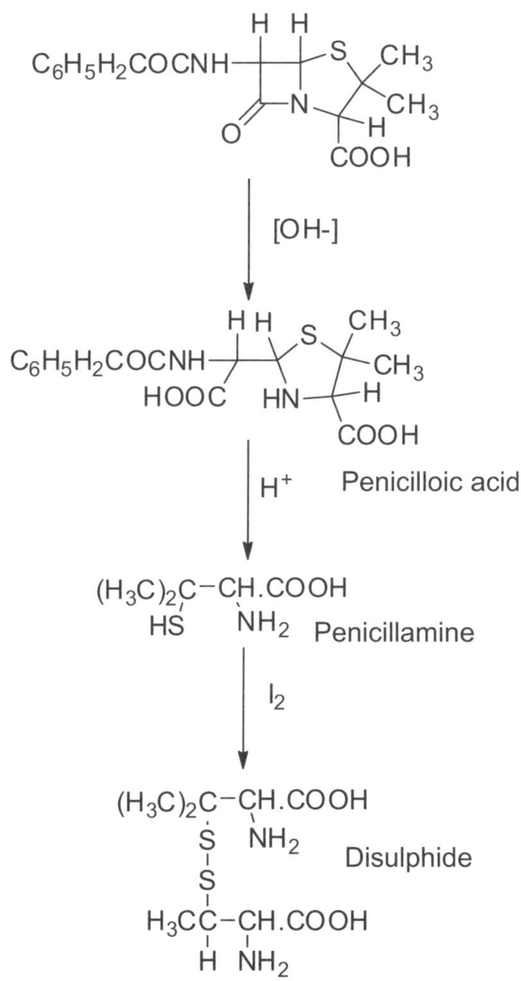

Fig. 6. Back -titration of Ampicillin by iodiometry

Benzyl penicillin sodium is standardized against a chemical reference substance of known potency.

The above method has two principal advantages over spectrophotometric method. First, it has a high degree specificity since non-penicillin impurities are to a large extent allowed for in the blank determination and secondly the relatively large iodine absorption makes the method highly sensitive.

Apart from the titrimetric methods of assay reported above, an attempt was made to assay penicillin by spectrophotornetric method. Beckett and Stenlake,1976 described the spectrophotometric method by the use of imidazole mercury reagent. Mercuric chloride is known to attack the sulphur atom of the penicillin leading to a rearrangement involving penicillin side chain. The rearrangement is catalyzed by imidazole. The product III formed is a penicillenic acid mercuric mercaptide which absorbs between 325 - 345nm. The equation of reaction is shown in figure 7.

By using a reference and test samples, the absorbance given by the penicillenic acid mercuric mercaptide formed from a known weight of the reference is compared with that of the test. From the comparison of the absorbance given by a reference as compared to that of the test sample, the actual penicillin content can be computed. The method is official in British Pharmacopoeia (1993) and India Pharmacopoeia (1985) for assay of cloxacillin, fluocloxacilin and ampicillin.

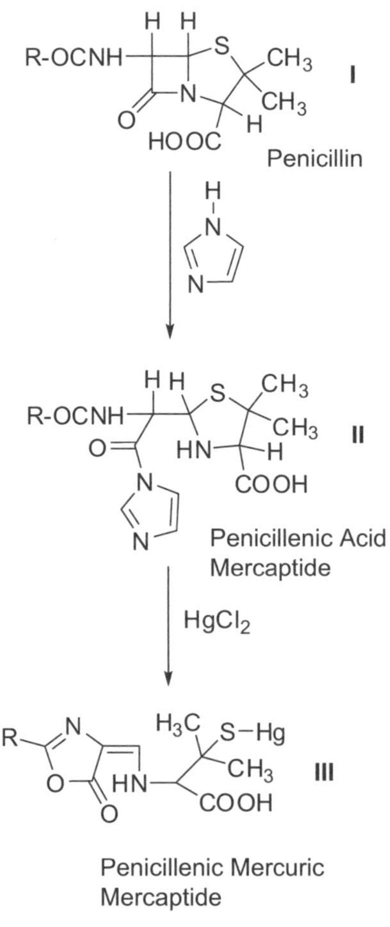

Fig. 7. Spectrophotometric analysis of Penicillin

The second instrumental method for the assay of penicillin was reported by Bungaard and Larsen,(1983) . They made use of sorbitol reagent in the analysis of Ampicillin, Amoxycillin and Cyclacillin. Sorbitol reagent is an hydroxyl compound, it was used to form a penicilloyl ester intermediate II, which rearranges to a corresponding piperizinedione derivative III. Treatment of the derivative with 1M NaOH gives a highly absorbing product with absorption maximum at 322nm. The equation of reaction is shown in figure 8.
The reaction is catalysed by the reaction in catalyzed by metal ions. Zinc ions were found to be most effective catalyst. Aminopenicillin such as Ampicillin, Amoxycillin and Cyclacillin can be determined quantitatively by this method.
Most official books such as, British Pharmacopoeia (1993), United states pharmacopoeia (1990) and International Pharmacopoeia (1979) described the analysis of tetracycline HCl using microbiological, non-aqueous and spectrophotometric methods.

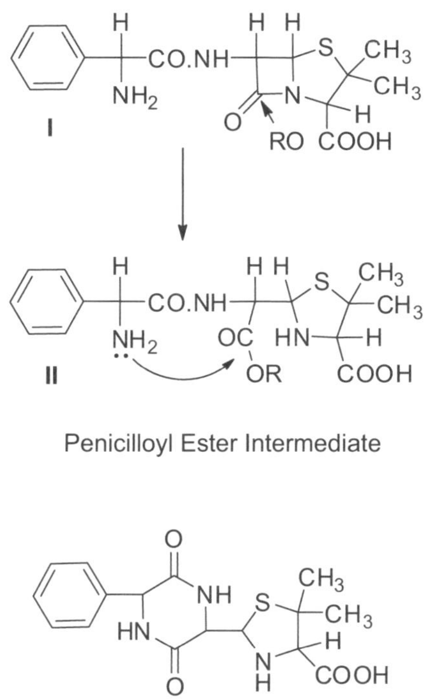

Penicilloyl Ester Intermediate

Piperizinedione derivative

Fig. 8. Spectrophotometric analysis of Ampicillin

As far back as 1956, Woolford and Chiccarelli proposed first spectrophotometric method of analysis of tetracylines, which is based on the formation of yellow colour with an absorption maximum at 320nm when tetracycline HCl is dissolved in 0.2M sodium hydroxide. The disadvantage of the method is that it is not applicable to oxytetracycline and chlortetracycline. Effort was made to develop analytical method that will be applicable to all three common tetracyclines. Monastero *et al.* (1951) were able to come up with a method which applied to all three tetracyclines. The method involves mixing of dilute hydrochloric acid solution with ferric chloride solution to form orange-brown colour which gives maximum absorption at 490nm. This method has been adopted as official method for the analysis of oxytetracylines in Indian Pharmacopoeia., 1985). Phosphates, fluorides, thiocyanates and other substances that combine with iron (III) interfere and therefore if present must be removed before carrying out determination.

Yokohoma and Chatten, (1958) reported non-aqueous method for tetracycline. It was shown that tetracycline hydrochloride may be titrated non-aqueously with perchloric acid in dioxan by addition of mercuric acetate. Tetracycline which is halide salt is treated with mercuric acetate to precipitate the halide ion as undissociated mercuric (II) halide and form the acetate salts which reacts quantitatively with perchloric acid thus preventing interferences of halogen.The methods are not selective for separation and determination of tetracycline degradation products in tetracycline HC1 powder and capsules. In order to find solution to this problem, Omer *et al*, (1958) used thin layer chromatography for the separation and semiquantitative determination of tetracycline degradation products in tetracycline Hydrochloride powders and capsules.

It failed in an attempt to fully quantify the degradation products and content of residual drug. This led to the development of more sensitive technique that quantified and differentiate between the degradation products and intact molecule of drug by Kabela in 1982 using HPLC. He was able to determine the amount of degradation products (4-epianhydrotetracycline, Anhydrotetracyline and 4 - epitetracycline) present in tetracycline Hydrochloride.

British Pharmacopoeia (1980) described a spectrophotornetric method to measure the total absorbance of degradation products in tetracycline hydrochloride. The absorbance limit set by the British Pharmacopoeia (1980) is twice that which has been set for tetracycline hydrochloride capsules and powder. The samples of tetracycline hydrochloride capsules and standard were examined at 430nm.

Some official books British Pharmacopoeia (1993), International Pharmacopoeia (1979) described spectrophotometric method for the analysis of chloramphenicol, by taking the absorbance of chloramphenicol solution at 278nm.

Chloramphenicol eye drops, ear - drops and capsules can also be determined by colorimetric method based upon reduction of the nitro group with zinc, stannous chloride or better sodium dithionate followed by diazotization and coupling with N-(-1-naphthyl) ethylene diamine.

Most assay methods for the drug are based upon its functional groups and will not differentiate between chloramphenicol and its degradation products. It is known that chloramphenicol loses its antibiotic activity by the hydrolysis of the amide bond to give 2-amino-1-(4-nitrophenyl) propane-1,3-diol.

This latter compound commonly occur in pharmaceutical preparation and because of the structural similarity of the chloramphenicol and 2-amino-1-(4-nitrophenyl) propane-1,3-diol, the specific determination of chloramphenicol by spectrophotometer is unsuitable. The absorption is due principally to the nitrophenyl functional group and hence degraded chloramphenicol will also absorb at 278nm.

To ensure freedom from interference due to degradation of the chloramphenicol molecule, the chromatographic separation procedure is suggested preliminary to all analyses. Higuchi *et al.* (1954) developed a simple partition chromatographic procedure which separated degradation products of chloramphenicol from chloramphenicol molecule. Many other interferences are also removed, since partition method is extremely sensitive to differences in molecular structure. The eluate obtained containing chloramphenicol is then determined by spectrophotometer at 272nm.

Piergiorgio, 1979 proposed the application of simplified TLC method for the simultaneous determination of chloramphenicol and 2-amino-1-(4-nitrophenyl) propane-1,3-diol. The principle of this method consists of spotting the solution on a Thin layer chromatography plate together with standard solution and developing the plate with an ethyl acetate-formic Acid - water (10:2:8, upper plate) solvent. The intensities of the various spots are then measured by a densitometer and the peak height of the standards are used to calculate the concentration of chloramphenicol and 2-amino-1-(4-nitrophenyl) propane-1,3-diol. in unknown sample. This procedure is too laborious, not precise and tedious.

A more rapid, precise and accurate method was developed by Belle and Young,(1979). The method was based on the use of High performance Liquid chromatography for determination of chloramphenicol and 2-amino-1-(4-nitrophenyl) propane-1,3-diol. in Pharmaceutical formulations.

The method possesses a distinct advantage because the method does not require extraction nor derivatization for the determination of both chloramphenicol and 2-amino-1-(4-nitrophenyl) propane-1,3-diol. in capsule, ophthalmic solution and ophthalmic ointment formulation.

The solvent extraction method has been described in British Pharmacopoeia (1980) for the determination of 2-amino-1-(4-nitrophenyl) propane-1,3-diol. Albendazole is a broad anthelmintic (figure 9). It is used for the treatment of threadworm, hookworm and tapeworm.

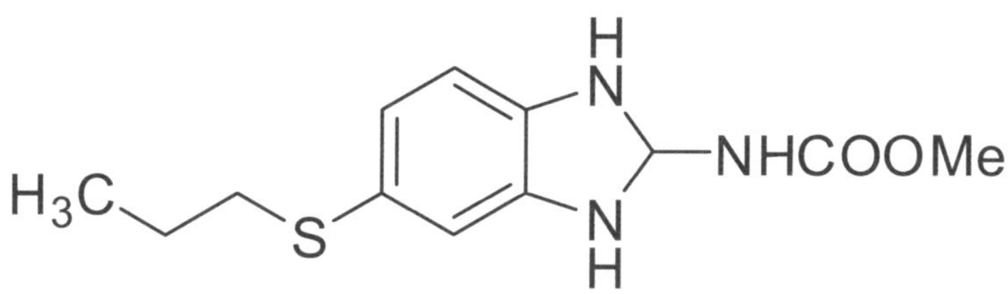

Fig. 9. Structure of Albendazole

Extensive literature survey reveals that the estimation of Albendazole in dosage and suspension forms are not available in pharmacopoeia and therefore require much investigation. The drug is readily available in Nigeria market in tablet, bolus (veterinary prepararion) and suspensions. The need to come up with a simple and sensitive method of analysis for the estimation of drug in pharmaceutical preparations therefore arises. Spectrophotometric method for the estimation of Albendazole in both solid and liquid preparation was developed (Tella *et al,*2010). Treatment of Albendazole with methanolic-glacial acetic acid gives a highly absorbing product with absorption maximum at 235nm. Beer law was obeyed in the concentration range 2.5-20µg/ml

Brands	Forms	Label claim(mg,mg/ml)	Found(mg,mg/ml)
A	Tablet	200	194.24±0.553
B	Bolus	250	243.07± 0.512
C	Bolus	500	497.6 ± 0.272
D	Suspension	25	25.11± 0.386
E	Suspension	25	24.60± 0.246

Table 7. Results of assay of Albendazole of different brands in solid dosage and suspension form

The results obtained from analysis of different brands of Albendazole tablets, bolus and suspension were in good agreement with the the label claims as shown in Table 7.
The little difference might be due to batch variation of medicaments in tablets, bolus and suspension, instrumental errors or degradation of active ingredients with time. The method could be considered for the determination of Albendazole in quality control laboratories.
Another alternative method developed for estimation of sulphadimidine in tablets is simple and rapid titrimetric method (Back titration). Sulphadimidine is bacteriostatic. It is an antibacterial (intestinal drug) (Figure 10).

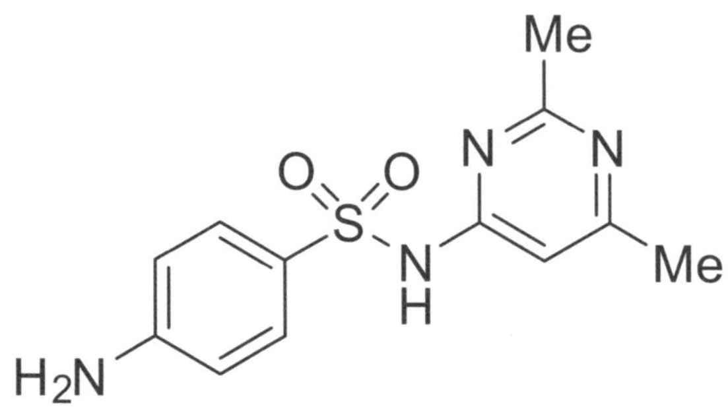

Fig. 10. Structure of Sulphadimidine

The drug is presented in tablets, injections, suspension and veterinary tablets.
Official methods for the analysis of the drug in pharmaceutical preparations are the potentiometric and Nitrite Titrations. The end point of Nitrite titration method is detected either electrometrically or by using an external indicator. By streaking a few drops of the titrated solution upon starch iodide paper or paste, a dark blue colour is obtained. Excess nitrous acid oxidizes the iodide in the indicator to give iodine which gives blue color with starch. The disadvantage of this method is that the visual end point with external starch iodide indicator may seem somewhat indefinite and difficult to determine. The great disadvantage of the electrometric end point and the potentiometric titration is lack of specificity. They also require expensive equipments, some level of expertise and are time - consuming. The average time to complete the reaction requires about 2 hours. However , the proposed method can be completed within 20minutes (Tella *et al,* 2010). The reagents are also available.
The procedure is described as follows: Standard drug solution was prepared by dissolving 250mg of Sulphadimidine B. P. in 30ml ethanol (96%) in a 250ml conical flask. 40ml of 0.1M sodium hydroxide solution was then added.
The content of the flask was mixed well and warmed on water bath for 5 minutes. It was allowed to cool. The excess alkali was then titrated with standardized 0.1M hydrochloric acid using 3 drops of Phenolphthalein solution as indicator. The operation was repeated without the substance being examined.
The difference between the titrations represents the amount of 0.1M sodium hydroxide required by the Sulphadimidine. Each milliliter of 0.1M sodium hydroxide is equivalent to 0.02783g of $C_{12}H_{14}N_4O_{25}S$ (Sulphadimidine)
Equation of the reaction is shown below (Figure 11)
The result obtained by the proposed and reported methods for the weighed Sulphadimidine B. P. and commercial samples of Sulphadimidine tablets are given in Table 8. The percentage recoveries show that the proposed method can be adopted for routine analysis of

Sulphadimidine tablets. The results obtained by the proposed method were in good agreement with the labeled amount.

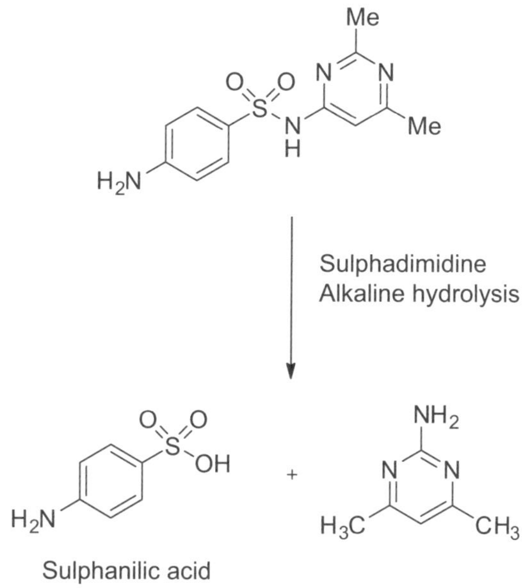

Fig. 11. Titrimetric analysis of Sulphadimidine

Titrimetric method (back titration)				Reported method (nitrite titration)		
Labeled Amount (mg)	Quantity Found (mg)	Recovery (%)	Standard Deviation	Recovery (mg)	Recovery (%)	Standard Deviation
500	498.98	99.80	± 0.06	496.24	99.25	± 0.07
500	495.43	99.09	± 0.09	490.83	98.17	± 0.05
500	493.43	98.69	± 0.10	500.00	100.00	± 0.06
500	500.18	100.04	± 0.08	494.56	98.91	± 0.03
500	499.24	99.85	± 0.03	492.48	98.50	± 0.02

(Average of 10 determinations)

Table 8. Average recoveries from the various commercial samples of Sulphadimidine tablets.

4. Adsorption of drugs on pharmaceutical exicipents

It has been established that the presence of adsorbent, such as activated charcoal interferes with the drug adsorption process resulting in a decrease bioavailability of some drugs. The interference in the systematic availability of drug is brought about by its adsorption on the activated surface of the solid adsorbent, thus preventing the adsorbed fraction of the drug from permeating through the gastro- intestinal mucosa into the blood stream.

Some of these drugs may be lost when adsorbent are administered concomitantly with the drugs.

Furthermore, in sub-Saharan Africa, the abuse of various drugs has increased considerably in the last decades. Many drugs used in treatment of tropical diseases have been implicated in various intentional and accidental poisoning. Adsorption and interaction of chlorapheneramine and chloroquine phosphate on pharmaceutical materials like magnesium trisilicate, Activated charcoal, magnesium carbonate and magnesium stearate was investigated by our research team. Freudlich Adsorption isotherm was adopted to evaluate adsorption capacity of each adsorbent on chloroquine phosphate. The freudlich parameter kf which is adsorption capacity obtained for the adsorbents are 0.053, 0.145, 0.131 and 0.173mg/g for magnesium carbonate, magnesium stearate, magnesium trisilicate and activated charcoal respectively showed that these adsorbents have ability to adsorb or remove chloroquine phosphate molecules from solution at PH 5.0 (Adediran *et al*,2006)

The extent of adsorption of chloroquine phosphate by the adsorbents followed the sequence; Activated charcoal > magnesium trisilicate > magnesium stearate > magnesium carbonate. Differences in surface characteristics and chemical structure of adsorbent may be responsible for the trend observed above.

Activated charcoal has the highest adsorption capacity which may be due to its organic nature and presence of phenolics and carboxyl moieties.

Magnesium trisilicate (Antacid) adsorbed Chloroquine better than magnesium stearate, because there is chemisorptions interaction between the negative charge of the adsorbent and positive charge of the drug molecule. The presence of small amount of oleate molecules in magnesium stearate enhances adsorption over magnesium carbonate. The findings are in agreement with the work of Mcginity and Lach, 1976, Cooney 1977 and Guay et al, 1984. Our investigation revealed that concurrent administration of these pharmaceutical adsorbents and chloroquine drug might interfere with chloroquine adsorption. Furthermore, these adsorbents can serve as alternative antidote for chloroquine poisoning. We also investigated the in-vitro absorption of chlor pheniramine maleate on these adsorbents. Chlorapheniramine maleate is an antihistamine which reliefs red, itchy and watery running nose. The study was carried out at P^H = 5.0 and 37°C using Batch method. Freudlich parameters were determined for each adsorbents as shown in Table 9). The freudlich parameter (kf) are 4.68, 4.47, 4.80 and 1.91 for activated charcoal, magnesium trisilicate, magnesium stearate and talcum powder (Tella and Owalude, 2007). The adsorbents have ability to adsorb or remove chlorapheniramine maleate from solution at 3.0 - 5.0mg/l adsorbate. The drug was mostly adsorbed by the activated charcoal and least absorbed by talcum powder.

We concluded that concurrent administration of these pharmaceutical adsorbents and chlorapheniramine maleate might induce interference between them thereby affecting the bioavailability of the drug to the system. There is possibility of using these adsorbents as antidote in case of Chlorapheniramine maleate over dose or poisoning.

Absorption	**1/n**	**Kg x 10^{-3}mg/g**
Activated charcoal	0.65	4.68
Mg Si O_3.	0.66	4.47
Magnesium stearate	0.77	3.80
Talcum powder	0.99	1.91

Table 9. Freudlich adsorption parameters of CPM on Adsorbents

5. Intravenous fluids

An intravenous fluid is a sterile, pyrogen-free, particle-free solution used for therapeutic purposes by infusion through the veins.
Intravenous fluids (I.V. Fluids) are solutions sometimes containing electrolytes such as sodium chloride, potassium chloride and calcium chloride; energy-giving compounds like dextrose and other ion-balancing solutions such as compound of sodium lactate (Hartman's and Ringer Lactate Solutions).
Examples of I.V. Fluids are:

- Normal Saline (0.9%$^w/_v$ Sodium Chloride in water)
- Dextrose 5% $^w/_v$ Saline (containing g/Litre Sodium Chloride and 50g/Litre dextrose anhydrous).
- Dextrose 5% $^w/_v$ (containing 50g/litre dextrose anhydrous).
- Dextrose 4.3% $^w/_v$ + 0.18% Saline (containing 43g/Litre dextrose anhydrous + 18g/Litre Sodium Chloride).
- Dextrose 50% $^w/_v$ Solution (containing 50g/100ml Dextrose anhydrous)
- Dextrose 10% $^w/_v$ Solution (containing 100g/Litre dextrose anhydrous).
- Metronidazole Injection - 0.5% $^w/_v$ (containing 0.5g metronidazole / 100ml).
- Hartman's Solution
- Darrow's Solution - Full strength and ½ Strength.
- Plasma expanders such as 4% polyvinyl pyrollidone (povidone k30 - in water).

5.1 Uses / functions of I.V. Fluids

I.V. Fluids are normally infused into ambulatory patients - usually very weak, unable to eat or drink, or totally of unconscious, in shock or acetate coma. I.V. Fluids are therefore, a life saving device for critical care of patients. I.V. Fluids have constituents that are used selectively to correct certain imbalances in the body fluids of patients and to supply, the required energy by directly infusing the metabolisable carbohydrate monomer - D-glucose in the various concentrations, depending on the specific requirement of the patient.
I.V. Fluids essentially do the following:

a. Rehydrate patients
b. Replace lost ions such as sodium ion, chloride ion from normal saline (0.9% sodium chloride I.V. Solution). potassium, calcium and chloride ions from Darrow's solutions full strength and half strength. Calcium, sodium, potassium and chloride ions. Lactates from ringers (Hartman's solution).
c. Increase total blood volume in short time (in cases of server blood loss) for accident victims. Plasma expanders such as Isoplasma (4%$^w/_v$ polyvinyl pyrollidone in 0.78%$^w/_v$ saline) he as to replace blood volume without affecting ion - balance in the patients.
d. Supply energy in the form of dextrose anhydrous. All dextrose containing I.V. Solutions are energy sources for ambulatory patients. The specific need of each patient must be ascertained to determine what to give him/her.
e. Lactate - containing products help to correct low pH in the blood by metabolizing lactate to release bicarbonate ions (HCO_3^-) into the blood and hence neutralize the excess hydrogen ions in the blood.
f. Amino Acid, fatty acids, mineral and vitamin nutritional supplements are nowadays available as intravenous infusions

Intravenous fluids belong to a group of pharmaceuticals called parenterals. i.e. medications that are administered by other routes than through the intestinal absorption into the blood.

Other parenteral preparations include irrigation solutions, Peritoneal Dialysis Solution, Heamodialysis Concentrates e.t.c.

5.2 Quality of intravenous fluids

Intravenous fluids are administered directly into the blood stream through the veins. The veins empty it through the heart, which pumps it round the body. Hence it is very easy to deliver proper medication and hence therapy through I.V. Fluids or poison contaminations or germs through the same route if the I.V. Fluid is not of the right quality. I.V. Fluids must be sterile, pyrogen-free, particle - free and contain the right quantity of constituents as per the labelled amount of the product. The acceptable limit of the constituent throughout the shelf life of the product must remain between 95% and 105% of the label claim and in some case 90% to 110% at most.

5.3 Critical quality of I.V. Fluids

Sterility

I.V. Fluids must be free of viable organisms be it bacteria, fungi, algae or any microbe. If the I.V Fluid is not sterile after preparation it may remain clear for a while and later turn cloudy or show massive macroscopic growth. A seemingly clean pouch may actually not be sterile. But such contaminated pouch will later turn cloudy. A non-sterile material when infused poses dangers of sepsis (heavy blood contamination by germs) to the patient and resultant adverse reaction and death. Therefore, sterility is a critical quality of I.V. Fluids.

Pyrogen - Free status

Pyrogen simply means a substance which when injected elicits adverse reactions such as fever, rigours, palpitations and restlessness in the patients that receive it. Pyrogens are endotoxin produced by Gram negative bacteria. The bacteria may be killed (destroyed) by sterilization but the endotoxin present in them is released into the fluid medium. The pyrogenic solution when injected cause adverse reactions in the patient. Therefore, pyrogen-free status is a critical, acceptable quality of I.V. Fluids. I.V. Fluids must be free of solid or suspended particles I.V. Fluids packaged must remain intact. A broken package that lets in air becomes contaminated and loses its sterile status.

Bacterial endotoxin as impurity in sterile pharmaceuticals

Gram negative bacteria produce bacterial endotoxin. They are made up of the lipopolysaccharide (LPS) that constitute the cell walls of Gram negative bacteria. They are called endotoxin because they are not released to the outside environment of the bacteria until the cells die. They are released after cells disruption. Bacterial endotoxin abounds everywhere. The Gram negative bacteria exist in particulate matter, in air, water and soil (Schaumann, *et al.*, 2008).

Endotoxin is detectable in ambient aerosols and it is an important component of tobacco smoke. (Larson *et al.*, 2004) It has been reported that early life exposure to endotoxin protects against the development of allergies. (Braun-Farhlander, *et al.*, 2002). Exposure to household endotoxin is a significant risk factor for increased asthma prevalence in adults. Higher levels of exposure to endotoxin were significantly associated with asthma diagnosis (Schaumann, *et al.*, 2008). It is a known fact that in asthma patients' inhalation of endotoxin causes a significant decrease in lung functions with enhanced airway hyperactivity (AHR). (Schaumann *et al*,. 2008). Endotoxin is also an impurity in sterile pharmaceuticals especially

Large Volume Parenterals (LVPs) and it has to be tested for in the products meant for intravenous administration (Radhakrishnan, 2010).

6. Current methods and manufacturers (users) experience

The test for pyrogens in LVPs was recognized during the 1940's in the US when the Food and Drug Administration, the National Institutes of Health and fourteen pharmaceutical manufacturers, undertook a collaborated study. This study led to the adoption of the procedure, which first appeared in the XII Edition of The United States Pharmacopoeia and was the only official test for the detection of bacterial endotoxin until the discovery of LAL.

6.1 Limitations to the rabbit test of pyrogen (bacterial endotoxin)

Rabbit test is limited by the elaborate nature of the test. It is expensive, time-consuming and subject to the variability of animal test. Rabbit test can detect endotoxin but cannot determine the actual concentration or endotoxin present in a solution. The Limulus Amebocyte Lysate (LAL) test had been described in literature as the most sensitive convenient method currently available for detecting bacterial endotoxin. (Bergheim, 1978) LAL being an *in vitro* test is useful in In-process detection, an important practice in In-process quality control. This is a quantitative determination of the negative side or the limit. In-process material cannot be injected into rabbits since final sterilization had not been done on the product. An un-sterilized product portends greater risks to the animals. Hence, LAL has an edge over the Rabbit test of pyrogen in this regard.

In 1973, Travenol laboratory developed its own in-house LAL test which measured the activated amounts of protein precipitated. In the LAL gelation reaction, samples were tested for the presence of protein using the Lowry protein assay and resulting differentials were read on spectrophotometer. This eliminates the problem of subjective reading the gel-clot endpoint (Bergheim, 1978).

6.2 The Nigerian experience

LAL in this part of the world (Nigeria) is not readily in use because the kits have to be imported. In the US, a laboratory will charge up to $140 per sample to run LAL test.

There are about six LVP - manufacturing plants in Nigeria as at 2010. None of the plants used LAL to test for pyrogen, perhaps due to non-availability of the material locally. There is need to develop other in-vitro tests similar to LAL, but using extracts from animals readily available in the tropics.

In an on-going research, Salawu *et al.,* (2010) have demonstrated that delay in sterilization of parenteral solutions of up to 48 hrs could lead to production of highly pyrogenic solutions, provided the solution had been contaminated with Gram negative organism like *Escherichia coli* before the delayed sterilization. In their report the resultant increase in the population of the contaminating bacteria before sterilization caused an intolerable rise in pyrogen level even after sterilization. Such a product in real production must be discarded after the production cycle had been completed. This was because only sterilzed product can be admisnistered to rabbit for pyrogen tests.

6.3 Investigation of endotoxin-induced protein coagulation in *Archachatina marginata*

Archachatina marginata is a gasropod, found in the forest and savannah zones of West Africa. In Nigeria, it is a source of dietary protein, eaten in stews and soups. In traditional practice, the haemolymph of the snail is applied as disinfectants to baldes and fresh cuts of circumscicion. This was believed to prevent sepsis of the wound and speed of healing of the fresh cuts of circumscision. Endotoxin-binding properties of the snail's haemolymph fraction was first reported by Salawu *et al.* (2011).

Fig. 12. *Archachatina marginata* (Source: Salawu, 2011)

In the research, the haemolymphs of the snails were collected by the apical cracking method (Ogunsanmi *et al.*, 2003). The haemolymph was mixed with anticoagulant and plasma was obtained by centrifugation. The pellets was washed with anticoagulant, followed by 0.1 M $CaCl_2$ and the pellet containing the hemocytes (amebocytes) were homogenised and suspended in buffer. Exposure of the fractions from the hemocytes: hemocyte lysate (HL), hemocyte lysate supernatant (HLS) and hemocyte lysate debris (HLD) and the plasma were respectively incubated at 37°C for 1 h with endotoxin (1EU/ml) and calcium ions. Controls were set up with the fractions exposed to endotoxin-free water (<0.025 EU/ml) and calcium ions. The fraction exposed to endotoxin produced coagulates which had higher protein content than those exposed to endotoxin-free water. Further investigation reveaealed that combination of plasma and HL of the snail in various ratios produced optimal protein coagulation at a plasma: HL ratio of 1:1. Exposure of the mixture producing the optimal coagulation to varied concentrations of endotoxin ranging from 1 to 5.0 EU/ml, followed by incubation at 37 °C for 1h produced protein coagulation in the mixture which was linear up to a concentration of 1EU/ml. Further increase in endotoxin did not elicit icrease in protein coagulation. There was a drop in coagulation at endotoxin concentrations above 1EU/ml. From this study, it was concluded that the haemolymph of A. marginata contained endotoxin-binding proteins. It was suggested that the haemolymph may serve as a souce of endotoxin detection and quantification kit for testing parenteral solutions in the future (Salawu *et al.*, 2011).

The choice of *Archachatina marginata* was inspired by the traditional medicine practice which had no scientific backing. *A. marginata* moves by creeping on soil, wood and rock surfaces and produces slime from its foot which binds dirt and possibly entraps microbes found along its path. Such an immunological adaptation suggests a very strong defence againt pathogens which was thought to be worthy of study in respect of endotoxin. This effort has opened more investigation and a possibility for development of '*Archachatina Amebocyte* Lysate' (AAL) kit for testing endotoxin.

This on-going research in the University of Ilorin, Nigeria, is promising in terms of having a tropical source of test kit for pyrogen status of parenterals and hence more affordable and safer, locally produced intravenuous fluids in Nigeria. A success of this research will be a great contributon to delivery of critical care in the developing countries, especially Nigeria.

7. Conclusion

Emphasis should be placed on degradation/stability studies of drugs because improper storage and distribution of pharmaceuticals can lead to their physical deterioration and chemical decomposition resulting in reduced activity and occasionally, in the formation of toxic degradation products.

The increasing rate of introduction of fake and adulterated drugs into sub-saharan Africa countries markets makes development of alternative analytical methods a necessity due to lack of reagent and unavailability of equipments required in official books

Studies of Adsorption of pharmaceuticals to excipients and additives are needed in order to investigate their interaction which may affect bioavailaibility of the drug. The clinical usefulness of these additives and excipients in the management of acute toxicity in drug overdose patients can be discovered from in-vitro adsorption study.

8. References

[1] Adediran, G.O. and Tella, A.C(2000). Biosciences Research Communication, 12,4, 457-465.

[2] Adediran G.O., Tella, A.C., Nwosu, F.O. and Ologe, M.O. (2006). Centre point(SCience Edition) 14, 1 and 2, 31-38.

[3] Adediran, G.O., Tella, A.C. and Olabemiwo O.M. (2003). Science focus 3, 112-115

[4] Alicino, J.F. (1946). Ind. Eng. Chem. Anal. Ed., 18,619.

[5] Beckett, A.H. and Stenlake, J.B. (1976). Practical Pharmaceutical Chemistry, 3rd Ed. (Part One), The Antjone Press, London,10-15.

[6] Bergheim O.B. (1978). Limulus Amebocyte Lysate (LAL) Tests for detecting pyrogens in parenteral products and Medical devices- current method and manufacturers' experience. In: Large Volume Parenterals - proceedings of a Seminar held in Oslo, June 6-8, 1978.

[7] Braun-Fahrlander, C., Riedler J. H., Eder W., Waser M., Arize L., Maisch S., Carr D., Gerlarch F., Buffe A., (2002). N. Eng. J. Med. 347: 869- 877.

[8] British Pharmacopoeia(1980), Her Majesty stationery office, London, Vol.1 and 2., 447-448

[9] British Pharmacopoeia (1993). Her Majesty stationery office, London, Vol.1 and 2., 131-132,661-662

[10] Bungard, A and Larsen, E (1983). J.Pharm. and Biomed, Analyst,1,29.

[11] Christie, W.W (2008). Lipopolysaccarides In: The Lipid Library. Eds., searched on 25 October 2008.

[12] Cooney, D.O. (1976). J. Pharm. Sci.67,426-428

[13] Fadiran, E.O and Grudzinski, S.K. (1987). The Nig. J.Pharm. 50, 219-221.

[14] Gornall AG, Bardawill CJ, David MM. (1949). Determination of serum protein by means of Biuret reaction. *J Biol Chem*, 177, 751–756.

[15] Guay, D.R, Meatherall, R.C, Macaulay, P.A and Yeug, C. (1984). Int. J. Clin. Pharmacol. THer. Toxicol. 22, 395-400
[16] Hamlin, W.E., Chulski, T., Johnson, R.H and Wagner, J.G. (1960). J. Am. Pharm. Ass. Sci. Ed., 49,253.
[17] Higuchi, T., Marcus, A.O. and Bias , C.D. (1954). J. Am. Pharm. Assoc. Sci. Ed. 43,135
[18] Hippenmier F., (1978). 'A plant for the production of Large Volume Parenterals in Wintherthur' In: Large Volume Parenterals - Proceedings of a Seminar held in Oslo from June 6 -8, 1978.
[19] Hvalka, P.A. (1989). J.Agric and Food Chem., 37,221-231.
[20] International Pharmacopoeia (1979). 3rd Edition, Vol 2, World Health Organization, Geneva, 65-66, 277-269
[21] Kabela, A.E. (1982). Influence of Temperature on stability of solid tetracycline hydrochloride measured by HPLC . J. Chromatogr.246, (2), 350-355
[22] Kaplan, M.A., Cappola, W.P., Nunning, B.G and Granate. K. A. (1976). Current Therapeutic Research 20,352.
[23] Kornblum, S.S and Zoglio, M.A. (1967). J. Pharm. Sci. 56, 1569.
[24] Le- Belle, M.J. and Young, D.C. (1979). J. Chromatogr. 170,282-287.
[25] Leeson, L.G. and Mattocks, A.M. (1958). J.Am. Pharm. Ass. Sci. Ed., 47,329-332
[26] Matsui, F., Roberton, D.L., Lafontaine, P., Kolasinski, H. and Lavering, E.G. (1978). J. Pharm. Sci.67, 646.
[27] Mc- ginity J.W. and Lach J.L. (1976). J. Pharm. Sci. 65,899-902.
[28] Monastero, F., Means, J.A., Grenfell, T.C. and Hedger, F.H. (1951). J. Am. Pharm. Ass. Sci. Ed. 40,241.
[29] Ogunsanmi, A.O.; Taiwo, V.O. and Akintomide, T.O. (2003). *Tropical Veterinarian,* 21 (2) 43-48.
[30] Omer, A.I., Gad. Karem, E.A. and Salama, R.B. (1981). J. Chromatogr. 205, 456-489.
[31] Owoyale, J.A. and Elmarkby, Z.S. (1989). Int. J. Pharm. 50, 219-221.
[32] Piergiogio, P. (1979). J. Chromatogr. 177, 177-179
[33] Radhakrishna S.T. (2010). Rabbit Pyrogen test; United States Pharmacopoeia XXIX, USP 29-NF24 p. 2546; available online at http://www.pharmacopeia.cn/v29240/usp29nf24s0_c151.html
[34] Salawu M.O., Oloyede O.B., Oladiji A.T., Muhammad N.O., Yakubu M.T. (2011). Pharmaceutical Biology (0,0) :1–5. Posted online on 23 Mar 2011, can be found at: http://informahealthcare.com/doi/pdf/10.3109/13880209.2011.560952
[35] Salawu M.O., Oloyede O.B., Oladiji A.T., Yakubu M.T., Atata R.F. (2010). *Afr J Biotech* 9, 6948–6951.
[36] Schaumann, F. Meike, M. Braun, A., David L., Peden, A., Hoh, J.M. (2008). American Journal of Respiratory and Critical Care Medicine searched on 13th September, 2010.
[37] Tella, A.C., Adediran, G.O. and Awolola, G.V. (2008). Research Journal of Pharmacology, 2(3), 28-30
[38] Tella, A.C. and Owalude S.O. (2007). Research Journal of Applied Sciences, 2,8,875-878.
[39] Tella, A.C., Olabemiwo, O.M., Salawu, M.O and Obiyenwa, G.K. (2010). International Journal of Physical Sciences, 5(4), 379-382.
[40] Tella, A.C., Olabemiwo, O.M., Salawu, M.O and Obiyenwa, G.K. (2010). African Scientist, 11,3,189-193.
[41] United States Pharmacopoeia (1990). Mack Publishing Co. New York, 18th Edition, 83-93.
[42] Woolford, M.H. and Chiccarelli, F.S. (1956). J. Am. PHarm. Ass. Sci. Ed., 45,400
[43] Yokohoma, F. and Chatten, L.G. (1958) J. Am. Pharm. Ass. Sci. Ed. 47,548.

10

Need for Quality Assurance Program of Donor Screening Tests

Young Joo Cha
Chung-Ang University College of Medicine
Republic of Korea

1. Introduction

Transfusion of blood and blood preparations is indispensible in modern medicine, and the processes of delivering a transfusion to a patient provide additional opportunity for risk, despite the remarkable progress. A spectrum of blood-borne infectious agents is transmitted through transfusion of infected blood donated by apparently healthy and asymptomatic blood donors. The diversity of infectious agents includes hepatitis B virus (HBV), hepatitis C virus (HCV), human immunodeficiency viruses (HIV-1/2), human T-cell lymphotropic viruses (HTLV-I/II), Cytomegalovirus (CMV), Parvovirus B19, West Nile Virus (WNV), Dengue virus, trypanosomiasis, malaria, and variant CJD[1]. Post-transfusion hepatitis caused by HBV or HCV make up the major problems of blood-transmitted infections. Clinical characteristics, such as pathophysiology and clinical progress, of post-transfusion hepatitis are the same as those of hepatitis by other causes, except of transmission route.

HBV presents a higher residual risk of transmission by transfusion than HCV or HIV. While most infectious blood units are removed by new testing methods such as chemiluminescent serologic assays for hepatitis B surface antigen (HBsAg), there is clear evidence that transmission by HBsAg-negative components occurs, in part, during the serologically negative window period, but more so during the late stages of chronic infection that HBV DNA could be detected despite HBsAg seronegativity defined as occult HBV infection (OBI). OBI is a challenging clinical entity, recognized by two main characteristics: absence of HBsAg, and low viral replication. The frequency of OBI depends on the relative sensitivity of both HBsAg and HBV DNA assays. It also depends on the prevalence of HBV infection in the population. OBI may follow recovery from infection, displaying antibody to hepatitis B surface antigen (anti-HBs) and persistent low-level viraemia, escape mutants undetected by currently available HBsAg assays, or healthy carriage with antibodies to hepatitis B e antigen (anti-HBe) and to hepatitis B core antigen (anti-HBc)[2]. Over time, in the latter situation, anti-HBe and, later, anti-HBc may become undetectable. Blood donated in the stage of so-called 'window period' after exposure is more infectious than that of OBI. It is reported that blood from donors in window period can infect, even if there might be only 10 virus particles because of its high infectivity. On the other hand, in case of chronic HBV infections in which HBsAg is negative or carriers lasting proliferation of HBV, Dane particles have been developing immune complexes with antibodies like anti-HBs, so infectivity is weaker than acute window period. By look-back study[3] reported in Japan, serological responses showing acute infection have been observed in 12 (19%) among 158 patients transfused with HBV-infected blood. Among them, serological responses showing acute infection have been observed in 11 (50%) among 22 patients transfused with blood donated from HBV-infected window period, on the other hands, observed in only 1 (3%) among 33 patients transfused with blood donated from OBI. However, all forms have been shown to be infectious in immunocompromised individuals, such as organ- or bone marrow-transplant recipients.

HBsAg become positive 50-60 days after infection, preceded by a prolonged phase (up to 40 days) of low-level viraemia. NAT pooling will only detect a small proportion of this pre-HBsAg window period (Fig. 1). Unlike HBV, the risk of HCV transmission by transfusion reduced by introducing HCV nucleic acid testing (NAT) and that of HIV transmission by transfusion also reduced by usage of HIV combined antibody-antigen tests and of HIV NAT. Window period of 16 days (p24 antigen) may be reduced to 11 days by NAT (Fig. 2) and HCV NAT theoretically reduce the window period by 41-60 days (Fig. 3).

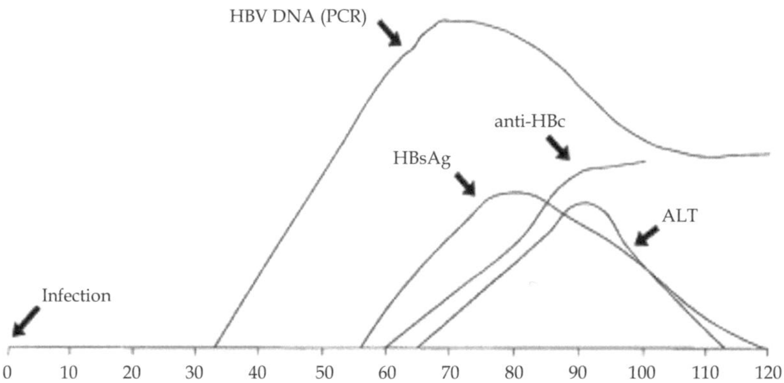

Infection	Day 0
HBV DNA	Variyble up to 23 days prior to HbsAg (average, 6-15 days)
HBsAg	Day 56; disappears Day 120

Fig. 1. Estimated window period in each HBV test

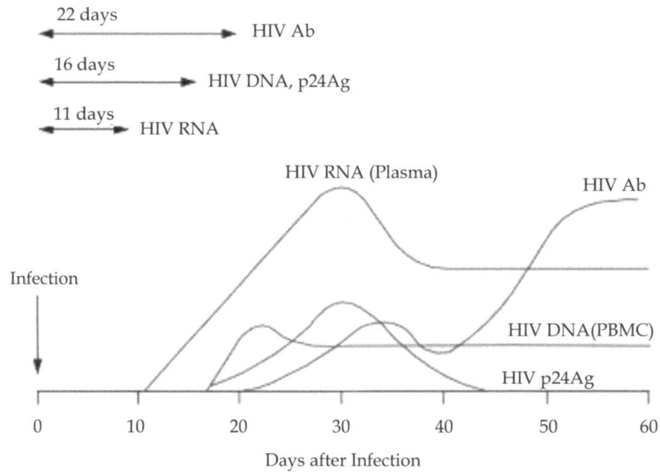

Fig. 2. Estimated window period in each HIV test

The risk of transfusion-transmitted infection (or "residual risk") refers to the chance that an infected donation escapes detection because of a laboratory test's window period (i.e., the time between infection and detection of the virus by that test). The residual risk depends on the prevalence of viremia in the population, especially in blood donors and the sensitivity of the donor screening tests. Prevalence of viremia in blood donors is much less than that of general population. The window period risk can be estimated using the incidence of infection in donors and the length of the window period for tests in use, with an adjustment for atypical inter-donation intervals in seroconverting donors.

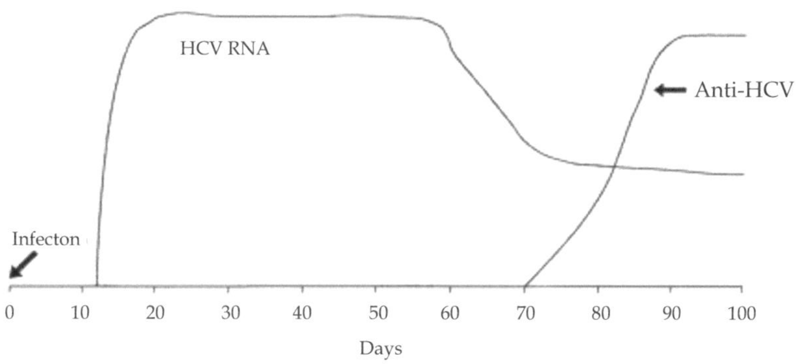

Infection	Day 0
HCV RNA	Day 12
HCV Antibody	Day 70

Fig. 3. Estimated window period of each HCV test

Following the introduction of NAT for HIV and HCV, the American Red Cross estimates[4] the risk of transfusion-transmitted human immunodeficiency virus to be 1:1,215,000 (per unit transfused) and 1:1,935,000 for transfusion-transmitted hepatitis C virus. Hepatitis B virus nucleic acid testing has not been implemented, and the risk of transfusion-transmitted hepatitis B virus in the United States remains relatively high at an estimated 1:205,000. The risk of transfusion-transmitted human T-cell leukemia virus I/II is 1:2,993,000, based on Red Cross estimates. The residual risk per million donations was 0.10 for HIV, 0.35 for HCV, 13.88 for HBV and 0.95 for HTLV reported by the the Canadian Red Cross Society and Canadian Blood Services in 2003[5]. The estimated frequency of infectious donations entering the blood supply during 1996-2003 was 1.66, 0.80 and 0.14 per million for HBV, HCV and HIV respectively, in the United Kingdom[6].

Virus	Risk of transfusion-transmitted infection per 1,000,000 donations		
	USA	Canada	UK
HBV	4.88	13.88	1.66
HCV	0.52	0.35	0.80
HIV	0.82	0.10	0.14
HTLV	0.33	0.95	

Table 1. Incidence and estimated rates of residual risk for HIV, HCV, HBV and HTLV in blood donors

2. Donor screening tests

Donor screening tests to prevent blood-borne virus infections include tests for HBV, HCV, HIV, and HTLV. Accurate detection of HBsAg is an important aid in successful screening blood donors infected with the HBV. Prevention of transfusion-transmitted HBV has historically relied on serological screening of blood donors using progressively more sensitive HBsAg assays; in some countries anti-HBc assays have also been employed to detect chronic carriers with low-level viremia who lack detectable HBsAg. According to the study conducted by the International Consortium for Blood Safety (ICBS) to identify high-quality test kits for detection of HBsAg, seventeen HBsAg enzyme immunoassay (EIA) kits among the 70 HBsAg test kits from around the world had high analytical sensitivity <0.13 IU/ml, showed 100% diagnostic sensitivity, and were even sensitive for the various HBV variants tested[7]. An additional six test kits had high sensitivity (<0.13 IU/ml) but missed HBsAg mutants and/or showed reduced sensitivity to certain HBV genotypes. As regards the sensitivity of HBsAg assays, diagnostic efficacy of the evaluated HBsAg test kits differed substantially, and the analytical sensitivity of HBsAg assays may be dependent on the genetic variability of HBV. Laboratories should therefore be aware of the analytical sensitivity for HBsAg and check for the relevant HBV variants circulating in the relevant population[8]. HBV mutants are stable over time and can be transmitted horizontally or vertically. The sensitivity of HBsAg assays for mutant detection is continuously improved. Immunoassays based on polyclonal capture antibody show the highest sensitivity for the recognition of recombinant mutants or serum samples harboring mutant forms of HBsAg. However, they do not guarantee full sensitivity. Detection of HBsAg needs to be improved by the introduction of new HBsAg assays able to recognize so far described S-gene mutants and with a lower detection threshold than current immunoassays in order to detect smallest amounts of HBsAg in low level carriers. There is also a need for more complete epidemiological data on the prevalence of HBsAg mutants and strategies for the (differential) screening of mutants need to be developed and evaluated[9].

NAT for HCV and HIV has been successfully introduced to screen donors in many developed countries over the past several years. HCV/HIV NAT screening has been applied to mini-pools (MP) of eight to 96 donor specimens, with only minimal impact of MP dilutions on clinical sensitivity for interdiction of window period donations. HBV NAT was only recently introduced in several countries (e.g., Japan and Germany), to detect HBsAg-negative, anti-HBc-negative blood units donated during early acute infection or from OBI[10], although many countries including England and France are still difficult to introduce HBV NAT because of the cost. HBV NAT in donor screening has been introduced in the Finland and Netherland since 2009 and in Korea since 2011.

Although theoretical benefits of HBV NAT relative to HBsAg has been proven through comparison data on seroconversion panels as been using HBsAg assays of varying sensitivities, benefit of pooled-sample NAT is relatively small in areas of low endemicity, with greater yields in areas highly endemic for HBV[11]. Japan is the first country introducing HBV NAT as a donor screening test in 1999, now using 20-MP since 2004. In Japan, frequency of OBI from donors was 1 in 107,000 donations, on the other hands, frequency of OBI from donors in Europe was 1 in 7500~63,000, because of using 6~8 MP. Frequency of OBI is differ from country to country, depending on the prevalence and the number of MP. Frequency of OBI detection in Japan is lower than Europe, so the number of MP should be reduced to increase efficiency of OBI detection.

Single-sample NAT would offer more significant early window period closure and could prevent a moderate number of residual HBV transmissions not detected by HBsAg assays. Although the major vendors of NAT systems (Roche and Chiron/Gen-Probe) have been developing triplex assays that include HBV DNA detection capacity without compromising HIV or HCV detection, there is controversy over the magnitude of the incremental yield and clinical benefit of HBV MP-NAT over serological screening strategies, as well as the impact of implementation of HBV NAT on need for retention of HBsAg and anti-HBc screening. Fully automated, high through-put single-sample HBV NAT systems are needed for blood donor screening, now being developed in Korea.

Each country will need to develop its blood screening strategy based on HBV endemicity, yields of infectious units detected by different serologic/NAT screening methods, and cost effectiveness of test methods in ensuring blood safety.

3. Need for quality assurance program of donor screening tests

Serological tests and NAT implemented as donor screening tests for transfusion-transmitted viruses should be most accurately performed, because their false positive results might hinder the effective use of blood and their false negative results might cause the risk of blood-transmitted infections[12,13]. Therefore, systematic quality assurance program is required to minimize false positive or false negative results, keeping the accuracy of donor screening tests strictly.

Quality Assurance program for donor screening tests is composed of 4 steps. The first step for quality assurance is in registration/licensing step of in vitro diagnostic reagents for donor screening tests. In the US or Europe, special licensing is required after validating safety and clinical effectiveness in order to be used as donor screening tests, even if it might be the same virus markers as those for diagnostic purpose. The second step for quality assurance is in production/distribution process of in vitro diagnostics for donor screening tests. There is a system verifying each lot of products for donor screening tests in the US or Europe. The third step for quality assurance is to monitor the quality of carrying out donor screening tests. For this, each process should be performed according to standard operating procedures (SOP) and accredited by inspecting institution or society. The last step for quality assurance is to conduct the external proficiency program verifying the accuracy of results of donor screening tests.

The external proficiency program for donor screening tests should be operated to verify the ability detecting low level of viral antibodies or antigens including genetic variability. To do this, wide range of the quality control specimens, including standard serum panels or low titer panels made from patients' sera, should be used for the external proficiency program. Ability for detecting low titer of antibodies or mutant viral antigens should be also confirmed, because blood transfusion by low titer or variant virus has been reported[14,15] all over the world. Two blood donors with mutant HBsAg have been also reported in Korea. World Health Organization recommends each country to develop national standard materials for donor screening tests for its people and make use of them for quality evaluation, if possible. In England and Australia, national standard materials of biological medicines have been established at national level, being used for the external proficiency program. These standard materials can be also provided to other countries asking for. Singapore enforces outside and inside quality assurance by using national standard materials made by National Standard Reference Laboratories in Australia. Each country should develop its quality assurance program for donor screening tests.

4. References

[1] Allain JP, Stramer SL, Carneiro-Proietti AB, Martins ML, Lopes da Silva SN, Ribeiro M, Proietti FA, Reesink HW. Transfusion-transmitted infectious diseases. Biologicals. 2009;37:71-7.

[2] Allain JP. Occult hepatitis B virus infection: implications in transfusion. Vox Sang. 2004;86:83-91.

[3] Satake M, Taira R, Yugi H, Hino S, Kanemitsu K, Ikeda H, Tadokoro K. Infectivity of blood components with low hepatitis B virus DNA levels identified in a lookback program. Transfusion. 2007;47:1197-1205.

[4] Pomper GJ, Wu Y, Snyder EL. Risks of transfusion-transmitted infections: 2003. Curr Opin Hematol. 2003;10:412-8.

[5] Chiavetta JA, Escobar M, Newman A, He Y, Driezen P, Deeks S, Hone DE, O'Brien SF, Sher G. Incidence and estimated rates of residual risk for HIV, hepatitis C, hepatitis B and human T-cell lymphotropic viruses in blood donors in Canada, 1990-2000. CMAJ. 2003;169:767-73.

[6] Soldan K, Davison K, Dow B. Estimates of the frequency of HBV, HCV, and HIV infectious donations entering the blood supply in the United Kingdom, 1996 to 2003. Euro Surveill. 2005;10:17-9.

[7] Scheiblauer H, El-Nageh M, Diaz S, Nick S, Zeichhardt H, Grunert HP, Prince A. Performance evaluation of 70 hepatitis B virus (HBV) surface antigen (HBsAg) assays from around the world by a geographically diverse panel with an array of HBV genotypes and HBsAg subtypes. Vox Sang. 2010;98(3 Pt 2):403-14.

[8] Huh HJ, Chae SL, Cha YJ. Comparison study with enzyme immunoassay and chemiluminescence immunoassay for hepatitis B virus surface antigen detection. Korean J Lab Med. 2007;27:355-9

[9] Weber B. Genetic variability of the S gene of hepatitis B virus: clinical and diagnostic impact. J Clin Virol. 2005;32:102-12.

[10] Busch MP. Should HBV DNA NAT replace HBsAg and/or anti-HBc screening of blood donors? Transfus Clin Biol. 2004;11:26-32

[11] Kuhns MC, Busch MP. New strategies for blood donor screening for hepatitis B virus: nucleic acid testing versus immunoassay methods. Mol Diagn Ther. 2006;10:77-91.

[12] Oh DJ, Cho YJ, Kwon SY, Cho NS, Kwon SW, Um TH, et al. A proposal for developing a national quality assurance program for donor blood assays. Korean J Blood Transfus 2008;19: 197-206

[13] Cha YJ. The results of external proficiency tests to prevent transfusion-transmitted virus infection: there is a need for a quality assurance program for donor screening tests to prevent blood-borne virus infections. Korean J Blood Transfus 2010;21:25-35

[14] Hou J, Wang Z, Cheng J, Lin Y, Lau GK, Sun J, et al. Prevalence of naturally occurring surface gene variants of hepatitis B virus in nonimmunized surface antigen-negative Chinese carriers. Hepatology 2001;34:1027-34

[15] Levicnik-Stezinar S. Hepatitis B surface antigen escape mutant in a first time blood donor potentially missed by a routine screening assay. Clin Lab 2004;50:49-51

11

Quality Control in Pharmaceuticals: Residual Solvents Testing and Analysis

Changqin Hu and Ying Liu
National Institutes for Food and Drug Control, Beijing
China

1. Introduction

Organic solvents are constantly present in the pharmaceutical production processes. They are usually used at any step of the synthesis pathway during the drug product formulation process. Organic solvents play an important role in the pharmaceutical industry, and appropriate selection of the solvents for the synthesis of drug substance may enhance the yield, or determine characteristics such as crystal form, purity, and solubility. Because of some physical and chemical property, the solvents are not completely removed by practical manufacturing techniques. Usually some small amounts of solvents may remain in the final drug product. They are called as residual solvents. Thus, residual solvents in pharmaceuticals are defined as organic volatile chemicals that are used or produced in the manufacture of drug substances or excipients, or in the preparation of drug products (International Conference on Harmonisation of Technical Requirement for Registration of Pharmaceuticals for Human Use [ICH], 2009). Since there is no therapeutic benefit from residual solvents, all residual solvents should be removed to the extent possible to meet product specifications, good manufacturing practices, or other quality-based requirements. If the presence of residual solvents in pharmaceuticals exceeds tolerance limits as suggested by safety data, they may be harmful to the human health and to the environment. That's the reason that residual solvents testing become one of the important parts of quality control in pharmaceuticals. This chapter will review the regulation of residual solvents and methods for residual solvents testing and analysis. Special emphasis will be given to the recent progress of residual solvents analysis and systematic study on residual solvents analysis in pharmaceuticals.

2. Regulation of residual solvents testing

The toxicity of residual solvents was recognized by the regulatory agency in the world in 90's. The United States Pharmacopeia was the first one that adopted residual solvent testing in 22 th edition 3 rd supplement in 1990 (The United States Pharmacopoeia [USP], 1990) British Pharmacopeia (1993 edition supplement) (British Pharmacopoeia [BP], 1996), European Pharmacopeia (3 rd edition) (European Pharmacopoeia [EP], 1997) and Chinese Pharmacopeia (1995 edition) (Pharmacopoeia of the People's Republic of China [ChP], 1995) subsequently adopted residual solvent testing, but only 6-8 residual solvents were controlled at that time. (Table 1)

Organic volatile impurities	Limit (ppm)			
	USP 22 edition 3rd supplement	BP(1993) supplement	EP 3rd	ChP 1995 edition
Benzene	100	100	100	100
Chloroform	50	50	50	50
1,4-Dioxane	100	100	100	100
Ethylene oxide	10	-	-	10
Dichloromethane	100	100	100	100
Trichloroethene	100	100	100	100
Acetonitrile	-	50	50	-
Pyridine	-	100	100	100
Toluene	-	-	-	100

Table 1. Categories and limits of residual solvents initially controlled in each pharmacopoeia

At that time, each pharmacopeia used various guidelines for residual solvents control in pharmaceutical products with different categories and acceptance limits. Moreover, only 6-8 residual solvents were controlled, which was far behind from the categories that were really used in pharmaceutical industry. Internationally, a standard guideline for control of residual solvents is needed to be established. Efforts were made to harmonize the guideline for residual solvents by ICH. On 17 July 1997, the Q3C parent guideline on residual solvent guidelines and limits was approved by the Steering Committee under Step 4 and recommended for adoption the three ICH regulatory bodies. 69 organic solvents that are commonly used in pharmaceutical industry were classified in 4 categories by ICH guideline (Table 2). Solvents in Class 1 are known carcinogens and should not be employed in the manufacture of drug substances, excipients, and drug products because of their unacceptable toxicity or their deleterious environmental effect. However, if their use is unavoidable in order to produce a drug product with a significant therapeutic advance, then their levels should be restricted as shown in Table 2, unless otherwise justified. The limits of Class l solvents are usually between 2-8 ppm except 1,1,1-trichloroethane is 1500 ppm, which is an environmental hazard. Class 2 solvents are nongenotoxic animal carcinogens. Solvents of this class should be limited in pharmaceutical products because of their inherent toxicity. The concentration limits of these solvents are in the range of 50 ~ 3880 ppm. Class 3 solvents have less toxic and lower risk to human health. Class 3 includes no solvent known as a human health hazard at levels normally accepted in pharmaceuticals. However, there are no long-term toxicity or carcinogenicity studies for many of the solvents in Class 3. They are less toxic in acute or short-term studies and negative in genotoxicity studies. The concentration limits of these solvents are 5000 ppm. Class 4 solvents are the solvents that may also be of interest to manufacturers of excipients, drug substances, or drug products. However, no adequate toxicological data was found. Manufacturers should supply justification for residual levels of these solvents in pharmaceutical products.

Solvent	Concentration limit (ppm)
Class 1 solvents (solvents to be avoided)	
Benzene	2
Carbon tetrachloride	4
1,2-Dicloroethane	5
1,1-Dichloroethene	8
1,1,1-Trichloroethane	1500
Class 2 solvents (solvents to be limited)	
Acetonitrile	410
Chlorobenzene	360
Chloroform	60
Cyclohexane	3880
1,2-Dichloroethene	1870
Dichloromethane	600
1,2-Dimethoxyethane	100
N,N-Dimethylacetamide	1090
N,N-Dimethylformamide	880
1,4-Dioxane	380
2-Ethoxyethanol	160
Ethyleneglycol	62
Formamide	220
Hexane	290
Methanol	3000
2-Methoxyethanol	50
Methylbutyl ketone	50
Methylcyclohexane	1180
N-Methylpyrrolidone	4840
Nitromethane	50
Pyridine	200
Sulfolane	160
Tetralin	100
Toluene	890
1,1,2-Trichloroethene	80
Xylene	2170
Class 3 solvents (solvents which should be limited by GMP or other qualitybased requirements)	
Acetic acid	5000
Acetone	5000
Anisole	5000
1-Butanol	5000
2-Butanol	5000
Butyl acetate	5000
tert-Butylmethyl ether	5000
Cumene	5000
Dimethyl sulfoxide	5000

Solvent	Concentration limit (ppm)
Ethanol	5000
Ethyl acetate	5000
Ethyl ether	5000
Ethyl formate	5000
Formic acid	5000
Heptane	5000
Isobutyl acetate	5000
Isopropyl acetate	5000
Methyl acetate	5000
3-Methyl-1-butanol	5000
Methylethyl ketone	5000
Methylisobutyl ketone	5000
2-Methyl-1-propanol	5000
Pentane	5000
1-Pentanol	5000
1-Propanol	5000
2-Propanol	5000
Propyl acetate	5000
Tetrahydrofuran	5000
Class 4 solvents (solvents for which no adequate toxicological data was found)	
1,1-Diethoxypropane	Methylisopropyl ketone
1,1-Dimethoxymethane	Methyltetrahydrofuran
2,2-Dimethoxypropane	Petroleum ether
Isooctane	Trichloroacetic acid
Isopropyl ether	Trifluoroacetic acid

Table 2. List of solvents included in the guideline of ICH

After the ICH guideline regarding residual solvents in pharmaceuticals became official in 1997, consequently, pharmacopeias of different countries have adopted it and have revised their general methods to reflect it. EP (3rd edition) was the first one that accepted ICH guideline with the same categories and limits of residual solvents. In general chapter: Identification and control of residual solvents, general methods for residual solvent determination were described. Gas chromatography (GC) with headspace injection is proposed in both systems. Two procedures (systems), A and B, are presented, and System A is preferred whilst System B is employed normally for confirmation of identity (EP, 1999). Japanese Pharmacopoeia accepted ICH guideline in 14th edition (Japanese Pharmacopoeia [JP], 2001). ICH guideline was accepted by Chp in 2005 edition (Chp, 2005). Three methods were used to screening and analysis residual solvents in pharmaceuticals: Isothermal temperature HS-GC method, Programmed temperature HS-GC method, and direct injection method. Until USP 28, residual solvents testing was finally updated to comply with ICH guideline. Current official methods for residual solvent determination are described in <467> chapter Organic Volatile Impurities. Three procedures (A, B, C) for water-soluble and water-insoluble articles, are available. Procedures A and B are useful to identify and quantify residual solvents, when the information regarding which solvents are likely to be present in the material is not available. In cases when we have information about residues of solvents that may be expected in the tested material, only procedure C is needed for quantification of the amount of residual solvents (USP, 2005).

3. Methods for residual solvents analysis

In the early stage, one of the simplest methods for determining the content of volatile residues consists in measuring the weight loss of a sample during heating. However, this method suffers the great disadvantages of being totally non-specific (multicomponent solvent blends cannot be analysed and there will always be a doubt on humidity contamination) and of needing several grams of product to achieve a detection limit of about 0.1% (Benoit, 1986; Dubernet, 1990; Guimbard, 1991). Nevertheless, when carried out by thermogravimetry, the limit can be lowered to 100 ppm using only a few milligrams of substance (Guimbard, 1991). Infrared spectroscopy (IR) (Osawa & Aiba, 1982) and Fourier Transform Infrared Spectrometry (FTIR) (Vachon & Nairn, 1995) were used to determine residual Tetrahydrofuran (THF), dichloroethane and methylene chloride in polymer samples by measuring the characteristic solvent bands in the spectra. The most common limiting factors in these methods are possible interferences of solvent and matrix peaks and, in the case of IR, the high detection limit (above 100 ppm) and a lack of accuracy at low concentrations (Weitkamp & Barth, 1976). Avdovich et al. determined benzene, toluene, acetone, methyl ethyl ketone and ethyl ether (in a few samples also methylene chloride and ethyl acetate) in cocaine samples by NMR, which allowed a quantification down to 100 ppm, with possibly detection or identification problems in the case of ethyl ether and methyl ethyl ketone at these low levels (Avdovich, 1991). However, these detection limits are too high to satisfy the requirements relating to residual solvents determination, especially for the most toxic solvents.

The methods mentioned above were replaced by GC. GC is the natural choice for residual solvent analysis. Firstly, because of its excellent separation ability, according to the chromatographic conditions and the column and, secondly, because of its low detection limits and the possibility of analysing liquid or solid samples of a complex nature. Modern capillary-column GC can separate a large number of volatile components, permitting identification through retention characteristics and detection at ppm levels using a broad range of detectors. The most popular detectors are: the flame ionization detector (FID), which is a rather universal detector for organic volatile compounds; and, the electron capture detector (ECD), which is especially suited to detection of halogenated compounds. However, FID is by far the most preferred for release-related tasks because of its low detection limits, wide linear dynamic range, robustness, ease of operation, and general reliability and utility, especially for trace organic compounds. There are three type of GC classed by different sample preparation procedures: direct-injection GC, headspace (HS) GC and solid-phase microextraction (SPME) GC. Application of these three GCs in residual solvent analysis will be reviewed below.

3.1 Direct-injection GC

Residual solvent determination using direct-injection sample preparation is the oldest technique, and, historically, it was preferred because of its simplicity, reliability, ease of operation and throughput (Witschi & Doelker, 1997). The drug substance or the formulation is dissolved in or extracted with a high-boiling-point solvent, such as water, dimethylsulfoxide (DMSO), dimethylformamide (DMF), dimethylacetamide (DMA), benzyl alcohol (BA) or ethylene glycol. Using high-boiling-point solvents has the advantage that the diluent solvent peak will elute later, thus not interfering with the earlier eluting analyte peaks. However, it has the big disadvantage that non-volatile components, such as the drug substance or the formulation components, are also injected, and that leads to injector contamination, column contamination and deterioration, together with unavoidable matrix effects. Furthermore, as the matrix is also injected onto the column, this must be eluted prior to beginning the next injection, and that has the effect of prolonging the analytical run. From Witschi and Doelker (Witschi & Doelker, 1997) and Hymer's (Hymer, 2003) reviews, the data in the literature on direct injection applications was summarized up to 2003. It was evident from the trend that, in more recent times, based on the number of publications, the interest of industry-research groups has shifted to other sample-preparation techniques, such as static headspace and sorbent-based approaches.

3.2 Headspace GC

Two types of HS sampling are available: dynamic HS analysis (also called purge-and-trap); and static HS analysis. The theory of static headspace is thoroughly described in three books, by Hachenberg and Schmidt (Hachenberg & Schmidt, 1977), Loffe and Vitenberg (Loffe & Vitenberg, 1984), and Kolb and Ettre (Kolb & Ettre, 2006). It was summarized by Snow and Bullock as below (Snow & Bullock, 2010). In HS extraction, the vapor phase directly above and in contact with a liquid or solid sample in a sealed container is sampled and an aliquot is transferred to a GC for separation on a column, detection and quantitation. The ability to determine the amount of a substance within a liquid or solid sample by analyzing the headspace vapor above it in a closed vessel derives from three critical fundamental principles: Dalton's Law, Raoult's Law and Henry's Law. Generally, static HS sampling is the most widely used technique for residual solvent determination in pharmaceuticals. This fact comes from some of the advantages of this technique, mainly that only volatile substances and dissolution medium can be injected onto the column. Also HS systems are fully automated, in addition, a sample preparation is easy, and the sensitivity of analysis is sufficient for the majority of solvents mentioned in ICH guidelines. Static HS sampling is based on thermostatic partitioning of volatile compounds in a sealed vial between the sample diluent and the gas phase. Sample diluent is a critical factor affecting HS-GC method sample load, sensitivity, equilibration temperature and time. A good sample diluent for analyzing residual solvents in pharmaceutical products should have a high capability for dissolving a large amount of samples, a high boiling point and a good stability. There are a number of commonly used sample diluents for HS analysis, such as water, DMSO, DMF, DMA, BA, 1,3-dimethyl-2-imidazolidinone (DMI), and mixtures of water-DMF or water-DMSO. For water-soluble samples, water is the choice of diluent. The influence of the matrix medium used for the determination of residual solvents in pharmaceuticals was investigated by Urakami et al (Urakami et al, 2004). A guide for the choice of a matrix medium suitable for the determination of residual solvents was proposed. Water, DMSO, DMF, DMA, BA, DMI were studied as matrix media, and seventeen solvents were used as target analytes. The peak shapes of each analytes were not affected by the matrix medium, whereas the peak intensities for all solvents were strongly affected by the matrix medium. Otero et al established a static HS GC method for quantitative determination of residual solvents in a drug substance according to European Pharmacopoeia general procedure. A water-dimethylformamide mixture is proposed as sample solvent to obtain good sensitivity and recovery (Otero et al,

2004). Recently, ion liquid was used as matrix medium in HS analysis in residual solvent analysis. Liu et al used a new solvent room temperature ionic liquid (1-butyl-3-methylimidazolium teterafluoroborate) as matrix medium in static HS to determine residual solvents in pharmaceutical. Six residual solvents were analyzed and better sensitivities were gained with it as diluent comparing with DMSO (Liu & Jiang, 2007). Laus et al reported that 1-n-Butyl-3-methylimidazolium dimethyl phosphate (BMIM DMP) was identified as the most suitable ionic liquid as solvent for the HS-GC analysis of solvents with very low vapor pressure such as dimethylsulfoxide, N-methylpyrrolidone, sulfolane, tetralin, and ethylene glycol (Laus et al, 2009). The main drawback of static HS is the lower detection limit compared to dynamic HS. Partition Coefficient (K) is the key factor that affects the sensitivity of HS analysis, which represented the concentration ratio of a volatile in the liquid and gas phase at a defined temperature and pressure at equilibrium stage. Substance with low partition coefficient ($K < 10\text{-}100$) is easier to go to the gas phase, and is considered to suitable for HS analysis. Several methods are available for reducing the partition coefficient of volatiles, in particular in aqueous systems, and thus to improve the HS sensitivity, such as salting-out, pH adjustment or increasing the equilibration temperature of the sample. Dynamic headspace sampling technique involves the passing of carrier gas through a liquid sample, followed by trapping of the volatile analytes on a sorbent and desorption onto a GC. A major advantage of this technique is that a thermodynamic equilibrium is not necessarily needed, and the sensitivity of the method is increased by enrichment of the anlaytes on the trap. Consequently, limit of detection reported for dynamic headspace are lower (pg/ml) than those obtained with static headspace (ng/ml) (Arthur & Pawliszyn, 1990). Therefore, the automation of the instrument and reproducibility of the results are not as good as static headspace, so the application of purge and trap in residual solvent analysis was not popular.
Dynamic headspace analysis is particular suited for the determination of volatile residual solvents at very low concentrations. Recently, Lakatos reported that four Class 1 solvents were analyzed in a water-soluble drug using dynamic headspace technique. The results show that the Purge and trap technique is more sensitive than the static headspace. Repeatability, accuracy and the linearity were examined, and these characteristics of the method were proved to be suitable for residual solvent analysis. It was found that the Purge and trap could be an alternative sample preparation method besides the static headspace method (Lakatos, 2008).

3.3 Solid-phase microextraction GC

SPME, in which a small amount of extracting phase, a stationary phase is coated on a support. Commonly, a fused silica fiber is used. The extracting phase is placed in contact with the sample matrix for a predetermined amount of time. If the time is long enough, a concentration equilibrium of the volatile analyte is established between the sample matrix and the extraction phase, then the analytes adsorbed on the fiber are thermally desorbed in the injector of the GC. In general, two types of SPME extractions can be performed. The first type, "Direct extraction" or "immersion" involves bringing the SPME fiber in contact with the sample matrix. The second type of SPME is headspace SPME, in which, the volatile analytes need to be transported through the barrier of air above the sample before they can reach the SPME extracting phase. It helps to protect the fiber coating from damage by high molecular-mass and other non-volatile interferers present in the sample matrix. Since the headspace SPME was developed in 1993 and has experienced the strongest growth in research interest over the past decade. Advantages of SPME include simplicity of execution, low cost of the instrument and less solvent consume. Headspace SPME attracted more attention in residual solvent testing area due to it can avoid the interference from the non-volatile pharmaceuticals. Camarasu et al used two types of SPME methods to determine residual solvents in pharmaceuticals. Three fibers with different polymer films were compared and the polydimethylsiloxane/divinylbenzene coated fiber was found to be the most sensitive one for the analyzed analytes. Bewteen the investigated sample preparation techniques, gastight-SPME proved to be the most sensitive one. Headspace SPME is more precise. Compared with the static headspace technique, SPME method showed superior results (Camarasu et al, 1998). Another paper from Camarasu reported that an SPME method has been developed and optimized for the polar residual solvents determination in pharmaceutical products. The headspace SPME from aqueous solutions was found to be ten times more sensitive than Immersion SPME and Headspace SPME from organic solutions (Camarasu, 2000)

3.4 Recent progress

A new method for direct determination of residual solvents in solid drug product using multiple headspace sing-drop microextraction (MHS-SDME) was reported by Yu et al. The MHS-SDME technique is based on extrapolation to an exhaustive extraction of consecutive extractions from the same sample which eliminates the matrix effect on the quantitative analysis of solid samples. Factors affecting the performance of MHS-SDME including extraction solvent, microdrop volume, extraction time, sample amount, thermostatting temperature and incubation time were studied. Experimentally, a model drug powder was chosen and the amounts of residues of two solvents, methanol and ethanol were investigated. Quantitative results of the proposed method showed good agreement with the traditional dissolution method. Compared with the conventional method for determination of residual solvents, the MHS-SDME technique can eliminate possible memory effects with less organic solvents. The results also indicated that MHS-SDME had a great potential for the quantitative determination of residual solvents directly from the solid drug products due to its low cost, ease of operation, sensitivity, reliability and environmental protection (Yu et al, 2010).

A novel on-line solvent drying technique has been described that is capable of simultaneously measuring the solvent end point in vapor phase and maintaining high accuracy with precision. The technique used non-contact infrared sensor for monitoring the solvent vapors during the pharmaceutical solvent drying process. The data presented demonstrated that on-line combined with non-contact sensor method had high degree of precision and accuracy for monitoring the end point of the solvent drying (Tewari et al, 2010).

4. Systematic study of analysis residual solvents in pharmaceuticals-database

Analysis of residual solvent is known to be one of the most challenging analytical tasks in pharmaceutical analysis and control. The challenge is due to the different manufacturer produce the same pharmaceutical products using different manufacturing processes. Unknown peaks are often detected during routine quality control testing using GC. When this happened, the only thing we can do is to try different solvent standards to find out which has the same retention time with the unknown peak. It is a time consuming work, sometimes the unknown peak is not a residual solvent, but an interference peak. To address this problem, a systematic study was conducted by our laboratory; three databases were established for fast screening, confirmation and method optimization in the analysis of residual solvents in pharmaceuticals. These three databases were published separately (Liu & Hu, 2006, 2007, 2009) and were combined here for a better understanding purpose since they are three parts of the intact database for residual solvent analysis.

4.1 Screening database

4.1.1 Establishment of screening database

When analysis residual solvent using GC, unknown peaks often show up. It is hard to tell the unknown peak is another residual solvent or interference peak. Moreover, some organic solvents controlled by ICH have the same retention time on a GC column. To solve these problems, a database for preliminary screening of residual solvents in pharmaceuticals has been established using the parallel dual-column system. The basic principle is that different compounds may have the same retention times on one column, but it is highly unlikely that different compounds will have the same retention times on another column with opposite polarities. So if an organic solvent is present in both columns in the screening procedure, then it is a suspect residual solvent in pharmaceutical. The establishment and application of the screening database were described in one of article published by our lab (Liu & Hu, 2007). Two columns with different polarities, SPB-1 and HP-INNOWAX, connected with a 'Y' splitter, constituted the dual pathways system. Fifty-two solvents that suitable for static headspace analysis were studied according to the guidelines for residual solvents regulated by ICH on this system. The retention times of 52 organic solvents in both systems were recorded under the above conditions. The dead time was determined using methane, and the adjusted retention times of each solvent were calculated. The relative retention times (RRTs) of each solvent in both systems were then calculated as follows, using methyl ethyl ketone (MEK) as the reference standard.

$$RRT= [t_R(\text{compound})-t_0]/[t_R(\text{MEK})-t_0] \quad (1)$$

Where t_R is the retention time of the compound, and t_0 is the retention time of methane. The RRT was selected as the basis of identification. The RRTs of the 52 organic solvents in both systems constituted the database (Table 3).

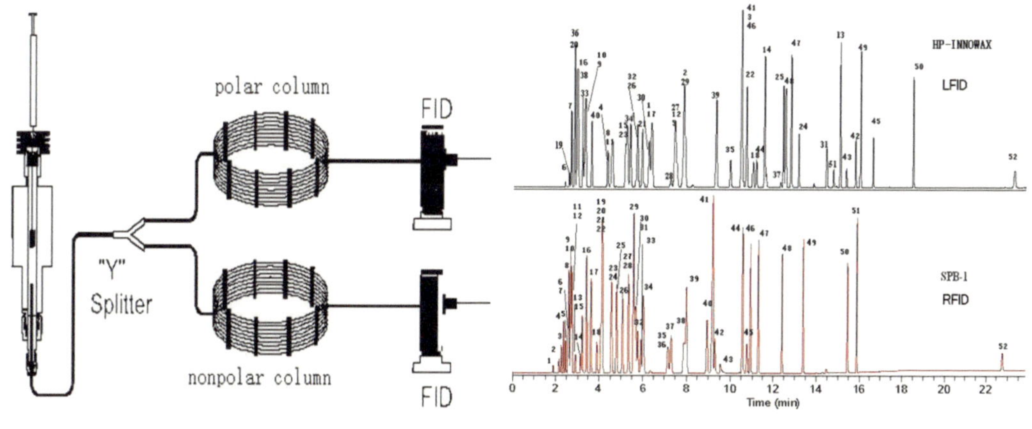

Fig. 1. Schematic diagram of parallel-column system and two chromatograms of all of 52 organic solvents obtained from the system in a single run

Non-polar system SPB - 1				Polar system HP - INNOWAX			
Order	Organic solvent	t_R(min)	RRT	Order	Organic solvent	t_R(min)	RRT
1	methanol	1.872	0.135	1	pentane	2.432	0.038
2	ethanol	2.155	0.274	2	hexane	2.607	0.081
3	acetonitrile	2.237	0.315	3	ethyl ether	2.675	0.098
4	acetone	2.345	0.368	4	isooctane	2.848	0.141
5	2-propanol	2.447	0.419	5	isopropyl ether	2.850	0.141
6	pentane	2.557	0.473	6	tert-butyl methyl ether	2.928	0.161
7	ethyl ether	2.568	0.479	7	heptane	2.987	0.175
8	ethyl formate	2.600	0.495	8	cyclohexane	3.232	0.236
9	1,1-dimethoxymethane	2.672	0.530	9	1,1-dichloroethene	3.277	0.247
10	1,1-dichloroethene	2.687	0.538	10	1, 1,1-dimethoxymethane	3.348	0.264
11	methyl acetate	2.730	0.559	11	methylcyclohexane	3.652	0.339
12	dichloromethane	2.733	0.560	12	acetone	4.378	0.518
13	nitromethane	2.903	0.644	13	ethyl formate	4.492	0.547
14	1-propanol	3.135	0.759	14	methyl acetate	4.562	0.564
15	1,2-dichloroethene	3.222	0.802	15	1,2-dichloroethene	5.190	0.719
16	tert-butyl methyl ether	3.407	0.894	16	tetrahydrofuran	5.217	0.725
17	methyl ethyl ketone	3.622	1.000	17	methyl tetrahydrofuran	5.378	0.765
18	2-butanol	3.892	1.134	18	1,1,1-trichloroethane	5.692	0.843
19	hexane	4.072	1.223	19	carbon tetrachloride	5.693	0.843
20	isopropyl ether	4.103	1.238	20	ethyl acetate	5.893	0.892
21	ethyl acetate	4.122	1.247	21	isopropyl acetate	6.250	0.980
22	chloroform	4.127	1.250	22	methyl ethyl ketone	6.330	1.000
23	tetrahydrofuran	4.537	1.453	23	methanol	6.358	1.007
24	2-methyl-1-propanol	4.560	1.464	24	1,2-dimethoxyethane	7.270	1.232
25	1,2-dichloroethane	4.788	1.577	25	2-propanol	7.390	1.262
26	1,1,1-trichloroethane	5.047	1.705	26	methyl isopropyl ketone	7.400	1.264
27	methyl isopropyl ketone	5.310	1.835	27	dichloromethane	7.470	1.281
28	1,2-dimethoxyethane	5.348	1.854	28	ethanol	7.802	1.363
29	benzene	5.563	1.960	29	benzene	7.827	1.369
30	isopropyl acetate	5.652	2.004	30	propyl acetate	9.355	1.746
31	1-butanol	5.718	2.037	31	1,1,2-trichloroethene	9.937	1.890
32	carbon tetrachloride	5.743	2.049	32	methyl isobutyl ketone	10.495	2.028
33	cyclohexane	5.903	2.128	33	acetonitrile	10.503	2.030
34	methyl tetrahydrofuran	5.997	2.175	34	isobutyl acetate	10.655	2.067
35	1,1,2-trichloroethene	7.143	2.741	35	chloroform	10.980	2.147
36	isooctane	7.278	2.808	36	2-butanol	11.182	2.197
37	1,4-dioxane	7.337	2.837	37	toluene	11.568	2.292
38	heptane	7.883	3.107	38	1-propanol	11.610	2.303
39	propyl acetate	7.997	3.164	39	1,4-dioxane	12.258	2.463
40	methylcyclohexane	8.933	3.627	40	1,2-dichloroethane	12.463	2.513
41	methyl isobutyl ketone	9.177	3.747	41	butyl acetate	12.540	2.532
42	3-methyl-1-butanol	9.270	3.793	42	methyl butyl ketone	12.800	2.596
43	pyridine	9.652	3.982	43	2-methyl-1-propanol	13.170	2.688
44	toluene	10.548	4.425	44	1-butanol	14.355	2.980
45	1-pentanol	10.737	4.519	45	cumene	15.030	3.147
46	isobutyl acetate	10.932	4.615	46	nitromethane	15.065	3.155
47	methyl butyl ketone	11.278	4.786	47	pyridine	15.357	3.227
48	butyl acetate	12.428	5.355	48	3-methyl-1-butanol	15.747	3.323
49	chlorobenzene	13.375	5.823	49	chlorobenzene	16.015	3.390
50	anisole	15.443	6.846	50	1-pentanol	16.618	3.538

Non-polar system SPB - 1				Polar system HP - INNOWAX			
Order	Organic solvent	t_R(min)	RRT	Order	Organic solvent	t_R(min)	RRT
51	cumene	15.887	7.066	51	anisole	18.523	4.008
52	tetralin	22.778	10.474	52	tetralin	23.303	5.188
	methane	1.600			methane	2.277	

Table 3. The relative retention times of 52 organic solvents on non-polar system and polar system

4.1.2 Applications of the database

4.1.2.1 Screening the residual solvents in parmacuticals in a single run

Amoxicillin sodium and clavulanate potassium (5:1), an antibacterial drug registered by a foreign company in China, was analyzed. The preliminary screening results (Table 4) were obtained simultaneously in a single run. According to Table 4, the solvents that appeared on both column systems simultaneously may be the residual solvents in the pharmaceuticals. The possible residual solvents were acetone, methyl acetate, ethyl acetate and 2-propanol in this case. All of these solvents were mentioned by the manufacturer, except for methyl acetate. It was confirmed by the reference standard. The confirmation database was used to give further identification of this peak, and the results indicated that the peak was indeed methyl acetate (4.2.3.1). Finally, the manufacturer admitted that methyl acetate was actually used in the manufacturing process, but for some reason it was not disclosed in the manufacturer's product information sheet. In addition, although only 4 out of the 8 impurities detected in Table 4 could be identified as residual solvents, it showed that the database could eliminate the interference of thermal degradation products or other volatile impurities (which were not the 52 residual solvents we concerned), which was one of the advantages of the database.

4.1.2.2 Eliminating the interference of co-elution

Potassium clavulanate and cellulose microcrystallistate (1:1), an enzyme inhibitor of β-lactamase, was registered by a foreign company in China. The content of methanol was reported much higher than the limit specified by the ICH in the routine residual solvent test. The database was used to check this result. The preliminary screening results are given in Table 5.

According to Table 5, the solvents that appeared on both column systems simultaneously may be the residual solvents in the pharmaceutical product. The possible residual solvents in the drugs were acetone and 2-propanol without methanol. If the peak whose RRT was 0.129 was judged only according to the results of SPB-1, it would definitely be identified as methanol, but on the HP-INNOWAX there was no peak with the RRT of methanol. Therefore, this peak was not methanol and was not included in the 52 residual solvents; it might be a degradation product from the headspace process. The database can eliminate the interference of co-elution and avoid false positive result.

4.2 Confirmation database

Mass spectrometry (MS) and FTIR are powerful tools for identification of organic compounds. GC is the most common technique for separation of volatile and semi-volatile mixtures. It is well accepted that when GC is coupled with spectral detection methods, such as FTIR or MS that the resulting combination is a powerful tool for the separation and

Column	Number	t_R(min)	RRT	Data in database	Possible residual solvents
SPB-1	1	2.352	0.369	0.368	acetone*
	2	2.445	0.416	0.419	2-propanol*
	3	2.727	0.556	0.538 0.559 0.560	1,1-dichloroethene methyl acetate* dichloromethane
	4	3.807	1.092	1.134	2-butanol
	5	4.135	1.255	1.223 1.238 1.247 1.250	hexane isopropyl ether ethyl acetate* chloroform
	6	5.863	2.113	2.037 2.049 2.128 2.175	1-butanol carbon tetrachloride cyclohexane methyl tetrahydrofuran
	7	8.697	3.520	3.627	methylcyclohexane
	8	17.988	8.133	no corresponding data	
HP-INNOWAX	1	3.080	0.197	no corresponding data	
	2	4.347	0.510	0.518	acetone*
	3	4.563	0.563	0.547 0.564	ethyl formate methyl acetate*
	4	5.913	0.897	0.892	ethyl acetate*
	5	6.308	0.995	0.980 1.000 1.007	isopropyl acetate methyl ethyl ketone methanol
	6	7.425	1.271	1.232 1.262 1.264	1,2-dimethoxyethane 2-propanol* methyl isopropyl ketone
	7	7.702	1.339	1.363 1.369	ethanol benzene
	8	16.808	5.590	no corresponding data	

Note: The organic solvents appeared in both columns were marked with *.

Table 4. The preliminary screening results of the SPB-1 column and the HP-INNOWAX column

Column	Number	t_R (min)	RRT	Data in database	Possible residual solvents
SPB-1	1	1.878	0.129	0.135	methanol
	2	2.363	0.368	0.368	acetone*
	3	2.458	0.415	0.419	2-propanol*
	4	4.832	1.588	1.577	1,2-dichloroethane
HP-INNOWAX	1	3.037	0.194	no corresponding data	
	2	4.287	0.505	0.518	acetone*
	3	7.310	1.258	1.232	1,2-dimethoxyethane
				1.262	2-propanol*
				1.264	methyl isopropyl ketone
	4	11.543	2.311	2.292	toluene
				2.303	1-propanol

Note: The organic solvents appeared in both columns were marked with *.

Table 5. The preliminary screening results of the SPB-1 column and the HP-INNOWAX column

identification of components in complex mixtures. Gas chromatography-mass spectrometry (GC-MS) has superior detection limits and is widely used in qualitation of volatile organic compound. Gas chromatography-fourier transform infrared spectrometry (GC-FTIR) also has applications in the identification for compound. The combination application of mass spectra and FTIR spectra is a very powerful coupling because of the complementary nature of the data acquired, which will make the confirmation more confident. Another problem is that the residual solvents testing is trace analysis, usually the concentration of residual solvent in the drugs is very low. So it was hard to get good results using the commercial MS spectra library when the analytes at low concentration. To address this issue, 60 organic solvents introduced by ICH were studied using GC-MS and GC-FTIR. The standard mass spectra library, limit of detection (LOD) mass spectra library, standard vapor-phase infrared spectra library and limit of detection (LOD) vapor-phase infrared spectra library were obtained to establish a confirmation database for determining residual solvents in pharmaceuticals. The confirmation database can be used to identify the unknown residual solvents without using reference organic solvents.

4.2.1 Establishment of the confirmation database

One microliter of each stock standard solution was injected into the GC-MS system and the mass spectra and the retention time of the organic solvents were recorded. The limit of detection was considered as the quantity of analyte that generated a response three times greater than the noise level at the retention time by diluting the stock standard solutions as required, and the mass spectra of organic solvents were recorded. The mass spectra library was established with Xcalibur software by exporting to the Library Brower a spectrum that had background subtracted and then attaching the chemical structure, compound name, molecular weight and molecular formula among other standard characteristics.

One microliter of each stock standard solution was injected into the GC-FTIR system and the vapor-phase infrared spectra and the retention time of organic solvents were recorded. The limit of detection was considered as the quantity of analyte that generated a response ten times greater than the noise lever at the retention time in the Gram-Schmidt chromatogram. This limit was achieved by diluting the stock standard solutions as required, and the vapor-phase infrared spectra of organic solvents were recorded. The vapor-phase infrared spectra library was established with OPUS software by exporting to the library a spectrum that had background subtracted and attaching an information mask that included compound name, molecular weight, molecular formula, melting point, boiling point and other standard characteristics.

4.2.2 Mass spectra library and vapor-phase infrared spectra library can verify and complement each other

The advantages of mass spectra in compound identification include the ability to give the molecular weight of compound, the ability to distinguish homologues, and superior detection limits. The LODs of organic solvents are usually in the picogram range. The main limitations of mass spectra include the inability to give the intact information of compound and the inability to distinguish closely related isomers. The advantages of FTIR spectra in compound identification are that it can give information about the intact molecule, and similar structures such as isomers can be distinguished. The main limitation of infrared is lower sensitivity. Obviously, the combination application of mass spectra and FTIR spectra is a very powerful coupling because of the complementary nature of the data acquired. If mass spectra and infrared spectra give the same result, then the result can be considered accurate with greater confidence.

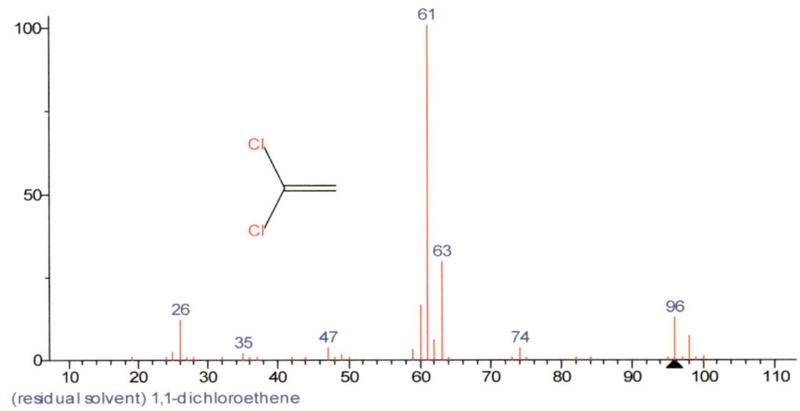

(a1)

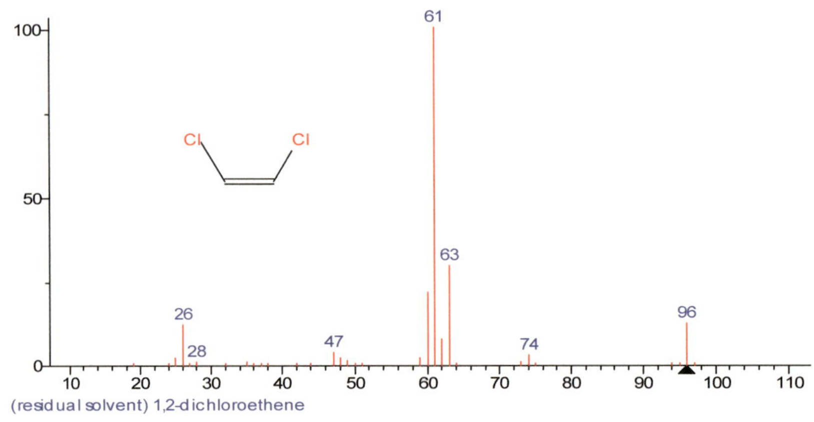

(a2)

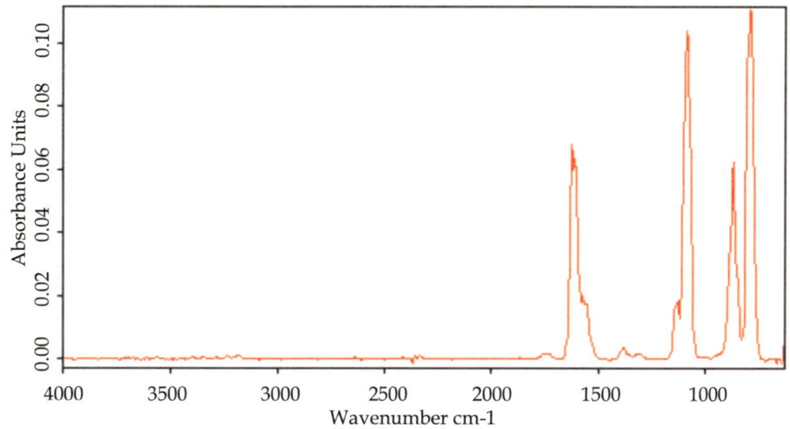

(b1)

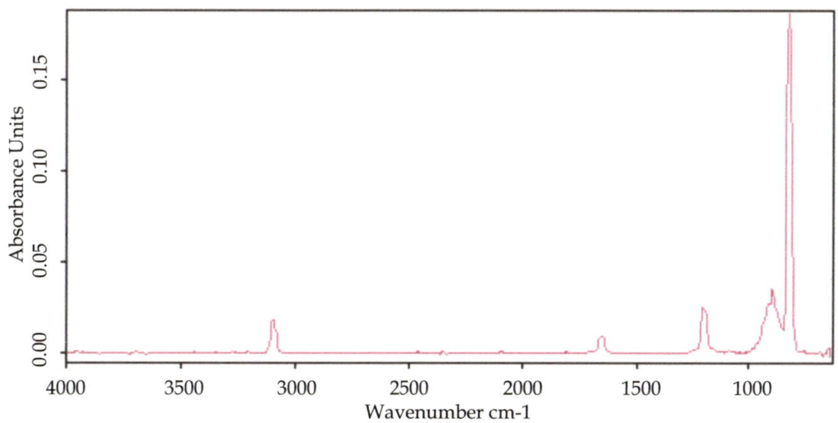

(b2)

Fig. 2. (a) Mass spectra of (1)1,1-Dichloroethene and (2)1,2-Dichloroethene. (b) Vapor-phase infrared spectra of (1)1,1-Dichloroethene and (2)1,2-Dichloroethene

Of the 60 organic solvents were determined, 1,1-Dichloroethene and 1,2-Dichloroethene were isomers. They had very similar mass spectra (Fig. 2.a), and they were difficult to distinguish in the mass spectra library search. But their vapor-phase infrared spectra showed a significant difference (Fig. 2.b).

Isomers that had very similar mass spectra were suited for determination by a vapor-phase infrared spectra library; the normal alkanes(homologs) which had simple vapor-phase infrared spectra (Fig. 3.a) were suited for determination by mass spectra library (Fig. 3.b).

4.2.3 Applications of the confirmation database

4.2.3.1 Confirmation for the residual solvents that were preliminarily identified in pharmaceuticals

Amoxicillin sodium and clavulanate potassium (5:1), an antibacterial medicine registered by a foreign company, was analyzed by the screening database. According to the screening

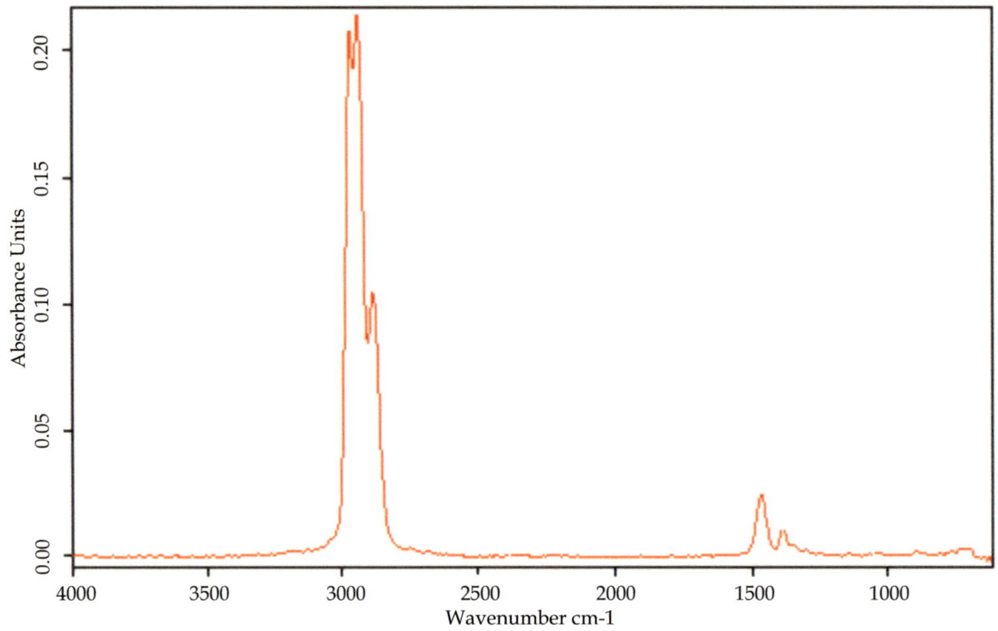
Absorbance Units
0.00
0.05
0.10
0.15
0.20
4000
3500
3000
2500
2000
1500
1000
Wavenumber cm-1

(a1)

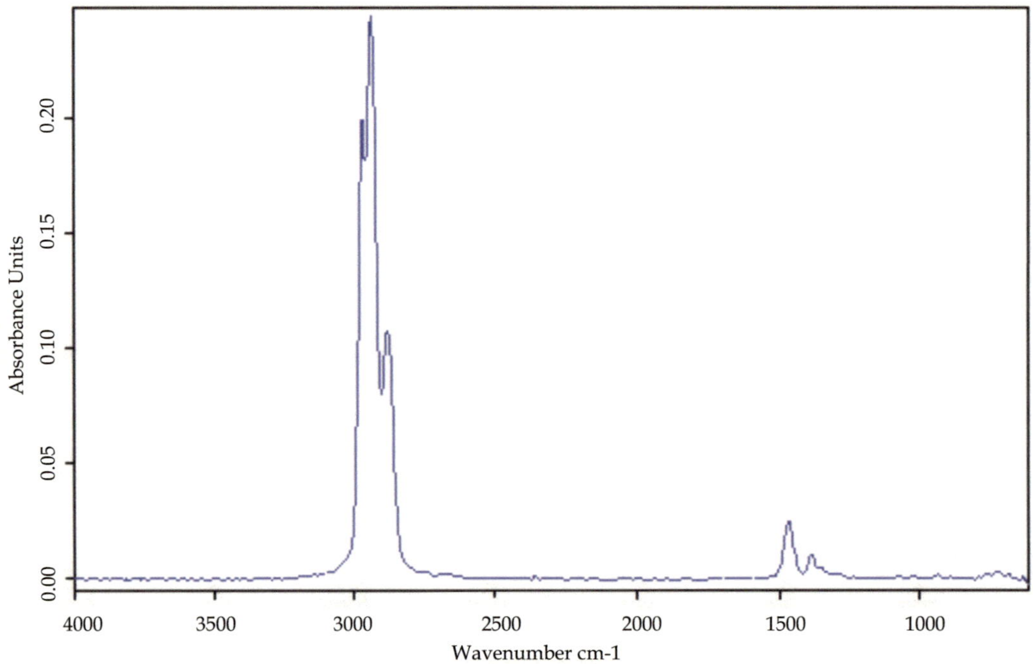
Absorbance Units
0.00
0.05
0.10
0.15
0.20
4000
3500
3000
2500
2000
1500
1000
Wavenumber cm-1

(a2)

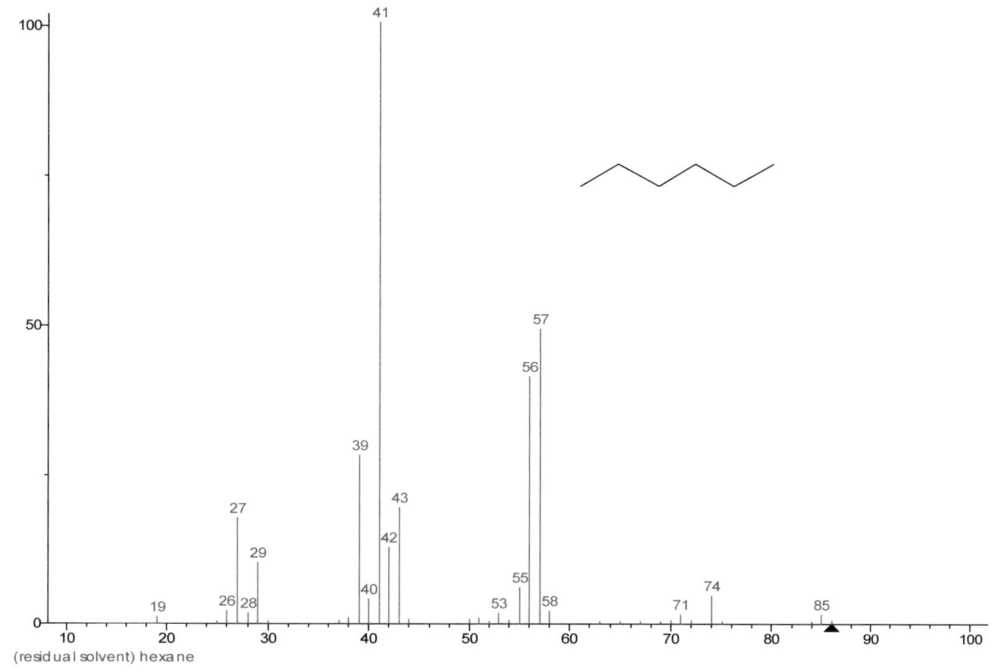

(b1)

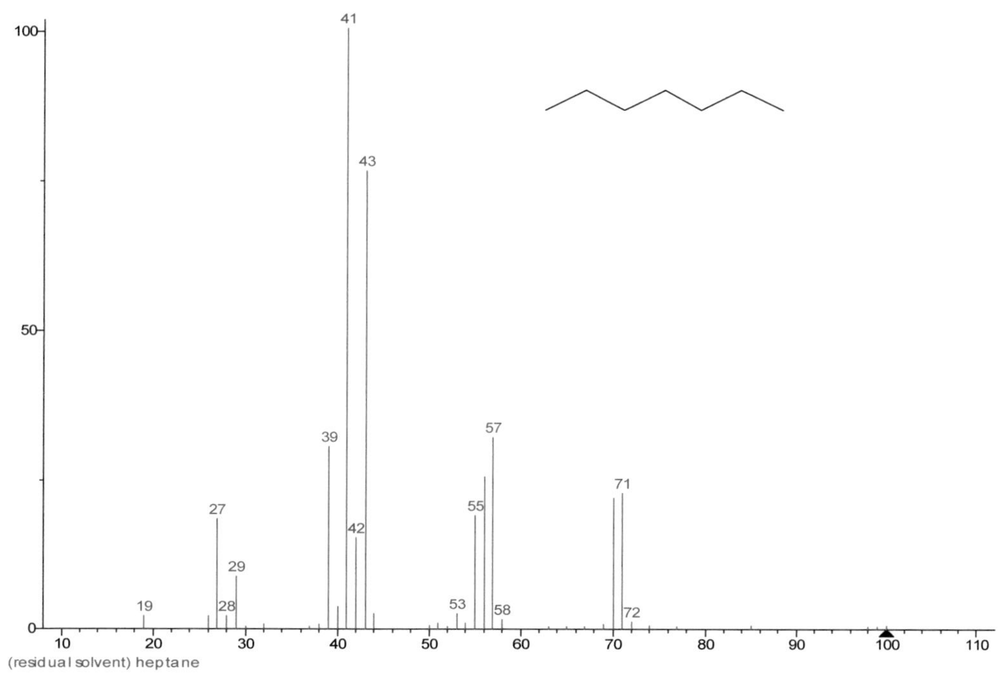

(b2)

Fig. 3. (a) Vapor-phase infrared spectra of (1) hexane and (2) heptane. (b) Mass spectra of (1) hexane and (2) heptane

results, acetone, isopropanol and methyl acetate were found in the product. Besides acetone and isopropanol were used in the synthesis, methyl acetate was not included. The confirmation database was used to confirm the screening results. According to the result from GC-MS, Ethyl acetate was the rank 1 compound according to the standard mass spectra library, and the similarity value was 913 (Fig. 4.a). The sample was analyzed by GC-FTIR using the standard vapor-phase infrared spectra library. Methyl acetate was also the rank 1 compound, and the similarity value was 983 (Fig. 4.b). The screening result was confirmed by the confirmation database, and methyl acetate was confirmed in the product.

4.3 Method optimization database

After the databases for screening and confirmation of residual solvents in pharmaceuticals were established, our next challenge is to focus on systematic method development and optimization, such as the fast selection of appropriate columns and optimization of chromatographic conditions. The solvation parameter model was applied in the development of a method for the analysis of residual solvents in pharmaceuticals. The interactions between organic solvents and six different stationary phases were studied using gas chromatography. The retention times of the organic solvents on these columns could be predicted under isothermal or temperature-programmed conditions using the established solvation parameter models. The predicted retention times helped in column selection and in optimizing chromatographic conditions during method development, and will form the basis for the development of a computer-aided method.

The solvation parameter model, first introduced by Abraham (Abraham, 1994a, 1994b, 1997), is a useful tool for delineating the contribution of defined intermolecular interactions to the retention of neutral molecules in separation systems based on a solute equilibrium between a gas mobile phase and a liquid stationary phase. The solvation parameter model in a form suitable for characterizing the retention properties of stationary phases in gas-liquid chromatography is shown below (Abraham, 2004):

$$\mathrm{SP} = c + eE + sS + aA + bB + lL \quad (2)$$

Where SP, is the gas chromatography retention data for a series of solutes. c is the model intercept, the lower case letters (e, s, a, b, l) are the system constants representing the stationary phase contribution to intermolecular interactions. l, *for the* contribution from cavity formation and solute-stationary phase dispersion interactions; e, *for the capacity of the phase to interact with* n- and π-electrons present in the solute; s, *for the ability to interact with dipoles of the solute*; a and b for the facility to interact with basic or acid solutes through hydrogen-bond forces, respectively.

The capital letters (E, S, A, B, L) are the solute descriptors for the complementary interactions with the system constants of the stationary phase. L being the gas-hexadecane partition coefficient; E, the molar refraction excess; S, the effective dipolarity/polarizability of the solute; A, the hydrogen-bond effective acidity of the solute; B, the hydrogen-bond effective basicity of the solute.

4.3.1 Prediction of retention time under isothermal conditions

The chromatographic columns used in this work were: SPB-1 (100% dimethyl siloxane, 30.0 m×0.32 mm×1 μm); HP-5 (5% diphenyl, 95% dimethyl siloxane, 30.0 m×0.53 mm×1.5 μm, used in Table 2); HP-5 (5% diphenyl, 95% dimethyl siloxane, 30.0 m×0.32 mm×0.25 μm);

HP-35 (35% diphenyl, 65% dimethyl siloxane, 30.0 m×0.53 mm×1 μm); DB-624 (6% cyanopropylphenyl, 94% dimethyl siloxane, 30.0 m×0.53 mm×3 μm); AT-225 (50% cyanopropylphenyl, 50% dimethyl siloxane, 30.0 m×0.32 mm×0.25 μm); ZB-WAX (100% polyethylene glycol, 30.0 m×0.32 mm×1 μm). The retention times of 39 organic solvents were determined on six columns at 40°C, 60°C, 80°C and 100°C. The dead time was determined using methane, and the RRTs of each organic solvent on each column were calculated using Eq. (1).

The system constants of these columns were obtained using Eq.(2) by multiple linear regression analysis. *SP* in this case was RART. The solute descriptors were taken from the literature (Kiridena, 2001; Abraham, 1993; Poole, 2002)], and are listed in Table 6. Multiple linear regression and statistical calculations were performed using SPSS software.

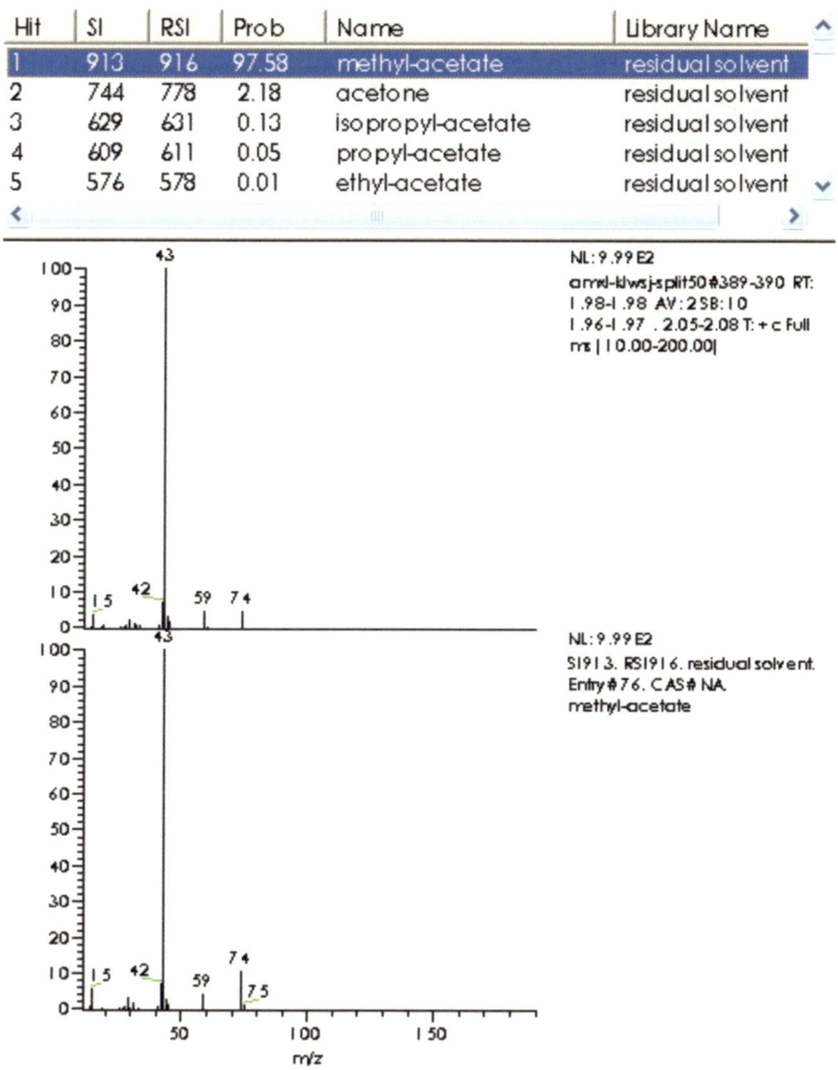

(a)

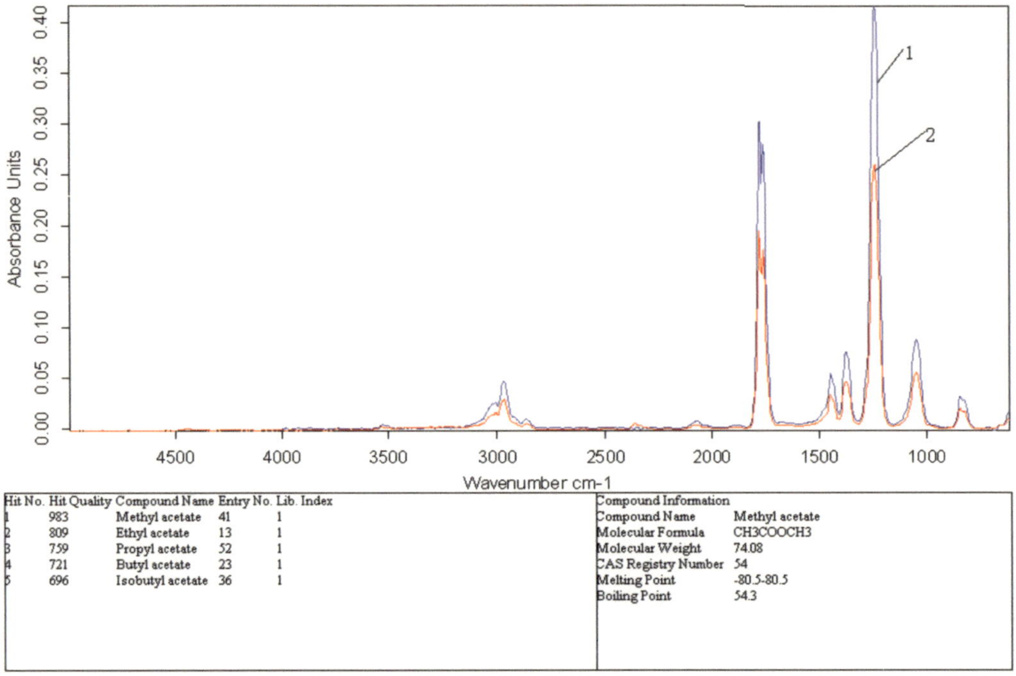

(b)

Fig. 4. Search result from (a) the standard mass spectra library and (b) the standard vapor-phase infrared spectra library (1) Spectrum of methyl acetate in the standard vapor-phase infrared spectra library (2) Spectrum of the residual solvent to be determined

The procedure for predicting retention time under isothermal conditions included the following steps:

i. The column t_0 is determined using methane, and t_R is measured for the standard (MEK).
ii. The value of *LogRRT* is calculated using the solvation parameter model and the known system constants and solute descriptors (Abraham, 1999).
iii. The retention time of the residual solvent is calculated from Eq. (1).

4.3.2 Prediction of retention time under temperature-programmed conditions

According to Cavalli's theory (Cavalli & Guinchard, 1995, 1996), retention time under temperature-programmed conditions can be calculated using only a few sets of isothermal experiments. The hypothesis is that, in temperature-programmed gas chromatography, the column acts as a series of short elements undergoing a succession of isothermal stages. The retention factor of the solute (*k*) decreases with increased column temperature and the logarithm of retention factor (*ln k*) has a linear correlation with the reciprocal of column temperature *(T)*. A and B can easily be determined experimentally from the linear regression using the following formula:

$$\ln k = \ln\left(\frac{t_R}{t_0} - 1\right) = \frac{A}{T} + B \quad (3)$$

where T is the oven temperature, A and B are fitting coefficients.

	Organic solvents	Solute descriptors				
		E	*S*	*A*	*L*	*B*
1	1,1,1-Trichloroethane	0.369	0.41	0	2.733	0.09
2	1,1,2-Trichloroethene	0.524	0.4	0.08	2.997	0.03
3	1,1-Dichloroethene	0.362	0.34	0	2.11	0.05
4	1,1-Dimethoxymethane	0.099	0.46	0	1.894	0.52
5	1,2-Dichloroethene	0.425	0.41	0.09	2.278	0.05
6	1,2-Dimethoxyethane	0.116	0.67	0	2.654	0.68
7	1-Butanol	0.224	0.42	0.37	2.601	0.48
8	1-Propanol	0.236	0.42	0.37	2.031	0.48
9	2-Butanol	0.217	0.36	0.33	2.338	0.56
10	2-Methyl-1-propanol	0.217	0.39	0.37	2.413	0.48
11	2-Propanol	0.212	0.36	0.33	1.764	0.56
12	Acetone	0.179	0.7	0.04	1.696	0.49
13	Acetonitrile	0.237	0.9	0.07	1.739	0.32
14	Benzene	0.61	0.52	0	2.786	0.14
15	Carbon tetrachloride	0.458	0.38	0	2.823	0
16	Chloroform	0.425	0.49	0.15	2.48	0.02
17	Cyclohexane	0.305	0.1	0	2.964	0
18	Dichloromethane	0.387	0.57	0.1	2.019	0.05
19	Ethanol	0.246	0.42	0.37	1.485	0.48
20	Ethyl acetate	0.106	0.62	0	2.314	0.45
21	Ethyl ether	0.041	0.25	0	2.015	0.45
22	Ethyl formate	0.146	0.66	0	1.845	0.38
23	Heptane	0	0	0	3.173	0
24	Hexane	0	0	0	2.668	0
25	Isooctane	0	0	0	3.106	0
26	Isopropyl acetate	0.055	0.57	0	2.546	0.47
27	Isopropyl ether	0	0.19	0	2.482	0.45
28	Methanol	0.278	0.44	0.43	0.97	0.47
29	Methyl acetate	0.142	0.64	0	1.911	0.45
30	Methyl ethyl ketone	0.166	0.7	0	2.287	0.51
31	Methyl isobutyl ketone	0.111	0.65	0	3.089	0.51
32	Methyl isopropyl ketone	0.134	0.65	0	2.692	0.51
33	Methyl tetrahydrofuran	0.241	0.48	0	2.82	0.53
34	Methylcyclohexane	0.244	0.1	0	3.323	0
35	Nitromethane	0.313	0.95	0.06	1.892	0.31
36	Pentane	0	0	0	2.162	0
37	Propyl acetate	0.092	0.6	0	2.819	0.45
38	Tetrahydrofuran	0.289	0.52	0	2.636	0.48
39	Toluene	0.601	0.52	0	3.325	0.14

Table 6. Solute descriptors of organic solvents

The prediction of the retention times of residual solvents under temperature-programmed conditions involves three steps:

i. The retention times of four different temperatures within the range of the temperature-programmed conditions, such as 40°C, 60°C, 80°C and 100°C is predicted using the solvation parameter model.
ii. The values of A and B is calculated using Eq.(3) and the retention times obtained from step (i).
iii. The retention time of residual solvent under temperature-programmed conditions is calculated according to Cavelli's theory.

Column	T (°C)	System constant (b=0 in all cases)					Statistics		
		r	s	a	l	c	ρ	SE	F
SPB-1	40	-0.162	0.297	0.355	0.766	-1.916	0.992	0.050	511
	60	-0.108	0.254	0.270	0.692	-1.730	0.993	0.043	582
	80	-0.065	0.223	0.210	0.628	-1.570	0.994	0.036	685
	100	-0.024	0.190	0.162	0.569	-1.425	0.994	0.032	759
HP-5	40	-0.155	0.435	0.385	0.769	-2.021	0.993	0.045	602
	60	-0.094	0.373	0.301	0.696	-1.825	0.994	0.039	695
	80	-0.045	0.324	0.235	0.629	-1.649	0.995	0.033	785
	100	-0.009	0.276	0.185	0.572	-1.493	0.995	0.029	858
HP-35	40	-0.057	0.926	0.544	0.760	-2.359	0.993	0.045	600
	60	0.009	0.809	0.487	0.690	-2.134	0.994	0.038	678
	80	0.067	0.710	0.376	0.618	-1.912	0.995	0.032	810
	100	0.108	0.627	0.313	0.560	-1.713	0.995	0.029	849
DB-624	40	-0.245	0.689	0.815	0.765	-2.193	0.993	0.041	637
	60	-0.173	0.601	0.653	0.687	-1.967	0.994	0.035	710
	80	-0.114	0.529	0.531	0.621	-1.777	0.995	0.031	773
	100	-0.068	0.471	0.433	0.563	-1.611	0.994	0.029	758
AT-225	40	-0.178	1.680	1.878	0.707	-2.803	0.994	0.047	682
	60	-0.098	1.530	1.627	0.630	-2.533	0.994	0.044	657
	80	-0.040	1.397	1.415	0.564	-2.299	0.993	0.041	615
	100	0.009	1.293	1.254	0.512	-2.115	0.992	0.041	534
ZB-WAX	40	0.401	2.007	3.045	0.575	-2.712	0.991	0.080	448
	60	0.388	1.801	2.698	0.517	-2.448	0.992	0.068	504
	80	0.384	1.617	2.378	0.463	-2.205	0.992	0.058	542
	100	0.373	1.467	2.126	0.421	-2.011	0.992	0.052	558

ρ= Overall multiple linear regression correlation coefficient; SE= standard error in the estimate; F = Fischer statistic; n = 39 in all cases.

Table 7. System constants for six columns at different temperatures

4.3.3 Prediction of system constants at different temperatures

The system constants (Eq. (2)) were summarized in Table 7. The overall multiple linear regression coefficients (ρ) of the solvation parameter models were all above 0.990 which indicated that the solvation parameter models could predict the retention times of the organic solvents.

The relationship between system constant and temperature was also studied. The system constants were reversely correlated with temperatures as indicated in the following equation:

$$y = \frac{m}{T} + n \quad (4)$$

where y is a system constant, T is the column temperature, and m and n are coefficient obtained by linear regression (Table 8).

Column	System constant	m	n	r^2
SPB-1	r	-267.12	0.6928	0.9996
	s	205.75	-0.3614	0.9985
	a	374.78	-0.8481	0.9938
	l	382.6	-0.4565	1.0000
	c	-954.11	1.1333	1.0000
HP-5	r	-323.08	0.852	0.9981
	s	320.86	-0.5702	0.9995
	a	455.2	-1.0223	0.9935
	l	389.59	-0.4709	0.9999
	c	-1044.4	1.2913	0.9998
HP-35	r	-323.84	0.9799	0.9973
	s	582.54	-0.9376	0.9994
	a	452.13	-0.9015	0.9994
	l	392.27	-0.4915	0.9992
	c	-1260.1	1.6599	0.9992
DB-624	r	-345.47	0.8615	0.9979
	s	424.98	-0.6718	0.9984
	a	743.05	-1.5676	0.9963
	l	392.84	-0.4912	0.9998
	c	-1131.9	1.4272	0.9997
AT-225	r	-362.72	0.9853	0.9961
	s	756.94	-0.7413	0.9991
	a	1220.1	-2.029	0.9980
	l	380.94	-0.5121	0.9988
	c	-1344.5	1.4992	0.9990
ZB-WAX	r	53.664	0.2285	0.9892
	s	1054.7	-1.3651	0.9996
	a	1798.9	-2.7054	0.9995
	l	301.68	-0.3893	0.9994
	c	-1371.4	1.6713	0.9996

Table 8. Fitted regression coefficients for Eq. (4)

These coefficients were used to further predict the retention at any temperature in the studied range.

For instance, the system constants of SPB-1 column were predicted at 50°C using Eq. (4) as follows: r = -0.134, s = 0.276, a = 0.312, l = 0.728, and c = -1.821. Meanwhile the system constants of this column were determined under 50°C and r = -0.145, s = 0.282, a = 0.326, l = 0.734, and c = -1.837. The results showed that the differences between predicted and experimental values were very small, and the system constants can be well predicted at any temperature within the ranges of 40°C to 100°C.

4.3.4 Application in the process of method development

The control of 8 residual solvents (methanol, ethanol, dichloromethane, chloroform, hexane, benzene, methyl isobutyl ketone and toluene) was evaluated in rabeprazole sodium formulations. Methyl ethyl ketone was used as internal standard (IS). The solvation parameter models were used to select columns under isothermal conditions and to optimize chromatographic conditions under temperature-programmed conditions in the analysis of residual solvents in rabeprazole sodium.

4.3.4.1 Column selection under isothermal conditions

The retention times of these solvents were predicted on SPB-1 (non polar), ZB-WAX (polar) and DB-624 (moderately polar) columns at 40°C using the solvation parameter model. The optimum column was selected according to the results shown in Table 9. Hexane and chloroform could not be separated on the SPB-1 column. On the HP-INNOWAX column, the predicted retention time of methanol was close to that of methyl ethyl ketone, as were ethanol and benzene. On the DB-624 column, all the residual solvents could be separated according to the predicted retention times, therefore the DB-624 column was selected in this experiment. The residual solvents were determined on the DB-624 column, and the results were compared with the predicted results shown in Table 10. These findings indicated that the predicted results were consistent with the experimental results, and that the 8 residual solvents could be separated on this column.

Organic solvent	Predicted t_R (min)		
	SPB-1	ZB-WAX	DB-624
Methanol	1.838	5.098	2.551
Ethanol	2.157	5.320	3.606
Dichloromethane	2.800	4.398	5.179
Methyl ethyl ketone (IS)	3.704	5.142	8.172
Chloroform	4.228	6.832	9.167
Hexane	4.315	1.766	6.271
Benzene	5.398	5.336	10.836
Methyl isobutyl ketone	10.130	8.016	25.493
Toluene	11.457	9.161	27.114

Table 9. Predicted retention times of residual solvents in rabeprazole sodium on 3 different columns at 40°C using Eqs. (1) and (2)

Organic solvent	t_R (min) Predicted	Experimental	Δt_R
Methanol	2.551	2.606	0.055
Ethanol	3.606	3.539	-0.067
Dichloromethane	5.179	4.928	-0.251
Hexane	6.271	6.296	0.025
Methyl ethyl ketone (IS)	8.172	8.199	0.027
Chloroform	9.167	9.190	0.023
Benzene	10.836	10.833	-0.003
Methyl isobutyl ketone	25.493	25.016	-0.477
Toluene	27.114	27.409	0.295

Table 10. Comparison between the predicted and experimental retention time of residual solvents in rabeprazole sodium on DB-624 column at 40°C using Eqs. (1) and (2)

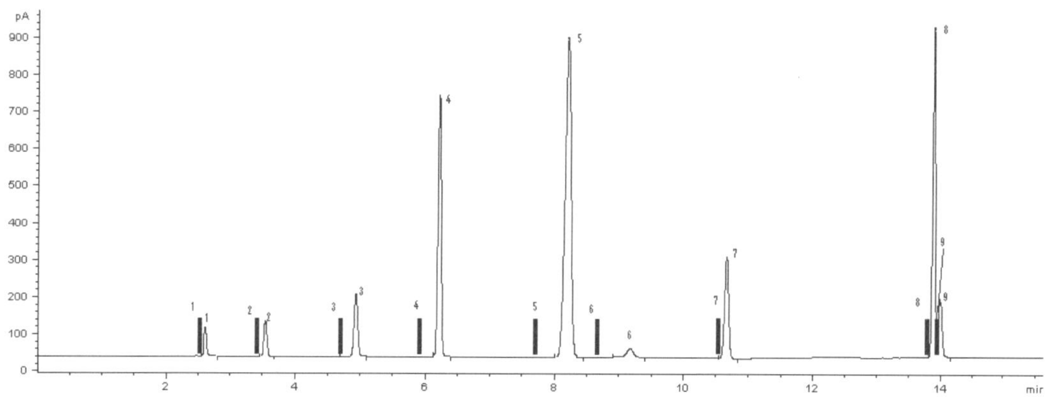

1-Methanol; 2-Ethanol; 3-Dichloromethane; 4-Hexane; 5-Methyl ethyl ketone (IS); 6-Chloroform; 7-Benzene; 8-Methyl isobutyl ketone; 9-Toluene;

Note: Predicted retention times of each organic compound were indicated by the vertical bars inserted in the chromatogram

Fig. 5. Chromatogram of 8 organic solvents under temperature-programmed conditions on DB-624 column

4.3.4.2 Optimization of chromatographic conditions under temperature-programmed conditions

From Table 10, it can be seen that the separation of these 8 residual solvents on the DB-624 column at 40°C took approximately 30 min, and no peak was eluted between 10 and 25 min, therefore temperature-programmed conditions can be used to shorten the analysis time. The method for predicting retention time under temperature-programmed conditions can be used to optimize the chromatographic conditions. The retention times of the solvents under designated temperature-programmed conditions were first calculated, and according to the predicted retention times, separations among the solvents were evaluated. If some of the solvents could not be separated under that condition, the temperature program was revised and the retention times were recalculated. This process was repeated until optimal chromatographic conditions were found under which all the solvents could be separated. In this case, the temperature-programmed conditions were as follows: oven temperature was maintained at 40°C for 10 min, and then raised to 120°C by a rate of 20°C/min for 2 min. These 8 residual solvents were determined under the optimized conditions, and the results were compared with the predicted results (Fig. 5). These findings indicated that the predicted results were consistent with the experimental results, and that the 8 residual solvents were separated within 15 min. The analysis time was decreased by 15 min compared to the analysis time under isothermal conditions. Therefore workload and time were dramatically decreased following the process of method optimization using the proposed approach.

5. Conclusion

Residual solvents from the processes in the manufacture of pharmaceuticals are a problem and must be removed. The ICH guideline is already accepted by different pharmacopeias. GC analysis is the ideal methodology for residual solvent analysis. Now the official method for sample preparation is still static headspace analysis, which gives a high level of automation from the instrumentation currently available and has a low impact on GC column life. Other methods such as SPME, MHS-SDME are useful alternative methods for residual solvents testing.

From the regulatory perspective, each pharmacopoeia focused on comprehensive analysis of residual solvents in pharmaceuticals. The official methods in USP and EP use two system and all the organic solvent reference standards to screening residual solvents. The established database for residual solvents analysis was adopted by ChP. Different from USP and EP, reference standards were not required for all organic solvents. Organic solvents having the same or similar retention times on one column usually have quite different retention times on the column with opposite polarity. The nature of the organic solvents can be identified using the two columns. The screening database was used to make a full-scale screening of the residual solvents in the pharmaceuticals. Only a few organic solvent reference standards were needed to confirm the screening result. If there are residual solvents that were not mentioned in the specification or production process, first class solvents or unknown solvents were found, that can be analyzed by GC-MS and GC-FTIR, using the confirmation database to make a confirmation. The dababase system can solve the difficult problem of unknown residual solvents determination, making it a powerful tool for determining residual solvents in pharmaceuticals.

6. References

Abraham, M. H., (1993). Scales of solute hydrogen-bonding: their construction and application to physicochemical and biochemical processes. *Chem. Soc. Rev.* 22, 73.

Abraham, M.H., Chadha, H. S., Leo, A. J., (1994). Hydrogen bonding: XXXV. Relationship between high-performance liquid chromatography capacity factors and water-octanol partition coefficients. *J. Chromatogr. A.* 685, 203-211.

Abraham, M. H., Roses, M., (1994). Hydrogen bonding. 38. Effect of solute structure and mobile phase composition on reversed-phase high-performance liquid chromatographic capacity factors. *J. Phys. Org. Chem.* 7, 672-684.

Abraham, M. H., Roses, M., Poole, C. F., Poole, S. K., (1997). Hydrogen bonding. 42. characterization of reversed-phase high-performance liquid chromatographic c18 stationary phase. *J. Phys. Org. Chem.* 10, 358-368.

Abraham, M. H., Poole, C. F., Poole, S. K., (1999). Classification of stationary phases and other materials by gas chromatography. *J Chromatogr. A* 842, 79-114.

Abraham, M. H., Ibrahim, A., Zissimos, A. M., (2004) Determination of sets of solute descriptors from chromatographic measurements *J. Chromatogr. A* 1037, 29-47.

Arthur C. L., Pawliszyn J. (1990). Solid phase microextraction with thermal desorption using fused silica optical fibers *Anal. Chem.* 62, 2145-2148.

Avdovich, H.W., Lebelle, M.J., Savard, C., Wilson, W.L., (1991). Nuclear magnetic resonance identification and estimation of solvent residues in cocaine. *Forensic Sci. Int.* 49, 225–235.

Benoit, J.P., Courteille, F., Thies, C., (1986). A physicochemical study of the morphology of progesterone-loaded poly(d,l-lactide) microspheres. *Int. J. Pharm.* 29, 95–102.

British Pharmacopoeia [S] 1993 edition supplement, The Stationery Office, London, 1996.

Camarasu, C.C., Mezei-Szuts, M., Varga, G.B., (1998). Residual solvent determination in pharmaceutical products by GC-HS and GC-MS-SPME. *J. Pharm. Biomed. Anal.* 18, 623-638.

Camarasu, C.C., (2000) Headspace SPME method development for the analysis of volatile polar residual solvents by GC-MS. *J. Pharm. Biomed. Anal.* 23, 197-210.

Cavalli, E. J., Guinchard, C., (1995). Forcasting retention times in temperature-programmed gas chromatography. *J. Chromatogr. Sci.* 33, 370-376.

Cavalli, E. J., Guinchard, C., (1996). Forecasting retention times in temperature-programmed gas chromatography: Experimental verification of the hypothesis on compound behavior. *J. Chromatogr. Sci.* 34, 547-549.

Dubernet, C., Rouland, J.C., Benoit, J.P., (1990). Comparative study of two ethylcellulose forms (raw material and microspheres) carried out through thermal analysis. *Int. J. Pharm.* 64, 99–107.

European Pharmacopoeia [S] 3rd Edition, Council of Europe, Strasbourg, 1997

European Pharmacopoeia [S] 3rd Edition supplement, Council of Europe, Strasbourg, 1999.

Guimbard, J.P., Besson, J., Beaufort, S., Pittie, J., Gachon, M., (1991). Evaluation des solvant résiduels. *S.T.P. Pharma Pratiques*, 1, 272–277.

Hachenberg, H., Schmidt, A.P., Gas Chromatographic Headspace Analysis, Hayden and Son, London, 1977

Hymer, C. B., (2003). Residual Solvent Testing: A Review of Gas-Chromatographic and Alternative Techniques. *Pharm. Res.* 20, 337-344.

International Conference on Harmonization of Technical Requirements for the Registration of Pharmaceuticals for Human Use, Q3C (R4) Impurities: Guideline for Residual Solvents, 2009

Japanese Pharmacopoeia [S]14 th edition, Shibuya,Tokyo, 2001.

Kiridena, W., Koziol, W. W., Poole, C. F., (2001). Selectivity assessment of DB-200 and DB-VRX open-tubular acaillary columns *J. Chromatogr. A* 932, 171-177.

Kolb, B., Ettre (Eds.), L.S., Static Headspace–Gas Chromatography: Theory and Practice (2nd edition), John Wiley and Sons, New York, 2006.

Laus, G., Andre, M., Bentivoglio, G., Schottenberger, H., (2009). Ionic liquids as superior solvents for headspace gas chromatography of residual solvents with very low vapor pressure, relevant for pharmaceutical final dosage forms. *J. Chromatogr. A.* 1216, 6020-6023

Liu, F., Jiang, Y., Claramunt, J., (2007). Room temperature ionic liquid as matrix medium for the determination of residual solvents in pharmaceuticals by static headspace gas chromatography. *J. Chromatogr. A.* 1167, 116-119.

Liu, Y., Hu, C. Q., (2006). Establishment of a knowledge base for identification of residual solvents in pharmaceuticals. *Anal. Chim. Acta.* 575, 246-254.

Liu, Y., Hu, C. Q., (2007). Preliminary identification and quantification of residual solvents in pharmaceuticals using the parallel dual-column system. *J. Chromatogr. A.* 1175, 259-266.

Liu, Y., Hu, C. Q., (2009). Application of the solvation parameter model in method development for analysis of residual solvents in pharmaceuticals. *J. Chromatogr. A.* 1216, 86-91.

Loffe, B.V., Vitenberg, A. G., Head-space Analysis and Related Methods in Gas Chromatography. John Wiley and Sons. New York, 1984

Osawa, Z., Aiba, M., (1982). Effect of residual solvent on the photodegradation of poly(vinyl chloride). *Poly. Photochem.* 2, 339–348.

Otero, R., Carrera, G. Dulsat, J. F., Fabregas, J., Claramunt, J., (2004). Static headspace gas chromatographic method for quantitative determination of residual solvents in pharmaceutical drug substances according to European Pharmacopoeia requirements. *J. Chromatogr. A.* 1057, 193-201.

Pharmacopoeia of the People's Republic of China [S]. 1995 ed. Part II. Beijing: Chemical Industry Press, 1995.

Pharmacopoeia of the People's Republic of China [S]. 2005 ed. Part II. Beijing: Chemical Industry Press, 2005.

Poole, C. F., Kiridema, W., Nanas, M. I., Koziol, W. W., (2002). Influence of composition and temperature on the selectivity of stationary phases containing either mixtures of poly(ethylene glycol) and poly(dimethylsiloxane) or copolymers of cyanopropylphenylsiloxane and dimethylsiloxane for open-tubular column gas chromatography. *J. Sep. Sci.* 25, 749-759.

Snow, N. H., Bullock, G. P., (2010). Novel techniques for enhancing sensitivity in static headspace extraction -gas chromatography, *J. Chromatogr. A.* 1217, 2726-2735

Tewari, J., Dixit, V., Malik, K., (2010). On-line monitoring of residual solvent during the pharmaceutical drying process using non-contact infrared sensor: A process analytical technology (PAT) approach. *Sensors and actuators B,* 144, 104-111.

The United states Pharmacopoeia [S] 22th edition 3rd supplement, The United States Pharmacopeial Convention, Inc., Rochville , 1990.

The United states Pharmacopoeia [S] 28th edition, The United States Pharmacopeial Convention, Inc., Rochville, 2005.

Urakami, K., Higashi, A. Umemoto, K. Godo, M., (2004). Matrix media selection for the determination of residual solvents in pharmaceuticals by static headspace gas chromatography. *J. Chromatogr. A.* 1057, 203-210.

Vachon, M.G., Nairn, J.G., (1995). Physico-chemical evaluation of acetylsalicylic acid-Eudragit RS 100 microspheres prepared using a solvent partition method. *J. Microencapsul.* 12, 287–305.

Weitkamp, H., Barth, R., (1976). Bestimmung kleiner Gehaltswerte nach dem Aufstockverfahren. In: H. Weitkamp, R. Barth, Einführung in die quantitative Infrarot-Spektrophotometrie. Georg Thieme Verlag, Stuttgart, pp. 58–67.

Witschi, C., Doelker, E., (1997). Residual solvents in pharmaceutical products: acceptable limits, influences on physicochemical properties, *Eur. J. Pharm. Biopharm.* 43, 215-242.

Yu, Y. J, Chen, B., Shen, C., Cai, Y., Xie, M. F., Zhou, W., Chen, Y., Li, Y., Duan, G. Li, (2010) Multiple headspace single-drop microextraction coupled with gas chromatography for direct determination of residual solvents in solid drug. *J. Chromatogr. A.* 1217, 5158-5164.

12

The Application of the Potentiometric Stripping Analysis to Determine Traces of M(II) Metals (Cu, Zn, Pb and Cd) in Bioinorganic and Similar Materials

Biljana Kaličanin[1] and Ružica Nikolić[2]
[1]*University of Nis/Faculty of Medicine, Department of Pharmacy,*
[2]*University of Nis/Faculty of Sciences, Department of Chemistry,*
Serbia

1. Introduction

The development and application of new technologies in all spheres of life and work carries with it the ever-increasing pollution of the environment through harmful and toxic substances. Pesticides and heavy metals are among some of the more prominent pollutants of the environment. Heavy metals significantly contribute to human environment pollution due to the impossibility of their biodegradation, and because some of them have cumulative toxic properties. Sources of contamination by means of metals are numerous, the most important ones being combustion products in the chemical industry and metallurgy, industrial waste waters and landfills, agrochemicals, and exhaust gasses of motor vehicles. People are, therefore, exposed to toxic metals that act both directly through the contaminated air and drinking water, and indirectly through the soil, underground waters and poisoned plants and animals found in food, the pharmaceutical and cosmetic industry.

Copper and zinc are essential bioelements which, in addition to their biological role and their importance for the development of the human body, also have a toxic effect when found in amounts higher than normal in the human body. Lead and cadmium are highly toxic metals, even when found only in traces (Goyer, 1997; Goyer & Klaassen, 1995).

Copper is one of the essential biometals necessary for the growth, development and normal functioning of the human body, for the synthesis of hemoglobin, melanin, and the mineralization and development of bones. The lack of copper can lead to serious illnesses. Nevertheless, its presence in the human body in values greater than 10^{-6} mol/dm^3 inhibits certain enzymes, which hinders the bonding of other essential microelements, or even leads to bonding with certain cofactors. The increased content of copper in the human body leads to coronary and vascular disease, arteriosclerosis, hypertension and various forms of damage to the central nervous system (Uauy, et al., 1998; Hart, et al. 1928; Chapman, 2008).

Zinc is an essential oligoelement which is found in significant amounts in the human body (0.02 - 0.03 g/kg of body weight). It is necessary for the synthesis of proteins and nucleic acids, DNA replication, the human reproductive ability, and maintaining high level healthy immune function. A shortage of zinc in the human body can lead to the harmful effect of pancreatic enzymes, anemia, pulmonary disease, neurological disorders and the occurrence of certain types of cancer (Walsh et al., 1994).

The necessary amounts of these elements for the normal functioning of the human body are introduced through water and food of plant or animal origin. Recommended amounts of zinc in various products range from 0.1 to 80 mg/kg, and of copper from 2 to 100 mg/kg (Goyer & Klaassen, 1995).

Lead is a toxic metal with a cumulative effect, which competes with the essential metals in the human body (Ca, Fe, Cu, Zn). A relatively low content of lead has a negative effect on the heart, blood vessels, kidneys, liver, and respiratory system. Based on its physical-chemical characteristics, Pb(II)- ions can replace Ca (II) - ions isomorphically as part of hydroxyapatite, which leads to the accumulation of this metal in mineral tissue - the teeth and bones. During physiological processes of bone tissue remodeling, part of the Pb (II)- ions, by migration through the oral and other biological fluids, reach other remote organs - the brain, kidneys, and the liver (Pocock et al., 1994; Banks et al., 1997; Vig & Hu, 2000).

Cadmium is considered one of the most dangerous occupational and environmental poisons. It is presumed that excessive amounts of this metal in the human body are undesirable. The basis of cadmium toxicity is its negative influence on the enzymatic systems of cells, owing to the substitution of other metal ions (mainly Zn^{2+} and Cu^{2+}) in metalloenzymes and its very strong affinity to biological structures containing –SH groups. Excessive Cd exposure may give rise to renal, pulmonary, hepatic, skeletal, reproductive effects and cancer. The major effects of this type of metal poisoning are found in the lungs, kidneys and bones. Obviously, the monitoring of the cadmium level at trace level in different environment matrices which are directly related with human health is of great importance. The World Health Organization (WHO, 1996) reported tolerable weekly intakes of cadmium of 0.007 mg/kg body weight, for all groups of humans. Briefly, it is considered that this metal can have a dangerous effect human health even at ultra trace concentrations.

Due to the harmful and toxic effects of copper, zinc, lead and cadmium, it is necessary to determine and monitor their content in water, soil, food, pharmaceutical and cosmetic products, packaging. For medicinal-diagnostic purposes it is sometimes necessary to monitor the contents of these metals in clinical-biological material. Data regarding the deposits and transport mechanisms of Cu, Zn, Pb and Cd in the body can be obtained through an analysis of biopsy material both of human and animal origin (Brzoska & Moniuszko-Jakoniuk, 1998; Florianezyk, 1995).

Due to the high toxicity and stability of Pb, Cd, Zn and Cu it is necessarity to determinate their content in materials, food, water and other samples.

In order to determine the content of the aforementioned metals in the analyzed samples, an electroanalytic technique was used - the potentiometric stripping analysis (PSA). The PSA is a highly-sensitive, selective microanalytic technique for determining heavy metal traces, including metals such as lead, cadmium, copper and zinc (Vydra et al., 1976; Suturović, 2003). The advantage of this technique in relation to other current, more unavailable and costly techniques is also its low exploitation and instrumentation cost, ease of use, the ability to simultaneously determine a greater number of metals in the same sample, as well as the infinite number of analyses of the same sample, even though it has previously been analyzed (Kaličanin, 2006).

The results involved in determining micro amounts of Cu, Zn, Pb and Cd within samples of various types and origin (water, soil, packaging, dental-prosthetic material, beauty products, teas, biopsy material) by using the PSA method have been outlined in this paper, and are in agreement with the data found in the literature in regards to the detection limits of other analytic techniques. This technique can successfully be used in the quality control of bioinorganic and similar material and the analysis of biopsy material for the presence of heavy metals, considering the high values of result reproduction (Danielsson et al., 1981).

2. The electrochemical stripping analysis (ESA)

In order to determine the content of toxic heavy metals in real samples, where even element amounts lower than 1 μg/dm^3 can be significant, the proper selection of the appropriate analysis techniques is also necessary. The analytical methods used for measuring concentrations of traces of M (II) metals (Cu, Zn, Pb and Cd) in bioinorganic and similar materials include atomic absorption spectrometry (AAS), neutron-activation analysis (NAA), inductively coupled plasma atomic emission spectroscopy (ICP-AES), inductively coupled plasma optic emission spectroscopy (ICP-OES) and electrochemical stripping analysis (ESA). The success as well as the frequency of the abovementioned techniques is different; they depend on the detection limit, selectivity and reproducibility of the given technique, the rapidity and simplicity of the method as well as the price of the device and its exploitation (Vydra et al., 1976; Jagner, 1979; McKenzie, 1988; Brainina & Neyman, 1993).
The electrochemical stripping analysis (ESA) has the greatest sensitivity (10^{-11}mol/dm^3) coming second to the neutron activation analysis (10^{-21} mol/dm^3). Besides, the cost of its application and exploitation is much lower than with the other above-mentioned techniques while the procedure for carrying out the analysis is relatively simple and fast (Suturović, 2003; Kaličanin et al., 2002).

2.1 Characteristics of the ESA

The electrochemical stripping analysis (ESA) as a highly sensitive and selective instrumental microanalytic technique is used for the quantitative determination of metals, that is metal ions, but in the last few years it has increasingly been used to determine micro-amounts of organic compounds and anions. Bearing in mind the possibilities and the demands of the ESA, we could say that it can fulfill the very rigorous general and specific micro-analytical demands to a significant extent. The most significant features of this technique include, in addition to exceptional sensitivity, very good analytical selectivity:

- The ability to determine a great number of elements simultaneously,
- The ability of unlimited repeated analyses of the same solution,
- The small size of the instrumentation,
- The ability of carrying out analyses outside of the laboratory, "on the spot".

The sample being analyzed with the help of the ESA has to be in a re-solvent condition. If the sample is in liquid form and if its content (matrix) is not complex (as is the case with water, for example), the preparation of the sample usually requires only the addition of an auxiliary electrolyte which primarily provides the necessary conditions for the ESA, but is often used as a de-complexing agent for the studied substance. When the liquid sample has a more complex matrix, the interfering influence of the matrix can significantly be reduced by means of the dilution of the sample, with the addition of the auxiliary electrolyte. This type of preparation is possible due to the high sensitivity of the ESA.
If the sample is in solid form, it has to be dissolved or extracted. Samples in liquid and solid form, which contain high amounts of organic substances, must be prepared for analysis by means of some of the procedures for the destruction of organic matter (Bock, 1979).

The analysis of gaseous samples requires a previous concentration (adsorption) of the analyzed material on suitable filters, and then their degradation by means of concentrated acids or though annealing, with subsequent dissolution. The preparation of the sample (solution) for the ESA includes the addition of the so-called auxiliary, indifferent electrolyte, with a concentration of 0.1-0.5 mol/dm^3. Most often these include salt solutions (chlorides, nitrates), mineral acids, bases or buffer systems (acetate or citrate). The role of the auxiliary electrolyte is to enable the maximum utilization of the electrochemical depositing, the conductivity of the solution, and to minimize the electrical current which enables migration and set the appropriate pH value (Vydra et al., 1976).
The analysis of the samples, which was carried out by means of the electrochemical stripping analysis, takes place in an electrochemical cell, which is made up of three electrodes (a working electrode, a reference electrode and an auxiliary electrode), a reaction zone and a solution mixer. The working electrode is a thin-layer mercury electrode which is obtained by adding a thin layer of mercury, 10 to 1000 nm thick (Jagner, 1982; Suturović, 2003), to the surface of the inert carrier made of vitreous carbon. Electrodes made of vitreous carbon are especially suitable due to their chemical inertness and relatively wide interval of varying potentials, ranging from -0.75 V to + 1.0 V (Konvalina et al., 2000). Electrodes made of vitreous carbon, despite their high chemical inertness, should not be exposed to the effects of concentrated solutions of potent oxidizing agents and acids, such as hydrogen-peroxide, nitric acid and sulfuric acid. In addition, these electrodes must not be exposed to overly positive potentials $E > 2$ V (ZKE), since this could lead to their irreversible destruction (Kaličanin, 2006).
Thin layers of mercury are deposited using a special mercury (II)- ion solution, whose value must be smaller than 2, at a constant electrical current (galvanostatic conditions), since it enables one to obtain the required thickness of the layer of mercury, irrespective of the resistance in the electrochemical cell. The reference electrode in the ESA is a silver-silver chloride electrode, while the auxiliary (counter) electrode is solely a platinum electrode.
The sensitivity of the ESA is in great part dependent on the intensity of the mixing. The mixing is usually done with a stirring stick mechanical mixer with good control of the number of rpms, constant geometry of the electrochemical cell and other conditions necessary for the analysis (Wang, 1985).

2.2 Processes within the ESA

The electrochemical stripping analysis is a specific analytical technique which is carried out in four successive steps. The first step is the culonometric process during which the determined material is concentrated either on or in the working electrode, under controlled hydrodynamic conditions and during a precisely specified period of time. The concentration of analytes can be carried out by means of an electrolysis, which is most often the case, or by unspecific adsorption or specific chemical reactions. The sensitivity of the ESA is directly dependent on the effectiveness of this step, while the precision is dependent on the determination of the degree of reproducibility of the conditions under which it takes place (Wang, 1985).
The factors which affect the effectiveness of the concentration of the analytes by means of electrolysis include: the potential of the electrolysis, the duration of the concentration, the value of the pH, the conditions of mass transport and the features of the amalgam.
The potential of the electrolysis is the most important factor in electrolysis concentration, as it affects the amount of the separated deposit, and thus the sensitivity and reproducibility of

the determination. The value of the electrolysis potential must as a rule range from 300 to 500mV more negative than the polarographic half-wave potential of the determined element (Bard et al., 1985).

The time needed for the analyte electrolysis depends on its concentration. In the case of more diluted solutions, longer electrolysis time is needed, while for solutions of greater concentrations, the duration of the electrolysis is shorter. During the ESA, the usual duration of the electrolysis ranges from 60 to 900 s, where the utilization is from 5-10 %.

The pH value of the analyzed solution affects the chemical state of the analyte, that is, the amount of electroactive ions and in general the analyte ion form in the solution. If the environment is alcaline, the determination of most of the metals is not possible due to the deposition of metal ions in the form of their hydroxides. An overly acidic environment influences the creation of chemical disturbances and also does not enable the determination of certain analytes (organic compounds, elements with a negative redox potential), due to the extraction of hydrogen to the surface of the electrode (Suturović, 2003). The optimum pH values, auxiliary electrolytes and dilution potentials of certain metals in the ESA are shown in Table 1.

Element	pH	Auxiliary electrolyte	Dilution potential (V)
Cu	1-2	HCl	-0.15
Zn	4.6	acetate buffer	-1.00
Pb	4.6	acetate buffer	-0.46
Cd	4.6	acetate buffer	-0.57

Table 1. Optimum pH values, auxiliary electrolytes and dilution potentials of certain metals in the ESA

The effectiveness of the concentration of the studied solution in the ESA improves considerably due to mass transfer through convection. Mass transport through convection is achieved by mixing the solution, using a stirring stick or the rotation of a reaction vessel (Jagner, 1982).

The viscosity of the amalgam influences the diffusion of the metals dissolved in the mercury, or the speed of the electrochemical reactions on the amalgam electrodes.

The second phase of the ESA is the reduction in the velocity of the flow of the solution which lasts from 15 to 30 s and which provides the necessary conditions for the diffusion mass transfer in the next step, and the homogenization of the deposit in the working electrode.

The first two steps are identical for all the "stripping" techniques, while the techniques differ in the third analytic step.

The techniques which include the use of chemical oxidation or less reductive means in the analytic step of the ESA are **potentiometric techniques**.

The fourth step is necessary in the case of the repeated analysis of the same solution, and requires the stirring of the solution for a period of 5 s, so that it could become homogenous and the deposit could dissolve completely.

The content of the analyte is determined through some of the relative methods, most commonly through the method of standard addition or the calibration curve. This is why it is important that the experimental conditions in all four steps of the ESA be very reproducible.

2.3 The potentiometric stripping analysis

The potentiometric stripping analysis (PSA) is the youngest of all the stripping techniques, and was first presented in 1976 (Jagner, 1976). Due to its simplicity and the work of Jagner et al., the PSA very soon gained substantial practical significance (Jagner, 1979, 1982, 1993).
The most frequently used oxidizing agents in the PSA are Hg^{2+}- ions, which are used to determine elements with a more negative redox potential than mercury, or the MnO_4^- or $Cr_2O_7^{2-}$ ions to determine mercury or more precious metals (Jagner, 1979). The concentration of the oxidizing agent, which is added to the sample, must always be high enough to enable the oxidation of all of the analytes, but not too high, as in that case it speeds up the oxidation (it reduces the sensitivity of the determination) and increases the contamination of the sample. (Jagner et al., 1981).
In 1979 Jagner proved the possible uses of the diluted oxygen as an oxidizing agent in the PSA (Jagner, 1979). This specific modification of the PSA with oxygen as the oxidizing agent has a great advantage, since oxygen is already present in the solution and does not require the use of deaeration, which significantly lengthens the duration of the analysis and represents a risk in the sense of the contamination of the solution, damage to the thin-layer electrode and the blocking of the active surface of all three electrodes.
As part of this technique, great care must be taken regarding the negative influence of some of the products of its reduction, as well as in terms of its direct (non-electrical) influence. Thus, due to the reduction of oxygen during electrolysis, the hydroxyl ions will be separated in the vicinity of the working electrode, which in turn can cause the hydrolysis of certain metal ions, while the effect is more pronounced the closer the pH value is to neutral. This effect can be prevented by adjusting the pH value with the help of the appropriate buffers. In addition, the presence of oxygen in this modification of the PSA can contribute to the separation of calomel on the mercury electrode if the solution contains chloride ions. In order to avoid this problem, it is necessary to separate the working electrode and the analyzed solution at the end of the analysis.
The formation of analyte deposits, in the PSA with oxygen as the oxidizing agent, is carried out during the electrolysis procedure. Once the electrolysis is completed, the potentiostatic control is discontinued and a change in the potential of the working electrode is registered, which occurs during the chemical oxidation of the formed deposit. Thus, due to the oxidizing effect of the oxygen, the potential of the working electrode increases and shifts towards the more positive values. The potential increases until it becomes the same as the potential for the dissolution of the most negative of the deposited metals (E_R) and remains constant until the metal with the smallest redox potential is oxidized. i Following that, there is a sudden increase in potential, until the oxidation potential of the following positively deposited metal is reached. The moment the most positive deposited metal on the electrode that can be oxidized by means of oxygen is dissolved, the potential of the working electrode increases to a borderline value, which is in the function of the pH solution, and then remains constant (at around +0,1 V, ZKE) (Fig. 1). This is, at the same time, an indication of the completion of the analytic step of the PSA (Suturović, 2003).

2.3.1 Interference in the potentiometric stripping analysis

The most frequently occurring interference which can take place during stripping analysis is a result of the presence of organic compounds, due to the formation of intermetal compounds and the overlap between the dilution potential of the determined elements (Suturović, 2003).

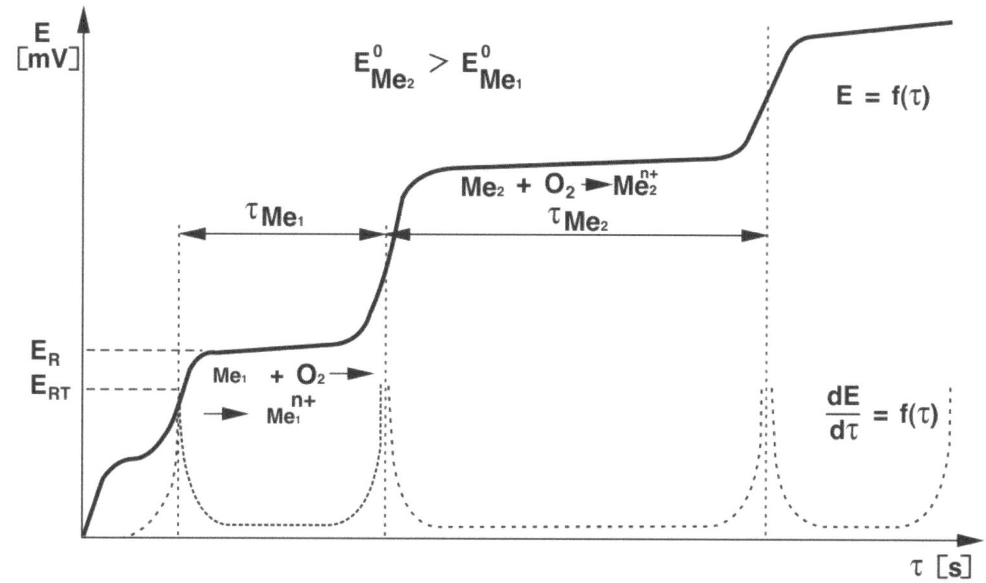

Fig. 1. Response signals in the PSA

Surface active agents such as polysaccharides, alcohols, salts of fatty acids, proteins and the like, cause interference in the PSA by adsorbing to the surface of the working electrode and influencing the mechanism of mass transfer, increasing the value of the dilution potential and decreasing the sensitivity of the determination. These difficulties can be eliminated through the use of a destruction procedure of the organic material, such as microwave radiation, UV radiation, ozone oxidation, dry or wet destruction procedures and the like (Suturović, 2003).

Humic and fulvic acid, in addition to being surface active agents, are also complexing agents, and so their presence can doubly interfere with the electrolytic and analytic steps of the PSA (Suturović, 2003). In this case, the pH value of the studied solution must be adjusted, so that the increase in the pH value will replace the degree of adsorption, but on the other hand, hinder the de-complexing of the analyte and facilitate the creation of organic-metal complexes.

Intermetal compounds can be formed between the metals concentrated in the mercury electrodes or between the electrode material (solid electrodes) and the deposited metals. Intermetal compounds are usually formed with copper and zinc, considering the fact that both metals are present in most of the real samples. These disturbances are mostly eliminated by adding a third element which builds more stable intermetal compounds with the elements which cause an interference, or with the addition of the auxiliary electrolyte, which will form a complex with the *interfering* element (Tyszczuk et al., 2006; Wang, 1985).

The overlap between the dilution potential of certain elements can have a significant influence on the PSA only in the case of elements which have similar dilution potentials (Sn and Pb, Cd and Tl, Bi and Sb). These interferences can be eliminated through the selected communication of selectively statement the potential of the deposit or the use of computerized equipment, which enables the reduction of the analytical signal of the interfering element from the overall signal.

3. Heavy metals

Heavy metals significantly contribute to the increase in the pollution of the human environment, due to the fact that they cannot be biologically decomposed, and the cumulative-toxic effect that some of them have. The sources of metal contamination are numerous, while the most significant are the products of combustion in the chemical industry and metallurgy, industrial waste waters and landfills, agrochemical products and exhaust fumes from motor vehicles. Special attention needs to be focused on contamination due to lead, cadmium, mercury and thallium, since these metals have a cumulative-toxic effect, and are also deposited in the human body (Goyer & Klaassen, 1995).

Essential metals, such as copper, zinc, nickel and others can lead to serious illnesses when they are present in the human body in insufficient amounts, but can also have a harmful and toxic effect if found in doses higher than the recommended ones.

Metals, like all the chemical substances which make their way into the human body, with their physical-chemical features can cause changes and numerous structural or functional damage to one or more of the organs or system of organs.

The penetration of metals from the outside environment into places where they could have a negative effect and the manifestation of toxic effects represents a process which includes the exposition phase (contact in the outside environment), the toxicokinetic phase (absorption, distribution, deposition, disintegration, transformation and elimination) as well as the toxicodynamic phase.

Metals display an affinity towards various organic molecule ligands, especially those that contain S, N and O and which are electron donors: -OH, -COOH, -SH, $-NH_2$ and others.

The affinity, and thus the toxicity of divalent metal ions, according to the appropriate ligands is:

- for the –SH group: Hg > Ag > Pb > Cd > Zn;
- for the –COOH group: Cu > Ni > Co >Mn;
- for the $-NH_2$ group: Hg > Cu > Ni >Pb > Zn > Co > Cd >Mn > Mg

3.1 Copper

Copper is an essential biometal necessary for the proper growth, development and normal functioning of the human body. It takes part in the metabolic acceleration, the increase in the oxidation of glucose, the strengthening of tissue respiration, the mineralization and development of bones, contributes to the resorption of iron in the digestive tract, and catalyzes the biosynthesis of hemoglobin by aiding the inclusion of iron into the hem. Along with calcium, copper takes part in the metabolism of phosphorus (Uauy, et al., 1998).

The biological significance of copper for humans was practically discovered in the work of Hart et al. (Hart, et al. 1928), who have shown that copper plays a very important role in the process of erythropoiesis, that is, the production of red blood cells.

The afore mentioned various roles of copper in the human body are made possible due to its polyvalence on the one hand, and propensity for the formation of stable complex compounds, on the other. The copper (I)- ion bonds with lingands via the –SH groups of proteins. The copper (II)-ion reacts with aminoacids and amino groups of proteins, and with nitrogen in the DNA and RNA molecules. Metabolic disorders involving copper are related to many illnesses, such as diabetes, Wilson's disease, acute and chronic hepatitis, cirrhosis of the liver, cardiovascular disease, osteoporosis.

The daily requirements of an adult range from 2-3 mg of copper (Goyer & Klaassen, 1995).

3.1.1 The toxicity of copper

Copper is a biometal essential for human life. Nevertheless, at concentrations of 10^{-6} mol/dm^3 copper inhibits certain enzymes (acid phosphatase), preventing the bonding of other essential microelements or bonds to certain cofactors, such as glutation (Ahasan et al., 1994).

The world health organization (WHO, 1996) stated that 10-12 mg/per day can be the minimum amount of safe daily intake. Nevertheless, if approximately 2 mg of copper salts are introduced into the body, copper-induced hemolytic anemia and kidney failure could ensue.

Copper found in drinking water or beverages, in amounts of 8 ppm $CuSO_4$ (0,022 mg Cu/kg) (Gotteland et al., 2001) causes nausea, vomiting, abdominal pain, and diarrhea. Amounts of 1-2 g can cause severe poisoning symptoms, and lead to hemolysis, destructive changes to the brain tissue and liver, which could be terminal (Ahasan et al., 1994).

The use of agrochemical substances based on copper, can also lead to an increased intake of copper via food which is produced on soil that has been exposed to it.

The increased content of copper in the body can stem from food or drinking water, which are in immediate contact with copper. Acidic foods or drink can dissolve milligrams of copper, which are sufficient enough to cause acute toxicity and symptoms (Goyer & Klaassen, 1995). Increased contents of copper in the human body can have a harmful effect on the cardiovascular system, leading to coronary disease, and high blood pressure. The toxicity of copper is usually a consequence of excessive intake or small amounts of other necessary nutrients. Small amounts of copper in the food lead to an increase in the content of copper, as they compete for absorption in the gastrointestinal tract (Uauy, et al., 1998). This indicates the necessity of monitoring copper content in various samples, including soil, water, food, and air since copper and other metals are involved in the circulation of matter in nature.

The content of copper in the human body can be determined on the basis of blood work, urine samples, nails or teeth.

3.2 Zinc

Zinc is an essential oligoelement, which can be found in significant amounts in the human body, approximately 0.02-0.03 g/kg of body weight of it. Of the overall amount of zinc in the human body, 20 % of it can be found in the skin, and it can be found in significant amounts in the pancreas, teeth, bones, blood, liver, kidneys, and nervous system (Walsh et al., 1994). Zinc is a necessary microelement for the lives of humans, animals, plants and microorganisms. It influences growth and development, bone formation, blood, the metabolism of nucleic acids, proteins, and carbohydrates. Participation in these processes is bound to the effect of enzymes, of which zinc is an important component or activator.

Among patients with diabetes, the amount of zinc in the pancreas is approximately half of what is found in healthy people.

Zinc also plays a certain role in the stabilization of the cell membrane, as well as in the functioning of certain nervous structures under whose control we find the senses of taste and sight. Zinc has an extremely significant, but insufficiently studied function in immunological processes. Zinc ions, along with Cu(II) and Co(II) ions improve the body's immune system.

Through a normal, varied diet, man daily intakes from 10 to 15 mg of zinc. The human body *stores* zinc and in the case of a lack of zinc, its excretion is reduced (Walsh et al., 1994).

3.2.1 The harmful effect of zinc on the human health

The lack of zinc in the human body can occur due to reduced intake, imbalanced absorption, increased excretion and the body's increased need for zinc. This lack of zinc leads to pathological states which are manifested in the occurrence of dermatitis, diarrhea, alopecia, mental disturbances, mental lethargy, stunted growth and development, loss of appetite, reduced neuropsychological functions, the occurrence of infantilism and slow wound healing.

On the other hand, the significant concentrations of zinc salts, such as chlorides, can have a harmful effect on the human body, to the extent that they could even damage tissue epithelials. A high content of zinc can have a harmful effect on the storage of iron (Walsh et al., 1994). Toxic amounts of zinc in the food can reduce the life span of red blood cells and can lead to anemia, since the use of iron is much faster. (Goyer, 1997).

Zinc as an element is necessary for the normal exocrine and endocrine functioning of the pancreas. Its concentration in this tissue is many times greater than in the plasma and is an important means of zinc elimination. Studies carried out *in vivo* (Chobanian, 1981) have shown that a high zinc content (800 mg/per day) causes a considerable increase in amylase and lipase in the serum, and an increase in blood sugar levels.

The increased zinc content can be connected to the occurrence and development of neurological disease. The significant increase of zinc in the human body can lead to a disturbance in neurological functions and the occurrence of multiple sclerosis among workers involved in production processes where zinc is the basic ingredient. Zinc contents ranging from 6.54 to 16.35 mg/dm^3 lead to minor damage, while amounts exceeding 16.35 mg/dm^3 are neurotoxic (Choi et al., 1988).

The recommendation is that zinc should be taken with dairy products since milk contains picolinic acid with which zinc builds chelating complex compounds which are best absorbed in the intestines.

3.3 Lead toxicity

Lead belongs to a group of the most toxic of elements, with a cumulative-toxic effect. (Vig & Hu, 2000). Lead is not an essential metal, but is present in all of the tissues and organs of mammals, and can mostly be found in mineral tissue - bones and teeth (over 90 % of the overall amount of this element) (Gulson, & Gillings,1997). If it is constantly introduced into the human body, even in small amounts, lead partially replaces calcium in the tertiary calcium-phosphate bone skeleton, where its toxic effect is gradually increased.

Daily amounts of lead which a human normally absorbs mostly through food and drink, can range up to 0,3 mg, but this amount does not cause poisoning, since lead is excreted in approximately the same amount daily from the human body (Goyer, 1997).

Lead intake can occur in different ways. Lead bound in tetraethyl lead, as an addition of gasoline, through its combustion is transferred into the atmosphere and reaches the human respiratory system. Part of the lead is absorbed by plants and animals alike, so that it is introduced into the human body by means of food. The innards used in the human diet (especially the liver and kidneys) contain high concentrations of lead. Nevertheless, it has been proven that only 3 % the lead in the innards is absorbed into the human body.

Besides food, lead can be introduced into the body by mans of water, which lead reaches via the air, soil or pipelines. Studies have shown that lead from the water or other beverages is reabsorbed to a greater extent than that from food. In addition, lead introduced into the body between meals is absorbed to a greater extent than the lead introduced during a meal, while the greater frequency of food intake minimizes the absorption of lead. It has been proven that 50 % of lead is absorbed from water, following an overnight fast (Vig & Hu, 2000). Professional exposure to lead, in factories and workshops leads to severe and prolonged illnesses as the gravest of professional illnesses.

According to the American Center for Disease Control (CDC) (Goyer & Klaassen, 1995) a lead content in the blood of less than 240 μg/dm^3 is *normal*, 250-490 μg/dm^3 belongs to the *moderate risk* category, 500-690 μg/dm^3 to the *high risk* category and contents of over and amounts that exceed 700 μg/dm^3 fall into the *urgent risk* category. Nevertheless, prolonged exposure to low-level toxicity (< 240 μg/dm^3) can lead to various psychological disorders and learning disabilities among children. Naturally, these symptoms can occur among children even in the case of lead amounts of less than 50 μg/dm^3 (Banks et al., 1997).

Lead competes with essential metals (Ca, Fe, Zn, Cu) for various important functions in the human body.

The intake of smaller amounts of iron and vitamins C and D, as part of our diet, can increase the lead content in the blood, absorption from the intestines and can lead to the accumulation of lead in the body (Pocock et al., 1994).

Lead competes with iron for the binding spots in the ferritin transport protein Fe^{3+}-ion and can block the active center of the ferritin, building lead-sulfide which can only be dissolved with difficulty.

According to some of its physical-chemical features, lead is chemically similar to calcium, and so in the body it behaves like calcium and can be found in calcified tissue (bones and teeth), in blood plasma bound to proteins or in an ionized form, as well as in the form of compounds with various biomolecules (such as citrates) (Kaličanin et al., 2004).

Naturally, the very important affinity of lead towards mineral tissue (bones, teeth) where it accumulates over time, has been confirmed in the works of many authors. Lead in the bones contributes to the development of osteoporosis, the reduction of bone tissue, changes in the structure of bone structure and increases the resorption of bones tissue among the elderly (Gulson, & Gillings, 1997).

3.4 Cadmium toxicity

Cadmium is one of the most dangerous poisons of the working and living environment, which can be introduced into the body by means of air, food or drinking water. Thus, the lack of iron can significantly increase the accumulation of cadmium, and sufficient amounts of iron in the blood inhibit the accumulation of cadmium. In addition, increased doses of vitamin D act as an antidote to cadmium poisoning. The cadmium content in the human body has a value of $1 \cdot 10^{-4}$ % of the overall body mass (Danielsson et al., 1981).

Cadmium poisoning can be acute and chronic. Acute poisoning occurs due to inhalation of the fumes of particles of cadmium salts (oxides, chlorides, sulfides, sulfates, carbonates and acetates) whose concentration in the air is approximately 1mg/m^3 (Goyer & Klaassen, 1995).

The toxic effect of cadmium to a great extent depends on the intake of calcium. Low calcium intake leads to higher cadmium absorption, the retention, accumulation and increased toxic effect of this metal. The consequences of this include kidney and bone damage (osteomalacia) as well as hypertension and anemia. (Brzoska & Moniuszko-Jakoniuk, 1998).

The presence of cadmium in the air originates from the combustion of oil derivatives, coal and plastic mass, and is found in cigarette smoke. The absorption of cadmium from the air mostly takes place through breathing, and to a lesser extent through the gastrointestinal tract, and in trace amounts via the skin.

Once it enters the body, cadmium is transported into the blood by means of red blood cells and a highly molecular blood protein - albumin. The normal level of cadmium in the blood of adults is less than 1 μg/dm^3. Even though cadmium circulates via the blood throughout the entire body, the greatest accumulation (from 50 to 60 % of the body's cadmium load) can be found in the kidneys and liver (Florianezyk, 1995).

As is the case with other metals, cadmium also participates very little or not at all in the direct metabolic exchange, but is bound to various biological components, such as proteins, thiol (-SH) groups and anion groups of various macromolecules. The basis of the toxicity of cadmium is its negative influence on the enzyme system of the cells, due to the exchange of other metal ions (mostly Zn^{2+} and Cu^{2+}) (Kaličanin, 2006).

The toxic effect of cadmium to a significant extent depends on nutritive factors. A protein-free diet with insufficient amounts of calcium, vitamin D along with a zinc, manganese, copper and selenium deficiency in the body increases while vitamins C and E reduce the toxicity of cadmium. (Deng et al., 2004).

4. Experimental conditions for the determination of Cu, Zn, Pb and Cd by using the PSA

As a carrier of the layered mercury electrode, during PSA, a disc electrode made of vitreous carbon was used in a standard electrochemical cell (Suturović, 2003, Kaličanin, 2006).

In all of the analyses, the layered mercury electrode was created from a special solution of mercury(II)-ions (with a content of 100 μg/dm^3), which was made acidic y means of hydrochloric acid (pH ~ 1.65). The mercury was deposited by means of electrolysis at to a constant electrical current of -50 μA, for a period of 240 s. The thickness of the level of mercury formed in this way was approximately 130 nm.

The initial volume of all of the analyzed solution models and samples was constant (25 cm^3). Hydrochloride acid was used as an auxiliary electrolyte, due to the fact that HCl is suitable for de-complexing most of the metals in real samples (Kaličanin, 2001a). In some cases 4 % CH_3COOH was used as the auxiliary electrolyte, which is at the same time in some of the experiments it was used for the extraction of Cu, Zn, Pb and Cd from various samples.

In order to determine the content of the soluble metals (Pb, Cu, Zn and Cd) the simplest modification of the PSA was used, the PSA with oxygen as the oxidizing agent.

4.1 Optimization of the conditions for the PSA of Cu, Zn, Pb and Cd

PSA as an exceptionally sensitive and selective microanalytic technique enables us to concurrently determine a large number of elements. Nevertheless, considering the fact that in the samples we determined the contents of Cu, Zn, Pb and Cd, due to the possible interferences in the PSA as a result of the formation of intermetal compounds of Cu-Zn, Cu and Pb from the same solution were usually determined together, while Zn and Cd were determined together in a different series of analyses, with the addition of a complexing agent, most often galium, which would form complexes with the Cu and enable the unhindered determination of Zn.

For the PSA of the cited metals, optimization was carried out, and the most optimum values of the parameters of analysis were selected: the potential of the electrolysis, the duration of the electrolysis, and solution stirring speed. The value of the pH model solution and samples ranged around 2.6.

The optimized experimental conditions for determining Cu, Pb, Cd and Zn are shown in Table 2. The duration of the electrolysis depends on the contents of the determined metal in the studied analyte and varied from 60 to 600, that is 900 s.

Parameters	Pb, Cu	Cd , Zn
Electrolysis potential Ag/AgCl/KCl (3.5 mol/dm³) (V)	-0.962	-1.522
Final potential Ag/AgCl/KCl (3.5 mol/dm³) (V)	0	-0.15
Sample volume (dm³)	0.025	0.025
Duration of the break (s)	15	15
Solution mixing speed (rpm)	4000	4000
Electrolysis duration (s)	60-600	60-900

Table 2. The conditions for determining Pb, Cu, Cd and Zn with the help of PSA

4.2 The linearity of the analytical signal in the PSA of PSA Cu, Zn, Pb and Cd

After the optimization of the determination conditions, both the linearity and the reproducibility of the analytical signal were defined.

The linearity of the analytical signal of each of the heavy metals (Cu, Zn, Pb and Cd) was studied on model dilutions of specific mass concentrations.

Under the defined optimum values of the electrolysis potential (-0.702 V for Cu, -1.31 V for Zn, -0.962 V for Pb i -1.10 V for Cd), the stirring speed (4000 min^{-1}) and the duration of the electrolysis (300 s, 480 s, 360 s i 600 s), the linearity of the analytical signal in the PSA was studied for the cited metals.

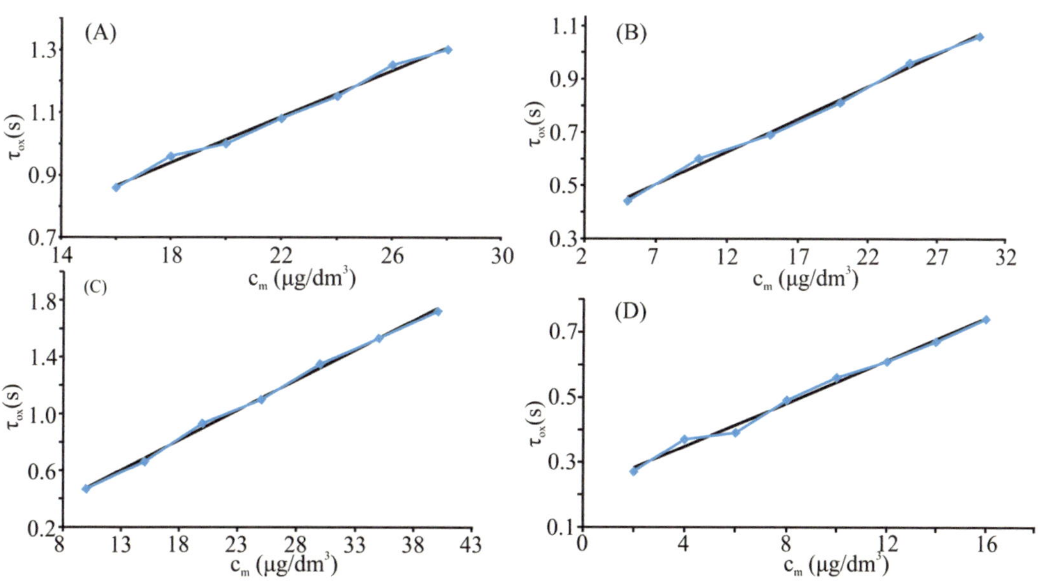

Fig. 2. The linearity of the analytical signal in PSA: A) Cu; $\tau_{ox} = 0.2804 + 0.0366 \cdot c_m$; r = 0.9964; B) Zn; $\tau_{ox} = 0.33 + 0.0246 \cdot c_m$; r = 0.9976; C) Pb; $\tau_{ox} = 0.0532 + 0.0422 \cdot c_m$; r = 0.9987; D) Cd; $\tau_{ox} = 0.2168 + 0.0329 \cdot c_m$; r = 0.9957

Using the method of the smallest squares , for each of the studied metals, the values were calculated for the intercept size (a), the direction coefficient (b) and correlation coefficient (r) in the linear function $\tau_{ox} = a + b \cdot c_m$. The results of the study are given in figure 2, as the dependence of the duration of oxidation in the function of the concentration.

5. Application of PSA for the determination of Cu, Zn, Pb and Cd in different samples

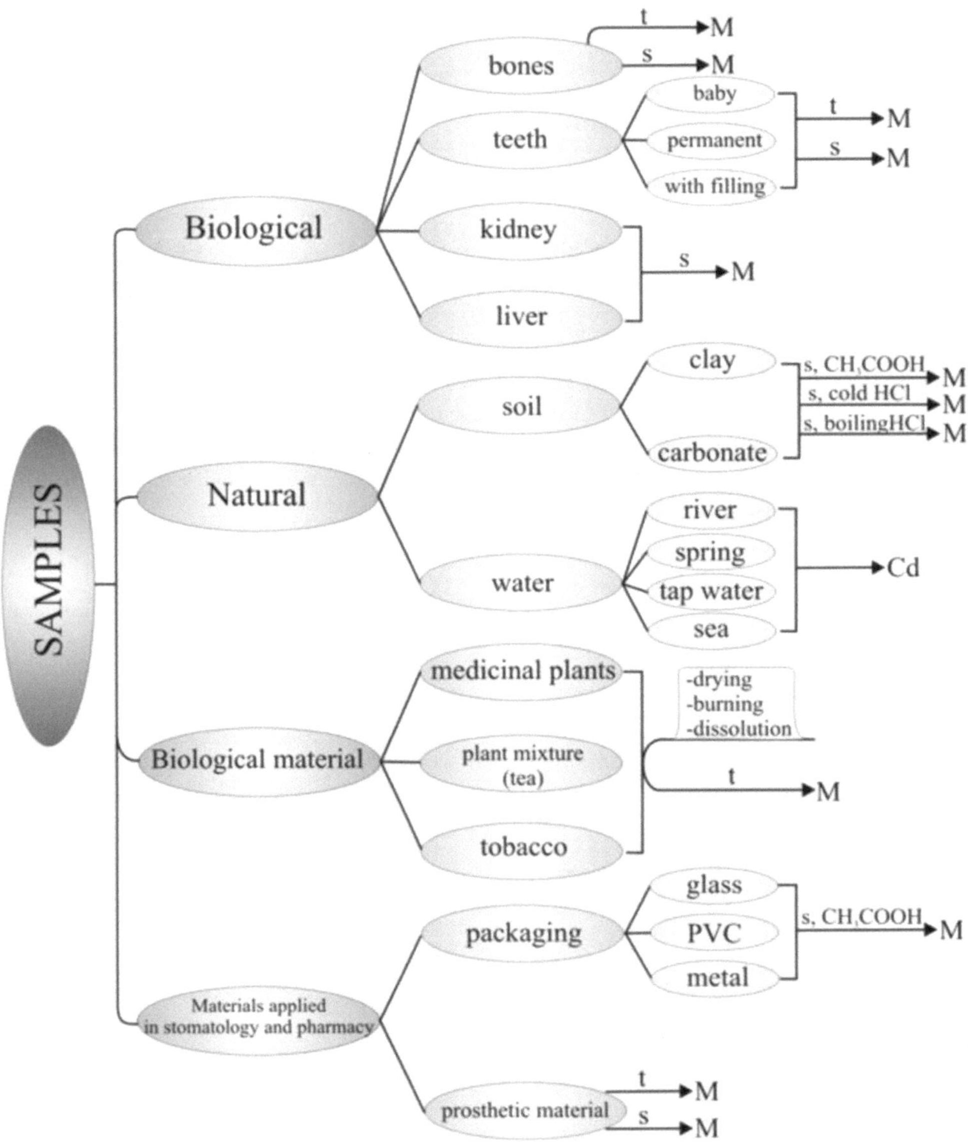

Fig. 3. The experimental design of the PSA analysis of heavy metals in various sample types: M - Cu, Zn, Pb and Cd; t- the total content of M in a sample dissolved in HNO_3, HCl and H_2O, at 130°C for a period of 2 hours; s – the content of soluble metals in the given mediums

5.1 Determining the content of Cu, Zn and Pb in natural human teeth

The overall content of Cu, Zn and Pb was determined in baby teeth, permanent intact teeth and teeth from which the amalgam filling, which had been a part of the teeth for years, had been mechanically extracted. These teeth were also used to determine the content of metal which is leached in a 4 % CH_3COOH solution, over a period of 24 hours.

Metals	Content (μg/g)	Sample		
		Permanent tooth	Baby tooth	Tooth with removed filling
Cu	t*	1.44 - 3.20	2.75 - 4.07	18.20 - 87.26
	s*	1.01 - 2.27	2.50 - 3.76	11.60 - 66.27
Cu leached	(%)	70.14- 73.93	87.74 – 92.38	53.72 – 75.94
Zn	t*	116.16 - 153.20	128.50 - 142.32	1505.20 - 5684.74
	s*	63.70 - 84.20	67.53 - 72.23	565.64 - 919.92
Zn leached	(%)	54.84 – 58.27	50.75 – 52.55	11.38 – 37.58
Pb	t*	2.20 - 5.41	1.52 - 2.17	2.70 - 6.73
	s*	0.82 - 1.99	0.97 - 1.40	1.52 - 3.90
Pb leached	(%)	35.26 – 37.27	63.81 – 64.52	56.13 – 57.94

Table 3. The overall and diluted content of copper, zinc and lead from human teeth

Over time heavy metals accumulate in the mineral tissue of the teeth, including Cu ranging from 1 to do 4 μg/g, Pb ranging from 1.5 to do 5.5 μg/g and Zn ranging from approximately 110 to do 150 μg/g. From the teeth which had an amalgam filling, a significantly higher content of the cited metals was noted. It was determined that the mobility of these metals in the 4 % CH_3COOH, over a period of 24 is almost identical to that in the permanent and baby teeth in the case of zinc, while there is significantly more copper and lead in the baby teeth. The greater mobility of lead from these teeth is probably a consequence of imperfect structure, low hardness and the frequent occurrence of caries (Kaličanin & Nikolić, 2008, 2010; Kaličanin et al., 2003, 2004; Nikolić et al., 2004).
The lead ions are, due to their size (r = 0.132 nm), more dominant over the smaller ions, and from the surface of the dentin, where they arrive from food, water, air and amalgam fillings, they transfer to an acidic solution, which increases their migration (Oehme et al., 1978; Gulson & Gillings, 1997).

5.1.1 The release of toxic metals from teeth under the influence of soft drinks

In today's day and age more and more refreshing soft drinks and energy drinks are being consumed by the younger population. Most of them contain carbon-dioxide, artificial sweeteners, preservatives and other additives and supplements, which enhance their taste. Frequent and constant consumption of these drinks, as well as of sour-tasting food, can have harmful effects on the mineral structure of teeth (Cairns et al., 2002; Attin et al., 2003; Ehlen et al., 2008). This analysis was carried out with the aim of identifying the effects of consuming sour-tasting food and refreshing drinks on the bone tissue of teeth among teenagers. The cumulative effect of a year-long exposure of teeth bone tissue to the erosive effects of a model system of acidic mediums (citric acid, lactic acid, acetic acid, apple vinegar, lemonade, the soft drink Sprite, mineral water) was studied. The effects were registered by means of the amount of released biometal ions, of zinc and copper, and toxic lead, during a period of 24 hours at room temperature, using the potentiometric stripping analysis.

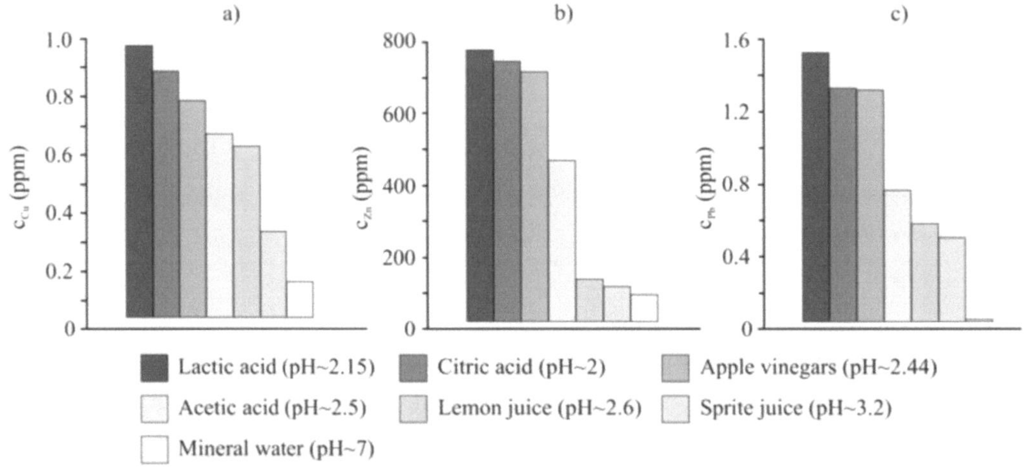

Fig. 4. The graphic dependence of the contents of released a) copper; b) zinc and c) lead (in ppm) depending on the studied system

In the given time span, 75 to 750 ppm of zinc, from 0.1 to 1.0 ppm of copper, and up to 1.5 ppm of lead was released from the dental matrix, while the release of cadmium was below the level of detection.

5.2 The application of the PSA in the analysis of pathohystological material (the kidneys and the liver)

The PSA method has been used to monitor the toxic effect of cadmium and lead from prosthetic implants, to which these metals have been added during their manufacture (5 μg/g of body weight), in the experiment model involving rats of the Spague Dolly kind, all of them female, 6-8 weeks old (Kaličanin et al., 2008). The content of Pb and Cd was determined in the kidney tissue and liver tissue, after a period 12 and 24 weeks during the experiment.

Sample	Content of Pb (μg/g)		Content of Cd (μg/g)	
	12 weeks	24 weeks	12 weeks	24 weeks
Kidneys	27.52	48.06	11.50	30.60
Liver	5.06	10.30	1.70	2.50

Table 4. The content of Pb and Cd in the kidneys and livers of the rats

Our results indicated that lead content is higher in the kidneys than in the, liver which corresponds to the results obtained by other authors, by means of other analysis techniques.

5.3 The use of the PSA in the analysis of soil for heavy metal content

As part of the Earth's crust, we find many silicate minerals which contain Al, Fe, Ca, Mg, K, Na, and Ti. Elements such as Zn and Cu are not present to such an extent. Nevertheless, toxic metals such as Pb and Cd both in their mineral form and as "additions" of silicate can also be found.

The results of the PSA analysis of the contents (mg/kg) of toxic metals Pb, Cd, Zn and Cu in various phases of demineralization of the soil samples, show that they primarily contain kaolinite as their main component, $Al_2(Si_2O_5)(OH)_4$, as shown in Table 5.

Metal	I (CH_3COOH)	II (HCl, cold)	III (HCl, boiling)	Average soil sample content
Pb	4.20	1.14	2.20	0.5-30
Cd	-	4.57	5.15	<1
Zn	-	3.24	4.60	10-300
Cu	-	17.21	20.13	2-10

Table 5. The mean values of the toxic metal content in naturaly occurring kaolinite

Lead can be found in the soil in the form of carbonite, phosphate or sulphate. It is easily adsorbed to other minerals and so its mobility is relatively small. When the soil becomes more acidic, Pb is released and then becomes available to the plants (Kaličanin et al., 2005; Mendil et al., 2009).

The increased presence of cadmium in the soil occurs due to the use of phosphate fertilizer, and fungicides. It is found in motor oil and tires, which leads to the increased content in the soil near roads and landfills. The geo-chemical relation between Cd and Zn leads to both of these elements being present in the soil at the same time. Zinc actually has a positive effect on cadmium absorption.

Zinc can be found in the soil in approximately 10-300 ppm, depending on the soil composition, pH values, and the presence of other metal ions. The Zn^{2+} ion is easily adsorbed onto the clay minerals, but also onto organic matter (Premović et al., 2001). Cadmium builds complexes well with ligand anions, which suits its solubility and higher mobility in the environment (Anđelković et al., 2006).

Copper can be found in the soil in very large extents (2-100 ppm). It is not a very mobile ion in the soil because it forms very good complexes with organic matter. High levels of copper in the soil can cause disturbances in the absorption of other microelements (Fe, Mo and Zn). What also has an effect on the toxicity and tolerance to copper, in addition to its content, is also the presence of organic matter, the level of phosphates in the soil and pH values.

5.4 Using PSA to analyze water

A potentiometric stripping method for a direct measurement of cadmium in various natural water samples using a Stripping analyzer M1 unit has been developed (Jagner et al., 1981). In order to ascertain optimum conditions for the determination of the effect of the mercury time electro-deposition, the electrolysis potential and the solution stirring rate of the cadmium analytical signal have been examined. Linearity of the cadmium analytical signal was achieved within the wide range of the mass concentrations, from 1 $\mu g/dm^3$ to 15 $\mu g/dm^3$. A detection limit of 0.10 $\mu g/dm^3$ was obtained, with a reproducibility of 3.55% expressed as the coefficient of variation. The accuracy of the method was confirmed by parallel analyses by flameless atomic absorption spectrophotometry as the reference method (Kaličanin, 2009).

Source of water	Content of cadmium ($\mu g/dm^3$)		
	Calibration curve method (min-max)	Standard addition method (min-max)	FAAS (min-max)
Tap	0.12 - 0.36	0.13 - 0.38	0.11 - 0.38
Mineral	0.15 - 0.21	0.17 - 0.20	0.14 - 0.23
River	1.25 - 3.52	1.31 - 3.60	1.37 - 3.60
Sea	6.67 - 13.25	6.80 - 13.60	6.50 - 12.95

Table 6. Cadmium contents in the water samples determined by PSA and FAAS methods

The results of the comparative analysis showed a very good agreement between the PSA and the FAAS methods. The best results agreement was achieved with the samples with micro cadmium contents as achieved, thanks to the great sensitivity of the given method in analyzing the cadmium in the water samples. On the basis of the values of the coefficient of variation, it can be concluded that it is possible to apply the calibration curve method for obtaining more reproducible values than by the standard addition method where the deviations are somewhat greater. Hence, the calibration curve method is proposed as the standard method for the determination of cadmium in the various natural water samples.

5.5 Determining Cu, Pb and Cd in herbs and herbal tea mixtures

Mineral matter has several roles in the development and growth of plants. Microelements take part in the metabolism of plants, have an effect in very small concentrations, in a highly specific manner and in certain phases of growth and development. Cu and Zn are necessary microelements for certain metabolic processes. Nevertheless, in higher concentrations, irrespective of the physiological role, they have a toxic effect. The heavy metal content of Pb, Cd, Cr and others in herbs can be an indicator of the contamination of the area in which they are growing. The accumulation of heavy metals in certain parts depends on the plant life, mobility of the metal, soil composition, pH values, precipitation, and the presence of other ions and molecule types in the soil.

We analyzed the plan species *Thymus serpyllumm,* in order to determine its Cu, Pb and Cd content. This plant is used to make tea, and as an aromatic spice in aromatherapy (Jeftović, 2001).

Location	Pb		Cd		Cu	
	t	s	t	s	t	s
I	6.62	6.60	0.42	0.40	3.20	3.18
II	15.80	14.90	1.06	1.04	3.80	3.60
III	2.42	2.50	0.25	0.30	1.06	1.10
IV	0.84	1.00	0.06	<0.50	0.84	0.80

Table 7. The toxic metal content in the plant species *Thymus serpyllumm* from various location in the vicinity of Niš ($\mu g/kg$). I – a location near a well-known local landfill; II – a meadow off the Niš-Belgrade highway; III – a mountain meadow; IV – a field in the mountain

Our findings showed especially high levels of lead in the flower of the plants which grew 20-30 m off the highway (15.80 ppm) and near the local landfill where waste is burned

(6.60 ppm). In the soil, lead turns into relatively soluble compounds, carbonate and phosphate, from which it is released due to acidification. When present in higher concentrations it causes numerous physiological, anatomic, morphological and chemical changes (Deng et al., 2004).

Cd reaches plants through their roots, but also from the air and above-ground parts of the plant. Its ability to form complexes with Cl^- and OH^- ions also contributes to this, which leads to greater mobility in the environment and increases the possibility of altering adsorption to cations (Ca^{2+} and Zn^{2+}). The increased level of Cd in the plants which were collected near the landfill where waste is burned and next to the highway is probably the result of accumulation which was made possible through the above ground parts of the plants.

The content of copper in the flower of the plant *Thymus serpyllumm* is within the normal limits for herbaceous annual plants and vegetables. Copper can be found in the soil up to 20mg/kg, but it is not a very mobile cation, so it is easily bonded to clay minerals, adsorbed, to form complexes and so is not readily available to plants and so is more frequent in the soil.

The results of the determination of the overall content of lead in commercial plant drugs, as well as the content of lead in teas prepared according to the recommendations of the manufacturers, by means of the PSA are shown in Table 8.

Sample	Content of Pb (μg/g)		Pb leached (%)
	t	s	
Chamomillae flos	0.73-0.81	0.32-0.35	43.20-44.00
Senae folium	1.21-1.64	0.58-0.74	45.12-47.93
Theae folium	1.74-2.37	0.27-0.39	14.28-16.46
Menthae folium	1.18-1.34	0.42-0.48	35.59-36.59
Uvae ursi folium	0.92-1.13	0.27-0.30	25.66-29.34

Table 8. The overall and soluble contents of Pb in herbal drugs and tea mixtures

The obtained results indicate that plant-based drugs contain a certain amount of lead, but that the obtained contents are within the limits prescribed for this metal. The content of lead which is released from the plant-based drug into the tea was 3 to 5 times lower than the overall content of this metal. The smallest percentage of the leached lead was found in green tea, which indicates that the migration of lead from this plant-based drug into the tea is the smallest. The lead content in teas depends on how they are prepared and is higher in teas which are prepared as a decoction, and the lowest in those prepared as macerate.

5.6 Using the PSA in the quality control of glass packaging for the food and pharmaceutical industry

Glass is a high quality packaging material, which is used in the food and pharmaceutical industry. Packaging, in addition to its basic components, can also contain metals (lead, zinc) as pollutants or as components used to achieve higher quality packaging. Considering the fact that products over a longer period of time, from packaging to use, are in contact with the packaging material, there is the possibility of ion metal migration from the packaging into the product. The international standard (ISO 7086/2) prescribes that the content of the

leached lead from glass packaging cannot exceed the prescribed limits of 5 mg/dm³, for small hollow glass and 2.5 mg/dm³, for large glass, under prescribed conditions.
Table 9 compares the results from a measuring of the contents of soluble lead from glass packaging for the food industry, under prescribed conditions, using different stripping analysis techniques and the AAS technique, as the referential technique prescribed by the standard (Kaličanin et al., 2001a, 2001b, 2001c; 2002).

Sample	C_{Pb} (μg/dm³)		
	Analytical methods		
	PSA	PSA-i_R	AAS
Bottle for fruit juice, 1 dm³ of volume, colorless	2.20	2.30	1.80
Jar, 1.5 dm³ of volume, colorless	2.80	2.91	1.80
Jar, 0.72 dm³ of volume, colorless	1.70	1.82	0.80
Bottle for strong alcoholic drinks, 0.7 dm³ of volume, of green color	2.70	2.86	2.80

Table 9. Lead contents in the glassware for the food industry extracts by applying the PSA, PSA-i_R (potentiometric stripping analysis with constant inverse current in the analytic step) and AAS

These results indicate that there is proper agreement between the contents obtained through the stripping techniques and AAS technique as the referential one.
Figure 5, shows the content of the leached Pb from various packaging material, which is used in the pharmaceutical industry, during a period of 1, 5 and 7 days. Most of the lead is leached from glass packaging of brown color and of greater volume. According to our research, plastic packaging is more durable to the effects of an acidic medium.

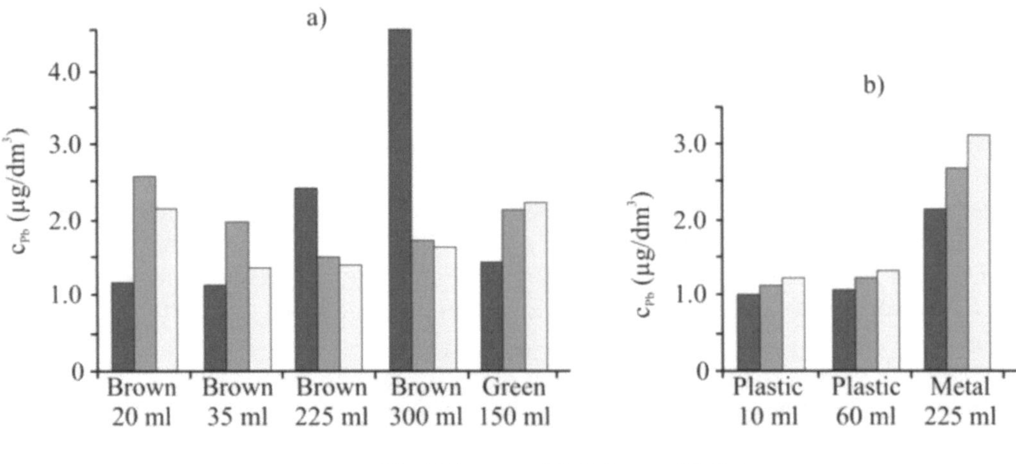

Fig. 5. The content of leached lead from packaging for the pharmaceutical industry a) glassware b) plastic and metal, depending on the duration exposure, volume and color of the packaging

5.7 The use of the PSA in the quality control of dental-prosthetic material

Dental-prosthetic material is very pure material of varying compositions. The same materials can consist of toxic heavy metals (Pb, Cd, Zn, Cu) which can be released under the influence of the corrosive effect of the oral medium or food with a high acidic taste (O'Brien, 2002). During the production phase of prosthetic implants, due to various physical-chemical processes, they are transformed into more stable units, they become less mobile, so that the finished product (metalceramic crown) limited release (Kaličanin & Ajduković, 2008; Kaličanin et al., 2007; Nikolić et al., 2001; Kaličanin & Nikolić, 2008, 2010).The results shown in table 10 indicate that these materials also contain Cu, Zn, Pb and that their traces can also be determined by means of the PSA technique.

Sample	Cu	RSD (%)	Zn	RSD (%)	Pb	RSD (%)
Dental ceramic	1.98	3.87	1.20	2.83	104.50	1.30
Ceramic color	2.05	1.93	0 .55	4.70	1.25	3.30
Cast alloys	2.60	2.14	215.95	4.39	0.33	10.60
Metal-ceramic crown	3.40	3.13	6.30	7.15	0.65	4.80
Zinc-phosphate cement	1.20	2.80	-	-	2.65	1.90
Glass- ionomer cement	0.76	1.50	1.98	2.15	0.35	1.80
Acrylic materials	n.d.	-	53.05	12.50	4.50	3.80
Hydroxyapatite	33.05	8.50	116.10	15.50	5.70	3.50

Table 10. The content (μg/g) of released copper, zinc and lead from various dental prosthetic materials under the effect of 4% acetic acid, over a period of 24 hours

6. Conclusion

The potentiometric stripping analysis is a highly sensitive and highly selective instrumental microanalytic technique for the quantitative determination of the metal ions. This technique can be used to determine low contents of heavy metals (Cu, Zn) and highly toxic metals (Pb and Cd) in samples of various origins Lead can be determined up to levels of 0.65 μg/dm^3, and cadmium up to 0.10 μg/dm^3, under the prescribed optimal conditions.
PSA fulfills very strict general and specific microanalytic demands:

- High sensitivity and proper analytical selectivity
- The possibility of determining a large number of elements at the same time
- The possibility of the unlimited re-analyses of the same solution
- The relatively small instrumentation and the possibility of "on-the-spot" analyses
- Lower cost of the instrumentation and exploitation in relation to other techniques.

The results of the determination of the content of Cu, Zn, Pb and Cd in the samples of bioinorganic and similar origin, have shown that the PSA technique with oxygen as an oxidant, as the simplest modification that can be done to this technique, can be used in the analysis and quality control of various samples with success. This technique can also be used to analyze:

- Clinical-biological material (mineral and soft tissue *in vivo* and *in vitro* analyses)
- Samples significant in quality control of the environment (water, soil)

- Plant material samples (herbs and aromatic plants, tea mixtures and spices)
- Packaging material and packaging for food and pharmaceutical products (glass, ceramics, plastic, metal)
- Highly pure bioinorganic material (dental-prosthetic materials)
- Beauty products.

7. Acknowledgment

Some results presented here are part of projects N° 45017 and 41018, which have been realized with partial financial support of the Republic of Serbia Ministry of Science and Environmental Protection.
We would also like to thank Marta Dimitrijevic for translating the original paper from Serbian into English.

8. References

Ahasan, H.A.M.N.; Chowdhury, M.A.J. & Azhar, M.A. (1994). Copper sulphate poisoning. *Tropical Docor,* Vol. 24, pp. 52-53, ISSN 0049-4755

Anđelković, T.; Perović, J., Purenović, M., Blagojević, S., Nikolić, R., Anđelković, D. & Bojić, A. (2006). Spectroscopic and potentiometric studies on derivatized natural humic acid. *Analytical Sciences*, Vol. 22, pp. 1553-1558, ISSN 0910-6340

Attin,T.; Meyer, K., Hellwig, E., Buchalla, W. & Lennon, A.M. (2003). Effect of mineral supplements to citric acid on enamel erosion. *Archives of Oral Biology,* Vol. 48, pp. 753 – 759, ISSN 0003-9969

Banks, E.C.; Ferretti, L.E. & Shucard, D.W. (1997). Effects of low level lead exposure on cognitive function in children: a review of behavioral, neuropsychological and biological evidence. *Neurotoxicology,* Vol. 18, pp. 237-281, ISSN 0161-813X

Bard, A.J.; Parsons, R. & Jordan, J. (1985). *Standard potentials in aqueous solution,* pp. 576-602, Marcel Dekker, Inc., ISBN 0824772911, New York-Basel

Bock, R. (1979). *A handbook of decomposition methods in analytical chemistry,* pp. 122-138, International texbook company, ISBN 0470265019, Edinburgh

Brainina, K. & Neyman, E. (1993). *Electroanalytical Stripping Methods (Chemical Analysis: A Series of Monographs on Analytical Chemistry and Its Applications),* pp. 1-190, Wiley-Interscience, ISBN 9780471595069, New York

Brzoska, M.M. & Moniuszko-Jakoniuk, J. (1998). The influence of calcium content in diet on cumulation and toxicity of cadmium in the organism. *Archives of Toxicology,* Vol. 72, pp. 63-73, ISSN 0340-5761

Cairns, A.M.; Watson, M., Creanor, S.L. & Foye, R.H., (2002). The pH and titrable acidity of a range of diluting drinks and their potential effect on dental erosion. *Journal of Dentistry,* Vol. 30, pp. 313-7, ISSN 0300-5712

Chapman, T.L. (2008). *Genetic Heavy Metal Toxicity: Explaining SIDS, Autism, Tourette's, Alzheimer's and Other Epidemics,* pp. 1-119, Universe Inc. Lincoln, ISBN 978-0-595-48056-2, New York

Chobanian, S.J. (1981). Accidental ingestion of liquid zinc chloride: local and systemic effects. *Annals of Emergency Medicine,* Vol. 10, pp. 91-93, ISSN 0196-0644

Choi, D.; Yokoyama, M. & Koh, J. (1988). Zinc neurotoxicity in cortical cell culture. *Neuroscience,* Vol. 24, pp. 67-79, ISSN 0306-4522

Danielsson, L.; Jagner, D., Josefson, M. & Westerlund, S. (1981). Computerized potentiometric stripping analysis for the determination of cadmium, lead, copper and zinc in biological materials. *Analytica Chimica Acta,* Vol. 127, pp. 147-156, ISSN 0003-2670

Deng, H., Ye, Z.H. & Wong, M.H. (2004). Accumulation of lead, zinc, copper and cadmium by 12 wetland species thriwing in metal-contaminated sites in China. *Environmental Pollution,* Vol. 132, pp. 29-40, ISSN 0269-7491

Ehlen, L.A.; Marshall, T.A., Qian, F., Wefel, J.S. & Warren, J.J. (2008). Acidic beverages increase the risk of in vitro tooth erosion. *Nutrition Research* Vol. 28, pp. 299–03, ISSN 0271-5317

Florianezyk, B. (1995). Toxic and cancerogenic properties of cadmium. *Nowiny Lekarskie,* Vol. 64, pp. 737-745, ISSN 0860-7397

Glassware and Glass Ceramic Ware in Contact with Food; Release of Lead and Cadmium, Part. 2: Permissible limits, ISO 7086/2

Gotteland, M.; Araya, M. & Pizarro, F. (2001). Effect of acute copper exposure on gastrointestinal permeability in health volunteers. *Digestive Diseases and Sciences,* Vol. 46, pp. 1909-1914, ISSN 0163-2116

Goyer, R.A. (1997). Toxic and essential metal interactions. *Annual Review of Nutrition,* Vol 17, pp. 37-50, ISSN 0199-9885

Goyer, R. & Klaassen, C.D. (1995). *Metal Toxicology,* pp. 31-45, Academic Press, ISBN 0122943759, San Diego

Gulson, B.L. & Gillings, B.R. (1997). Lead exchange in teeth and bone-a pilot study using stable lead isotopes. *Environmental Health Perspec*tives, Vol. 105, pp. 820-824, ISSN0091-6765

Hart, E.B.; Steenbock, H., Waddell, J. & Elvehjem, C.A. (1928). Iron nutrition. VII. Copper as a supplement to iron for hemoglobin building in the rat. *Journal of Biological Chemistry,* Vol. 77, pp. 797-812, ISSN 0021-9258

Jagner, D. (1979). Potentiometric stripping analysis in non-deaerated samples. *Analytical Chemistry,* Vol.51, pp. 324-345, ISSN 0003-2700

Jagner, D. (1982). Potentiometric stripping analysis. *Analyst,* Vol. 107, pp. 593-599, ISSN0003-2654

Jagner, D. & Graneli, A. (1976). Potentiometric stripping analysis. *Analytica Chimica Acta,* Vol. 83, pp. 19-26, ISSN 0003-2670

Jagner, D.; Josefson, M. & Westerlund, S. (1981). Determination of zinc, cadmium, lead and copper in sea water by means of computerized potentiometric stripping analysis. *Analytica Chimica Acta,* Vol. 129, pp. 153-161, ISSN 0003-2670

Jagner, D.; Sahlin, E., Ratana-Ohpas & Axelsson, R. B. (1993). Rapid method for the determination of copper(II) and lead (II) in tap water using a portable potentiometric stripping analyser. *Analytica Chimica Acta,* Vol. 278, pp. 237-242, ISSN 0003-2670

Jeftović, M. (2001). *Zelena riznica,* pp. 135-137, Zadužbina Andrejević, ISBN 86-7244-250-4, Beograd

Kaličanin, B. (2006). *Zubi, dentalni materijali i teški metali,* pp. 7-66, Zadužbina Andrejević, ISBN 86-7244-590-2, Beograd

Kaličanin, B. (2009). Determination of very toxic metal – Cadmium in natural water samples. *Desalination,* Vol. 249, pp. 58-62, ISSN 0011-9164

Kaličanin, B. & Ajduković, Z. (2008). Influence of saliva medium on feeing heavy metal ion from fixed dentures. *Science of the Total Environment,* Vol. 397, pp. 41-45, ISSN 0048-9697

Kaličanin, B.; Ajduković, Z., Kostić, M., Najman, S., Savić, V. & Ignjatović, N. (2007). The role of synthetic biomaterials in resorptive alveolar bone regeneration. *Hemijska Industrija,* Vol. 61, pp. 96-100, ISSN 0367-598X

Kaličanin, B.M., Marjanović, N.J. & Suturović, Z.J. (2002). Application of potentiometric stripping analysis with constant inverse curent in the analitic step for determining lead in glassware. *Journal of the Serbian Chemical Society,* Vol. 67, pp. 213-220, ISSN 0352-5139

Kaličanin, B.M., Marjanović, N.J. & Suturović, Z.J. (2001a). Determination of the soluble lead in the glassware by the potentiometric stripping analysis. *Acta Periodica Technologica,* Vol. 32, pp. 61-70 ISSN 1450-7188

Kaličanin, B.M., Marjanović, N.J. & Suturović, Z.J. (2001b). The development of a high sensitivity method for the electrochemical determination of soluble lead in glassware. *Hemijska industrija,* Vol. 55, pp. 407-410, ISSN 0367-598X

Kaličanin, B.M. & Nikolić, R. (2010). Copper release from dental prosthetic crowns, dental materials, and human teeth into acetic acid. *Connective Tissue Research,* Vol. 51, pp. 31-35, ISSN 0300-8207

Kaličanin, B.M. & Nikolić, R. (2008). Potentiometric stripping analysis of zinc and copper in human teeth and dental materials. *Journal of Trace Elements in Medicine and Biology,* Vol. 22, pp. 93-99, ISSN 0946-672X

Kaličanin, B.M., Nikolić, R.S. & Marjanović, N.J. (2004). Application of potentiometric stripping analysis with constant inverse current in the analytic step for determining soluble lead in human teeth. *Analytica Chimica Acta,* Vol. 525, pp. 111-114, ISSN 0003-2670

Kaličanin, B.M., Nikolić, R.S. & Nikolić, G. (2003). The analysis of migration property of lead from primary matrix tooth and metal ceramic crowns in less acid medium, *Acta Stomatologica NAISSI,* Vol. 44, pp. 151-158, ISSN 0352-5252

Kaličanin, B.M., Nikolić, G. & Nikolić, R. (2005). Metal ion (Cu, Zn) migration in the agrochemicals-soil-food-man chain, *Ecologica,* Vol. 12, pp. 297-300 ISSN 0354-3285

Kaličanin, B.M.; Todorović, Z.B., Marjanović, N.J. & Suturović, Z.J. (2001c). Potentiometric stripping analysis of soluble lead in the glassware used for needs of food and pharmaceutical industry. *Research Journal of Chemistry and Environment,* Vol. 5, pp. 7-12, ISSN 0972-0626

Kaličanin, B.; Velimirović, D. & Ajduković, Z. (2008). Ispitivanje uticaja Cd i Pb na efikasnost regeneracije kostiju pacova sa implantiranim CP/PLGA kompozitom, *Tehnika, Novi materijali,* Vol. 17, pp. 3-8, ISSN 0040-2176

Konvalina, J.; Khaled, E. & Vytras, K. (2000). Carbon paste electrode as a support for mercury film in potentiometric stripping determination of heavy metals. *Collection of Czeshosovak Chemical Communications,* Vol. 65, pp. 1047-1054, ISSN 0010-0765

Labar, Ch. & Lamberts, L. (1994). Determination of metals in animal tissues by potentiometric stripping analysis without chemical destruction of organic matter. *Electrochimica Acta,* Vol. 39, pp. 317-325, ISSN 0013-4686

McKenzie, H.A. (1988). *Quantitative Trace Analysis of Biological Materials: Principles and Methods for Determination of Trace Elements and Trace Amounts of Some Macro Elemements,* pp. 567-785, Elsevier Science Ltd, ISBN 0444809589

Mendil, D., Celik, F., Tuzen, M. & Soylak, M. (2009). Assessment of trace metal levels in some moss and lichen samples collected from near the motorway in Turkey, *Journal of Hazardous Materials,* Vol. 166, pp. 1344-1350, ISSN 0304-3894

Munoz, E. & Palmero, S. (2006). Determination of heavy metals in honey by potentiometric stripping analysis and using a continuous flow methodology. *Food Chemistry,* Vol. 94, pp. 478–483, ISSN 0308-8146

Munoz, E. & Palmero, S. (2004). Determination of heavy metals in milk by potentiometric stripping analysis using a home-made flow cell. *Food Control,* Vol. 15, pp. 635–641, ISSN0956-7135

Nikolić, R., Kaličanin, B. & Nikolić, G. (2004). The determination of soluble lead and soluble cadmium in dental prosthetic materials and teeth by potentiometric stripping analysis. *Journal of the Serbian Chemical Society,* Vol. 69, pp. 575-580, ISSN 0352-5139

Nikolić, R., Kaličanin, B. & Marjanović, N.J. (2001). Potentiometric stripping analysis of the soluble lead released from dental ceramic materials. *Facta Universitatis, series: Physics, Chemistry and Technology,* Vol. 2, pp. 159-163, ISSN 0354-4656

O'Brien, W.J. (2002). *Dental Materials and Their Selection (Third Edition),* pp. 8-76, Quintessence Pub Co, ISBN 0867154063, New York

Oehme, M.,; Lund, W. & Jonsen, J. (1978). The determination of copper, cadmium and zinc in human teeth by anodic stripping voltammetry. *Analytica Chimica Acta,* Vol. 100, pp. 389-398, ISSN 0003-2670

Pocock, S.; Smith, M & Baghurst, P. (1994). Environmental lead and children's intelligence: A systematic review of the epidemiological evidence. *British Medical Journal,* Vol. 309, pp. 1189–1196, ISSN 0959-535X

Premović, P.I., Nikolić, N.D., Pavlović, M.S., Todorović, B.Ž., Đorđević, D.M., Nikolić, R.S., Lopez, L. & Lomonaco, S. (2001). The Cretaceous-Tertiary boundary Fiskeler at Stevns Klint, Denmark: the geochemistry of the major trace metals. *Journal of the Serbian Chemical Society,* Vol. 66, pp. 647-670, ISSN 0352-5139

Suturović, Z. (2003). *Elektrohemijska striping analiza,* pp. 1-120, Tehnološki fakultet, ISBN 86-80995-40-1, Novi Sad

Tyszczuk, K.; Korolczuk, M. & Grabarczyk, M. (2006). Application of gallium film electrode for elimination of copper interference in anodic stripping voltammetry of zinc. *Talanta,* Vol. 71, pp. 2098-2101, ISSN 0039-9140

Uauy, R., Oliveras, M. & Gonzales, M. (1998). Essentiality of copper in humans. *American Journal of Clinical Nutrition,* Vol. 67, pp. 952S-959S, ISSN 0002-9165

Vig, E.K. & Hu, H. (2000). Lead toxicity in older adults. *Journal of the American Geriatrics Society,* Vol. 48, pp. 1501-1506, ISSN 0002-8614

Vydra, F., Stulik, K. & Julakova, E. (1976). *Electrochemical Stripping Analysis,* pp. 1-260, Ellis Horwood Ltd, ISBN 085312017X

Wang, J. (1985). *Stripping Analysis,* pp. 1-147, VCH Publishers, ISBN: 0895731436, Florida

Walsh, C. T., Sandstead, H.H., Prasad, A.S., Newberne, P.M. & Fraker, P.J. (1994). Zinc: Health effects and research priorities for the 1990s. *Environmental Health,* Vol. 102, pp. 5-46, ISSN 0091-6765

World Health Organization. Trace elements in human nutrition and helth, Geneva, 1996.

13

Near Infra Red Spectroscopy

Ahmed Badr Eldin
Sigma Pharmaceutical Corp.,
Egypt

1. Introduction

NIR stands for Near Infrared and refers to the region of light immediately adjacent to the visible range, falling between 750 and 3,000 nanometers (nm = nanometers or 1/1000000000 of a meter) in wavelength. Most organic materials have well defined reflectance or transmittance features at these wavelengths. According to the principles of quantum physics, molecules may only assume discrete energy levels. Similar to the vibrating string of a musical instrument, the vibration of a molecule has a fundamental frequency, or wavelength, as well as a series of overtones. For molecules, the fundamental vibrations involve no change in the center of gravity of the molecule. The spectrum shape for any material is the result of these characteristic fundamentals and overtones. Near-infrared spectra are primarily the result of overtones, whereas there are many fundamentals in the mid and far infrared regions. Since the molecular structure of most compounds is very complex, the resulting spectra are actually the result of many overlapping peaks and valleys. Generally speaking, persons performing NIR analysis must then identify and characterize specific features in the spectra by means of statistical methods. Chemometrics software is designed to accomplish this task.

The absorption of NIR radiation by organic molecules is due to overtone and combination bands primarily of O-H, C-H, N-H and C=O groups whose fundamental molecular stretching and bending absorb in the mid-IR region. These overtones are anharmonic, i.e., they do not behave in a simple fashion, making NIR spectra complex and not directly interpretable as in other spectral regions. Below is a graph depicting the prominent absorption bands as they relate to the overtone and combination bands of the fundamental vibrations occurring in the Mid IR region.

To understand the types of measurements possible using NIR light, it is useful to understand several general properties of electromagnetic waves, as well as basics of classical molecular and atomic structure. EM radiation, is in the form of waves, and as such, has all the properties of a wave; including wavelength. Figure 1 graph is a typical wave.

Wavelength is a distance between two points. Wavelength is particularly important to our discussion as it is closely connected to energy. Wavelength and energy are readily convertible from one to the other when speaking of EM waves. *See figure 2 below*

They are related in the following manner

$$E = \frac{hc}{\lambda} \tag{1}$$

E = energy, *h* = *Planck's* constant (6.626 x 10-27), *c* = *speed* of light (2.998 x 1010dm/s), and l = wavelength.

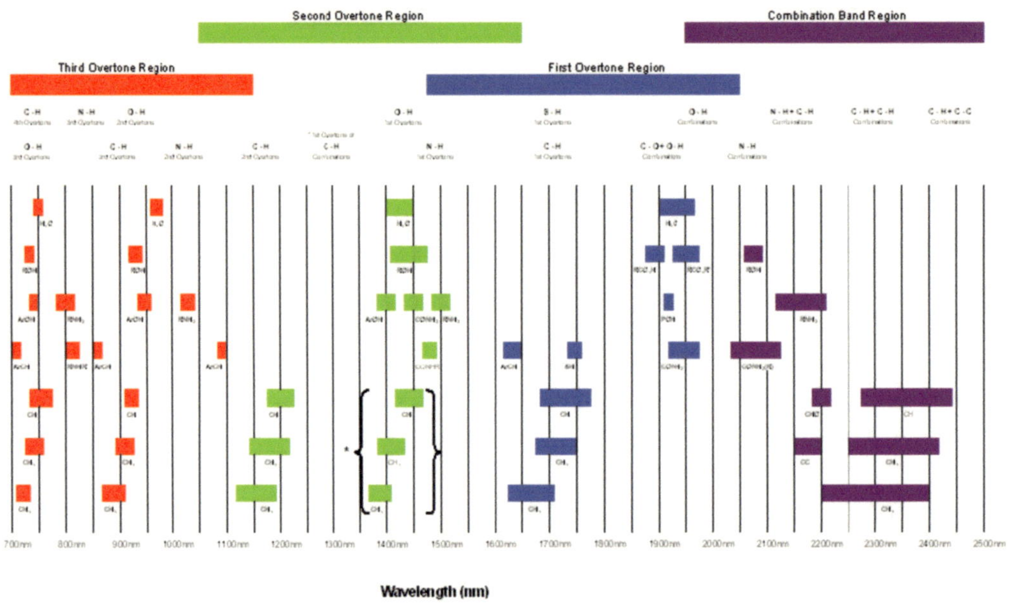

Fig. 1. Graph of near-infrared overtone absorptions, peaks and positions

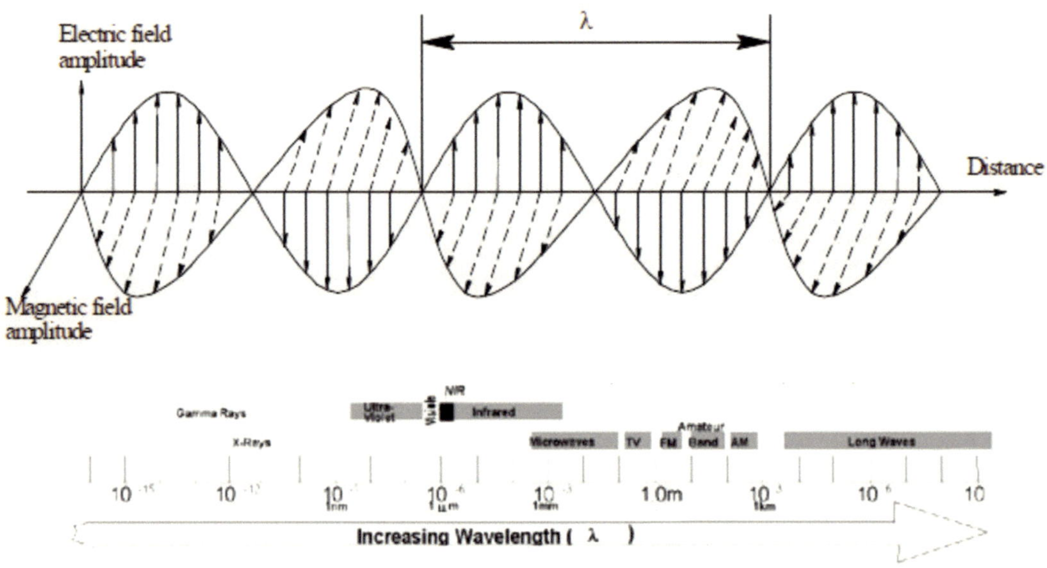

Fig. 2. Electromagnetic spectrum

It is the energy or wavelength that gives a wave its particular properties, and it is the amount of energy an EM wave carries (its wavelength) that determines whether or not a wave (radiation) is harmful. Waves with different wavelengths (energies) act differently. Wavelengths with certain energies will produce the effects associated with an x-ray to microwaves. The general properties of waves of certain energies allow us to classify them across the full EM spectrum. Another property of light is the manner in which energy is transferred from itself to whatever it may encounter. Light, as well as being a wave, consists of photons. Photons have properties of both waves and particles. For this discussion, we will think of photons as the "carriers" and "transferers" of energy. Now that we have discussed light and its properties, it is appropriate to talk about matter. Matter is defined as anything which has mass and takes up space. Matter (pen, paper, ink) is made up of atoms. Atoms are made up of smaller constituents known as neutrons, protons, and electrons. Protons are charged electrically positive, neutrons have no charge, and electrons are negatively charged. This means that protons and electrons are attracted to one another in a similar manner as are magnets of differing polarities. This also means that protons are repelled by other protons, and electrons are repelled by other electrons. These small particles can be arranged in many different ways. The simplest model is shown in *Figure 3*.

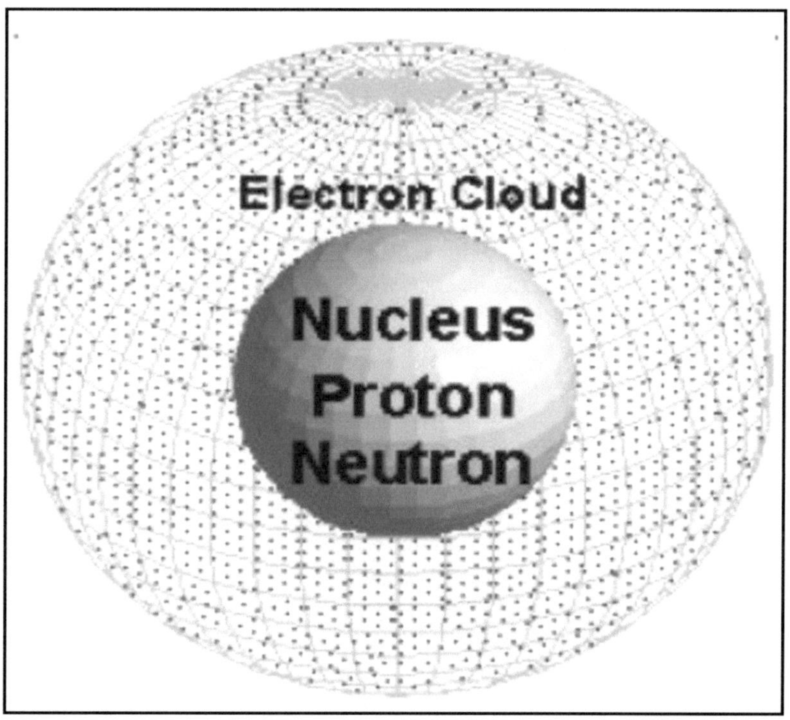

Fig. 3. Hydrogen atom

The center area, where the neutrons and protons are located, is referred to as the nucleus. Around the nucleus is the space in which the electrons reside and is knows as an orbital. Orbitals are distinct areas where an electron can exist. Orbitals also have distinct energies with which they are associated. Continuing addition of protons, neutrons, and electrons would produce atoms in the numeric sequence listed in the periodic chart of the elements.

Molecules are a group of atoms which have combined together to form a chemical compound. Molecules are simply substances made of several atoms of similar or different elements. Chemicals made of different types of atoms may have completely different properties than the properties exhibited by the individual atoms of which they are made. The interactions of protons and electrons help to hold the molecules together by producing bonds between the different atoms. Different arrangements of different numbers and kinds of atoms produce different properties and characteristics.
With NIR we will only deal with organic molecules (generally water, H_2O, is an exception). This will limit the types of molecules we will observe with NIR, since organic molecules are classified as molecules that contain carbon. Every living thing on earth is made up of thousands upon thousands of different organic molecules. Generally speaking, the interactions of EM waves with matter will simply involve the transfer of energy. The type of interaction we will observe and use is absorption of EM radiation by molecules. Actually, only a small portion of the molecule is involved in the absorption process—the electrons. As stated before, we know electrons exist in orbitals around the nuclei of atoms. Orbitals are also energy levels and if the electron is orbiting about, at a particular distance from the nucleus and with a particular speed, it will have a particular energy. Because of quantum mechanics, scientists now know that electrons can exist only in specified energy states; in other words, specific orbitals. Electrons cannot exist in between energy states (orbitals). This means electrons can only absorb discrete amounts (packages) of energy as the next orbital is a specific amount of energy away. *Figures 4a* and *4b* illustrate the process of light being absorbed by an electron.
The photon is absorbed by an electron causing the electron to jump up to a higher energy level. Electrons in differing original orbits will absorb different amounts of energy. Remember that energy and wavelength are closely related (see *Equation 1*) so if electrons absorb differing energies, this also translates into different wavelengths.
Molecules' atoms are built of electrons, protons, and neutrons in different configurations. Similarly, the electrons, protons, and neutrons in water have different characteristics than those in protein. This also means varying substances absorb different wavelengths of light. This type of absorption is considered an *electronic* absorption. The absorptions in the NIR are slightly more complicated though they still involve the absorption of energy (light) by electrons. Remember molecules consist of atoms bonded together. Bonds are produced by atoms sharing and/or giving up electrons to another atom. These bonds actually act similar to little springs (see *Figure 4c*). As an electron moves about the atom(s), the bonded atom is drawn or repulsed from the atom to which it is bonded, creating a vibrating motion. Whenever something moves consistently (vibrates) in time in this manner, it is said to have a frequency (*n*=frequency). The frequency is the number of times the atom vibrates in a second. The absorptions occurring in the NIR region will therefore be considered *vibrational* absorptions. These possible absorptions are also quantum mechanical in nature; only discrete energy amounts can be absorbed. These levels can be roughly calculated using *Equation 2*

$$En = \left(n + \frac{1}{2}\right)\frac{h}{2\pi}\sqrt{\frac{k}{\mu}} \tag{2}$$

Where En = the molecule vibrational energy, n = (0,1,2,3 ...), h = Plank's constant, k = the force = the reduced mass.

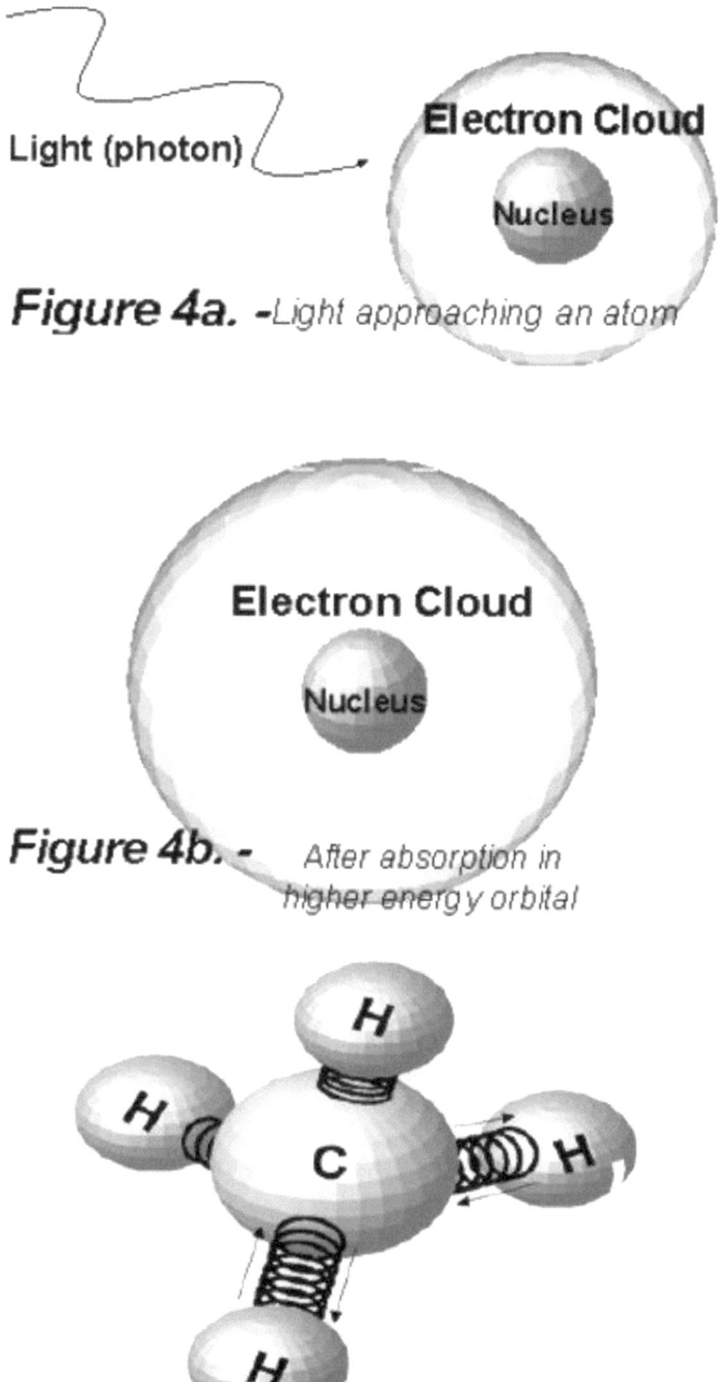

Fig. 4. (a) Light approaching an atom. (b) After absorption in higher energy orbital
(c) Vibrating methan molecul

N is considered a quantum number and can be constant and take on only whole integer values. A transition where *n=1* is known as a fundamental absorption. These fundamental absorptions are about 100 times less energetic in the NIR region and less energetic means longer wavelength. When *n* is greater than 1, the transition is known as an overtone. By looking at *Equation 2*, it is evident that as *n* increases, the energy to be absorbed also increases. This in turn indicates that shorter wavelengths will need to be absorbed. These absorptions generally occur in the NIR region. *Equation 2* predicts fairly well the absorptions of two atoms bonded together (called diatomic molecules), but does not take into account all of the surrounding effects for polyatomic (many atom) molecules, such as overlapping absorption bands or hydrogen bonding. Organic molecules exist in energy states that absorb NIR wavelengths (energies). Metals, such as silver, lead, and most inorganics, cannot absorb NIR light because they have electrons incapable of absorbing NIR wavelengths, therefore there is no interaction to measure. Generally, only organic molecules can absorb wavelengths in the NIR region. It is actually the energy state of a molecule which allows us to perform a measurement with NIR.

Now imagine a sample made up of many, many electrons, protons and neutrons. These particles are arranged into atoms, and further into molecules. The sample can be made of different types of molecules, meaning there can be water molecules, protein molecules and so on. When they take on these arrangements, they also take on different properties such as the ability to absorb different wavelengths of light, therefore, quite a few different energies might be absorbed. When a measurement is performed on this sample, what the instrument is measuring is the number of photons which undergo the absorption process for a particular wavelength. The number of photons absorbed is proportional to the amount of particular type of molecule present in the sample. This statement is more or less *Beer's Law* which states that "absorption is proportional to concentration." In principle, that is what is occurring and is the basis for an NIR measurement.

Bouguer-Lambert-Beer Law (BLB Law, 'Beer's Law)[1]

$$\log(1 / Transmittance_{\lambda}) = \alpha_{\lambda} l c$$

where α_{λ} is the molar absorption coefficient, l is the path length, and c is the analyte concentration. This equation is called the BLB Law and the quantity $\log(1 / Transmittance_{\lambda})$ is called 'absorbance'. Absorbance is a unitless quantity, however, the term absorbance units (AU) is often used to indicate this type of measurement. BLB is valid only for transmittance measurements and much has been written on the mathematics and physics of this law. There is no rigorous derivation of a similar law that relates reflectance to analyte concentration (see *Log(1/Reflectance)*). Absorbance cannot be measured directly since there is no way to directly count the number of photons as they disappear one-by-one. Therefore, what is being measured is actually transmittance.

1.1 Chemometric models[2]

The single step in NIR analysis requiring the most planning preparation is the assembly of the samples, often called the training set to be used for the development of calibrations. A crucial step in achieving success is ensuring the samples have been analyzed as accurately and precisely as conventional techniques allow. These analyses are termed reference analyses. In order for any NIR analyzer to make quantitative measurements or qualitative discriminations, the controlling computer must have access to one or more chemometrics models which represent the type of material being tested. The model is a mathematical construct developed using samples of the same product or class of products. The controlling computer applies the model(s) to the target spectrum and returns a model result. A chemometrics model is developed by collecting spectral readings from a group of samples that display (a) the maximum variability of the characteristic of interest, and (b) non-correlating or random variability in all other characteristics. The same samples are submitted for independent testing to measure the characteristic of interest by a standard analytical method. The spectral data and independent test data are then analyzed using commercially available chemometrics software. The statistical processes used in quantitative spectral analysis include multiple linear regression, classical least squares, inverse least squares, and principal component regression. The statistical processes used in qualitative spectral analysis include K-nearest neighbors, SIMCA and others.

When a sufficient number of samples have been collected and properly analyzed, a mathematical model is constructed that describes the relationship between specific spectral features and the sample characteristic of interest. Thereafter, a chemist or technician may quickly measure that same characteristic in a new target sample by applying the chemometrics model to the spectrum of the target sample. Essentially a calibration is interpreting the information coming from the instrument. If the instrument is taught (calibrated) properly, it will predict the correct amount of parameter in our sample. Once calibrations are obtained, they are entered into the NIR spectrophotometer. Following the scanning of unknowns, requiring a few seconds per sample, numerous constituents or parameters of interest are simultaneously predicted. In this mode, NIR is a rapid, cost-effective, non-destructive, accurate and efficient analytical method.

1.2 Advantages

The biggest advantage of NIR over Mid-IR and Far-IR is little or no sample preparation, and near real-time analysis. Unlike most conventional analytical methods, NIRS is rapid, non-destructive, does not use chemicals, or generate chemical wastes requiring disposal, simultaneously determines numerous constituents or parameters, and can be transported to nearly any environment, or true portable for field work. NIR instrumentation is simple to operate by non-chemists, and operates without fume hoods, drains, or other installations. NIR is not a stand-alone technology. Its accuracy is dependent upon the accuracy of the reference method used for training, however, the data from the NIR method has better reproducibility than the primary method.

Another advantage of NIR over Mid-IR and Far-IR is 'thermal' noise. All internal electronic components are a source of thermal noise in the Mid-IR and Far-IR. However, internal sources of IR are either insignificant to NIR detectors or can be made insignificant by minor shielding.

With NIR analysis most of the useful features in a spectrum consist of overtones, or combinations of overtones, which are more subtle than the fundamentals found in Mid-IR and Far-IR spectra. However, recent developments in off-the-shelf chemometrics software and powerful PC's have made NIR analysis the practical choice for most applications. Because the absorbances in the NIR region are lower than in neighboring regions and generally obey the Beer/Lambert Law, i.e., absorbance increases linearly with concentration, it is possible to analyze bulk samples without the need for dilution or other elaborate sample preparation. Thus, the results provided by NIR are typically more representative than that provided by other analytical means.

1.3 Disadvantage

NIR is not a stand-alone technology. Separate calibrations are required for each constituent or parameter and a portion of unknown samples must periodically be analyzed by the reference method to ensure that calibrations remain reliable. It may be necessary to update calibrations several times during the initial phases of use to incorporate "outlying" samples, until the calibration is acceptable. Despite the intuitive disadvantage of broad and overlapping absorption bands, sophisticated chemometric techniques can extract meaningful information from the complex NIR spectra.

2. Making an NIR analyzer work for you

2.1 Abstract [3]

In recent years, NIR analysis has steadily grown in popularity because of its ability to quickly provide qualitative and quantitative information on many products, especially raw materials. To determine if NIR spectroscopy is a reasonable alternative to more traditional methods, many factors must be considered. These factors include sample characteristics, experiment configuration, and data analysis.

2.2 Sample consideration

The chemical constituents and physical phenomena of interest should have direct or indirect absorbance in the NIR region. Virtually all organic compounds do, particularly those with functional groups like hydroxyl, carboxyl, amine and carbon-hydrogen. A good reference for researching near infrared spectra is *The Atlas of Near Infrared Spectra*, Bio-Rad Sadtler Division, Philadelphia, Pennsylvania. For calibration samples, the amount of analyte in the sample set should be above the detection limit and have sufficient variability. Some analytes, e.g. water, are detectable at the ppm level. For most analytes, the nominal detection limit is 1% or above. The analytical chemist must have an accurate independent method for measurement of the properties and must know the level of error in the reference methods. Errors in NIR prediction most often arise from errors in the reference methods, instability of the NIR spectrometer, and/or inappropriate choice of the calibration model method. The samples used in the development of calibration sets must be representative. All the variations in the future unknown samples should be covered in the "training" calibration sets—for example, sample composition and particle size, homogeneity, and temperature variation at the working environment. As a rule of thumb, the more samples you have for the training set, the more reliable the calibration model.

2.3 Experiment configuration

When using an NIR analyzer, instrument characteristics such as sensitivity, resolution, and signal-to-noise ratio parameters need to be evaluated. The quality of these values is a function of the light source stability, optics throughput, dispersion/filter element accuracy, and detector sensitivity in the instrument. The choice of accessories is application dependent. For liquid samples, transmission and transflectance modes are commonly employed using fiber optic probes or cuvettes. The optimum path length is sample.

dependent, usually ranging from 0.1 to 1 cm. The advantage of using a fiber-optic probe is that sample preparation is significantly reduced, and noninvasive or nondestructive measurements are possible. For solid samples, diffuse-reflectance spectra collected by a reflectance probe will provide information for analytes. Diffuse reflectance should be measured without interference from specular reflectance. The setup configuration, such as the angle of incident light and the distance of light illumination/collection ought to be consistent throughout all the measurements, including those taken in developing the calibration set and for predicting the future unknowns. For solid samples, the sample should be rotated and measurements done on different spots of the sample to average out surface effects and sampling error. A group of spectra may be averaged to increase the signal-to-noise ratio. Random noise is reduced by the factor square root of the number of spectra averaged. For ASD's NIR spectrometer, it takes 0.1 seconds to acquire one spectrum. Therefore, a 10 second measurement reduces the random noise by a factor of 10.

2.4 Data analysis

NIR spectroscopy is an extremely rapid method of measurement, capable of performing an analysis in under a minute. The time-consuming part of NIR work is the data analysis phase, where chemists try to find the correlation between near-infrared spectral characteristics and the property, or properties, of interest as measured by more traditional methods. There are several commercially available software packages for accomplishing this task. Data analysis involves the following steps.

Data preprocessing

When the spectral data plots are presented, first determine if there is any baseline drift or slope in the spectra, which often occurs in diffuse-reflectance measurements. If necessary, baseline subtraction, first derivative and second derivative transformations may be performed to reduce these effects. There is a trade-off though as each successive degree of derivative taken introduces additional noise into the spectral data.

Outlier detection

An outlier is a data point that falls well outside the main population. Outliers result from lab measurement errors, samples from different categories, and instrument error. It is important to check for, and remove, outliers in both the training set and the set of unknowns on which calibration testing will occur (see "validation" and "prediction").

Building a good calibration model

This is one of the most important steps in NIR analysis. Developing a calibration model involves calculating the regression equation based on the NIR spectra and the known analyte information. The model is then used to predict the future unknowns. Multiple Linear Regression (MLR), Principal Component Regression (PCR) and Partial Least Squares (PLS) are commonly used linear calibration methods, along with Locally Weighted Regression (LWR) for nonlinear models. In developing a calibration model, several parameters are evaluated: factors, loadings, and scores. When choosing the number of factors, one should try to avoid under-fitting, i.e. too few factors, and over-fitting, i.e. too many factors. If an insufficient number of factors are chosen, the prediction is not reliable because useful information has been omitted. If too many factors are chosen, however, more uncertainty is included in the calibration set which results in errors in prediction. Scores are used to check the sample homogeneity and possible clusters, while loadings are used to interpret how the variables are weighted in principal component space.

Approximate [4] Wavelengths of Some Common Functional Groups Functional Group	Wave-length (nm)	Functional Group	Wave-length (nm)
C-H second overtone	1143	O-H stretch/C-O stretch second overtone combination	1820
C=O stretch fourth overtone	1160	C-Cl stretch sixth overtone	1860
C-H second overtone	1170	C=O stretch second overtone	1900
C-H second overtone	1195	O-H stretch first overtone	1908
C-H second overtone	1215	C=O stretch second overtone	1920
C-H second overtone	1225	O-H stretch/HOH deformation combination	1930
C-H combination	1360	O-H bend second overtone	1940
C-H combination	1395	C=O stretch second overtone	1950
O-H first overtone	1410	O-H stretch/O-H bend combination	1960
C-H combination	1415	Asym N-H stretch/N-H in-plane bend; C-N stretch combination	1980
C-H combination	1417	N-H stretch/N-H bend combination	1990
O-H first overtone	1420	C=O stretch second overtone	2030
C-H combination	1440	N-H/N-H in-plane bend; C-N stretch or N-H/C-N stretch; N-H in-plane bend or combination	2050
C-H combination	1446	Sym N-H stretch/C=O stretch combination	2060
O-H stretch first overtone	1450	N-H bend second overtone or N-H bend/N-H stretch combination	2060
C=O stretch third overtone	1450	N-H deformation overtone	2070
Sym N-H stretch first overtone	1460	O-H combination	2070
N-H stretch first overtone	1471	C-H combination	2090
N-H stretch first overtone	1483	O-H bend/C-O stretch combination	2100
N-H stretch first overtone	1490	Asym C-O-O stretch third overtone	2100
O-H stretch first overtone	1490	C-H stretch/C=O stretch combination or sym C-H deformation	2140
Sym N-H stretch first overtone	1490	Asym C-H stretch/C-H deformation combination	2170
N-H stretch first overtone	1492	N-H bend second overtone or C-H stretch/C=O stretch combination, or C=O stretch C-N stretch; N-H in-plane bend. Combination	2180
N-H stretch first overtone	1500	C-H stretch/C=O stretch combination	2200
N-H stretch first overtone	1510	O-H stretch/C-O stretch combination	2270
N-H stretch first overtone	1520	C-H stretch/CH2 deformation	2280
N-H stretch first overtone	1530	C-H bend second overtone	2300
O-H stretch first overtone	1540	C-H bend second overtone	2310
N-H stretch first overtone	1570	C-H stretch/CH2 deformation combination	2322
C-H stretch first overtone	1620	C-H stretch/CH2 deformation combination	2330
C-H stretch first overtone	1685	C-H stretch/C-H deformation	2335
C-H stretch first overtone	1695	CH2 bend second overtone	2352
C-H stretch first overtone	1705	C-H stretch/C-C stretch combination	2380
C-H stretch first overtone	1725	C-H combination	2470
S-H stretch first overtone	1740	Sym C-N-C stretch overtone	2470

Table 1.

Validation

The validity of the model must be tested. This is usually done by splitting the sample set into two sets; one set for calibration and the other for validation. If there are not enough samples, "leave-one-out" cross validation can be performed. This means leaving one sample out, using the rest of the samples to build a calibration model and then using the model to predict the one left out. The advantage of doing cross validation is that unlike calibration with a full data set, the sample being predicted is not included in the calibration model. Thus, the model can be tested independently.

2.5 Prediction

Finally, the calibration can be used to predict future unknowns, assuming the unknowns are in the same sample population as those used in the calibration set. Whether the unknown is an outlier needs to be tested.

2.6 Summary

Applying an NIR analyzer to a particular application requires the development of a reliable calibration model. The most important steps involve a thorough consideration of experimental design and multivariate calibration. Once this is established, one can enjoy the advantages of the NIR analysis. The speed of the analysis will save time and avoid mistakes instantaneously. The speed advantage is so valuable to engineers involved with on-line process monitoring that instruments are routinely installed in or near the process line with feedback loops. With an NIR analyzer such as QualitySpec® Pro spectrometer, samples can be non-invasively analyzed on-the-spot, dramatically reducing costly and time consuming laboratory analysis as well as preventing unnecessary product waste and/or downtime. The low absorptivity in the NIR region allows measurements to be taken on raw materials, in process and finshed product without elaborate sample preparation. In the food, agricultural, pharmaceutical, polymer, cosmetics, environmental, textile, and medical fields, NIR analysis serves a wide range of applications, with still many unknown applications waiting to be discovered. With the maturity of this technique, more and more people will use NIR analysis for convenience and flexibility.

2.7 Wavenumber and wavelength

$$Y = 10^7 / X$$

where
Y = the number of nanometers (nm)
X = the number of wavenumbers (cm^{-1})

$$Y = 10^7 / 28571 = 350$$

Resolution in cm^{-1}, ($R_{cm^{-1}}$) is dependent upon wavelength position.
So, Resolution in nanometers, (R_{nm}) is calculated as follows:

$$\begin{aligned} R_{nm} &= +/- [Y - Y'] \\ &= +/- \{[10^7 / X] - [10^7 / X']\} \\ &= +/- 10^7 * \{[1 / X] - [1 / X']\} \\ &= +/- 10^7 * \{[1 / X] - [1 / (X - R_{cm^{-1}})]\} \end{aligned}$$

3. References

[1] Burns D. A. and E. W. Ciurczak (Eds.), Handbook of Near-Infrared Analysis, (Volume 13 in Practical Spectroscopy Series), Marcel Dekker, Inc., New York, 1992

[2] Hildrum K. I., T. Isaksson, T. Naes and A. Tandberg (Eds.), Near Infra-red Spectroscopy, (Ellis Horwood Series in Analytical Chemistry), Ellis Horwood, Ltd., England, 1992

[3] Murray I. and I. A. Cowe (Eds.), Making Light Work: Advances in Near Infrared Spectroscopy, 4th International Conference on Near Infrared Spectroscopy, Aberdeen, Scotland, August 19-23, 1991, Weinheim, New York, Basel, Cambridge, VCH, 1992

[4] Burns D. A. and E. W. Ciurczak (Eds.), Handbook of Near-Infrared Analysis, (Volume 13 in Practical Spectroscopy Series), Marcel Dekker, Inc., New York, 1992, pgs 393-395.

Part 3

Quality Control in Clinics

14

Quality Control in Hospital Bone Banking

Eline W. Zwitser and Barend J. van Royen
Department of Orthopaedic Surgery
VU Medical Center Amsterdam
The Netherlands

1. Introduction

The use of allogenic bone transplantation is nowadays a standard orthopaedic procedure. It is widely used for reconstruction of bone defects that arise from trauma (Friedlaender 1987), infection, resection of bone tumours (Mankin et al. 1996) or it is used in spinal fusion (Raizman et al. 2009) and as impaction grafting in revision of total joint arthroplasty (Slooff et al. 1996). Although autologous bone is generally preferred because of its osteoconductive and osteoinductive activity, autologous bone is often not sufficiently available and comes with donor site morbidity (Summers & Eisenstein 1989). Therefore allogenic bone grafts are often used in orthopaedic procedures. These bone allografts are provided by an orthopaedic bone bank. It might be financially attractive for a hospital to manage its own local hospital bone bank, especially if they perform many procedures in which bone allograft is used. The main advantage of managing a hospital bone bank however, is the easy accessibility to and availability of bone allograft. The bone allografts in a hospital bone bank are femoral heads obtained from suitable patients who underwent total hip replacement surgery. Management of an orthopaedic bone bank is a complex process. The bone bank procedure has to meet the requirements of the national law and European guidelines 2004/23/EC and 2006/86/EC. This law states the technical requirements for coding, processing, preserving, storing, and distributing of human tissue and cells. Human tissue should be traceable and serious side effects and incidents with human tissue and cells should be reported. The bone bank procedure should be carefully described in an extensive protocol. Neither in the Netherlands, nor in any other European country, there are official guidelines for the organization and management of an orthopaedic bone bank. Our bone banking procedure protocol is based on guidelines of The American Association of Tissue Banks (AATB 1993), the criteria of the Council for Blood Transfusion of the Netherlands Red Cross (Richtlijn Bloedtransfusie 2004), the recently merged Netherlands Bone Bank Foundation (NBF) and Bio Implant Services (BIS); (NBF-BIS Foundation 2010) and the guidelines of the European Association of Musculoskeletal Transplantation (EAMST). The latter has been discontinued because of diverging European legislation. This bone bank protocol extensively describes the procedure, which includes a thorough questionnaire for donor selection, extensive serological, bacteriological and histopathological examination, as well as standard procedures for registration, processing, preservation, storage and distribution of bone allografts (Zwitser et al. 2010). In this chapter we describe our local hospital bone banking procedure and protocol. This constantly updated protocol is of the utmost importance in order to prevent the transmission of infectious diseases. Because of the potential risk of transmission of diseases from donor to recipient we performed routine histological examination in the screening protocol. We found a relatively high percentage of pathological conditions in retrieved femoral heads (Zwitser et al. 2009; Sugihara et al. 1999). Therefore, we recommend the routine histopathological evaluation of all femoral heads removed during elective total hip arthroplasty as a tool for quality control. The cost-performance ratio of routine histopathological evaluation is discussed in literature (Kocher et al. 2000; Meding et al. 2000; Lawrence et al. 1999; Campbell et al. 1997). Therefore we performed an intern evaluation of the bone banking process. We compared the costs made to harvest, store and implant one bone allograft of our bone bank and the costs of one allograft obtained from the central bone bank. Furthermore this evaluation brought valuable information of improvements to be made. We describe these conclusions and have suggestions how to further improve the quality and cost effectiveness of the bone banking process in the near future.

2. History of bone transplantation

The history of bone transplantation can be traced back to the seventeenth century. In 1668 Job van Meekeren, a Dutch surgeon was the first to perform bone transplantation (Schweiberer et al. 1990). He repaired a skull defect in a soldier with part of a skull from a dog. As soon as the soldier was informed about the transplant, he requested immediate removal of the dog's skull. This was not possible because the xenograft was already fully incorporated in the man's skull. The first human allograft ever reported describes a case of a bone transplant in a young male who suffered an osteomyelitis of the entire humeral shaft (MacEwen 1881). In 1881, the treatment for osteomyelitis was surgical debridement or resection of the affected bone. After this surgical procedure, it took several years for the infection to extinguish, where after the surgeon could replace the bony defect with fresh allografts from diaphysis of tibial shaft. In the following seven years these allografts slowly but successfully incorporated in the recipient humeral shaft (MacEwen 1909). In the following decades the technique of transplantation of large allografts in septic arthritis and osteomyelitis was further developed and popularised by a German surgeon. He used fresh long bones of amputated limbs and used them as osteoarticular allografts with a reported success rate of 50% (Lexer 1908, 1925). Later, in 1929 Alexander Fleming discovered the antibacterial properties of penicillium and treatment modalities of osteomyelitis changed (Fleming 1929). In the subsequent years antibiotics were further developed and introduced for clinical medical use in the 1950's. From now on, the treatment of choice for osteomyelitis consisted of appliance of antibiotics in stead of surgery.

Transplantation of large allografts was applied as a limb-saving treatment in high grade malignant bone tumours of the lower extremity (Parrish 1973, Mankin et al. 1976). In a series of 19 allograft replacements for osseous malignancies satisfactory results were reported in 75% of patients. In the largest series of two hundred lower extremity osteoarticular allografts performed between 1976 and 1997 for malignant bone tumours, results were diminished by radiation and chemotherapy (Hazan et al. 2001).

In the first century of bone transplantation the greatest impediment to the use of allografts was availability of fresh bone grafts, because there were no means for preservation.

Only fresh amputated limbs could be used as a donor allograft. For this reason, autografts were used much more frequently than allografts. In the 1940's storage methods were developed for preservation longer than a few days or hours by refrigeration or freezing.

In 1949 in Bethesda, Maryland the first United States Navy Tissue Bank was established, because of a military need for bone allografts (Hyatt 1950). In that Tissue Bank allografts were obtained from the nearby National Naval Medical Centre. In addition, they developed a method for freeze-drying of allografts by lyophilisation, which made it possible to store and preserve allografts for several years without the need for refrigeration or freezing (Kreuz et al. 1951). In the first century of bone transplantation disease transmission was not of great concern. In addition, serological tests for transmittable diseases other than syphilis were not available. The first case ever of the transmission of viral disease by frozen bone was reported by Shutkin in 1954. The donor had undergone an above the knee amputation. The allograft bone was cut into portions under aseptic conditions, placed in double sterile containers and frozen at a temperature of -10 to -20 °C. Five months later, the bone was implanted into a medical student and transmitted hepatitis B. In the early 1980's the first publications concerning a new disease AIDS were published (Centers for Disease Control [CDC] 1981, 1982). Only a few years later in 1984 the first transmission of disease by bone allograft was reported (CDC 1988). A serological test was not yet available at the time of transmission. Even with the first serological tests used for screening purposes another transmission occurred in 1985, due to a very recent donor infection in the so-called "window period" of testing, with a less accurate test. Therefore, in an expert conference guidelines and recommendations were developed (CDC 1988). The most important conclusion drawn was that the disease was transmitted by blood and bone marrow containing allografts. Recommendations concerned donor screening, testing and re-testing of living donors after 6 months. The constant update for screening of donors for infectious diseases proved to be important in the subsequent years with two reports on transmission of hepatitis C virus by bone allografts (Eggen & Nordbø 1992; Conrad et al 1995). In the following years tissue banks developed better techniques for processing and preparation of bone allografts and more reliable blood tests came to market (Busch 1991, 1994; Alter et al. 1990). However, donor screening methods are constantly updated and revised with the introduction of new infectious diseases, like SARS (Lam et al. 2004). The safety of allograft bone transplants can never be taken for granted, but recent safety records for bone allografts are excellent.

As more complex orthopaedic surgical procedures are performed nowadays the need for (safe) bone allografts has increased (Nielsen et al. 2001).

3. Indications for the use of bone allograft

Massive bone defects can arise from trauma, infection, osteolysis after arthroplasty or resection of bone tumours and are a challenging problem in orthopaedic practice. These bone defects can be filled with either autograft or allograft bone transplants. Ideally, autograft is preferred because of its osteoconductive and osteoinductive activity. However, autografts are available in limited number and size and therefore not sufficiently available. In addition harvesting is associated with extended surgical time and involves donor site morbidity (Aro&Aho 1993; Summers and Eisenstein 1989). Therefore allografts supplied by a bone bank are commonly used instead. Allogenic bone exclusively has osteoconductive activity; it serves as an acellular mineralized frame against which newly formed bone gets deposited (Elves and Pratt 1975; Urist 1953). Indications for the use of allografts are wide and include treatment of bone defects as a result from trauma, tumour surgery and infection (Friedlaender 1987; Mankin et al. 1976, 1996; Finkemeier 2002; Jupiter et al. 1987). In addition it is used in spinal fusion (Raizman et al. 2009; Takaso et al. 2011). However the most applied indication is the use of morselized allograft femoral heads as impaction grafting in revision or primary total joint arthroplasty. Aseptic loosening of the acetabular or femoral component of a total hip prosthesis is becoming an increasingly significant problem in orthopaedic surgery (Slooff et al. 1996). The migration of implants during loosening and procedures to remove the prosthesis and cement during revision induce significant bone destruction, resulting in enlargement of the acetabulum and widening of the femoral medullary cavity. Because of the magnitude of the loss of bone, allograft bone is often needed, at revision, to provide stability for the new socket or stem. Using a technique for revision of the acetabular component with impacted morselized cancellous bone grafting and a cemented or uncemented acetabular component, excellent long term results were described, with a survival of the revised hip prostheses of 87% at 20 years. (Schreurs et al. 2009; Garcia-Cimbrelo 2010; Paxton 2011). In different subgroups of patients this technique was applied with good results: patients under the age of 50, with rheumatoid arthritis, with acetabular fracture and dysplastic hips. In addition femoral revision with use of an impaction bone-grafting technique and a cemented polished stem resulted in an excellent prosthetic survival rate at eight to thirteen years postoperatively. However several studies report on high complication rates on this procedure, mostly femoral fractures and authors refined their indications (Leopold et al. 1999; Meding et al. 1997; Toms et al. 2004; Sierra et al. 2008). Because of the use of femoral head allograft bone in revision hip surgery, which is a common orthopaedic operation, the need for these allografts is increased. The retrieval of these femoral heads in daily orthopaedic practice is simple: the procedure for implantation of a total hip always requires removal of the femoral head. However the bone banking procedure for safe donor selection, retrieval, documentation and storage is a complex process. Therefore in many countries central bone banks were founded, containing bone- and tendon allografts of various sizes and origin, obtained from living and deceased donor patients. These central bone banks also provide femoral head allografts to other hospitals, which can be ordered at a cost price in advance of a planned operation.

Hospitals that perform a high quantity of orthopaedic procedures that require the use of bone allografts (total hip revisions, spine surgery and tumour surgery) might consider foundation of a hospital owned bone bank. In the Netherlands these hospital bone banks contain only allografts of femoral heads of living donors removed at the time of hip replacement surgery. Main advantages for the hospital and orthopaedic surgeons consist of easy accessibility to - and availability of allografts. In addition there might be a financial advantage, depending on the quantity of allografts needed per year. However the break-even point is not clear and dependant on local costs which differ between hospitals, countries and bone banking protocols. In the next paragraph a bone banking protocol is described which meets the requirements of Dutch national law and European guidelines 2004/23/EC and 2006/86/EC and is based on several guidelines: The American Association of Tissue Banks (AATB 1993), the criteria of the Council for Blood Transfusion of the Netherlands Red Cross (Richtlijn Bloedtransfusie 2004), the central bone bank in the Netherlands NBF-BIS (NBF-BIS Foundation 2010) and the guidelines of the European Association of Musculoskeletal Transplantation (EAMST), which has been discontinued because of diverging European legislation.

4. Hospital bone bank protocol

A bone bank procedure should be extensively described in a protocol concerning the five components: organization, donor selection, documentation, storage and processing, and implementation. The Head of Department (HOD). of the Department of Orthopaedics and the bone bank administrator compose this protocol.

4.1 Organization

In an organization chart we describe the responsibilities of different stakeholders. One of the orthopaedic departmental members, preferably the HOD of Orthopaedic surgery, has general responsibility for the bone bank. The HOD should have appropriate technical support by a bone bank administrator, who is responsible for the daily management of the bank. This bone bank administrator can be a paramedic with appropriate training for the required tasks. The bone bank administrators' responsibilities include administration as well as storage and allocation of donor bone. Additionally, the administrator takes care of the maintenance and cleaning of the storage facilities (freezers, etc.), and verifies the registration forms of femoral heads meeting the requirements for storage in the bone bank. Furthermore the team consists of a theatre nurse, a medical microbiologist, an anatomic pathologist, a clinical chemical analyst, a haematological laboratory technician, and a trainer. The knowledge and skills concerning surgical techniques and clinical hygiene are guaranteed by the orthopaedic surgeon and theatre nurse. The bone bank administrator and the trainer are responsible for training of bone bank employees. Apart from an orientation module for new employees, the training program consists of regular refresher courses for all members of the staff, in order to keep the knowledge of the procedures updated.

4.2 Donor selection

Preceding the hip replacement procedure, the attending orthopaedic surgeon requests the patient for his permission to store any removed tissue for donation. It concerns patients whose femoral head grafts will be retrieved in order to be replaced by a total hip prosthesis. The quantity of corticospongious bone removed during knee or shoulder arthroplasty is not sufficient for donor purposes; therefore patients undergoing such procedures cannot be taken into consideration for donation of bone tissue.

The attending orthopaedic surgeon informs the patient both orally and written. In case the patient grants permission he or she signs the consent forms, and fills out a standard survey (see Table 1). The orthopedic surgeon now decides whether the patient is suitable for being a donor; he uses general and specific exclusion criteria (see Tables 2, 3).

All criteria must be met; if not, exclusion necessarily follows. The orthopedic surgeon examines the patient thoroughly: blood samples are collected to determine blood type, Rhesus-factor and Erythrocyte Sedimentation Rate (ESR). (Tables 4, 5).

During surgery, bacterial culture swab samples from the hip capsule are collected and a biopsy of 1 cm^3 corticospongious bone is sent for histopathological analysis. Serological screening for infectious diseases is performed six months after surgery. Once all requirements are met (Tables 1, 2, 3, 4, 5), a femoral head can be released for donation:

- approval donor
- signed consent forms of donor
- completed survey; all questions should yield a negative answer
- preoperative ESR rates within criteria
- no abnormal bacteriological values in derived tissue

- no abnormal histopathological structures in derived tissue
- no abnormal serological values six months after surgery

4.3 Documentation

Accurate documentation and coding are of the utmost importance for a well functioning bone bank. A unique registration code is allocated to each femoral head. Only the bone bank administrator is able to trace the donor based on this code. Of every registered femoral head, a file, containing the consent forms and results of ESR, bacteriological and histopathological examination, is kept updated. Other relevant data, such as the size of the femoral head and the allocation date are also documented and stored in this file. When the file is completed (which takes at least 6 months due to the serological examination), and no abnormalities are recorded, both bone bank administrator and the responsible orthopedic surgeon sign the forms. The femoral head is now available for transplantation. In case a file cannot be completed in full, or any abnormal values are recorded, the femoral head will be destroyed according to hospitals' protocol.

In the past 3 months, did you suffer any infection? If so, what infection?
In the past 3 months, did you have any vaccination or inoculation, or have you been injected with narcotic drugs?
In the past 6 months, did you have a malaria attack or did you use anti-malarial medication?
Have you ever been infected with a sexually transmitted disease?
Have you ever been diagnosed with jaundice or liver illness?
In the past 6 months, have you been in contact with patients diagnosed with jaundice/hepatitis?
In the past 6 months, have you been in contact with patients diagnosed with AIDS, or individuals at risk to AIDS? If yes, how and when?
Have you ever been tested for HIV/AIDS?
Have you had homosexual intercourse after 1977? (Males only).
Have you emigrated after 1977? If so, to what country?
Are you diagnosed with hemophilia? If yes, are you using anticoagulants?
Are you a sexual partner of an individual for which any of the abovementioned questions can be answered with 'yes'?
Have you been actively involved in prostitution after 1977, or have you been a sexual partner of a person involved in prostitution in the past 6 months?
Have you ever been diagnosed with a hematological disease or any malignant disorder?
Have you ever been treated for diabetes mellitus?
Have you ever been treated for chronic brain- or neurological diseases?
Have you ever received radiation therapy?
Have you ever been diagnosed with rheumatoid arthritis?
Have you ever been diagnosed with tuberculosis?
Have you ever been diagnosed with any disease, other than the abovementioned?
Have you ever received hormonal treatment?
Do you use any prescribed medication?
Have you ever used any narcotic drugs?
Have you recently been exposed to hazardous or toxic materials? If yes, please specify.
What is your alcohol consumption per week?
Have you recently been in surgery? If so, when? Did you receive blood from a blood transfusion?
In the past 14 days, have you been traveling through or staying in a region exposed to a SARS epidemic, or have you been in contact with patients infected with SARS?
In the past 6 months, have you tattooed yourself or did you get a piercing?

Table 1. Questionnaire patient for orthopaedic bone donation

4.4 Storage and processing

Retrieval of the femoral head is performed at the time of routine orthopaedic total hip replacement under aseptic conditions. The removed femoral head is inspected and capsule and synovial tissue are cultured on aerobic and anaerobic bacteria. In order to exclude malignancies, auto-immune processes, or infections, a biopsy of 1 cm3 corticospongious bone and capsule is collected for histopathological examination. After determining its size, the femoral head is wrapped in three layers of sterile packing material, labeled and stored in the freezer within 30 min. The freezer has a temperature of -80°C, and has a continuous temperature registration device installed. Should the temperature fall outside the acceptable range of -90 and -70°C, an alarm system gives off a warning signal to the Technical Service, guaranteeing a 24- hours security against temperature-induced damage to the tissue. A nitrogen tank is fitted onto the freezer, as a backup cooling mechanism in case of mechanical breakdown of the freezer. In deep frozen condition, the allogenic bone tissue can be preserved for a maximum of 5 years. The temperature data is stored and managed by the bone bank administrator for a period of at least 5 years.

No permission from patient
Under aged donor (<18 years).
Active or recent systemic infection/sepsis
Active infection of transplantation tissue (especially coxitis/osteomyelitis).
Previously infected with tuberculosis
Active "slow-virus" infection or anamnesis in the past
Anamnesis of previous infection with hepatitis B or C, AIDS or AIDS related complex, or tested positive for HIV
Active or past syphilis infection
Recent (<4 weeks). vaccination with live vaccine (measles, yellow fever, mumps, polio, oral typhoid, rubella).
Rheumatoid arthritis
Diffuse connective tissue disorders/autoimmune diseases
Metabolic disorders
Existing insulin dependent diabetes mellitus
Treatment with growth hormones
Chronic medication (especially corticosteroids).
Recent exposure to toxic substances
Malignancies
Donor location has been exposed to radiation
Chronic neurological disorders
Dementia
Language barrier or when patient does not understand the information for any reason (e.g. psychiatric patients).

Table 2. General exclusion criteria

4.5 Allocation and implementation

If during surgery an orthopaedic surgeon decides to use a femoral head as an allograft, a femoral head from the freezer together with its documents are handed over to the orthopedic surgeon and surgery team. The orthopedic surgeon and theatre nurse verify the file and expiration date of the femoral head. The femoral head is thawed in physiological saline; after being defrosted the theatre nurse takes a bacterial culture swab.

The hospital or care institution warrants fulfillment of the traceability requirements, which implies storing the file of the femoral head and records of the receiving patient for 30 years post implantation.

A clinically proven HIV infection
Men having homosexual intercourse after 1977
Intravenous medication/narcotics use, currently or in the past
Immigrants (after 1977). from countries of which it is known that heterosexual intercourse is an important factor for HIV transmission
Hemophilic patients administered with clotting factors concentrate
Sexual partners from abovementioned individuals
Men and women active in prostitution since 1977, and individuals being their partner in the past 6 months
Individuals who recently (past 6 months). placed a tattoo or piercing
Individuals who have had a blood transfusion before 1980
Individuals who have had a blood transfusion outside Europe or North America
Individuals who stayed in a SARS epidemic area or individuals who had face-to-face contact with a SARS patient

Table 3. Specific exclusion criteria

Blood type and Rhesus factor
Erythrocyte sedimentation rate (ESR), age and sex dependant
Normal values
Male
<50 years: 0–15
>50 years: 0–20
Female
<50 years: 0–20
>50 years: 0–30

Table 4. Hematological examination before surgery

Viral (hepatitis).
Hepatitis B antigen
Hepatitis B antibody
Hepatitis C antibody
Viral nucleic acid
Viral (additional).
HIV
HTLV
Bacterial syphilis
TPHA (Treponema Pallidum, Lues).

Table 5. Serological examination 6 months postoperative

5. Quality control in hospital bone banking

The bone bank protocol contains several procedures to secure the retrieval, storage and implementation of predictable, safe femoral head allografts. To prevent transmission of infectious diseases and malignancies not only thorough donor screening, but also histopathological screening, culture swabs and serological screening are performed as a tool for quality control. In literature these complementary screening methods are discussed.

5.1 Histopathological screening

There is no uniform practice concerning histopathological evaluation of femoral heads retrieved from living donors. Some hospitals routinely examine all removed tissue specimens histopathological, suitable for bone banking or not. At some only gross examination is performed and at some neither evaluation is done. The utility and cost-performance ratio of routine histopathological evaluation is discussed in literature. Several studies doubt the cost effectiveness and necessity of routine histopathological examination (Campbell et al. 1997; Kocher et al. 2000; Lawrence et al. 1999; Meding et al. 2000; Raab et al. 1998). Campbell et al. performed a retrospective review of 715 pathologic findings of specimens obtained at total joint arthroplasties (283 hips and 432 knees). Of 715 specimens 3 were suggestive for low grade lympho-proliferative disorder and 3 for rheumatoid arthritis. In none of these cases this resulted in alteration in patient care. Follow up by telephone at 35 months revealed no signs of systemic disease. Therefore they conclude that routine histo-pathologic evaluation is not cost effective (Campbell et al. 1997). Kocher et al. performed 1234 arthroplasties (471 hips 763 knees) and found discrepant diagnoses in 28 patients, including (pseudo)gout, pigmented villonodular synovitis, osteonecrosis, granulomatous disease, hemochromatosis and rheumatoid arthritis. In two patients they found hypercellular marrow suggestive for lympho- proliferative disease. They performed no formal hematological work-up. They conclude that because of low prevalence of findings that altered patient management, routine pathological examination has limited cost effectiveness (Kocher et al. 1998). Meding et al. drawed similar conclusions. They found 27 discrepancies in pathology in 951 cases, however no neoplasia was noted. Lawrence et al. found malignancies in 11 of 1388 patients, including three cases of lympho- proliferative disease. They stated that the surgeon was able to identify malignancy with a sensitivity of 100 % and a specificity of 99.9%. They conclude that the surgeon preoperatively can exclude the possibility of malignancy and should reserve histopathological analysis for cases where the diagnosis is suspect or in case of unsuspected intra-operative findings (Lawrence et al. 1999). Raab et al. found two malignancies in a group of 168 patients, one of which had findings suggestive for lympho-proliferative disorder, however patient management did not alter. One patient showed osteomyelitis and this had clinical significance for patient treatment. In retrospect this turned out to be a misdiagnosis (Raab et al. 1998). The authors of these studies state that omission of routine pathology would bring about substantial decreases in charges. Campbell et al. estimated that the potential annual cost savings would be between $43 million to $61 million if all hip and knee arthroplasties in the USA were considered. Kocher et al. reported on $122,000 in total costs for them to discover only one unexpected case of granulomatous inflammation. Total reimbursements were only slightly more than $106,000, leaving them with a $16,000 loss for the hospital. Clearly, desisting with routine pathology for presumed osteoarthritis in elective total joint cases could substantially decrease expenditures. The earlier described studies make compelling arguments for omission of routine pathologic examination after elective joint arthroplasty for presumed osteoarthritis. The cost savings they estimate are enormous and patient management did not change because of an unexpected histologic diagnosis.

On the other hand some studies find higher percentages of unexpected pathological diagnoses and therefore advise routine histopathological examination (Billings et al. 2000; Clark et al. 2000; DiCarlo et al. 1994; Lauder et al. 2004; Palmer et al. 1999; Sugihara et al. 1999; Zwitser et al. 2009). DiCarlo et al. documented a 5.4% disagreement between the clinical and pathologic diagnoses of 1794 femoral heads from total joint arthroplasties , with a large-cell lymphoma, myeloma, sarcoma, ochronosis, Gaucher's disease, Paget's disease, and enchondroma. The authors recommended routine pathology for elective arthroplasty to both verify the diagnosis and to serve as a measure of quality control. Billings et al. described the unexpected finding of an occult primary bone sarcoma in two patients with otherwise benign clinical findings and therefore underscore the necessity for routine pathological examination of femoral head specimens from patients who are at risk for the development of a secondary malignant tumor (Billings et al. 2000).

Lauder et al. describe a patient with histopathological findings indicative for a low grade B cell lymphoma who developed systemic disease after 8 months. They underscore the fact that without routine pathologic examination, neoplasms could still be missed, even in patients who lack risk factors for malignancy and despite of a thorough preoperative evaluation. Furthermore they discuss the fact that studies in which low grade malignancies were found, suggested that their patients had been free of signs or symptoms for underlying disease, but did not perform a formal hematological evaluation for malignancy (Kocher et al. 2000; Campbell et al. 1997). They ask what the cost-benefit is when one is able to diagnose an unsuspected occult malignancy (Lauder et al. 2004). Clark et al. conclude that the routine histological evaluation of tissue excised from patients with an uncomplicated case of osteoarthritis may not be necessary at all hospitals, but when a patient has suspection of another disorder then osteoarthritis and when gross examination suggests an unexpected finding, or when the results of such analysis are used for ongoing quality-assurance studies, histological examination is warranted. Other studies performed analysis of histopathological screening of femoral heads for bone banking screening purposes.

Palmer et al. analysed the histological findings in 1146 osteoarthritic femoral heads which would have been considered suitable for bone bank donation to determine presence of pathological lesions and found that 91 femoral heads (8%) showed evidence of disease. The most common benign conditions were chondrocalcinosis (63), avascular necrosis (13), osteomas (6), metabolic bone disease (2) and rheumatoid arthritis (4). Three cases of malignant tumour were described (one case of low-grade chondrosarcoma and two of well-differentiated lymphocytic lymphoma). They conclude that occult pathological conditions are common and recommend that histopathological screening should be included as part of the screening protocol for bone bank collection (Palmer et al. 1999). Sugihara et al. describe similar histological findings in routine bone bank screening of 137 femoral heads and found abnormal histopathological findings in five femoral heads: three were highly suspicious of low-grade B-cell lymphoma, one of monoclonal plasmacytosis and the other of non-specific inflammation of bone marrow (Sugihara et al.,1999). In routine histopathological screening the subsequent years this group found variable numbers of low-grade B-cell lymphoma, even in a group of femoral heads that were eligible for bone transplantation. In a long term follow up of these patients, with serendipitously found low grade B cell lymphoma on routine histologic examination, two developed systemic disease. Therefore we recommend and perform routine histopathological screening as part of the bone banking protocol (Zwitser et al. 2009).

5.2 Culture swabs

In order to prevent transplantation of infected bone allografts we routinely perform culture swabs of the femoral head and synovium. If these culture swabs are positive for bacterial contamination the femoral head is discarded. Because femoral heads are readily available, any suspicion of contamination is respected, regardless of the source of the organism. At the time of implementation another two culture swabs from the thawed femoral head are performed. A culture of a specimen at the time of use of the femoral head serves two purposes: it is a quality-control check on the banking procedure, without the risk of additional contamination by separate culturing and handling, and it also allows the surgeon to administer an appropriate antibiotic should the culture be positive, especially if an infection occurs postoperatively (Tomford et al. 1986). However literature suggests that routine culture swabs are not always able to detect bacterial contamination.
Veen et al (1994) describe analysis of 75 fibular specimens obtained from cadaver donors under sterile conditions. All specimens were culture swabbed as routinely performed for retrieved allografts. Of these allografts 92 % were contaminated when cultured entirely but swab cultures were positive in only 45% After swabbing, all specimens were placed in BHI-culture medium. Three different protocols were subsequently followed: 1) culture of the entire bone specimen in BHI-culture medium, 2) culture of the swab incubated on blood agar and chocolate plates, and 3) culture of the swab in BHI-culture medium. A control group included 20 sterilized bone specimens that were cultured entirely according to Protocol 1. The negative predictive value and sensitivity and were found to be 9% and 10% in Protocol 2 and 13% in Protocol 3. Therefore they conclude that swab cultures are inadequate to detect bacterial contamination of bone allografts in all cases. However, because of an acceptable infection rate after transplantation of the allografts that does not exceed those reported in other similar series, there is suggestion of an acceptable bioburden.
Vehmeyer et al (2002) analyzed the bacterial contamination of 106 allografts of femoral heads obtained from living donors. From 15 initially swab positive grafts only five grafts were contaminated when cultured entirely. From 10 of 91 initially swab culture negative allografts microorganisms could be isolated when cultered in their entirety. They conclude that the routine swab culture technique seems to be less suitable for assessing the bacterial load of femoral heads obtained from living donors. Therefore they advise to routinely perform antibacterial processing before releasing an allograft for transplantation.
Antibiotic rinsing of the allograft seems not to be an effective decontamination method in allografts obtained from post-mortem donors (Deijkers et al. 1997).
James et al. (2002; 2004) determined whether the swab culture results had any clinical implication on wound problems or infections in the donor. In performed studies the rate of contamination was 9%, which is consistent with other studies. There was no difference in the complication rate of patients with a positive culture swab compared to those with a negative culture swab and therefore they conclude that positive culture swabs have no clinical implications for the donor.

5.3 Immunogenic screening

A question of interest to all bone banks was raised by a case report of a young Rhesus-negative female patient in whom antibodies to a Rhesus antigen developed after she received a femoral-head bone allograft that had been stored by freezing. The graft was procured from a Rhesus positive donor, and the recipient had no other sources of sensitization (Johnson 1985). The immunogenic reaction of allografts is well known and extensively described in literature (Stevenson & Horowitz 1992). This immunogenic response is a reaction on the blood and bone marrow in the allograft. Fresh allografts, which have not been frozen, cause a massive vascular reaction, as has been recently proven by a CAM model (Holzmann et al. 2010). However, freezing of the femoral heads to - 80°C for only three days caused a significant reduction of early vascularisation. Keeping the allografts frozen for longer than one month minimizes the angiogenic potential. Therefore a transfusion reaction after transplantation is unlikely. However, as shown by that case report, sensitization is possible, particularly to the Rhesus (D) (Rh-positive) antigen, which is highly immunogenic. We currently record blood type of all donors, however we only provide Rhesus-compatible grafts to Rhesus-negative women of child-bearing age, in an attempt to prevent problems with future pregnancies or transfusions.

5.4 Audit of a bone bank; further improvements

As a tool for quality control we performed an intern audit of our hospital bone bank, containing only femoral head allografts from living donors. For this audit we assessed all data from the bone bank registry from November 1994 and March 2010. We also included data from potential allografts which eventually pointed out to be not suitable for transplantation as determined by the aforementioned in-and exclusion bone banking criteria. We retrieved 643 femoral heads as potential allografts from 550 donors. Of 643 harvested femoral heads 242 (38%) were discarded. Based on one or more exclusion criteria 123 grafts were excluded based on the questionnaire or due to incomplete pre-operative donor data or tests. Furthermore, 34 grafts were discarded based on positive microbiological, histopathological or serological examination. In total, 64 grafts were discarded due to missing microbiological, histopathological or serological test after at least 6 months. The rest had to be excluded because of tears in the package, loosening of labels, discovery of malignancies in the donor patient and deceased donors in which serological examination could not take place. We calculated the costs associated with complete testing of one femoral head as potential graft which includes all laboratory, histopathological and bacteriological tests. If all completely tested femoral head allografts would be suitable for donation this bone bank would be financially advantageous, even with routinely performed histopathological assessment. It is never possible that all potential donor allografts are suitable for bone banking. However in our bone bank the major loss of potential allografts is mainly due to managing, administrative and logistic omissions. Therefore currently managing our own hospital bone bank offered no financial benefits. We provide safe and reliable allografts with good accessibility. We calculated that hospital bone banking can be a financially viable strategy, when logistic procedures are more accurate. We made improvements in the logistic procedure of testing and retesting and expect future improvements of our financial bone banking balance.

6. Conclusions

There are no uniform guidelines for management of a bone bank. The bone bank protocol should meet national law. The described bone bank protocol from our hospital provides for safe and easy accessible allografts. We routinely perform histopathological screening, this practice is extensively discussed on in literature. At this moment we have no financially viable bone bank. This is due to organisational and logistical problems, which have our attention in order to further improve the bone banking process in the near future.

7. References

Alter HJ, Epstein JS, Swenson SG, VanRaden MJ, Ward JW, Kaslow RA, Menitove JE, Klein HG, Sandler SG, Sayers MH, et al. Prevalence of human immunodeficiency virus type 1 p24 antigen in U.S. blood donors--an assessment of the efficacy of testing in donor screening. The HIV-Antigen Study Group. *N Engl J Med.* 1990 Nov 8;323(19).:1312-7.

American Association of Tissue Banks (AATB). (1993). In: *Standards for tissue banking.* Available from: http://www.aatb.org

Aro HT & Aho AJ. Clinical use of bone allografts. *Ann Med.* 1993 Aug;25(4).:403-12.

Billings S.D, Wurtz L.D, Tejada E & Henley J.D (2000). Occult sarcoma of the femoral head in patients undergoing total hip arthroplasty: A report of two cases. *J Bone Joint Surg* 2000; 82A : 1536-9

Busch MP, Eble BE, Khayam-Bashi H, Heilbron D, Murphy EL, Kwok S, Sninsky J, Perkins HA & Vyas GN (1991). Evaluation of screened blood donations for human immunodeficiency virus type 1 infection by culture and DNA amplification of pooled cells. *N Engl J Med.* 1991 Jul 4; 325(1).:1-5.

Busch MP. HIV and blood transfusions: focus on seroconversion. *Vox Sang.* 1994; 67 Suppl 3:13-8. Review.

Campbell ML, Gregory AM & Mauerhan DR (1997). Collection of surgical specimens in total joint arthroplasty. Is routine pathology cost effective? *J Arthroplasty* 1997; 12:60-63.

Centers for Disease Control (CDC 1981). Pneumocystis pneumonia--Los Angeles. *MMWR Morb Mortal Wkly Rep.* 1981 Jun 5; 30(21).:250-2.

Centers for Disease Control (CDC 1982). Update on acquired immune deficiency syndrome (AIDS).-United States.*MMWR Morb Mortal Wkly Rep.*1982 Sep 24; 31(37).:507-8,513-4

Centers for Disease Control (CDC). Transmission of HIV through bone transplantation: case report and public health recommendations. *MMWR Morb Mortal Wkly Rep.*1988 Oct 7; 37(39).:597-9

CDC, no authors listed. Leads from the MMWR. Transmission of HIV through bone transplantation: case report and public health recommendations. *JAMA.* 1988 Nov 4; 260(17).:2487-8.

Clark C.R. &Bauer T. Routine pathological examination of operative specimens from primary total hip and total knee replacement: another look. *J Bone Joint Surg* 2000;82A : 1529.

Conrad EU, Gretch DR, Obermeyer KR, Moogk MS, Sayers M, Wilson JJ & Strong DM (1995). Transmission of the hepatitis-C virus by tissue transplantation. *J Bone Joint Surg Am* 77:214–224

Deijkers RL, Bloem RM, Petit PL, Brand R, Vehmeyer SB & Veen MR (1997). Contamination of bone allografts: analysis of incidence and predisposing factors. *J Bone Joint Surg Br.* 1997 Jan;79(1).:161-6.

DiCarlo E.F, Bullough P.G , Steiner G, Bansal M & Kambolis C (1994). Pathological examination of the femoral head. *Mod Pathol* 1994; 7(1).: 6A.

Eggen BM & Nordbø SA. Transmission of HCV by organ transplantation. *N Engl J Med.* 1992 Feb 6; 326(6).:411

Elves MW, Pratt LM. The pattern of new bone formation in isografts of bone. *Acta Orthop Scand.* 1975 Sep; 46(4).:549-60.

Fleming A (1929). On the antibacterial action of cultures of a penicillium, with special reference to their use in the isolation of B. influenzae. *British Journal of Experimental Pathology* 10: 226-236

Finkemeier CG (2002). Bone-grafting and bone-graft substitutes. *J Bone Joint Surg Am.* 2002 Mar; 84-A (3).:454-64.

Friedlaender GE (1987). Bone grafts. The basic science rationale for clinical applications. *J Bone Joint Surg Am* 69:786–790

Garcia-Cimbrelo E, Cruz-Pardos A, Garcia-Rey E, Ortega-Chamarro J (2010). The survival and fate of acetabular reconstruction with impaction grafting for large defects. *Clin Orthop Relat Res.* 2010 Dec;468(12).:3304-13.

Hazan EJ, Hornicek FJ, Tomford W, Gebhardt MC& Mankin HJ. The effect of adjuvant chemotherapy on osteoarticular allografts. *Clin Orthop Relat Res.* 2001 Apr; (385).:176-81.

Holzmann P, Niculescu-Morzsa E, Zwickl H, Halbwirth F, Pichler M, Matzner M, Gottsauner-Wolf F, Nehrer S.(2010). Investigation of bone allografts representing different steps of the bone bank procedure using the CAM-model. *ALTEX.* 2010;27(2).:97-103.

Hyatt GW (1950). Fundamentals in the use and preservation of homogenous bone. *US Armed Forces Med J* 1950 Aug; 1(8).:841-52

James LA & Gower A (2002).The clinical significance of femoral head culture results in donors after hip arthroplasty: a preliminary report. *J Arthroplasty.* 2002 Apr;17(3).:355-8.

James LA, Ibrahim T & Esler CN.(2004). Microbiological culture results for the femoral head. Are they important to the donor? *J Bone Joint Surg Br.* 2004 Aug;86(6).:797-800.

Johnson CA, Brown BA & Lasky LC. Rhesus Immunization Caused by Osseous Allograft (letter). *New England J. Med,* 312: 121- 122, 1985.

Jupiter JB, Bour CJ & May JW Jr (1987). The reconstruction of defects in the femoral shaft with vascularized transfers of fibular bone. *J Bone Joint Surg Am.* 1987 mar; 69(3).:365-74.

Kocher MS, Erens G, Thornhill TS & Ready JE: Cost and effectiveness of routine pathological examination of operative specimens obtained during primary total hip and knee replacement in patients with osteoarthritis. *J Bone Joint Surg Am* 2000, 82-A: 1531-1535.

Kreuz FP, Hyatt GW Turner TC & Bassett AL (1951). The preservation and clinical use of freeze-dried bone. *J Bone Joint Surg Am.* 1951 Oct; 33-A (4).:863-72

Kwaliteitsinstituut voor de Gezondheidszorg CBO(2004). In: *Richtlijn Bloedtransfusie* Available from: http://www.cbo.nl/thema/Richtlijnen/Overzicht- richtlijnen/Overig/ (in Dutch).

Lam MF, Ooi GC, Lam B, Ho JC, Seto WH, Ho PL, Wong PC, Liang R, Lam WK & Tsang KW(2004).An indolent case of severe acute respiratory syndrome. *Am J Respir Crit Care Med.* 2004 Jan 1; 169(1).:125-8.

Lauder AJ, Cheatham SA & Garvin KL: Unsuspected non-Hodgkin's lymphoma discovered with routine histopathology after elective total hip arthroplasty. *J Arthroplasty* 2004; 19: 1055-60

Lawrence T, Moskal JT & Diduch DR: Analysis of routine histological evaluation of tissues removed during primary hip and knee arthroplasty. *J Bone Joint Surg Am* 1999; 81:926-931.

Leopold SS, Berger RA, Rosenberg AG, Jacobs JJ, Quigley LR & Galante JO. Impaction allografting with cement for revision of the femoral component. A minimum four-year follow-up study with use of a precoated femoral stem. *J Bone Joint Surg Am.* 1999 Aug; 81(8).:1080-92.

Lexer E (1908). Die verwendung der freien Knochenplastik nebst Versuchen uber Gelenkversteifung und Gelenktransplantation. *Archiv für Klinsche chirugie* 1908; 68: 940-48

Lexer E (1925). Joint transplantation and arthroplasty. *Surg Cynec Obstet* 1925; 40 782-88

Mac Ewen W (1909). Intrahuman bone grafting and reimplantation of bone. *Ann Surg* 1909; 50:959-63

MacEwen W. Observations concerning Transplantation of Bone: illustrated by a case of inter-human osseous transplantation, whereby over two thirds of the shaft of a humerus was restored. *Proc Royal Soc (London).* 1881, 232-247.

Mankin HJ, Fogelson FS, Thrasher AZ & Jaffer F (1976). Massive resection and allograft transplantation in the treatment of malignant bone tumors. *N Engl J Med.* 1976 Jun 3; 294(23).:1247-55.

Mankin HJ, Gebhardt MC, Jennings LC, Springfield DS & Tomford WW (1996). Long-term results of allograft replacement in the management of bone tumors. *Clin Orthop Relat Res* 324:86–97

Meding JB, Ritter MA, Keating EM & Faris PM (1997). Impaction bone-grafting before insertion of a femoral stem with cement in revision total hip arthroplasty. A minimum two-year follow-up study. *J Bone Joint Surg Am.* 1997 Dec; 79(12).:1834-41.

Meding JB, Ritter MA, Jones NL, Keating EM & Faris PM: Determining the necessity for routine pathologic examinations in uncomplicated total hip and total knee arthroplasties. *J Arthroplasty* 2000, 15:69-71.

Nielsen HT, Larsen S, Andersen M & Ovesen O (2001). Bone bank service in Odense, Denmark. Audit of the first ten years with bone banking at the Department of Orthopaedics, Odense University Hospital. *Cell Tissue Bank* 2:179-183

NBF-BIS Foundation (2010). Available from: http://www.bisfoundation.org

Palmer S.H, Gibbons C.L. & Athanasou N.A, (1999). The pathology of bone allograft. *J Bone Joint Surg* 1999; 81B: 333-5.

Parrish FF (1973). Allograft replacement of all or part of the end of a long bone following excision of a tumor: report of twenty-one cases. *J Bone Joint Surg* 1973; 55A:1-22

Paxton ES Jr, Keeney JA, Maloney WJ & Clohisy JC (2011). Large acetabular defects can be managed with cementless revision components. *Clin Orthop Relat Res.* 2011 Feb; 469(2).:483-93.

Raab SS, Slagel DD, Robinson RA. The utility of histological examination of tissue removed during elective joint replacement. A preliminary assessment. *J Bone Joint Surg Am.* 1998 Mar;80(3).:331-5.

Raizman NM, O'Brien JR, Poehling-Monaghan KL & Yu WD (2009). Pseudarthrosis of the spine. *J Am Acad OrthopSurg* 17:494–503

Schreurs BW, Busch VJ, Welten ML, Verdonschot N, Slooff TJ & Gardeniers JW(2004). Acetabular reconstruction with impaction bone-grafting and a cemented cup in patients younger than fifty years old. *J Bone Joint Surg Am.* 2004 Nov; 86-A (11).:2385-92.

Schreurs BW, Zengerink M, Welten ML, van Kampen A & Slooff TJ.(2005). Bone impaction grafting and a cemented cup after acetabular fracture at 3-18 years. *Clin Orthop Relat Res.* 2005 Aug;(437).:145-51.

Schreurs BW, Luttjeboer J, Thien TM, de Waal Malefijt MC, Buma P, Veth RP& Slooff TJ (2009). Acetabular revision with impacted morselized cancellous bone graft and a cemented cup in patients with rheumatoid arthritis. A concise follow-up, at eight to nineteen years, of a previous report. *J Bone Joint Surg Am.* 2009 Mar 1; 91(3).:646-51.

Schreurs BW, Keurentjes JC, Gardeniers JW, Verdonschot N, Slooff TJ & Veth RP(2009).Acetabular revision with impacted morselized cancellous bone grafting and a cemented acetabular component: a 20- to 25-year follow-up. *J Bone Joint Surg Br.* 2009 Sep; 91(9).:1148-53.

Schweiberer L, Stutzle H & Mandelkow HK. Bone transplantation. *Arch Orthop Trauma Surg* 1990; 109(1).:1-8.

Shutkin NM (1954). Homologous-serum hepatitis following the use of refrigerated bone-bank bone. *J Bone Joint Surg Am* 1954, 36-A: 160-162.

Sierra RJ, Charity J, Tsiridis E, Timperley JA & Gie GA. The use of long cemented stems for femoral impaction grafting in revision total hip arthroplasty. *J Bone Joint Surg Am.* 2008 Jun; 90(6).:1330-6.

Slooff TJ, Buma P, Schreurs BW, Schimmel JW, Huiskes R & Gardeniers J (1996). Acetabular and femoral reconstruction with impacted graft and cement. *Clin Orthop Relat Res* 324:108–115

Somford MP, Bolder SB, Gardeniers JW, Slooff TJ & Schreurs BW (2008). Favorable survival of acetabular reconstruction with bone impaction grafting in dysplastic hips. *Clin Orthop Relat Res.* 2008 Feb;466(2).:359-65.

Stevenson S & Horowitz M (1992). The response to bone allografts. *J Bone Joint Surg Am.* 1992 Jul;74(6).:939-50.

Sugihara S, van Ginkel AD, Jiya TU, van Royen BJ, van Diest PJ & Wuisman PI (1999). Histopathology of retrieved allografts of the femoral head. *J Bone Joint Surg Br* 81:336–341

Summers BN & Eisenstein SM (1989). Donor site pain from the ilium. A complication of lumbar spine fusion. *J Bone Joint Surg Br* 71:677–680

Takaso M, Nakazawa T, Imura T, Ueno M, Saito W, Shintani R, Fukushima K, Toyama M, Sukegawa K, Okada T & Fukuda M (2011). Surgical treatment of scoliosis using allograft bone from a regional bone bank. *Arch Orthop Trauma Surg.* 2011 Feb; 131(2).:149-55.

Tomford WW, Ploetz JE & Mankin HJ. Bone allografts of femoral heads: procurement and storage. *J Bone Joint Surg Am.* 1986 Apr;68(4).:534-7.

Toms AD, Barker RL, Jones RS & Kuiper JH(2004). Impaction bone-grafting in revision joint replacement surgery. *J Bone Joint Surg Am.* 2004 Sep; 86-A (9).:2050-60.

Urist MR & Maclean FC (1953). The local physiology of bone repair with particular reference to the process of new bone formation by induction. *Am J Surg.* 1953 Mar;85(3).:444-9

Veen MR, Bloem RM & Petit PL.(1994).Sensitivity and negative predictive value of swab cultures in musculoskeletal allograft procurement. *Clin Orthop Relat Res.* 1994 Mar;(300).:259-63.

Vehmeyer SB, Slooff AR, Bloem RM & Petit PL (2002). Bacterial contamination of femoral head allografts from living donors. *Acta Orthop Scand.* 2002 Apr;73(2).:165-9.

Zwitser EW, de Gast A, Basie MJ, van Kemenade FJ & van Royen BJ (2009). B-cell lymphoma in retrieved femoral heads: a long term follow up. *BMC Musculoskelet Disord* 2009; 10:53

Zwitser EW, Jiya TU, George Licher H & van Royen BJ. Design and management of an orthopaedic bone bank in the Netherlands. *Cell Tissue Bank.* 2010 Nov 13. [Epub ahead of print]

15

Future Applications of Electronic-Nose Technologies in Healthcare and Biomedicine

Alphus Dan Wilson
USDA Forest Service, Southern Hardwoods Laboratory
United States of America

1. Introduction

The development and utilization of many new electronic-nose (e-nose) applications in the healthcare and biomedical fields have continued to rapidly accelerate over the past 20 years. Innovative e-nose technologies are providing unique solutions to a diversity of complex problems in biomedicine that are now coming to fruition. A wide range of electronic-nose instrument types, based on different operating principles and mechanisms, has facilitated the creation of different types and categories of medical applications that take advantage of the unique strengths and advantages of specific sensor types and sensor arrays of different individual instruments. Electronic-nose applications have been developed for a wide range of healthcare sectors including diagnostics, immunology, pathology, patient recovery, pharmacology, physical therapy, physiology, preventative medicine, remote healthcare, and wound and graft healing. E-nose biomedical applications range from biochemical testing, blood compatibility, disease diagnoses, drug purity, monitoring metabolic levels, organ dysfunction, and telemedicine. This review summarizes some of the key technological developments of electronic-nose technologies, arising from past and recent biomedical research, and identifies a variety of future e-nose applications currently under development which have great potential to advance the effectiveness and efficiency of biomedical treatments and healthcare services for many years. A concise synthesis of the major electronic-nose technologies developed for healthcare and medical applications since the 1980s is provided along with a detailed assessment and analysis of future potential advances in electronic aroma detection (EAD) technologies that will provide effective solutions to newly-emerging problems in the healthcare industry. These new e-nose solutions will provide greatly improved quality controls for healthcare decisions and diagnoses as well as badly needed final confirmations of appropriate patient treatments. The purpose of this chapter is to provide some detailed insights into current and future e-nose applications that will yield a variety of new solutions to detection-related tasks and difficult problems in the fields of healthcare and biomedicine. The uses of electronic-noses for quality control (QC) and quality assurance (QA) issues, associated with numerous diagnostic-testing activities conducted within the medical field, also are discussed.

2. History of Electronic-nose developments for biomedical applications

Use of the olfactory sense (of smell) as an indicator of disease probably originated with Hippocrates around 400 BC. Observations that unusual human odors or aromas provided some indication of human ailments caused early medical practitioners to recognize that the presence of human diseases changed the odor of bodily excretions that could be used to diagnose certain common diseases.

2.1 Early use of aroma-detection in evaluating health conditions

Medical doctors have utilized the sense of smell to facilitate determinations of the physical state and general health of their patients for centuries. The application of smell as useful sensory clues used by physicians to identify the causes of human ailments resulted in the development of qualitatively descriptive odors (or aromas) and specialized terms used to describe and identify odors associated with specific human diseases and physiological disorders. Some descriptive aromas found to be associated with some common human diseases are presented in Table 1. The use of olfactory information provided valuable additional information for physicians in assessing patient conditions and formulating accurate diagnoses before modern analytical equipment and chemical-detection devices became available for this purpose. Notice that in some cases the same term, such as "amine-like" for bacterial vaginosis and bladder infections, occasionally was used to describe common

Disease / Disorder	Body source	Descriptive aroma	References
Anaerobic infection	Skin, sweat	Rotten apples	Pavlou & Turner, 2000
Bacterial vaginosis	Vaginal fluid	Amine-like	Pavlou & Turner, 2000
Bladder infection	Urine	Amine-like	Pavlou & Turner, 2000
Congestive heart failure	Heart	Dimethyl sulfide	Smith, 1982
Fetor hepaticus	Breath	Newly-mown clover	Hayden, 1980
Gout	Skin	Gouty odor	Liddell, 1976
Hyperaminoaciduria	Infant skin	Dried malt or hops	Liddell, 1976
Hypermethioninemia	Infant breath	Sweet, fruity, fishy	Liddell, 1976; Hayden, 1980
Isovaleric acidemia	Skin, breath	Sweaty, cheesy	Hayden, 1980; Pavlou & Turner, 2000
Ketoacidosis	Breath	Acetone-like	Hayden, 1980
Liver failure	Breath	Musty fish, feculent	Hayden, 1980; Smith, 1982
Maple syrup disease	Sweat, urine	Maple syrup	Liddell, 1976; Pavlou & Turner, 2000
Pseudomonas infection	Skin, sweat	Grape	Pavlou & Turner, 2000
Scrofula	Body	Stale beer	Liddell, 1976
Smallpox	Skin	Pox stench	Liddell, 1976
Trimethylaminuria	Skin, urine	Fishy	Pavlou & Turner, 2000
Typhoid	Skin	Freshly-baked bread	Liddell, 1976; Hayden, 1980
Uremia	Breath	Fishy, ammonia	Hayden, 1980
Yellow fever	Skin	Butcher's shop	Liddell, 1976; Hayden, 1980

Table 1. Descriptive aromas previously used for diagnosing human diseases

odors associated with completely different diseases. This occurred because different diseases can result in the production of very similar compounds even though the mechanism of disease is quite different. In other cases such as for use of the term "fishy" for hypermethioninemia and uremia, both of these diseases cause the buildup of similar or identical compounds in the blood due to similar physiological processes that are often referred to as in-born genetic or metabolic diseases resulting from the absence of certain enzymes or the failure of certain organs. Many other metabolic diseases caused by genetic enzyme deficiencies are associated with various distinctive odors due to the accumulation of undecomposed metabolites in the body.

Some descriptive aromas, such as maple syrup and pox stench, are so diagnostic that the aroma was named after the specific disease referred to by name. Other diagnostic terms for descriptive aromas include fetor hepaticus, diabetic breath, and uremic breath which have been included in common medical vocabulary and continue to be used to some extent even in contemporary vernacular. Once modern analytical instrumentation became available in the twentieth century, the actual volatile compounds responsible for these characteristic smells began to be identified. Probably the first such identification was done by Linus Pauling, the noted chemist who was able to freeze out and identify some of the volatiles in urine using cold traps, followed by gas chromatography (Pauling et al., 1971). Many other discoveries of VOCs associated with specific human smells related to particular diseases followed in subsequent years leading up to the identification of diagnostic bioindicators of disease. These compounds are highly correlated with the presence of specific diseases in the body as discussed in the following section.

2.2 Discovery of bioindicators of disease

The discovery and recognition of particular volatile organic compounds (VOCs), released from various diseased human body parts or fluids derived from these tissues, have been found to be associated with specific human diseases through the use of specialized modern analytical instruments. These instruments have included such analytical machines as gas chromatographs working in tandem with mass spectrometers (GC-MS) and other such technical instruments used in analytical chemistry. The results of intense chemical analyses from numerous research studies have been the identification of many volatile biomarkers of disease and their associated chemical structures. The identification of unique molecular markers (volatile metabolites) associated with particular diseases has become an extremely effective and powerful tool for the early detection of diseased tissues and infectious agents in the human body. For example, the analysis of patients' breath odors has had a long history of application for the detection of various human diseases, not only respiratory diseases. Even though the human breath contains hundreds of volatile organic compounds at low concentrations, relatively few (less than fifty) of these are detected in the majority of healthy humans under normal physiological conditions (Phillips et al., 1999a). However, a much smaller number of aberrant VOCs are often found only in patients when disease is present somewhere in their bodies. Thus, the association of specific volatile metabolites, released within the expired human breath of patients, not only provides indicators of particular diseases, but also reflect the overall physiological state as an indication of general health and a useful index of disease (Jacoby, 2004). These volatile markers of disease often are released several hours to several days before outwardly-noticeable physical symptoms of illness appear and thus provide early indicators of disease or physiological disorders. New molecular markers that are indicators of specific diseases, both infectious and noninfectious, are being increasingly revealed by new scientific research. Some examples of these volatile molecular biomarkers (or bioindicators) of disease and physiological disorders, reported hitherto by various researchers, are summarized in Table 2.

Disease / Disorder	**Volatile chemical biomarkers**	**References**
Allograft rejection	Carbonyl sulfide	Studer et al., 2001
Breast cancer	C4-C20 alkanes	Phillips et al., 2003b
Cholera	p-menth-1-en-8-ol, dimethyl disulphide	Garner et al., 2009
Chronic hepatitis	Methyl-mercaptan, dimethyl sulfide	Kaji et al., 1978
Cirrhosis	Dimethyl sulfide, mercaptans	Chen et al., 1970
Cystic fibrosis	Leukotriene B4, interleukin-6, carbonyl sulfide, alkanes	Carpagnano et al., 2003; Phillips et al., 2004
Diabetes	Acetone, ethanol, methyl nitrate	Rooth & Ostenson, 1966; Crofford et al., 1997; Ping et al. 1997; Novak et al., 2007
Halitosis	Methanethiol, Hydrogen sulfide, methyl mercaptan, dimethyl sulfide	Kaizu, 1976; Van den Velde et al., 2009
Hepatic encephalopathy	3-methylbutanal	Goldberg, 1981
Histidinemia	2-imidazolepyruvic acid, 2-imidazolelactic acid, 2-imidazoleacetic acid	Bondy & Rosenberg, 1980
Liver cancer	Hexanal, 1-octen-3-ol, octane	Xue et al., 2008
Lung cancer	Alkanes, ketones, specific aromatic hydrocarbons (benzene derivatives)	Manolis, 1983; Gordon et al., 1985; Preti et al., 1988; Phillips et al., 1999b, 2003a
Maple syrup disease	2-oxoisocaproic acid	Bondy & Rosenberg, 1980
Necrotizing enterocolitis	2-Ethyl-1-hexanol	De Lacy Costello et al., 2008
Oxidative stress	8-isoprostane	Montuschi et al., 1999
Periodontal disease	Pyridine, picolines	Kostelc et al., 1981
Phenylketonuria	Phenylpyruvic acid, phenyllactic acid, phenylacetic acid	Bondy & Rosenberg, 1980
Schizophrenia	Pentane, carbon disulfide	Smith & Sines, 1960; Smith et al., 1969; Phillips et al., 1993
Tyrosinemia	p-hydroxyphenylpyruvic acid	Bondy & Rosenberg, 1980
Trimethylaminuria	Trimethylamine	Pavlou & Turner, 2000
Uremia	Dimethylamine, trimethylamine	Simenhoff et al., 1977

Table 2. Molecular biomarker VOCs of specific human diseases and disorders

Analysis of expired human breath is considered particularly valuable because it can be monitored noninvasively (without causing physical damage to patients), yet provide information about the chemical and physiological state of the entire body. The reason that information about the physical health of the entire body is possible by the analysis of expired breath is because most volatile metabolites of infectious agents of disease, or those produced from abnormal tissues, are eventually eliminated from the body through the lungs, often soon after being formed within diseased tissues. Alternatively, other less volatile abnormal metabolites are eliminated through the urine which may be similarly analyzed using aroma-sensing instruments such as electronic noses.

Cao and Duan (2006) summarized some of the advantages and disadvantages of breath analysis for clinical practice and diagnosis. They found breath tests were noninvasive, easily repeated, and caused less discomfort and embarrassment to patients than blood and urine tests. Breath samples closely reflected arterial concentrations and provided much less complicated mixtures than serum or urine analyses and more direct information on respiratory function than by other means. They listed limitations of breath testing for clinical practice to include the lack of standardization of analytical methods, the high water content of breath samples affecting detection, relatively expensive costs compared to simple chemical tests (but much less time-consuming for results), and the lack of well-established links between breath VOCs and certain kinds of diseases. Biomarkers in chronic obstructive pulmonary disease (COPD) also may be useful in aiding diagnosis, monitoring exacerbations, evaluating effects of drugs, and defining specific phenotypes of disease (Borrill et al., 2008).

Frey & Suki (2008) found risk assessments, disease progression, and control of asthma and COPD required multidimensional fluctuation analysis of the dynamics of lung-function parameters that needed to be quantified and monitoring via precise biomarkers of these diseases using instruments capable of direct, electronic monitoring of these biomarkers.

The importance of the use of biomarkers for the detection of disease has become so prominent that Bentham Science, a leading international publisher of high quality scientific journals, decided to launch a new journal called *Recent Patents on Biomarkers* in January 2011 to publish reviews and research articles written by experts on recent patents and research relating to biomarkers in basic and applied, medical, environmental, and pharmaceutical research, and including patent biomarker applications, clinical development, and molecular diagnostics.

3. Current e-nose technologies utilized in healthcare and biomedicine

Electronic-noses are ideal instruments for biomedical uses because of their versatility, low-cost, rapid output of results, capabilities of continuous operation (for physiological-monitoring purposes), and the wide range of VOCs and other cellular chemical constituents that may be analyzed. The potential for miniaturization of electronic-nose devices also is great due to their microcircuitry and microsensor components. Some key ways in which e-noses have been particularly useful in various sectors of the healthcare industry are discussed in the following sections.

3.1 Electronic-nose technology types and applications

A variety of different types of e-noses, based on different working principles, have been used for biomedical tasks including conductive polymers (CP), metal-oxide semiconductor (MOS), quartz crystal microbalance (QCM), and surface acoustic waves (SAW) among others. Each e-nose technology has different advantages, disadvantages, and limitations that largely determine what types of medical applications that individual e-nose sensor types are best suited for in practical clinical settings.

3.2 Point-of-care medicine

Point-of-care testing (POCT) may be defined as diagnostic testing at or near the site of patient care (Kost, 2002). The objective of POCT is to bring the test conveniently and immediately to the patient. The POCT approach to diagnostic testing increases the likelihood that the patient will receive the results and treatment in a timely manner. POCT is accomplished through the use of transportable, portable, and handheld instruments and test kits. The use of cheaper, smaller, faster, and smarter POCT devices, such as e-noses, has increased the use of POCT approaches by making diagnostic tests more cost-effective for many diseases.

3.3 Working e-nose applications in current medical practice

E-noses in general have the advantages of providing patient laboratory results much faster than standard cultures or wet chemistry tests and the capability of providing early detections of diseases before symptoms appear. These characteristics have been compelling reasons for the development of e-nose systems for clinical medicine. Some recent uses of electronic noses in hospitals and universities around the world are presented in Table 3.

The development of new e-nose applications for POC treatments will no doubt continue to increase as the breadth of existing e-nose systems is expanded with new capabilities and practical e-nose uses are discovered and implemented through more extensive empirical testing. This work will require extensive trials in hospitals and clinics as well as in the field (for portable units) to determine the range of multiple tasks that individual e-nose systems can perform for various types of detection and diagnostic testing needs. The cooperation of many levels of healthcare professions working in cooperation with e-nose manufacturers, clinical technicians and medical research scientists will be required to accomplish these tasks. This effort is quite a challenge in many situations because of the limited time available to physicians for testing new experimental equipment.

3.3.1 Health monitoring

Continuous monitoring of the physiological states of patients is essential to determine the current physiological condition of patients and whether treatment and recovery is progressing favorably. For example, the continuous monitoring of serum glucose levels, particularly with the aid of sophisticated algorithms, provides a means of generating alerts when glucose concentration exceeds the normal high and low threshold ranges (Sparacino et al., 2010). Monitoring exhaled VOC biomarkers of endogenous metabolic processes using electronic noses is an ideal means of detecting altered metabolic pathways resulting from diseases such as diabetes. The use of e-nose sensors for continuous glucose monitoring requires accurate calibration, filtering of data to enhance the signal-to-noise ratio, and effective predictions of future glucose concentration in order to generate alerts with minimal risk of causing false alarms or missing entirely the occurrence of life-risking events. Electronic-nose devices also might be used to facilitate the study of transcriptional gene regulation of glucose sensors in pancreatic β-cells and liver by monitoring changes in breath volatiles (primarily ethanol, acetone, and methyl nitrate) associated with hyperglycemia in type 2 diabetes patients (Bae et al., 2010; Lee et al., 2009).

Country	Hospital, University or Research Facility	E-nose utilized	Application	References
USA	University of Pennsylvania	Experimental model	Distinguish cerebrospinal fluid	Thaler et al., 2000
USA	Merck Research Laboratories	Fox 4000	Flavor analysis for drug formulation	Zhu et al., 2004
United Kingdom	Birmingham Heartlands Hospital	Cyranose 320	Identify *Staphalococcus*	Dutta et al., 2005
Germany	University of Applied Sciences	DE 101	Detect renal dysfunction	Voss et al., 2005
USA	Cleveland Clinic	unspecified	Diagnose lung cancer	Erzurum et al., 2005
Belgium	University of Antwerp	PEN 2	Clinical diagnoses of bacteria	Moens et al., 2006
United Kingdom	South Manchester University Hospital	experimental model	Burn and wound infection types	Persaud, 2006
USA	University of Pennsylvania	unspecified	Diagnosis of diseases via breath	Anthes, 2008
USA	California Institute of Technology	JPL ENose	Detect & differentiate brain cancers	Kateb et al., 2009
Australia	Prince Charles Hospital	unspecified	Detect chronic lung disease	Dent, 2010
Netherlands	Amsterdam Academic Medical Center	Cyranose 320	Discriminate inflammation airway diseases	Lazar et al., 2010
Italy	Catholic University	experimental model	Asthma detection	Montuschi, 2010
Tanzania	National Institute of Medical Research	Bloodhound EN	Diagnosis of Tuberculosis	Kolk et al., 2010
United Kingdom	Gloucestershire Royal Hospital	NST 3320	Diagnosis of ventilator-associated pneumonia	Humphreys et al., 2011

Table 3. Electronic-nose uses in hospitals and universities around the world

Monitoring inorganic anions and cations in the body are equally important for maintaining proper electrolyte levels and water balance in tissues. Thus, routine clinical assays of electrolyte levels (such as chloride, sodium, and potassium) in biological samples like serum, blood, plasma, and urine provide useful information about the proper functioning of organ systems and regulatory hormones of patients receiving treatments. Assessing urinary chloride concentration helps in the diagnostic evaluation of metabolic alkalosis and other physiological conditions caused by improper osmotic pressure, water imbalances within extracellular spaces, and acid-base imbalance. The Microcontroller P89C668 is an instrument that measures urinary chloride concentration to determine body electrolyte levels based on a mercuric thiocyanate colorimetric principle (Vasumathi & Neelamegam, 2010). This instrument works by measuring color intensity of a colored complex formed between chloride ions and mercuric thiocyanate which is proportional to the chloride concentration in the sample. Colorimetric e-noses operate similarly by measuring changes in absorbance caused by color changes resulting from interactions of the target analyte with an organic dye.

3.3.2 Infection and disease detection

Electronic-nose systems probably were first tested for disease detection in the biomedical field through the discrimination of pathogenic microbes in pure cultures (Gardner et al., 1998). Microbial identification is an integral part of infectious disease diagnosis and the subsequent determination of proper treatments as a consequence of the wide range of disease mechanisms associated with pathogenesis generated by various microbial agents. Dutta et al. (2002) used a portable Cyranose 320 e-nose, consisting of 32 polymer carbon black composite sensors, to identify six bacterial species responsible for eye infections. The bacteria were cultured at various concentrations in a saline solution and the VOCs from the headspace were analyzed using linear PCA and other data-clustering algorithms. The Self Organizing Map (SOM) network provided an accuracy of 96% for bacterial classification, but the Radial basis function network (RBF) allowed identifications with up to 98% accuracy. Most laboratory-grade instruments such as the Cyranose 320 are now being replaced with simpler and cheaper e-noses that are easier to use by trained clinical technicians. Many new types of experimental e-noses, based on different operating principles, currently are being tested for numerous healthcare applications.

Microbial biosensors are being employed increasingly to detect human diseases. These sensors, like e-noses, consist of a transducer that converts biochemical signals into a quantifiable electronic response, but instead of utilizing electronic sensors, the transducer is used in conjunction with either viable or unviable microbial cells. A variety of different transducers may be used such as acoustic, electrochemical, electric, or optical types. D'Souza (2001) did an early review of applications of microbial biosensors and gave some advantages and limitations of various types. Biosensors will be discussed in greater detail in section 4.2.1.

3.3.3 Detecting exposure to toxins and hazardous chemicals

Food safety and exposure to toxic substances in the environment has become of greater concern to man in the world today as a result of the acceleration and increasing frequency of bioterrorism and the growing susceptibility of world crops to toxic sprays and disease due to the planting of crop monocultures and the application of agricultural chemicals from the air. Toxic volatile solvents also are found in the air within certain areas of hospitals despite the filtering of air. All of these opportunities for incidental human exposures to toxic substances necessitate the monitoring of food supplies and ambient air to assure that levels of harmful substances are below damaging levels. The occurrence of various toxins in food is potentially very harmful to human health. Sensor technologies such as electronic noses have been recognized as possible useful tools for determinations of the geographical origin of food products, now quite important for the identification of food lots that have become contaminated by toxins or other harmful substances in order to remove these specific food sources from grocery shelves (Luykx & van Ruth, 2008). Other examples include the occurrence of mycotoxins, toxic secondary metabolites (e.g. aflatoxin and ochratoxin A) produced by fungi such as *Aspergillus* and *Fusarium* species that commonly grow on agricultural products in the field or in storage (Huang et al., 2006). Cheli et al. (2009) very effectively utilized the PEN2 e-nose with principal component analysis (PCA) to detect the presence of aflatoxin in maize samples at a high level of confidence. This method was potentially useful for screening maize food lots for aflatoxin contamination prior to marketing.

Mujahid et al. (2010) used cholesteric liquid crystals (CLCs) as sensitive coatings on acoustic devices such as QCM e-noses for the detection of organic solvent vapors of both polar and non-polar compounds by the frequency shift of analyte samples. They were able to gain mechanical stability by combining CLCs with imprinted polymers. This e-nose application would be useful for detection of pharmaceutical preparations requiring solvent extraction or delivery, and for detection of potential patient exposures to hazardous chemical solvents in the hospital environment.

3.4 Quality control

There are many potential uses for e-nose instruments in quality control (QC) applications in medicine. These machines can be used to quickly double check diagnoses to help assure that patients are receiving the correct and precise treatments prescribed by physicians. Another possible related application is the e-nose evaluation of food quality and control measures to assure that food contaminants and toxins, that can adversely affect food safety and human health, are not present. Improved QC has been accomplished through use of specialized algorithms that increase analyte discriminations and confirm the results.

3.4.1 Electronic-nose algorithms

The efficiency with which electronic-nose systems are able to identify and discriminate VOCs associated with analyte mixtures largely depends on the effectiveness of discriminating algorithms used during headspace analysis. Pattern-recognition algorithms are heavily used for integrating signal outputs of sensor arrays and comparing such outputs to patterns of known analyte standards held in recognition reference libraries. This discrimination process is very similar to those used in GC-MS analyses that use reference libraries. New gas-recognition algorithms have provided a means of improving the effectiveness, robustness, and accuracy of gas detection and identification for the medical industry.

Flitti et al. (2008) developed a gas-recognition algorithm for an on-chip Complementary Metal-Oxide Semiconductor (CMOS) tin-oxide (SnO_2) gas sensor array that operates at high temperature (typically 300°C) with the advantages of cost effectiveness and high sensitivity to various gases, but the disadvantages of low selectivity, high sensitivity to humidity, nonlinearities of sensor-response, and drift in signal output. Many pattern-recognition algorithms have attempted to correct for low selectivity of sensors, yet most do not address the problem of drift which was largely corrected according to experimental results in this study, indicating that more than 98% correct recognition was obtained using this robust method. Polat and Güneş (2006) proposed a decision-tree classifier system using fuzzy weighted preprocessing methods for the diagnosis of erythemato-squamous diseases. They used twelve clinical-evaluation criteria and twenty-two histopathological features in the diagnostic analysis. Similar fuzzy-reasoning methods have been used in e-nose algorithms to discriminate sensor-array patterns produced from headspace volatiles. Thus, many different types of diagnostic information may be used in these decision-tree classifiers. Seising (2006) created a similar model using fuzzy reasoning to address the phenomenon of vagueness in a physician's style of thinking concerning reasoning used to make clinical diagnoses.

3.4.2 Drug development, purity, and delivery

Spin-offs of electronic-nose technologies similar to conductive polymer (CP) e-noses, but with single sensors instead of an array, are being developed to work in aqueous solutions for the detection of drugs and other chemicals used in pharmaceutical preparations. Manganese (III) porphyrins are particularly useful for the construction of polymeric membranes. Vlascici et al. (2010) developed ion-selective electrode sensors composed of two types of manganese (III) porphorins, high molecular weight polyvinyl chloride (PVC) and sol-gel, for the determination of diclofenac in pharmaceutical preparations by direct potentiometry. Diclofenac is a nonsteroidal drug used in the treatment of ankylosing spondylitis, osteoarthritis, and rheumatoid arthritis due to its antipyretic, anti-inflammatory and analgesic properties. Their best results were obtained with PVC membrane plasticized with dioctylphtalate and incorporated with sodium tetraphenylborate as a lipophilic anionic additive. Electrode response to diclofenac was linear in the concentration range of 3×10^{-6} to 1×10^{-2} M and in good agreement with a High Pressure Liquid Chromagraphy (HPLC) reference method.

Continuous glucose monitoring systems (CGM) may soon offer the possibility of continuous dynamic assessment and control of daily fluctuations in blood glucose concentration for diabetes treatment. The emergence of a new generation of open-loop and closed-loop subcutaneous insulin-infusion devices that are controlled by continuous glucose-monitoring sensors will soon make glycemic control and insulin treatment more reliable (Torres et al., 2010). New smart machines are on the horizon to simplify diurnal treatments, allowing diabetics to be less attentive to their daily insulin needs.

4. Future potential medical applications of electronic noses

The potential applications of electronic-nose devices in the healthcare and biomedical industries will continue to expand with greater research and in-hospital testing as new ways of using these chemical-detection machines are discovered, and the breadth of capabilities widened, particularly in the area of coordinated uses in combination with other medical devices. The combined uses of e-noses with other electronic medical instruments will facilitate the development and availability of improved real-time information of patient conditions, leading to even more effective decisions and treatments by physicians in hospitals and POCT clinics. The future potential of combining the capabilities of e-nose devices with other types of detection technologies are examined here in light of new technological discoveries in chemical sensor-detection that are currently emerging.

4.1 Emerging e-nose biomedical developments

Electronic noses have even greater potential synergistic capabilities when used cooperatively in combination with many other electronic medical devices. The potential advantages of combining their use are enormous considering the possible permutations of combinations in which these analytical devices may be combined for cooperative tasks. Sometimes these advantages are so useful that e-noses are often combined with other technologies to produce compound e-nose instruments. Both theoretical and practical aspects of these conceptual instrument mergers are discussed in greater detail in section 4.3.

4.2 E-nose uses in cooperative combination with other electronic devices

Synergistic applications of e-nose technologies, used in combination with other medical devices, are receiving increasing attention in the healthcare industry because these instrument-combinations are viewed as ways of achieving greater cooperative effectiveness in improving clinical services to patients. Complimentary information obtained in this way leads to better diagnoses and prognoses. The ultimate results of synergistic uses of instrument combinations are better, more detailed and quality information for medical decisions and thus more effective treatments leading to faster patient recoveries.

One key area where electronic noses are effectively used in combination with other medical instruments is in the application of e-nose information on various physiological conditions of patients toward more effective treatments for particular ailments. E-nose information may be used to confirm the physiological states or functions in patients that are identified in pre-scanning and preliminary assessments of patient conditions during initial examinations. Medical infrared thermography (MIT) is a non-invasive, non-radiating thermal imaging method used to analyze physiological functions based on localized thermal abnormalities characterized by increases or decreases in skin temperature. MIT involves detection of infrared radiation usually related to variations in blood flow that affect skin temperature. Reduced muscular activity or degeneration leads to dermal hyperthermia whereas inflammation causes a hyperthermic pattern. Use of a MIT detection tool has been particularly useful in sports medicine for pre-screening athletes for injuries or muscular inflammation and degeneration (Hildebrandt et al., 2010). E-noses also might be used in combination with wearable motion-sensing sensor technologies for confirming physiological activities after monitoring mobility-related activities in individuals with chronic disease conditions (Allet et al., 2010). Electronic-noses could be used in combination with drug-delivery devices to monitor physiological responses and provide feedback to these devices during or following the administration of drugs. The feedback would then adjust the rate of drug-delivery to ease physiological stress of adverse reactions and thus regulate release rates of drug payloads and resorption rates (Anglin et al., 2008). Similar systems are possible using fiber-optic sensors such as the Sencil system for continuous monitoring of glucose (Liao et al., 2010). Other potential applications include uses in combination with associated cerebrospinal fluid (CSF) tests for analysis and monitoring of specific CSF constituents associated with specific diseases (Di Terlizzi & Platt, 2009), and in combination with the Liver Disease Quality of Life (LDQOL) instrument for liver transplantation evaluation in ambulatory adults with advance, chronic lung disease (Gralnek, 2000).

4.2.1 Biosensors

Biosensors are analytical devices that combine a biological-sensing element with a chemical or physical transducer to quantitatively and selectively detect the presence of specific compounds in a given external environment (Vo-Dinh and Cullum, 2000). Chaubey and Malhotra (2002) summarized the commercialization and applications of four different types of mediated biosensors based on the type of transducer used to convert the physico-chemical change in the selected biologically-active material, resulting from interactions with the analyte to produce the output signal. Biosensor technologies previously have been divided into optical, calorimetric, piezoelectric, and electrochemical biosensors. Optical sensors as based on the measurement of light absorbed or emitted from a biochemical reaction and guided with optical fibers into the sensor. Calorimetric biosensors detect the analyte by the heat released from the biochemical reaction of the analyte with a suitable enzyme. Piezoelectric biosensors operate by generating electrical dipoles through the subjection of anisotropic natural crystals to mechanical stress. The adsorption of an analyte to the sensing crystal increases the mass of the crystal which alters its frequency of oscillation that is recorded in the instrument output. QMB e-nose sensors essentially operate by this same principle. Electrochemical (EC) biosensors measure the generation or consumption of electrons during a bio-interaction process. EC biosensors are the most commonly used class of biosensors and are further subdivided into amperometric, conductometric, and potentiometric sensor types depending on the electrochemical property to be measured by the detector system. Specific EC biosensors such as the Ion selective electrodes (ISE), ion selective field effect transistors (ISFET), and pH electrodes usually measure the oxidation of specific substrates to produce an oxidized product. Two mediated biosensors were previously commercialized early on in biosensor development, including the lactate analyzer (LA 640) in 1976 and a glucose analyzer in 1987. The LAPS (light addressable potentiometric sensor) optical biosensor was commercialized in 1993.

New types of biosensor technologies have been tested and developed recently. For example, Thanyani et al. (2008) examined an affinity biosensor technology to detect antibodies to mycolic acid in tuberculosis patients. Mycolic acids are useful detection targets for tuberculosis because each *Mycobacterium* species produces unique types of mycolic acids in chemical structure and in association with specific liposomes. Komaitis et al. (2010) developed a fully-automated flow-injection bioluminescent biosensor for the assessment of water toxicity, particularly heavy metal toxicity. Kumar & Kumar (2008) analyzed a DNA biosensor for selective detection of target genes responsible for diseases using DNA hybridization with a specific probe. PCR-free DNA biochips are emerging new tools in the field of diagnosis because of the greater advantages of electrochemical biosensors due to the electrochemical behavior of labels associated with hybridization.

There are several notable recent reviews on the development of biosensor applications within the biomedical field. Yoo & Lee (2010) recently reviewed the present status and use of glucose biosensors in the management of diabetes in clinical practice. Dzyadevych et al. (2008) discussed the advantages and disadvantages of amperometric enzyme biosensors for medical diagnostics and other potential healthcare applications. Gomila et al. (2006) described some advances in the development of methods and techniques for the production, mobilization, electrical characterization, and development of olfactory nanobiosensors.

Implantable short-term and long-term biosensors offer utility for a plethora of clinical applications, particularly in the areas of point-of-care medicine, intensive care, and surgery (Guiseppi-Elie, 2010). Biosensors provide invaluable real-time data on the metabolic and physiological status of patients that are required by clinicians and physicians to make medically-important, informed decisions that impact the short- and long-term outcome of patients. These devices potentially could save countless lives in the emergency room or in triage on the battle field where patient mortality is high due to trama-induced hemorrhaging and rapid decisions concerning patient status are essential to provide immediate care to individuals based on their current condition.

4.2.2 BioMEMS and MIP sensors

Biological Micro-Electro-Mechanical Systems (BioMEMS), also known as BioChips, are micro- or nano-scale devices that detect biochemical entities by either mechanical, electrical, or optical means. Mechanical BioMEMS use cantilever sensors on a chip that operate in either stress-sensing or mass-sensing mode. In stress-mode sensing, biochemical reactions cause changes in surface free energy resulting in stress and bending of the cantilever. In mass-mode sensing, the cantilever is excited mechanically so that it vibrates as a certain resonant frequency. A change in mass due to adsorption of chemical species on the sensor is detected by shifts in the resonant frequency. BioMEMS have a wide variety of important biomedical applications including the processing, delivery, manipulation, analysis, and construction of biological and chemical entities (Bashir, 2004). Some important major areas of research and applications range from diagnostic detections (e.g. DNA and protein micro-assays), micro-fluidics, and tissue engineering to surface modification, drug preparation and delivery, cell lysing, mixing, separation, implantable monitoring and sensing. Diagnostics probably represents the largest segment of applications because a very large number of BioMEM devices have been developed for diagnostic applications. Diagnostic detections of pathogenic viruses, bacteria, and fungi as well as small molecular components produced by these microbes may be detected. The advantages of using micro- and nano-scale detection technologies are greater portability through miniaturization, higher sensitivity, reduced reagent volumes with lower associated costs, and perhaps most useful is reduced time to results due to smaller volumes and higher effective concentrations (Bashir, 2004). Aponte et al. (2006) summarized the potential uses of BioMEMS devises to detect the presence of molecular markers in body fluids as indicators of immune system responses. The reviewed research focused on candidate biomarkers that could be useful for in-flight monitoring of astronaut immune status using MEMS and Nano-Electro-Mechanical System (NEMS) devices. They found cytokine levels were significantly affected by space flight conditions. Cytokines are chemical messengers directly related to immune responses and various diseases. They are classified as chemokines, colony-stimulating factors, growth factors, interleukins, interferons, lymphokines, stress proteins, and tumor necrosis factors (Stvrtinova et al., 1995).

Molecular Imprinted Polymer (MIP) microsensors utilize polymeric materials for the recognition of particular chemical substances that are complementary to a specific receptor cavity. MIP materials usually consist of a copolymerized monomer matrix cross-linked to a template molecule that creates a receptor cavity complimentary to the template molecule when the template is removed from the polymer matrix (Tokonami et al., 2009). These nanostructured MIP objects may be used to develop micro-and nano-sized sensors or sensor arrays for chemical sensing and detection. The small size of MIP materials provides the advantages of faster equilibrium with the analyte, increased number of accessible complementary cavities per material weight, and enhanced catalytic activity of the sensor surface. Large-scale sensor array systems utilizing MIP sensors are capable of handling large sample throughput as high density detection for primarily biochemically-related substances such as enzymes, antibodies, and DNA (Tokonami et al., 2009).

4.2.3 Electroconductive hydrogels

Electroconductive hydrogels (ECH) are composite biomaterials made of polymeric blends that combine conductive electroactive polymers (CEPs) with highly hydrated hydrogels. They bring together the redox-switching and electrical properties of conductive electroactive polymers (CEPs) with the small-molecule transport and compatibility of cross-linked hyrogels (Guiseppi-Elie, 2010). CEPs are incorporated into biosensors for the detection of chemical species (e.g., antigens, drug metabolites, enzyme substrates, neurotransmitters, and ssDNA fragments) of medical importance. Biosensors based on CEPs operate either with electrochemical, gravimetric, or optical detectors. They are used for measurements of constituents in low-volume samples with continuous-flow systems and fast response times, high sensitivities, and detection limits in the µM range for enzyme substrates, and even lower detection ranges for DNA fragments. CEPs do have some serious limitations including slow switching speeds in bio-electronic applications, formation of reactive species due to over-oxidation, and time-temperature drift. ECH-based sensors are a new class of devices with potential for *in vivo* biocompatibility in human-implantable biosensors, low voltage actuation for electrically-stimulated drug release devices, and with low interfacial impedances suitable for neural prosthetic devices such as deep-brain stimulation electrodes (Guiseppi-Elie, 2010). ECH characteristics of soft elastic nature, low interfacial tension, and high swelling capacity results in low tissue irrigation and high permeability to low molecular weight drugs and metabolites (Li et al., 2004). These characteristics have allowed hydrogels to be used in biosensors, catheters, contact lenses, wound dressings, and tourniquets. Hydrogels can be designed to possess hydration characteristics and mechanical properties similar to that of human tissue. Thus, uses of ECH as a biorecognition membrane layer in biosensors has extended potential applications to clinically important biomedical diagnoses (using analyte-specific enzymes), neural prosthetic and recording devices (NDPs and NRDs), electro-stimulated drug-release devices (ESDRDs) and implantable electrochemical biosensors. A hydrogel synthesized from a poly(HEMA)-based hydrogel and poly(aniline was fashioned into a biosensor (by incorporation into recombinant cytochrome P450-2D6) that was responsive to the drug fluoxetine, the active ingredient in Prozac (Iwuoha et al., 2004). These polymeric materials provide a non-cytotoxic interface between the biosensor device and native living tissue or cell culture medium (Fonner et al., 2008).

Gawel et al. (2010) reviewed the various principles involved in the design of biospecific hydrogels acting through various molecular mechanisms to transduce the recognition of label-free analytes. The range of different responsive characteristics displayed by hydrogels include changes in equilibrium swelling volume in response to various changes in solution parameters such as solvent pH, ionic strength, temperature, electrical fields, and presence of surfactants.

4.2.4 Porous polymers and resins

Porous polymers and resins provide applications as enantio-selective catalysts, artificial antibodies, and sensors in electro-optical and micro-electronic devices. Unlike inorganic porous gels such as silica gel carriers, porous polymers have unique properties such as flexibility, ductility, and capability to incorporate a wide range of organic functional groups useful for biotechnical and biomedical sensor applications (Hentze & Antonietti, 2002). Initial applications of porous polymers have included uses as insulators and ion exchange resins, employed in the field of column chromatography for separation and purification of organic compounds. Applications of porous polymers have now been extended into sensor development. Some potential pharmaceutical applications of template porous polymer gels are in the development of controlled drug-delivery devices, drug-monitoring devices, and for biological receptor mimetics. These materials have become particularly useful as active components in optical sensors.

4.3 Compound electronic-nose devices

E-nose hybrid devices are created by combining e-nose technologies with other types of sensors into one instrument. This is different from instrument systems such as GC-MS or HPLC-MS instruments used in tandem. In an e-nose hybrid device, different sensor types are found within the same instrument not in separate instruments combined in tandem. There are a number of different combined-technology commercial e-noses available with various types of e-nose sensors combined with other sensor types. The e-nose components of such compound-sensor devices usually contain MOS, SAW, QMB, or CB electronic-nose sensors with different combinations of electron capture (EC), ion mobility spectrometer (IMS), photoionization (PI), mass spectrometer (MS), oxygen (O_2), carbon dioxide (CO_2), and humidity sensors. The sensing range and capabilities of these compound e-noses are considerably greater, but also generally more expensive than typical e-nose devices alone. The efficacy and justification of expense depends on the particular combination of sensing needs that are required for specific medical applications.

Other possibilities exist for integrating e-nose components with DNA probes within a microarray. One such possibility might be the integration of the CombiMatrix microarray system with 12,544 electrodes in which multiplexed CMOS microarray DNA probes are on individual electrodes coated with electro-polymerized polypyrrole (PPY) that is a common material used in many conductive polymers e-noses (Maurer et al., 2010). The possibility of combining PPY sensors for detecting DNA as well as other similar sensors for VOCs within the same instrument is theoretically possible. Lorenzelli et al. (2005) have integrated a MOS detector with a microcapillary GC silicon-based system for clinical diagnostics and other biomedical applications. Initial planned future work are to test this biosensor-based e-nose micro-GC system for determining and monitoring hamovanillic acid (HVA) and vanillylmandelic acid (VMA) catecholamine metabolite concentrations, end-products of dopamine and norepinephrine metabolism, in urine samples as well as for oncological (cancer) diagnoses.

5. Conclusions

Many research and development (R&D) feasibility studies have demonstrated the effectiveness of electronic-nose technologies for detection-type applications in many diverse areas of the healthcare and biomedical industries. Electronic noses have proven to be very competent and effective in discriminating between VOCs and other cellular biochemical constituents, showing great potential for improving and speeding up detections for a myriad of applications. Most of this feasibility work has been done with expensive laboratory-grade instruments designed to allow maximum discriminations and sample-sensitivity for rigorous scientific testing. Consequently, a number of major problems have resulted from attempts by commercial e-noses manufacturers to use laboratory-grade instruments for practical clinical POCT applications. Laboratory-grade instruments generally are too expensive, too complicated for operation by industry technicians, require extensive training (for operation, maintenance, and data-interpretation), and are too versatile in terms of numbers and permutations of control settings that are possible (adjustable) which complicates repeatability (precision and accuracy) within the normal range needed for diagnostic testing. All of these problems have contributed to the failure of applying laboratory-grade e-nose instruments to practical applications. The common mistake and practice of skipping the additional needed steps of customizing e-noses (in both design and operation) for specific biomedical applications has been costly, causing some potential end-users to lose faith in e-nose technologies, and has resulted in the business failures of some e-nose instrument manufacturers as a result of marketing instruments that are not simplified, adapted, and customized to the specific uses required by healthcare professionals.

Now, the electronic-nose industry is at the stage where lessons of design and manufacture have been learned and the path forward has shifted to designing e-noses that are smaller, less expensive, more application-specific (specialized), easier to use by operators, and produce results that are easily interpreted by the user due to limited data outputs. The only final steps left to be completed today for e-nose development for practical uses in many modern-day applications are largely limited to efficacy testing to determine such things as the range and breadth of applications of individual instruments, procedural uses that are possible in combination with other medical instruments or diagnostic tests, quality control between individual instruments (calibration concerns), and developing specialized aroma libraries, software and algorithms for specific medical applications. Once these tasks are completed, use of electronic noses should accelerate in diagnostic laboratories and POCT clinics, replacing many conventional time-consuming methods and instruments used in diagnostics and providing fast, reliable information useful for speeding up effective patient care with the most appropriate treatments.

6. Acknowledgments

The author wishes to thank Drs. Manuela Baietto, Francesco Ferrini (Dipartimento di Ortoflorofrutticoltura, Università di Firenze, Sesto Fiorentino, Italy), and Daniele Bassi (Dipartimento di Produzione Vegetale, Università degli Studi di Milano, University of Milano, Milan, Italy) for previous cooperative international research studies that led to new advancements in electronic-nose applications. These prior collaborative studies ultimately made this review possible through additional contacts and interactions with other e-nose research scientists throughout the world. The author also acknowledges the efforts of Mrs. Charisse Oberle who compiled and collated the references used in producing this review.

7. References

Allet, L., Knols, R.H., Shirato, K. & de Bruin, E.D. (2010). Wearable Systems for Monitoring Mobility-Related Activities in Chronic Disease: A Systematic Review. *Sensors*, Vol.10, No.10, (October 2010), pp. 9026-9052, ISSN 1424-8220

Anglin, E.J., Cheng, L., Freeman, W.R. & Sailor, M.J. (2008). Porous Silicon in Drug Delivery Devices and Materials. *Advanced Drug Delivery Reviews,* Vol.60, No.11, (August 2008), pp. 1266–1277, ISSN 0169-409X

Anthes, E. (January 11, 2008). E-noses Could Make Diseases Something to Sniff at, In: *Scientific American,* Health News, Date of access April 6, 2011, Available from: http://www.scientificamerican.com/article.cfm?id=electronic-noses-could-make-diseases-something-to-sniff-at

Aponte, V.M., Finch, D.S. & Klaus, D.M. (2006). Considerations for Non-invasive In-flight Monitoring of Astronaut Immune Status with Potential use of MEMS and NEMS devices. *Life Sciences,* Vol.79, No.14, (August 2006), pp. 1317–1333, ISSN 0024-3205

Bae, J.-S., Kim, T.-H., Kim, M.-Y., Park, J.-M. & Ahn, Y.H. (2010). Transcriptional Regulation of Glucose Sensors in Pancreatic β-cells and Liver: An Update. *Sensors,* Vol.10, No.5, (May 2010), pp. 5031-5053, ISSN 1424-8220

Bashir, R. (2004). BioMEMS: State-of-the-art in Detection, Opportunities and Prospects. *Advanced Drug Delivery Reviews,* Vol.56, No.11, (September 2004), pp. 1565–1586, ISSN 0169-409X

Bondy, P.K. & Rosenberg, L.E. (1980). Histidinemia, In: *Metabolic Control and Disease, 8th ed.,* Bondy, P.K. & Rosenberg, L.E., eds., pp. 1010-1014, W.B. Sanders, ISBN 0721618448, Philadelphia, PA, USA.

Borrill, Z.L., Roy, K. & Singh, D. (2008). Exhaled Breath Condensate Biomarkers in COPD. *European Respiratory Journal,* Vol.32, No.2, (February 2008), pp. 472–486, ISSN 0903-1936

Cao, W. & Duan, Y. (2006). Breath Analysis: Potential for Clinical Diagnosis and Exposure Assessment. *Clinical Chemistry,* Vol.52, No.5, (March 2006), pp. 800–811, ISSN 0009-9147

Carpagnano, G.E., Barnes, P.J., Geddes, D.M., Hodson, M.E. & Kharitonov, S.A. (2003). Increased Leukotriene B4 and Interleukin-6 in Exhaled Breath Condensate in Cystic Fibrosis. *American Journal of Respiratory and Critical Care Medicine,* Vol.167, No.8, (April 2003), pp. 1109-1112, ISSN 1073-449X

Chaubey, A. & Malhotra, B. (2002). Mediated Biosensors. *Biosensors & Bioelectronics,* Vol.17, No.6, (June-July 2002), pp. 441–456, ISSN 0956-5663

Cheli, F., Campagnoli, A., Pinotti, L., Savoini, G. & Dell'Orto, V. (2009). Electronic Nose for Determination of Aflatoxins in Maize. *Biotechnology, Agronomy, Society and Environment,* Vol.13, No.1, (January 2009), pp. 39-43, ISSN 1370-6233

Chen, S., Zieve, L. & Mahadevan, V. (1970). Mercaptans and Dimethyl Ssulfide in the Breath of Patients with Cirrhosis of the Liver. *Journal of Laboratory and Clinical Medicine,* Vol. 75, No.4, (April 1970), pp. 628-635, ISSN 0022-2143

Crofford, O.B., Mallard, R.E., Winton, R.E., Rogers, N.L., Jackson, J.C. & Keller, U. (1977). Acetone in Breath and Blood. *Transactions of the American Clinical and Climatological Association,* Vol.88, No.1, (October 1996), pp. 128-139, ISSN 0065-7778

De Lacy Costello, B., Ewer, A.K., Garner, C.E., Probert, C.S.J., Ratcliffe, N.M. & Smith, S. (2008). An Analysis of Volatiles in the Headspace of the Faeces of Neonates. *Journal of Breath Research,* Vol.2, No.3, (September 2008), pp. 1-8, ISSN 1752-7155

Dent, A. (2010) The Use of an Electronic Nose to Detect Chronic Lung Disease, In: *The Prince Charles Hospital Foundation,* Current Research Projects: Thoracic Research, Date of access April 6, 2011, Available from: http://www.tpchfoundation.org.au/getdoc/82c41668-103c-4739-8a2f-1b72c6e157ca/The-use-of-an-electronic-nose-to-detect-chronic-lu.aspx

Di Terlizzi, R. & Platt, S. (2009). The Function, Composition and Analysis of Cerebrospinal Fluid in Companion Animals: Part II - Analysis. *The Veterinary Journal,* Vol.180, No.1, (April 2009), pp. 15–32, ISSN 1090-0233

D'Souza, S. (2001). Microbial Biosensors. *Biosensors and Bioelectronics,* Vol.16, No.6, (August 2001), pp. 337–353, ISSN 0956-5663

Dutta, R., Hines, E.L., Gardner, J.L. & Boilot, P. (2002). Bacteria classification using Cyranose 320 electronic nose. *BioMedical Engineering OnLine,* Vol.1, No.4, (April 2002), pp. 1-7, ISSN 1475-925X

Dutta, R., Morgan, D., Baker, N., Gardner, J.W. & Hines, E.L. (2005). Identification of *Staphylococcus aureus* Infections in Hospital Environment: Electronic Nose Based Approach. *Sensors and Actuators B,* Vol.109, No.2, (September 2005), pp. 335-362, ISSN 0925-4005.

Dzyadevych, S.V., Arkhypova, V.N., Soldatkin, A.P., El'skaya, A.V., Martelet, C. & Jaffrezic-Renault, N. (2008). Amperometric Enzyme Biosensors: Past, Present and Future. *Innovation and Technology in Biological Medicine – Review of Biological Medicine,* Vol.29, No.2, (April 2008), pp. 171–180, ISSN 1297-9562

Erzurum, S.C., Burch, T., Laskowski, D., Mazzone, P.J., Mekhail, T., Jennings, C., Stoller, J.K., Machado, R.F., Pyle, J., Deffenderfer, O. & Dweik, R.A. (2005). Can the Electronic Nose Really Sniff out Lung Cancer? *American Journal of Respiratory and Critical Care Medicine,* Vol.172, No.8, (October 2005), pp. 1060-1061, ISSN 1073-449X

Flitti, F., Guo, B., Far, A.B., Bermak, A. (2008). A Robust and Low-complexity Gas Recognition Technique for On-chip Tin-oxide Gas Sensor Array. *Journal of Sensors,* Vol. 2008, No.1, (January 2008), pp. 1-6, ISSN 1687-725X

Fonner, J.M., Forciniti, L., Nguyen, H., Byrne, J.D., Kou, Y-F., Syeda-Nawaz, J. & Schmidt, C.E. (2008). Biocompatibility Implications of Polypyrrole Synthesis Techniques. *Biomedical Materials,* Vol.3, No.3, (September 2008), pp. 034124, ISSN 0955-7717

Frey, U. & Suki, B. (2008). Complexity of Chronic Asthma and Chronic Obstructive Pulmonary Disease: Implications for Risk Assessment, and Disease Progression and Control. *Lancet,* Vol. 372, No.9643, (September 2008), pp. 1088–1099, ISSN 0140-6736

Gardner, J.W., Craven, M., Dow, C. & Hines, E.L. (1998). The Prediction of Bacteria Type and Culture Growth Phase by an Electronic Nose with a Multi-layer Perceptron Network. *Measurement Science and Technology,* Vol.9, No.1, (January 1998), pp. 120-127, ISSN 0957-0233

Garner, C.E., Smith, S., Bardhan, P.K., Ratcliffe, N.M. & Probert, C.S. (2009). A Pilot Study of Fecal Volatile Organic Compounds in Feces from Cholera Patients in Bangladesh to Determine their Utility in Disease Diagnosis. *Transactions of the Royal Society of Tropical Medicine and Hygiene,* Vol.103, No.11, (November 2009), pp. 1171-1173, ISSN 0035-9203

Gawel, K., Barriet, D., Sletmoen, M. & Stokke, B.T. (2010). Responsive Hydrogels for Label-Free Signal Transduction within Biosensors. *Sensors*, Vol.10, No.5, (May 2010), pp. 4381-4409, ISSN 1424-8220

Goldberg, E.M. (1981). A Gas Chromatographic-Mass Spectrometric Study of Profiles of Volatile Metabolites in Hepatic Encephalopathy. *Journal of Chromatography B*, Vol.226, No.2, (December 1981), pp. 291-299, ISSN 1570-0232

Gomila, G., Casuso, I., Errachid, A., Ruiz, A., Pajot, E., Minic, J., Gorojankina, T., Persuy, M.A., Aioun, J., Salesse, R., Bausells, J., Villanueva, G., Rius, G., Hou, Y., Jaffrezic, N., Pennetta, C., Alfinito, E., Akimov, V., Reggiani, L., Ferrari, G., Fumagalli, L., Sampietro, M. & Samitier, J. (2006). Advances in the Production, Immobilization, and Electrical Characterization of Olfactory Receptors for Olfactory Nanobiosensor Development. *Sensors and Actuators B*, Vol.116, No.1, (July 2006), pp. 66–71, ISSN 0925-4005

Gordon, S.M., Szidon, J.P., Krotoszynski, B.K., Gibbons, R.D. & O'Neill, H.J. (1985). Volatile Organic Compounds in Exhaled Air from Patients with Lung Cancer. *Clinical Chemistry*, Vol.31, No.8, (August 1985), pp. 1278-1282, ISSN 0009-9147

Gralnek, I.M., Hays, R.D., Kilbourne, A., Rosen, H.R., Keeffe, E.B., Artinian, L., Kim, S., Lazarovici, D., Jensen, D.M., Busuttil, R.W. & Martin, P.M. (2000). Development and Evaluation of the Liver Disease Quality of Life Instrument in Persons with Advanced, Chronic Liver Disease—The LDQOL 1.0. *The American Journal of Gastroenterology*, Vol.95, No.12, (December 2000), pp. 3552-3565, ISSN 0002-9270

Guiseppi-Elie, A. (2010). Electroconductive Hydrogels: Synthesis, Characterization and Biomedical Applications. *Biomaterials*, Vol.31, No.10, (April 2010), pp 2701–2716, ISSN 0142-9612

Hayden, G.F. (1980). Olfactory Diagnosis in Medicine. *Postgraduate Medicine*, Vol.67, No.4, (April 1980), pp. 110-118, ISSN 1597-1627

Hentze, H.-P. & Antonietti, M. (2002). Porous Polymers and Resins for Biotechnological and Biomedical Applications. *Reviews in Molecular Biotechnology*, Vol. 90, No.1, (March 2002), pp. 27-53, ISSN 1389-0352

Hildebrandt, C., Raschner, C. & Ammer, K. (2010). An Overview of Recent Application of Medical Infrared Thermography in Sports Medicine in Austria. *Sensors*, Vol.10, No.5, (May 2010), pp. 4700-4715, ISSN 1424-8220

Huang, B., Tao, W., Shi, J., Tang, L. & Jin, J. (2006). Determination of Ochratoxin A by Polyclonal Antibodies based Sensitive Time-resolved Fluoroimmunoassay. *Archives of Toxicology*, Vol.80, No.8, (May 2006), pp. 481–485, ISSN 0340-5761

Humphreys, L., Orme, R.M.L., Moore, P., Charaklias, N., Sahgal, N., Planas, P.N., Magan, N., Stone, N. & Kendall, C.A. (2011). Electronic Nose Analysis of Bronchoalveolar Lavage Fluid. *European Journal of Clinical Investigation*, Vol.41, No.1, (January 2011), pp. 52-58, ISSN 0014-2972

Iwuoha, E.I., Wilson, A., Howel, M., Mathebe, N.G.R., Montane-Jaime, K., Narinesingh, D. & Guiseppi-Elie, A. (2004). Cytochrome P450 2DG (CYP2DG) Bioelectrode for Fluoxetine. *Analytical Letters*, Vol.37, No.5, (March 2004), pp. 943-956, ISSN 0094-8276

Jacoby, M. (2004). Breath Analysis for Medical Diagnosis. *Chemical and Engineering News*, Vol.82,No.13, (March 2004), pp. 29-31, ISSN 0009-2347

Kaji, H., Hisamura, M., Saito, N. & Murao, M. (1978). Gas Chromatographic Determination of Volatile Sulphur Compounds in Expired Alveolar Air in Hepatopathic Patients. *Journal of Chromatography B*, Vol.145, No.3, (May 1978), pp. 464-468, ISSN 1570-0232

Kaizu, T. (1976). Analysis of Volatile Sulphur Compounds in Mouth Air by Gas Chromatography. *Nippon Shishubyo Gakkai Kaishi (Journal of Clinical Periodontology)*, Vol.18, No.1, (March 1976), pp. 1-12, ISSN 1600-051X

Kateb, B., Ryan, M.A., Homer, M.L., Lara, L.M., Yin, Y., Higa, K. & Chen, M.Y. (2009). Sniffing Out Cancer Using the JPL Electronic Nose: A Pilot Study of a Novel Approach to Detection and Differentiation of Brain Cancer. *NeuroImage*, Vol. 47, Suppl.2, (August 2009), pp. T5-T9, ISSN 1053-8119

Kolk, A., Hoelscher, M., Maboko, L., Jung, J., Kuijper, S., Cauchi, M., Bessant, C., van Beers, S., Dutta, R., Gibson, T. & Reither, K. (2010). Electronic-Nose Technology Using Sputum Samples in Diagnosis of Patients with Tuberculosis. *Journal of Clinical Microbiology*, Vol.48, No.11, (November 2010), pp. 4235-4238, ISSN 0095-1137

Komaitis E., Vasiliou E. & Kremmydas, G. (2010). Development of a Fully Automated Flow Injection Analyzer Implementing Bioluminescent Biosensors for Water Toxicity Assessment. *Sensors,* Vol. 10, No.8, (August 2010), pp. 7089-7098, ISSN 1424-8220

Kost, G.J. (2002). Goals, guidelines and principles for point-of-care testing, In: *Principles & practice of point-of-care testing*. pp. 3–12, Lippincott Williams & Wilkins, ISBN 0-7817-3156-9, Hagerstwon, Maryland

Kostelc, J.G., Zelson, P.R., Preti, G. & Tonzetich, J. (1981). Quantitative Differences in Volatiles from Healthy Mouths and Mouths with Periodontitis. *Clinical Chemistry*, Vol.27, No.6, (June 1981), pp. 842-845, ISSN 0009-9147

Kumar, S. & Kumar, A. (2008). Recent Advances in DNA Biosensor. *Sensors and Transducers Journal,* Vol. 92, No.5, (May 2008), pp. 122-133, ISSN 1726-5479

Lazar, Z., Fens, N., van der Maten, J., van der Schee, M.P., Wagener, A.H., de Nijs, S.B., Dijkers, E. & Sterk, P.J. (2010). Electronic Nose Breathprints are Independent of Acute Changes in Airway Caliber in Asthma. *Sensors,* Vol.10, No.10, (October 2010), pp. 9127-9138, ISSN 1424-8220

Lee, J., Ngo, J., Blake, D., Meinardi, S., Pontello, A.M., Newcomb, R. & Galassetti, P.R. (2009). Improved Predictive Models for Plasma Glucose Estimation from Multi-linear Regression Analysis of Exhaled Volatile Organic Compounds. *Journal of Applied Physiology*, Vol.107, No.1, (May 2009), pp. 155-160, ISSN 8750-7587

Li, H., Wang, D.Q., Liu, B.L. & Gao, L.Z. (2004). Synthesis of a Novel Gelatin-carbon Nanotubes Hybrid Hydrogel. *Colloids and Surfaces B: Biointerfaces*, Vol.33,No.2, (February 2004), pp. 85-88, ISSN 0927-7765

Liao, K.-C., Chang , S.-C., Chiu, C.-Y. & Chou, Y.-H. (2010). Acute Response in vivo of a Fiber-Optic Sensor for Continuous Glucose Monitoring from Canine Studies on Point Accuracy. *Sensors*, Vol.10, No.8, (August 2010), pp. 7789-7802, ISSN 1424-8220

Liddell, K. (1976). Smell as a Diagnostic Marker. *Postgraduate Medicine,* Vol.52, No.605, (March 1976), pp. 136-138, ISSN 1597-1627

Lorenzelli, L., Benvenuto, A., Adami, A., Guarnieri, V., Margesin, B., Mulloni, V. & Vincenzi, D. (2005). Development of a Gas Chromatography Silicon-based

Microsystem in Clinical Diagnostics. *Biosensors and Bioelectronics*, Vol.20, No.10, (April 2005), pp. 1968–1976, ISSN 0956-5663

Luykx, D. & van Ruth, S.M. (2008). An Overview of Analytical Methods for Determining the Geographical Origin of Food Products. *Food Chemistry*, Vol.107, No.2, (March 2008), pp. 897-911, ISSN 0308-8146

Manolis, A. (1983). The Diagnostic Potential of Breath Analysis. *Clinical Chemistry*, Vol.29, No.1, (January 1983), pp. 5-15, ISSN 0009-9147

Maurer, K., Yazvenko, N., Wilmoth, J., Cooper, J., Lyon, W. & Danley, D. (2010). Use of a Multiplexed CMOS Microarray to Optimize and Compare Oligonucleotide Binding to DNA Probes Synthesized or Immobilized on Individual Electrodes. *Sensors*, Vol.10, No.8, (August 2010), pp. 7371-7385, ISSN 1424-8220

Moens, M., Smet, A., Naudts, B., Verhoeven, J., Ieven, M., Jorens, P., Geise, H. J. & Blockhuys, F. Fast Identification of Ten Clinically Important Micro-organisms Using an Electronic Nose. *Letters in Applied Microbiology*, Vol. 42, No.2, (February 2006), pp. 121-126, ISSN 0266-8254

Montuschi, P., Corradi, M., Ciabattoni, G., Nightingale, J., Kharitonov, S.A. & Barnes, P.J. (1999). Increased 8-isoprostane, a Marker of Oxidative Stress, in Exhaled Condensate of Asthma Patients. *American Journal of Respiratory and Critical Care Medicine*, Vol.160, No.1, (July 1999), pp. 216-220, ISSN 1073-449X

Montuschi, P., Santonico, M., Mondino, C., Pennazza, G., Mantini, G., Martinelli, E., Capuano, R., Ciabattoni, G., Paolesse, R., Di Natale, C., Barnes, P.J. & D-Amico, A. (2010). Diagnostic Performance of an Electronic Nose, Fractional Exhaled Nitric Oxide, and Lung Function Testing in Asthma. *Chest*, Vol. 137, No.4, (January 2010), pp. 790-796, ISSN 0012-3692

Mujahid, A., Stathopulos, H., Lieberzeit, P. & Dickert, F.L. (2010). Solvent Vapour Detection with Cholesteric Liquid Crystals - Optical and Mass-Sensitive Evaluation of the Sensor Mechanism. *Sensors*, Vol.10, No.5, (May 2010), pp. 4887-4897, ISSN 1424-8220

Novak, B.J., Blake, D.R., Meinardi, S., Rowland, F.S., Pontello, A., Cooper, D.M. & Galassetti, P.R. (2007). Exhaled Methyl Nitrate as a Noninvasive Marker of Hyperglycemia in Type 1 Diabetes. *Proceedings of the National Academy of Science*, Vol.104, No.40, (August 2007), pp. 15613-15618, ISSN 0027-8424

Pauling, L., Robinson, A.B., Teranishi, R. & Cary. P. (1971).Quantitative Analysis of Urine Vapor and Breath by Gas-Liquid Partition Chromatography. *Proceedings of the National Academy of Science USA*, Vol.68, No.10, (October 1971), pp. 2374-2376, ISSN 0027-8424

Persaud, K. (January 1, 2006). Wound Monitor - Mobile system for Non-invasive Wound State Monitoring, In: *The University of Manchester*, Date of access April 6, 2011, Available from: http://www.woundmonitor.manchester.ac.uk/

Pavlou, A.K. &Turner, A.P.F. (2000). Sniffing out the Truth: Clinical Diagnosis Using the Electronic Nose. *Clinical Chemistry and Laboratory Medicine*, Vol.38, No.2, (February 2000), pp. 99-112, ISSN 1434-6621

Phillips, M., Sabas, M. & Greenberg, J. (1993). Increased Pentane and Carbon Disulphide in the Breath of Patients with Schizophrenia. *Journal of Clinical Pathology*, Vol.46, No.9, (September 1993), pp. 861-864, ISSN 0021-9746

Phillips, M., Herrera, J., Krishnan, S., Zain, M., Greenberg, J. & Cataneo, R.N. (1999a). Variation in Volatile Organic Compounds in the Breath of Normal Humans. *Journal of Chromatography*, Vol.729,No.1-2, (June 1999), pp. 75-88, ISSN 1570-0232

Phillips, M., Gleeson, K., Huges, J.M.B., Greenberg, J., Cataneo, R.N., Baker, L. & McYay, W.P. (1999b). Volatile Organic Compounds in Breath as Markers of Lung Cancer: A Cross-sectional Study. *Lancet*, Vol.353, No.9168, (June 1999), pp. 1930-1933, ISSN 0140-6736

Phillips, M., Cataneo, R.N., Cummin, A.R., Gagliardi, A.J., Gleeson, K., Greenberg, J., Maxfield, R.A. & Rom, W.N. (2003a). Detection of Lung Cancer with Volatile Markers in the Breath. *Chest*, Vol.123, No.6, (June 2003), pp. 1788-1792, ISSN 0012-3692

Phillips, M., Cataneo, R.N., Ditkoff, B.A., Fisher, P., Greenberg, J., Gunawardena, R., Kwon, C.S., Rahbari-Oskoui, F. & Wong, C. (2003b). Volatiles Markers of Breast Cancer in the Breath. *The Breast Journal*, Vol.9, No.3, (May 2003), pp. 184-191, ISSN 1075-122X

Phillips, M.P., Boehmer, J.P., Cataneo, R.N., Cheema, T., Eisen, H.J., Fallon, J.T., Fisher, P.E., Gass, A., Greenberg, J., Kobashigawa, J., Mancini, D., Rayburn, B. & Zucker, M.J. (2004). Prediction of Heart Transplant Rejection with a Breath Test for Markers of Oxidative Stress. *American Journal of Cardiology*, Vol.94, No.12, (December 2004), pp. 1593-1594, ISSN 0002-9149

Ping, W.; Yi, P.; Haibao, X. & Farange, S. (1997). A Novel Method for Diabetes Diagnosis based on Electronic Nose. *Biosensors and Bioelectronics*, Vol.12, No.9, (November 1997), pp. 1031-1036, ISSN 0956-5663

Polat, K. & Güneş, S. (2006). The Effect to Diagnostic Accuracy of Decision Tree Classifier of Fuzzy and *k*-NN Based Weighted Pre-processing Methods to Diagnosis of Erythemato-squamous Diseases. *Digital Signal Processing*, Vol.16, No.6, (November 2006), pp. 922–930, ISSN 1051-2004

Preti, G., Lobows, J.N., Kostelc, J.G., Aldinger, S. & Daniele, R. (1988). Analysis of Lung Air from Patients with Bronchogenic Carcinoma and Controls using Gas Chromatography-Mass Spectroscopy. *Journal of Chromatography A*, Vol.432, No.11, (November 1988), pp. 1-11, ISSN 0021-9673

Rooth, G. & Ostenson, S. (1966). Acetone in Alveolar Air, and the Control of Diabetes. *Lancet*, Vol.2, No.7473, (November 1966), pp. 1102-1105, ISSN 0140-6736

Seising, R. (2006). From Vagueness in Medical Thought to the Foundations of Fuzzy Reasoning in Medical Diagnosis. *Artificial Intelligence in Medicine*, Vol.38, No.3, (November 2006), pp. 237-256, ISSN 0933-3657

Simenhoff, M.L., Burke, J.F., Saukkonen, J.J., Ordinaria, A.T. & Doty, R. (1977). Biochemical Profile of Uremic Breath. *The New England Journal of Medicine*, Vol.297, No.3, (July 1977), pp. 132-135, ISSN 0028-4793

Smith, K. & Sines, J. (1960). Demonstration of a Peculiar Odor in the Sweat of Schizophrenic Patients. *A.M.A. Archives of General Psychiatry*, Vol.2, No.1, (February 1960), pp. 184-188, ISSN 0003-990x

Smith, K., Thompson, G.F. & Koster, H.D. (1969). Sweat in Schizophrenic Patients: Identification of the Odorous Substance. *Science*, Vol.166, No.3903, (October 1969), pp. 398-399, ISSN 0036-8075

Smith, M. The Use of Smell in Differential Diagnosis. (1982). *Lancet*, Vol.320, No.8313, (December 1982), pp. 1452-1453, ISSN 0140-6736

Sparacino, G., Facchinetti, A. & Cobelli, C. (2010). "Smart" Continuous Glucose Monitoring Sensors: On-line Signal Processing Issues. *Sensors*, Vol. 10, No.7, (July 2010), pp. 6751-6772, ISSN 1424-8220

Studer, S.M., Orens, J.B., Rosas, I., Krishnan, J.A., Cope, K.A., Yang, S., Conte, J.V., Becker, P.B. & Risby, T.H. (2001). Patterns and Significance of Exhaled-breath Biomarkers in Lung Transplant Recipients with Acute Allograft Rejection. *The Journal of Heart and Lung Transplantation*, Vol.20, No.11, (November 2001), pp. 1158-1166, ISSN 1053-2498

Stvrtinova, V., Jakubovsky, J. & Hulin, I. (1995). *Inflammation and Fever Pathophysiology: Principles and Diseases*. Academic Electronic Press, ISBN 80-967366-1-2, Bratislava, Slovak Republic.

Thaler, E.R., Bruney, F.C., Kennedy, D.W. & Hanson, C.W. (2000). Use of an Electronic Nose to Distinguish Cerebrospinal Fluid from Serum. *Archives of Otolaryngology Head & Neck Surgery*, Vol.126, No.1, (January 2000), pp. 71-74, ISSN 0886-4470

Thanyani, S.T., Roberts, V., Siko, D.G., Vrey, P. & Verschoor, J.A. (2008) A Novel Application of Affinity Biosensor Technology to Detect Antibodies to Mycolic acid in Tuberculosis Patients. *Journal of Immunological Methods*, Vol. 332, No.1, (March 2008), pp. 61–72, ISSN 0022-1759

Tokonami, S., Shiigi, H. & Nagaoka, T. (2009). Review: Micro- and Nanosized Molecularly Imprinted Polymers for High-throughput Analytical Applications. *Analytica Chimica Acta*, Vol. 641, No. 1, (May 2009), pp. 7–13, ISSN 0003-2670

Torres, I., Baena, M.G., Cayon, M., Ortego-Rojo, J. & Aguilar-Diosdado, M. (2010). Use of Sensors in the Treatment and Follow-up of Patients with Diabetes Mellitus. *Sensors*, Vol.10, No.8, (August 2010), pp. 7404-7420, ISSN 1424-8220

Van den Velde, S., van Steenberghe, D., Van Hee, P. & Quirynen, M. (2009). Detection of Odorous Compounds in Breath. *Journal of Dental Research*, Vol.88, No.3, (March 2009), pp. 285-289, ISSN 0022-0345

Vasumathi , R. & Neelamegam, P. (2010). Development of Bio-analyzer for the Determination of Urinary Chloride. *Sensors & Transducers Journal*, Vol. 119, No.8, (August 2010) pp. 142-150, ISSN 1726-5479

Vlascici, D., Pruneanu, S., Olenic, L., Pogacean, F., Ostafe, V., Chiriac, V., Pica, E., Bolundut, L., Nica, L. & Fagadar-Cosma, E. (2010). Manganese(III) Porphyrin-based Potentiometric Sensors for Diclofenac Assay in Pharmaceutical Preparations. *Sensors*, Vol.10, No.10, (October 2010), pp. 8850-8864, ISSN 1424-8220

Vo-Dinh, T. & Cullum, B. (2000). Biosensors and Biochips: Advances in Biological and Medical Diagnostics. *Fresenius' Journal of Analytical Chemistry*, Vol. 366,No.6, (March 2000), pp. 540-551, ISSN 0937-0633

Voss, A., Baier, V., Reisch, R., Von Roda, K., Elsner, P., Ahlers, H. & Stein, G. (2005). Smelling Renal Dysfunction via Electronic Nose. *Annals of Biomedical Engineering*, Vol.33, No.5, (May 2005), pp. 656-660, ISSN 0090-6964

Xue, R., Dong, L., Zhang, S., Deng, C., Liu, T., Wang, J. & Shen, X. (2008). Investigation of Volatile Biomarkers in Liver Cancer Blood using Solid-Phase Microextraction and

Gas Chromatography/Mass Spectrometry. *Rapid Communications in Mass Spectrometry*, Vol.22, No.8, (April 2008), pp. 1181-1186, ISSN 1097-0231

Yoo, E.-H. & Lee, S.-Y. (2010). Glucose Biosensors: An Overview of Use in Clinical Practice. *Sensors*, Vol. 10, No.5, (May 2010), pp. 4558-4576, ISSN 1424-8220

Zhu, L., Seburg, R.A., Tsai, E., Puech, S. & Mifsud, J.C. (2004). Flavor Analysis in a Pharmaceutical Oral Solution Formulation Using an Electronic Nose. *Journal of Pharmaceutical and Biomedical Analysis*, Vol.34, No.3, (February 2004), pp. 453-461, ISSN 0731-7085

16

Quality Assurance and Quality Control of Equipment in Diagnostic Radiology Practice - The Ghanaian Experience

Stephen Inkoom[1], Cyril Schandorf[2],
Geoffrey Emi-Reynolds[1] and John Justice Fletcher[2]
[1]Radiation Protection Institute, Ghana Atomic Energy Commission, Accra;
[2]School of Nuclear and Allied Sciences, University of Ghana Atomic Campus, Accra;
Ghana

1. Introduction

The World Health Organization (WHO) defines a quality assurance (QA) programme in diagnostic radiology as an organized effort by the staff operating a facility to ensure that the diagnostic images produced are of sufficiently high quality so that they consistently provide adequate diagnostic information at the lowest possible cost and with the least possible exposure of the patient to radiation: (World Health Organization [WHO], 1982). The nature and extent of this programme will vary with the size and type of the facility, the type of examinations conducted, and other factors. The determination of what constitutes high quality in any QA programme will be made by the diagnostic radiology facility producing the images. The QA programme must cover the entire X-ray system from machine, to processor, to view box.

Quality assurance actions include both quality control (QC) techniques and quality administration procedures. QC is normally part of the QA programme and quality control techniques are those techniques used in the monitoring (or testing) and maintenance of the technical elements or components of an X-ray system. The quality control techniques thus are concerned directly with the equipment that can affect the quality of the image i.e. the part of the QA programme that deals with instrumentation and equipment. An X-ray system refers to an assemblage of components for the controlled production of diagnostic images with X-rays. It includes minimally an X-ray high voltage generator, an X-ray control device, a tube-housing assembly, a beam-limiting device and the necessary supporting structures. Other components that function with the system, such as image receptors, image processors, automatic exposure control devices, view boxes and darkrooms, are also parts of the system. The main goal of a QC programme is to ensure the accuracy of the diagnosis or the intervention (optimising the outcome) while minimising the radiation dose to achieve that objective

In a typical diagnostic radiology facility, QC procedures may include the following:

a. Acceptance test and commissioning

Acceptance test is performed on new equipment to demonstrate that it is performing within the manufacturer's specifications and criteria (and also to confirm that the equipment meets the purchaser's specifications i.e. the requirements of the tender). Commissioning is the process of acquiring all the data from equipment that is required to make it clinically useable in a specific department. This commissioning test will give the baseline values for the QC procedures.

b. Constancy tests

Constancy tests are performed at specific intervals to check on the performance of some key parameters. The frequencies reported for the control of constancy may be with a tolerance of ±30 days.

c. Status tests

Stautus tests are normally performed with full testing at longer periods, e.g. annually.

d. Performance test

Performance tests are specific tests performed on an X-ray system after a pre-determined period of time.

e. Verification of radiation protection (RP) and QC equipment and material
f. Follow up of necessary corrective actions taken in response from previous results of QC procedures. This is important because simply performing QC measurements without documentation of corrective actions and a follow ups are not sufficient.

On the other hand, quality administration procedures are those management actions intended to guarantee that monitoring techniques are properly performed and evaluated and that necessary corrective measures are taken in response to monitoring results. These procedures provide the organizational framework for the quality assurance programme. A diagnostic radiology facility as used in this sense refers to any facility in which an X-ray system(s) is used in any procedure that involves irradiation of any part of the human or animal body for the purpose of diagnosis or visualisation. Offices of individual physicians, dentists, podiatrists, chiropractors, and veterinarians as well as mobile laboratories, clinics, and hospitals are examples of diagnostic radiology facilities.

A quality assurance programme should contain the following elements listed below:

1. Responsibility.

There must be a clear assignment of responsibility and authority for the overall quality assurance programme as well as for monitoring, evaluation, and corrective measures. Responsibilities for certain quality control techniques and corrective measures may also be assigned to personnel qualified through training and experience, such as qualified experts or representatives from maintenance personnel outside the facility.These should be specified and written in a quality assurance manual.

2. Purchase specifications.

The purchasing specifications for diagnostic radiology equipment should be in writing and should include performance specifications. Staff of the diagnostic radiology facility should determine the desired performance specifications for the equipment.

3. Monitoring and maintenance.

A routine quality control monitoring and preventive maintenance system incorporating state of the art procedures should be established. This should be performed properly and according to a planned timetable.

4. Standards for image quality.

Standards of acceptable image quality which are diagnostic enough should be established. This should be comparable to International Standards such as the quality criteria established by the European Commission (European Commission 1996a, 1996b, 1999 & Bongartz et al., 2004). Ideally these should be objective as much as possible, e.g., acceptability limits for the variations of parameter values, but they may be subjective, e.g. the opinions of professional personnel, in cases where adequate objective standards cannot be adequately defined. These standards should be routinely reviewed and redefined as and when the need arises.

5. Evaluation.

The facility quality assurance programme should make provisions for results of monitoring procedures to evaluate the performance of the X-ray system(s) to determine whether corrective actions are needed to adjust the equipment so that the image quality consistently meets the standards for image quality. Additionally, the facility quality assurance programme should also include means for evaluating the effectiveness of the programme itself.

6. Records.

The programme should include provisions for the keeping of records on the results of the monitoring techniques, any difficulties detected, the corrective measures applied to these difficulties, and the effectiveness of these measures. Typically, records should contain the following:

- Results of the calibration and verification of the measurement instruments,
- Results of acceptance and quality control tests,
- Patient dosimetry results and comparison with guidance or diagnostic reference levels (DRLs),
- Inventory of X-ray systems.

7. Manual

A quality assurance manual should be written in a format which permits convenient revision as needed and should be made readily available to all personnel.

8. Education and training.

A quality assurance programme should make provisions for adequate training for all personnel with quality assurance responsibilities. The training should be specific to the facility and the equipment in use.

9. Setting up of committee.

Large facilities such as teaching or referral or specialist hospitals should consider the establishment of a quality assurance committee whose primary function would be to maintain lines of communication among all groups with quality assurance and/or image production or interpretation responsibilities.

The extent to which each of these elements of the quality assurance programme is implemented should be determined by an analysis of the facility's objectives and resources conducted by its qualified staff or by qualified outside consultants. Implementation should also be based on Regulatory requirements (Regulations, Codes or Guides), Health Service Policy as well as the Hospital's Local Rules on the application of ionising radiation. The expected benefits from any additional actions should be evaluated by comparing to the resources required for the programme.

Several studies have indicated that many diagnostic radiological facilities produce poor quality images and give unnecessary radiation exposure to patients. Inkoom et al. recommends for the institution of regular assessment of QC parameters that affect patient dose and image quality at diagnostic facilities, since patient protection is an essential element for the overall management of patient undergoing X-ray examination (Inkoom et al., 2009).

A QA programme should also address issues of radiation protection in the diagnostic radiology. This will ensure that the image quality of radiographs meet minimum quality criteria for confident diagnosis, patient doses are as low as reasonable achievable (ALARA) and exploration of optimisation options. For instance, the International Basic Safety Standards (BSS) (BSS, 1996) requires Licensee / Registrant to;

- establish the Radiation Protection Programme (RPP),
- provide the necessary resources to properly apply the RPP,
- ensure that the RPP addresses all phases of diagnostic and interventional radiology from purchase, installation, maintenance, qualifications and training of users. etc. and
- ensure appropriate protection for patients, staff and members of the public.

This paper reviews the current QA programme and QC for diagnostic radiology practice in Ghana. The state of equipment in clinical use, QC measurements that are done, Regulatory Guidelines for QA/QC and what holds for the future are presented.

2. Equipment used in diagnostic radiology practice in Ghana

The inventory of number of items of diagnostic X-ray equipment in Ghana is compared with Health-care level III category of Zimbabwe (UNSCAER 2008 Report, 2010) as shown in Table 1.

Country	X-ray generators							
	Medical (General)	Mammo-graphy	Dental	Interven-tional	General fluoroscopy	Angio-graphy	Bone densito-metry	CT scanners
Health-care level III								
Zimbabwe*	250	2	200	2	30	15	-	8
Ghana+	230	8	17	-	9	1	2	11

* (UNSCEAR 2008 Report, 2010)
+ (Regulatory Authority Information System [RAIS], 2010).

Table 1. Comparison of number of items of diagnostic X-ray equipment between Ghana and Zimbabwe

2.1 Human resource present

As a third world country, a major challenge confronting diagnostic radiology practice is the availability of the requisite human resources. The various categories of Radiographic Staff available in Ghana is shown in Table 2.

For instance, earlier Consultant Radiologists were trained overseas until the last five years when training of Radiologists started in Ghana and the accreditation is given by either the Ghana College of Surgeons or the West African College of Physicians and Surgeons. The School of Allied Health Sciences (SAHS), College of Health Sciences (CHS) of the University of Ghana (UG) came into being in the year 2001, after an initiative from Ghana's Ministry of Health to produce medical and dental technical graduates in physiotherapy, medical laboratory science and radiography. Since its inception, SAHS has trained more than 200 radiographers.

Similarly, most Medical Physicists in Ghana were trained abroad, until 2004 when the School of Allied Health Sciences began training Medical Physicists after it admitted the first batch of six students to pursue the M.Phil degree in Medical Physics. Subsequently training of eight more Medical Physicist has been taken over from SAHS by a Post-Graduate School of Nuclear and Allied Sciences (SNAS). Currently, there are four students in training.

As part of measures aimed at training the requisite human resoursce in nuclear science applications, a Post-Graduate School of Nuclear and Allied Sciences has been established jointly by the Ghana Atomic Energy Commission and University of Ghana, in co-operation with the International Atomic Energy Agency (IAEA). The SNAS has been designated by the IAEA as African Regional Cooperative Agreement for Research, Development and Training Related to Nuclear Science and Technology (AFRA) Centre to assist in training engineers and scientists from neighbouring countries and the sub-region.

Radiographic Staff category	Number
Population estimate, 2010 [a]	24,000,000
GDP per capita, 2010 [b]	US$ 1,609
Radiologists [c]	25
Radiographers (B. Sc.) [d]	104
Radiographers (Diploma) [d]	97
X-ray Technicians [e]	149
Medical Physicists [f]	26

[a] (Wikipaedia, 2010); [b] (International Monitory Fund [IMF], 2010); [c] (Ghana Association of Radiologist, 2011); [d] (School of Allied Health Sciences, University of Ghana, 2010); [e] (Korle-Bu Teaching Hospital, 2006); [f] (International Organisation for Medical Physicist [IOMP], 2009)

Table 2. Categories of radiographic staff in Ghana

The number of physicians and health care professionals in Ghana is also compared with that of Health-care level III category under UNSCEAR 2008 Report and WHO Health Statistics for 2010, which is shown in Table 3.

3. Advances in technology

The transition of film screen radiography to computed radiography (CR) and digital radiography (DR) is anticipated to increase in Ghana. Currently, DR and CR systems account for about 4% of conventional X-ray machines in Ghana. With the introduction of digital X-ray systems in medical imaging, QC is becoming increasingly more important. One of the reasons is that overexposed detectors, which provided a natural dose limitation for conventional image receptor systems are no longer observed in digital systems (Zoetelief et al., 2008). Also, such new technology brings with it new challenges in terms of its control and quality assurance management. In view of this, KCARE (KCARE 2005a, 2005b) have developed protocols for both CR and DR receptors; Institute of Physics and Engineers in Medicine [IPEM], (2005) have expanded their X-ray system tests to encompass digital technologies; American Association of Physicist in Medicine (AAPM) have also published a protocol for CR QA (AAPM, 2006).

The generators and X-ray tubes that are used in the radiographic systems for both CR and DR remain the same as their film screen system counterparts and QA of the X-ray tube and generators in digital systems follows the standard methods (IPEM, 2005). However, it must be noted that whenever automatic exposure control (AEC) system is selected, the X-ray output is linked (directly or indirectly) to the detector performance and this demands consideration. This can lead to an increase or decrease in patient dose when the X-ray system becomes faulty or changes in the output consistency occurs. The detectors that are currently available in CR and DR have a wide exposure dynamic range which means there is significant potential for the initial setup of such systems not to be optimised (Medicines and Healthcare products Regulatory Agency [MHRA], 2010).

Number	Health-care level III	Country	
		Zimbabwe*	Ghana
Population (thousand)		12 000	24 000
All physicians		13	1(2587)^
Physicians conducting radiological procedures		15	25 [c]
Radiology technicians		180	350 [d, e]
Medical Physicists		4	26 [f]
Interventional Cardiologists		-	-
Other Physicians performing radiology		-	-
Dentists		200	0.5

* (UNSCEAR 2008 Report, 2010); [c] (Ghana Association of Radiologist, 2011); [d] (School of Allied Health Sciences, University of Ghana, 2010); [e] (Korle-Bu Teaching Hospital, 2006); [f] (International Organisation for Medical Physicist [IOMP], 2009); ^ World Health Organization, World Health Statistics, ISBN 978 92 4 156397 7, France Note: the values in the bracket represent the actual numbers.

Table 3. Comparison of physicians and health care professionals with UNSCEAR 2008 Report and WHO 2010 Health Statistics

Another part of the radiographic chain which is often neglected is the performance of monitors. Subjective evaluations of image quality assessment are made at a workstation/review monitor and as such this must be part of the QA programme. In the era of CTs, there has also been a transition from single slice to multi-slice CT and Ghana's first 64 multi-slice CT together with other accessories like cardiac monitor and automatic contrast agent injector has been installed recently. Indications are that the transition from film screen technology to digital technology is expected to be very rapid in Ghana. This calls for re-organisation and re-alignment of current structures by all relevant stakeholders of the diagnostic imaging community so as to face the challenges that this new technology offers.

4. Regulatory guidelines for quality assurance/quality control measurements

In Ghana, the National Competent Regulatory Authority charged with the responsibility for Authorisation and Inspection of practices using radiation sources and radioactive materials is the Radiation Protection Board (RPB) (Radiation Protection Instrument LI 1559, 1993). The Regulatory Authority was established in 1993 by the Provisional National Defence Council (PNDC) Law 308. The PNDC law 308 was an amendment of the Atomic Energy Act 204 of 1963 (Atomic Energy Act 204, 1963), which has been superseded by the Atomic Energy Act 588 of 2000 (Atomic Energy Act 588, 2000). However, before the inception of RPB, the Health Physics Department of the Ghana Atomic Energy Commission (GAEC) was providing QC and other services like environmental monitoring and film badge services in Ghana. RPB now has a memorandum of understanding with the National Health Service in order to address issues of ionizing radiation in the health delivery sector.

Just as acceptance testing and routine quality control testing of diagnostic imaging equipment are the requirements of European (Council Directive 97/43/ EURATOM, 1997) and many other national legislations, the LI 1559 of 1993 also requires Registrants and Licensees to establish a comprehensive QA programme for medical exposures with the participation of appropriate qualified experts in radiation physics taking into account the principles established by the WHO and the Pan American Health Organization (PAHO).
The operational functions of the RPB are carried out by the Radiation Protection Institute (RPI), which was established by the Ghana Atomic Energy Commission in 2000 to provide scientific and technical support for the enforcement of the legislative instrument, LI 1559. Some major activities that are undertaken by RPI include:

- conducting regulatory inspections and safety assessments for purposes of authorisation and enforcement of the requirements of the LI 1559 of 1993,
- promoting human resource development in radiation protection, safety and nuclear security by promoting training of regulatory staff and organising courses for registrants and licensees,
- carrying out radiation and waste safety services, and
- carrying out relevant research to enhance protection of workers, patients, the public and the environment from the harmful effects of ionising radiation and the safety and security of radiation sources.

In exercise of the powers conferred by regulations 8 (2) and 11 (c & e) of the Legislative Instrument LI 1559 of 1993, RPB has issued the following Guides to ensure compliance with the Regulations intended to protect patients, workers and the general public from the risks associated with exposure to ionising radiation in the course of operating a practice in Ghana. In all, it has issued ten Guides which are listed below:

1. Radiation Protection and Safety Guide No. GRPB-G1-Qualificaiton and Certification of Radiation Protection Personnel (Schandorf et al., 1995).
2. Radiation Protection and Safety Guide No. GRPB-G2-Notificaiton and Authorisation by Registration or Licensing, (Schandorf et al., 1995).
3. Radiation Protection and Safety Guide No. GRPB-G3-Dose Limits, (Schandorf et al., 1995).
4. Radiation Protection and Safety Guide No. GRPB-G4-Inspection, (Schandorf et al., 1995).
5. Radiation Protection and Safety Guide No. GRPB-G5-Safe Use of X-Rays, (Schandorf et al., 1998).
6. Radiation Protection and Safety Guide No. GRPB-G6-Safe Transport of Radioactive Material, (Schandorf et al., 2000).
7. Radiation Protection and Safety Guide No. GRPB-G7-Enforcement, (Schandorf et al., 2000).
8. Radiation Protection and Safety Guide No. GRPB-G8-Occupational Radiation Protection, (Schandorf et al., 2000).
9. Radiation Protection and Safety Guide No. GRPB-G9-Medical Exposure, (Schandorf et al., 2003).
10. Radiation Protection and Safety Guide No. GRPB-G10-Safe Application of Industrial Radiography, (Schandorf et al., 2003).

Currently there are Institutional reforms to establish an independent Regulatory Body to regulate the peaceful uses of nuclear energy which will be known as Ghana Nuclear Regulatory Authority (GNRA), independent from Ghana Atomic Energy Commission as it is currently. The current Regulatory functions of RPB will then be transferred to the new Regulatory Authority (GNRA).

5. Present trend of quality assurance/quality control of diagnostic radiology in Ghana

For the present trend, the Regulatory Authority is still largely in charge of QA/QC of diagnostic radiology in Ghana, which ideally is supposed to be an external audit. This practice has been so due to the non-availability of qualified personnel (medical physicists, radiation protection experts, health physicists, etc.) to man diagnostic facilities, and also this requirement not being a major one for granting of authorisation as is in radiotherapy practice in which qualified personnel availability is mandatory.

The QA/QC is done through Regulatory inspections that are undertaken by the Radiation Protection Institute to conduct safety assessment for the issuance of authorisations. The safety assessment includes detailed inventory of X-ray equipment, availability of skilled and trained operators, adequacy of personal monitoring, health status and structural shielding adequacy with respect to actual practice, usage of personal protective devices for staff and comforters, usage of radiation protection devices for patients, etc. All these parameters which are related to radiation protection are verified and checked.

The inspections are conducted every one to three years depending upon the risk classification of practice and also, whenever there is a major maintenance or change of some key components of the X-ray system.

Some quality control measurements that are supposed to be done (because not all parameters listed under each measurement is currently carried out) to monitor the following key components of the X-ray system are:

a. Film-processing.
b. Basic performance characteristics of the X-ray unit.
c. Cassettes and grids.
d. Darkroom.
e. For specialised equipment.
f. View boxes.

Some parameters of the above-named components and of more specialised equipment that are supposed to be monitored are as follows:

a. For film processing:
   An index of speed.
   An index of contrast.
   Base plus fog.
   Darkroom and solution temperatures.
   Processor condition, film artifact identification.
   Cassettes, intensifying screens, film, etc.
b. For basic performance characteristics of the X-ray unit:
   1. For fluoroscopic X-ray units:
      Tabletop exposure rates.
      Centering alignment.
      Collimation.
      kVp accuracy and reproducibility.
      mA accuracy and reproducibility.
      Exposure time accuracy and reproducibility.
      Reproducibility of X-ray output.
      Focal spot size consistency.

Half-value layer.
Max air kerma rate and air kerma rate at the entrance of patient.
Calibration of kerma area product (KAP) meter.
Radiation leakage.
Relationship between current and voltage stabilising.

2. For image-intensified systems, the following tests are required in addition to (1) above:
   Focusing.
   Distortion.
   Glare.
   Low contrast resolution.
   Spatial resolution with high contrast
   Physical alignment of camera and collimating lens.
   Air kerma rate at the entrance of image rececptor.
   Distance from focus to Image receptor.
3. For radiographic X-ray units with screen-film:
   Reproducibility of X-ray output.
   Linearity and reproducibility of mA/mAs.
   Reproducibility and accuracy of timer.
   Reproducibility and accuracy of kVp.
   Accuracy of source-to-film distance indicators.
   Light/X-ray field congruence.
   Half-value layer.
   Focal spot size consistency.
   Representative ESAK
   X-ray tube housing leakage
4. For radiographic X-ray units with CR and DR:
   In addition to the tests in (3), the following tests are needed.
   Detector dose indicator consistency/sensitivity (for 1 plate of each size)
   Uniformity
   Dark noise
   Threshold contrast detail detectability
   Limiting spatial resolution (in one quadrant at 45^0 only)
   Erasure cycle efficiency
   Scaling errors
   Blurring and stiching artefacts
   Dosimetry (receptor doses)
5. For mammographic X-ray units with screen-film
   Reproducibility of X-ray output.
   Linearity and reproducibility of mAs.
   Reproducibility and accuracy of timer.
   Reproducibility and accuracy of kVp.
   Accuracy of source-to-film distance indicators.
   Light/X-ray field congruence.
   Half-value layer.
   Focal spot size consistency.
   X-ray tube housing leakage
   Mean glandular dose.

6. For mammographic X-ray units with CR and DR
   Reproducibility of X-ray output.
   Linearity and reproducibility of mAs.
   Reproducibility and accuracy of timer.
   Reproducibility and accuracy of kVp.
   Accuracy of source-to-film distance indicators.
   Half-value layer.
   Light/X-ray field congruence.
   Focal spot size consistency.
   X-ray tube housing leakage
   Mean glandular dose.
7. For dental X-ray units
   Reproducibility of X-ray output.
   Linearity and reproducibility of mAs.
   Reproducibility and accuracy of kVp.
   Accuracy of source-to-film distance indicators.
   Half-value layer.
   Focal spot size consistency.
   Representative ESAK.
8. For automatic exposure control devices:
   Reproducibility.
   kVp compensation.
   Field sensitivity matching.
   Minimum response time.
   Backup timer verification.

c. For cassettes and grids:
   1. For cassettes:
      Film/screen contact.
      Screen condition.
      Light leaks.
      Artefact identification.
   2. For grids:
      Alignment and focal distance.
      Artefact identification.

d. For darkroom:
   Darkroom integrity.
   Safe light conditions.

e. For specialised equipment:
   1. For tomographic systems:
      Accuracy of depth and cut indication.
      Thickness of cut plane.
      Exposure angle.
      Completeness of tomographic motion.
      Flatness of tomographic field.
      Resolution.
      Continuity of exposure.
      Flatness of cassette.
      Computed tomography dose index.

2. For computerised tomography:
   Precision (noise).
   Linearity and contrast scale.
   Spatial resolution with high contrast.
   Low contrast resolution.
   Alignment light/slice congruence.
   Mean CT Number.
   Slice thickness.
   Computed tomography dose index.
   Positioning the patient support.
   Sensitivity profile of slices.
   Coronal and Saggital resolution.

f. View boxes
   Consistency of light output with time.
   Consistency of light output from one box to another.
   View box surface conditions.

5.1 Ghana's participation in IAEA project

Ghana is involved in several IAEA Technical Cooperation Projects, but one of significant importance to the subject matter under discussion is on Strengthening Radiological Protection of the Patient and Medical Exposure Control. The main objectives of this Project are to upgrade / strengthen radiological protection of the patient in medical exposures due to:

i. Diagnostic Radiology and Interventional Radiological procedures,
ii. Nuclear Medicine procedures and
iii. Radiotherapy practice.

Ghana is participating in four tasks of the Project which are:

1. Surveys of image quality and patient doses in simple radiographic examinations; establishing guidance levels and comparison with international standards.
2. Survey of mammography practice from the optimisation of radiation protection view point.
3. Patient dose management in computed tomography with special emphasis to paediatric patients.
4. Taking steps to avoiding accidental exposure in radiotherapy.

For task (1) above, the entrance surface air kerma (ESAK) in some selected X-ray rooms were estimated from output data of the X-ray machine. A calibrated Ionisation chamber was used to measure air kerma (in mGy) at 1 m focus-detector-distance for different kVp settings. The values of X-ray tube output (in mGy/mAs) were plotted against tube potential (kVp) and the resulting output-kVp curve fitted to a square function. Then at the indicted kVp, the analytical equation (1) was used to evaluate the ESAK.

$$ESAK = Y(kVp, FFD) * mAs * \left[\frac{100}{FSD}\right] * BSF \quad \text{(mGy)} \qquad (19)$$

where

Y(kVp, FFD) is tube output for actual kVp used during examination (derived from mGy/mAs-kVp curve) at 1 m, mAs is actual tube current-time product used during

examination, FSD is the difference between the focus-to-film distance (FFD) and patient thickness (in m) in the anatomic region of interest, BSF is the backscatter factor.
The mean entrance surface air kerma estimates from six X-ray rooms from Ghana and other African countries that participated in the IAEA project is shown in Table 4 (Muhogora et al., 2008).

Radiographic Projection	Entrance surface air kerma (mGy)							
	Congo	Ghana	Madagascar	Sudan	Tanzania	Zimbabwe	Diagnostic Reference Level (Rehani, 2001)	
							400 Film-Screen	200 Film-Screen
Chest, posteroanterior	0.3	0.1	0.29	0.21	0.3	0.2	0.2	0.4
Lumbar spine, anteroposterior	0.4	8.3	3.92	1.63	2.1	0.7	5.0	10.0
Lumbar spine, lateral	-	14.4	6.61	3.29	4.7	2.0	15.0	30.0
Abdomen, anteroposterior	0.3	10.3	3.92	1.5	0.9	0.6	5.0	10.0
Pelvis, anteroposterior	0.1	7.0	3.92	0.9	1.5	1.1	5.0	10.0
Skull, anteroposterior	-	-	2.95	1.02	-	0.8	2.5	5.0

Dash (-) indicates that data not available.

Table 4. Mean entrance surface air kerma to adult patients before implementing a quality control program in participating centers in Africa (Muhogora et al., 2008)

Data on technique factors used for most computed tomography (CT) examinations (head, chest & abdomen) and the frequency of examinations / year for both adult and paediatric patients were collected from four hospitals, which is shown in Table 5.

Hospital	Examination	Number / year	
		Adult	Paediatric
A	Head	2080	780
	Chest	520	520
	Abdomen	520	520
B	Head	1820	520
	Chest	520	-
	Abdomen	1040	-
C	Head	1300	260
	Chest	520	520
	Abdomen	780	260
D	Head	5200	520
	Chest	780	260
	Abdomen	1300	260

Dash (-) indicates that no data was available at the time of the survey

Table 5. Frequency of CT examinations surveyed in each hospital

For Task (3) above, the CT dose descriptors that were used were weighted and volume computed tomography dose index ($CTDI_w$, $CTDI_{vol}$) and dose length product (DLP). Computed Tomography Dose Index (CTDI) is the patient CT dose defined as the integrated dose profile (in z-direction) for a single slice, normalised to the nominal slice thickness and the DLP for a complete examination. The DLP takes into account the scan length and number of sequences. Standard methods were used to determine the CT dose descriptors [European Commission 1999, McNitt-Gray 2002, Wall 2004].
The summary of the mean $CTDI_w$ values for adults from four participating hospitals in Ghana for each CT procedure is shown in Table 6 together with other countries that participated in the project (Muhogora et al., 2009).

By country	Method	Mean $CTDI_w$ (mGy)[a]					
			Chest	Chest HR	Lumbar spine	Abdomen	Pelvis
		DRL	(30)	(35)	(35)	(35)	(35)
Algeria	P		9.2	6.8	16.2	15.4	19.1
Ghana	P or I		17.1	17.2	20.4	20.4	20.4
Kenya	P		20	-	-	13	20
Morocco	P		10	25.8	11.9	11.9	10.6
Sudan	P, I or C		19.2	14.1	-	20.5	7.3
Tanzania	I		16.8	13.9	38.8	22.7	26
Tunisia	C		24.3	-	-	-	25.4
Japan	C		14	15	19.3	19.3	19.3
Kuwait	C		12	18.2	-	11.7	-
Syria	P		18.6	24.3	-	21.6	28.4
Thailand	C or P		15.3	14.4	19.5	18.5	16.8
Bulgaria	P		16.7	14.7	20.7	16.3	18.2
Czech Republic	P, I or C		21.3	16.9	23.9	18.4	20.3
Bosnia & Herz.	P		13.5	20.6	21.2	21.2	20.1
Srpska B & H	C		6.9	-	22.8	10.2	8.3
Estonia	C		15.7	-	-	19	14.5
FYROM	I		11.4	-	-	13	11.4
Malta	C		11.5	10	15.4	14.8	21.8
Serbia	C		20.1	-	12.3	12.3	14.1

Dash (-) indicates that data not available.
The Federation of Bosnia and Herzegovina is stated as Bosnia & Herz, Republic of Srpska as Srpska B&H and the former Yugoslav Republic of Macedonia as FYROM.
[a]For examinations of the trunk, calculated values of $CTDI_w$ relate to the 32 cm diameter CT dosimetry phantom (Shrimpton et al. 2006).

Table 6. Mean $CTDI_w$ values for adult patients in different countries. The determination method is indicated as based on phantom measurements (P), calculation by Internet data (I) or display of console (C). The DRL (European Commission, 1999) is shown in brackets (Muhogora et al., 2009)

The summary of the mean DLP values for adults from four participating hospitals in Ghana for each CT procedure is shown in Table 7 together with other countries that participated in the project (Muhogora et al., 2009).

By country	Mean DLP (mGy.cm)[a]					
	DRL	Chest	Chest HR	Lumbar spine	Abdomen	Pelvis
		(650)	(280)	(780)	(780)	(570)
Algeria		347	194	646	554	604
Ghana		396	348	523	496	415
Kenya		933	-	-	1314	837
Morocco		256	121	341	341	271
Sudan		423	171	-	725	163
Tanzania		382	366	363	602	494
Tunisia		874	-	-	-	599
Japan		564	404	513	513	513
Kuwait		223	561	-	552	-
Syria		416	103	-	638	545
Thailand		301	99	720	574	390
Bulgaria		512	-	-	435	322
Czech Republic		341	-	507	444	466
Bosnia & Herz.		437	330	460	460	323
Srpska B & H		246	-	541	448	231
Estonia		833	-	-	910	698
FYROM		342	-	-	526	416
Malta		296	117	289	480	268
Serbia		148	-	512	512	305

Dash (-) indicates that data not available.
The Federation of Bosnia and Herzegovina is stated as Bosnia & Herz, Republic of Srpska as Srpska B&H and the former Yugoslav Republic of Macedonia as FYROM.
[a] For examinations of the trunk, calculated values of DLP relate to the 32 cm diameter CT dosimetry phantom (Shrimpton et al. 2006)

Table 7. Mean DLP values for adult patients in different countries. The DRL (European Commission, 1999) is shown in brackets (Muhogora et al., 2009)

The results of the $CTDI_w$ and DLP show some wide variations, with some CT centres recording values greater than diagnostic refrence levels. This calls for some optimisation studies in order to reduce patient dose without a compromise in image quality.

6. Future of quality assurance/quality control

Optimisation of patient dose and image quality is of primary concern in the field of diagnostic imaging. It is recognised that comprehensive quality assurance programmes are a vital component of the optimisation process. Due to the importance of quality control in diagnostic imaging, it is recommended that the appropriate facility personnel review the control tests, data and images periodically (eg. quarterly reviews).

With the availability of training institutions like the School of Allied Health Sciences and the Post-Graduate School of Nuclear and Allied Sciences, more radiologic staff are expected to be churned out to meet the manpower needs of the diagnostic imaging community. For instance, the next 10-15 years, it is projected that about 100 Medical Physicists / Engineers are expected to be trained.

There are also plans for the registration of Ghana Society for Medical Physics (GSMP) association. GSMP will draw out necessary modalities to streamline the Education and Training of Medical Physicists and other professionals since Medical Physics experts are identified as one of the professional groups for whom training is mandatory. GSMP will also work on the accreditation and recognition of Medical Physics Profession in Ghana and job placement of Medical Physicists in Hospitals in Ghana, starting with the Teaching Hospitals.

It is expected that the human resources trained locally will be employed to establish Physics Units or Departments in the hospitals for the establishment of quality assurance programmes and quality control services that meets regulatory requirements and international best practices. The Medical Physicists will take charge of the routine QC procedures at their departments, undertake periodic dose audits and assist in the establishment of local reference levels and national guidance levels. These levels are to be compared with diagnostic reference levels and other international recommendations which are internationally recognised as a practical tool in the optimisation of radiological protection.

The independent GNRA when it becomes operational will put in place regulatory control system including authorisation, inspection and enforcement for the beneficial and peaceful uses of nuclear energy for all practices in Ghana. The GRNA is expected to revise/update the protocols that are currently being used to conduct safety assessment to authorise diagnostic radiology departments in order to keep pace with the emergence of modern medical equipment, and also due to the transition from screen-film technology to digital technology in the country. Additional equipment and test protocols will be needed in this regard. When the country attains the necessary critical mass of expertises, the RA may have to consider licensing some Technical Support Organizations (TSO) including Radiation Protection Institute, which will undertake some of the regulatory inspections of facilities on behalf of the Authority, and submit reports to the RA to issue the necessary authorisation.

A comprehensive review of all the RPB Guides that have been issued since 1995 to 2003 is necessary. This will address current challenges of diagnostic radiology practice due to rapid advances in technology. For instance current regulatory guidelines do not cover the application of non-ionising radiaion such as ultrasound and magnetic resonance imaging (MRI).

Quality control for view boxes conditions must be incorporated in the QA programme as this is also part of the radiographic chain. Parameters such as consistency of light output with time, consistency of light output from one box to another and view box surface conditions can be incorporated in the QC measurements.

When all appropriate QA programmes are put in place, these will enable the facility to recognise when parameters are out of limits, which could result in poor quality images and can increase the radiation exposure to patients (Compliance Guidance of Radiographic Quality Control, 2003).

7. Conclusion

It has been increasingly recognised that quality assurance programmes directed at equipment and operator performance can be of great value in improving the diagnostic information content, reducing radiation exposure, reducing medical costs, and improving departmental management. Quality assurance programmes thus contribute to the provision of high quality health care.

There are strong indications that access to diagnostic radiological services will increase in Ghana in the near future. This comes with complex challenges of QA, QC, radiation protection and patient dose management. In all this, the ultimate goal should aim at achieving a diagnostic image that meets clinical requirements with doses to patients as low as possible. Now is the time for all stakeholders (Regulatory Authority, Heath Authorities, Universities and other Training Institutions, Physicists, Hospital or Biomedical Engineers, Radiologists, General Physicians, etc.) to work together to improve the quality of patient protection and management.

8. Acknowledgement

The authors are grateful for the support received from the Radiation Protection Institute of Ghana Atomic Energy Commission and the Graduate School of Nuclear and Allied Sciences, University of Ghana.

9. References

American Association of Physicist in Medicine [AAPM] (2006). *Acceptance Testing and Quality Control of Photostimulable Storage Phosphor Imaging Systems*, Report of AAPM Task Group 10, AAPM Report No. 93.

Bongartz, S.J., Golding, A.G., Jurik, M., Leonardi, E., van Persijn van Meerten, R., Rodríguez, K., Schneider, A., Calzado, J., Geleijns, K.A., Jessen, W. Panzer, P. C. Shrimpton and Tosi G. (2004), *European Guidelines for Multislice Computed Tomography*, Funded by the European Commission. Contract number FIGM-CT2000-20078-CT-TIP

Council Directive 97/43/EURATOM of 30 June 1997 on health protection of individuals against the dangers of ionizing radiation in relation to medical exposure. Official Journal L 180, p. 22, 9.7.1997. European Union. (Repealing Directive 84/466/EURATOM, O.J. no. L265, p. 1, 5.10.1984),

Compliance Guidance of Radiographic Quality Control (2003), New Jersey Department of Environmental Protection, Bureau of Radiological Health, 4th Edition, Trenton NJ 08625,

European Commission (1996a). *European Guidelines on Quality Criteria for Diagnostic Radiographic Images*. Report EUR 16260 EN, Office for Official Publications of the European Communities, ISBN 92-827-7248-5, Brussels, Belgium.

European Commission (1996b). *European Guidelines on Quality Criteria for Diagnostic Radiographic Images in Paediatrics*. Report EUR 16261 EN, Office for Official Publications of the European Communities, ISBN 92-827-7248-6, Brussels, Belgium.

European Commission (1999). *European Guidelines on Quality Criteria for Computed Tomography*. Report EUR 16262 EN, Office for Official Publications of the European Communities, ISBN 92-828-7478-8, Brussels, Belgium.

Food and Agriculture Organization of the United Nations, International Atomic Energy Agency, International Labour Organisation, OECD Nuclear Energy Agency, Pan American Health Organization, World Health Organization (1996). *International Basic Safety Standards for Protection against Ionizing Radiation and for the Safety of Radiation Sources*, Safety Series No. 115, IAEA, Vienna

Ghana Association of Radiologist (2011), Accra, Ghana.

Inkoom, S., Schandorf, C. & Fletcher, J. J., (2009), Optimisation of patient radiation protection in conventional X-ray imaging procedures using film reject analysis: a demonstration of the importance of rare earth screen-film systems *Radiation Protection Dosimetry*, Vol. 136, No. 3 (September 2009), pp. 196–202

Institute of Physics and Engineers in Medicine [IPEM] (2005). *Recommended Standards for Routine Performance Testing of Diagnostic X-ray Imaging Systems*, IPEM Report 91.

International Monetary Fund (2010). World Economic Outlook Database-October 2010 (Data refer to the year 2009), Date of access: December 15, 2010. Available from: http://en.wikipedia.org/wiki/List_of_countries_by_GDP_%28PPP%29_per_c apita

International Organisation for Medical Physics [IOMP[, (2009).

KCARE (01/06/2005). *Protocol for the QA of Computed Radiography Systems, Commissioning and Annual QA Tests*, Date of access: February 15 2011, Available from: http://www.kcare.co.uk/content.php?page=protocols.htm&folder=Education

KCARE. (01/06/2005). *Protocol for the QA of Direct Digital Radiography Systems, Commissioning and Annual QA Tests*. Date of access: February 15 2011, Available from: http://www.kcare.co.uk/content.php?page=protocols.htm&folder=Education

McNitt-Gray M. F. (2002), Radiation Dose in CT, *Radiographics*, Vol. 22, pp 1541-1553.

Medicines and products Regulatory Agency [MHRA] (29 October 2010), *Radiation dose issues in digital radiography systems*, Date of access: February 21 2011, Available from: http://www.mhra.gov.uk/home/idcplg?IdcService=SS_GET_PAGE&nodeId=263

Muhogora W. E., Ahmed N. A., Almosabihi A., Alsuwaidi J. S., Beganovic A., Ciraj-Bjelac O., Kabuya F. K., Krisanachinda A., Milakovic M., Mukwada G., Ramanandraibe M. J., Rehani M. M., Rouzitalab J. & Shandorf C.,(2008), *Patient Doses in Radiographic Examinations in 12 Countries in Asia, Africa, and Eastern Europe: Initial Results from IAEA Projects*, Am. J. of Roentgenol, Vol. 190 (June 2008) pp. 1453–1461

Radiation Protection Instrument LI 1559 (1993), Provisional National Defence Council Law 308, Accra, Ghana. Date of Gazette Notification: 2nd April, 1993.

Radiology Department, Korle-Bu Teaching Hospital (2006), Accra, Ghana.

Regulatory Authority Information System [RAIS] (2010), Radiation Protection Institute, Ghana Atomic Energy Commission, Accra, Ghana.

Rehani M. M., (2001*), Protection of patients in general radiography. In: International Atomic Energy Agency. Proceedings of the international conference*, Malaga, March 2001, Vienna, Austria: IAEA. pp. 169–180

Schandorf, C., Darko, E. O., Yeboah, J., Osei E. K., and Asiamah S. D., (1995). *Qualification and Certification of Radiation Protection Personnel*, Radiation Protection and Safety Guide No. GRPB-G1, Radiation Protection Board, Ghana Atomic Energy Commission, Accra, Ghana.

Schandorf, C., Darko, E. O., Yeboah, J., Osei E. K., and Asiamah S. D., (1995). *Notification and Authorization by Registration or Licensing*, Radiation Protection and Safety Guide No.

GRPB-G2, Radiation Protection Board, Ghana Atomic Energy Commission, Accra, Ghana.

Schandorf, C., Darko, E. O., Yeboah, J., Osei E. K., and Asiamah S. D., (1995). *Dose Limits,* Radiation Protection and Safety Guide No. GRPB-G3, Radiation Protection Board, Ghana Atomic Energy Commission, Accra, Ghana.

Schandorf, C., Darko, E. O., Yeboah, J., Osei E. K., and Asiamah S. D., (1995). Inspection, Radiation Protection and Safety Guide No. GRPB-G4, Radiation Protection Board, Ghana Atomic Energy Commission, Accra, Ghana.

Schandorf, C., Darko, E. O., Yeboah, J., Osei E. K., and Asiamah S. D., (1998). *Safe Use of X-Rays,* Radiation Protection and Safety Guide No. GRPB-G5, Radiation Protection Board, Ghana Atomic Energy Commission, Accra, Ghana.

Schandorf, C., Darko, E. O., Yeboah, J., Osei E. K., and Asiamah S. D., (2000). *Safe Transport of Radioactive Material,* Radiation Protection and Safety Guide No. GRPB-G6, Radiation Protection Board, Ghana Atomic Energy Commission, Accra, Ghana.

Schandorf, C., Darko, E. O., Yeboah, J., Osei E. K., and Asiamah S. D., (2000). *Enforcement,* Radiation Protection and Safety Guide No. GRPB-G7, Radiation Protection Board, Ghana Atomic Energy Commission, Accra, Ghana.

Schandorf, C., Darko, E. O., Yeboah, J., Osei E. K., and Asiamah S. D., (2000). *Occupational Radiation Protection,* Radiation Protection and Safety Guide No. GRPB-G8, Radiation Protection Board, Ghana Atomic Energy Commission, Accra, Ghana.

Schandorf, C., Darko, E. O., Yeboah, J., Osei E. K., and Asiamah S. D., (2003). *Medical Exposure,* Radiation Protection and Safety Guide No. GRPB-G9, Radiation Protection Board, Ghana Atomic Energy Commission, Accra, Ghana.

Schandorf, C., Darko, E. O., Yeboah, J., Osei E. K., and Asiamah S. D., (2003). *Safe Application of Industrial Radiography,* Radiation Protection and Safety Guide No. GRPB-G10, Radiation Protection Board, Ghana Atomic Energy Commission, Accra, Ghana.

School of Allied Health Sciences (2010), College of Health Sciences, University of Ghana, Legon, Accra, Ghana.

Shrimpton P. C., Hillier M. C., Lewis M. A. and Dunn M., (2006), National survey of doses from CT in the UK-2003, *Br J Radiol,* Vol. 79, pp 968-980.

The Two Hundred And Four Act of the Parliament of the Republic of Ghana (1963). Atomic Energy Commission Act, Accra, Ghana.

The Five Hundred And Fifty-Eight Act of the Parliament of the Republic of Ghana (2000). Atomic Energy Commission Act, Accra, Ghana.

United Nations Scientific Committee on the Effects of Atomic Radiation [UNSCEAR] (2010) *2008 UNSCEAR Report to the General Assembly Annex A Medical Radiation Exposures* Vol. 1, ISBN 978-92-1-142274-0,New York NY United Nations

Wall B. F., (2004), Radiation protection dosimetry for diagnostic radiology patients, *Rad Prot Dosim,* Vol. 109, pp 409-419.

World Health Organization (1982). *Quality Assurance in Diagnostic Radiology,* Macmillan Procrom, ISBN 92 4 154164 4, Geneva.

World Health Organization (2010), World Health Statistics, ISBN 978 92 4 156397 7, France.

Wikipaedia (2010), Date of access: January 15 2010, Available from: http://en.wikipedia.org/wiki/Ghana

Zoetelief, J., Schultz, F. W., Kottou, S., Gray, L., O'Connor, U., Salat, D., Kepler, K., Kaplanis, P., Jankowski, J., Schreiner, A. and Vassileva J.(2008). *Quality Control Measurements for Fluoroscopy Systems in Eight Countries Participating in the Sentinel EU Coordination Action,* (Invited Paper) Radiation Protection Dosimetry, Vol. 129, No. 1-3, (March 2008) pp. 237-243

17

Pressure-Sensitive Adhesives for Medical Applications

Zbigniew Czech, Agnieszka Kowalczyk and Jolanta Swiderska
West Pomeranian University of Technology, Szczecin
Poland

1. Introduction

Since their introduction half a century ago, pressure-sensitive acrylic adhesive has been successfully applied in many fields. They are used in self-adhesive tapes, labels, sign and marking films and protective films as well as in dermal dosage systems for pharmaceutical applications, in biomedical electrodes, plasters, the assembly of automotive parts, toys, and electronic circuits and keyboards.

In the last sixty years or so, pressure-sensitive adhesive (PSA) acrylics have made tremendous strides from what was virtually a black art to what is now a sophisticated science. So much so that both the few larger manufacturers of pressure-sensitive adhesive articles and their even larger suppliers now use very expensive equipment to study pressure-sensitive adhesive behavior: tack, adhesion and cohesion.

Three properties which are useful in characterizing the nature of pressure-sensitive adhesives are tack, peel (adhesion) and shear (cohesion). The first measures the adhesive's ability to adhere quickly, the second its ability to resist removal by peeling, and the third its ability to hold in position when shearing forces are exerted. Generally speaking the first two are directly related to each other but are inversely related to the third.

1.1 Definition

Pressure-sensitive adhesives (PSA) are nonmetallic materials used to bond other materials, mainly on their surfaces through adhesion and cohesion. Adhesion and cohesion are phenomena, which may be described thermodynamically and chemically, but actually they cannot be measured precisely. It was shown that the most important bonding processes are bonding by adhesion and bonding with pressure-sensitive adhesives.

In the long history of this technology, pressure-sensitive adhesives and tapes as we know them are a fairly recent concept. However, to trace their origins, one needs to study the history of adhesives as a whole, including the many failures and near misses along the way, as well as the fusion of various technologies, which eventually led to their development.

Since the dawn of history, people learned of the healing powers of certain leaves and plants. There is archaeological evidence indicating that adhesives have indeed been found on primitive tools. More than 6000 years ago, on the arrival of the Egyptian Civilization, the art of healing was already a profession. A primitive tape concept used by Egyptians was the use of a paste of starch in water applied to cloth strips. It indicates that surgical bandages, made of a mixture of fat and honey, were in use. There is very little known of the other raw materials used in Egyptian/Greek times for surgical dressings. It is known, though, that resins, pitches and so on, were in common use in other trades and professions, for instance in the ship-building industry, and such resins would no doubt work well as tackifying resins in pressure-sensitive adhesive systems.

Pressure-sensitive adhesives were in wide use since the late 19th century, starting with medical tapes and dressings. The earliest was awarded in 1845. This was for a surgical pressure-sensitive adhesive that used natural rubber as the base, and pine gum as the tackifier, with balsam of Peru, turpentine, and spirits of turpentine also being added. Ninety years later Stanton Avery developed and introduced the self-adhesive label. Two major industries resulted from these innovations: pressure-sensitive tapes and labels.

In the late 1800s and early 1900s, the development first of the bicycle and then the automobile and their need for tires, allowed the rubber industry to flourish. Greater demands were placed on the industry to develop improved rubber-based products, and this improved technology naturally filtered into the existing adhesive tape industry. Industrial tapes were introduced in the 1920s and 1930s followed by self-adhesive labels in 1935. While various materials in roll form were available early in the 20th century that could have been used as adhesive tape backings, cotton cloth remained the backing of choice, with manufacturing geared to producing surgical tape. The history of PSAs was described by Villa.

Minnesota Mining and Manufacturing Company, popularly known as 3M, was the supplier of sandpaper to the automobile industry in the 1920s, their brand being known as "Wetordry". Richard Drew, then a laboratory technician for 3M would occasionally call at the automobile plants and body repair shops to take developmental samples of sandpaper for testing. There followed a whole series of patents by 3M on pressure-sensitive adhesive tapes, which laid the cornerstone of the industrial adhesive tape industry. The patents were awarded in 1933 for the transfer of his masking tape know-how to cellophane film, making the first pressure-sensitive film tape, giving the world the generic name of "Scotch" tape.

The major raw materials for pressure-sensitive adhesives in the mid-thirties were natural rubber, either as pale crepe, smoked sheet rubber, or wild rubber, with reclaim rubber for primer formulations, coumarone gum resin, burgundly pitch, pine oil, wood resin and gum olibanum as tackifiers, liquid paraffin, or mineral oil, lanolin and beeswax as softeners, zinc oxide as filler, with whiting as filler for prime coats, and benzole or low-boiling-point aliphatic petroleum hydrocarbons as solvent. There was little else available.

The elastomers in common use were polyisobutylene, or Oppanol B, polyvinyl isobutyl ether or Oppanol C, and some styrene butadiene, or Buna S. It is significant to note that as early as 1941 a polypropyl acrylic ester, known as Acronal 4, from I.G.Farben, was being used as a one component pressure-sensitive adhesive, the first indication of an acrylic pressure-sensitive adhesive system.

In the 1940s hot-melt adhesives were introduced. The post-war times brought with them exploration, and initial investigations began with balloons being sent into the stratosphere. It was soon learned that adhesives would be needed that were capable of functioning at extremely low temperatures. A research contract was to develop such an adhesive, and from it came Dow Corning`s silicone pressure-sensitive adhesive, which could perform in the range from -62°C to +260°C, the forerunner of other low/high temperature silicone pressure-sensitive adhesive systems.

However the fifties brought with them an acceleration of research work to convert the practice from an art to a science, and the mystery of tack and adhesion was explored in depth. Also industrial adhesive tape companies began to communicate with one another for the common good of the industry. Acrylic pressure-sensitive systems, although still more than twice the cost of rubber-based systems, were not viable. For most companies this meant buying a commercial pressure-sensitive adhesive produced by someone other than themselves, so they lost the ability to manipulate by formula adjustment.

On the arrival of the 1970s, a very large proportion of the raw materials used by the adhesive tape industry were petroleum derived. The 1980s continued to bring raw material upgrades and new products, particularly in the area of alternate hot-melt elastomers, but little in the way of changes in pressure-sensitive adhesive technology. Product and process development in the industry continues in its upward spiral as can be seen by the number of related patents, which are granted every week throughout the world. Continuing environmental concerns now force the industry to look to other coating techniques than solvent-based systems, with calendaring the original technique still holding its own as a 100% solids system capable of laying down a heavy coat of adhesive at reasonably high speeds. Work continues to develop effective crosslinked hot-melt adhesive systems to replace those based on natural rubber, and water-based adhesive systems now becoming more viable, with a greater selection of raw materials to choose from, and with an improvement in economics. But the number and uses of pressure-sensitive adhesive and tape products continue to grow as the capability of the pressure-sensitive adhesive system improves, and as the user continues to be educated as to their potential.

At the end of the 1980s and during the early 1990s 3M, Beiersdorf, BASF and Lohmann presented the first solvent-free pressure-sensitive adhesive acrylics crosslinked with UV-radiation. Six years later 3M Company presented a new adhesive tape with pressure-sensitive thermosetting adhesives, the semi-structural adhesive tape.

The term PSA has a very precise technical definition and was dealt with extensively in the chemical literature. The function of PSAs is to ensure instantaneous adhesion upon application of a light pressure. Most applications further require that they can be easily removed from the surface to which they were applied, through a light pulling force. Thus PSAs are characterized by a built-in capacity to achieve this instantaneous adhesion to a surface without activation, such as a treatment with solvents or heat, and also by having sufficient internal strength so that the adhesive material will not break up before the bond between the adhesive material and the surface ruptures. The bonding and the debonding of PSAs are energy driven phenomena. Pressure-sensitive adhesives must possess viscous properties in order to flow and to be able to dissipate energy during the adhesive bonding process.

Polymers employed as PSAs have to fulfill partially contradictory requirements; they need to adhere to substrates, to display high shear strength and peel adhesion, and not leave any residue on the substrate upon debonding. In order to meet all these requirements, a compromise is needed. When using PSAs there appears another difference with wet adhesives, namely the adhesive does not change its physical state because film forming is inherent to PSAs.

Thus PSAs used in self-adhesive tapes are adhesives which through their viscoelastic fluid state, can build up the joint without the need to change this flow state during or after application. On the other hand, their fluid state allows controlled debonding giving a temporary character to the bond. Because of the fluid character of the bonded adhesive, the amount of adhesive (i.e., the dimensions of the adhesive layer) is limited; the joint works as a thin-layer tape, laminate or composite. The solid state components of the tape exert a

strong influence on the properties of the adhesive in the composite. Therefore, there exists a difference between the measured properties of the pristine adhesive, and of the adhesive enclosed within the laminate.
The properties, which are essential in characterizing the nature of PSAs comprise: tack, peel adhesion, and shear. The first measures the adhesive's ability to adhere quickly, the second its ability to resist removal through peeling, and the third its ability to hold in position when shear forces are applied.

1.2 Kinds of pressure-sensitive adhesives according to polymer groups

The pressure sensitive adhesive market includes a number of polymeric raw materials. There are used natural rubber, various types of synthetic rubber, such as styrene-butadiene and-ethylene co-polymers, polyvinyl ether, polyurethane, acrylic, silicones and etylene-vinyl acetate-copolymers. However, basic pressure-sensitive adhesive formulations are acrylics, rubbers and silicones (Fig. 1)

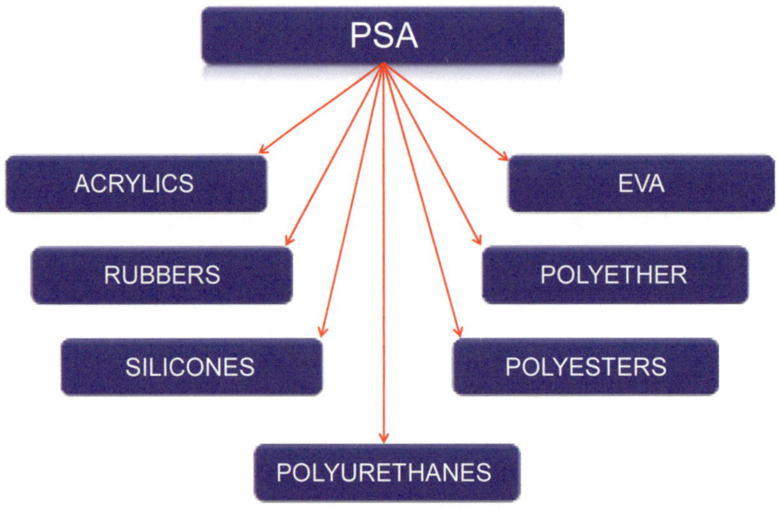

Fig. 1. Polymer classes as potential raw materials for the manufacture of PSAs

Typical performances of those three groups, such as tack, peel adhesion, shear strength, UV-resistance, solvent, chemical, plasticizer and thermal resistance, colour, costs and other important properties are presented in Table 1.
Rubber-based pressure-sensitive adhesives consist of natural or synthetic rubber, various resins, oils, and antioxidants. Blending rubber with tackifiers produces high quality PSAs, the properties of which are determined by tackifier. Because natural rubber is expensive addition of fillers modify properties and reduce costs as well. Other components, such as antioxidants, included protecting the unsaturated backbone polymer from degradation, pigments, plasticizers, and fillers are also added. Rubber-based adhesives are said to be the most cost-effective PSA system. However, long-term aging stability is low. Most of rubbers PSAs are produced as 35 % solution in hexane, similar petroleum fraction, or toluene. Synthetic polyisoprene use in PSA has poorer cohesive strength and his production costs are higher, then natural rubbers PSA.

Property	Acrylic	Rubber	Silicone	Polyurethane	Polyester	Polyether
Tack	Low to high	Typically high	Typically low	Typically low	Medium to high	Medium
Peel adhesion	Medium to high	Moderate to high	Low to moderate	Low to medium	Medium to high	Low to medium
Cohesion	Low to high	Moderate to high	High	Low to medium	Low to medium	Low to medium
UV resistance	Excellent	Low	Excellent	Excellent	Excellent	Excellent
Solvent/chemical resistance	High	Good	Excellent	High	Medium	Excellent
Plasticizer resistance	Low to medium	Generally low	Good	Medium	Medium	Low to medium
Humidity resistance	Excellent	Excellent	Excellent	Excellent	Medium	Excellent
Temperature range	−40 to 160°C	−40 to 70°C	−50 to 260°C	−30 to 120°C	−30 to 140°C	−40 to 120
Adhesive colour	Clear to straw	Yellow (more with time)	Clear	Clear to straw	Clear	Clear
Cost	Medium	Low	High	High	Medium to high	Medium
Other characteristics	Good hydrolysis resistance	Good adherence to low and high energy surfaces	Good adherence to low and high energy surfaces	Good removability	Good flexibility	Excellent flexibility
	Easy to apply	Good flexibility	Good oxidation resistance			

Table 1. Typical performance of the basic adhesive formulations

Acrylic-based pressure-sensitive adhesives are made from higher alkyl esters of acrylic acid without need of tackifiers and provide excellent physical properties. Monomer composition and molecular weight of the polymer determine most of the adhesive's properties. Because acrylic PSAs can be free of those additions, they are less irritating to skin and often preferred to medical applications. Nevertheless, many commercial acrylic PSAs are formulated with other components such as tackifiers, antioxidants, pigments, and fillers. Modified acrylic adhesives contain tackifiers that improve initial tack and adhesion levels while decreasing resistance to solvents, plasticizers, and high temperatures. Acrylic PSAs have superior environmental stability, adhesion to high surface energy materials and greater resistance to oxidation when compared with rubber-based PSAs. In addition, they are more stable to light and heat. The dominate raw materials used for production PSAs are acrylic esters of C_4-C_{12} alcohols, from which the most commonly use are butyl acrylate and 2-ethylhexyl acrylate. Acrylic PSAs are applied in solution, water dispersion or in form of 100 % systems, which are hot-melts or LVS-low viscosity systems also known as room-temperature RT-coatable PSAs. They are used for a multitude of types, labels, protective films, sign and market films, medical products.

Property	Solvent–borne PSAs	Water–borne PSAs	Solvent–free PSAs
Solid content, %	20–50	40–69	100
Drying rate	Usually fast, drying time adjustable by choice of solvents	Slow to dry; however, high solids content meant less to dry	No volatiles to evaporate
Surface wet–out	Excellent on most surfaces, even without minor contamination	Good on porous substrates, poor to fair on non polar surfaces like plastics	Good on most surfaces, may require heating the surface before application
Water resistance	Usually good to excellent	Fair to good	Excellent
Temperature range	−55 to 160°C	−40 to 120°C	−45 to 140°C
Surface attack	May cause degradation of some plastic and paint	Causes shrinkage and wrinkling of fabric and paper; corrosive to metal	Usually no problem; heat may deform thin plastics
Flexibility	Excellent	Excellent	Excellent
Hazards	Requires vapour control; many are flammable and many require emission control	Usually no problem; some have trace levels of chemicals that require control of vapours	Few problems other than working with hot dispensing equipment
Cleanup	Solvent	Water when wet; solvent when dry	Solvent
Cost	Low to moderate cost per gallon; *high cost* per dry unit of weight	Moderate cost per gallon; *low cost* per dry unit of weight	*Moderate cost* per dry unit of weight

Table 2. Properties and characteristic of pressure-sensitive adhesives

Silicone-based pressure-sensitive adhesives are made from two basic components, which are resin and a gum. The resin is obtained from reaction of silicic or polysilicic hydrosol with trimethylchlorosilane. The gums are high molecular weight linear polysiloxane polymers. The simplest method of PSA preparation is mixing resin and gum in a suitable solvent, typically toluene. The physical properties of the silicone PSA depend on resin/gum ratio, type of gum and the heating of the blends in toluene. Silicone PSAs are the only PSA that consistently bond to silicone substrates, also they have excellent high temperature stability over 500°C, flexibility over a wide temperature range, unique adhesion to surfaces of both high and low surface energy. Despite of those superb properties, silicone–based adhesives are expensive in relation to other types of systems, and they have low initial tack and adhesion.

Another group of PSAs is poly(vinyl alkyl ether) blends. That polymers group are mostly use in medical application, especially poly(vinyl ethyl ether) because of his high moisture vapour permeability, valuable for patient comfort during long-term application to the skin. Also is used poly(vinyl isobutyl ether). The production involves blending high and low molecular weight polymers, from which low MW polymer provides spreading on the adherent surface and the high MW polymer determine cohesive strength.

1.3 Kinds of pressure-sensitive adhesives according to synthesis methods

Pressure-sensitive adhesives are applied from solvent solutions, water dispersions or solvent-free systems. After that, they form a permanent tacky film after the evaporation of the solvent or weather phase or after cooling of a hot-melt. Bonding is affected by slightly pressing the adhesive surface onto the adherend. These three groups can demonstrate different properties (Table 2).

Solvent-borne adhesives are synthesized in solvent and are coated onto a web. Following coating, the solvent evaporates, leaving a functional adhesive. Solvent-borne PSA acrylics offer several advantages such as excellent aging characteristics and resistance to elevated temperatures and plasticizers, exceptional optical clarity due to the polymer compatibility and non-yellowing. They also have the highest balance of adhesion and cohesion and an excellent water resistance. The numerous advantages of solvent-based acrylic PSA have led to their wide use in the manufacture of self–adhesive products.

Solvent-free system and emulsion are two additional types of adhesives. Emulsion adhesive ingredients are polymerized in water, applied to the web and dried to create a functional adhesive. Solvent-free as hot-melt adhesives are made from thermoplastic rubbers that formulate with tackifying resins, oils, and antioxidants to achieve coating on the web at high temperatures. Solvent-free pressure-sensitive adhesives are relatively new group of self–adhesive technical and medical products and demonstrate many advantages in opposite to solvent-based pressure-sensitive adhesives. The main aspect is reduction of environmental impact to a minimum during production and exploitation.

Advantages and disadvantages of those three PSAs group are presented in Table 3.

1.4 Main properties of pressure-sensitive adhesives

The term pressure-sensitive describes adhesives, which in the dry form are aggressively and permanently tacky at room temperature and firmly adhere to a variety of dissimilar surfaces upon mere contact, without the need of more than finger or hand pressure. PSAs possess adhesion, required for bonding and debonding, and cohesion necessary against debonding. Adhesion is characterized by tack and peel, whereas cohesion is described by shear resistance, and partially by peel. The special balance of these properties, the adhesion/cohesion balance, embodies the pressure-sensitive character of the adhesive. The efficiency of the bonding process is related to the adhesive's ability to exhibit viscous flow. In order to achieve peel adhesion the bonding stage involves some dwell time. During this time the adhesive must flow in the absence of any externally applied forces. The more liquid-like the behavior of the polymer under these conditions, the more pronounced the degree of bond formation. The debonding process involves a more rapid deformation of the adhesive mass. The polymer's resistance to deformation at higher strain rates becomes very important; the higher this resistance, the higher the force which must be applied to separate the adhesive from the adherent (i.e. the peel resistance). Therefore, high tack, high peel strength adhesives should exhibit good flow at low strain rates, but good resistance to flow at higher strain rates. A proper balance between high tack, peel adhesion, and high cohesion is necessary in most cases. The behavior of any pressure-sensitive adhesive can be reduced to three fundamental and interconnected physical properties: tack, adhesion (peel adhesion), and shear resistance (cohesion). A clear understanding of each property and term is essential.

Pressure−sensitive adhesive	Advantage	Limitation
Solvent-borne	Quick drying	Flammability
	Form homogenous films	Toxicity
	Good adhesion to non polar substrate	Relatively low solid content
	Good key on certain plastics	Difficult cleaning
Water-borne	Easy cleaning	Slow drying
	Good adhesion to polar substrates	Require heat to dry
	Good heat and ageing resistance	Poor adhesion on non polar substrates
	Environment friendly	Presence of surfactant
	High solid content	
Hot-melt	100% active	High equipment costs
	Environment friendly	Require heat
	Very fast setting	Thermal degradation
		Difficult to clean
		Can melt the substrate

Table 3. The advantages and limitations of solvent-borne, water-borne and hot-melt PSA

1.4.1 Tack (initial adhesion)

Tack has been one of the favorite subjects of theorists over many years, often resulting in the derivation of complex formulae in an attempt to explain the property. Nevertheless, tack is still considered and rated by many as how well a pressure-sensitive adhesive sticks to the finger following only slight pressure and short dwell time. While in many cases this can be an approximate measure, this method is badly flawed in that it is highly subjective.

When a pressure-sensitive adhesive is applied to a surface, it takes time for that adhesive to wet out the surface until optimum contact area and adhesion is achieved. This time may be a small fraction of second, or may take days or even weeks. The wetting process can be aided by the degree of pressure applied and the length of time given to that pressure. The rate at which wetting take place varies inversely with the amount of surface still available for wetting. Because the rate of wettability can be considerably accelerated by applied pressure, in a number of tests this application pressure is reduced to a minimum to increase the sensitivity of the test. In summation, the tack of a pressure-sensitive adhesive can be considered to be primarily a measure of the wettability of that adhesive under controlled application conditions, with due regard for its optimum adhesion value. Tack increases continuously upon adding soft, viscous components to the formulation.

1.4.2 Peel adhesion (adhesion)

High peel adhesion requires certain tack level for bonding and certain cohesion for debonding. The dependence of the peel on the ratio of elastic/viscous components is more complex, going through a maximum as a function of the level of the soft component. There are two meanings of the term "adhesion". On the one hand, adhesion is understood as the process through which two bodies are attached to each other when brought together. In this sense adhesion characterizes the sum of all intermolecular and electrostatic forces acting across the interface. On the other hand, we may examine the process of breaking the already adhesive in contact. In this case adhesion is the force, or the energy, required to separate the two bodies, often called "practical adhesion" or "adherence".

One would believe that in describing the adhesion of a pressure-sensitive adhesive in the form of tapes it would be a measure of the force that holds that pressure-sensitive adhesive tape to an applied surface. In fact, though, it is actually a measure of the force required to remove it from that surface. Removal involves work done in extension of the adhesive, the work done in distorting the backing during the stripping action, and the work done in separating the adhesive/surface interface, the last being the smallest of the three. According to ASTM (American Society for Testing Materials), adhesion is "a state in which two surfaces are held together by interfacial forces which may consist of valence forces or interlocking action, or both".

Many theoretical models of adhesion have been proposed, which together are both complementary and contradictory: mechanical interlocking, adsorption (or thermodynamic) theory, electrostatic theory, chemical bonding theory, diffusion theory of adhesion, adhesive effect of thin liquid films and theory of weak boundary layers. Each of these theories of adhesion is supported by experimental analysis but for each there are also convincing counter arguments.

1.4.3 Shear strength (cohesion)

According to ASTM definitions of cohesion include "The propensity of a single substance to adhere to itself, the internal attraction of molecules towards each other; the ability to resist partition from the mass; internal adhesion; the force holding a single substance together". The most important means to influence the cohesion of PSAs are tackification and crosslinking. PSAs possess typical viscoelastic properties, which allow them to respond to both a bonding and a debonding step. For permanent adhesives the most important step is the debonding one; the adhesive should not break under debonding (mainly shear and peel) forces (i.e., permanent adhesives must provide a higher level of cohesive or shear strength than removable adhesives).

At one end of the spectrum, with a high stress or a rapidly increasing stress, the behavior will be largely elastic and the adhesive will separate at the interface leaving a trace of adhesive residue, or the tape backing will break. At the other end of the spectrum, the liquid component of the adhesion can respond fully, allowing molecular disentanglement within the adhesive resulting in cohesive failure.

Typical shear resistance testing is performed with a controlled area of adhesive tape (pressure-sensitive adhesive layer) applied to a standard test surface. Because shear failure is the inability of the pressure-sensitive adhesive to resist a continuous stress, any task that is a measure of stress relaxation within the adhesive gives meaningful data. A high shear resistant adhesive will maintain the stress, while a poor shear resistant adhesive will relieve the stress quite rapidly.

1.5 Characteristic of medical quality of pressure-sensitive adhesives

The focus of the development of self-adhesive medical products is on one hand directed towards customer-oriented requirements such as adhesion, biocompatibility and permeability for water vapor or air. The customer wants highly tolerable, breathable products which are also characterized by very good skin and optimal release. On the other hand, the economic targets of medical-products manufactures must be considered.

The three domains: kind of raw material, technologies and application, supply the basis for the trends of the development of adhesives for medical products. The use of highly tolerable substances with minimal allergenic potential is the primary factor with regard to raw materials. Additionally, the choice is limited by other external influences too.

Pressure-sensitive adhesives (PSA) for stick-to-skin medical products are required to perform a wide variety of tasks under a range of conditions on a complex and highly variable substrate. The origins of something like PSA go back to ancient Egypt, and the first modern adhesive was patented in 1845. Although, the materials have evolved, they had the elements that we recognize today, including an elastic polymer, an adhesion resin, filler and plasticizer.

Nature rubber became a standard base for adhesives, and in 1899, Johnson and Johnson introduced a zinc oxide-containing cloth-backed tape. Shortages of materials during World War II led to a proliferation of synthetic polymers, and by the 1960s, when polar comonomers were used to give cohesive strength to the polymer, acrylic adhesives began to dominate the medical field because of their inherent tack, good oxidative stability and easy adaptability to a variety of needs.

Today, medical devices are called upon to meet a multitude of needs, and are formulated accordingly. Tapes are used to secure dressing and devices, with gentleness, low cost, security, breathability or other characteristics being of primary importance, depending on the particular application. A so-called "paper" tape, which has a non-woven backing coated with a thin and often porous coating of adhesive [5], is chosen for low-stress applications such as securing a surgical dressing, where low cost is important and moderate adhesion is sufficient.

Transparent dressings are widely used to cover and protect superficial wounds. They consist of thin, breathable polyurethane films coated with acrylic adhesives. Since they function as bacterial barriers, their generally high-moisture vapor transmission is accomplished by diffusion of water through the adhesive and backing layers. The more effective adhesives have hydrophilic character and are lightly crosslinked by gamma radiation during the sterilization process.

As a surface to which medical adhesives must stick, skin presents a substantial challenge. It is the largest organ of the body, averaging 3 000 square inches and seven pounds in an adult. As the interface between the body and the outside world, it provides protection against microbial invasion, controls perspiration for temperature regulation, limits transpiration for moisture and provides sensory information.

Skin is a very complex organ. It is made up of two layers. The lower layer is the dermis, whose two major proteins, collagen and elastin, form a matrix that supports the outer layer called the epidermis. Skin is highly variable with gender, age, ethnicity, location on the body and ambient conditions. It is also a structurally weak surface. The top layer, called the stratum corneum, is made of cells that migrated from the base of the epidermis and are being sloughed off as the skin renews and replenishes itself.

The body sheds roughly 10 million cells per day, or about 10 thousand per minute of walking. Skin turns over completely in about 20 to 30 days. So medical adhesive devices must stick to layer that is being shed. It is also a very rough surface, with hair, folds, creases, pores for sweat and oil glands, and wrinkles. The surface energy of skin is also low and adhesion is further compromised by contamination with water, oils, salts and loose debris. As a consequence of these characteristics, there is an upper limit on how much adhesion can be achieved [6], since surface contact is limited and failure of the adhesion bond occurs mostly in the stratum corneum.

Material	Critical surface-energy [mN/m]	Polar [%]
Skin	25-29	40
Polyethylene	31	3
Teflon	18	0,2
Water	72	68

Table 4. Skin as low-energy surface

Pressure-sensitive adhesives for skin need to have a combination of traits that are unique compared to general-purpose adhesive products. Because a medical device is applied to a body that is compromised by acute, chronic, or systemic conditions, it is important that no component in the adhesive aggravate the system further. Obviously, there should be no toxic components that can be absorbed through broken or compromised skin but also there should be nothing that causes an allergic sensitization, response or an acute chemical irritation. Common sources of concern are unreacted free monomers, stabilizers, crosslinking agents, residues from initiators surfactants and processing aids.

In use, an adhesive needs to have good water resistance, both to applied water from outside, such as from bathing or leaking tubes, and to water from under a tape or dressing in the form of perspiration, blood or wound exudates. The self-adhesives need to retain sufficient cohesive strength when wet so the device can be removed cleanly without leaving any adhesive residue and without relevant changes in removal force. In addition, because skin transpires water vapor into the air, an adhesive device also needs to allow enough water diffusion so the skin tissue does macerate and eventually break down.

Quality control of PSAs for medical applications requires, in addition to a simply identity check, also more detailed analyses like proof of homogeneity or quantitative composition. Commonly used methods for quality control of PSAs medical grade are GC-chromatography (residue monomers), IR-spectroscopy (presence of additional derivatives) and typical rheological methods as K-value, peel adhesion on defined substrates and holding power tests. Rheological test methods offer, due to specific geometry and stress, the potential to analyze of viscoelastic properties of PSAs, important at application on skin and independently of test equipment.

2. Acrylic pressure-sensitive adhesives

The composition of acrylic polymers that are inherently pressure-sensitive is a combination in the polymer chain of soft (low glass transition temperature T_g), hard (high glass transition temperature T_g) and functional monomers (Fig. 2). Normally acrylic PSAs contains typical hard and soft monomers, as well as the types of functionality that can be incorporated into the polymer chain.

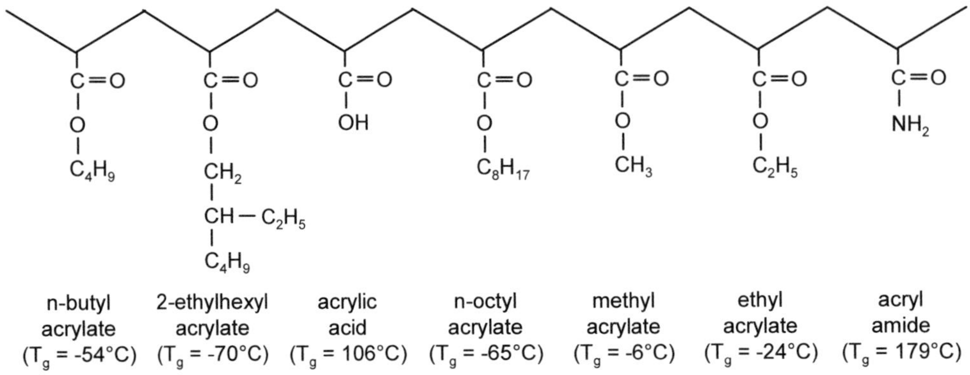

Fig. 2. Typical chain of acrylic PSA copolymer

Solvent-borne, water-borne and solvent-free acrylic PSAs are nowadays predominantly manufactured by polymerization from a wide selection of acrylic, methacrylic and other monomers, often with low levels of monomers having pendant functional groups in a refluxing organic solvent in the present of an initiator, such as organic peroxides or azo compounds: Solvent-borne PSA acrylics offer several advantages such as excellent aging characteristics and resistance to elevated temperatures and plasticizers, exceptional optical clarity due to the polymer compatibility and non-yellowing. They also have the highest balance of adhesion and cohesion and an excellent water resistance. Lower adhesion to non-polar polyolefins is caused by the polar chemistry of acrylics. Acrylics polymer chemistry is expanding through the introduction and utilization of new raw materials, new polymerization process, new modification methods, new crosslinking agents and new crosslinking kind and technology.

2.1 General properties

Although the pressure-sensitive acrylic adhesives may be dwarfs in terms of quantity, they are giants when considered from the quality point of view. Only by means of these acrylic specialties was it possible to succeed in drafting the present surprisingly efficient generation of medical pressure-sensitive adhesive tapes and other self-adhesive materials medical grade for prominent assembly projects at justifiable cost for medical applications.

The most important requirements for a pressure-sensitive adhesive, such as high tackiness (adhesion by the touch), high cohesion (inner stability), high stickiness (adhesion), UV, solvent and temperature stability are fulfilled by acrylics in an outstanding way.

Solvent-borne, water-borne or solvent-free acrylic PSAs offer several advantages such as excellent aging characteristics and resistance to elevated temperatures and plasticizers, exceptional optical clarity due to the polymer compatibility, non-yellowing and free of residual monomers. They also have the highest balance of adhesion and cohesion and an excellent water resistance. Acrylics are harder than rubbers. This can be seen in a less aggressive tack and slower build-up of peel strength. Lower adhesion to non-polar polyolefins is caused by the polar chemistry of acrylics.

2.2 Special properties for medical quality

Acrylic pressure-sensitive adhesives are available on the market as the major types in form of solvent-born, water-borne or 100% polymer systems, which can be tailor-made for defined product purposes.

The target function of adhesives, especially acrylic PSAs, which can be used for skin adhesion, can be concentrated on three basic characteristics. The fast skin wetting during initial adhesion and the secure adhesion on skin within the application time as well as the complete removability after application. A balanced relationship between these three basic characteristic is the primary aim of the formulation of pressure-sensitive adhesives for skin application.
Nowadays, the medical self-adhesive products represent a vast part of the total adhesive materials on the medical market. It all started a long time ago already two centuries ago, in Europe, a druggist issued the first patent on a medical tape. In 1882 Paul Beiersdorf claimed a patent for a "medical plaster". Since then, the production of PSA products started for hospital and first-aid applications. It took until the 1920`s before the benefits of PSA products were introduced in industrial applications. Today, medical products do more than merely fixing medical dressings to the body. Over the years, the investigations in medical PSA technology have concentrated on a wide range of formulations to tailor adhesive properties to meet specific needs, resulting in the development of PSA`s that form a vital part of the modern wound care dressings.
Although the medical self-adhesive materials can be classified in similar categories as typical technical industrial products, their performance and composition differs significantly for similar technologies. Medical self-adhesive products are mainly applied to human skin. It is this complex substrate which requires a unique approach for the formulation and production of medical PSAs. In order to develop a suitable medical skin adhesive, it is important to understand the basics of skin anatomy and physiology.

2.3 Medical applications

The focus of the development of self-adhesive medical products is on one hand directed towards customer-oriented requirements such as adhesion, biocompatibility and permeability for water vapor or air. The customer wants highly tolerable, breathable products which are also characterized by very good skin adhesion and optimal release. On the other hand, the economic target of medical-product manufactures must be considered. Typical aspects would be an increase in machine speed and the reduction of manufacturing costs as well as the corresponding environmental aspects concerning both product and process.
The three domains, namely raw materials, technologies and application, supply the basis for the trends of the development of adhesives for medical products. The use of highly tolerable substances with minimal allergenic potential is the primary factor with regard to raw materials. Additionally, the choice is limited by other external influences.
One example of this is discussions concerning the integration of animal-derived raw material for the manufacturing of medical products. Furthermore, the requirements of the raw material with regard to the characteristics of the finished products and easy processability are continuously increasing. During the development of the process, those technologies are to be preferred where critical products such as organic solvents are avoided. Typical examples are hot-melt systems, water-borne adhesives and solvent-free acrylic systems. Also, those systems where serious savings can be achieved with regard to process time and investments are focused upon as a major point of interest.
New applications of self-adhesive acrylic medical products are aimed at developing easier handling or other additional unique selling propositions. There are medical systems, for example, where medication is achieved my means of drug supply through the simple applications of an island dressing. Typical medical application of acrylic pressure-sensitive adhesive medical grade extend over plaster and pads, transdermal drug delivery systems (TDDS), OP-tapes, biomedical electrodes, self-adhesive hydrogels and surgical drapes.

2.3.1 Plaster and pads

Medicinal plasters (Fig. 3) and pads have been utilized around the globe for centuries to treat multiple ailments both topical and systemic. History teaches us that as far back as 14th century China, certain plants were being ground and placed on the skin for the purpose of malady containment and cure. As an evolutionary step, ointments, creams and gels have been developed over the years to treat everything from toothaches and mosquito bites to rheumatoid arthritis and melanomas, thus attesting to the viability of the skin as a delivery portal for topical and systemic drugs.

Fig. 3. Self-sticking plasters

2.3.2 Transdermal drug delivery systems (TDDS)

Physicians and hospitals make every effort to ensure that patients actually complete the course of medication therapy that is prescribed. If the medication is in the form of self-administered oral drugs or injections it is not easy to monitor compliance with the prescribed course. The development of systems that allow the controlled delivery of drugs through the skin a "therapeutic patch" was therefore welcomed by the medical profession with enthusiasm. Transdermal drug delivery systems (TDDS) offer real, practical advantages to the patient by releasing precise amounts of medication through the skin directly into the blood stream. Once the patch is applied to the skin no further action is required of the recipient-the patch conzinues to administer a uniform dosage over an extended period of time.

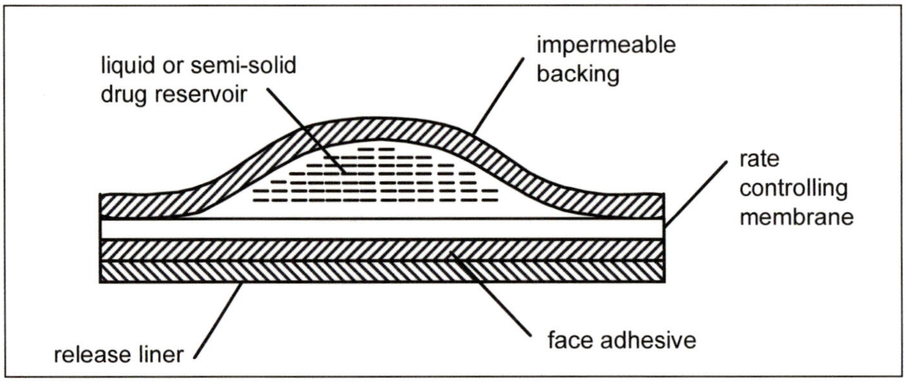

Fig. 4. Reservoir transdermal system with face adhesive

Transdermal drug delivery systems (TDDS) constitute evolutionary step in the passage of active agents through the skin. Transdermal drug delivery is complex but essentially comprises a drug reservoir with a protective outer cover, a permeable membrane (sometimes), a self-adhesive and a release liner. Figures 4 to 8 represent designs of the commercially available transdermal drug delivery systems.

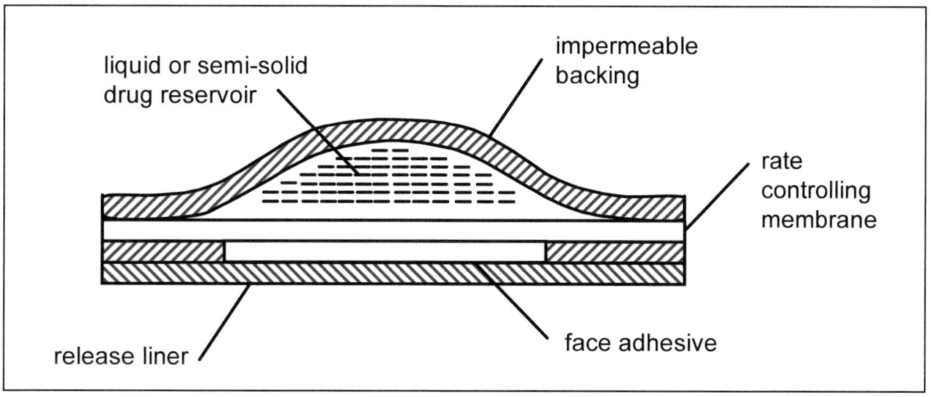

Fig. 5. Reservoir transdermal system with perimeter adhesive

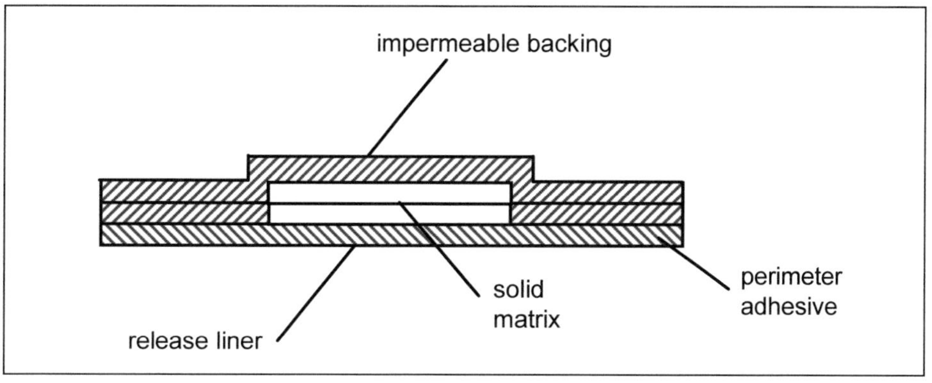

Fig. 6. Solid matrix transdermal systems with perimeter adhesive

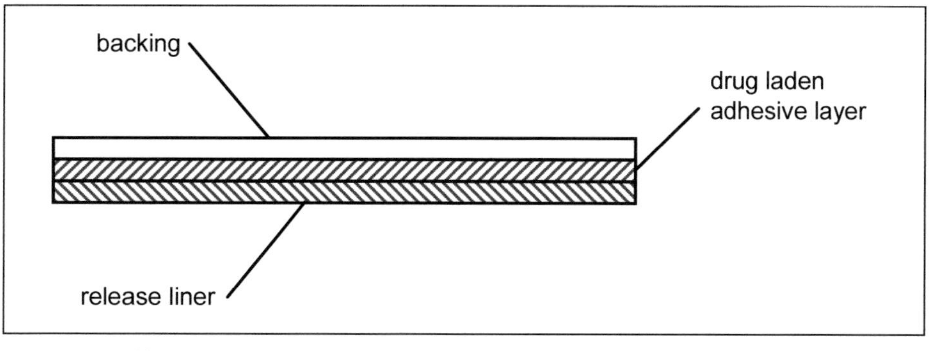

Fig. 7. Drug-in-adhesive transdermal systems

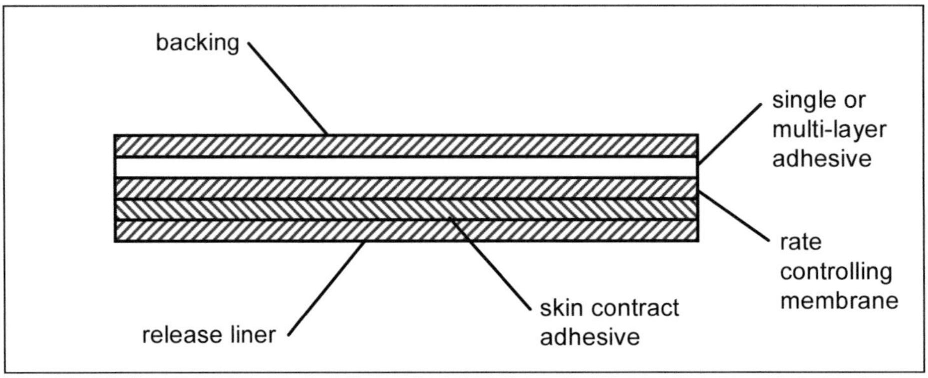

Fig. 8. Multilayer drug-in-adhesive transdermal systems

Figure 9 shows the typical TDDS construction for diverse drugs usable in medical applications.

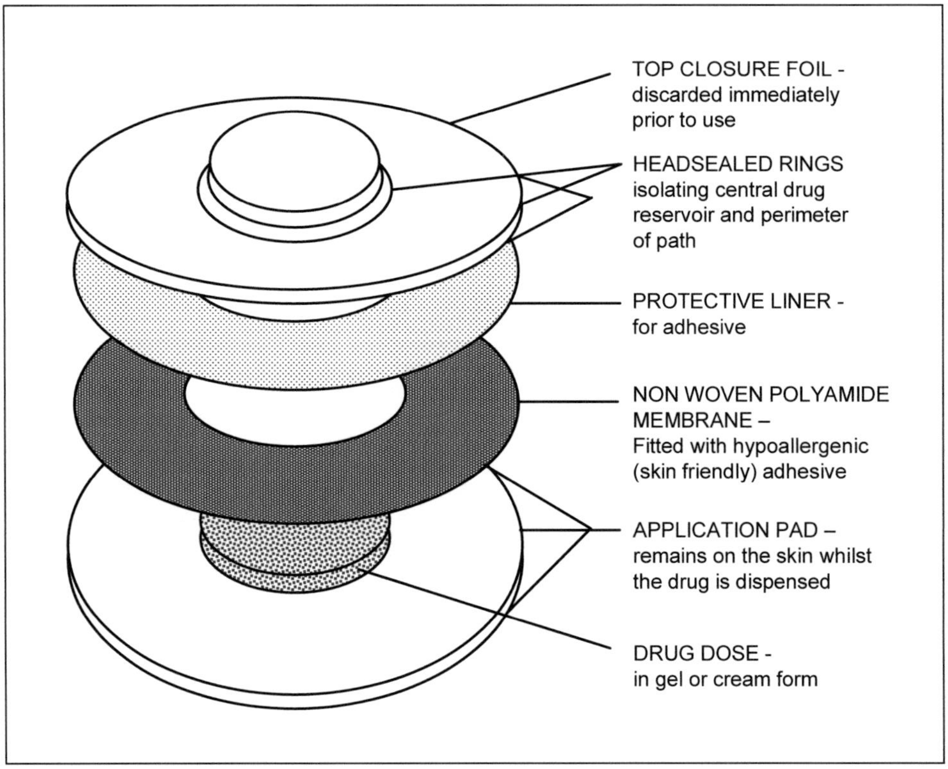

Fig. 9. TDDS construction

Health practitioners make every effort to ensure that patients actually complete the course of the prescribed medication therapy. If the treatment is in the form of self-administered oral drugs or injections it is difficult to monitor compliance with the prescribed course. Therefore the development of systems that allow the controlled delivery of drugs through the skin using a "therapeutic patch" was warmly welcomed by the medical profession.

Following the pioneering work by the California-based Alza Corp. with Ciba-Geigy at 80-ties, the first commercial TDDS products were patches containing scopolamine for motion sickness and nitroglycerine (NTG) for angina sufferers. NTG TDDS significantly reduced the risk of myocardial infarction. This success stimulated the search for other drugs suitable for sustained transdermal delivery. At least 30 projects has been now known to be under development, including patches to treat sexual dysfunction, depression, Parkinson and even Alzheimer disease. Beside the ongoing research the following drugs are available in TDDS form: scopolamine, NTG, clonidine, nicotine, estradiol, testosterone, norethindrone acetate, fentanyl, lidocaine.

The benefits of transdermal route of drug delivery may be best seen in gynecology. This includes hormone replacement therapy and contraception. Oral administration of estradiol derivatives is associated with a significant risk of vascular complications: thromboembolism and myocardial infarction. Women at major risk are smokers, patients with atherosclerosis and thrombophilias (carriers of clotting factor mutations), with a history of deep venous thrombosis or coronary heart disease. Oral administration of estrogens changes metabolism of liver dramatically. Some metabolic pathways are stimulated while others are partially blocked. While administered orally, the first pass effect modulates synthesis of important clotting factors, which results in altered activity of factors II,VII, IX and X as well as proteins S and C. This pathological state of "hypercoagulation" may lead to formation of thrombi and clinical complications like DVT, pulmonary embolism, myocardial infarctions or cerbrovascular accidents.

The risk of oral hormone replacement therapy (HRT) among menopausal women has its reflection in the results of women health initiative (WHI) study. Since the results have been published, the demand for the oral HRT has declined dramatically. On the other hand the transdermal administration of estradiol and progestins avoids liver portal circulation, thus at least theoretically decreases the risks related to the hormonal treatment. However the evaluation of the true safety of transdermal route of hormone administration awaits further meticulous research.

American data derived from women and health care practitioners indicate that women desire user-friendly contraception simplifying their lives. Despite enormous progress made in the field of contraception, in fact there are only 2 effective methods: hormonal contraception and intrauterine devices. The latter method bears significant risks therefore administration is narrowed to limit group of patients. On the other hand hormonal contraception, also not completely free from potential complications, requires patients to be very compliant.

Data from clinical studies are surprising. At least 1/3 of pregnancies are unplanned. Among this, 2/3 happen in women using contraceptive methods. It has been established that among women taking combined oral contraceptive pills, at least 60% of unintended pregnancies resulted from errors of daily drug administration. Efficacy of contraception is measured by Pearl index (PI). PI is determined by the number of unintentional pregnancies related to 100 women years. For instance, a hundred women can use contraception for a year, each with a method that is going to be examined. If five pregnancies occur during this time in this studied group, the Pearl index is 5.0. The theoretical PI for oral combined contraceptives is 0,3 which reflects perfect use of the method. However the practical PI may reach even 8,0, which reflects common errors made by pill users. About 10 to 30 % of women forget up to 3 pills in the cycle. This observation helps explain the differences between theoretical and practical values. Figure 10 clarifies the detrimental consequences of non-compliance in pill users, and shows the benefits of transdermal administration of the hormones.

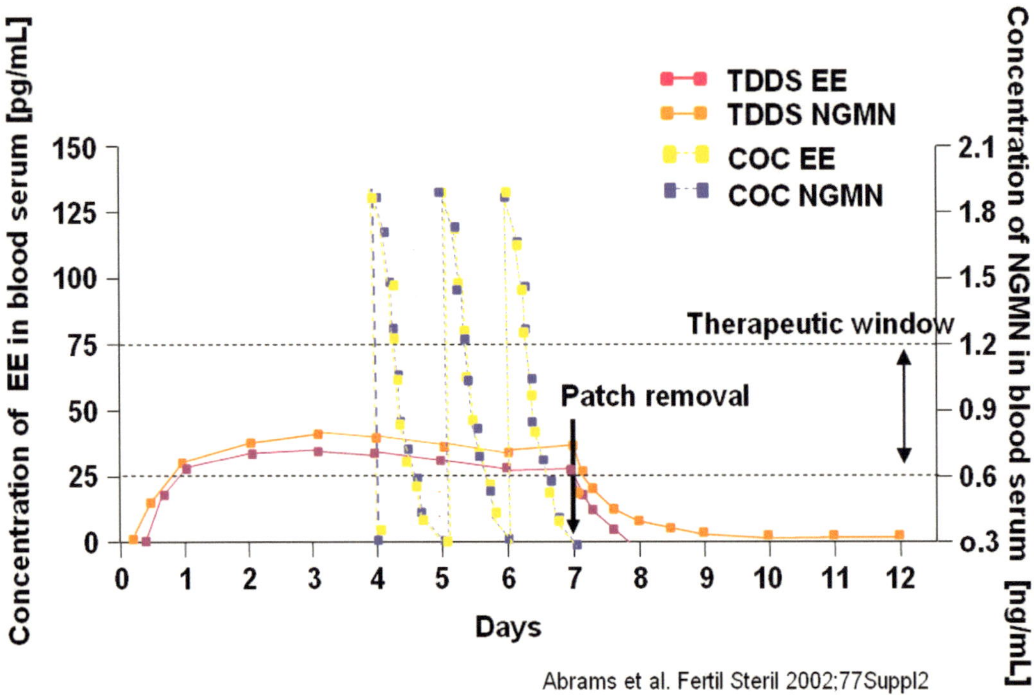

Fig. 10. Pharmacokinetics of hormonal contraception in relation to the way of administration. TDDS - transdermal delivery systems, COC - combined oral contraceptives, EE - ethynyl estradiol, NGMN - norelgestromin

In TDDS, effective levels of serum hormones are reached a day following patch application and maintained within the therapeutic window throughout the seven days of wear. When the patch is then removed on day 7, hormone levels decline, however are negligible only by day 10. This profile of steady levels of EE and NGMN throughout the seven days of patch-wear, stays in contrast to the daily peaks and troughs seen with a pill taken once a day. In pill users, the levels of hormones drop fast below the therapeutic level, which may lead to unintended pregnancy. This may not happen in TDDS users, who if make errors, usually forget to REPLACE the patch. In such situations, blood serum hormone levels are found to remain within the therapeutic window least two extra days. Sustained concentrations of EE and NGMN suggest that clinical efficacy can be maintained even if scheduled patch change is missed. This gives an extra time of two days of relative contraceptive safety, compared with 12 hours given by the pill.

Pressure-sensitive adhesives used for delivery systems are primarily based on acrylics, polyisobutylens and silicones, although the most important role plays acrylics. Close cooperation between pharmaceutical and companies which produce PSAs is essential for the successful development of such systems, for example, has led to the development of permeation enhancers that temporarily modify the permeability of skin, allowing drugs with larger molecular structures to be considered for TDDS therapy.

One problem area requiring further work is the limited solubility of drugs in adhesives. The solubility of same drug molecules is less than 1 wt. % in polyisobutylens and only slightly better in acrylics. The using of co-solvents allows solving this problem. To ensure the finished product has adequate cohesive strength, high molecular weight adhesives with high shear resistance need to be used if non-volatile co-solvents are blended in to reach the necessary drug solubility levels.

Another difficulty is that dispersions adhesives cannot easily be used in TDDS applications because they tend to re-emulsify when exposed to perspiration. Transdermal patches are increasingly worn over several days, so that the adhesion must be maintained in the presence of wet skin. For this reason transdermals have had to rely on solvent-based and low-temperature hot-melt adhesives.

2.3.3 OP-tapes

An interesting application of acrylic pressure-sensitive adhesives medical grade is for securing sheets and other covering materials used in hospital operation theatres. The special composition of the acrylic water-soluble PSAs allows such pressure-sensitive adhesive tapes to be used even on hydrophobic, low surface energy cotton cloths coated with polyfluorocarbon resins. The acrylic adhesives used for securing operation theatre linen must be largely moisture-resistant, insoluble in cold water and must have a removable adhesion to the skin as well as hypoallergenic properties. The target in this case is a complete water-solubility of the adhesive, and thus a complete dispersibility of the OP-tape, is reached above 60 to 70°C and in the pH range (pH > 9). For this application, the availability of water-soluble carrier materials is also demanded.

Due to the growing environmental problems, reusable operation theatre linen is becoming increasingly important throughout Europe. The number of hospitals which employ reusable linen is rising continuously. The market for reusable medical systems of this kind is expanding very rapidly. The textile materials with polyfluorocarbon resins are used with OP-tapes especially developed for the medical sector and fixed after steam sterilisation for 20 min at 121°C on the skin of the patient. After use, the operation-tape is removed without residue during the washing process (pH range > 9 at 65°C) from the textile, the pressure-sensitive adhesive is dissolved and the carrier completely dispersed.

In view of the current situation regarding disposal of hospital waste, top priority must be given to waste prevention. Therefore, products which can be reused several times are of fundamental importance, especially textiles because, by their nature, they are designed for long-term use. A double-sided medical tape is applied to the textile sheets, which are then be secured to the patient's skin. The quality of such tapes must be such that they have sufficient adhesion during use but can be removed completely from the textile cotton cloths after use, i.e., during washing (Fig. 11).

2.3.4 Biomedical electrodes

The term "biomedical electrode" as used here means an electrode for establishing an electrical connection between the skin of a living body and an electromedical apparatus.

Scientists have conducted their own developments, which is based on water-soluble, ionically conductive pressure-sensitive adhesives. The adhesive used in layers allows important biological processes to be stimulated by bioelectrical signals. The outstanding features of such adhesive electrodes are their resistance to skin moisture, resistance to drying out, and their safe use on the skin.

Many biomedical electrodes are known in the art, they use electro-conductive materials such as conductive creams, pastes, and gels that incorporate natural polymers such as karaya

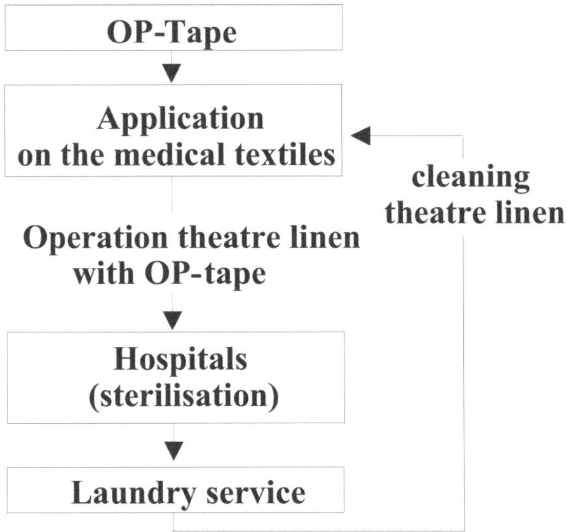

Fig. 11. Recycling of reusable operation theatre linen

gum, so as to provide good contact between the skin surface and the electrode and reduce electrical resistance across the skin-electrode interface. Such conductive materials are placed between the skin and the electrode plate so as to ensure good electrical connection of the skin surface to biomedical diagnostic equipment such as high-impedance electro-myographs and electrocardiographs. Conductive creams and pastes are unpleasant to use, are sloppy and will often foul the skin surface. Adhesion to the skin must be adjustable within a certain range and removal of the electrodes must be gentle and should cause no discomfort.

The following kinds of large-area biomedical electrodes are used:

TENS (Transcutaneous Electrical Nerve Stimulation) electrode coupling media is produced from low to medium concentration of sodium chloride in the hydrogel sheets.

ESU (Electro-Surgical Unit) electrode is produced from a low ionic hydrogel sheet.

EKG (electrocardiogram) electrodes are the poly(ethylene oxide) hydrogel-based electrodes. They have a variety of specific use applications, made possible by the ability to produce hydrogels of specific ionic strengths.

DEFIBRILLATION pad. The defibrillator pad is produced from a sheet containing conductive ions. The pad is usually used as a conductive medium for the application of large amounts of electricity (voltage and current) and also is used as a sensing electrolyte for EKG monitoring through the same electrodes.

BIO-FEEDBACK electrodes. Bio-Feedback electrodes are produced from a high ionic concentration gel sheet. They are used with a wide variety of clinical electrodes, and permits immediate signal reception.

New biomedical electrodes (Fig. 12) consist of an electrically conductive foil (1), a contact (2) and an electrically conductive pressure-sensitive adhesive (3), which is applied over the surface of the electrically conductive foil.

Biomedical electrodes are applied in the following medical areas:

- stimulation of biological processes
- percutaneous administration of medicines
- discharge of currents from surgical high-frequency cutting instruments

- pain relief by means of appropriate electrical signals
- monitoring the patient's state of health in the intensive care and as well as in the operation theatre

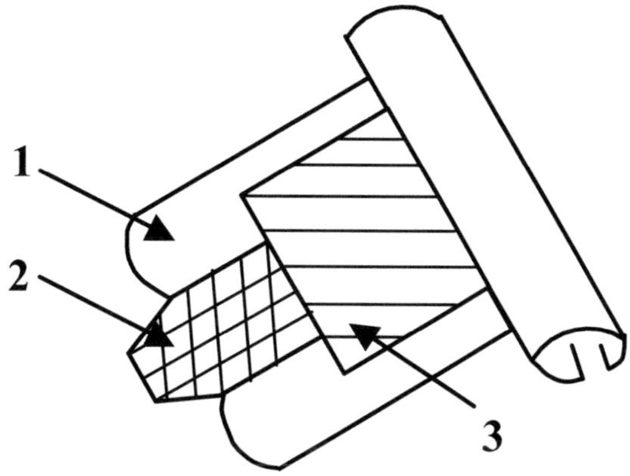

Fig. 12. Design of a new biomedical electrode

New biomedical electrodes have a transparent, electrically conductive, highly elastic and hypoallergenic layer of pressure-sensitive adhesive based on acrylic, silicone or poly(vinyl pyrrolidone) (PVP). The gentle adhesion of the hydrophilic layer to the skin is not impaired by the skin moisture or sweat. Principally, both the adhesive and cohesive strengths of the electrically conductive adhesive layer are also sufficient to ensure that the electrodes remain fully functional throughout the prescribed duration of application. The acrylic adhesive layers, which crosslink at room temperature, contain a soft resin, a (poly)electrolyte and a moisturiser.

Biomedical electrodes are monitored constantly with respect to electrical conductivity and resistance to skin moisture and drying out. In addition, the electrically conductive adhesive layers are tested for their hypoallergenicity and resistance to ageing. As confirmed by trials with biomedical electrodes, the newly developed biological electrodes have outstanding resistance to skin moisture and retain their most important properties, such as electrical conductivity and resistance to drying out during storage.

2.3.5 Self-adhesive hydrogels

Self-adhesive hydrogels are three-dimensional hydrophilic water-swellable polymeric materials in form of polymeric films characterized in dry and in water-swollen state by tack, peel adhesion and shear strength. They are crosslinked polymeric structures containing either covalent bonds produced by the simple reaction of one or more comonomers, chemically crosslinked using conventional at room temperature reactive crosslinking agents, thermal reactive crosslinkers or crosslinked by the use of UV radiation in the area between 200 and 400 nm. The hydrogen bonds, based on a dipole-dipole attraction of polar groups such as -COOH, $-CONH_2$ or -OH and association bonds such as van der Waals forces between polymer chains are not enough strong for excellent structure of hydrogels. Hydrogen bonds are fully reversible and three to four times as strong as van der Waals forces. Consequently PSAs with H-bonds are significantly stronger than those, which are interconnected only by van der Waals forces. Secondary valence crosslinkings are in general distinguished from primary valence crosslinkings by their thermoreversible nature. When warmed up the crosslinking is lost and during cooling it is formed again.

The physical behavior of hydrogels is dependent on their equilibrium and dynamic swelling behavior in water, since upon preparation they must be brought in contact with water to yield the final, swollen network structure. Hydrogels are a unique class of macromolecular networks that may contain a large fraction of aqueous solvent within their structure. The hydrophilic-hydrophobic balance of the hydrogels, the degree of crosslinking, and especially, the degree of ionization and its interaction with counterions are the important parameters which control the equilibrium swelling, dimensional change and the release patterns of drugs from these carries. Hence, mathematical modeling of hydrogel swelling and predictability of swelling behavior has gained considerable attention during past decades. The favorable property of hydrogels is their ability to swell, when put in contact with a thermodynamically compatible solvent, in this case water. When a hydrogel in its initial state is in contact with water molecules, the latter attacks the hydrogel surface and penetrates into the polymeric network. In this case, the unsolvated glassy phase is separated from rubbery hydrogel region with a moving boundary. Regularly the meshes of the network in the rubbery phase will start expanding, allowing other water molecules to penetrate within the hydrogel network.

Self-adhesive hydrogels are characterized by permanent adhesiveness performance before, during and after water absorption (Fig. 13). They are characterized by fast swelling, excellent mechanical properties, and high transparency after water absorption and good elastic properties. In the lab of Westpomeranian University of Technology in Szczecin has been developed a new generation of self-adhesive hydrogels based on acrylic polymers.

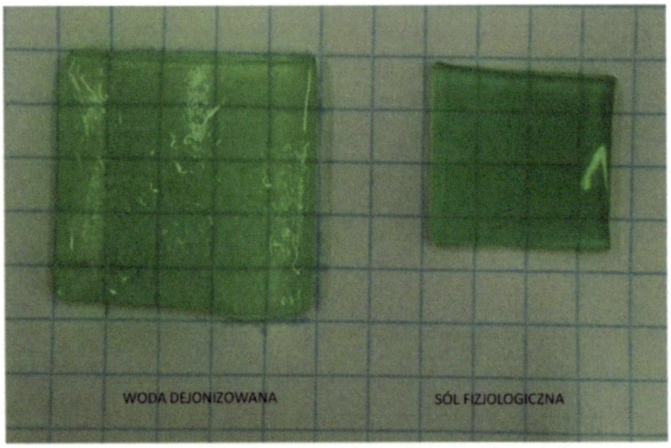

Fig. 13. Medical self-adhesive hydrogel after (right) and before (left) absorption of water

The physical properties of self-adhesive hydrogels make them attractive for a variety of technical, biomedical and pharmaceutical applications. The applications of hydrogels are grown extensively. They are currently used as scaffolds in tissue engineering where they may contain human cells in order to repair tissue. Environmental sensitive hydrogels have the ability to sense environmental stimuli, such as changes of pH, temperature, or the concentration of metabolite and then release their load as a result of such a change. Self-

adhesive hydrogels can be used as biosensors as well as in drug delivery systems (DDS). These kinds of hydrogels are also used as controlled-release delivery devices for bioactive agents and agrochemicals. Contact lenses are also based on hydrogels.
Hydrogels are formulated to absorb moisture resulting in a skin compatible system as well. The absorption of trans epidermal water (TEW) can enhance the wetting and the adhesion on the skin of these adhesive systems, resulting in a gradual increasing adhesion during water. The acceptance of moist wound healing as being most appropriate for rapid healing has lead to the further development of hydrogels in occlusive dressings capable of maintaining the correct wound environment. Hydrogels are very suitable in achieving a balance between exudates handling and maintaining a moist environment. The absorption capacity of these adhesive systems is measured by fluid handling capacity in g/m^2h. This intrinsic property of hydrogels allows them to be introduced in medical tapes suitable for damaged and even enhancing the skin healing process.

2.3.6 Surgical drapes

Medical surgical drapes are carrier-based products, which are generally manufactured by adhesive coating. Because of the absorbtion, dosage, and storage function of such tapes, they are coated on special porous carrier materials with high coating weight of special acrylic adhesives. These adhesives do not contain volatiles. Acrylic pressure-sensitive adhesives in medical drapes may display adhesion buildup in the time, or weakening of the cohesive strength due to migration of oils. These disadvantages are avoided by crosslinking with built-in special monomers or special crosslinking agents. Surgical drapes as medical products allow using of diverse raw materials in form of hotmelts acrylic PSAs, UV-crosslinked acrylic prepolymer, solvent-borne acrylic PSAs and other non acrylic polymers. Commercial acrylic PSAs used for skin application are easier to remove and cause less skin irritation than other kinds of adhesives. Acrylic adhesive compositions for medical surgical drapes which do not leave adhesive residues on skin contain modified copolymers having a molecular weight of 2500 to 3000 dalton. Cotton cloth and hydrophobic textile materials coated with low energy polyfluorocarbon resins are used as carrier materials for surgical tapes. Because the skin is part of the systemic and immune system, medical surgical drapes require testing to indicate the suitability on human skin. As part of medical devices, medical tapes are tested in accordance with FDA guidance and ISO 10993 standards. The safety evaluation involves advanced biocompatibility testing appropriate to the intended use of the component material. Medical surgical drapes applied on healthy skin are categorized as skin surface devices and tested on skin irritation and sensitization as detailed in the ISO guidelines.

3. References

[1] D. Satas, "Handbook of Pressure Sensitive Technology", Van Nostrand-Rheinhold Co, New York, USA (1982)
[2] Z. Czech, Crosslinking of Pressure-Sensitive Adhesives based on acrylics, Szczecin University of Technology, Szczecin (1999)
[3] I. Benedek, Developments in Pressure-Sensitive Products, Taylor & Francis a CRC Press Book, New York (2006

[4] Z. Czech, R. Kurzawa, "Acrylic Pressure-Sensitive Adhesives for Transdermal Drug Delivery Systems", Journal of Applied Polymer Science" 106 (2007) 2398-2404

[5] Z. Czech, A. Wróblewska, E. Meissner, R. Kurzawa, „Pressure-Sensitive Adhesives for Transdermal Drug Delivery Systems", Conference", SURUZ 2007, (2007) 443-446

Part 4

Quality Control in Food Sector

18

The Role of Empirical Rheology in Flour Quality Control

Tamara Dapčević Hadnađev, Milica Pojić,
Miroslav Hadnađev and Aleksandra Torbica
Institute for Food Technology, University of Novi Sad
Serbia

1. Introduction

Rheology, as a branch of physics, studies the deformation and flow of matter in response to an applied stress or strain. According to the materials' behaviour, they can be classified as Newtonian or non-Newtonian (Steffe, 1996; Schramm, 2004). The most of the foodstuffs exhibit properties of non-Newtonian viscoelastic systems (Abang Zaidel et al., 2010). Among them, the dough can be considered as the most unique system from the point of material science. It is viscoelastic system which exhibits shear-thinning and thixotropic behaviour (Weipert, 1990). This behaviour is the consequence of dough complex structure in which starch granules (75-80%) are surrounded by three-dimensional protein (20-25%) network (Bloksma, 1990, as cited in Weipert, 2006). Wheat proteins are consisted of gluten proteins (80-85% of total wheat protein) which comprise of prolamins (in wheat - gliadins) and glutelins (in wheat - glutenins) and non gluten proteins (15-20% of the total wheat proteins) such as albumins and globulins (Veraverbeke & Delcour, 2002). Gluten complex is a viscoelastic protein responsible for dough structure formation.

Among the cereal technologists, rheology is widely recognized as a valuable tool in quality assessment of flour. Hence, in the cereal scientific community, rheological measurements are generally employed throughout the whole processing chain in order to monitor the mechanical properties, molecular structure and composition of the material, to imitate materials' behaviour during processing and to anticipate the quality of the final product (Dobraszczyk & Morgenstern, 2003). Rheology is particularly important technique in revealing the influence of flour constituents and additives on dough behaviour during breadmaking. There are many test methods available to measure rheological properties, which are commonly divided into empirical (descriptive, imitative) and fundamental (basic) (Scott Blair, 1958 as cited in Weipert, 1990). Although being criticized due to their shortcomings concerning inflexibility in defining the level of deforming force, usage of strong deformation forces, interpretation of results in relative non-SI units, large sample requirements and its impossibility to define rheological parameters such as stress, strain, modulus or viscosity (Weipert, 1990; Dobraszczyk & Morgenstern, 2003), empirical rheological measurements are still indispensable in the cereal quality laboratories. According to the empirical rheological parameters it is possible to determine the optimal flour quality for a particular purpose. The empirical techniques used for dough quality control are generally recognized as standard methods by ICC, AACC, ISO and different national standards.

In comparison to rheological methods generally applicable in food quality control, dough rheological tests are probably the most diverse. The devices suitable to monitor the dough behaviour during different processing operations such as mixing, kneading, moulding, fermentation and baking have been developed. For example, Farinograph and Mixograph provide information about mixing properties of flour, whilst the rheological properties of the dough during moulding are assessed using Extensograph, Alveograph and recently introduced Kieffer dough and gluten extensibility rig, which measure rheological properties of dough in extension (e.g. dough strength and resistance to extension). Proving properties of dough (gas production and retention) can be monitor by Rheofermentometer and Maturograph. For monitoring of starch gelatinization properties and/or indirect determination of α-amylase activity, Amylograph and Falling Number techniques are employed.

The most of the instruments listed above, developed in the early days of dough rheology, have remained their original principle (Weipert, 2006) and thus their shortcomings. One of the newest rheological devices, called Mixolab, has overcome some of the problems encountered with other empirical rheological instruments. The Mixolab System measures dough behaviour during mixing and heating, which enables determination of both protein and starch contribution to dough rheological properties in a single test. Therefore, it is able to perform continuous measurement throughout a simulated baking process, which means that one can use the same instrument for several applications.

The aim of this study is to give a review on the empirical rheological instruments, their principles and techniques and interpretation of results by comparing various international and national standards. Moreover, the demands to change the parameters interpretation along the changes in wheat cultivars over time and varieties in different regions will be discussed. The special emphases will be given on the influence of climatic changes on the rheological quality parameters of Serbian wheat flour. The past and current studies employing empirical rheological tests, the correlation between different empirical rheological parameters, as well as their correlation to sensory attributes will be also presented.

2. Mixing and kneading devices

Mixing is very important operation in wheat flour processing. During this phase, hydratation of flour particles and formation of three-dimensional viscoelastic gluten network from glutenin and gliadin components occur. The rheological changes, which occur in gluten structure during mixing, greatly determine the final product quality (Dobraszczyk & Morgenstern, 2003).

The most important empirical rheological devices used to monitor the behaviour of dough during mixing and kneading are Farinograph and Mixograph.

2.1 Farinograph

The most popular and accepted device for measuring dough physical properties is Brabender Farinograph. It measures and records the mechanical resistance of the dough during mixing and kneading. Physical properties of dough are measured by placing a defined mass of flour in a tempered (30°C) mixing bowl equipped with two Z type kneaders. Depending on the available quantity of flour, tests can be performed in 300g, 50g and 10g mixing bowls. In order to obtain the dough, which rheological properties are actually measured, water is added to the flour in amount which ensures the dough consistency of 500 BU (arbitrary Brabender units). The working principle of Farinograph, as well as, the interpretation of resulting curves is described in details in different official metods (ICC 115/1, AACC 54-21, ISO 5530-1). Also, it is within the scope of different national standards, where Serbian national standard, which had previously been taken from Hungarian, is one of them. The main difference between all of them lies in the interpretation of resulting diagram, i.e. in definition of obtained values.

The Farinogram parameter which has the greatest practical value is the water absorption. Water absorption is directly related to the yield of finished bakery product and it is one of the most important parameters in assessing the "flour strength" and in product price calculations. According to ICC method, dough development time is the elapsed time from the beginning of the kneading until maximum consistency is achieved. Dough stability represents the time during which the maximum dough consistency does not change or changes very little. The degree of softening can be described as the distance between the centre of the curve at the end of analysis time and the central line which passes through the maximum of the curve.

Different approaches of the evaluation of the obtained Farinograms suggested by ICC (ICC Standards, 1996), AACC (AACC methods, 2000) and by actual Serbian method (Serbian official methods, 1988) are shown in Figure 1.

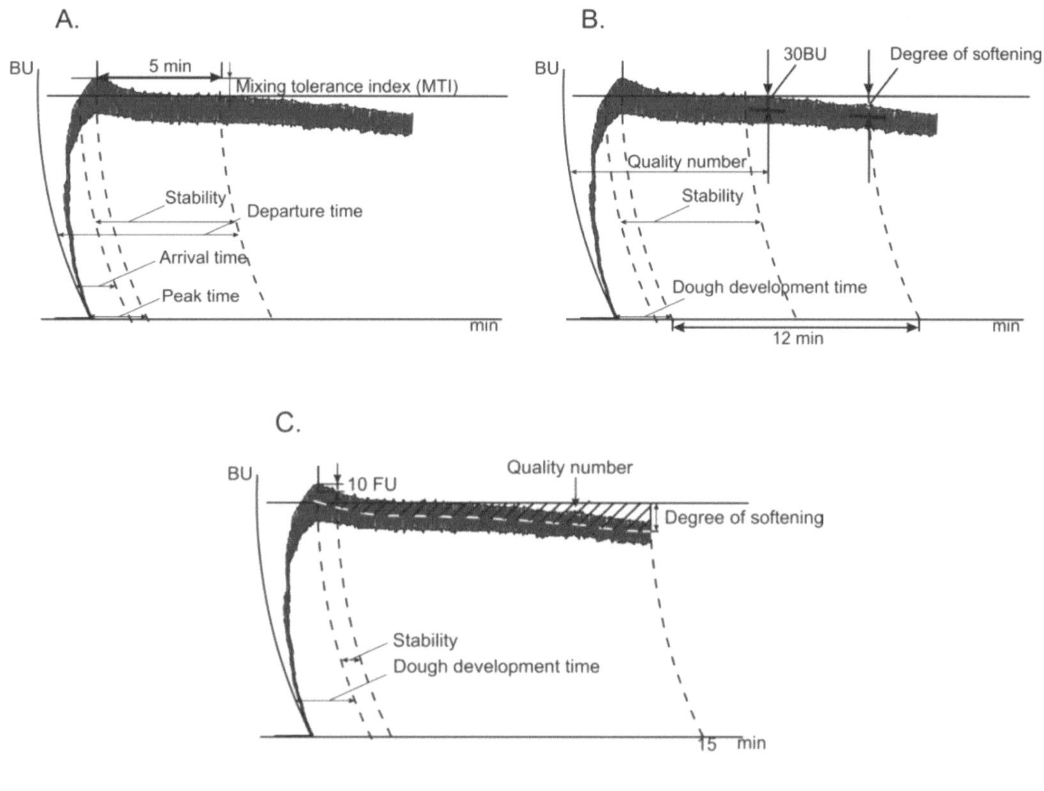

Fig. 1. Farinograph parameters according to: A.) AACC, B.) ICC and C.) Serbian method

The different duration of measurement makes the first difference between the methods. According to ICC method (ICC standards, 1996) measurement lasts 12 minutes from the end of development time. However, according to AACC method Farinograph measurement lasts 5 minutes after reaching the maximum consistency of the dough (peak time) in the case of dough with weak gluten, or until the consistency of the dough falls below 500 BU (departure time). Serbian method proposes that Farinograph measurements are performed during the same time interval regardless of flour quality, i.e. 15 min from the water addition. Water absorption is measured identically in all methods. Dough development time, stability and degree of softening are evaluated differently as it can be seen in Figure 1.

Farinograph water absorption is mainly influenced by the properties of flour main components: gluten and starch. In order to be properly interpreted, it must be compared to the other Farinograph parameters. Thus, high water absorption, combined with low degree of softening indicates good quality flour, whereas a high water absorption combined with a high degree of softening indicates poor quality flour. Dough development time depends on the gluten quality, starch granule size and degree of starch damage. Furthermore, dough development time increases with the increase in the proteolytical degradation of protein. It also increases with a decrease in the size of starch granule and the increase in the content of damaged starch due to the increase in specific surface area which absorbs water. The stability and the degree of softening are the gluten quality parameters which describe the viscoelastic properties of formed gluten complex. In practice, higher stability and lower degree of softening indicate that dough will be more able to sustain long mechanical processing treatments. Increased degree of softening is particularly important indicator of proteolytic degradation of gluten.

Farinograph also enables monitoring the influence of additives, and thus allows optimization of flour processing in terms of standardization of flour quality produced from raw materials of variable quality.

Flour quality is defined and classified differently in European countries depending on its end-use purpose. In Serbian method, the quality number (Figure 1) represents the area enclosed by line passing through the centre of the Farinograph curve and the central line which passes through the maximum of the curve (500±10 FU). According to the value of the area, wheat flours are classified into six quality classes: A1 (0-1.4 cm^2), A2 (1.5-5.5 cm^2), B1 (5.6-12.1 cm^2), B2 (12.2-17.6 cm^2), C1 (17.7-27.4 cm^2) and C2 (27.5-50 cm^2).

However, due to the breeding process and the development of new varieties, the existing ranges of Farinograph parameter values which classified the flour into good or bad, have to be redefined to accurately access the quality of flour.

According to the results of the research performed during the past decade in our laboratory (Torbica et al., 2010a), the quality of wheat varieties cultivated in Serbia was strongly affected by climatic changes and global warming. Namely, frequent occurrence of heat stress changed the course of biosynthesis of gluten complex proteins in the direction of synthesis of larger amounts of gliadin in relation to glutenin. Moreover, concerning the starch component, large quantities of larger starch granules (A granules) were synthesized rather than small starch granules (B granules).

Figure 2 illustrates trends in values of the most important Farinograph parameters of the flour from the Serbian wheat varieties harvested over the past ten years.

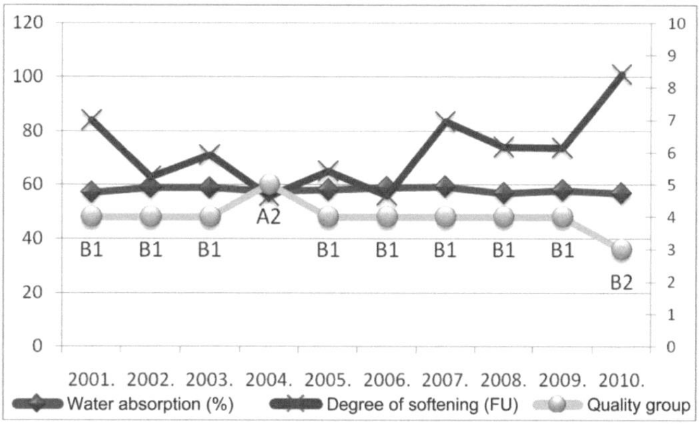

Fig. 2. Farinograph values of average wheat flour samples harvested in Serbia during ten years

As it can be seen in Figure 2, the average value of the water absorption was almost unchanged in the examined period of time, while in the last five years the average values of the degree of softening were increased. This resulted in reduction of the flour quality from B1 to B2 quality class (Torbica et al., 2011).

Although the Farinograph has been the most commonly used device for monitoring the physical properties of dough in order to assess flour quality, its implementation has been improved. Newly introduced the Farinograph-AT allows determination of dough properties by the mixer blades of different profiles, has the ability to change the mixing speed and temperature of mixing bowl. Due to this fact, it has already found its place in research and development laboratories.

2.2 Mixograph

Similary to Farinograph, Mixograph is a rheological device that measures the dough resistance during kneading. However, these two instruments differ in kneading process and in intensity of mechanical stress applied on dough during the analysis. The Farinograph method requires dough kneading until consistency of 500 BU is reached, while Mixograph always operates with a constant amount of water which resulted in dough of different consistency (Weipert, 1990; Mann et al., 2008). The difference in the dough mechanical treatments in Mixograph compared to Farinograph measurements is reflected in the different curve profile i.e. the obtained curves are characterized by different curve width. Although Mixograph monitors identical properties of dough as Farinogram does, the obtained parameters are not equivalent. A major drawback for the wider Mixograph application is its impossibility to determine water absorption due to Mixograph principle to operate with constant amount of water, as reported in AACC 54-40A (AACC methods, 2000). The major advantage of the Mixograph method is that it is not time consuming and it requires small amount of flour sample (2g, 10 g and 35 g depending on mixing bowls). Therefore, this method remained traditional among breeders, who handle with small amounts of sample, and for whom this method was initially designed (Graybosch et al., 2011).

3. Fermentation recording devices

Most rheological tests are performed on non-yeasted dough. These results are relevant in assessing the quality of cookie, cake and other bakery products which do not contain yeast.

However, for a bread dough, fermentation is an important step in processing chain where the expansion of air bubbles incorporated during mixing led to formation of aerated crumb structure which appearance greatly contribute to sensory assessment and consumers acceptability of bread (Dobraszczyk et al., 2000).
There are different types of devices which measure dough fermentation parameters. The changes in the dough volume are monitored by Brabender Maturograph and Oven Rise Recorder, while the formation of CO_2 in dough is recorded using Brabender Fermentograph. Unlike the listed equipment, Chopin Rheofermentometer allows simultaneous measurement of dough height during fermentation and interaction of CO_2 development and retention.

3.1 Rheofermentometer

Chopin Rheofermentometer is the unique device which provides information about dough properties that traditionally have been obtained by employing several different tests, i.e. by combination of at least two analyses such are Maturograph and Fermentograph. Moreover, it indirectly correlates with the Farinograph measurements. Rheofermentometer allows evaluation of flour fermentation capacity, yeast activity and indirectly indicates the quality of gluten complex proteins.
The parameters are measured in real time (3 h) during which two curves are simultaneously generated: the first one that describes the dough development and the second one describing the volume of CO_2 which retains in the dough as well as the volume and the time of CO_2 release, which also represents the appearance time of dough porosity (T_x parameter). At the end of the Rheofermentometer analysis all the values are automatically calculated by microprocessor (Lallemand, 1996; Ktenioudaki et al., 2011).
The basic parameters of Rheofermentometer curves are shown in Figure 3.

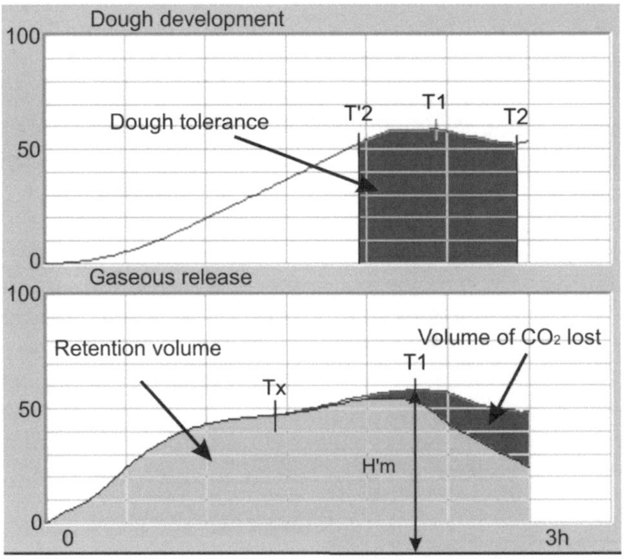

Fig. 3. Rheofermentometer curves consisted of dough development time curve and gaseous release curve

Rheofermetometer analysis of flour and dough enables accurate simulation of processing conditions during production of baked goods containing yeast. Moreover, it is possible to precisely calculate the amount of necessary additives (oxidizing and reducing agents, emulsifiers, enzymes) in order to optimize production processes.

During the past decade, our researches indicated that prediction of the quality of the final product, based only on Extensograph and Alveograph measurements was pretty uncertain. The harvest of wheat crop in 2008 in Serbia showed that the energy values estimated by Extensograph were very heterogeneous. Therefore, for the research purposes two flour mixtures were formed. The first flour mixture had the Extensograph energy value of 58 cm^2 and Alveograph deformation energy W=159 x 10^{-4} J. Amylograph peak viscosity was 285 BU; flour mixture 1 was estimated as A2 Farinograph quality group, with a favourable ratio of dough development and stability value and the degree of softening of 55 BU. Rheofermentometer curve showed that the dough after the fermentation completely retained 87% of the total CO_2 produced. The maximum dough volume of 1186 ml was reached after 58.9 min, dough tolerance during fermentation was 58 min and 30 s, and porosity was estimated after 1 hour, 25 minutes and 30 seconds. Another flour mixture had the Extensograph energy value of 26 cm^2 and Alveograph deformation energy W=115x10^{-4} J. Amylograph peak viscosity was 315 BU and a flour mixture was estimated as B1 Farinograph quality group with a favourable dough development and stability value and the degree of softening of 60 BU. Rheofermentometer curve indicated that the dough after the fermentation completely retained 88% of the total CO_2 produced. The maximum volume of 1112 ml was reached after 54.9 min, dough tolerance during fermentation was 34 min and 30 s, and porosity was estimated after 1 hour 52 minutes and 30 seconds. Performed experiments have shown the importance of flour or dough characterization by Rheofermentometer. Namely, it is evident that the two examined flour mixtures tested by Rheofermentometer showed similar properties although they possessed different parameters obtained by Extensograph and Alveograph.

However, the sensory analyses of baked products confirmed the results obtained by Rheofermentometer. Bread made of flour mixture 1 had a specific volume of 4.63 ml/g, good elasticity, somewhat rough pores and light dark, shiny crust. The addition of improvers in the flour mixture resulted in a product having slightly higher specific volume while the other quality parameters were rated with the highest score (5). Bread made of flour mixture 2 had a specific volume of 4.22 ml/g, poor elasticity, rough pores and light dark, shiny crust. The addition of improvers in the flour mixture resulted in a bread having much higher specific volume while the other parameters of sensory evaluation were scored with 4.3 points. Similar sensory quality of bread samples produced from flour mixtures 1 and 2 confirmed that the assessment of flour quality using Rheofermentometer was reliable and necessary to precisely evaluate the flour quality.

4. Extensional techniques

Extensibility represents the most unique property of wheat dough, which enables getting characteristic structure and volume of the baked products. This property is enabled by the presence of gluten complex proteins (Kieffer, 2006).

Extensibility tests are typically conducted on wheat dough to evaluate its tensile strength and extensibility characteristics which are heavily dependent on the protein quality

(Dobraszczyk & Morgenstern, 2003). Also, a great concern for the extensibility of wheat dough has been influenced by the relevance of the extensibility to baking performance and the final product quality (Cauvain & Young, 1998; Grausgruber et al., 2002; Sahin & Sumnu, 2006). In this regard, the baking performance is influenced by the interrelation between the maximum resistance and extensibility, since it is indirectly responsible for the extent of the expansion during the fermentation process (Anderssen et al., 2004). In extensibility tests, a shaped dough piece is submitted to large deformations until rupture occurs. Simultaneously, the resistance that occurs in a dough during stretching is recorded, providing the data relevant to the assessment of dough handling behavior and baking performance (e.g. resistance to large deformations and stretching suitability) (Grausgruber et al., 2002; Vergnes et al., 2003; Nash et al., 2006).

To measure the extensional properties of dough, two types of extensional test are generally applied:

- The measurement of uniaxial extension, where dough is stretched in one direction, and
- The measurement of biaxial extension, where the dough is stretched in two opposing directions, which can be achieved either by compression between lubricated surfaces or by bubble inflation (Dobraszczyk & Morgenstern, 2003; Abang Zaidel et al., 2010)

which both measure power input during dough development caused by extensional deformation (Sahin & Sumnu, 2006).

The measurement of uniaxial extension is one of the most widely used test principle to measure materials properties. The methods are performed by clamping a strip of material at both ends, its stretching at a fixed rate in a suitable testing device, and by recording the force-extension curve. The most commmon methods used for measurement of the uniaxial extensional properties of doughs are Brabender Extensograph and Stable Micro Systems Kieffer dough and gluten extensibility rig (Dobraszczyk, 2004). The measurement of biaxial extension implies stretching a dough piece at equal rates in two perpendicular directions in one plane. The most widely used principle in the measurement of biaxial extension properties of dough is based on inflation technique (e.g. bubble expanding) as in Alveograph method (Dobraszczyk, 2004; Sahin & Sumnu, 2006). The Alveograph method measures the resistance to biaxial extension of a thin sheet of dough prepared from flour, water and salt, generally at a constant hydration level, although the measurements could be performed at adapted hydration as it is the case with the Alveoconsistograph method. From the above mentioned it follows that the extensibility tests that are in common use are carried out by the fundamentally different measuring equipment, although the resulting curves equally describe the extensional work, resistance to extension and extensibility of the tested dough (Weipert, 2006).

4.1 Extensograph

The Brabender Extensograph is an internationally accepted standard method that is in compliance with ISO 5530-2, ICC 114/1, AACC 54-10. It is applicable for measurement of physical properties of dough subjected to mechanical handling and resting. Precisely, an Extensograph provides information about dough resistance to stretching and extensibility by measuring the force to pull a hook through a cylindrically shaped piece of dough. During the measurement the resistance of dough to stretching and the distance the dough stretches before breaking is recorded in the form of diagram extensogram. Measurement procedure comprises of several steps:

- Preparation of dough (with 2% salt based on flour weight) in the Brabender Farinograph mixer, usually at 2% less than its optimum absorption to compensate the salt addition.
- Moulding of dough pieces on the Extensograph into a cylindrically shaped dough pieces
- Resting of the dough pieces for a fixed period of time (45, 90, 135 min)
- Stretching the dough pieces until they rupture and recording the extensibility of the dough and its resistance to stretching (Kent & Evers, 1994; Rasper & Walker, 2000; Sahin & Sumnu, 2006).

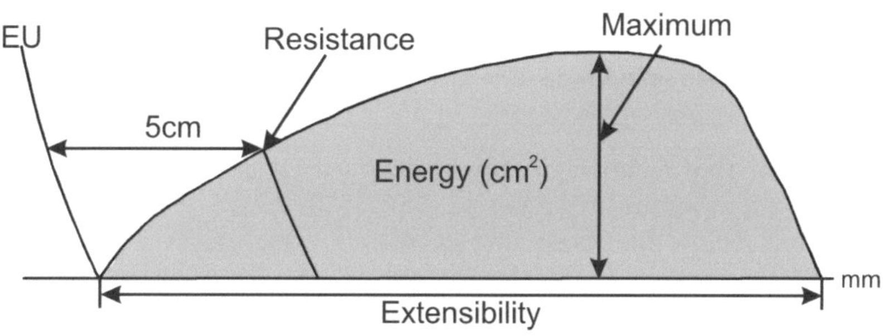

Fig. 4. Extensograph curve

The data obtained from the extensogram (Figure 4) include:

1. The maximum resistance (R_{max}), or the resistance at constant deformation usually corresponds to the height of the curve at 50 mm from the beginning of stretching (R_{50}). The latter is preferably expressed within the cereal testing laboratories since it represents the resistance at a fixed extension for all tested doughs. This parameter is expressed in Brabender units.
2. The dough extensibility (E) expressed in mm, which represents the distance of stretching before rupture.
3. The ratio of resistance to extensibility. High ratio indicates the short gluten properties resulting in low volume of baked products.
4. The area under the curve, which is proportional to the energy required to stretch the test piece to its rupture point. This parameter, expressed in cm^2, is a convenient single figure for characterizing flour strength. The stronger the flour, the more energy is required to stretch the dough.

The shape of extensogram curve gives an indication of results that could be expected for baking performance (Freund & Kim, 2006). For example the shape of the extensogram curve gives an indication of the appearance of the cross section of the loaf of bread. Curves characterized by low resistance to extension indicates the small baking volume and vice versa. Hence, the dough with the balanced ratio between the resistance and extensibility is considered as a raw material of a suitable quality for baking production.

It can be also used, in the same way as Farinograph, for monitoring the influence of additives on the physical properties of dough.

During the last decade, the average value of extensibility parameters of wheat dough have been significantly deteriorated and at the same time being characterized by wide ranges of

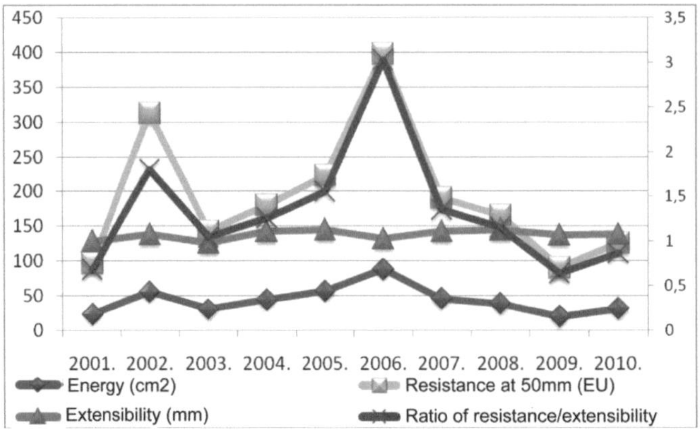

Fig. 5. Extensograph values of average wheat flour samples harvested in Serbia during ten years

values indicating variable technological quality of wheat in Serbia. Figure 5 shows the average values of extensogram in the period 2001-2010. Average energy values (except in 2006) were generally below the acceptable values for bread making (Torbica et al., 2011). Moreover, in 2004 there were significant amount of samples with the energy values that could not be registered. On the basis of common interpretation of extensogram, very bad baking performance could be expected, which was only partially confirmed. However, the samples with destroyed protein structure as a result of wheat bug damage were characterized by very bad baking performance that could not be improved by using baking improvers. On the other hand, there were flour samples damaged by inappropriate climate conditions during growth (e.g. heat stress) that demonstrated quite satisfactory baking performance. These samples were characterized by higher gliadin content in relation to common values. Altered ratio of gliadin and glutenin caused extremely increased extensibility of dough, but with the use of baking improvers satisfactory baking performance was enabled (Torbica et al., 2007). Based on the experience of examined time period, and the fact that wheat breeders have developed varieties with lower but more focused protein content we would like to point out that the energy values of the extensogram can not be interpreted as in previous decades in the evaluation of technological wheat quality (Anderssen et al., 2004; Torbica et al., 2011). Also, it stresses the need to increase the understanding of the parameters obtained in extensibility tests and its relevance to baking perfomance, and on the other hand to relate the gliadin content and ratio of gliadin and glutenin to extensional properties of wheat dough.

4.2 Alveograph

The Chopin Alveograph is an internationally accepted standard method that is in compliance with ISO 27971, 5530-2, ICC 121, AACC 54-30A. An Alveograph, provides similar information as Extensograph does, by measuring the pressure required to blow a bubble in a sheeted piece of dough. The phases of the Alveograph method simulate the sheeting, rounding and molding of the dough pieces during baking process. The five dough pieces are prepared by mixing and extruding, followed by shaping into small discs which are then left for resting during 20 minutes. After that, air is blown under the disk at a constant rate creating a bubble. The pressure inside the bubble is recorded until the bubble ruptures creating the alveogram (Figure 6) giving the data relating to the dough's resistance to deformation such as (Vergnes et al., 2003; Sahin & Sumnu, 2006; Dubois et al., 2008):

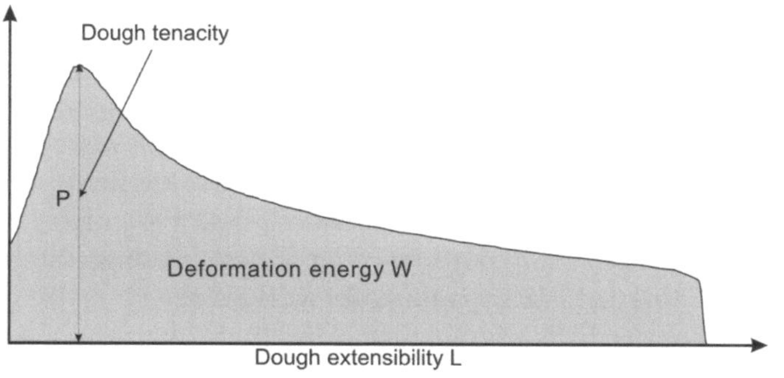

Fig. 6. Alveograph curve

1. Overpressure (P) is measured as the maximum height of the alveogram multiplied by the correction of 1.1 for the difference between the geometry of old water manometer and inner vertical tube of the manometer. This parameter is expressed in mm H_2O and is commonly used in the interpretation of the results of Alveograph method. Generally, it is considered as an indicator of the dough's tenacity.
2. Average abscissa to rupture (L) represents the average length of the alveogram from the point in which the bubble starts to inflate to the point in which the bubble ruptures. The L value is generally related to dough extensibility and predicts the handling properties of the dough. It is expressed in mm
3. Configuration ratio (P/L) indicates the shape of alveogram
4. Swelling index (G) represents the square root of the air volume needed to rupture the dough bubble. This value is the measure of dough extensibility.
5. Elasticity index (Ie)
6. Deformation energy (W) represents the energy needed for the dough piece inflation until rupture, expressed in 10^{-4} J. The W value is related to the baking "strength" of the flour. This parameter has the uppermost significance in the interpretation of alveogram since it summaries all the others (Bordes et al., 2008).

In general, strong flours are characterized by high H and W and low or medium L values. Particularly, the suitability of certain wheat/flour for intended purpose on the basis of alveogram values depends between different countries and is influenced by the availability of raw material of certain quality (Vergnes et al., 2003; Dubois et al., 2008).

Thus, by performing the quality testing of bread wheat cultivars from worldwide core collection, Bordes at al. (2008) classified available wheat cultivars on the basis of alveogram parameters. It was also previously done by Aldovrandi & Vitali (1995). P values for standard wheat quality range 60-80 mm H_2O, very good wheat quality 80-100 mm H_2O, whilst extra strong wheats are characterized by P value higher than 100 mm H_2O. Moreover, the L of 100 mm is generally considered as good, although some applications require higher values (e.g. biscuit production). The configuration ratio is commonly used in wheat trade where value of 0.50 indicates either resistant and very extensible dough or moderately extensible less resistant dough. Furthermore, value of 1.50 indicates very strong and moderately extensible dough, whilst raw material with P/L value in the range 0.40-0.80 is suitable for bakery production. Wheat suitable for confectionary products should exhibit P/L value lower than 0.50. Concerning the interpretation of W value for confectionary wheat is characterized by W value lower than 115 (* 10^{-4} J), standard quality wheat is characterized by W value in range 160-200, whilst good quality wheat and improving wheat are characterized by W value in the range 220-300 and higher than 300, respectively (Bordes et al., 2008). Traditional measurements of the properties of durum dough for pasta production include the application of the Alveograph method. Indicators of suitable durum wheat quality for pasta production determined by Alveograph include P/L values in the range 1.5-2.5 and W values in the range 200-250, or even higher than 300 for superior pasta quality (Marchylo & Dexter, 2001). It is common that parameters derived from the Alveogram are used for selection purposes and wheat grading throughout the wheat storage and milling sectors and bakery companies worldwide.

Physical dough tests such as Extensograph and Alveograph are used in various parts of the world to give an indication of dough mixing characteristics. Although the application of Extensograph and Alveograph in flour quality control is more culture related, the data they produce can equally be used to help the baker better utilize the tested flour. They are primarily used in Germany, France, and the British Isles, and their respective former colonies (De Bry, 2006; Khan & Nygard, 2006). The Alveograph is mainly used in the Mediterranean countries, certain parts of South America and former French colonies in Africa.

The main difference in the Extensograph and Alveograph method is that the Alveograph assumes the constant hydratation resulting in dough of different consistency. Unlike the Alveograph, the Extensograph method assumes the adjusted hydratation resulting in dough of equal consistency. The main limitation on the use of Extensograph and Alveograph method is the large amount of sample that is required which is limiting factor in application of these methods in breeding programs. This situation has initiated the attempts to develop the micro-extensograph methods that proved to be significantly correlated with standard Extensograph method despite the differences in the mass and shape of dough pieces and hooked speed used for stretching (Grausgruber et al., 2002; Anderssen et al., 2004; Uthayakumaran et al., 2004) Also, none of these give rheological data in fundamental units of stress and strain, because the sample geometry is not defined, dimensions change extensively and nonuniformly during testing, which makes impossible to define any rheological parameters such as stress, strain, strain rate, modulus, or viscosity (Dobraszczyk, 2004).

4.3 Kieffer dough and gluten extensibility rig

Another piece of equipment developed for the measurement of uniaxal extension of dough is Kieffer extensibility rig for the TA.TX2i texture analyzer. It uses the same principle as Extensograph, except that the sample is stretched upwards. It enables extensibility measurement of both dough and gluten which represents its additional advantage. Apart from that, it require small quantities of sample (cca 0.8 g), the force is expressed in Newton and the speed of the hook is adjustable (Nash et al., 2006; Abang Zaidel et al., 2008; Abang Zaidel et al., 2010). However, in comparison to the Extensograph, the Kieffer rig is characterized by worse reproducibility (coefficient of variation extensibility in the range of 2-12% compared to coefficient of variation 2-7% obtained by Extensograph (Ktenioudaki et al., 2011).

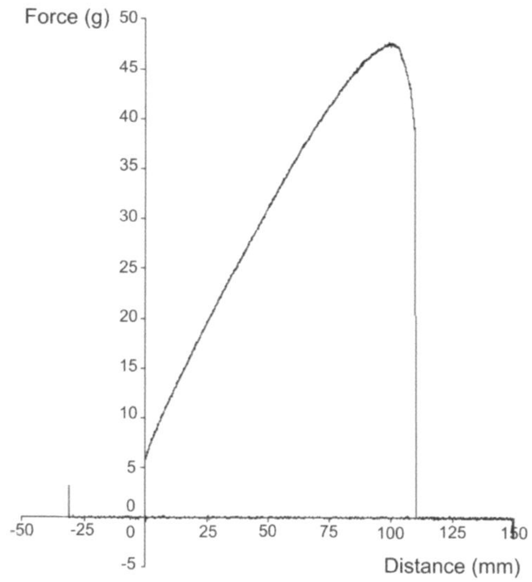

Fig. 7. Kieffer force-distance curve

The data obtained from the Kieffer force-distance curve (Figure 7) include:

1. Maximum force (g)
2. Maximum extensibility (mm)
3. Area under the force vs. distance curve (mm^2) (Dobraszczyk & Salmanowicz, 2008).

It was reported that the parameters obtained by using the Kieffer extensibility rig were significantly correlated with the Extensograph parameters especially those related to the strength of dough (Suchy et al., 2000; Grausgruber et al., 2002; Mann et al., 2005; Ktenioudaki et al., 2011). The reason why the extensibility parameters were not exhibit such a high correlation as strength parameters was explained by the differences in the sample size and strain rate between the two methods (Ktenioudaki et al., 2011).

5. Devices for monitoring the properties of carbohydrate complex

The term dough rheology is mostly connected to protein component of dough. However, starch, as the major component (75-80%), also contributes to the formation of texture and quality of bakery products, since it dilutes the gluten, absorbs water from the gluten during gelatinization, and thus provides a bread structure permeable to gas which will not collapse while cooling (Miyazaki et al., 2006). The properties of cereal starches, as well as the α-amylase activity at high temperatures are mostly examined by Brabender Amylograph and Hagberg-Perten Falling Number instruments. Besides that, in order to determine starch properties during heating and subsequent cooling Viscoamylograph and Rapid Visco Analyser are employed.

5.1 Amylograph

Pasting properties of wheat and rye flour starch, starch properties solely as well as α-enzymatic activity can be determined using Brabender Amylograph (Duisburg, Germany). The measurement principle consists of heating the flour (starch) suspension at constant speed rate of 1.5 °C/min to 95 °C or until the significant decrease of measured torque after

the pasting peak is reached. The obtained results are influenced by different starch flour properties, milling conditions, α-enzymatic activity, different wheat or rye varieties and botanical source if starch is tested material.
Generally, three parameters are recorded: gelatinization temperature, peak viscosity and temperature at peak viscosity that is presented in Figure 8.

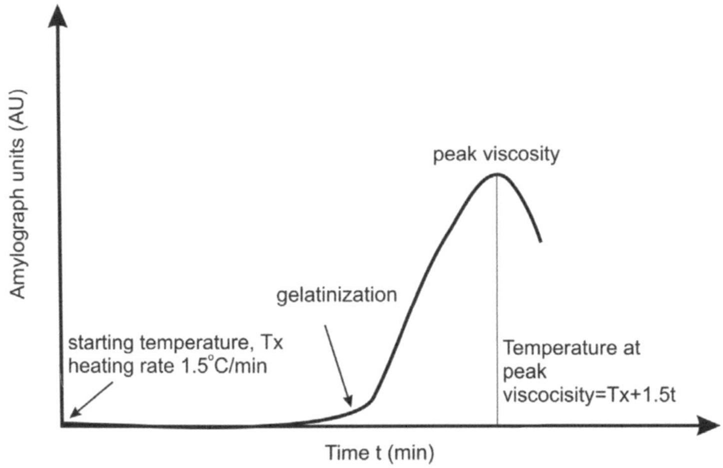

Fig. 8. Amylograph curve

Brabender Amylograph is approved and standardised method by ICC (ICC 126/1), AACC (AACC 22-10.01; old number AACC 22-10) and ISO (ISO 7973:1992). Although the operational principle is basically similar, there are still minor differences between these methods, such as different concentration of flour suspension. α-enzymatic activity which can be especially found in wheat flour obtained from sprouted wheat grains and normally found in rye flour has large influences on liquefying starch paste during heating expressing consequently lower peak viscosity. Therefore, low peak viscosity can be ascribed to α-enzymatic activity and to starch properties as well. Wheat varieties, climate conditions and wheat milling process influence the pasting properties of tested flours. In order to study starch pasting properties solely, excluding the α-enzymatic effect, the Amylograph procedure can be modified. According to Mariotti et al. (2005) 1mM solution of $AgNO_3$ was used instead of distilled water which has been proved to be efficient alpha-amylase inhibitor. The obtained results showed that system prepared with $AgNO_3$ solution revealed significant increase in peak viscosity in comparison to system prepared with distilled water. On the other hand, milling conditions can cause mechanical damage of starch granules, especially during the milling of hard wheat varieties (Hoseney, 1994). Damaged starch granules can absorb more water and are more susceptible to amylase attack which results in decrease in the peak viscosity (Léon et al., 2006). Starch granule size and structure, amylase/amylopectin ratio and molecular weight also influence pasting properties of flours (Thomas & Atwell, 1999). For example, large granule size influence higher viscosity but this effect is limited, because large granules of starch are more sensitive to shear and more easily destructible, especially if lower heating rates are applied. Heating rate of 1.5°C/min simulates the baking procedure (Mariotti et al., 2005). Time-consuming measurement is one of the major drawbacks of the Amylograph method (cca. 45 min). Therefore, the Rapid Visco

Analyser (RVA) as a rapid method for measuring degree of sprout damage in wheat was introduced (Ross et al., 1987). However, the effect of α-amylase decreases as heating rate increase due to the shorter time available for enzyme that can act on starch granules which resulted in higher peak viscosities (Collado & Corke, 1999). Also, if the applied shear rate is increased, starch granules are more easily destructible and final peak viscosities are lowered (Suh & Jane, 2003). The results obtained by Amylograph are in arbitrary i.e. Brabender Units (BU) because precise measurements could not be achieved due to complex system geometry that makes the determination of flow field inside the cup very difficult. Also, shear rate is not uniform within the sample. Therefore, viscosity cannot be calculated from torque measurements (Lagarrigue & Alvarez, 2001). Consequently, in order to keep conditions of shear rate and temperature conditions well defined and in order to obtain results expressed in viscosity units, rotational rheometers equipped with different measuring geometries were introduced. According to measurement performed in our laboratory significant correlation of peak viscosities was obtained by Brabender Amilograph and Haake Mars rheometer (Thermo Scientific, Germany). The Haake Mars measuring geometry consisted of Z40 measuring cup (43.4 mm diameter, 8 mm gap) and FL2B propeller shaped rotor with 2 blades. The concentration of flour-water suspension and heating rate was the same as it was in Amylograph measurements. However, significantly lower amount of wheat flour (cca 12.4g) was needed in comparison to ICC 126/1 Amylograph measurements (cca. 80g) (ICC Standards, 1996). Twenty flour samples having different pasting properties were examined by Brabender Amylograph and Haake Mars rheometers and the correlation of 0.99 was obtained.

5.2 Falling number

The Falling number (FN) is a method used to determine α-amylase activity in meal, wholemeal and flour of wheat, rye, barley, as well as other cereals, starch containing and malted products. It measures time (expressed in seconds) required to stir and allow stirrer to fall through a hot aqueous flour or meal gel or starch gel undergoing liquefaction influenced by α-amylase activity. Pre-harvest sprouting represents one of the major reasons for high α-amylase activity. Therefore, lower falling number values - shorter time needed for stirrer to fall distance through a liquefied paste indicates higher α-amylase activity and vice versa. Falling number is a method which indicates the effect of a present α-amylase rather than the actual enzyme amount, i.e. any other parameter which could impact the viscosity can affect the falling number, too. Falling number test is also affected by endosperm quality; susceptibility of starch to α-amylase i.e. the ease of starch gelatinization and accessibility to enzyme activity (Best & Muller, 1990). Increased α-amylase activity could cause problems in breadmaking which can be manifested in sticky dough, lower water absorption, crumb discolouration, mechanical handling breakdowns (Dimmock & Gooding, 2002, as cited in Chamberlain et al., 1982; Gooding & Davies, 1997; Sorenson, 2006). Also, pasta and noodle products are highly affected by α-amylase activity. Pasta/noodle produced from raw material of low FN is soft and mushy, cooking loss is increased and there are production problems regarding uneven extrusion, strand stretching and irregularities in drying (Kweon, 2010).

Falling number method is recognized by ICC (ICC 107/1), AACC (AACC 56-81.03) and ISO (ISO 3093).

There are different factors affecting the falling number value: altitude, climate conditions, late maturity alpha amylase, fungicidal treatment, waxy wheat, fusarium infection, nitrogen fertilization rate. Therefore, altitude correction was incorporated in revised AACC method in 1982 (Lorenz & Wolt, 1981). It was estimated that an average relative humidity of over 80% and a maximum daily temperature of below 13°C during grain filling affected decrease in the falling number to below 120 s (commercially acceptable starch quality). Also, average relative humidity fell below 70% and average maximum temperature above 16°C during grain filling affected increase in falling number over 230 s (bread wheat quality) (Karvonen et al., 1991). Kettlewell (1999) proved that application of nitrogen fertilization affected the increase of Falling number in the absence of sprouting. In addition, it was estimated that the use of fungicides may reduce falling number (Ruske et al., 2004), but this effect is cultivar dependent (Wang et al., 2004). Falling number test can be also influenced by genotype variation. One of the extreme examples of genotype variation is implementation of waxy wheats that are characterized by lower amylose content (Graybosch et al., 2000). Beside the pre-harvest sprouting which is known to affect low falling number, there are also a number of additional causes of low falling number such as late maturity α-amylase (Mares & Mrva, 2008) or prematurity α-amylase and retained pericarp α-amylase (Lunn et al., 2001).

6. Determination of mixing and heating properties of dough in one test - Mixolab

Although it is a relatively new device, introduced in 2004 by Chopin Technologies (Villeneuve la Garenne, France), it has already been within the scope of many scientific papers dealing with the assessment of dough rheological behaviour (Rosell et al., 2007; Collar et al., 2007; Kahraman et al., 2008). Mixolab working principle comprises the combination of Farinograph and Amylograph methods (described earlier in the text). Moreover, Mixolab system offers additional application called Mixolab Simulator whose results correspond to values and units obtained by Farinograph. However, in contrast to Farinograph which works with the constant flour mass (50 or 300 g), Mixolab flour mass depends on a flour water absorption, where the parameter which is fixed is the dough mass (75 g). The difference between Amylograph measurements, which are performed using flour-water suspension, is that Mixolab monitors starch gelationization in water-limited dough system resembling the real baking conditions. The development of a Mixolab also represents a step toward expression of the consistency (measured as a torque) in a real SI unit (Nm), unlike arbitrary Brebender units. Namely, usage of arbitrary units is one of the major drawbacks of empirical rheological methods over the fundamental ones (Weipert, 1990; Dobraszczyk & Morgenstern, 2003).

Regardless the existing differences between the Mixolab and Farinograph, significant correlation was found between the obtained parameters (Dapčević et al., 2009), e.g. $r = 0.98$ for water absorption, $r = 0.97$ for dough development time. A significant correlation coefficient ($r = 0.88$) was determined between Amylograph peak viscosity and Mixolab C3 torque. Significant correlations were also found with parameters derived from Alveoconsistograph, Zeleny sedimentation and baking test (Kahraman et al., 2008).

Ţăin et al. (2008) determined that the bread's volume was significantly negatively correlated with C2 value ($r = -0.76$) and with C5-C4 value ($r = -0.73$). According to Kahraman et al. (2008) most of the Mixolab parameters (C2, C3, C4 and C5) were significantly correlated with cake volume index.

In order to simulate the phases of the breadmaking process and thus to investigate the thermo-mechanical behaviour of the dough, Chopin+ protocol is generally employed. This protocol is integrated into Mixolab software and it is standardize as ICC 173, as well as AACC 54-60.01 method. It is very easy to operate with, since the software is guiding the operator through all the necessary steps. The first step is the determination of flour water absorption. For that purpose nearly 50 g of flour, of known moisture content, is placed into Mixolab bowl and kneaded between the two kneading arms in order to achieve a consistency of 1.1 Nm. Since the necessary consistency is rarely achieved in the first step, the correction has to be made with the new mass of flour, in order to obtain 75 g of dough of consistency of 1.1 Nm. Subsequently, the following procedure is performed: mixing the dough under controlled temperature of 30 °C during 8 minutes, followed by temperature sweep until 90 °C and a cooling step to 50 °C. Total duration of the second step is 45 min.

Since, during 45 min the dough is subjected to mechanical and thermal constraints, the data concerning the quality of the protein network and the starch changes during heating and cooling can be obtained in a single test. A typical Mixolab profile is shown in Figure 9. It can be divided into five different stages, depending on physicochemical phenomena which occur during that processing condition and which determine the rheological properties of the system.

The first stage starts with an initial mixing (8 min) when the hydration of the flour compounds occurs, followed by the stretching and alignment of the proteins which led to the formation of a three-dimensional viscoelastic dough structure (Rosell et al., 2007; Huang et al., 2010). During the first stage, an increase in the torque is observed until a maximum consistency (C1 = 1.1 Nm) at 30 °C is reached. After that the dough is able to resist the deformation for some time, which is related to the dough stability.

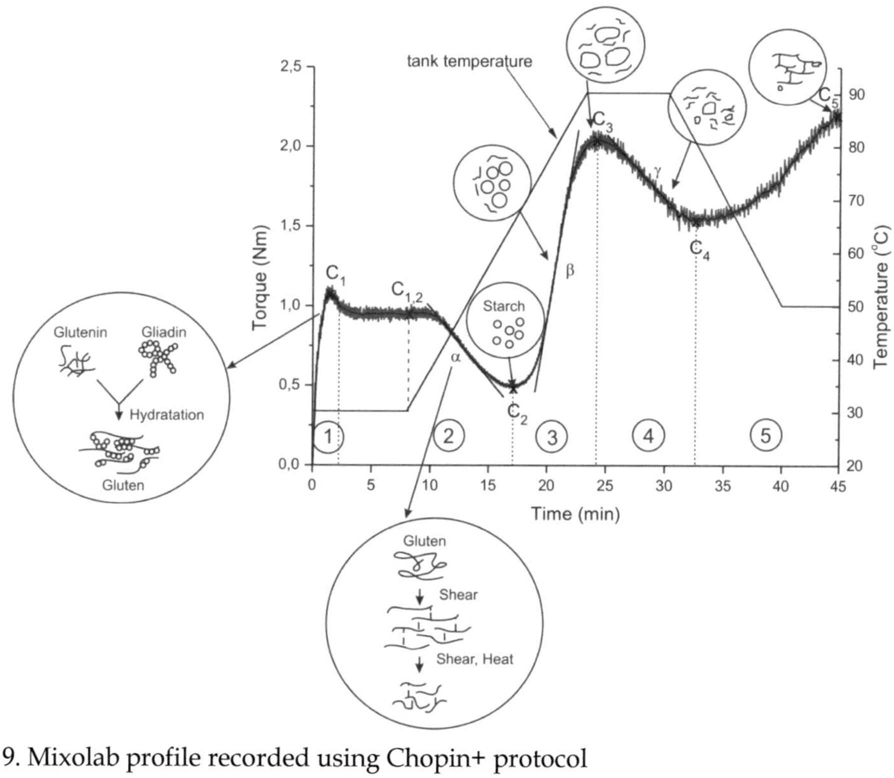

Fig. 9. Mixolab profile recorded using Chopin+ protocol

The parameters obtained during the first stage are thus related to dough mixing characteristics and are listed below:

1. Initial maximum consistency (Nm), C1 - used to determine the water absorption
2. Water absorption (%), WA - the percentage of water required for the dough to produce a torque of 1.1 Nm
3. Dough development time (min), DDT - the time to reach the maximum torque at 30 °C
4. Stability (min) - time until the loss of consistency is lower than 11% of the maximum consistency reached during the mixing
5. Amplitude (Nm) – refers to dough elasticity
6. Torque at the end of the holding time at 30 °C (Nm), C1.2 - used to determine the mechanical weakening

After the dough's stability period, which indicates the end of the first stage and the beginning of the second stage, a torque decrease is registered. Depending on a flour quality, the second stage can start within the initial mixing period or later. Namely, the longer the stability period is, the better the protein quality is. During the second stage, the protein weakening occurs. The weakening is firstly the consequence of a mechanical shear stress, which is subsequently followed by temperature increase. The resulting torque decrease is related to the native protein structure destabilization and unfolding (Rosell et al., 2007; Huang et al., 2010). The rise of the dough temperature led to the protein denaturation involving the release of a large quantity of water. Moreover, within the temperature range of second stage, the proteolytic enzymes have an optimal activity (Stoenescu et al., 2010), represents in the Mixolab curve by the α slope.

The parameters obtained during the second stage include:

1. Minimum consistency (Nm), C2 - the minimum value of torque produced by dough passage while being subjected to mechanical and thermal constraints
2. Thermal weakening (Nm) - the difference between the C1.2 and C2 torques
3. Protein network weakening rate (Nm/min), α

Further protein changes during heating are minor and the torque variations during the last three stages is governed by the modification of the physico-chemical properties of the starch (Rosell et al., 2007; Huang et al., 2010). In the third stage the dough heating and the water available from the thermally denaturated proteins causes the starch gelatinization. Namely, during this stage, starch granules absorb the water, they swell and amylose chains leach out into the aqueous intergranular phase (Thomas & Atwell, 1999) resulting in the increase in the dough consistency and thus the increase in the torque. The maximum consistency of the dough in the third stage will be higher as the starch's gelling power increases and the α-amylase activity decreases. The starch gelatinization rate recorded in the third stage is defined by the β slope.

The parameters obtained during the third stage are the following:

1. Pasting temperature (° C) - the temperature at the onset of the rise in viscosity
2. Peak torque (Nm), C3 - the maximum torque produced during the heating stage
3. Peak temperature (° C) - the temperature at the peak viscosity
4. Gelatinization rate (Nm/min), β

At the fourth stage, consistency decreases as a result of physical breakdown of the starch granules due to mechanical shear stress and the temperature constraint (Rosell et al., 2007). The rate of dough consistency decrease is given by the γ slope, which refer to cooking stability rate (Rosell et al., 2007).

The parameters obtained during the forth stage includes:
1. Minimum torque (Nm), C4 - minimum torque reached during cooling to 50°C
2. Breakdown torque (Nm) - calculated as the difference between C3 and C4
3. Cooking stability rate (Nm/min), γ

During the final stage registered at the Mixolab profile, the decrease in the temperature causes an increase in the consistency of dough. That increase is referred to as setback and corresponds to the gelation process of the starch, when starch molecules (especially amylose) comprising gelatinized starch begin to reassociate in an ordered structure, which results in an increase in crystalline order (Thomas & Atwell, 1999). This stage is related to the retrogradation of starch molecules. Since retrogradation is one of the causes for staling of bread (Ross, 2003), the difference between C5 and C4 value can be the indicator of bread shelf life.

The following parameters can thus be recorded:
1. Final torque (Nm), C5 - the torque after cooling at 50°C
2. Setback torque (Nm) - the difference between C5 and C4 torque

Most of the parameters listed above are extracted from the curve legend. However, since Mixolab is highly versatile device, it enables manual reading of some extra parameters (such as C1.2) from Mixolab curve. Moreover, there is a possibility to create your own protocol that differs from Chopin+, e.g. for evaluation of the thermomechanical properties of gluten-free flours Torbica et al. (2010b) have established the dough mass of 90 g instead of 75 g as listed in Chopin+ protocol.

Although being a highly scientificly utilized, Mixolab can also be used as a quality control tool either in accredited laboratory or in flour and cereal processing industry. Namely, using the Mixolab Profile option, it is possible to simplify the interpretation of the results obtained by Chopin+ protocol. The Mixolab Profiler converts the Mixolab Standard curve into six flour quality factor indexes (water absorption, mixing behaviour, gluten strength, maximum viscosity, amylase resistance and retrogradation) graduated from 0 to 9. The meaning of the parameters is the following (Chopin Technologies Application Team, 2009):
1. Absorption stands for water absorption and as it is well known it is mainly influenced by the moisture content, protein content and level of damaged starch in the flour
2. Mixing index represents the resistance of the flour to kneading and it is used as an indicator of overall flour protein quality
3. Gluten+ index represents the behaviour of the gluten when heating the dough and it is therefore the measure of protein strength. It has to be pointed out that Gluten+ index is not the measure of gluten content
4. Viscosity represents the maximum viscosity during heating. It depends on both amylase activity and starch quality
5. Amylase stands for resistance of starch component to α-amylase and a high value of index corresponds to low amylase activity
6. Retrogradation index provides information about final product staling rate, where a high value indicates a poor staling rate of the final product

For example, the quality of the average wheat flour sample harvested in Serbia in 2008 and 2010 is presented in Figure 10.

Year 2008 was characterized with high temperatures during the harvest, while in 2010 there were extremely large amounts of rain which interrupted the harvest. Rain conditions, during the ripening stage of the crop 2010, increased sprouting and thus α-amylase activity

(Morris & Paulsen, 1985) which resulted in low Amylase index. This also affected the low Viscosity index. On contrary, low Viscosity index of sample 2008 was not the consequence of increased amylase activity, as it can be seen from high Amylase index value, but it was caused by a heat stress. Concerning the protein quality, both samples have shown low gluten strength as expressed in low values of Gluten+ index. Sample 2010 even exhibited very low Mixing index due to destroyed proteins structure as a result of the attacks of wheat bugs. Namely, sample 2010 contained 2% bug-damaged kernels where bug's proteolytic enzymes caused the breakdown of the gluten proteins during the breadmaking process (Olanca & Sivri, 2004).

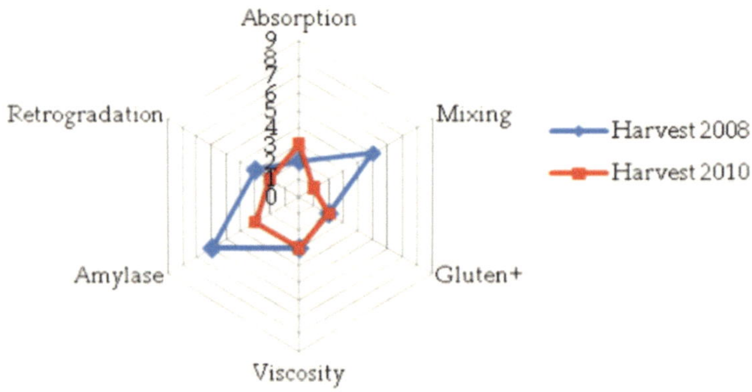

Fig. 10. Mixolab Profiler values of average wheat flour sample harvested in Serbia in 2008 and 2010

7. Conclusion

In order to get more comprehensive insight into the structural changes during the dough processing, fundamental rheology has the greater advantages over the empirical rheology. Therefore, the basic rheometry is an important tool among cereal scientists.

On contrary, ease in the interpretation and application of the result obtained by empirical rheology methods, as well as their high correlation with dough processing behaviour and end product quality, has made the descriptive rheological devices indispensable in cereal quality control laboratories and among cereal technologists.

However, in order to get complete picture of dough behaviour during all breadmaking stages, one have to employ a wide range of different empirical rheological devices, which is very time consuming and requires large amount of sample. Therefore, the future trends in development of new dough empirical rheological instruments or attachments to existing devices would be the combination of different devices and principles in one instrument and reduction of the sample amount to a quantity which will still be able to imitate real processing and baking conditions.

8. Acknowledgment

The financial support of Brabender® GmbH & Co. KG (Duisburg, Germany) and Chopin Technologies (Villeneuve-la-Garenne Cedex, France) towards this study is hereby gratefully acknowledged.

The results expressed and conclusions arrived at are the part of the project (project number TR-31007) funded by Ministry of Science and Technological Development, Republic of Serbia.

9. References

AACC methods. (2000). Approved methods of the American Association of Cereal Chemists, 10th edition, American Association of Cereal Chemists, St. Paul, MN, USA

Abang Zaidel, D. N., Chin, N. L., Abdul Rahman, R. & Karim R. (2008). Rheological Characterisation of Gluten from Extensibility Measurement. *Journal of Food Engineering*, Vol. 86, No. 4, (June 2008), pp. 549-556, ISSN 0260-8774

Abang Zaidel, D. N., Chin, N. L. & Yusof, Y. A. (2010). A Review on Rheological Properties and Measurements of Dough and Gluten. *Journal of Applied Sciences*, Vol. 10, No. 20, pp. (2478-2490), ISSN 18125654

Aldovrandi L. & Vitali F. (1995). Criteri di valutazione del frumento tenero da parte dell'industria molitoria. *Molini d'Italia*, Vol. 46, No. 3, pp. 16-19

Anderssen, R. S., Bekes, F., Gras, P.W., Nikolov, A. & Wood, J. T. (2004). Wheat-Flour Dough Extensibility As a Discriminator for Wheat Varieties. Journal *of Cereal Science*, Vol. 39, No. 2, (March 2004), pp. 195–203, ISSN 0733-5210

Best, S., Muller, R. (1990). Use of the Hagberg Falling Number Apparatus to Determine Malt and Barley Quality. *Journal of the Institute of Brewing*, Vol. 97, No 4, (July-August 1991), pp. 273-278, ISSN 0046-9750

Bordes, J. Branlard, G., Oury, F. X., Charmet, G. & Balfourier, F. (2008). Agronomic Characteristics, Grain Quality and Flour Rheology of 372 Bread Wheats in a Worldwide Core Collection. *Journal of Cereal Science*, Vol. 48, No. 3, (November 2008), pp. (569-579), ISSN 0733-5210

Cauvain, S. P. (1998). Breadmaking Process, In: *Technology of Breadmaking*, S. P. Cauvain & L. S. Young, (Eds.), 18-44, Blackie Academic and Professional, ISBN 9780834216853, London, UK

Chopin Technologies Application Team. (2009). *Mixolab applications handbook, Rheological and Enzymatic Analysis* (No. 28), Chopin Applications Laboratory, Villeneuve la Garenne, France

Collado, L. S., Corke, H. (1999). Accurate Estimation of Sweetpotato Amylase Activity by Flour Viscosity Analysis. *Journal of Agricultural and Food Chemistry*, Vol. 47, No. 3, (February 1999), pp. 832–835, ISSN 1520-5118

Collar, C., Bollain, C. & Rosell, C. M. (2007). Rheological Behaviour of Formulated Bread Doughs During Mixing and Heating. *Food Science and Technology International*, Vol. 13, No. 2, (April 2007), pp. 99–107, ISSN 1082-0132

Dapčević, T. Hadnađev, M. & Pojić, M. (2009). Evaluation of the Possibility to Replace Conventional Rheological Wheat Flour Quality Control Instruments with the New Measurement Tool - Mixolab. *Agriculturae Conspectus Scientificus*, Vol. 74, No. 3, (September 2009), pp. 169-174, ISSN 1331-7776

De Bry, L. (2006). Wheat Flour of the Third Millennium, In: *Bakery Products: Science and Technology*, Y.H. Hui, (Ed.), 87-96, Blackwell Publishing, ISBN-13 978-0-8138-0187-2, Oxford, UK

Dimmock, J. P. R. E. & Gooding M. J. (2002). The Effects of Fungicides on Hagberg Falling Number and Blackpoint in Winter Wheat. *Crop Protection*, Vol. 21, No. 6, (July 2002), pp. 475-487, ISSN 0261-2194

Dobraszczyk, B.J., Campbell, G.M. & Gan, Z. (2000). Bread—a unique food, In: *Cereals and Cereal Products: Technology and Chemistry*, B.J. Dobraszczyk, D.A.V. Dendy, (Ed.), pp. 182-232, Aspen Publishers, ISBN 0-8342-1767-8, Gaithersburg, Maryland, USA

Dobraszczyk, B. J. & Morgenstern, M. P. (2003). Rheology and the Breadmaking Process. *Journal of Cereal Science*, Vol. 38, No. 3, (November 2003), pp. 229-245, ISSN 0733-5210

Dobraszczyk, B. J. (2004). Dough Rheology. In: *Encyclopedia of Grain Science*, C. Wrigley, H. Corke & C. E. Walker, (Eds.), 400-416, Elsevier Ltd, ISBN 978-0-12-765490-4, Oxford, UK

Dobraszczyk, B. J. & Salmanowicz, B. P (2008). Comparison of Prediction of Baking Volume Using Large Deformation Rheological Properties. *Journal of Cereal Science*, Vol. 47, No. 2, (March 2008), pp. 292-301, ISSN 0733-5210

Dubois, M. Dubat, A. & Launay, B. (2008). *The AlveoConsistograph Handbook* (2nd edition), AACC International, ISBN 978-1-891127-56-4, St. Paul, Minnesota, USA

Freund, W. & Kim, M.-Y. (2006). Determining the Baking Quality of Wheat and Rye Flour. In: *Future of Flour – A Compendium of Flour Improvement*, L. Popper, W. Schäfer & W. Freund, (Eds.), 101-116, Verlag Agrimedia, ISBN 978-3-86037-309-5, Clenze, Germany

Gooding, M. J. & Davies, W. P. (1997). *Wheat Production and Utilization: Systems, Quality and the Environment*, CAB International, ISBN 0851991556, Wallingford, UK

Grausgruber, H, Schöggl, G. & Ruckenbauer, P. (2002). Investigations on the Validity of the Micro-Extensigraph Method to Measure Rheological Properties of Wheat Doughs. *European Food Research and Technology*, Vol. 214, No. 1, (January 2002), pp. 79-82, ISSN 1438-2385

Graybosch, R. A., Guo, G. & Shelton, D. R. (2000). Aberrant Falling Numbers of Waxy Wheats Independent of α-amylase activity. *Cereal Chemistry*, Vol. 77, No. 1, (January/February 2000), pp. 1-3, ISSN 0009-0352

Graybosch, A. R., Seabourn, B., Chen, R. Y. & Blechl, E. A. (2011). Quality and Agronomic Effects of Three High-Molecular-Weight Glutenin Subunit Transgenic Events in Winter Wheat. *Cereal Chemistry*, Vol. 88, No.1, (January-Ferbruary 2011), pp. 95–102, ISSN 0009-0352

Hoseney, R. C. (1994). *Principles of cereal science and technology* (2nd edition), American Association of Cereal Chemists, ISBN 0913250791, St. Paul, MN, USA

Huang, W., Li, L., Wang F., Wan J., Tilley, M., Ren C & Wu, S. (2010). Effects of Transglutaminase on the Rheological and Mixolab Thermomechanical Characteristics of Oat Dough. *Food Chemistry*, Vol. 121, No. 4, (August 2010), pp. 934-939, ISSN 0308-8146

ICC Standards. (1996). Standard Methods of the International Association for Cereal Science and Technology, International Association for Cereal Science and Technology, Vienna, Austria

Kahraman, K., Sakıyan, O., Ozturk, S., Koksel, H., Sumnu, G. & Dubat, A. (2008). Utilization of Mixolab® to Predict the Suitability of Flours in Terms of Cake Quality. *European Food Research and Technology*, Vol. 227, No. 2, (October 2008), pp. 565-570, ISSN 1438-2385

Karvonen, T. Peltonen, J., Kivi, E. (1991). The Effect of Northern Climatic Conditions on Sprouting Damage of Wheat Grains. *Acta Agriculturae Scandinavica*, Vol. 41, No. 1, pp. 55-64, ISSN 0001-5121

Kent, N. L. & Evers, A. D. (1994). *Technology of Cereals* (4th edition), Elsevier Science Ltd., ISBN 0 08 040833 8, Oxford, UK

Kettlewell, P. S. (1999). The Response of Alpha-Amylase Activity During Wheat Grain Development to Nitrogen Fertiliser. *Annals of Applied Biology*, Vol. 134, No. 2, (Jun 1999), pp. 241-249, ISSN 1744-7348

Khan, K. & Nygard, G. (2006). Gluten, In: *Bakery Products: Science and Technology*, Y.H. Hui, (Ed.), 97-107, Blackwell Publishing, ISBN-13 978-0-8138-0187-2, Oxford, UK

Kieffer, R. (2006). The Role of Gluten Elasticity in the Baking Quality of Wheat, In: *Future of Flour – A Compendium of Flour Improvement*, L. Popper, W. Schäfer & W. Freund, (Eds.), 169-178, Verlag Agrimedia, ISBN 978-3-86037-309-5, Clenze, Germany

Ktenioudaki, A., Butler, F. & Gallagher, E. (2011). Dough Characteristics of Irish Wheat Varieties I. Rheological Properties and Prediction of Baking Volume. *LWT - Food Science and Technology*, Vol. 44, No. 3, (April 2011), pp. 594-601, ISSN 0023-6438

Kweon, M. (2010). Falling Number in Wheat - How is it Calculated and What Does it Mean to Producers?, In: *2010 Research* Review *USDA-ARS Soft Wheat Quality Laboratory*, 2010, Available from: http://www.ars.usda.gov/SP2UserFiles/Place/36070500/InfoDianehasuploaded/2010ResearchReviewAnnualReport/MKweon-FN-012810.pdf

Lagarrigue, S. Alvarez, G. (2001). The Rheology of Starch Dispersions at High Temperatures and High Shear Rates: A Review. *Journal of Food Engineering*, Vol. 50, No. 4, (December 2001), pp. 189-202, ISSN 0260-8774

Lallemand. (1996). Using the Right Amount of Yeast, In: *Lallemand Baking Update*, Vol. 1, No. 4, 30.03.2011., Available from http://www.lallemand.com/BakerYeastNA/eng/PDFs/LBU%20PDF%20FILES/1_4DOSE.PDF

Léon, A. E, Barrera, G. N., Pérez, G. T., Ribotta P. D. & Rosell, C. M. (2006). Effect of Damaged Starch Levels on Flour-Thermal Behaviour and Bread Staling. *European Food Research and Techology*, Vol. 224, No. 2, (March 2006), pp. 187–192, ISSN 1438-2385

Lorenz, K. & Wolt, M. (1981). Effect of Altitude on Falling Number Values of Flours. *Cereal Chemistry*, Vol. 58, No. 2, (March-April 1981), pp. 80-82, ISSN 0009-0352

Lunn, G. D., Kettlewell, P. S., Major B. J. & Scott R. K. (2001). Effects of Pericarp Alpha Amylase Activity on Wheat (*Triticum Aestivum*) Hagberg Falling Number, *Annals of Applied Biology*, Vol. 138, No. 2, pp. 207-214, ISSN 1744-7348

Mann, G., Allen, H., Morell, M. K., Nath, Z., Martin, P., Oliver, J., Cullis, B. & Smith, A. (2005). Comparison of Small-Scale and Large-Scale Extensibility of Dough Produced from Wheat Flour. *Australian Journal of Agricultural Research*, Vol. 56, No. 12, (December 2005), pp. 1387–1394, ISSN 0004-9409

Mann, G., Diffey, S., Allen, H., Pumpa J., Nath, Z., Morell, M. K., Cullis, B. & Smith, A. (2008). Comparison of Small-Scale and Large-Scale Mixing Characteristics: Correlations Between Small-Scale and Large-Scale Mixing and Extensional Characteristics of Wheat Flour Dough. *Journal of Cereal Science*, Vol. 47, No. 1, (January 2008), pp. 90-100, ISSN 0733-5210

Marchylo, B. A. & Dexter, J. E. (2001). Pasta Production, In: *Cereals Processing Technology*, G. Owens, (Ed.), 109-130, Woodhead Publishing Limited, ISBN 0-8493-1219-1, Cambridge, UK

Mares, D. & Mrva, K. (2008). Late-Maturity α-Amylase: Low Falling Number in Wheat in the Absence of Preharvest Sprouting. *Journal of Cereal Science*, Vol. 47, No. 1, (January 2008) pp. 6–17, ISSN 0733-5210

Mariotti, M., Zardi, M., Lucisano, M. & Pagani, M. A. (2005). Influence of the Heating Rate on the Pasting Properties of Various Flours. *Starch – Stärke*, Vol. 57, No. 11, (November 2005), pp. 564–572, ISSN 1521-379X

Miyazaki, M., Hung, P. V., Maeda, T. & Morita, N. (2006). Recent Advances in Application of Modified Starches for Breadmaking. *Trends in Food Science & Technology*, Vol. 17, No. 11, (November 2006), pp. 591-599, ISSN 0924-2244

Morris, C. F. & Paulsen, G. M. (1985). Preharvest Sprouting of Hard Winter Wheat as Affected by Nitrogen Nutrition. *Crop Science*, Vol. 25, No. 6, pp. 1028-1031, ISSN 0011-183X

Nash, D., Lanning, S. P., Fox, P., Martin, J. M., Blake, N. K., Souza, E., Graybosch, R. A., Giroux, M. J. & Talbert, L. E. (2006). Relationship of Dough Extensibility to Dough Strength in a Spring Wheat Cross. *Cereal Chemistry*, Vol. 83, No. 3, (May/June 2006), pp. 255-258, ISSN 0009-0352

Olanca, B. & Sivri, D. (2004). Inhibition Effects of Plant Extracts on the Protease Activity in Bug (Eurygaster spp.) Damaged Wheats. Proceedings of 8th Gluten Workshop, pp. 417-420, ISBN 0-85404-633-X , Viterbo, Italy, September 8-10, 2003

Rasper, V. F. & Walker C. E. (2000). Quality Evaluation of Cereals and Cereal Products, In: *Handbook of Cereal Science and Technology*, K. Kulp & J.G. Ponte Jr., (Eds.), 505-538, Marcel Dekker, Inc., ISBN 0-8247-8294-1, New York, USA

Rosell, C. M., Collar, C. & Haros, M. (2007). Assessment of Hydrocolloid Effects on the Thermo-Mechanical Properties of Wheat Using the Mixolab. *Food Hydrocolloids*, Vol. 21, No. 3, (May 2007), pp. 452–462, ISSN 0268-005X

Ross, A. S., Walker, C. E., Booth, R. I., Orth, R. A. & Wrigley, C. W. (1987). The Rapid ViscoAnalyser: A New Technique for the Evaluation of Sprout Damage. *Cereal Foods World*, Vol. 32, No. 11, (November-December 1987), pp. 827–829, ISSN 0146-6283

Ross, Y. H. (2003). Molecular Mobility in Dough and Bread Quality, In: *Bread making: improving quality*, S. P. Cauvain, (Ed.), pp. 288-305, Woodhead Publishing Limited, ISBN 1 85573 553 9, Cambridge, England, UK

Ruske, R. E., Gooding, M. J. & Dobraszczyk, B. J. (2004). Effects of Triazole and Strobilurin Fungicide Programmes, With and Without Late-Season Nitrogen Fertiliser, on the Baking Quality of Malacca Winter Wheat. *Journal of Cereal Science*, Vol. 40, No. 1, (July 2004), pp. 1-8, ISSN 0733-5210

Sahin, S. & Sumnu, S. G. (2006). *Physical Properties of Foods*, Springer Science+Business Media, LLC, ISBN 978-0387-30780-0, New York, USA.

Schramm, G. (2004). *A Practical Approach to Rheology and Rheometry* (2nd edition), Thermo Electron (Karlsruhe) GmbH, Karlsruhe, Germany

Serbian official methods. (1988). Određivanje fizičkih osobina pšeničnog brašna Brabenderovim farinografom, In: Pravilnik o metodama fizičkih i hemijskih analiza za kontrolu kvaliteta žita, mlinskih i pekarskih proizvoda, testenina i brzo smrznutih testa (Regulation of methods of physical and chemical analysis for quality control of grain, milling and bakery products, pasta and quick frozen dough), Službeni list SFRJ 74/88, Serbia

Sorenson, B., (2006). Falling Numbers 101: Determining Sprout Damage in Grain, In: *Praire grains*, 2006, Available from: http://www.smallgrains.org/springwh/Mar06/falling/falling.html

Steffe, J. F. (1996). *Rheological Methods in Food Process Engineering* (2nd edition), Freeman Press, ISBN 0-9632036-1-4, East Lansing, MI, USA

Stoenescu, G., Ionescu, V., Vasilean, I., Aprodu, I. & Banu, I. (2010). Prediction the Quality of Industrial Flour Using the Mixolab Device, *BulletinoUniversity Agricultural Sciences and Veterinary Medicine Cluj-Napoca. Agriculture*, Vol. 67, No. 2, pp. 429- 434, ISSN 1843-5386

Suchy, J., Lukow, O. M. & Ingelin, M. E. (2000). Dough Microextensibility Method Using a 2-g Mixograph and a Texture Analyzer. *Cereal Chemistry*, Vol. 77, No. 1, (January/February 2000), pp. 39–43, ISSN 0009-0352

Suh, D. S. & Jane, J. J. (2003). Comparison of Starch Pasting Properties at Various Cooking Conditions Using the Micro Visco-Amylo-Graph and the Rapid Visco Analyse. *Cereal Chemistry*, Vol. 80, No. 6, (November/December 2003), pp. 745-749, ISSN 0009-0352

Ţăin, A.E., Zincă, G. & Banu, I. (2008). Studies about Obtaining Safe and Healthy Bakery Products Using the Beneficial Properties of Enzymes. *Chemical Bulletin of „POLITEHNICA" University of Timisoara*, Vol. 53(67), No. 1-2, pp. 110-114, ISSN 2069-6310

Thomas, D. J. & Atwell, W. A. (1999). *Starches*, Eagan Press Handbook series, ISBN 1891127012, St. Paul, MN, USA

Torbica, A., Antov, M., Mastilović, J., Knežević, D. (2007). The Influence of Changes in Gluten Complex Structure on Technological Quality of Wheat (*Triticum aestivum* L.). *Food Research International*, Vol. 40, No. 8, (October 2007), pp. 1038-1045, ISSN 0963-9969

Torbica, A., Živančev, D., Hadnađev, M. & Mastilović, J. (2010a). Influence of Heat Stress on Wheat Grain Quality, *Proceedings of 45th Croatian & 5th International Symposium on Agriculture*, pp. 940-944, ISBN 978-953-6331-79-6, Opatija, Hrvatska, February 15-19, 2010

Torbica, A., Hadnađev, M. & Dapčević, T. (2010). Rheological, Textural and Sensory Properties of Gluten-Free Bread Formulations Based on Rice and Buckwheat Flour, *Food Hydrocolloids*, Vol. 24, No. 6-7, (August-October 2010), pp. 626–632, ISSN 0268-005X

Torbica, A., Živančev, D., Mastilović, J., Knežević, D. & Bodroža-Solarov, M. (2011). Impact of Changes in Climate Conditions on the Technological Quality of Wheat, *Proceedings of 46th Croatian & 6th International Symposium on Agriculture*, pp. 617-621, ISBN 978-953-6135-71-4, Opatija, Hrvatska, February 14-18, 2011

Uthayakumaran, S., Lafiandra, D. & Gianibelli, M. C. (2004). Evaluation of Durum Wheat Quality Using Micro-Scale and Basic Rheological Tests, In: *The Gluten Proteins*, D. Lafiandra, S. Masci & R. D'Ovidio, (Eds.), 275-278, The Royal Society of Chemistry, ISBN 0-85404-633-X, Cambridge, UK

Veraverbeke, W. S. & Delcour, J. A. (2002). Wheat Protein Composition and Properties of Wheat Glutenin in Relation to Breadmaking Functionality. *Critical Reviews in Food Science and Nutrition*, Vol. 42, No. 3, (May 2002), pp. 179-208, ISSN 1549-7852

Vergnes, B., Della Valle, G. & Colonna P. (2003). Rheological Properties of Biopolymers and Applications to Cereal Processing, In: *Characterization of Cereals and Flours*, G. Kaletunç & K. J. Breslauer, (Eds.), 209-265, Marcel Dekker, Inc., ISBN 0-8247-0734-6, New York, USA

Wang, J., Pawelzik, E., Weinert, J., Zhao, Q. & Wolf, G. A. (2004). Effect of Fungicide Treatment on the Quality of Wheat Flour and Breadmaking. *Journal of Agricultural and Food Chemistry*, Vol. 52, No 25, (December 2004), pp. 7593-7600, ISSN 1520-5118

Weipert, D. (1990). The Benefits of Basic Rheometry in Studying Dough Rheology. *Cereal Chemistry*, Vol. 67, No. 4, (July-August 1990), pp. (311-317), ISSN 0009-0352

Weipert, D. (2006). Fundamentals of Rheology and Spectrometry. In: *Future of Flour – A Compendium of Flour Improvement*, L. Popper, W. Schäfer & W. Freund, (Eds.), 117-146, Verlag Agrimedia, ISBN 978-3-86037-309-5, Clenze, Germany

19

Sensory Analysis in Quality Control: The Gin as an Example

Montserrat Riu Aumatell
Universitat de Barcelona/ Departament Nutrició i Bromatologia
Spain

1. Introduction

The quality of a food product could be defined by different ways from a widely manner to a more detailed one. One of the most usual meanings is define the quality as "in conformity with consumer's requirements and acceptance, is determined by their sensory attributes, chemical composition, physical properties, and level of microbiological and toxicological contaminants, shelf-life, packaging and labelling". In order to manage the quality of a food product most industries have defined quality control and quality assurance programs. In the recent years, a lot of companies have established a quality control/sensory program especially the food industry. Frequently the quality control of a food needs some multidisciplinary approaches. In the last years, the advances in instrumental techniques have been enormous, increasingly the sensitivity and selectivity of the analytes detection so the control of chemical composition or toxicological contaminants must be easier. In spite of these the perception of flavour product usually must be measured by sensory analysis. But only some of the food industry use a sensory program compared to other disciplines (Muñoz, 2002). However some companies confirmed a relationship between instrumental and sensory measurements. The sensory analysis is a scientific discipline in which man is a measure instrument. It is often defined as "a discipline used to evoke, measure, analyse and interpret reactions to the characteristics of foods and similar materials as they are perceived by the sense of sight, smell, taste, touch and hearing" (Mc Ilveen & Armstrong, 1996; Piggott, et al., 1998). The latter has the same requirements as the chemical determinations, thus it means, it must be accurate, precise and valid. The discipline of sensory analysis use scientific principles drawn back from food science, physiology, psychology and statistics (Piggott, et al., 1998). The sensory quality is much difficult because it depends not only of food characteristics but of the consumer (Costell, 2002). Thus sensory quality could be product oriented or consumer oriented. Therefore, the role of sensory analysis in the food industry could be more important than it is actually. Sensory analysis have different approaches, requirements, and practical applicability and usually requires a lot of time, difficulties in analyzing data and the expertise are not always available. Is difficult organize a trained panel test, to have the adequate reference standards, and difficulties in focus the objective for the analysis so to perform the optimum sensorial test. If it's possible the sensory quality control must be applied to the ingredients or in-process. For this it's important that companies stipulate the specifications of the raw material in order to avoid the entrance of a defective ingredient in the product elaboration. This can suppose the detection of a defect in the finished product. Probably this kind of sensory evaluation will be more efficient. Sensory control is recommended only in critical steps while physical and chemical analyses are realized at different stages (Muñoz, 2002).

There are a great number of sensory methods. They can be divided in two groups' discriminant and descriptive methods (Piggott, et al., 1998). This chapter objective is to evaluate the role of the sensory quality control in the food industry. For this the most usual sensory methods were described and analyzed.

On the other hand, the industry of alcoholic beverages especially the spirit drinks is one of the most important of the world. Actually the improved communications and the expansion of travel have made the globalization a reality. Information about the sensory profile of alcoholic beverages could be interesting for the quality control of the worldwide beverage industry in order to obtain flavour integrity. Some alcoholic beverages as whiskey or brandy are widely studied. Other distilled beverages as gin in spite of they are widely consumed around the world there are few documented studies about this sensory profile (Piggott & Holm, 1983; Phelan et al., 2004; Riu-Aumatell et al., 2008). The descriptive analysis of gin is characterized by juniper and coriander preferably but other nuances could be detected when trained judges are used.

The sensory evaluation of gin as an alcoholic beverage example in the industry was studied. The references available about this topic were discussed.

2. Sensorial methods in food quality

Once the quality sensory standards were defined the optimum sensorial method was chosen. According to Costell, 2002, the choice of sensorial method depends of:

1. The objective of the quality control programme
2. The type of standard established
3. Whether or not the perceptible variability of a product can be defined by specific sensory attributes
4. The magnitude variability that must be detected
5. The level of quality to be assessed

The characteristics of a product are important to chosen the sensorial method. In order to perform a sensory quality control some preliminary steps must be taken into account, the first one the sensory quality specifications. Each company must define the quality standard of their products. The stability of a food product is an essential characteristic for a food quality. With foods it's very difficult to obtain products with uniform sensory characteristics during time. A definition of a descriptor should be given therefore a suitable stable reference should be assigned to a descriptor. The reference must be stable and reproducible with time. A standard for quality control is defined as "a representation for certain characteristics and a product that can be easily being obtained by, maintained or reproduced" (Costell, 2002). Some information about its variability and its influence on sensory attributes must be well defined. The variability of the standard must be quantified and also variation limits should be established.

Also, other factors that influence are the training of the panel, the conditions of the analysis, and the correct data analysis that are essential for the information obtained of the sensory analysis. Then to establish a quality program of sensory method also, it should be bear in

mind the training of the panellists when it was necessary, the type of established specifications and the use of controlled test conditions (Muñoz, 2002).
According to the authors considered, the sensorial methodology could be divided in different ways but the most usual and easy methods used in the quality control could be divided in discriminant and descriptive analysis. According to Muñoz et al., (1992), the sensory methods for quality control could be divided in eight types: overall difference test, difference from control, attribute or descriptive test, in/out of specifications, preference and other consumers test, typical measurements, qualitative description of typical production and quality grading. All of these methods present advantages and inconvenients. While according to Costell (2002), the most suitable test for the sensory quality control in the industry is that which make possible to measure a magnitude of variability between a product and a defined standard while a difference or acceptance test are not adequate. The difference test are too sensitive to small differences between products and do not determine the extent of a difference while the acceptance test with a small group of tasters not represent the consumer population. The most usual sensory methods for the sensory quality control are discussed below.
Sensory methods are usually classified in three categories: difference test (1), affective test (2) and descriptive test (3). Difference tests (1) are named of different manner but usually it could be divided in two ways: overall difference test and attribute difference test. The latter measures a single attribute of a sample which not imply that no overall difference exist between samples and includes the directional difference test named also paired comparison test or pair wise ranking test. While triangle, duo-trio, two-out-of-five and difference from control amongst others are test usually used by detect overall difference between samples. The most easily sensory methods for quality control are difference from control test. The aim of this test is to determine if a difference could be recognize between a sample and a control and to estimate the magnitude of the difference (Meilgaard, et al., 1999). Usually one sample is defined as control, standard or reference and the sample problem was evaluated with respect the control. The easier method should be the overall difference from standard. The judges rate the differences between a sample or samples and a control. Usually 20 or 50 presentations of the sample were needed. The judges must be semi trained. A more useful method should be to evaluate the difference between the sample and the standard but evaluating the differences of the most important attributes of the product (for example which sample of olive oil is more rancid). The latter should be more useful in order to apply corrections to the sample when it was necessary. When some change was applied to a food product it could be more useful use a scale with a control in the middle. This allows identifying the direction of a detected difference. It's no necessary that the subjects are trained only when the attribute is very important, for example a specific off-flavour, in this case the test requires high training judges.
The affective tests (2) evaluate the personal response (preference or acceptance) to a new product, or a single characteristic of a product. The affective tests involve the acceptance methods, the preference methods and the attribute diagnostics. The most usual test to evaluate a preference of a product includes paired preference, rank preference or multiple paired preferences. These test are based in arrange the food tested in the order of preference. The acceptance test is used by to rank the products in a scale of acceptability while attribute diagnostics consists in rank the principal attributes that determine the acceptance or the preference of the products. Some authors (as Costell, 2002) have the opinion that the affective test or the difference tests are not suitable for routine analysis. Probably, the

consumers are not prepared to identify flavours and sometimes they are not prepared to explain why they like or dislike a food sample. Affective test could provide a direct link between the consumer and the development and it could useful by marketing research (Sidel & Stone, 1993).

Probably, in quality control the most appropriate sensory methods are those that measure the magnitude of a variability of a sample between a standard. The objective of the methods involving the comparison to a standard evaluates the difference between a product and a standard. The standard must carry out the specifications of the method. The latter includes the methods cited in Table 1.

Comparison to a standard	Difference from a standard or a control product	
	Difference from a mental standard	In/out method
		Overall quality rating method
	Difference from a written standard	Quality grading method
Methods without standard	Descriptive methods	
	Other	

Table 1. The most usual methods used in sensory analysis according to Costell (2002)

The evaluation of difference between a sample and a control is useful when a food product have not very complex sensory characteristics. Generally one sample was considered the control or standard and the objective is the evaluation how different is the target sample from the control. Also, the magnitude of the difference was usually asked to the panellist. When a most important attributes were considered the analysis must be more useful because then the corrections necessary should be produced. Another possibility is use a scale which the control in the central point. This could be useful to understand the differences between the product and the control, for example when some change was produced in the formulation and it could be interesting to know the direction of the changes produced. The panel must be trained only when the difference between control and sample was taken into account. The use of a mental standard is not recommended but in some cases it could be useful for example in order to evaluate slight differences which modify the prize in foods as wine or olive oil or for example when the raw material or some ingredient was measured. Two methods evaluate the difference from a mental standard In/out method and overall quality rating method. In and out method is useful only when the differences were very clear or an off-flavour was considered. Another method is the Overall quality rating method. This method could be considered a mixture between an acceptance method and difference method. These cause that the results obtained are not easy to treat. When the method considered that a group of expert judges with a common mental standard could evaluate the overall quality of a food product. The results obtained do not translate into changes in food because it does not conclude in which direction and how the quality of the food differs from the mental standard. Therefore another method could be more useful as the quality grading method. This method analyses the basic attributes colour, flavour, texture and appearance. The attributes were evaluated by ordinary scales according to high quality, acceptable quality and rejected food. The judges must be very well trained in order to obtain significant results.

Finally it must be commented the methods which no need comparison to standard, basically Descriptive analysis. The descriptive methods (3) are the most sophisticated methods of sensory analysis and involve the detection and description of qualitative and quantitative attributes of a food product by a trained panel of judges. The qualitative attributes of a product involves aroma, appearance, flavour, texture, sound and aftertaste and the trained judges quantify these parameters to describe a target product (Murray, et al., 2001). Descriptive analysis is used in multiple ways as quality control, for comparison of product prototypes, for sensory mapping and product matching (Murray, et al., 2001). Descriptive analysis includes Flavour Profile Method, Texture Profile Method, Quantitative Descriptive Analysis ™, Spectrum ™ method, Quantitative Flavour Profiling and Generic Descriptive Analysis. The last one, combines different characteristics of the other methods and is usually used in descriptive analysis of food. Descriptive analysis could also be used to relate the results obtained with preference ratings and with instrumental data. The most important considerations in descriptive analysis are a strict list of terms and a highly trained panel of judges. The list of attributes must be consensued by the panel below the direction of the panel leader. Usually the training of the panel takes place with reference standard, with intensity ranking test and sometimes with food samples enriched with the descriptors identified in the samples. The results obtained were analysed with spider web diagrams, one-way ANOVA and multivariate methods as Principal Components Analysis (PCA). Numerous food products were analysed by descriptive analysis as alcoholic beverages gin, wine, cheese, meat or coffee.

3. Gin

Gin is a distilled beverage developed in the northern Europe in 17th century. Gin is one of the distilled beverages widely consumed around the world and it belongs to the juniper flavoured spirit drinks category according the European legislation (Regulation (EC), no. 110/2008). According to the European regulation juniper-flavoured spirit drinks could be divided in juniper-flavoured spirit drink, Gin, Distilled Gin and London gin. The last one is the one of the most popular spirit drinks with global sales adding up to approximately 50 million cases by volume (according to The Gin and Vodka Association). Gin is a colourless beverage with an alcoholic strength of at least 40% in the United States and 37.5% in the European Union (Greer, et al., 2008). Regardless of the elaboration process always the predominant flavour is Juniperus.

The production of Gin depends basically of the beverage type. The juniper flavoured spirit drink and gin are elaborated by flavouring ethyl alcohol of agricultural origin with *Juniperus* berries (*Juniperus communis* L. in Gin and *J. Communis* L. and/or *J. Oxicedrus* L. in juniper-flavoured spirit drinks). Most usual is the distilled Gin obtained by redistilling organoleptically suitable ethyl alcohol of agricultural origin of an appropriate quality with an initial alcoholic strength of at least 96% in stills traditionally used for gin in the presence of juniper berries and other natural species provided that the juniper taste is predominant. The stills traditionally used to obtain distilled gin are usually made of copper. The stills are heated using a steam jacket to remove the essential oil from the botanicals which provide the taste to the beverage. The early part (fore shots) usually abounding in fusel oil and the end of the run (feints) are of lower quality and to produce high quality distilled gin only the middle run is used. Moreover, the minimum alcoholic strength by volume shall be 37.5%. The flavouring ingredients of gin are all natural and are referred as botanicals. These botanicals are carefully selected qualitatively and quantitatively and vary according to the producer. Some authors talk about more than 100 botanicals added to a gin providing them its particular character. It could include mainly coriander seeds (*Coriander sativum* L.) but also other botanicals as orange peel (*Citrus sinensis*), purging cassia (*Cassia fistula*), orris root (*Iris florentina* L.), cardamom seeds (*Elettaria cardamomum* L.), angelica root (*Angelica archangelica* L.), cinnamon bark (*Cinnamomum zeylandicum*), calamus (*Acorus calamus* L.), fennel (*Foeniculum vulgare*), aniseed (*Pimpinella anisum*), lemon peel (*Citrus limon* L.), cumin (*Cuminum cynimum* L.), almond (*Prunus amygdalus* L.) and liquorice root (*Glycyrrhiza glaba*).

Finally, the production process of London Dry Gin is similar of that of distilled gin with high quality distillate (maximum methanol content of 5 grams per hectolitre of 100% vol. alcohol), and which not contain other added ingredients than water and with no colorants and artificial flavouring ingredients added. Gin and distillate gin no needs any period of maturation.

Moreover, when the production of gin takes place in a geographical area and accomplish some requirements according to its elaboration, composition and quality they could receive the denomination of geographical indication as Genièvre/Jenever/Genever (Belgium, the Netherlands, France and Germany), Jonge jenever, Oude jenever (Belgium, the Netherlands), Genièvre Flandres Artois (France), Ostfriesischer Korngenever and Steinhäger (Germany), Plymouth Gin (United Kingdom), Gin de Mahón (Spain), Vilniaus Džinas/Vilnius Gin (Lithuania) and Spišská borovička (Slovakia).

3.1 Chemical composition

The same as all the spirituous beverages the gin flavour is provided by several volatile and semivolatile compounds. The volatile composition of gin depends mainly of the volatile compounds of juniper berries and other botanicals added to the spirituous beverage. According to Barjaktarović et al., (2005) the composition of juniper essential oil was formed by monoterpenes (58-85%), sesquiterpenes (2-10.2%) and other minority compounds as aldehydes, alcohols and oxygenated compounds.

According to our own experience the volatile composition of 6 Gins were performed mainly of terpenoid compounds. The samples analysed were 4 London Dry Gins and the other two were gins with geographical indication (Gin de Mahón and Plymouth Gin). More than 60 volatile and semivolatile compounds were identified and quantified by Headspace/Solid Phase Microextraction coupled to Gas Chromatography/ Mass Spectrometry (HS/SPME-GC/MS). They belong mainly to the terpenoids family (monoterpenes, sesquiterpenes and they corresponding oxygenated compounds). Table 2 showed the mean of the concentration (mg/L) of the main compounds identified in gin. Samples were separated by London Dry Gin and geographical indication.

The volatile profile of London Dry Gins analysed differs from that of gins with geographical indication. Also, from juniper berries some of the principal volatile compounds come from other botanicals used in its elaboration and it's very different according to the gin considered. Gin with geographical indication 5 showed a higher values of limonene and γ-terpinene and the sesquiterpenes δ and γ-cadinene. Limonene and γ-terpinene could provide from citric fruit other than juniper berries. While the terpenoid compounds characteristics of juniper berries are found in samples with geographical indication 6. Also this last sample contains the highest values of oxygenated monoterpenes as verbenyl ethyl ether (for the first time identified in gin samples) and α-terpineol.

	London Dry Gins (n=4)	Geographical indication 5	Geographical indication 6
α-pinene	2.55	6.12	5.65
β-myrcene	4.01	6.17	11.09
Limonene	3.99	17.21	5.74
γ-terpinene	1.25	2.87	1.51
Linalool	22.37	16.83	1.93
Verbenyl ethyl ether	3.45	3.27	24.43
α-terpineol	1.51	3.80	9.03
Geranyl acetate	1.38	1.53	0.25
β-caryophyllene + β-elemene	0.54	0.77	0.93
α-humulene	0.33	0.47	0.90
δ and γ-cadinene	0.57	1.15	0.93
Caryophyllene oxide	0.23	0.09	5.13

Table 2. The main volatile compounds (mg/L) identified in gins

3.2 Sensory evaluation

Distilled beverages are complex mixtures of many individual compounds in an ethanol: water matrix. The composition of the distillate drinks depends of the raw material (grain or fruit) and also of the technology employed in its elaboration (mashing, fermentation, distillation and maturation). In some beverages as gin the composition depends of the flavouring agents added to a neutral alcohol. The unique recipe used by the producers implies a different sensory profile in these beverages. Particularly, in the gin technology it is well known that the distillation process at low temperatures (near 0°C) could benefits the retention of volatile compounds as oxygenated monoterpenes and a decrease of monoterpenes. This fact could imply high stability (Greer, et al., 2008).

Even though gin is one of the most consumed around the world, the knowledge about its sensory profile is limited. Only few studies exist about the sensory profile of gin, there are detailed in Table 3. Table 3 also showed the number of attributes identified in gin in the works published.

Piggott and Holm (1983)	21 attributes
Mc Donell et al., (2001)	2 attributes
Phelan (2004)	16 attributes
Riu-Aumatell et al., (2008)	5 attributes

Table 3. The studies founded in the literature about the sensory profile of gin

The quality of spirituous beverages could be defined by a small number of attributes (colour, aroma, taste and mouthfeel). Today, the tasting and nosing of distilled beverages remains very important in the distilleries. Actually a panel of trained judges are used replacing a single expert. The tasters have two clear objectives, for one hand the ability to identify each individual attribute in the whole flavour and at the same time develop a list of vocabulary. In order to obtain a consistent list of attributes, a reference material must be available which allow to the panellists to identify a reference material to a descriptor. As can be seen in Table 3, for juniper flavoured beverages doesn't exist a unified vocabulary and at the same time any sensory wheel. Except the juniper flavour also than citric the variability about the list of terms that defined gin is high. For this, each author elaborate its own list of vocabulary and they trained the panel according to the objective of the study. The variability about the training, or the botanicals or standards used in the training is high. Based in our experience, in order to obtain an optimum list of attributes a high training of the panellists is necessary. Moreover, it must be necessary to have materials, standards with optimum quality useful for the training of the panellists.

According to Simpson et al., (2004) there are three guides about the vocabulary development:

- Use one flavour for every word
- Use the smallest the sensory vocabulary consistent with sensory description task
- Avoid subjective terms (good/bad)

In the own study about sensory profile of gin we have a double objective. For one hand, establish a sensory characterization of gins and to do this was absolutely necessary firstly elaborate a vocabulary. To perform the work the lexicon development was performed according to the ISO 11035 and a Generic Descriptive Analysis was applied to 4 London Dry gins and two gins with geographical indication.

The panellists were 7 women and 7 men of the Nutrition and Food Science department of the University of Barcelona. All of them are selected according to the availability, health aspects and their experience on tasting food and alcoholic beverages. First a triangle test was performed to check differences between the gins tested. At the same time this test is useful to familiarize the panel with the gin samples. Then the generation of vocabulary was performed during four sessions detailed in Table 4.

1st session	**Intensity ranking test** Hydro alcoholic solution of myrcene, limonene, linalool and γ-terpinene
2nd session	**Description and recognition of orthonasal perception** Natural sensory references: juniper, coriander, aniseed, lemon peel
3rd session	**Vocabulary generation** Elaboration of a preliminary list with 44 terms Discussion with the panel leader Elaboration of a first list with 10 attributes (juniper, coriander, liquorice, spice, fruity, floral, citric peel, cardamom, aniseed/fennel, angelica root)
4th session	**Intensity Ranking test** Gin enriched with ethanolic extracts of juniper berries, aniseed and coriander seeds and angelica root

Table 4. The training sessions performed during the descriptive analysis of gin

The descriptors selection was performed according to the ISO 11035. The generic descriptive analysis (GDA) was performed in successive sessions to avoid the fatigue of the assessors. Also, the alcoholic strength was diminished in order to avoid fatigue and also, to equal the alcoholic strength of the samples. From the 10 first list of descriptors (Table 4) the final attributes were selected using geometric means and Principal Components Analysis (PCA).

The profile sheet used includes an unstructured scale from 0 to 5 (0 is the absence of perception) (Figure 1).

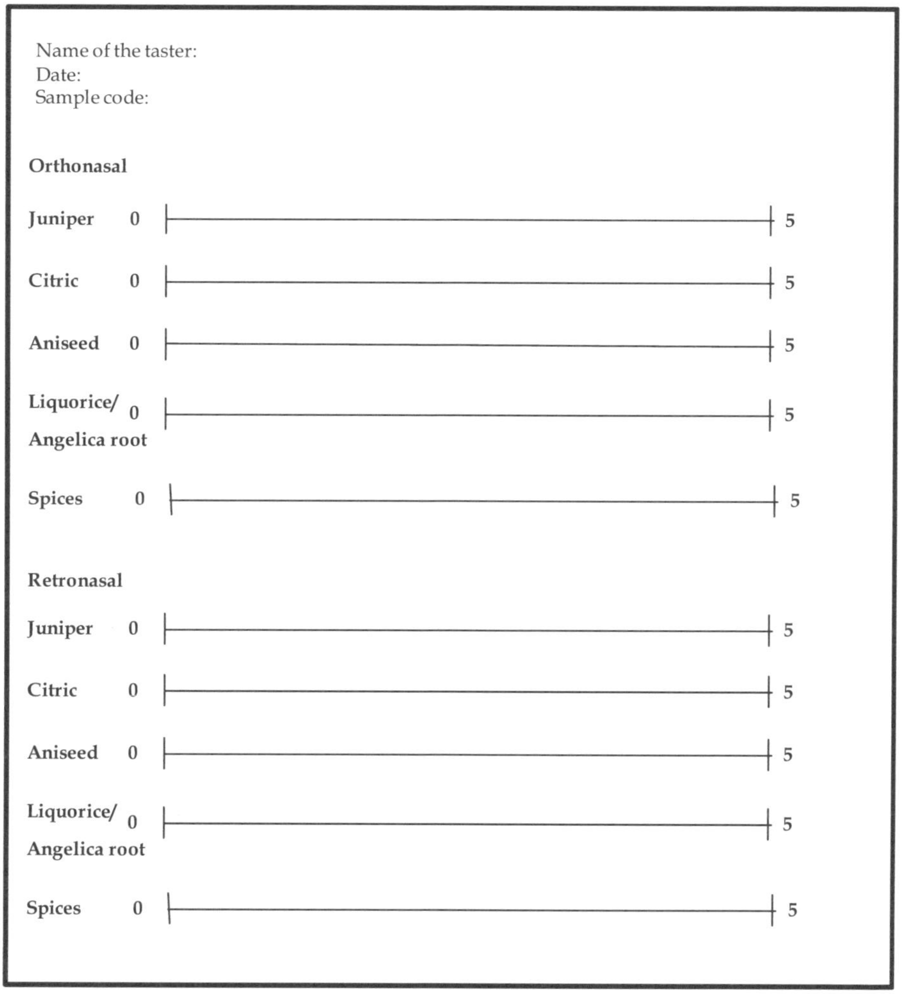

Fig. 1. Profile sheet of organoleptic assessment of gin

From a list of 10 attributes afterwards the vocabulary reduction a final list of 5 attributes was established (juniper, citric, aniseed, liquorice/angelica root, and spice). The generic descriptive analysis of gin samples was evaluated in duplicate in 3 sessions presented in randomized order and coded with three digit numbers. The results obtained are showed in Figure 2. G1-G4 is the mean of the results of 4 London Dry Gin while G5 and G6 are gins with geographic indication.

The results obtained by retronasal perception are similar to that of orthonasal perception (Data not shown). The sensory profile of London Dry gins was different of that of gins with geographical indication. The London Dry gins showed an equilibrated profile with intermediate values of all the descriptors evaluated. While gins with geographical indication were noticeable different. Gin G5 was characterized by citric attribute (probably because the species added were more citric as cardamom, coriander or citrus peel) while G6 was characterized by juniper descriptor. The high values of citric attribute in G5 and juniper in G6 could mask the detection of other attributes in this samples. The sensory profile could be related with sensory profile (Table 2). The results of sensory analysis are in accordance with that of chemical composition (Table 2).

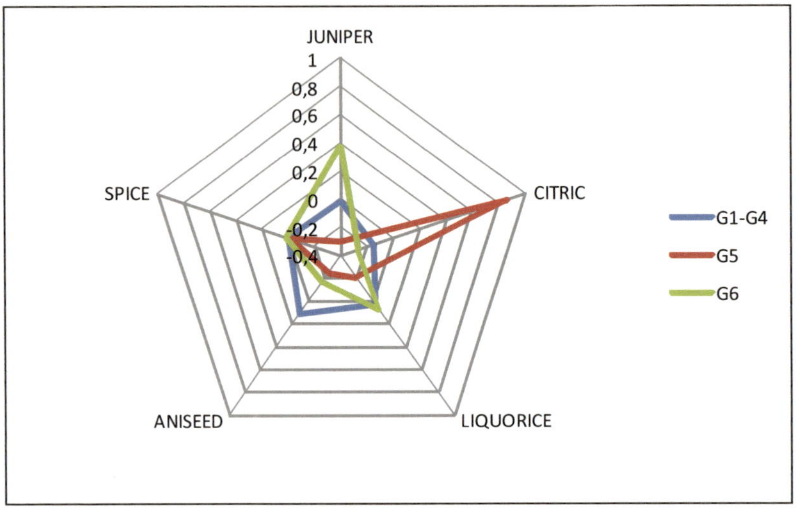

Fig. 2. Aroma profiles of gin samples obtained by orthonasal perception

4. Conclusions

In the last years the use of sensory analysis as quality control in the industry has increase its use. Some sensory methods as difference methods (particularly difference from standard) and descriptive methods are showed its usefulness in a wide range of industries. Nevertheless, is vital to continue the research in sensory analysis in order to ensure the optimum results. On the other hand methods as affective or preference methods are not useful to take part of a quality assurance program.

The results obtained by our group in the sensory analysis of gin showed as difference test and descriptive analysis could be a good method to evaluate the quality of a distilled beverage as gin. The variability of such samples makes them particularly important to obtain a consensus about the list of descriptors and therefore the training of the panel is especially important. This allows to evaluate the changes or alterations in the production process of alcoholic beverages and to make the appropriate modifications. Also, the results obtained from sensory analysis are in accordance with the results obtained in chemical analysis.

5. References

Barjaktarović, B., Sovili, M. & Knez, Ž. (2005). Chemical composition of Juniperus communis L. fruits supercritical CO_2 extracts: dependence on pressure and extraction time. *Journal of Agricultural and Food Chemistry*, 53, 7, (January 2005), pp. 2630-2636, ISSN 0021-856.

Costell, E. (2002). A comparison of sensory methods in quality control. *Food Quality and Preference*, 13, 6, (September 2002), pp. 341-353, ISSN 0950-3293.

Greer, D., Pfahl, L., Rieck, J., Daniels, J. & Garza, O. (2008). Comparison of a novel distillation method versus a traditional distillation method in a model gin system using a liquid/liquid extraction. *Journal of Agricultural and Food Chemistry*, 56, 19, (August 2008), pp. 9030-9036, ISSN 0021-856.

ISO 11035, (1994). Sensory analysis-identification and selection of descriptors for establishing a sensory profile by a multidimensional approach. International Organization for Standardization.

Mc Donnell, E., Hulin-Bertaud, S., Sheehan, E.M. & Delahunty, C.M. (2001). Development and learning process of a sensory vocabulary for the odor evaluation of selected distilled beverages using descriptive analysis. *Journal of Sensory Studies*, 16, 4, (August 2001), pp. 425-445, ISSN 1745-459X.

Mc Ilvenn, H. & Armstrong, G. (1996). Sensory analysis and the food industry: can computers improve credibility?. *Nutrition & Food Science*, 96, 1, (January/ February 1996), pp. 36-40, ISSN 0034-6659.

Muñoz, A.M., Civille, G.V. & Carr, B.T. (1992). *Sensory evaluation in quality control*. Van Nostrand Reinhold, ISBN 0442004591, New York.

Muñoz, A.M. (2002). Sensory evaluation in quality control: an overview, new developments and future opportunities. *Food Quality and Preference*, 13, 6, (September 2002), pp. 329-339, ISSN 0950-3293.

Murray, J.M., Delahunty, C.M. & Baxter, I.A. (2001). Descriptive sensory analysis: past, present and future. *Food Research International*, 34, 6, (December 2001), pp. 461-471, ISSN 0963-9969.

Piggott, J.R. & Holm, A.M. (1983). Descriptive sensory analysis of gin flavour, *in* Piggott, J.R. (ed), *Flavour of Distilled Beverages*, Ellis Horwood, Chichester, pp 145-153.

Piggott, J.R., Simpson, S.J. & Williams, S.A.R. (1998). Sensory analysis. *International Journal of Food Science and Technology*, 33, 1, (February 1998), pp. 7-18, ISSN 1365-2621.

Phelan, A. D., Jack, F. R., Conner, J. M., Reid K. J. G. & Priest F. G. (2004). Sensory assessment of gin flavour, *in* Bryce, J.H. and Stewart, G.G. (eds.), *Distilled spirits tradition and innovation*, Nottingham University Press, United Kingdom, ISBN 1-897676-39-5, pp. 53-58.

Regulation (EC) No 110/2008 of the European parliament and of the council of 15 January 2008 on the definition, description, presentation, labelling and the protection of geographical indications of spirit drinks.

Riu-Aumatell, M., Vichi, S., Mora-Pons, M., López-Tamames, E. & Buxaderas, S. (2008). Sensory characterization of dry gins with different volatile profiles. *Journal of Food Science*, 73, 6, (August 2008), pp. 286-293, ISSN 0022-1147.

Sidel, J.L. & Stone, H. (1993). The role of sensory evaluation in the food industry. *Food Quality and Preference*, 4, 1-2, (August 1993), pp. 65-73, ISSN 0950-3293.

Simpson, W. J., Boughton, R. A., & Hadman, S. I. (2004). Stabilised powder flavour standards for use in training and validation of tasters of distilled spirits, *in* Bryce, J.H. and Stewart, G.G. (ed.), *Distilled spirits tradition and innovation*, Nottingham University Press, United Kingdom, ISBN 1-897676-39-5, pp. 53-58.

Vichi, S., Riu-Aumatell, M., Mora-Pons, M., Buxaderas, S. & López-Tamames, E. (2005). Characterization of volatiles in different dry gins. *Journal of Agricultural and Food Chemistry*, 53, 26, (October 2005), pp. 10154-10160, ISSN 0021-856.

Vichi, S., Riu-Aumatell, M., Mora-Pons, M., Guadayol J.M., Buxaderas, S. & López-Tamames, E. (2007). HS-SPME coupled to GC/MS for quality control of Juniperus communis L. berries used for gin aromatization. *Food Chemistry*, 105, 4, (March 2007) pp. 1748-1754, ISSN 0308-8146.

20

Spectral Imaging as a Tool in Food Research and Quality Monitoring of Food Production

Stina Frosch, Bjørn Skovlund Dissing,
Jens Adler-Nissen and Michael Engelbrecht Nielsen
Technical University of Denmark, National Food Institute
Denmark

1. Introduction

A forward-looking food industry must obviously continue to develop its production technology to be able to produce foods that meet both present legislation and consumers' expectations and demands. Ensuring a healthy, secure and sensory food quality, as well as ensuring cost competitiveness / effectiveness is of high importance to survive the strong competition within the field. The high costs in many food-processing areas are primary caused by the extensive use of manual work e.g. for visual inspection of quality parameters and the subsequent sorting or removal of products. However, new production and / or distribution technologies in themselves neither create nor ensure high quality of products or the optimization of the production. This requires knowledge about both new opportunities to create specific production and distribution conditions in combination with knowledge about product response given the production and distribution conditions. Food and food production covers a broad variety of both raw materials and production processes. Therefore, application of new technology cannot be regarded as a simple procurement of accessible standard products but requires research and development including several tests to ensure optimal outcome.

The assessment of the visual appearance of food products from size and colour of the product to uniformity of packaging is an important part of the control system in the food supply chain. To ensure that the required standards are met, inspections at all stages from primary production to final retail distribution are needed. However, visual inspection of quality parameters and the subsequent sorting and sometimes also rejection of products by manual work are significant contributors to the total production costs in the food industry. To save costs and to enhance visual quality assessments, automatic vision systems are introduced and tested in many food-manufacturing operations. Vision systems are attractive for online quality assessment and process control, because the methods are rapid, contact free and non-destructive.

The introduction of vision systems for quality monitoring in the food manufacturing industry is challenged by the relatively harsh production environment, typically in the form of a high humidity, low / high temperatures, routine wash down and sanitation. The technology is maturing, however, and the special hardware requirements on system design are now met, so that computer-based vision systems are now gaining wider application for quality monitoring in food processing.

2. The present state-of-the art of vision systems for food quality control

The present state-of-the art vision systems for quality and process control in agriculture and the food processing industries are typically based on traditional trichromatic (Red Green Blue) or grey scale imaging, and related to visual quality characteristics such as size, shape, colour and appearance. Examples on industrial applications of vision systems are in relation to inspection and grading of fruit and vegetables (Chen et al., 2002). Within the fish industry examples include fish species sorting, length and colour measurement with subsequent grading and inspection ensuring right trimming and cutting of the fish (Erikson & Misimi, 2008).

It is, however, difficult with most trichromatic or grey-scale vision technologies to reliably distinguish small differences in the appearance of food products. To increase sensitivity and ability to discriminate, the combination of imaging with spectroscopy in the form of spectral imaging technology covering both multi- and hyperspectral imaging seems promising (Geladi et al., 2004). This technology makes use of light reflection at usually a large number of different wavelengths for producing a spectral image attaining both spatial and spectral information simultaneously of an object. Spectral imaging is particularly well suited for applications where it is crucial to detect small differences in texture, colour and surface chemistry (Hansen, 2005; Maletti, 2003; Gomez, 2005; Clemmensen & Ersbøll, 2006; Hansen et al., 2003; Sun, 2010).

The spectral images are analysed using advanced data analysis to extract the relevant information e.g. changes in both physical and chemical parameters relating to biological features or processing conditions. The use of spectral imaging is highly advantageous compared to traditional sensors based on point measurements when working with non-uniform material such as food; the full spectral signature for each spatial location (pixel) is obtained and thereby reflecting the natural diversity in the measured sample e.g. as astaxanthin in a fish fillet (Dissing et al., 2011; Frosch et al., 2011). In theory, this allows prediction of the specific quality attribute under consideration for each pixel in the image resulting in an image, which may also be considered as a chemical surface map (Burger & Geladi, 2006). An introduction of spectral image technology will enable measuring of more biological quality parameters at the same time and ensure highly specific information relating to the analysed product. An overview of the fundamentals and recently developed applications of spectral imaging for food quality analysis and control can be found in Sun (2010).

3. Recent developments in food related applications with spectral imaging

New cases from authors' on-going projects will be used in the following to elucidate the different applications where spectral imaging may play an important role in the future; Process control, process/quality optimization and documentation. Furthermore, the development within spectral imaging is also beneficial within research as illustrated in the last case described.

3.1 Control of extruded food and feed products

Extrusion is a commonly used processing technology both within food and feed production e.g. expanded snacks, breakfast cereals, pastas and fish feed pellets (Gonzalez et al., 2002). Food extrusion is a high-temperature short-time (HTST) technology (Frame, 1994) where the mixed, wet material is forced trough a die to obtain a product of definite shape, size and porosity (Rahman et al., 2002).

In fish feed production the pellet size is of outermost importance since a convincing correlation between suitable fish growth and pellet size exists; the pellet size is adapted to the size of the fish (Wankowski & Thorpe, 1979; Tabachek, 1988; Azaza et al., 2010). The pellet size i.e. the pellet diameter corresponding to the pellet size as declared by the feed industry, is determined by the size of the dyes. A set of rotating knifes placed on the other side of the disk containing the dyes cuts the extruded material into cylindrical shaped pellets. The length of the pellets is determined by a combination of the velocity of the knives and the pressure inside the extruder. However, when the extruder has been running for some time - depending on raw material composition and pellet size produced - the dyes starts to clot with raw material resulting in changes in dye size and number. Consequently, the pressure inside the extruder raise and thereby increase the speed of the feed material as it passes through the remaining dyes resulting in further changes in pellet size. The temperature, which is not constant during the production, might also have an impact on both material speed and pressure, and thereby affect the pellet size. As a consequence of the mentioned circumstances the main challenge when producing fish feed pellets by extrusion is the continuous drift in pellet size. To some degree, the size of the pellets can be adjusted by changing the production settings continuously. Because of the clotting a mandatory cleaning of the dyes will arise for a satisfying feed production meaning that the production needs to be stopped and the dyes cleaned before further production.

Today, monitoring of the pellet size is based on manual visual inspection, which are both labour demanding and relies on experienced assessors. The size of the pellets produced depends on the production specification. The product selection for fish feed producers typically covers pellets in the range from 0.5 millimetres to 9.0 millimetres. This, especially when producing the small size pellets, places great demand on the quality assessors. An automatic vision system for online size monitoring/control will be of great benefit to the fish feed industry. If such a system could indicate when the production is out of range of the quality in demand, the information could be used in a feed backward control system to adjust the settings of the extruder continuously. Such a system will improve the control of the pellet size and secure uniformity of pellets.

Foster et al. (1995) and Parsonage (2001) have both shown that it is possible to measure fish feed pellet size by image analysis. However, their work is only valid for measurements on single pellets, which limits the usability in relation to industrial production where the production process never leads to singlewise pellets. The pellets will always occur in bulks, which demand a measurement system capable of handling a pile of pellets at the same time. Recently, Ljungqvist et al. (Submitted to PloSOne) proposed an automatic size monitoring method based on image analysis working on a pile of pellets. The method is based on the mathematical morphological opening operation, meaning no image segmentation is needed. The outcome from the method is a general size distribution representing the general pellet area in the sample. Additionally, it was shown that the measurements on moist pellets could be related to dry pellets, which is essential since the pellets are moist right after extrusion. As the extrusion is the process in the production line, which is of most importance for the quality this is where the measurement should be performed in practice. Additionally, this will allow a fast process control, as measurements on the dried pellets will be to late in relation to use the information for process control. The new proposed method will therefor enable online size control by defining the smallest (upper specification limits) and largest (lower specification limits) allowable values. Based on the specification limits the information can be used for control charts in relation to statistical process control. Based on the information from the process the machine settings of the extruder can be adjusted to ensure that the right pellet size is produced and if necessary stop the production.

Another process in the fish feed production where image analysis can be applied for online measurement is in the control of the pellet coating procedure. Some feed-recipes intended for salmonids include coating of pellets with astaxanthin. Astaxanthin is a naturally occurring carotenoid with a high antioxidant activity essential for reproduction, growth and survival, and important for the development of the red colour in salmonid fishes (Owen, 1991). As fish cannot synthesize carotenoids de novo their intake rely on the content of carotenoids in the feed. Astaxanthin is highly expensive and even though it only constitutes less than 20% of the total fish feed costs, optimization of its use in fish feed production is of importance for a cost effective fish feed production. Today feed samples are taken from the production line on a regular basis and sent to laboratories for analysis. A fast but not necessarily online measurement system with pellet astaxanthin concentration as output will be a valuable tool in relation to process control and process optimization for the fish feed industry. Due to the coating procedure, that is a batchwise process not allowing any changes during the process, continuous automated monitoring will not be necessary. However directly after the batchwise coating process an automated system could ensure that the pellets contains the right astaxanthin concentration. If the outcome of the measurement gives a value outside the specified limits the operator can either coat the pellets again and / or change the settings for the next batch produced. In a preliminary study of Ljungqvist et al. (2010) it has been shown that it is possible to distinguish between feed pellets with and without astaxanthin coating using multispectral image analysis. The equipment used was a VideometerLab (Videometer A/S, Hørsholm, Denmark), which supports a multispectral resolution of up to 20 wavelengths distributed over the visible and first near infrared region, ranging from 385 nm to 1050 nm. The image resolution is 1280 x 960 pixels. Images of the feed pellets lying separately were captured, and the pellets were segmented from the background using a threshold. A canonical discriminant analysis on the mean pixel values for each pellet gave the best result with a classification correctness of 93%. Despite these promising results much research is still needed before the method is ready for use in the industry. A future challenge is how well the method will work for different concentrations of astaxanthin and different feed recipies.

In general, an important issue to take into consideration when evaluating these models is the criteria set by the industry - how precise do the results need to be to be satisfying compared with existing methods? Maybe three intervals e.g. low, middle, high, is enough and not necessarily exact numbers is mandatory.

3.2 Monitoring of frying processes

Fried food elements are used in various ready-made meals, which are sold in retail and served in canteens and fast food restaurants around the world. Fried minced meat is used in various production recipes, where the frying treatment gives the meat specific flavour and colour. The flavour and colour is essential for the recipe, and the end consumers high quality demands, which the industry needs to comply to. Frying of minced or small cut meat in large-scale batches is a difficult task, while still complying with consumer quality demands. Different industrial frying procedures used include batchwise frying on large pans or vessels, frying tables or continuous wok (Adler-Nissen, 2002). In order to meet the quality specifications, trained human operators are often judging the quality of the meat frying process by visual inspection.

Monitoring frying processes using automated methods poses different challenges than monitoring meat quality in general. When monitoring the meat frying process, the browning of the meat through mainly Maillard reactions must be closely followed through the whole process. The Maillard reactions create melanoidins giving the fried meat its distinctive brown colour and many of its organoleptic properties (Nursten, 2005). Controlling the browning process of the meat requires accurate control of the frying parameters. Continues sampling can be performed by non-contact methods as for example with an imaging device designed to measure reflected electromagnetic radiation from the meat. By simultaneously inspection of visual and near infrared wavelengths, disorders caused by thermal radiation may be avoided, without lose of ability to quantify colours and correlations with chemical compounds such as fat and water in the frying meat. The imaging technology enables the monitoring of a spatial distribution of water content and melanoidin production giving a dense surface sampling of large batches. A successive sampling may be used to create a feedback loop to control the frying media and thus create a fully automated frying system.
In an experiment described in Daugaard et al. (2010), minced meat has been monitored during frying. In the experiment meat was fried with variations both in frying time and temperature (see Table 1).

Time/Temp	120s	160s	200s	240s
200°C	I	I	I	I
225°C	I	A	A	A
250°C	A	A	A	A

I = The frying process have been inadequately processed; A = The frying process have been adequately processed

Table 1. Processing times and categorization of fried minced beef (From Daugaard et al., 2010)

Two skilled experts evaluated the state of the meat as being either inadequately or adequately processed. For each sample a multispectral image was acquired with spectral channels ranging from 430 nm to 970 nm. By training a canonical discriminant function (Fisher, 1936), the multispectral images were projected into a one-dimensional subspace ensuring the distance between the two classes; inadequate and adequate, was maximized. The subspace is used as a score space, termed "frying treatment score", which describes the browning of the meat. Utilizing this transformation it is possible to visualize the browning distribution as shown in Figure 1.
The irregular topology of the minced meat gives rise to scattering effects, which causes spectral differences between granules "peaks" and "valleys". Granule peaks have been preserved whereas remaining parts have been left out of the analysis using a grey scale morphology technique, known as the H-dome method (Vincent, 1993). A mean value of the frying treatment scores in an area may be used as an overall score of the area. The magnitude of the loadings of the canonical discriminant analysis, wich is illustrated in Figure 2, reveal which wavelengths are most significant for quantifying the browning of the meat. The loadings reveal that the browning is best described as a linear combination of recordings at 970 nm, 700 nm and 470 nm.
While the colour reveals the progression of the Maillard reactions, knowledge about future browning may be acquired by learning about the current water content (Nursten, 2005).

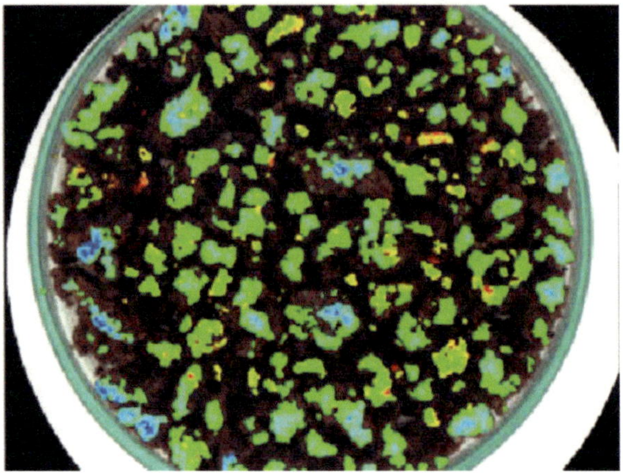

Fig. 1. Frying treatment scores superimposed on RGB (Red Green Blue) representation of multispectral image of minced meat, which was fried at 225°C for 200 seconds. Blue values represents less fried areas, red values represents more fried areas (From Daugaard et al., 2011)

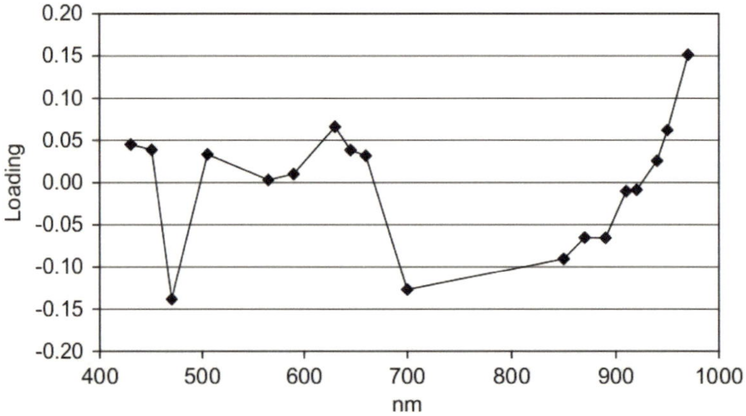

Fig. 2. Canonical Discriminant Function for fried meat samples (From Daugaard et al., 2010)

Because water has a large absorption peak in the near infrared area, quantification of the water content using a spectral camera recording electromagnetic waves in this area is possible as investigated in Dissing et al. (2009). Based on the same data as the quantification of the browning experiment, the water content of each batch was determined using a oven method for water removal.

Based on the near infrared spectral reflection values of the granular tops a large feature space was extracted. All pairwise ratios of the near infrared channels and following selection of percentiles in the resulting distributions were used as features. A variable selection procedure known as stepwise selection with a p-value of 0.03 was used to select the linear combination of features which best described the water content. The linear prediction model suggested by the stepwise selection is described in equation 1:

$$y = \alpha_1 q(\lambda_{970}, 50\%) + \alpha_2 q(\lambda_{890}, 99\%) + \alpha_1 q\left(\frac{\lambda_{920}}{\lambda_{910}}, 1\%\right) + \varepsilon \tag{1}$$

α_i denotes coefficient values, q(.) denotes the quantile function with two parameters; the wavelength ratio and the quantile-number, and ε is the error term. The model suggest that 970 nm together with 890 nm are the most significant wavelengths for predicting water content in fried minced meat. An absorption peak of water is sensitive to temperature, which means that the model might be different for other temperatures. The experiment was performed at 20°C. Figure 3 shows how the prediction model based on the feature set extracted from the multispectral images correlates very well with the actual water content with a root mean square error of prediction (RMSEP) of 0.50. It is very important to note that when having few data points as in this example, it is necessary to use cross validation to validate the prediction ability with the risk of obtaining an too optimistic model validation outcome.

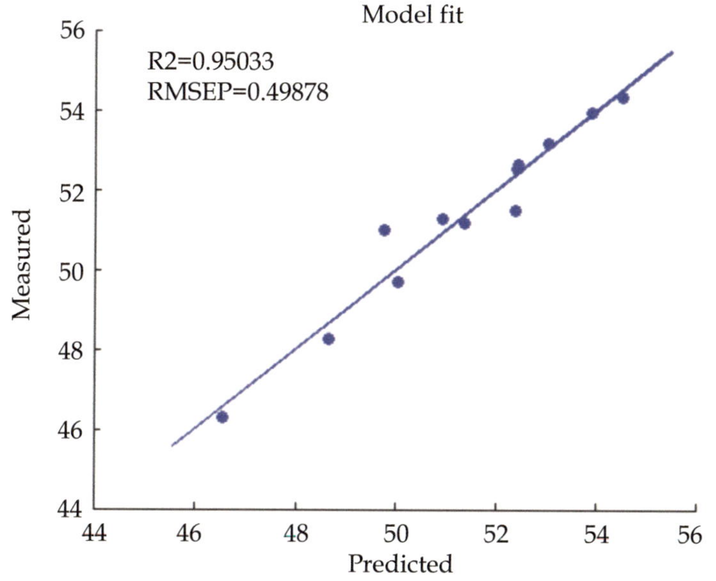

Fig. 3. Predicted versus measured water content in minced fried meat

Thus, by using a sensible combination of data acquisition techniques, image analysis and statistical methods a system for monitoring frying processes of minced meat may be designed. An advantage is the non-contact probing ensuring that the quality measurement will not cause contamination of the meat sample. Furthermore, the rapidness of the process monitoring and analysing as such will make it possible to conduct process adjustments very fast and could in the future be used to predict the process rather than monitor it directly, and thereby alter process parameters to avoid process drifts.

3.3 Spectral imaging as a method for measuring colouring compounds in salmonid fish

The high labour costs of fish processing are predominantly due to the extensive use of manual labour e.g. visual detection of quality parameters and removal of defect products. However, new processing concepts with automation represent a common strategy within the fish industry today (Erikson & Misimi, 2008). Only computer vision can ensure less fatigue, consistent accuracy, higher speed and adjustability, besides offering process-related feedback/ forward among handling and processing machines. Current state-of-the art vision systems for

quality and process control in the fish processing industries are typically based on traditional trichromatic (Red Green Blue) imaging. However, the aquaculture and fish processing industries are definitely areas where the added information in a spectral image can be exploited to improve the general quality and/or reduce the production cost e.g. assessment of fat and water content distribution in fillets (ElMasry & Wold, 2008), determination of freshness (Chau et al., 2009), detection of nematodes in cod fillets (Wold et al., 2001; Heia et al., 2007) and detection of melanin spots in Atlantic Salmon fillets (Mathiassen et al., 2007).
The consumers associate increased intensity of red in salmonid fishes with superior quality, the fish being fresher and having a better flavour (Gormley, 1992; Anderson, 2000). This means that colour is a highly important quality parameter in relation to the commercial production of salmonid fishes. To ensure that farmed salmon and trout achieve coloration that comply with the consumer preferences it is common practice within aquaculture to add astaxanthin to the fish feed. As previously mentioned astaxanthin is highly expensive, and control and optimization of the astaxanthin concentration from feed to fish is of paramount importance for a cost effective salmonid fish production. The traditional methods for determination of astaxanthin in fish meat are based on a chemical extraction of astaxanthin from a sample of minced fish meat, and subsequently determination by spectrophotometric analysis or high-performance liquid chromatography (HPLC) analysis. However, the methods are time consuming, labour demanding and sample destructive. Dissing et al. (2011) investigated the potential of using multispectral imaging for prediction of astaxanthin concentration in rainbow trout, *Oncorhynchus mykiss*. Fillets from 59 fish where imaged by a VideometerLab (Videometer A/S, Hørsholm, Denmark) prior to determination of the true concentration of astaxanthin by chemical extraction and HPLC analysis. The analysed samples ranged from 0.20 to 4.34 μg astaxanthin per g fish meat. The VideometerLab obtains multispectral images at 19 different wavelengths ranging from 385 to 970 nm (395 nm, 435 nm, 450 nm, 470 nm, 505 nm, 525 nm, 570 nm, 590 nm, 630 nm, 645 nm, 660 nm, 700 nm, 850 nm, 870 nm, 890 nm, 910 nm, 940 nm, 950 nm and 970 nm). After segmentation, removing unwanted information such as background material and arrears assumed to be fat and collagen, the spectra contained in the remaining part, the so-called region of interest (ROI) was transformed to a spectrum based on a mean calculation. Thus each image contributed with a single spectrum for the model calibration. Partial least squares regression [Brown, 1995] was then used to estimate a calibration model between the extracted spectra and reference values (the true values). The model was validated by a test set. The model showed good results with a prediction error expressed as root mean square error of prediction (RMSEP) of 0.27. For comparison a similar model where built for normal colour images, which yielded a RMSEP of 0.45. The results clearly illustrate that multispectral imaging is a promising method for rapid analysis of the astaxanthin concentration of rainbow trout, and thereby a qualified candidate for replacement of ordinary laborious and destructive sampling of astaxanthin for concentration prediction. Additionally, Dissing et al. (2011) and Frosch et al. (2011) illustrated how pixel-wise prediction can be used to get an impression of the spatial distribution of the astaxanthin concentration within the fillet. Figure 3 illustrates the pixel-wise prediction of astaxanthin within the analysed sample. The pixel-wise prediction is color-coded according to the amount of astaxanthin predicted in each pixel, so that pixels with high values of astaxanthin appears red, while low value astaxanthin pixels appear blue.

3.4 Spectral image analysis for monitoring biological process in research

The relationship between biological processes in the live organism and subsequent final food quality are an upcoming field of research. Such an understanding will have a huge

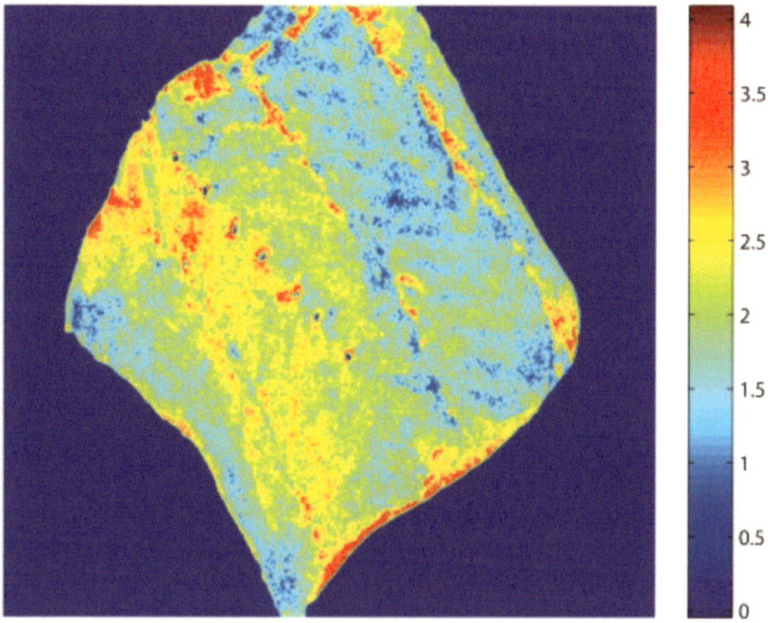

Fig. 3. Pixel-wise predicted astaxanthin image. The true astaxanthin concentration is determined to be 2.26 µg astaxanthin per g fish meat (From Dissing et al., 2011)

impact on the futures food production as knowing the history and thereby the capabilities of the raw materials can ensure its use in the most profitable way. A linkage between biological mechanisms related to tissue regeneration and texture changes have recently been established in salmonid fishes (Ingerslev et al., 2010a). Observed variations in the texture of the fish fillets relate to changes in structural components in the live organisms as a response to effects of production. It has further been found that changes in fillet texture due to bacterial infections is expressed in the eating quality of the steamed fish fillet whereas the effect is no longer expressed in smoked fillets (Ingerslev et al., 2010b). By knowing the history of the fish it will be possible to ensure that raw material is used in the most optimal way depending on whether the fish has been infected or not. Today these structural changes are monitored by laborious and expensive molecular methods (Ingerslev et al., 2010a). Additionally, the available methods are destructive meaning that it is not possible to follow the individual samples over time.

If structural changes in tissue can be examined using image analysis in a non-destructive manner and thus be used to follow impact of the biological process on the subsequent product quality this would be a valuable research tool. In general image analysis making non-destructive sampling possible, unlocks new possibilities to follow biological processes continuously, securing minimal usage of experimental animals and biological facilities. Furthermore the possibility of combining collection of spatial image information and specific individual sample recognition considerably reduces the impact of biological variance. Thus, for example enabling a visual detection of individual specimen's in a group and thereby overcome differences in experimental setup (Dissing et al., 2011; Schmidt et al., 2011).

The individual approach and the unique ability of vision / image analysis to compare on an individual level ensure the implementation of correlation between biological parameters and image data as a frontier in research for the future.

4. Perspectives for spectral image analysis

Spectral imaging systems are in the process of taking the step from being mainly an academic research tool for application oriented research in a variety of fields, to be developed into a tool for solving real life problems in a variety of applications ranging from airborne remote sensing to online quality control in e.g. the food industry. Image analysis of biological processes on individual level reveal new opportunities to the food processing industry as a unique possibility of single product handling can be the future. The rapid non-destructive measurement of raw material empowers differentiation between individual specimens / batches and thus enabling optimization of the use of raw material on an individual / batch level. An individual/batch sorting will ensure the best quality of the processed raw material where the knowledge from the raw product will be utilized based on known capacities. Ensuring a well designed and implemented traceability system in the production is therefore of significant importance (Frosch Møller, 2005; Frosch et al., 2008).

It is expected that vision systems based on spectral imaging will be employed to a much larger extent in the coming years, and that the food processing industry is an area where the added information in a spectral image can be exploited to improve the quality and or reduce the production cost.

Some of the challenges for the future implementation of image analysis in quality control of food are the rapid and reliable transfer of complex information to the operators in a logical way. The usage of image analysis is highly depend on which characteristics that explain the quality of a food product. Today the criteria for process control are often based on personal knowledge obtained during many years of practice of the operator. Thus further development within optimal automation of such operations will be dependent on close collaborative research between the food industry and the image analysis scientists to achieve a reliable transfer of the complex data from the image analysis into criteria for making automatic decisions to process control.

5. References

Adler-Nissen J. (2002). The continuous wok – A new unit operation in industrial food processes. *Journal of Food Process Engineering*, Vol. 25, pp. 435 – 453.

Anderson S. (2000). Salmon color and the consumer. *Proceedings of the Tenth Biennial Conference of the International Institute of Fisheries Economics and Trade Presentations*, ISBN 0-9763432-2-3, Corvallis: Oregon USA, July, 2000.

Azaza, M.; Dhraief, M.; Kraiem, M. & Baras, E. (2010). Influences of food particle size on growth, size heterogeneity, food intake and gastric evacuation in juvenile Nile tilapia, Oreochromis niloticus, L., 1758. *Aquaculture*, Vol. 309, pp. 193 - 202.

Burger, J. & Geladi, P. (2006). Hyperspectral NIR imaging for calibration and prediction: a comparison between image and spectrometer data for studying organic and biological samples. *The Analyst*, Vol. 131, pp. 1152 – 1160.

Chen, Y-R.; Chao, K. & Kim, M.S. (2002). Machine vision technology for agricultural applications. *Computers and Electronics in Agriculture*, Vol. 36, pp. 173 - 191.

Clemmensen, L. H. & Ersbøll, B. K. (2006). Multispectral recordings and analysis of psoriasis lesions. *9th MICCAI Conference Proceedings*, pages 15–18. MICCAI '06 - Workshop on Biophotonics Imaging for Diagnostics and Treatment, Copenhagen, Denmark, October 2006.

Daugaard, S. B.; Adler-Nissen, J. & Carstensen, J. M. (2010). New vision technology for multidimensional quality monitoring of continuous frying of meat. *Food Control*, Vol. 21, No. 5, pp.626 - 632.

Dissing, B. S.; Ersbøll, B. K. & Adler-Nissen, J. (2009). Monitoring water content using multispectral imaging and NIR in a minced meat preparation process. *The 14th international conference of Near Infrared Spectroscopy*, Bangkok, Thailand, November 2009.

Dissing, B. S.; Nielsen, M. E.; Ersbøll, B. K. & Frosch, S. (2011). Multispectral imaging for determination of astaxanthin concentration in salmonids. *PLoS One*,, Vol. 6, No. 5, e19032. Doi:10.1371/journal.pone.0019032.

Dissing, B. S.; Schmidt, J. G.; Ersbøll, B. K. & Nielsen, M. E. (2011). Visual effects of β-glucans on wound healing in fish. *Scandinavian workshop on imaging food quality (SWIFQ)*. Ystad, Sweden, May 2011.

ElMasry, G. & Wold, J. P. (2008). High-speed assessment of fat and water content distribution in fish fillets using online imaging spectroscopy. *Journal of Agricultural and Food Chemistry*, Vol. 56, No. 17, pp. 7672 - 7677.

Erikson, U. & Misimi, E. (2008). Atlantic salmon skin and fillet color changes as affected by perimortem handling stress, rigor mortis and ice storage. *Journal of Food Science*, Vol. 73, pp. 50-58.

Fisher, R. A. (1936). The use of multiple measurements in taxonomic problems. *Annals of Eugenics*, Vol. 7, pp. 179 - 188.

Foster, M.; Petrell, R.; Ito, M. R.; Ward, R. (1995). Detection and counting of uneaten food pellets in a sea cage using image analysis. *Aquacultural Engineering*, Vol. 14, pp. 251-269.

Frame, N. D. (1994). *The technology of extrusion cooking*. London: Blackie Academic and Professional.

Frosch Møller, S. (2005). The Importance of Data Quality and Traceability in Data Mining. Applications of Robust Methods for Multivariate Data Analysis. Ph.D. Thesis, BioCentrum, Technical University of Denmark.

Frosch S.; Dissing, B. S.; Nielsen, M. E. (2011). Determination of astaxanthin concentration in Rainbow trout (Oncorhynchus mykiss) by multispectral image analysis. 12th Scandinavian Symposium on Chemometrics, Billund, Denmark, 7 - 10 June 2011.

Frosch, S.; Randrup M. & Frederiksen M. T. (2008). Opportunities for the herring industry to optimize operations through information recording, effective traceability systems and use of advanced data analysis. Journal of Aquatic Food Product Technology, Vol. 17, No. 4, pp. 387 - 403.

Gomez, D.D. (2005). Development of an image based system to objectively score the severity of psoriasis. PhD thesis, Informatics and Mathematical Modelling, Technical University of Denmark, DTU, Richard Petersens Plads, Building 321, DK-2800 Kgs. Lyngby.

Gonzalez, R.; Torres, R. & Degreef, D. (2002). Extrusion-coccion de cereales. *Boletin*, Vol. 36, No. 2, 104e115, SBCTA, Campinas.

Gormley, T.R. (1992). A note on consumer preference of smoked salmon colour. *Irish Journal of Agricultural and Food Research*, Vol. 31, pp. 199 - 202.

Hansen, M. E.; Lund, F. & Carstensen, J. M. (2003). Visual clone identification of *Penicillium commune* isolates. *Journal of Microbiological Methods*, Vol. 52, pp. 221–229.

Hansen, M. E. (2005). Indexing and Analysis of Fungal Phenotypes Using Morphology and Spectrometry. PhD thesis, Informatics and Mathematical Modelling, Technical University of Denmark, DTU, Richard Petersens Plads, Building 321, DK-2800 Kgs. Lyngby.

Heia, K.; Sivertsen, A. H.; Stormo, S. K.; Elvevoll, E.; Wold, J. P. & Nielsen, H. (2007). Detection of nematodes in cod (Gadus morhua) fillets by imaging spectroscopy. *Journal of Food Science*, Vol. 72, No. 1, pp. E011 - E015.

Ingerslev, H. C. & Nielsen, M. E. (2010b). Infection of Atlantic Salmon (*Salmo salar*) by *Moritella viscosa* induces quality related markers in the musculature. *40th WEFTA Annual Meeting*, Izmir, Turkey, 2010.

Ingerslev, H. C.; Lunder, T. & Nielsen, M. E. (2010a). Inflammatory and regenerative responses in salmonids following mechanical tissue damage and natural infection. Fish and Shellfish Immunology, Vol. 29, No. 3, pp. 440-450.

Ljungqvist, M. G.; Frosch, S; Nielsen, M. E. & Ersbøll, B. K. (2010). Analysis of astaxanthin in fish feed pellets. Proceedings of West European Fish Technologists Association, Vol. 40, pp. 59–60, October, Izmir, Tyrkey 2010.

Ljungqvist, M. G.; Nielsen, M. E.; Ersbøll, B. K. & Frosch, S. Image analysis of pellet size for a control system in industrial feed production. Submitted to PLoS One.

Maletti, G. M.(2003). Novelty detection in dermatological images. PhD thesis, Informatics and Mathematical Modelling, Technical University of Denmark, DTU, Richard Petersens Plads, Building 321, DK-2800 Kgs. Lyngby.

Mathiassen, J. R.; Misimi, E. & Skavhaug, A. (2007). A Simple Computer Vision Method for Automatic Detection of Melanin Spots in Atlantic Salmon Fillets. Proceedings of the International Machine Vision and Image Processing Conference, IEEE Computer Society Washington, DC, USA.

Nursten, H. (2005). The Maillard reaction: Chemistry, biochemistry and implications. Culinary and Hospitality Industry Publications Services.

Owen, J. B. (1991). Genetic variation and nutrition - edited by A. P. Simopoulus and B. Childs. *Clinical Nutrition*, Vol. 10, No. 1, pp.: 61–62.

Parsonage, K. D. (2001). Detection of fish-food pellets in highly-cluttered underwater images with variable illumination. Master thesis, Department of Chemical and Biological Engineering, The University of British Columbia.

Geladi, P.; Burger, J. & Lestander, T. (2004). Hyperspectral imaging: calibration problems and solutions. *Chemometrics and Intelligent Laboratory Systems*, Vol. 72, pp. 209–217.

Rahman, L.; Rowe, P.; Cheyne, A. & Wilson, D. I. (2002). Ram extrusion of potato starch dough through multiholed dies. *Food and Bioproducts Processing*, Vol. 80, pp. 12 - 19.

Schmidt, J.; Dissing, B. S.; Ljungqvist, M. G.; Frosch, S.; Ersbøll, B. K. & Nielsen, M. E. (2011). Multispectral imaging of the effects of ß-glucans on wound healing in fish, 12th Scandinavian Symposium of Chemometrics, Billund, Denmark, June, 2011.

Sun, D. W. (2010). Hyperspectral imaging for food quality analysis and control. Academic Press, 2010.

Tabachek, J. L. (1988). The effect of feed particle size on the growth and feed efficiency of Arctic charr [Salvelinus alpinus (L.)]. *Aquaculture*, Vol. 71, pp. 319-330.

Vincent, L. (1993). Morphological grayscale reconstruction in image analysis: Applications and efficient algorithms. IEEE Transactions on Image Processing, Vol. 2, No. 2, pp. 176 - 201.

Wankowski, J. W. J. & Thorpe, J. E. (1979). The role of food particle size in the growth of juvenile Atlantic salmon (Salmo salar L.). *Journal of Fish Biology*, Vol.14, pp. 351-370.

Wold, J. P.; Westead, F. & Heia, K. (2001). Detection of parasites in cod fillets by using SIMCA classification in multispectral images in the visible and NIR region. *Applied Spectroscopy*, Vol. 75, No. 1, pp. 78 – 83.

Part 5

Quality Control in Environment

21

Mass Rearing and Quality Control Parameters for Tephritid Fruit Flies of Economic Importance in Africa

Sunday Ekesi and Samira A. Mohamed
International Centre of Insect Physiology and Ecology (ICIPE), PO Box 30772, Nairobi, Kenya

1. Introduction

Tephritid fruit flies are recognized worldwide as the most important threat to horticulture (Allwood and Drew, 1997; Barnes, 2004; Ekesi and Billah, 2007). Sub-Saharan Africa (SSA) is the aboriginal home to 915 fruit fly species from 148 genera, out of which 299 species develop in either wild or cultivated fruit. They belong mainly to four genera: *Bactrocera, Ceratitis, Dacus,* and *Trirhithrum* (White and Elson-Harris, 1992). Most of the fruit fly species are highly polyphagous attacking several important fruits and vegetables including mango, citrus, guava, avocado, tomatoes, pepper, cucurbits etc. Female fruit flies that lay eggs under the skin of fruits and vegetables cause direct losses. The eggs hatch into larvae that feed in the decaying flesh of the crop. Infested fruits and vegetables quickly rot and become inedible or drop to the ground. Beside the direct damage to fruits, indirect losses is associated with quarantine restrictions because infestation and sometimes mere presence of the flies in a particular country could also restrict the free trade and export of fresh horticultural produce to large lucrative markets abroad. The introduction of uniform and strict Maximum Residue Levels (MRL) across Europe compounds the problem and further jeopardizes export. Of greater concern is the fact that even in countries where fruit fly management methods are undertaken, rejection by European markets is on the increase largely because with global trade and passenger trafficking, they are easily translocated and the risk of majority of African fruit flies as key and potential quarantine pests is becoming increasingly realized (OleMoiYoi and Lux, 2004).

Although fruit flies attack a variety of horticultural crops in SSA, most of the fruit fly R&D in SSA has concentrated largely on mango infesting species largely because of the recognition of this crop as an important source of food and nutritional security. Globally, mango is the second most internationally traded crop after pineapple providing much needed income to poor rural households and the national economy. In Africa, assessments by the *icipe*-led African Fruit Fly program (AFFP) has revealed that out of the 1.9 million tons of mangoes produced annually across Africa about 40% is lost to fruit fly infestation (Lux et al., 2003). Majority of fruit fly species in Africa are little known but rank high on quarantine list worldwide. Some of the most important species attacking mango include the invasive fruit fly species *Bactrocera invadens* Drew Tsuruta & White and the native species; *Ceratitis cosyra* (Walker), *C. fasciventris* (Bezzi), and *C. rosa* (Karsch). The increasing awareness of the damage caused by these fruit flies to the fruit industry has created a demand for development of control measures against the fruit flies. Control methods to suppress or eradicate the insects are of utmost importance for the economy of developing countries whose income is highly dependent on agriculture. Pre-harvest control methods for fruit flies include methods such as the use of baiting and male annihilation techniques, Sterile Insect Technique (SIT), biological control (entomopathogens, parasitoids and predators), fruit bagging, early harvesting, and orchard sanitation (Allwood, 1997; Barnes, 2004; Mau, et al., 2007; Ekesi and Billah, 2007). To allow access to quarantine sensitive markets, pre-harvest methods are complemented with postharvest methods such as fumigation, heat treatment, cold treatments and irradiation (Armstrong, 1997; Ekesi and Billah, 2007). The success of majority of these control methods depends on the ability to establish a cost effective rearing methods. To conduct pre-requisite study on biology, response to attractants, and efficacy of the various biological control agents, a regular supply of good quality insects of pre-determined reproductive stages and age is necessary. Management methods that are especially based on the use of classical biological control, SIT, and post harvest treatment require the use of high quality developmental stages of various fruit fly species. To produce the required quality of flies, the average and variation in production, and quality control parameters must be established and closely monitored in mass rearing procedures. The quality of insect colonies also largely depends on the conditions under which they are maintained and the precision with which they are managed. Many aspects of mass-rearing environments directly affect the quality of fruit flies. Artificial larval and adult diets as well as rearing conditions (light intensity, photoperiod, insect density, temperature, humidity, etc.) can influence quality control parameters such as body size, survival, pupal weight, adult emergence, longevity, flight ability, fecundity, fertility and mating ability of the insects. Apart from *Ceratitis capitata* (Wiedemann), which has been reared for decades in different parts of the world (Vargas, 1989) with great levels of success, the mass rearing procedures for majority of economically important African fruit fly species such as *C. cosyra, C. fasciventris,* and *C. rosa* and the recently described invasive pest, *B. invadens* are little known and documented. Reliable methods for rearing large numbers of these insects have been developed by *icipe* scientists. In this chapter, the facilities and methodology for laboratory rearing of these species in relation to different quality control parameters is presented.

2. Rearing facilities

The fundamental and most essential requirement for sustainable fruit fly mass rearing is availability of well secured building. The entire building must be free from pesticide contamination and should be surrounded with a water trough at the ground level. The water trough prevents the entrance of crawling insects. Laboratory windows should be equipped with glass with the upper portion made up of fine wire mesh to ensure the provision of adequate natural light as well as to provide good ventilation. The fine mesh also safeguards against entry of lizard that may prey on the puparia and pupariating larvae. Proper lighting regime is vital for fruit fly behavior especially mating and additional artificial light which is controlled by a timer should be provided. For most of the mango infesting fruit fly species photoperiod of L12:D12 have been found to be adequate. Heaters and or cooling systems should also be installed to ensure that temperature is kept at the optimum. At the *icipe* rearing facility, temperature is maintained at 26-28°C, humidity is kept at the range of 60-70 RH%. The inside area of the laboratory should be spacious enough to allow movement of trolleys with trays and plastic boxes containing the successive stages of the flies. On the laboratory benches the rearing cages should be suspended on small stands within a plastic bowl (12 cm dia and 5.5 depth) containing water and few drops of liquid soap to prevent ant from climbing into the rearing cages. Also a layer of grease should be applied to the stand of the laboratory bench. McPhail traps baited with 3% NuLure should be hung at the ceiling of the rearing room to trap *Drosophila* species and any other flies that may escape during the handling. Trays and plastic boxes with diet containing developing larvae should be covered with fine netting to prevent contamination with *Drosophila* species.

3. Laboratory hygiene

High hygienic standards must be maintained in the laboratories at all production stages. Whole fruit, fruit domes, and important rearing materials such as Perspex cages, trays, sponges, racks etc must be frequently rinsed with 0.025% NaClO followed by several rinses in sterile distilled water to eliminate bacterial and fungal infection. The benches must be disinfected by wiping with 70% ethanol every morning to prevent infestations by mite which are common causes for drastic reduction in adult emergence. Diet ingredients should be purchased from well reputable sources and kept in very clean, cool and dry storage room.

4. Handling procedure for different developmental stages

4.1 Eggs

Eggs are harvested using plastic bottle (8 cm dia, 13 cm length) oviposition device that have been tested and validated for this purpose (see section 7 of this chapter). The plastic bottles are exposed to sexually matured flies after pre-oviposition period of 5 and 10 days for *Ceratitis* spp. and *B. invadens* respectively. Plastic bottles are perforated with 500 oviposition holes (at 1 cm apart). Before introduction of the bottle into the adult rearing cage, a 3 x 3 x 9 cm cellulose sponge saturated with fruit juice of the fly's preferred host is placed inside the bottle to provide the necessary ovipoistion stimulus and prevent egg dessication. After an exposure period of 10 hours, the bottles are removed from the cage. Fruit domes are also an excellent egg collection device. Pesticide free mango fruit should be surface sterilized with 0.025% sodium hypochloride (NaClO) and then cut into two equal halves. Thereafter, the mango flesh and seed are scooped out, and the domes are pierced several times with entomological pin (00). The domes are then placed on Petri dish (9 cm diameter) lined with moistened filter paper, and the edge of the dome along the Petri dish is sealed with cellotape (1 cm width) to prevent flies entering inside the dome. The flies are then allowed to oviposit for 10 hours. Egg collection from the oviposition devices is done by washing with water using a hand sprayer (0.5 l). Freshly collected eggs should be washed with 0.025% NaClO followed by several rinses with sterile distilled water. The number of eggs are then estimated volumetrically and recorded.

4.2 Larvae

Larvae are reared on artificial diets that normally include sources of protein, carbohydrate, lipids, microbial inhibitors, vitamins, salts, minerals, water, sterols, bulking agent and substances for adjusting pH. Several artificial diets have been evaluated for African fruit fly species and results are presented in Sections 9 and 10 of this chapter. Diet ingredients must be thoroughly mixed. A blender is most appropriate for mixing and especially ensuring that microbial inhibitor in the diet is evenly distributed. Prior to use, solid-based diet can be stored at 4°C for at least 24 h to allow the mixture to gel before decanting excess water. Diet pH should range between 4.0-5.5. If necessary, concentrated hydrochloric acid (HCl) (0.5-3.55%) or citric acid can be used to increase acidity. Whenever required, diet (500-800 g or ml) are transferred into shallow trays which have been shown to be the most appropriate for larval rearing (Chang et al., 2004). The thickness (solid diet) and volume (liquid diet) can be an important factor in reducing mortality as the greater the surface area to volume ratio the greater the likelihood of metabolic heat dissipation (Fay, 1989; Chang et al., 2006). Trays with diet and egg should be placed in plastic containers with screened tight fitting lids (to keep out *Drosophila* flies) above sterilized sand. During the first 3-4 days, the screen on the lid should be sealed to ensure high humidity for egg hatch and thereafter ventilation becomes important for efficient gas and heat exchange.

4.3 Puparia

Sand that is used as pupariation medium should be thoroughly washed and sterilized at 120°C for at least 2 hours and thereafter cooled at ambient temperature before use. Sterilized sand should be placed at the bottom of rearing cages at about 2 days before the larvae commence popping or jumping out of the diet trays to pupariate in the moist sterile sand at the bottom of the diet tray. After 7 d in the sand, puparia should be separated from the pupation medium by gentle sifting. For all mango-infesting fruit fly species, the puparia are usually held at 26-28°C and RH of 60-70% which is adequate to prevent pupal water loss and minimize development of mould. The puparia should be transferred into new cages at 2-3 days before emergence.

4.4 Adult

At the *icipe* rearing facilities, adult cages are made from Perspex cages (50 x 50 x 50 cm). One side of the cages is provided with a screen with mesh holes small enough to keep ants and *Drosophila* species out but ample for cross ventilation. Another side of the cage is provided with a sleeve to facilitate access to the flies. Colonies consist of 7 cages labeled cages 0-6 (cage 0 contains one week old flies and cage 6 contains older flies at 7 weeks). Cages are discarded at 8 weeks old. At one week old, the cages would normally hold between 5,000 to 6,000 flies and the numbers decrease successively with the age due to natural mortality. Fruit flies diet consist of a volumetric mixture of 3:1 sugar and enzymatic yeast hydrolysate ultrapure (USB Corporation, Cleveland, OH) in a Petri dish (9 cm). In each cage and depending on the number of flies, 2 - 6 Petri dishes containing the diet are placed inside the cage to minimize overcrowding. Water is provided in Petri dish with a layer of pumice granules and replaced daily. Egg production commences after a pre-oviposition period of 5-10 days depending on the species, and eggs are collected using oviposition devices as described in section 4.1. Egging is carried out three times a week and at the peak of egg production and depending on the species, a good cage can produce up to 50,000 eggs in 24 h.

5. Quality control parameters and recording

Boller and Chambers (1977) defined quality control of mass reared insects as "The degree to which a product meets the requirements of the objective or expected function". Quality control provides a means of optimizing insect mass rearing by identifying and gradually correcting deficient production processes, thereby preserving the genetic variability of the strain (Leppla and Ashley (1989). Quality control therefore integrates methods development, colonization, maintenance and arrays of process control details that affect the production and use of insects for pest management purposes. It helps in reducing cost of insect maintenance by improving rearing techniques through careful monitoring of protocols and a minimal number of sensitive parameters. Leppla and Ashley (1989) categorized quality control of mass reared insects into three main interrelated elements:

1. Production quality control which manages the consistency, reliability and timeliness of the production output. At the *icipe* facility, production quality control include standardization of diet ingredients and the production process by ensuring that all cages are appropriately labeled with species, dates, cage number, and diet type. The use and efficient storage of all materials away from pesticide contamination and any other pollutants forms part of efficient production quality control.
2. Process quality control assures the performance of the production process so that unacceptable deviations do not occur in product quality. Here, logistical consideration such as strict compliance to rearing protocols and maintenance schedules are crucial. In addition, environmental conditions such as temperature, RH, photoperiod, light and atmospheric quality are monitored constantly to determine the conditions of the rearing operations. Process control is recommended even when production is unstable although the diagnosis and correction of problems is more complicated.
3. Product quality control regulates the conformity of the product to acceptable standards of quality and predicts the effectiveness of the product in performing its intended function. In this regard, biological parameters are evaluated to identify possible deficiencies and to predict insect quality. Production can immediately be improved through testing of key and sensitive parameters and feedback mechanisms. Tests should be practical, uncomplicated, efficient and reproducible. A minimum number of parameters using the smallest sample size are recommended. In fruit fly mass rearing, important parameters include pupal recovery, pupal weight, percent adult emergence, percent survival, flight ability, percent fecundity and percent egg hatch (Calkins, 1989; Walker et al., 1997; FAO/IAEA/USDA, 2003). Walker et al., (1997) noted that there are variations between individual fruit fly species in any country and between the same species in different countries but there is a range that indicates that a colony is healthy. At the *icipe* facility, quality assurance are based on the following parameters: (1) pupal recovery from the number of eggs seeded should be >60%; (2) pupal weight, using 100 puparia should be regular for the same aged puparia; (3) adult emergence should be >70%; (4) percent fliers should be >80% and (5) percent egg hatch, using 100 eggs should >70%.

6. Colonization process

6.1 Fruit collection, handling, and processing

The evolutionary potential of a laboratory population is essentially determined when the breeding stock is isolated from the field (Mackauer, 1976). In this regard, all the mango infesting fruit fly species that are currently reared at the *icipe* facility were started by collecting several cultivated and wild host fruit species. For example, *C. capitata* originated from coffee berries, *Coffea arabica* L. (Rubiaceae) collected from Ruiru, central highlands of Kenya, *C. rosa* from *Monodora grandidieri* Welw (Annonaceae), a wild host from Mombasa, coastal Kenya, *C. fasciventris* from coffee in Rurima, central highlands of Kenya, *C cosyra* from mango *Mangifera indica* L. (Anacardiaceae) from Rurima and marula, *Sclerocarya birrea* (A. Rich.) Hochst. (Anacardiaceae) from Nguruman, Rift Valley Kenya, *C. anonae* from *Antiaris toxicara* (Pers) Lesh (Moraceae), a wild host from Kakamega forest in western Kenya and *B. invadens* from mango, Nairobi, Kenya. Leppla (1989) also recommended collections at different times of the day and year and this was also practiced in our collection. Fruit samples of the different plant species were placed separately in sisal or polypropylene woven sacks and then transported to the rearing facility. At the fruit incubation room, each fruit type is counted, weighed and later placed in secure well-aerated transparent plastic containers with a layer of sterilized moistened sand at the bottom. The rim of the containers are covered with a fine netting material held in place by the cover of the containers of which the greater part of its inner area is removed. The sand in the cage and the plastic containers served both as the pupariation medium for the larvae that exited the fruit in addition to soaking up fruit juices. The cage and the container holding the fruits should be labeled with information regarding fruit type, date of collection, GPS reading of location, number and weight of fruits. Fruits are held at ambient conditions for three weeks, after which they are dissected and the larvae that fail to pop out are assisted using a pair of soft forceps. Fruit holding cages should be checked daily and resultant puparia picked from the sand with a pair of soft forceps, counted and placed in Petri dishes with moistened filter paper. The Petri dishes with puparia should then be held in small, ventilated, transparent cylindrical plastic cages (5.5 by 12.5 cm) until adult eclosion. Emerging tephritid flies should be provided with diet that consists of a volumetric mixture of 1:3 enzymatic yeast hydrolysate and sugar, and water in Petri dish with a layer of pumice granules. When flies develop full coloration (after 24-48 hr) they are identified, and each species kept in a separate cage and provided with food and water as described above.

6.2 Colony establishment and maintenance

When initiating a colony from wild population, the first 3 generations are crucial for successful establishment and extreme care is necessary to minimize mortality by avoiding stress related mortality resulting from overcrowding and exposure to suboptimal temperatures (>30 or <20°C, for tropical species). The starter colony should also not be based on smaller number of flies with very narrow genetic diversity. Use smaller cages of 30 x 30 x 30 cm at the initial establishment with 2-3 cages each with ~ 50-100 flies, of similar age. As the number of flies increase, they should be transferred to larger cages of 50 x 50 x 50 cm. Within several generations, laboratory colonies of fruit flies can undergo, considerable changes with respect to the conditions under which they are reared, through selection (Tsitsipis, 1983) followed by rapid adaptation. Genetic changes can also occurr within only 4 generations of laboratory rearing in artificial larval diet flies. Maintenance of reared flies in such a way that laboratory colonies are genetically similar with wild populations is a criterion for ensuring quality of the reared insects. However, Leppla and Ashley (1989) opined that maintenance strategies are intuitive primarily because criteria for deciding to change strains in colony have not been quantified. They suggested various options for retaining a strain, these include continuous rearing without altering the population intentionally, rejuvenation by controlled selection, infusion with wild populations and hybridization of isolated lines. At the *icipe* facility, colonies are rejuvenated every 6-12 months by crossing wild males with laboratory reared females and at the same time crossing wild females with laboratory reared males. New cages are then set up with the progeny of the two cages. Therefore, the maintenance of vibrant colonies of fruit flies is a matter of careful rearing, diligent monitoring of quality control parameters and periodic strain restoration or replacement (Leppla, 1989). Although the process is laborious and costly, recent activities are exploring cryopreservation of embryos as described by Leopold (2007).

7. Evaluation of different oviposition devices for mass production of *Bactrocera invadens* and *Ceratitis cosyra*

Large quantities of eggs are usually required for mass production of fruit flies and an effective egg collection system is a prerequisite for mass rearing to reduce operational cost. In laboratory rearing, female fruit flies learn quickly to oviposit inside or through various substrates. To provide large number of flies for research and mass rearing of fruit fly parasitoids various egg collection systems were tested. The commonest egg collection device methods used for large-scale production of the most well known fruit fly, *C. Capitata,* is based on oviposition through screen into a trough containing water or moist blotter paper (Nadel, 1970). The process however requires genetic selection for screen-adapted flies which can take lengthy time period to adapt the insect to the laboratory procedure and Vargas (1984) opined that at mass production level, this may not be a straight forward process. Oviposition by *C. capitata* into perforated bottles and their subsequent removal from cages for egg collection has also been tested with significant levels of success (Tanaka 1965; Steiner and Mitchell 1966, Tanaka et al., 1970; Vargas 1984).

Using *B. invadens* and *C. cosyra,* we evaluated four egg collection devices namely mango dome, plastic bottle, funnel and plastic cup in order to select the most effective egg collection systems for mass rearing of these insects and other fruit fly species. Egg collection experiments were carried out using a 50 x 50 x 50 cm Plexiglas cages. Each device was perforated with 500 oviposition holes that were 1 cm apart. A 3 x 3 x 8 cm cellulose sponge saturated with 2% mango juice was placed in each perforated bottle to provide oviposition stimulus and prevent egg desiccation. Cages were then placed on a stand at ca. 1 m below two rows of 60-W fluorescent tubes. Oviposition cages were stocked with 500 ml of puparia. Four samples of 100 puparia were taken from each cage and held in small containers for determination of adult emergence and sex ratio. The results of which showed that in both species, adult emergence was > 85% and sex ratio was not different from the normal 1:1. The newly emerged adults were provided with water and a 3:1 volumetric mixture of sugar and enzymatic hydrolysate ultrapure. All cages were kept in a room maintained at 26- 28°C and 60 -70% RH under a photoperiod of L:D 12:12. The egging devices were exposed to sexually mature flies for a period of 4 hrs. Eggs were collected at 0900 h on days 8 through 17. The eggs were gently washed out of each egging device separately and the volume of eggs (ca. 12,000/ml for *B. invadens* and 13,000/ml for *C. cosyra*) obtained from each cage for each egging device was measured with a graduated cylinder. Three samples of 100 eggs were taken from the collection from each cage, spread with a Camel hair brush on moist blotting paper, and held in Petri dishes for determination of hatchability. This was replicated four times for each device. *Bactrocera invadens* readily accepted the 4 oviposition devices tested but to varying degrees. Over the 10-day standard egg collection period, the mean number of eggs collected was significantly higher in the bottle and mango dome compared with the funnel and plastic cup devices (Fig. 1A). The mean percentage egg hatch was also marginally higher in the bottle compared with the other devices (Fig. 1B). In *C. cosyra*, the performance of the various devices followed the same pattern as in *B. invadens* with the flies significantly laying more eggs in bottles and mango dome than the other two devices although egg hatch was not affected by devices (Fig. 2A & B). The bottle egg collection device is adopted in Hawaii fruit fly mass rearing facility because of the protection it provides to eggs inside the saturated atmosphere of plastic containers with wet sponges (Tanaka, 1965; Tanaka et al., 1970). Vargas (1984) demonstrated that oviposition by *C. capitata* and egg hatchability was significantly higher in bottles than on the screen device. The Oriental fruit fly *Bactrocera dorsalis* (a close relative of *B. invadens*); and *B. cucurbitae* are also mass reared in Hawaii by use of bottle egg collection devices. At the *icipe* rearing facility, bottles have been found to be the best egg collection system and presently in use for mass rearing of the different fruit fly species. Increase female acceptance of this oviposition device has led to substantial reduction in labor and cost of mass rearing of the various fruit fly species.

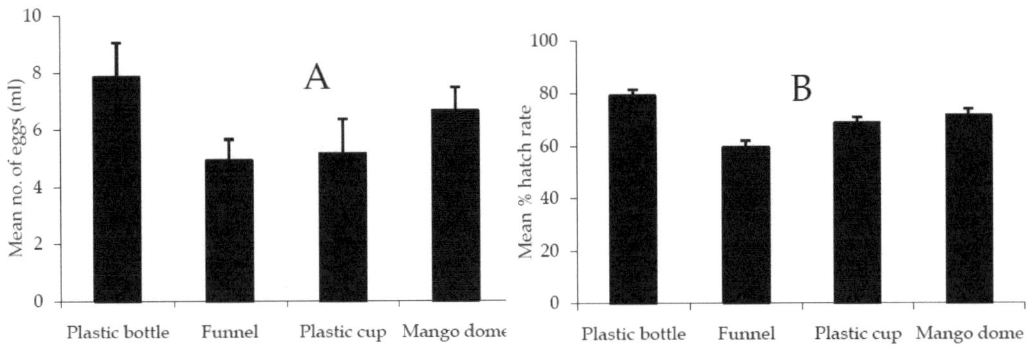

Fig. 1. Mean ± SE eggs collected (A) and percentage hatch rate (B) for four different devices in *Bactrocera invadens*

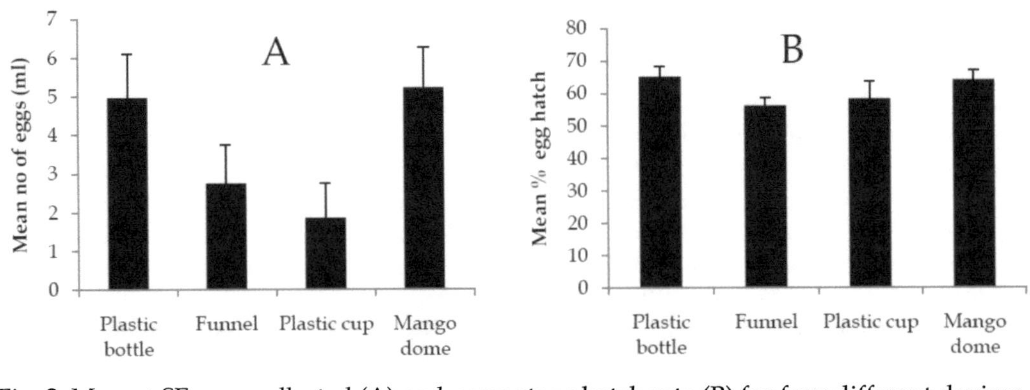

Fig. 2. Mean ± SE eggs collected (A) and percentage hatch rate (B) for four different devices in *Ceratitis cosyra*

8. Oviposition stimulant test

An important factor in the colonization process of insects is identification of a preferred oviposition stimulant which is important for egg deposition. Vargas et al., (1990) reported that fresh bell pepper juice stimulated oviposition by *Bactrocera latifrons*. We evaluated juices from different host plants of the target mango infesting fruit fly species for their ability to stimulate oviposition by the insects. Oviposition stimulant test was carried in a 50 x 50 x 50 cm Plexiglas cages. Plastic bottle similar to that used in the egg collection experiment above was used. A 3 x 3 x 8 cm cellulose sponge saturated with 100% juice of various hosts of the different fruit fly species as follows: *Bactrocera invadens* (mango, guava, *Terminalia catappa*, guava and marula), *C. cosyra* (mango, marula, guava and custard apple), *C. fasciventris* (mango, banana, guava, coffee, and citrus), and *C. rosa* (mango, *Monodora grandidieri, Uvaria lucida, Ludia mauritiana* and *Slacia elegans*). Sponges were placed in perforated bottle to provide oviposition stimulus and experimental procedures and egg collections were as previously described. In *B. invadens*, mean egg collection over a standard egg collection period of 14 days varied from 6.8-7.2 ml in mango, guava and marula and egg collection was 5.2 ml in *T. catappa*. When mango juice was compared at varying dilution rates of 1, 10, 50 and 100%, egg collection was at 3.5, 7.2, 5.8 and 7.0 ml, respectively. Vargas et al., (1993) showed that harvest of *B. latifrons* eggs from oviposition tubes that contained diluted papaya juices at 10% was 67% higher than from tubes that contained pure bell pepper juice which was the standard oviposition stimulant. They noted that with diluted juices, the humidity within the receptacle remained higher for a longer period of time compared with undiluted juices. In *C. cosyra*, juices of mango and marula at 100% elicited more egg laying (5.2 and 5.6 ml, respectively) compared with custard apple (2.1 ml) and dilution of custard apple juice had no effect on oviposition. Juices of mango, banana and coffee recorded 5.7, 6.1 and 5.8 ml of eggs, respectively in *C. fasciventris* compared with guava at 3.6 ml and citrus at 1.8 ml. Comparison of mango juice at varying dilution rates of 1, 10, 50 and 100%, resulted in *C. fascivntris* egg collection rates of 1.3, 5.8, 5.6 and 5.5 ml, respectively. Exposure of *C. rosa* adults to sponge cloth treated with juices of 4 different host plants in plastic bottles elicited oviposition but to varying degrees. Eggs laid in plastic bottles containing sponge cloth dipped in 100% juice of mango (6.2 ml), banana (5.8 ml) and *M. grandidieri* (3.9 ml) was higher followed by *U. lucida* (1.6 ml). Varying doses of mango had no effect on oviposition. In general, mango and banana juices seem to significantly elicit oviposition in all the target fruit flies perhaps not suprising given that these host plants are highly preferred by the target fruit fly species (Ekesi et al., 2006; Rwomushana et al., 2008; Mwatawala et al., 2010.). Fruits from different cultivars of both crops are also available all the year-round and banana is especially less expensive compared to mango and should serve as a readily available and convenient oviposition stimulants for mass production of the insects.

9. Adaptation of fruit flies to artificial rearing and quality control: case studies from *B. invadens* and other fruit flies

Conventional knowledge in fruit fly mass rearing demands that artificially reared flies produced for research should possess qualities and behavior nearly like insects in the wild (Calkins et al., 1994). However, laboratory colonization and mass production of fruit flies on artificial diet may require several generations for the insects to adapt to the artificial diet (Kamikado et al., 1987, Souza et al., 1988, Economopoulos 1992). In some cases, attempts to raise a colony from wild populations on artificial diet have completely failed (Rössler 1975), but, if successful, long-term rearing on artificial diet may improve insect performance, e.g., through reduction in the pre-oviposition period and an increase in egg production (Vargas and Carey 1989). Ekesi et al., (2007) assessed the adaptation process of *B. invadens* to artificial diet when the flies were moved from rearing on whole mango fruits to a wheat-based artificial diet.

The insects that were used for the adaptation studies had originated from rotten mangoes collected at a local market in Nairobi, Kenya and maintained on whole mango fruits for five generations. Eggs of *B. invadens* were collected using a mango dome. The dome was placed into fly stock colonies for 1 h and eggs were carefully removed from the underside of the dome with a fine camel hair brush and then placed on a 9-cm diameter moist blotting paper. After 36 h, 100 newly emerged larvae from the above lots of eggs were gently introduced with a fine camel's hair brush onto the surface of 100 g of wheat-based diet (Table 1) in open 150 ml plastic cups or into artificially drilled holes on whole mango fruit. Each hole was perforated with an entomological pin and measured ~ 1 mm in diameter and 1 cm in depth. After inoculation with larvae, the fruit were placed in 200 ml plastic cups over a 1 cm deep layer of sand and the cups were covered with muselin cloth. Mature larvae exited the fruit and pupariated in the sand. Each of the artificial diet cups was placed in larger 250 ml plastic cups that contained a layer of heat-sterilized sand on the bottom as pupariation medium. Mature larvae left the artificial diet and entered the larger cups to pupariate. An experiment consisted of 4 cups of diet containing 100 larvae, with each group of 4 cups replicated 3 times or 4 fruits with the same number of larvae and replications. At every generation tested, records were kept on the following quality control parameters (1) adult survival expressed as % insect-days of 20 d (total possible d), (2) larval stage duration, (3) % puparia recovered, (4) weight of puparia, (5) percentage adult emergence, (6) flight ability, (7) fecundity over a 10-d period and (8) fertility. Adult survival was tested in 15 x 15 x 15 cm Plexiglas cages. Percent pupal recovery was calculated based on the initial number of larvae introduced into each container of rearing medium and pupal weight was based on four lots of 20 puparia from each replicate. Adult emergence was based on four lots of 20 puparia from each replicate that were placed in screened 12 cm diameter plastic containers and observed for a period of 21 d. The flight-ability test was conducted using flies emerging from four lots of 20 puparia from each replicate using the method of Boller et al., (1981). Fecundity and fertility were based on daily egg collections from 10 pairs of flies held after a pre-oviposition period of 7 d. Eggs were collected using a mango dome and egg hatchability was assessed after 72 h. In all experiments adults were fed on a diet consisting of 3 parts sugar and 1 part enzymatic yeast hydrolysate ultrapure, and water on pumice granules. All experiments were carried out in a room maintained at 26-28°C, 60 ± 70% RH with a photoperiod of 12:12 (L:D). Results showed that there were significant differences between the two rearing media tested for all quality control parameters evaluated. During the first three generations, survival of *B. invadens* was significantly lower on artificial diet compared with flies reared on whole mango fruits (Fig. 3A). Thereafter, no significant difference was observed between the media in the other generations. In *C. capitata* Leppla et al., (1983) reported that adult survival was reduced during the first three generations of colonization on artificial diet when a colony was initially kept on host fruits for the first five generations and then switched to artificial diet with a continuous decline in the following four generations, but recovering to normal levels thereafter. Economopoulos (1992) also observed a reduction in adult survival of both sexes for three consecutive generations following colonization of new wild *C. capitata* colony with recovery to normal levels occurring in the fourth generation. Ekesi et al (2007) attributed the low survival to the reduced ability of flies to adapt to artificial diet and other rearing conditions, such as crowding. The authors did not observe any significant difference in larval developmental duration between the rearing media from the parent generation to generation F_3. However, in generation F_4 and F_5, development time was significantly shorter (7.8-8.2 d) on the artificial diet compared with that for the insects reared on whole mango (10.2-10.5 d) (Fig. 2B). With regard to pupal recovery, significantly less *B. invadens* were recovered on artificial diet between generations P to F_2 (Fig. 2C). From generation F_3 onward, the rate of pupal recovery was not significantly different between the two rearing media. At generations F_4 to F_5, pupal weight was higher on artificial diet compared with rearing on mango (Fig. 3D). Adult emergence and flight ability were not affected by the rearing media throughout the five generations tested (Fig. 3E &F). Over the generations, adult emergence ranged from 72-83% on artificial diet and 73 to 83% on mango. Flight ability varied from 75% to 85% on artificial diet and 72 to 82% on mango. Life history traits, including larval developmental period, are important quality control parameter in the evaluation of diet and the rearing process for fruit flies, and nutritional content of a diet can considerably affect development time, growth, and survival of fruit fly larvae (Krainacker et al., 1987, Vargas et al., 1994). With *B. invadens* Ekesi et al., (2007) observed a decrease in the length of larval development time in flies reared on artificial diet by generation F_4 and a corresponding increase in pupal weight at the same generation. Kaspi et al., (2002) demonstrated that the higher the protein level in larval diet the shorter the developmental time and the larger the size of individuals produced. The protein content in the artificial diet used in studies by Ekesi et al (2007) was higher than the quantity in the variety of mango used which could explain the shorter developmental time and higher pupal weight of *B. invadens.*

Bactrocera invadens fecundity over a 10-d period was found to be significantly lower (77 to 97 eggs) on artificial diet during generations P to F_2 compared with mango (101 to 124 eggs) (Fig. 3G). However, by generations F_3 to F_5, flies emerging from the artificial diet produced significantly more eggs (122 to 282 eggs) than insects reared on mango (109 to 117 eggs). During the first three generations, fertility was significantly reduced (34 to 61 %) on artificial diet compared with mango (70 to 84%). However, from generation F_3 onward, the egg hatchability did not differ significantly between the two rearing media (Fig. 3H). This study demonstrated that fecundity and fertility are affected during the adaptation process. It took four generations for the insects maintained on artificial diet to reach or surpass the level of fecundity recorded for flies reared on whole mango fruits. Economopoulos (1992) showed that even after nine generations, fecundity of wild *C. capitata* did not match the levels of laboratory-adapted flies. Ekesi et al., (2007) argued that the prolonged period of adaptation could be associated with poor mating, which the authors observed in flies maintained on artificial diet at the earlier stage of colonization. Previous studies with other fruit flies have demonstrated the importance of wheat and wheat-based products in fecundity and eclosion of larvae from eggs (Zumreoğlu et al., 1979, Vargas et al., 1994, Chang et al., 2006).

Leppla et al (1983) noted that the process of lab-adaptation of fruit flies is nonlinear with relative greater changes occurring initially and perhaps again later as the techniques are modified to increase rearing efficiency. Such 'crash-recovery cycle" was observed for *B. invadens* and it took three to four generations for the insect to adapt to artificial rearing medium. The laboratory adaptation of any insect species is a function of their inherited ability to adapt to the new rearing media as well as the quality of the rearing effort during the colonization.

Souza et al., (1988) observed that at least 10 generations were needed for adaptation of *C. capitata* to artificial diet. In the olive fruit fly *Bactrocera oleae* (Gmelin), about three to four generations were required to adapt (Tsitsipis 1983), while in *B. cucurbitae* Coquillett, it took 14 generations to reach a permanent plateau (Kamikado et al., 1987). In the olive fruit fly, Loukas et al., (1985) found that colonization resulted in significant ADH (alcohol dehydrogenase), 6-PGD (6-phosphogluconate dehydrogenase) and HK (hexokinase) allozymes frequency changes within three generations of artificial rearing. Additionally, Economopoulos and Loukas (1985) reported that the selection factor responsible for the genetic changes was the artificial medium. It is therefore evident that several genetic, physiological and behavioural changes occur during colonization of wild populations of fruit flies and depending on the species could take several generations for the insect to fully adapt to the rearing medium (Economopoulos 1992).

10. Comparison of different solid-based artificial diet for larval rearing and quality control for different fruit flies

The identification of dehydrated plant material (i.e. carrot) and dry yeast as important ingredients for larval development has led to the production of viable colonies of different fruit fly species across the globe (Hooper, 1978; Tsitsipis, 1989; Walker et al., 1997). Evidence suggest that carotene (from carrot) is an important feeding stimulant that promotes growth, particularly during the early larval development (Fay 1989). Before the development of artificial larval diet for the various African fruit fly species under discussion, all the target insects were mass reared on whole fruits mainly mango, banana and coffee. In the whole fruit rearing process, flies are allowed to oviposit in the target fruits that are placed in Perspex cages. Infested fruits are transferred into Styrofoam boxes above a layer of sterilized sand, and puparia are separated from the sand by sifting. Mass production of fruit flies on fruits is expensive in addition to being labor-intensive and require a lot of space. Moreover, fruits are impractical to handle, difficult to standardize with respect to larval density, and prone to inconsistent rates of decomposition (Debouzie, 1978). Futhermore, unpredictability in fruit availability due to seasonal fluctuations can affect the whole production process hence the need to develop, test and adapt various artificial diets for mass rearing of fruit flies.

The wheat millfeed diet has been the standard diet of choice for fruit fly mass rearing in different parts of the world (Tanaka et al., 1965; Vargas et al., 1993; Walker et al., 1997). In addition to the millfeed as the bulking agent other ingredients includes yeast-based products, sugar, antimicrobial agents (nipagen, sodium benzoate, and streptomycin) and water (Tanaka et al., 1969). Water is particularly important for minimizing the effect of metabolic heat build-up during the final stage of larval development (Hooper, 1978). Fruit fly rearing and quality control parameters from millfeed diet have been satisfactory although the quality of millfeed can sometimes vary and adversely affect rearing process. A variety of factors such as the cultivar of wheat, spring or winter season, grain hardness, blends of bran and germ, and impurities can affect the quality of millfeed (Orth and Shellenberger 1988, Bass, 1988). Therefore, the need to test various bulking compounds in the mass rearing of fruit flies on artificial diet becomes important.

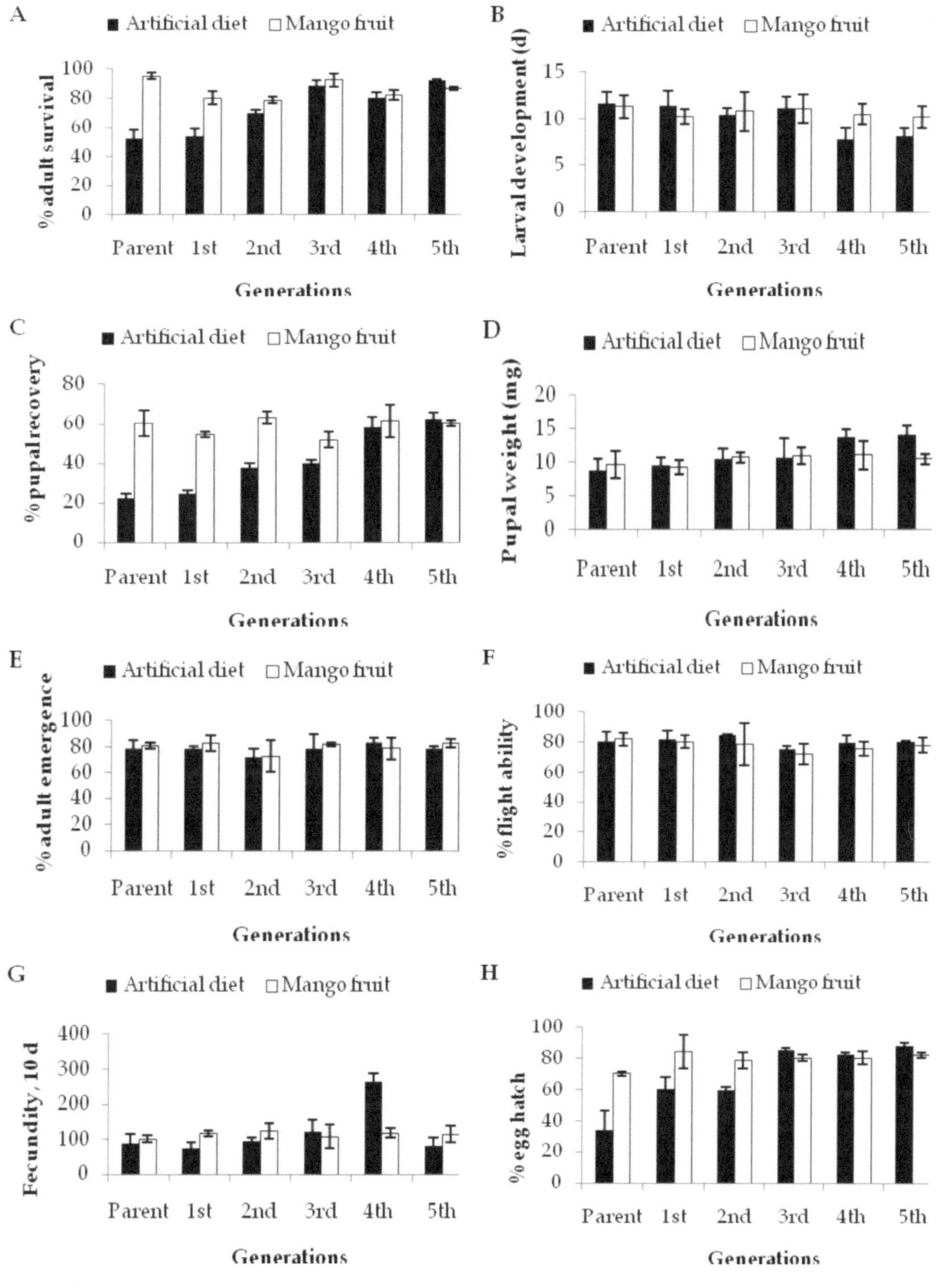

Fig. 3. Quality control parameters for *B. invadens* following adaptation to artificial rearing in the laboratory

Several bulking compounds including carrot, sugarcane bagasse, cassava, potato, corncob, paper, sawdust etc have been tested as alternative to millfeed in fruit fly mass rearing (Vargas et al., 1983; Minno and Holler, 1991; Stevens 1991, Vargas et al., 1994; Ekesi et al., 2007). These compounds absorb liquid when mixed in diets providing a uniform distribution of nutrients, physical consistency and texture, and serving as a substrate for larval feeding and development. In the Pacific Islands, the initial bulking agent used in the mass rearing of different *Bactrocera* species was cassava and sugarcane bagasse but fresh pawpaw has become the main ingredient in the mass production process because of its year-round availability (Walker et al., 1997). On the basis of these information, we explored different larval rearing media at the *icipe* mass rearing facility that are based on wheat, carrot, boiled cassava, sugarcane bagasse (*B. invadens, C. fasciventris* and *C. rosa*), and carrot supplemented with mango powder (*C. cosyra*) for mass rearing of the fruit fly species. The ingredients in the various diets are as presented in Table 1.

Ingredients	Solid-based						Liquid-based
	Wheat	Carrot alone	Carrot with mango A	Carrot with mango B	Sugarcane baggase	Cassava	
Torula yeast, g[a]	3.55	-	-	-	3.55	3.55	-
Brewer's yeast, g[b]	-	8.10	8.10	8.10	-	-	-
Lallemand yeast[c]	-	-	-	-	-	-	20.40
Methyl p-hydroxybenzoate, g[d]	0.11	0.20	0.20	0.20	0.20	0.20	0.20
Sodium benzoate, g[e]	0.11	-	-	-	-	-	0.20
Sugar, g[f]	7.83	16.20	16.20	16.20	16.20	16.20	12.18
Citric acid, g[g]	-	0.60	0.60	0.60	-	-	2.31
Mill feed, g[h]	31.20	-	-	-	-	-	-
Carrot, g[i]	-	24.20	21.20	18.20	-	-	-
Mango, g[i]	-	-	3.00	6.00	-	-	-
Sugarcane bagasse, g[f]	-	-	-	-	24.20	-	-
Cassava, g[i]	-	-	-	-	-	24.20	-
Wheat germ oil[j]	-	-	-	-	-	-	0.20
Water, ml	57.20	50.70	50.70	50.70	50.70	50.70	100

[a]ISCA Technologies, CA; [b]East African Breweries Limited, Nairobi; [c]Lallemand Inc. Bio-Ingredients (3 part whole cell yeast - LB12240 and 1 part hydrolyzed yeast -FNILS65); [d]Kobian Kenya Limited, Nairobi; [e]J.T. Baker Chemical Company, NJ; [f]Mumias Kenya Limited; [g]Chemoquip Limited, Nairobi; [h]Unga Limited, Nairobi; [i]Obtained as fresh fruit from hawkers market, Nairobi and then crushed, dried and blended into a powder; [i]Obtained as fresh roots from hawkers market, Nairobi and then boiled; [j]MP Biomedicals, Aurora, OH.

Table 1. Artificial larval diet ingredients used in the study

In this experiment, eggs (1 ml ~ 12, 000 eggs for *B. invadens* and ~ 13,000 eggs for *Ceratitis* spp.) collected within 1 hr after oviposition were seeded using a 1 ml transfer pipette onto a 5 x 5 cm strip of moist black cloth for each species separately. The cloth was placed on the center top of 1 kg of diet in 3 x 10 x 15 cm plastic trays. The experimental conditions were similar to the adaptation experiment described above. Upon hatching, larvae fed *ad libitum* and the subsequent larval and pupal handling was the same as that described in section 9. All the mentioned quality control parameter above were also evaluated in this experiment and were measured in the same manner as described above.

Quality control parameters for the 4 diets tested and 4 fruit fly species are presented in Tables 2 & 3. For *B. invadens* pupal recovery was generally highest in wheat- (65%) followed by sugarcane bagasse- (55%) and carrot-based (54%) diets and lowest in the cassava-based diet (32%). Pupal weight was not significantly different among carrot-, wheat- and sugarcane bagasse-based diet and ranged from 13.6 to 14.1 mg. However, the pupal weight for flies reared on the cassava-based diet were significantly lighter (12.5 mg) compared to the other three media. Adult emergence was significantly higher in carrot- and wheat-based diet (80-86%) compared with sugarcane bagasse-based diet (70-77%) (Table 2). Fecundity over ten days period was also significantly higher in the carrot-, wheat- and sugarcane bagasse- (342-366 eggs per female) than in the cassava-based diet (233 eggs per female) (Table 2). In *C. fasciventris* and *C. rosa*, all quality control parameters from carrot and wheat outperformed those of sugarcane bagasse and cassava (Table 2). The cassava-based diet particularly did not support pupal recovery and egg production in both fly species.

Parameter	Carrot	Wheat	S/cane baggase	Cassava
Bactrocera invadens				
Pupal recovery (%)	54.2 ± 2.1b	65.4 ± 1.5a	55.4 ± 1.1b	32.1 ± 1.4c
Pupal weight (mg)	13.8 ± 1.1a	14.1 ± 1.3a	13.6 ± 1.3a	12.5 ± 1.6b
Adult emergence (%)	80.4 ± 1.7a	86.3 ± 2.3a	76.5 ± 1.2b	70.3 ± 2.2b
Flier (%)	82.2 ± 2.3a	81.3 ± 2.5a	76.8 ± 2.6a	80.4 ± 1.8a
Fecundity, 10-day	366.2 ± 11.2a	354.4 ± 12.7a	342.4 ± 14.1a	233.2 ± 7.8b
F_1 egg hatch (%)	80.2 ± 3.6a	74.4 ± 1.4a	70.3 ± 1.6b	58.4 ± 2.2c
Ceratitis fasciventris				
Pupal recovery (%)	38.5 ± 3.1a	40.8 ± 2.8a	20.2 ± 1.3b	11.1 ± 1.3c
Pupal weight (mg)	8.1 ± 0.6a	8.5 ± 0.1a	5.3 ± 1.4b	4.8 ± 0.9b
Adult emergence (%)	76.2 ± 1.0a	81.2 ± 4.1a	60.8 ± 2.5b	58.4 ± 1.7b
Flier (%)	78.8 ± 1.2a	80.8 ± 2.3a	64.2 ± 1.8b	65.5 ± 2.5b
Fecundity, 10-day	222.3 ± 26.2a	212.2 ± 18.7a	101.4 ± 13.8b	76.5 ± 10.4c
F_1 egg hatch (%)	72.4 ± 2.5a	68.8 ± 1.2a	51.2 ± 3.2b	48.1 ± 2.5b
Ceratitis rosa				
Pupal recovery (%)	54.4 ± 2.3a	50.2 ± 2.6a	22.4 ± 1.8b	13.3 ± 1.5c
Pupal weight (mg)	8.8 ± 1.2a	8.1 ± 1.1a	5.7 ± 0.9b	5.1 ± 1.1b
Adult emergence (%)	74.2 ± 1.4b	84.4 ± 2.6a	58.5 ± 2.2b	55.2 ± 3.1b
Flier (%)	81.2 ± 2.5a	80.0 ± 2.2a	70.1 ± 2.4b	68.9 ± 2.4b
Fecundity, 10-day	115.1 ± 10.5a	101.8 ± 12.1a	98.9 ± 8.6b	80.1 ± 12.3c
F_1 egg hatch (%)	64.5 ± 3.2a	60.8 ± 2.2a	40.2 ± 1.1b	40.5 ± 1.5b

Table 2. Effect of four artificial larval diets on pupal recovery, pupal weight, adult emergence, flier, fecundity, and egg hatch of *Bactrocera invadens, Ceratitis fasciventris* and *Ceratitis rosa*. Means within a row followed by the same letter do not differ significantly by Tukey HSD test (P=0.05)

Diets of carrot, carrot supplemented with mango and sugarcane bagasse also supported growth and reproduction by *C. cosyra* (Table 3). However supplementation of carrot with mango powder significantly increased pupal weight, egg production and to some extent egg hatchability compared to the other diets. Pupal weight was 10.5-11.7 mg in carrot supplemented with mango powder, 9.2 mg in carrot alone and 7.2 mg in sugarcane bagasse (Table 3). In fruit fly mass rearing, high pupal weight is considered a desirable characteristic in the production process, as it is a good indicator of flies' body size. Churchill-Stanland *et al.*, (1986) found that size of *C. capitata* was important in mating success and noted that 8 and 9 mg insects experienced greatest mating success followed by those that weighed 6 and 4 mg. Fly size is also a determinant factor of insect fertility and fecundity.

Overall, quality control parameters from carrot, carrot supplemented with mango and wheat were superior for mass rearing of *B. invadens, C. fasciventris, C. rosa* and *C. cosyra* compared with diets based on sugarcane bagasse and boiled cassava. In *B. invadens* the wheat-based diet seem to have an upper edge in terms of pupal recovery. In mass rearing of *B. latifrons*, Vargas and Mitchell (1987) found carrot to be less nutritious than wheat on the basis of female reproductive parameters. In the current study, apart from pupal recovery, all quality control parameters on both diets were uniform for the target insects. Based on these findings, *B. invadens, C. fasciventris* and *C. rosa* have successfully been maintained on a carrot-based artificial diet but in *C. cosyra* carrot is supplemented with mango to obtain appreciable quality control parameters. Although wheat millfeed is an equally effective bulking agent, a number of factors had favored the use of carrot over wheat. Firstly, carrot fruits are usually available all the year although prices can be exorbitant during the off-season period. Secondly, inconsistencies in the quality of wheat-based bulking agent associated with chemical pesticide residues have been observed. In most of the African countries, carrots are not treated with pesticides, thus free of residues. Thirdly, ingredients in the wheat-based diet are more expensive than those found in the carrot-based diet. Table 4 shows the weekly production data on carrot-based artificial diet for the 4 fruit fly species, and using the methodology described here healthy culture have been maintained at the *icipe* rearing facility.

Parameter	Carrot	Carrot supplemented with 3% mango	Carrot supplemented with 6% mango	Sugarcane bagasse
Pupal recovery (%)	37.8 ± 5.2a	40.4 ± 2.4a	48.4 ± 2.4a	20.1 ± 1.8b
Pupal weight (mg)	9.2 ± 1.8b	10.5 ± 3.2a	11.7 ± 1.3a	7.2 ± 1.3c
Adult emergence (%)	86.4 ± 4.4a	82.2 ± 6.2a	80.2 ± 3.6a	72.4 ± 2.3b
Flier (%)	84.2 ± 1.4a	80.5 ± 2.0a	80.4 ± 2.5a	82.8 ± 2.6a
Fecundity, 10-day	201.4 ± 10.2c	276.4 ± 22.2a	256.2 ± 31.4b	101.5 ± 24.8d
F_1 egg hatch (%)	62.4 ± 2.2b	68.2 ± 2.4b	75.2 ± 1.4a	54.2 ± 3.2c

Table 3. Effect of four artificial larval diets on pupal recovery, pupal weight, adult emergence, flier, fecundity, and egg hatch of *C. cosyra*. Means within a row followed by the same letter do not differ significantly by Tukey (HSD) test (P=0.05)

Parameter	*B. invadensa*	*C. cosyra**	*C. fasciventris*	*C. rosa*
Puparia	3012.1 ± 605.4	1134.5 ± 141.4	2561.8 ± 214.8	2984.2 ± 162.2
Pupal recovery (%)	55.3 ± 7.2	38.2 ± 4.8	38.8 ± 2.4	55.1 ± 4.2
Pupal weight (mg)	13.7 ± 2.2	10.8 ± 1.4	8.2 ± 1.1	8.4 ± 1.4
Adult emergence (%)	84.0 ± 2.2	72.2 ± 2.5	76.4 ± 1.8	75.8 ± 2.4

Each tray (3 x 10 x 15 cm) contained 500- g diet. *Supplemented with mango.

Table 4. Mean ± SE weekly production (n=28) of four fruit fly species per tray on carrot - based diets

11. Liquid-based artificial diet for larval rearing and quality control for four fruit fly species

Although we have been able to successfully rear all target fruit fly species on the solid-based artificial diets for over 100 generations, in recent years, the importance of liquid diet as an excellent substitute to solid based diet for fruit fly rearing is being promoted (Chang, 2009). The liquid larval diet is reported to have the following advantages over solid larval diet: (1) total consumption by flies if accurate proportion of diet volume to egg density are established thereby leaving minimal amount of spent diet (2) spent diet is water soluble, and can be simply rinsed off with a water gun (3) liquid-diet reared fruit fly larvae can develop in the same humidity/temperature controlled room, without the necessity of moving trays around thus providing more space for rearing (4) flies can be reared in smaller and shallower trays than those currently used for solid diet with each tray generating an equal amount of pupal production (5) lastly the most convenient inert bulking agent in the liquid which is the sponge cloth is mainly composed of natural cellulose and fiber. The sponge cloth is light weight with high water absorbance, and reusable, recyclable, biodegradable and also environmentally friendly. We evaluated the performance of our four target fruit fly species on liquid diet in comparison to the conventional solid based diet.

Liquid diet ingredients used were similar in composition to that described by Chang et al., (2006) for *Bactrocera dorsalis* (Hendel) larval rearing (Table 1). Each diet mixture was formulated by weighing all the ingredients and blending for 5 min in 1 litre electronic blender with the appropriate amount of water to ensure a homogenous mix of all the components. In the case of the liquid diet, plastic containers (3.2 x 13.5 x 20.6 cm) were rinsed with sterile distilled water and a piece of garden net (15 x 10 cm) was placed inside the plastic container. Sponge cloth (4 x 6 cm), the primary support matrix for larval feeding was rinsed 3 times with sterile distilled water, squeezed dry and placed above the garden net. The liquid diet mixture (150 ml) was then poured over the sponge cloth. In the case of the control solid diet, 500 g of diet was scooped from the blender and transferred into plastic trays of similar dimension as above. The initial pH at the end of the mixing and transfer of both diets was at 3.5.

Eggs (0.5 ml, ~ 6000 eggs in the case of *B. invadens* and 8000 eggs in the case of *Ceratitis* species) collected within 1 hr after oviposition were seeded using a 1 ml transfer pipette onto a 5 x 5 cm strip of moist black cloth. The cloth with the egg was placed in a tray on the center top of either a sponge cloth in liquid diet or the 500 g carrot-based diet. Upon hatching the subsequent larval and pupal handling was the same as described in section 9 and quality control parameters measured as above.

Parameter	Carrot-based solid	Liquid-based
Bactrocera invadens		
Pupal recovery (%)	52.2 ± 2.1b	60.4 ± 7.3a
Pupal weight (mg)	12.8 ± 1.3b	13.8 ± 1.1a
Adult emergence (%)	76.5 ± 1.4b	92.4 ± 2.4a
Flier (%)	80.2 ± 2.5a	82.2 ± 1.2a
Fecundity, 10-day	209.8 ± 10.1b	214.2 ± 15.0a
F_1 egg hatch (%)	65.4 ± 1.4a	72.1 ± 4.1a
Ceratitis fasciventris		
Pupal recovery (%)	52.5 ± 5.1a	28.3 ± 2.4b
Pupal weight (mg)	7.3 ± 0.2a	6.3 ± 0.7b
Adult emergence (%)	70.8 ± 2.1a	70.2 ± 1.2a
Flier (%)	76.3 ± 2.4a	78.2 ± 2.1a
Fecundity, 10-day	98.5 ± 11.1a	65.8 ± 11.2b
F_1 egg hatch (%)	64.2 ± 1.6a	65.2 ± 2.2a
Ceratitis rosa		
Pupal recovery (%)	56.0 ± 2.8a	38.2 ± 1.2b
Pupal weight (mg)	8.5 ± 0.7a	6.8 ± 0.6b
Adult emergence (%)	72.2 ± 2.2a	65.6 ± 1.4b
Flier (%)	70.5 ± 1.3a	62.2 ± 1.4b
Fecundity, 10-day	95.6 ± 22.1a	58.4 ± 14.6b
F_1 egg hatch (%)	60.2 ± 1.4a	55.3 ± 1.8b
*Ceratitsi cosyra**		
Pupal recovery (%)	41.8 ± 1.2a	11.4 ± 1.4b
Pupal weight (mg)	11.7 ± 1.3a	5.5 ± 0.8b
Adult emergence (%)	76.4 ± 1.2a	56.5 ± 1.0b
Flier (%)	78.2 ± 1.4a	58.8 ± 2.6b
Fecundity, 10-day	142.2 ± 14.8a	77.9 ± 11.8b
F_1 egg hatch (%)	72.2 ± 1.6a	51.6 ± 1.4b

Table 5. Effect of solid and liquid artificial larval diets on pupal recovery, pupal weight, adult emergence, flier, fecundity, and egg hatch of *Bactrocera invadens, Ceratitis fasciventris* and *Ceratitis rosa.* Means within a row followed by the same letter do not differ significantly by *t* test (*P*=0.05). *Supplemented with 6% mango

Results showed that *B. invadens* reared on liquid diet had higher percent pupal recovery, heavier pupal weight, greater percent adult emergence and they were more fecund than those reared on solid carrot diet (Table 5). However, egg fertility and flight ability were not affected by diet. In the *Ceratitis* species in general, significant differences were observed between the liquid diet and the standard carrot-based solid diet especially with respect to pupal recovery, pupal weight and fecundity. *Ceratitis fasciventris* had a lower percent pupal recovery, lighter pupal weight, and lower fecundity when reared on liquid diet (Table 5). However, adult emergence, egg fertility and flight ability were not affected by diet type. In the other *Ceratitis* species, quality control parameters in the solid based diet were all superior to the liquid diet (Table 5). These results clearly indicate that the liquid diet can support the development of these insects but the insect probably require a longer period of adaptation to the diet to achieve the quality control parameters obtained from the solid diet. The rate of adaptation in laboratory rearing of insect depends on both the insect's ability to utilize the nutrients in the diet and also the quality of the rearing effort (Ekesi et al., 2007). The nutritional content of the liquid diet is quite high and has been found to be suitable for the development of other *Ceratitis* species such as *C. capitata* (Chang et al., 2007). Since fruit fly adaptation to artificial diet can vary with species (Souza et al., 1988; Tsisipis, 1983; Kamikado et al., 1987), it is likely that the *Ceratitis* species require a prolong period of adaptation to the liquid diet when compared with *B. invadens*.

12. Conclusions

Because artificially reared fruit flies used for research and management purposes must possess qualities and exhibit behaviours as close to that of wild insects as much as possible, quality control of mass reared insects is continuing to gain importance in mass rearing research. We have found that several oviposition stimulants are important for increasing egg deposition. Bottle egg collection system that have been used for producing related *Bactrocera* and *Ceratitis* species was identified as suitable for egg collection of all target fruit fly species. Using *B. invadens* as model insect to study adaptation of fruit flies to artificial diet, it was demonstrated that the process of adaptation for this insect, when moved from whole mango fruit rearing to artificial diet based on wheat-bran, took between three to five generations to reach the plateau of quality control parameters observed for rearing the insect on whole mango fruits. By comparing several solid-based artificial diets, it is concluded that diet based on carrot is the most suitable for mass rearing of all the target species based primarily on important quality control parameters (pupal recovery, pupal weight, adult emergence, flight ability and reproductive parameters), year-round availability of the bulking compound, absence of chemical impurities in carrot and low cost of other ingredients in the diet. The procedure allows for low-cost mass production system for all the target fruit fly species whereby 1135-3012 insects are produced per week for research and management purposes. *Bactrocera invadens* was also found to lend itself to mass rearing in liquid diet. Although the native *Ceratitis* species were able to successfully develop in the liquid diet, quality control parameters were consistently lower for all the species on liquid diet compared to the carrot-based solid diet. Long-term rearing on the diet may improve quality control parameters as the insect adapt to the liquid diet but additional research is required.

13. Acknowledgement

We thank John Kiilu, Peterson Nderitu and Martin Wanyonyi for assistance in the rearing procedures. We are also grateful to the International Atomic Energy Agency (IAEA), Vienna, Austria and the German Ministry of Economic Cooperation and Development for grant support for colony initiation and maintenance, and to Dr. N.K. Maniania for comments on an earlier draft of the manuscript. We are also grateful to the Entomological Society of America (ESA) for permission to reproduce published information.

14. References

Allwood, A. J. & Drew, R. A. I. (1997). *Management of Fruit Flies in the Pacific, ACIAR Proceedings No. 76, A regional Symposium,* Nadi, Fiji.

Allwood, A.J. (1997) Control strategies for fruit flies (Family Tephritidae) in the South Pacific, *ACIAR Proceedings No. 76: Management of Fruit Flies in the Pacific, A regional Symposium,* Allwood, A. J. and Drew, R. A. I. (Eds.), Nadi, Fiji, p. 171-178.

Armstrong J. W. and Jang E.B. (1997). An overview of present and future fruit fly research in Hawaii and the US mainland, *ACIAR Proceedings, No. 76, Management of Fruit Flies in the Pacific* (Edited by A. J. Allwood and R.A.I. Drew), pp. 30-42.

Bass, E.J. (1988) Wheat flour milling, pp. 1-68. In Y. Pomeranz (Ed.), Wheat: chemistry and technology, vol.2. American Association of Cereal Chemists, St. Paul, MN

Barnes B. (Ed.) (2004) *Proceedings of the 6th International Symposium on Fruit Flies of Economic Importance.* Ultra Litho (Pty), Johannesburg, 510 pp.

Boller, E. F., Katsoyannos, B. I. Remund, U. & Chambers, D. L. (1981). Measuring, monitoring, and improving the quality of mass-reared Mediterranean fruit flies, *Ceratitis capitata* (Wied).1. The RAPID quality control system for early warning. *Zeitschrift für Angewandte Entomologie,* 92: 67-83.

Boller, E.F. & Chambers, D.L. (1977). Concepts and approaches, *in* Boller, E.F. and. Chambers, D.L (Ed.), *Quality Control: An Idea Book for Fruit Fly Workers.* International Organization for Biological Control of Noxious Animals and Plants Bulletin SROP-WPRS, pp. 4-13.

Calkins, C.O. (1989). Quality control, *in*: Robinson, A.S. and Hooper, G. (ed.), *Fruit Flies: their Biology, Natural Enemies and Control,* Elsevier, Amsterdam, 3B:153-165.

Calkins, C. O., Bloem, K., Bloem, S. &. Chambers, D. L. (1994). Advances in measuring quality and assuring good field performance in mass reared fruit flies, *in* Calkins, C.O., Klassen W and. Liedo, P (ed.), *Fruit flies and the Sterile Insect Technique,* CRC Press, London, pp.85-96.

Chang C.L. (2009). Fruit fly liquid larval diet technology transfer and update, *Journal of Applied Entomology* 133, 164-173.

Chang, C. L., Caceres, C &. Jang, E. B. (2004). A novel liquid larval diet and its rearing system for melon fly, *Bactrocera cucurbitae* (Diptera: Tephritidae), *Annals of Entomological Society of America* 97: 524-528.

Chang, C. L., Vargas, R.I., Caceres, C., Jang, E. & Cho, I. K. (2006). Development and assessment of a liquid diet for *Bactrocera dorsalis* (Diptera: Tephritidae), *Annals of Entomological Society of America* 99: 1191-1198.

Chang, C.L, Caceres, C. & Ekesi, S. (2007). Life history parameters of *Ceratitis capitata* (Diptera: Tephritidae) reared on liquid diets, *Annals of Entomological Society of America* 100, 900-906.

Churchill-Stanland, C., Standland, R., Wong, T.T., Tanaka, N., McInnis, D.O. & Dowell, R.V. (1986) Size as a factor in the mating propensity of Mediterranean fruit flies, *Ceratitis capitata* (Diptera: Tephritidae), in the laboratory. *Journal of Economic Entomology* 79: 614-619.

Debouzie, D. (1978) Varibilité à lintérieur d'une population de *Ceratitis capitata* élevée sur un milieu naturel non renouvelé (Diptera: Trypidae), Annales De Zoologie Ecologie Animale. 10: 515-524.

Economopoulos, A. P. (1992). Adaptation of the Mediterranean fruit fly (Diptera: Tephritidae) to artificial diet, *Journal of Economic Entomology* 85: 753-758.

Ekesi, S., Nderitu, P.W. & Rwomushana, I (2006) Field infestation, life history and demographic parameters of the fruit fly *Bactrocera invadens* (Diptera: Tephritidae) in Africa. *Bulletin of Entomological Research* 96: 379-386

Ekesi, S., & Billah, M.K. (2007). *A field guide to the management of economically important tephritid fruit flies in Africa*. ICIPE Science Press, Nairobi, Kenya.

Ekesi, S.; Nderitu, P. W.; & Chang, C. L. (2007). Adaptation to and Small-Scale Rearing of Invasive Fruit Fly *Bactrocera invadens* (Diptera: Tephritidae) on Artificial Diet, *Annals of the Entomological Society of America*, 100: 562-567.

Fay, H.A.C. (1989). Multi-host species of Fruit Fly, *in* Robinson, A.S. and Hooper, G. (ed.), *Fruit Flies: their Biology, Natural Enemies and Control*, Elsevier, Amsterdam, 3B:129-140

[FAO/IAEA/USDA] Food and Agriculture Organization of the United Nations/International Atomic Energy Agency/United States Department of Agriculture (2003) FAO/IAEA/USDA manual for product quality control and shipping procedures for sterile mass-reared tephritid fruit flies. Version 5.0. IAEA, Vienna, Austria. http://www.iaea.org/programmes/nafa/d4/index.html

Hooper, G.H.S. (1978). Effects of larval rearing temperature on the development of the Mediterannean fruit fly, *Ceratitis capitata, Entomologia Experimentalis et Applicata* 23: 222-226

Kamikado, T., Chisaki, N., Kamiwada, H. &. Tanaka, A. (1987). Mass rearing of the melon fly, *Dacus cucurbitae* Coquillett, by the sterile insect release method. I. Changes in the amount of eggs laid and longevity of mass-reared insects, *Proceedings No 33 of the Association of Plant Protection of Kyushu, Kagoshima Agriculture*, Experiment Station, Naze, Kagoshima, Japan.

Krainacker, D.A., Carey, J. R & Vargas, R. I. (1987). Effect of larval host on life history traits of the Mediterranean fruit fly, *Ceratitis capitata, Oecologia* (Berlin) 73: 583-590.

Leopold (2007). Colony maintenance and mass-rearing: Using cold storage Technology for extending the shelf life of insect. *in* Vreysen, M.J.B., Robinson, A.S and Hendrichs, J. (ed.), *Area-Wide Control of Insect Pest: From Research to Field Implementation*, Springer, Dordrecht, The Netherlands, pp. 149-162.

Leppla, N. C. (1989). Laboratory colonization of fruit flies, *in* Robinson, A.S. and Hooper, G. (ed.), *Fruit Flies: their Biology, Natural Enemies and Control*, Elsevier, Amsterdam, 3B:91-104.

Leppla, N. C., Huettel, M. D., Chambers, D. L., Ashley, T. R., Miyashita, D. H., Wong, T.T.Y. & Harris. E. J. (1983). Strategies for colonization and maintenance of the Mediterranean fruit fly. *Entomologia Experimentalis et Applicata*. 33: 89-96

Leppla, N.C., & Ashley, T.R. (1989). Quality control in insect mass production: A review and model. *Bulletin of the Entomological Society of America* Winter 1989:33-44

Loukas, M., Economopoulos, A.P., Zouros, E., & Vergini (1985) Genetic changes in artificially reared colonies of the olive fruit fly, *Annals of Entomological Society of America*. 78: 159-165.

Lux, S. A., Copeland, R.S., White, I.M., Manrakhan, A. & Billah, M. K. (2003). A new invasive fruit fly species from the *Bactrocera dorsalis* (Hendel) group detected in East Africa, *Insect Science and its Application*, 32: 355-361.

Mackauer, M.(1976). Genetic problems in the production of biological control agents, *Annual Review of Entomology*, 21: 369-385

Mau, R.F.L., Jang, E.B. & Vargas, R.I. (2007). The Hawaii Area-Wide fruit fly pest management Programme: Influence of partnership and good education programme, *in* Vreysen, M.J.B., Robinson, A.S and Hendrichs, J. (ed.), *Area-Wide Control of Insect Pest: From Research to Field Implementation*, Springer, Dordrecht, The Netherlands, PP. 149-162

Minno, M.C. & Holler, T.C. (1991) Procedures manual for mass rearing the Caribbean fruit fly, Anastrepha suspense (loew) (Diptera: Tephritidae), Sterile Fly Laboratory, Florida department of Agriculture and Consumer Services, Division of Plant Industry, Gainville, FL.

Mwatawala, M.W.,. De Meyer, M,. Makundi, R.H &. Maerere. A.P. (2006). Seasonality and host utilization of the invasive fruit fly *Bactrocera invadens* (Dipt., Tephritidae) in Central Tanzania, *Journal of Applied Entomology* 130: 530-537.

Nadel, D.J. (1970) Current mass rearing techniques for the Mediterranean fruit fly, *Proceedings of a panel on sterile-male technique for control of fruit flies*, IAEA, Vienna, pp. 13-18

Ole-MoiYoi, O.K. and Lux, S.A. (2004) Fruit flies in sub-Saharan Africa: a long-neglected problem devastating local fruit production and a threat to horticulture beyond Africa, *Proceedings of the 6th International Symposium on Fruit Flies of Economic Importance* Barnes B. (Ed.), .Ultra Litho (Pty), Johannesburg, South Africa, pp. 5-10.

Orth, R.A. & Shellenberger, J.A. (1988) Origin production and utilization of wheat, pp. 1-4. In Y. Pomeranz (Ed). Wheat: chemistry and technology, vol.1. American Association of Cereal Chemists, St. Paul, MN

Rossler, Y. (1975). Reproductive differences between laboratory-reared and field-collected populations of the Mediterranean fruit fly *Ceratitis capitata*, *Journal of Economic Entomology* 68: 987-991.

Souza, H. M. L. de, Matioli, S. R. &. Souza, W. N. de. (1988). The adaptation process of *Ceratitis capitata* to the laboratory analysis of life history traits, *Entomologia Experimentalis et Applicata* 49: 195-201

Steiner, L.F. and Mitchell, S. (1966) Tephritid fruit flies, *in* Smith, C.N. (Ed), *Insect colonization and mass production*, Academic Press, New York, USA pp. 555-583

Stevens, L. (1991) Manual of standard operating procedures for the mass-rearing and sterilization of the Mexican fruit fly, Anastrepha ludens (Loew). USDA-APHIS, South Central Region, Mission, TX.

Tanaka N. (1965) Artificial egg receptacles for three species of tephritid flies, *Journal of Economic Entomology* 58: 177-178

Tanaka, N., Steiner, F., Ohinata, K., & Okamoto, R. (1965). Low-cost larval rearing medium for mass production of oriental and Mediterranean fruit flies, *Journal of Economic Entomology* 62: 967-968.

Tanaka, N., Okamoto, R & Chambers, D. L. (1969). Low costs larval rearing medium for mass-production of oriental and Mediterranean fruit flies, *Journal of Economic Entomology* 62: 967-968.

Tanaka, N., Okamoto, R. & Chambers, D.L. (1970). Methods for mass rearing of the Mediterranean fruit fly currently used by the U.S. Department of Agriculture, *Proceedings of a panel on sterile-male technique for control of fruit flies*, IAEA, Vienna, pp. 19-23

Tsitsipis, J. A. (1983). Changes of a wild ecotype of the olive fruit fly during adaptation to lab rearing. *Proceedings of the 1st International Symposium on Fruit Flies of Economic Importance*, R. Cvalloro (ed.), Athens, Greece, pp. 416-422.

Tsitsipis, J.A. (1989). Nutrition, *in* Robinson, A.S. and Hooper, G. (ed.), *Fruit Flies: their Biology, Natural Enemies and Control*, Elsevier, Amsterdam, 3A:103-119

Vargas, R.I., Chang, H., & Williamson, D.L. (1983). Evaluation of sugarcane bagasse larval diet for mass production of the Mediterranean fruit fly (Diptera Tephriridae) in Hawaii, *Journal of Economic Entomology* 76:1360-1362.

Vargas, R.I. (1984) Alternative egg collection system for mass production of Mediterranean fruit fly (Diptera: Tephritidae), *Journal of Economic Entomology* 77: 1064-1069

Vargas, R. I., &. Mitchell, S. (1987). Two artificial larval diets for rearing *Dacus latifrons* (Diptera: Tephritidae), *Journal of Economic Entomology*, 80: 1337-1339.

Vargas, R.I. (1989). Mass production of fruit flies, *in* Robinson, A.S. and Hooper, G. (ed.), *Fruit Flies: their Biology, Natural Enemies and Control*, Elsevier, Amsterdam, 3B:141-151

Vargas, R. I., &. Carey, J. R. (1989). Comparison of demographic parameters for wild and laboratory-adapted Mediterranean fruit fly (Diptera: Tephritidae), *Annals of Entomological Society of America*. 82: 55-59.

Vargas, R. I.,. Mitchell, S, Hsu, C-L, &. Walsh, W. (1993). Evaluation of mass-rearing procedure for *Bactrocera latifrons* (Diptera: Tephritidae), *Journal of Economic Entomology* 86: 1157-1161.

Vargas, R.I., Mitchell, S., Fujita, B. & Albrecht, C. (1990) Rearing techniques for *Dacus latifrons* (Diptera: Tephritidae), *Proceedings of Hawaii Entomological Society*, 30: 61-66.

Vargas, R. I., S. Mitchell, C-L. Hsu, & W. A. Walsh. (1994). Laboratory evaluation of diets of processed corncob, torula yeast, and wheat germ on four developmental stages of Mediterranean fruit fly (Diptera: Tephritidae), *Journal of Economic Entomology* 87: 91-95.

Walker, G.P., Tora Vueti, E., Hamacek, E.L. & Allwood, A.J. (1997). Laboratory-rearing techniques for Tephritid fruit flies in the South Pacific.

White, I. M.& Elson-Harris, M. M.. (1992). *Fruit Flies of Economic Importance: Their identification and bionomics*, CAB International, Wallingford, United Kingdom.

Zumreoğlu, A., Tanaka, N. & Harris, E. J. (1979). The need for wheat germ in larval diets of the Mediterranean fruit fly (Diptera: Trypetidae) of non-nutritive bulking material, *Turkish Journal of Entomology*, 3: 131-138.

22

Quality Control of Baculoviral Bioinsecticide Production

Solange Ana Belén Miele, Mariano Nicolás Belaich, Matías Javier Garavaglia and Pablo Daniel Ghiringhelli
LIGBCM-AVI (Laboratorio de Ingeniería Genética y Biología Celular y Molecular - Area Virosis de Insectos)
Universidad Nacional de Quilmes/Departamento de Ciencia y Tecnología
Argentina

1. Introduction

Agriculture is a discipline that has accompanied human beings since the beginning of civilization. The cultivation of different vegetables for centuries has allowed selecting varieties that far exceed the capabilities of many wild type plants originally used as a food source. That situation derived in the manipulation of natural ecosystems, transforming them into spaces where they can only grow and develop the desired species.

In our world, plants are the staple diet of many organisms including invertebrates like Lepidoptera. During the larval stage, these insects can consume a large amount of leaf tissue causing serious damage to the plant. If we think that most vegetables have insect predators, agricultural crops can be transformed into an inviting habitat, allowing the development of these animals. In conclusion, all crops have pests that threaten their productivity. Given this scenario, many pest control strategies have been used by human beings to protect the health of their crops: treatment with chemical insecticides, development of transgenic plants and biological control applications (Christou *et al*, 2006; Gilligan, 2008).

Baculovirus is a large family of insect pathogens that infect and kill different species of Lepidoptera, Hymenoptera and Diptera (Theilmann *et al*, 2005). In particular, many lepidopteron are pests in agriculture transforming these viruses in an important biocontrol tools for their natural hosts (Entwistle, 1998; Moscardi, 1999; Szewczyk *et al*, 2006). Baculoviruses have double-stranded circular DNA genomes of 80,000-180,000 bp, containing between 80 to 180 genes depending on the specie (van Oers & Vlak, 2007; Miele *et al*, 2011). In early stages of virus cycle, this pathogen is produced as Budded Viruses (BVs): the genome contained in a protein capsid (nucleocapsid), which is surrounded by a lipid membrane. In change, in the last phase of multiplication processes appear the Occluded Bodies (OBs): protein crystals (forming polyhedra or granules) containing nucleocapsids wrapped by a lipid membrane with a different composition (ODVs or Occluded Derived Viruses, with single or multiple nucleocapsids depending on the specie) (Rohrman, 2008). These two virus phenotypes have different biological properties; while OBs are specialists (infecting larvae by *per os* route with a narrow host range; responsible of primary infection in midgut cells), the BVs are generalists (infecting a wide range of different insect cells triggering their death; responsible for secondary infection). In the pest control strategies, baculoviruses (OBs) are introduced on the crops to infect and kill larvae through the production of an epizooty.

Genus	Name	Code	Accesion number	Genome (bp)	Total ORFs
Alphabaculovirus - Group I	*Antheraea pernyi* NPV-Z	APN	NC_008035	126629	145
	Antheraea pernyi NPV-L2	AP2	EF207986	126246	144
	Anticarsia gemmatalis MNPV-2D	AGN	NC_008520	132239	152
	Autographa californica MNPV-C6	ACN	NC_001623	133894	154
	Bombyx mori NPV	BMN	NC_001962	128413	137
	Bombyx mandarina NPV	BON	NC_012672	126770	141
	Choristoneura fumiferana DEF MNPV	CDN	NC_005137	131160	149
	Choristoneura fumiferana MNPV	CFN	NC_004778	129593	145
	Epiphyas postvittana NPV	EPN	NC_003083	118584	136
	Hyphantria cunea NPV	HCN	NC_007767	132959	148
	Maruca vitrata MNPV	MVN	NC_008725	111953	126
	Orgyia pseudotsugata MNPV	OPN	NC_001875	131995	152
	Plutella xylostella MNPV	PXN	NC_008349	134417	149
	Rachiplusia ou MNPV	RON	NC_004323	131526	146
Alphabaculovirus - Group II	*Adoxophyes honmai* NPV	AHN	NC_004690	113220	125
	Adoxophyes orana NPV	AON	NC_011423	111724	121
	Agrotis ipsilon NPV	AIN	NC_011345	155122	163
	Agrotis segetum NPV	ASN	NC_007921	147544	153
	Apocheima cinerarium NPV	APO	FJ914221	123876	118
	Chrysodeixis chalcites NPV	CCN	NC_007151	149622	151
	Clanis bilineata NPV	CBN	NC_008293	135454	129
	Ectropis obliqua NPV	EON	NC_008586	131204	126
	Euproctis pseudoconspersa NPV	EUN	NC_012639	141291	139
	Helicoverpa armigera NPV-C1	HA1	NC_003094	130759	135
	Helicoverpa armigera NPV-G4	HA4	NC_002654	131405	135
	Helicoverpa armigera MNPV	HAN	NC_011615	154196	162
	Helicoverpa armigera SNPV-NNg1	HAS	NC_011354	132425	143
	Helicoverpa zea SNPV	HZN	NC_003349	130869	139
	Leucania separata NPV-AH1	LSN	NC_008348	168041	169
	Lymantria dispar MNPV	LDN	NC_001973	161046	163
	Lymantria xylina MNPV	LXN	NC_013953	156344	157
	Mamestra configurata NPV-90-2	MCN	NC_003529	155060	169
	Mamestra configurata NPV-90-4	MC4	AF539999	153656	168
	Mamestra configurata NPV-B	MCB	NC_004117	158482	169
	Orgyia leucostigma NPV	OLN	NC_010276	156179	135
	Spodoptera exigua MNPV	SEN	NC_002169	135611	142
	Spodoptera frugiperda MNPV-3AP2	SF2	NC_009011	131330	143
	Spodoptera frugiperda MNPV-19	SF9	EU258200	132565	141
	Spodoptera litura NPV-II	SLN	NC_011616	148634	147
	Spodoptera litura NPV-G2	SL2	NC_003102	139342	141
	Trichoplusia ni SNPV	TNN	NC_007383	134394	144
Betabaculovirus	*Adoxophyes orana* GV	AOG	NC_005038	99657	119
	Agrotis segetum GV	ASG	NC_005839	131680	132
	Choristoneura occidentalis GV	COG	NC_008168	104710	116
	Cryptophlebia leucotreta GV	CLG	NC_005068	110907	129
	Cydia pomonella GV	CPG	NC_002816	123500	143
	Helicoverpa armigera GV	HAG	NC_010240	169794	179
	Phthorimea operculella GV	POG	NC_004062	119217	130
	Plutella xylostella GV	PXG	NC_002593	100999	120
	Pieris rapae GV	PRG	GQ884143	108592	120
	Pseudaletia unipuncta GV-Hawaiin	PUG	EU678671	176677	183
	Spodoptera litura GV-K1	SLG	NC_009503	124121	136
	Xestia c-nigrum GV	XCG	NC_002331	178733	181
Gamma	*Neodiprion abietis* NPV	NAN	NC_008252	84264	93
	Neodiprion lecontei NPV	NLN	NC_005906	81755	93
	Neodiprion sertifer NPV	NSN	NC_005905	86462	90
Delta	*Culex nigripalpus* NPV	CNN	NC_003084	108252	109

Table 1. Baculovirus complete genomes. Baculoviruses used in this study, sorted by genus (and within them by alphabetical order). MNPV is the abbreviation of multicapsid nucleopolyhedrovirus; NPV is the abbreviation of nucleopolyhedrovirus; SNPV is the abbreviation of single nucleopolyhedrovirus; GV is the abbreviation of granulovirus. The accession numbers are from National Center for Biotechnology Information (NCBI, http://www.ncbi.nlm.nih.gov) and correspond to the sequences of complete genomes. Code is an acronym used for practicity

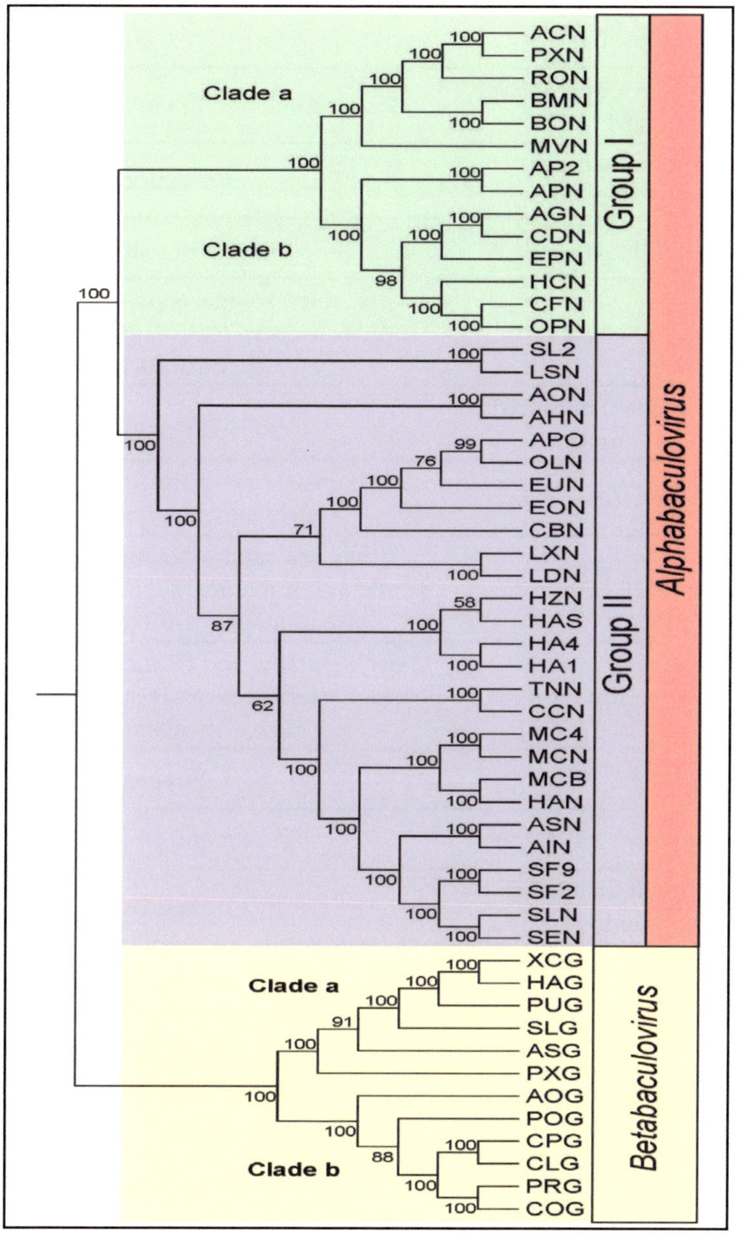

Fig. 1. Lepidopteron Baculovirus genome phylogeny. Cladogram based on amino acid sequence of 31 core genes. Core genes from Lepidopteron Baculoviridae family were independently aligned using MEGA 4 (GOP = 10, GEP = 1 and Dayhoff Matrix. Then, a concatemer was generated and phylogeny inferred using the same software [UPGMA; Bootstrap with 1000 replicates; gap/Missing data = complete deletion; Model = Amino (Dayhoff Matrix); patterns among sites = Same; rates among sites = Different (Gamma Distributed); gamma parameter = 2.25]. Baculoviruses are identified by the acronyms given in Table 1 and distribution in lineages and genera are also indicated. Clades proposed for Betabaculoviruses are shown in bold letters (Miele *et al*, 2011)

Virus code	Host (larvae)	Pest of…
ACN	Alfalfa looper, broad host range	Alfalfa and many other crops
AGN	Velvetbean caterpillar	Soybean crops
AHN	Smaller tea tortrix	Tea plants
AIN	Black cutworm	Vegetables, solanaceous, cucurbitaceous and industrial crops (cotton, essential-oil cultures, maize, tobacco, sunflower)
AOG	Summer fruit tortrix moth	Apples and pears
AON	Tea tree tortrix	Apple, pear, rose, plum, cherry, apricot, sweet cherry, currant, gooseberry, etc.
ASG	Black cutworm	Cotton, essential-oil cultures, maize, tobacco, sunflower, tomatoes, sugar beet and potato and also damage seedlings of tree species
ASN	Turnip moth	Many vegetable and field crops (corn, rape, beet, potatoes, cabbage, cereals, tobacco, vine and many others)
CBN	*Clanis bilineata*	Soybean
CCN	*Chrysodeixis chalcites*	Tomato and sweet pepper.
CDN, CFN	Eastern spruce budworm	Conifeorus trees
CLG	False codling moth, other Tortricid	Citrus, cotton, maize
COG	Western spruce budworm	Coniferous trees
CPG	Codling moth	Apples, pear and quince
EON	The tea looper caterpillar	Tea plants
EPN	Light brown apple moth	Apple, horticultural crops
HA1, HAN, HAS, HAG	Old world bollworm	Cotton, corn, baccy, tomato, maize, chick pea, alfalfa, soybean, pea, pumpkin
HCN	Fall webworm	Trees (cherry, plane, mulberry and persimmon)
LDN	Gypsy moth	Hardwoods

Virus code	Host (larvae)	Pest of…
LSN	Eastern armyworm	Many field crops in China
LXN	Casuarina moth	Casuarina, guava, longan, lychee, acacia
MCN, MC4, MCB	Bertha armyworm	Cruciferous oilseed crops in Canada.
MVN	Maruca pod borer	Leguminous crops (pigeon pea, cowpea, mung bean and soybean)
OLN	White-marked tussock moth	Wide variety of trees, deciduous and coniferous
POG	Potato tuber moth	Solanaceous cultures (potato, eggplant, tomato, pepper, and tobacco).
PRG	Small cabbage white	Cabbage, swede, turnip, radish, horseradish, garden radish, watercress, rape, turnip, and other cruciferous plants
PUG	Armyworm	Turfgrasses, small grains, corn, timothy, millet, and some legumes
PXG, PXN	Diamondback moth	Cruciferous crops
RON	Gray looper moth	Herbaceous plants
SEN	Beet armyworm	Asparagus, beans and peas, sugar and table beets, celery, cole crops, lettuce, potato, tomato, cotton, cereals, oilseeds, tobacco, etc.
SF2, SF9	Fall armyworm	Corn and small grain crops
SLN, SL2, SLG	Oriental leafworm moth	Wide range of plants, like cotton and tobacco.
TNN	Cabbage looper	Wide variety of cultivated plants and weeds (broccoli, cabbage, cauliflower, collards, kale, mustard, radish, rutabaga, turnip, snap bean, spinach, squash, sweet potato, tomato, watermelon, etc.)
XCG	Setaceous hebrew character	Huge variety of plants (tomato, tobacco, carrot, lettuce, alfalfa, potato, grape, maize, apple)

Table 2. Baculovirus and pest control. The table contains some Baculoviruses with their insect hosts, revealing their possible application as bioinsecticide

Actually, baculoviruses are classified in four genera according to their biological properties and gene content: *Alphabaculovirus*, polyhedroviruses that infect Lepidoptera (grouped into two lineages, Group I and Group II, according to their phylogenetic relationships and the identity of the fusogenic membrane protein presents in the BVs); *Betabaculovirus*,

granuloviruses that infect Lepidoptera; *Gammabaculovirus*, polyhedroviruses that infect Hymenoptera; and *Deltabaculovirus*, polyhedroviruses that infect Diptera (Table 1) (Jehle *et al*, 2006a).
Genomic sequence is known more than 50 different baculovirus species, being the recognized prototypes of each genus: AcMNPV, CpGV, NeleNPV and CuniNPV, respectively. Many of them have been used for biological pest control, being excellent biopesticides (Figure 1; Table 2).
However, most baculoviruses cannot efficiently compete with chemical insecticides, especially in the time of death. To overcome this problem, many researchers have been focused to introduce genetic modifications in order to accelerate the lethal effects of bioinsecticide or expand their host range. One strategy that has been explored is the introduction of genes encoding insect toxins, such as different neurotoxins from eukaryotic organism or the bacterial protein Cry (Inceoglu *et al*, 2006; Jinn et al, 2006; De Lima *et al*, 2007). Thus, these genetically modified viruses (GMV) would ensure better performance in biopesticide application.
Baculoviruses are produced by infection processes in susceptible larvae or in *in vitro* cell cultures. First approach is appropriate and inexpensive in small-scale, but big productions prefer the use of cell bioreactors(van Beek & Davis, 2007; Micheloud *et al*, 2009; Mengual Gómez *et al*, 2010). This technology would allow the standardization of production processes and achieve bioinsecticides with reproducible quality.
The main difference among these strategies consists in the starters used, being in one case OBs (in larvae) and BVs in the other (*in vitro* cell cultures); but always with the goal of producing OBs (infective phenotype in nature). Although the trend is moving toward baculovirus production in cell cultures, it is important to note some problems associated with that strategy. One of them is the genome stability. Because only the BVs infect cells growing in laboratory conditions, after successive rounds of infection tend to accumulate defective viral variants with smaller genomes (Lee & Krell, 1992). These quasispecies lose genomic segments encoding late proteins important for generating OBs, because there is no selection pressure associated with oral infection in larvae. Other problems are related to the composition of culture media and the availability of susceptible insect cell lines to each baculovirus. Actually, many researchers are working on the establishment of new cell lines or modifying existing ones to improve their performance, while others have focused on developing proper and cheaper formulations of growth media for cell propagation *in vitro* (Agathos, 2007; Micheloud *et al*, 2009).

2. Quality control assays

The production of baculoviruses for use as bioinsecticides required quality control processes to ensure their proper formulation. In either case above (wild type viruses or GMVs) or regardless of production method applied (larvae or *in vitro* cell cultures), is necessary to carry out a series of phenotypic and genotypic tests against which to assess the quality of each batch produced (Figure 2).
The formulation of one biological entity for some biotechnological application (e.g. baculovirus for agriculture pest control) requires its multiplication under controlled conditions and subsequent procedures for isolation and concentration. In this point, it is important to remember that all biological entities are object of evolution, natural phenomenon that can

influence and alter the biological properties of the product by the accumulation of point mutation or genome rearrangements.

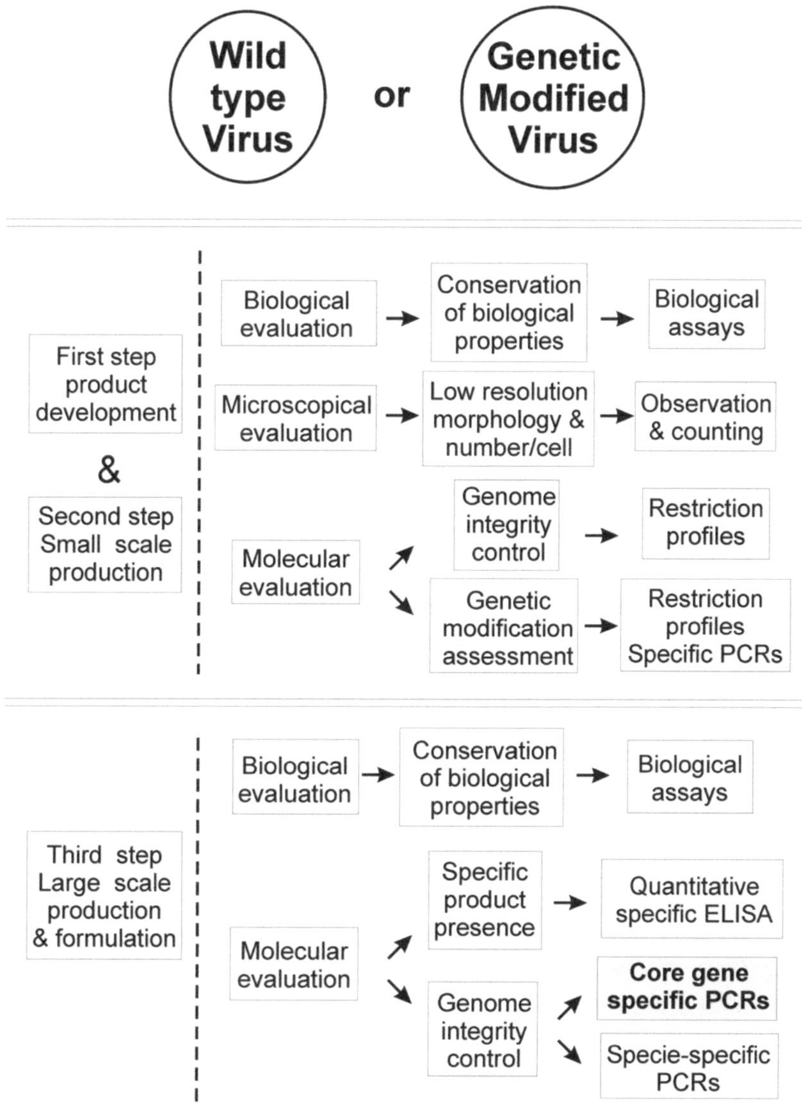

Fig. 2. Quality control scheme. A good quality control strategy is supported in the setting of and in the rigid adhesion to the procedures and protocols. These may include routine examinations of insect/cells stocks, microscopic examinations for infections, routine counting of ODVs, bioassays to assess bioinsecticide potency, restriction profiles of viral DNAs, and so on. First and second steps are developmental phases of the bioinsecticide production, in which the feasibility to obtain high amounts of good quality DNA is not an obstacle. In the third step, is of special importance the availability of sensitive molecular techniques to minimize the interference of formulation components

Thus, quality control assays emerge as central tools for verifying the baculovirus production in each of its stages allowing generating a product that can compete with chemical insecticides, whose production is highly optimized and controlled for years. Also, quality control strategies are useful to standardize the basic studies performed in laboratory scale, necessary for the generation of improved baculovirus.

2.1 Phenotype quality controls

First of all, it is important to have good methods to quantify the number of OBs produced and isolated from larvae or *in vitro* cell cultures. To fulfill this purpose, it is possible to make direct eye count using hemocytometer and optical microscopes. On the other hand, there are methodologies based on immunoassays or carried out by the use of flow cytometers. In the first case, the development of ELISA kits or other similar tests based on the immune detection of OBs (through the use of polyclonal or monoclonal antibodies against polyhedrin or granulin proteins) has standardized the quantitation of baculovirus allowing a more reliable measure (Parola *et al*, 2003). The use of flow cytometers also provides good results, but only so far for the quantification of BVs (Shen *et al*, 2002; Jorio *et al*, 2006).
Once quantified the production of OBs, should determine their biological activity. This involves setting parameters to estimate the ability of baculovirus to kill insect pests and control their population. In view of this, parameters like *median lethal time* (LT 50) and *median lethal dose* (LD 50) work as the best indicators to characterize the baculovirus activity (Li & Bonning, 2007; Lasa *et al*, 2008). These tests consist of exposing susceptible larvae reared in standardized conditions of temperature, light, moisture and food to the virus under evaluation. Then, through the register of deaths and the time in which they occur can be estimated both parameters.

2.2 Genotype quality controls

The production of baculoviruses for use as bioinsecticides requires accurate determination of the number of OBs and their biological activity expressed in LT 50 and LD 50 parameters. But it is also important to apply other methodologies that allow considering genotypic evaluations. As mentioned earlier, the processes of baculovirus production in insect cell lines growing in laboratory conditions may derived in problems with the integrity of their genomes. Consequently, the productivity of OBs can be seriously affected both in quantity and activity ruining the entire production. Of course, this is particularly relevant when dealing with GMVs. The stability of putative transgenes should be considered.
Most of baculoviruses applied as bioinsecticides derived from homogenous populations cloned or partially cloned by different procedures (Wang *et al*, 2003; Simón *et al*, 2004). This is a remarkable aspect since it allows establishing genotypic characteristic patterns that can be detected by different approaches. Among them, the visualization of RFLPs (*Restriction Fragment Lenght Polymorphism*) in agarose gel electrophoresis stained by different dyes and UV exposition is usually a good indicator of genome integrity, revealing the gain or loss of DNA (Simón *et al*, 2004; Eberle *et al*, 2009; Rowley *et al*, 2010). In fact, this is a classic approach to characterize genotypic variants of a viral species. The main problem that has this strategy is related to allocate part of baculovirus production to perform the isolation of viral genome, requiring high DNA masses to achieve reliable results. The complementation with hybridization assays solves part of that problem but requires the availability of suitable probes, adding experimental steps and costs of supplies and equipment.

In view of that, methodologies based on PCR (*Polymerase Chain Reaction*) are suitable and reproducible approaches to assess baculovirus productions because this technique can detect desired *locus* with high sensitivity and specificity. These characteristics transform the PCR in the best genotypic evaluation strategy due to its simple, fast and accessible properties for any laboratory production. Since the beginning of studies on the baculovirus genomes, many researchers have designed PCR tests to detect, identify and classify the different species of this virus Family. Thus, PCR assays based on *polyhedrin/granulin*, *p74*, *lef8*, *lef9* or *DNA polymerase* genes, among others, were used to describe new virus isolates which are candidates to bioinsecticide applications (Faktor *et al*, 1996; de Moraes *et al*, 1999; Wang *et al*, 2000; Rosisnki *et al*, 2002; Espinel-Correal *et al*, 2011; Rodríguez *et al*, 2011).
However, there are too many examples of the use of PCR as a technique for quality control in the production of a baculovirus, despite all the advantages mentioned above (Christian *et al*, 2001; Murillo *et al*, 2006).

2.2.1 MP-PCR to control baculovirus production

PCR amplification of several *loci* in the same reaction allows obtaining a profile of products that can be used for genome identification or control test in production processes. MP-PCR (*Multiplex* PCR) assays require the proper design of primers to amplify a set of fragments that are typical for a particular genome. This technique provides results composed of a set of enzymatic amplified fragments that are characteristic for a viral species (when primers were designed completely specific), or for a phylogenetic group (when primers derived from multiple alignments of orthologous sequences). With regard to trials designed to particular viruses, it should be noted the work developed for EpapGV (Manzán *et al*, 2008). Meanwhile, for the detection of groups of related viruses are not many references. Currently, the accepted practice to identify or preliminarily classify a new baculovirus is based on PCR amplification and subsequent sequencing of three genomic fragments corresponding to the *polyhedrin/granulin*, *lef9* and *lef8* genes (Jehle *et al*, 2006b). However, this approach is not itself an MP-PCR. In view of this, we propose an MP-PCR for *alpha* and *betabaculovirus* quality control based on universal primer designs.
Baculoviruses contain 31 core genes conserved by all known members (Miele *et al*, 2011). These orthologous sequences are present in each sequenced baculovirus, but their genomic distribution varies among species. From the analysis of gene distribution in genus prototypes *pif2*, *p49*, *p74*, *lef9*, *38k* genes were selected to primer design targets (Figure 3).
These sequences are properly distributed throughout the entire circular genome. Two genes (*pif2* and *p74*) encode *per os* infectivity factors essentials to the success of primary infection in midgut cells (Song *et al*, 2008; Peng *et al*, 2010). Other two genes (*p49* and *38 K*) encode proteins associated to packaging, assembly, and release of virions (Wu *et al*, 2008; Lin *et al*, 2010). Meanwhile, *lef9* gene encodes a polypeptide involved in virus transcription machinery (Crouch *et al*, 2007). Using multiple alignments derived from sequences corresponding to *P74*, *lef9* and *38k* genes from all *alpha* and *betabaculovirus* members were selected the two better regions of homology to design a set of primers (Figure 4). Thus, these three amplicons certified the presence of lepidopteron baculovirus DNA.
In change, because high divergence of *pif2* and *p49* sequences the primer design was conducted using multiple alignments derived from closest phylogenetic clades (Group I and Group II *alphabaculovirus*, and *betabaculovirus*). According to this, different pairs of primers were designed to generate amplicons from baculovirus genomes (Figure 5).

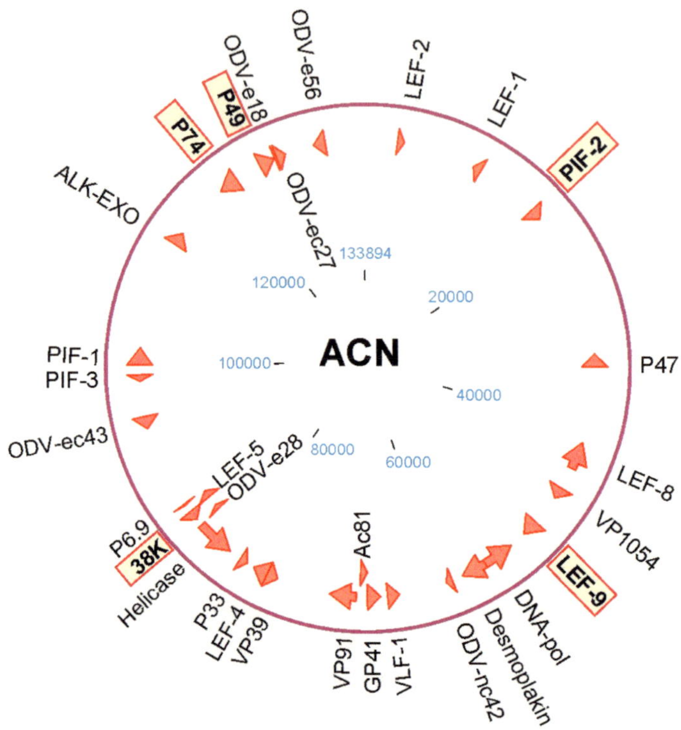
ACN
133894
20000
40000
60000
80000
100000
120000
ODV-e56
ODV-e18
P49
P74
LEF-2
LEF-1
PIF-2
ALK-EXO
ODV-ec27
PIF-1
PIF-3
P47
ODV-ec43
LEF-5
ODV-e28
LEF-8
VP1054
LEF-9
P6.9
38K
Helicase
Ac81
P33
LEF-4
VP39
VP91
GP41
VLF-1
ODV-nc42
Desmoplakin
DNA-pol

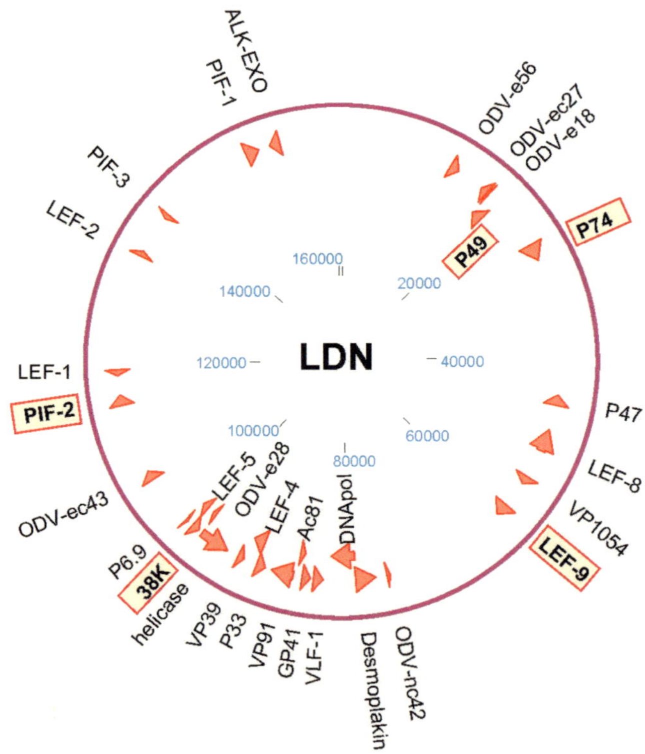
LDN
160000
20000
40000
60000
80000
100000
120000
140000
ALK-EXO
PIF-1
ODV-e56
ODV-ec27
ODV-e18
PIF-3
LEF-2
P49
P74
LEF-1
PIF-2
P47
LEF-8
VP1054
LEF-9
ODV-ec43
LEF-5
ODV-e28
LEF-4
Ac81
DNApol
P6.9
38K
helicase
VP39
P33
VP91
GP41
VLF-1
Desmoplakin
ODV-nc42

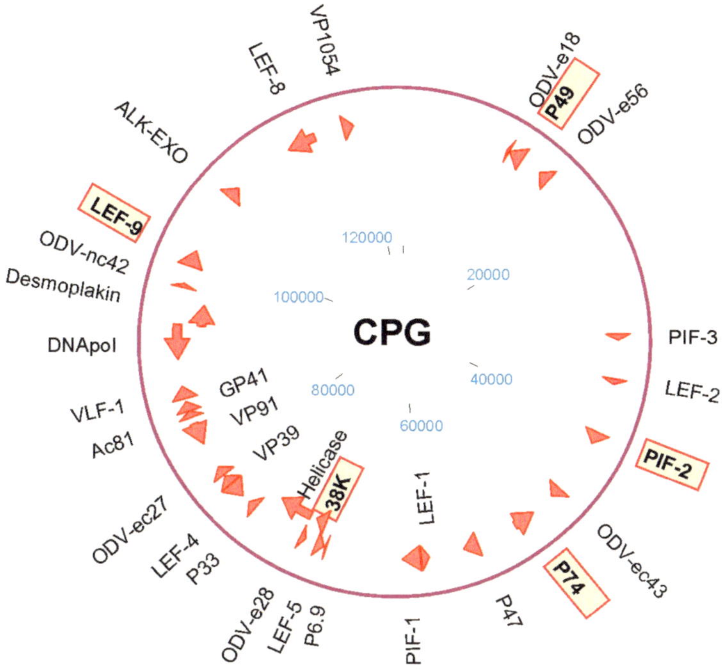

Fig. 3. Physical maps of ACN, LDN and CPG (Arrows shows the physical location of the 31 Core genes. The five selected Core genes for primer designs are highlighted in bold and red boxed.)

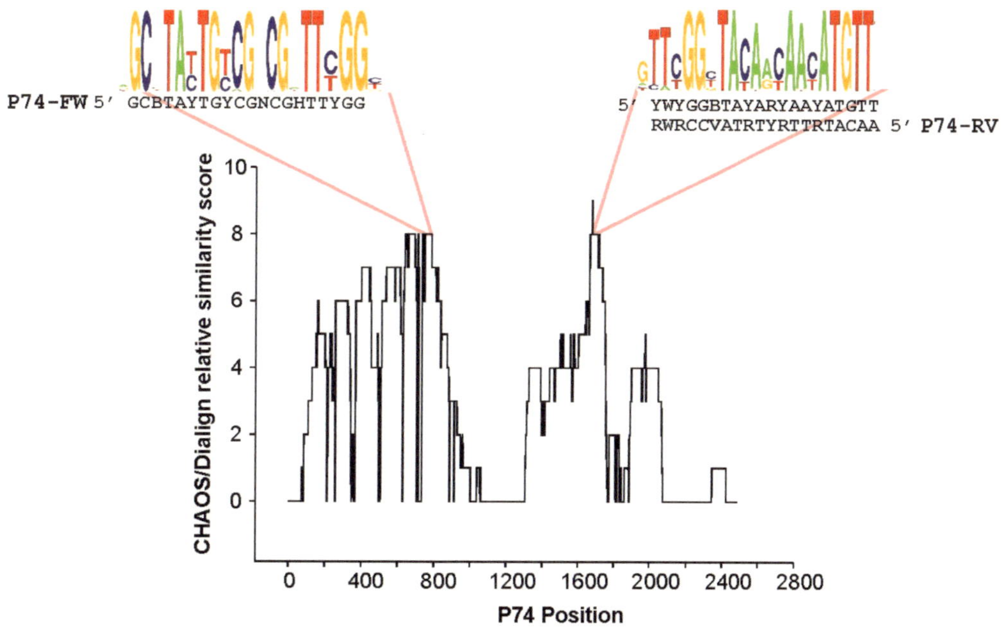

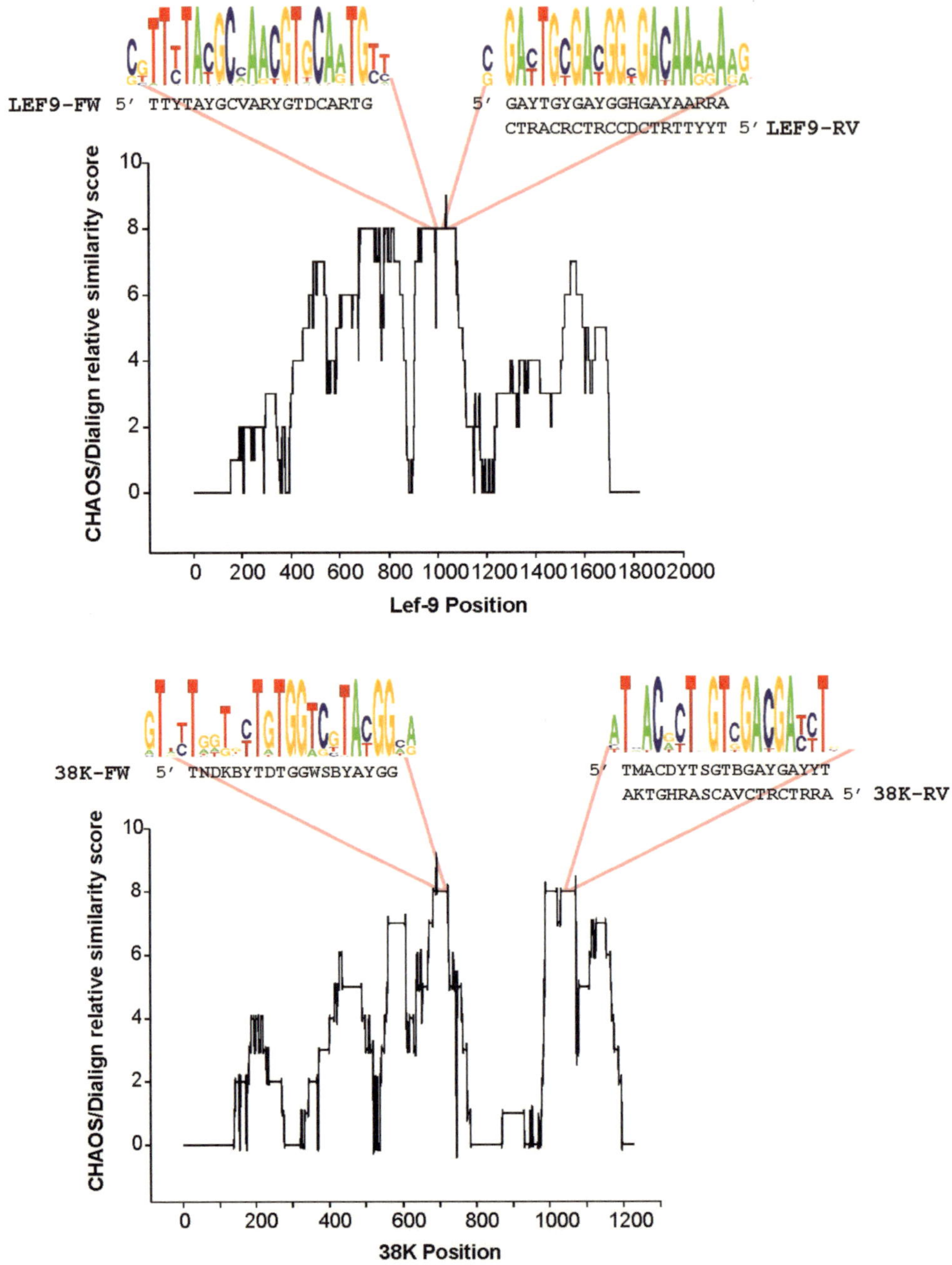

Fig. 4. Primer design for *p74*, *lef-9* and *38K* genes. The orthologous sequences of *p74*, *lef-9* and *38K* genes from Alpha and Betabaculovirus members were aligned by CHAOS/DIALIGN program (Brudno *et al*, 2004). A consensus line in the multiple alignment is a set of numbers (between 0-9) that roughly reflect the degree of local similarity among the sequences. These scores were used to generate plots. The regions with higher relative similarity were selected to design primers. These sequences are showed at the top in Sequence Logos (Crooks *et al*, 2004)

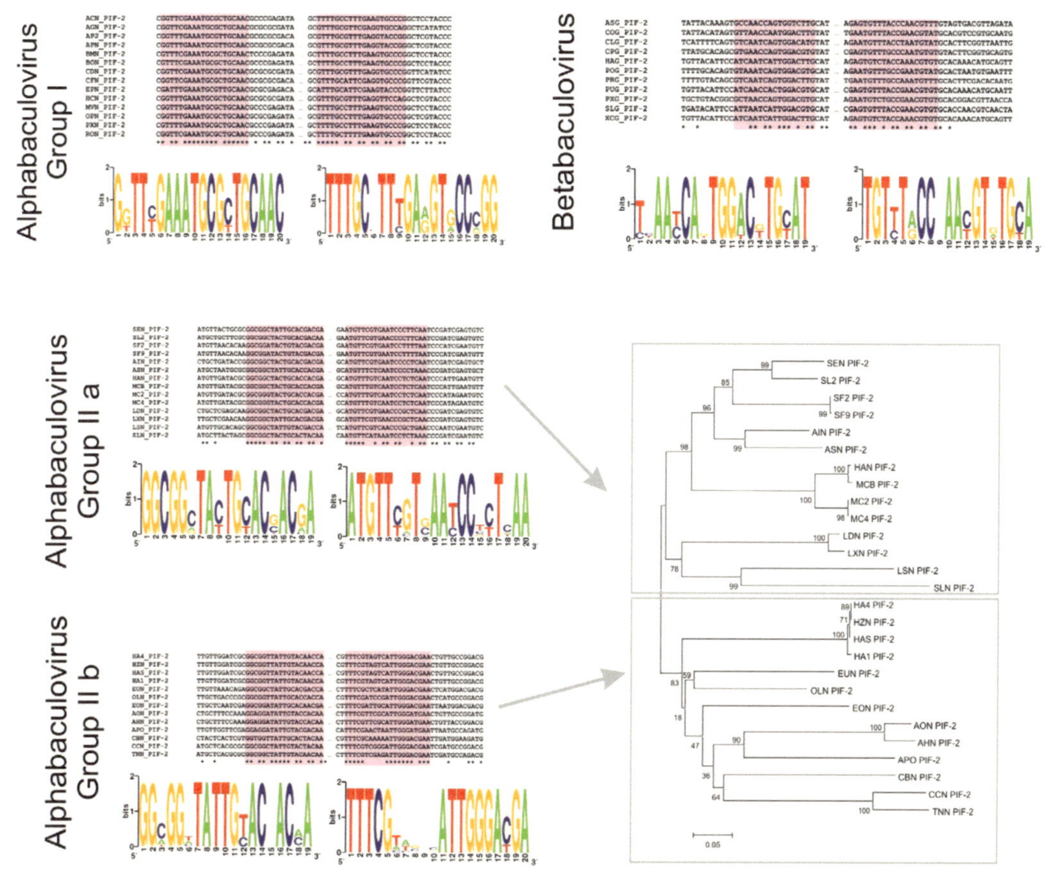

Fig. 5. Primer design for *pif-2* gene. The orthologous sequences of *pif-2* gene from Group I Alphabaculovirus or Group II Alphabaculovirus or Betabaculovirus members were aligned by T-Coffee program (Notredame *et al*, 2000; Poirot *et al*, 2003). The regions with higher similarity were selected to design primers. These sequences are showed at the bottom of each multiple alignment in Sequence Logos (Crooks *et al*, 2004). The cladogram was made with nucleotide sequences of *pif-2* Group II Alphabaculovirus using MEGA 4. It showed a significative grouping in two lineages (Group II a and Group II b), which were considered to design primers. For *p49* sequence analysis a similar approach was conducted (data not shown)

Sets of proposed primers for MP-PCR would allow to detect the proper integrity of genomes in a baculovirus production (Table 3).

Gene	Baculovirus		Primer sequence	Product (bp)		
			5´ to 3´	ACN	LDN	CPG
lef-9	Alpha + Beta	FW	TTYTAYGCVARYGTDCARTG	245	245	247
		REV	TYYTTRTCDCCRTCRCARTC			

Gene	Baculovirus		Primer sequence 5′ to 3′	Product (bp)		
				ACN	LDN	CPG
38K	Alpha + Beta	FW	TNDKBYTDTGGWSBYAYGG	247	260	218
		REV	ARRTCRTCVACSARHGTKA			
pif-2	Alpha Group I	FW	GDTTYGAAATGCGYTGCAAC	382	---	---
		REV	CCBGGHACYTCRAASGCAAA			
	Alpha Group IIa	FW	GGCGGVTAYTGYACBACVA	347	347	---
		REV	TTDARVGGRTTSACRAACAT			
	Alpha Group IIb	FW	GGMGGHTATTGYACNACVA	---	260	---
		REV	TCRTCCCAATBNSDDCGAAA			
	Beta	FW	YYAAYCAKTGGWCDTGYAT	306	306	306
		REV	TRCAHACRTTNGGYARACA			
p74	Alpha + Beta	FW	GCBTAYTGYCGNCGHTTYGG	824	830	935
		REV	AACATRTTRYTRTAVCCRWR			
p49	Alpha GroupI	FW	AGTYTATTTGAYYTRAAARA	1284	---	---
		REV	ACTTTCGTAATCACCTCTTA			
	Alpha GroupII	FW	TAYGCNACNAAYYTKTTYGT	---	970	---
		REV	AATCWCCTCTTATRAWWARAT			
	Beta	FW	CARMGVGAYTAYRTHTWYGA	---	---	596
		REV	AATAARYTYRVWAHVGTRTT			

Table 3. Primer sequences to perform a MP-PCR assay. The table contains all the primer sequences designed by two different approaches and the hypothetical length of amplified fragments using the genome prototypes as reaction template. The specificity of annealing and the size of the amplicons were verified using jPCR (Kalendar *et al*, 2009). FW: forward primer. REV: reverse primer. Ambiguities are indicated in IUPAC code, B=C,G,T; D=A,G,T; H=A,C,T; K=G,T; M=A,C; N=A,C,G,T; R=A,G; S=C,G; V=A,C,G; W=A,T; Y=C,T

3. Conclusion

Integrated control management of agricultural pests requires the combination of different insecticide strategies. Among them, the use of baculovirus is an excellent solution as biological control agent. There are many known members of this viral family, with dozens of sequenced genomes. Some of the limitations that exist in their massive application are given by their time of action and modes for their production. Regarding the latter, quality control methodologies are emerging as essential to ensure proper development and formulation. In view of that, in this work are proposed a series of primers for PCR assays

that would amplify a fragment profile appropriate to certify the genomic integrity and identity of batch production. Furthermore, adding other specific primers (e.g. specific of transgenes) could be confirmed genotypic stability of genetically modified viruses.
Also, the methodology here proposed could be used to characterize new baculoviral isolates, which could be used as bioinsecticides and produced and controlled without the knowledge of their genome sequences.

4. Acknowledgment

This work was supported by research funds from *Agencia Nacional de Promoción Científica y Técnica* (ANPCyT) and *Universidad Nacional de Quilmes*. PDG is member of the Research Career of CONICET (*Consejo Nacional de Ciencia y Tecnología*); MNB holds a postdoctoral fellowship of CONICET, SABM holds a fellowship of CONICET and MJG holds a fellowship of CICBA.

5. References

Agathos, S.N. (2007). *Development of serum-free media for lepidopteran insect cell lines*. Methods Mol Biol. Vol 388, pp. 155-186. ISSN: 1064-3745.

Brudno, M.; Steinkamp, R.; Morgenstern, B. (2004). *The CHAOS/DIALIGN WWW server for multiple alignment of genomic sequences*. Nucleic Acids Res., 2004 Jul 1, vol. 32, pp. 41-44. ISSN 0305-1048.

Christian, P.D.; Gibb, N.; Kasprzak, A.B.; Richards, A. (2001). *A rapid method for the identification and differentiation of Helicoverpa nucleopolyhedroviruses (NPV Baculoviridae) isolated from the environment*. J Virol Methods, 2001 Jul, vol 96, pp. 51-65. ISSN: 0166-0934.

Christou, P.; Capell, T.; Kohli, A.; Gatehouse, J.A.; Gatehouse, AM. (2006). *Recent developments and future prospects in insect pest control in transgenic crops*. Trends Plant Sci, 2006 Jun, vol 11, pp. 302-308. ISSN: 1360-1385.

Crouch, E.A.; Cox, L.T.; Morales, K.G.; Passarelli, A.L. (2007). *Inter-subunit interactions of the Autographa californica M nucleopolyhedrovirus RNA polymerase*. Virology, 2007 Oct 25, 367, pp. 265-274. ISSN: 0042-6822.

De Lima, M.E.; Figueiredo, S.G.; Pimenta, A.M.; Santos, D.M.; Borges, M.H.; Cordeiro, M.N.; Richardson, M.; Oliveira, L.C.; Stankiewicz, M.; Pelhate, M. (2007). *Peptides of arachnid venoms with insecticidal activity targeting sodium channels*. Comp Biochem Physiol C Toxicol Pharmacol. 2007 Jul-Aug, vol 146, pp. 264-79. ISSN: 1532-0456.

de Moraes, R.R.; Maruniak, J.E.; Funderburk, J.E. (1999). *Methods for detection of Anticarsia gemmatalis nucleopolyhedrovirus DNA in soil*. Appl Environ Microbiol, 1999 Jun, vol 65, pp. 2307-2311. ISSN: 0099-2240.

Eberle, K.E.; Sayed, S.; Rezapanah, M.; Shojai-Estabragh, S.; Jehle J.A. (2009). *Diversity and evolution of the Cydia pomonella granulovirus*. J Gen Virol, 2009 Mar, vol 90, pp. 662-671. ISSN: 0022-1317.

Entwistle, P.F. (1998). *A world survey of virus control of insect pests*, pp. 189-200. In F. R. Hunter-Fujita, P.F. Entwistle, H.F. Evans, and N. E. Crook (ed.), "Insect Viruses and Pest Management". John Wiley & Sons Ltd., Chichester, England. ISBN0-471-96878-1.

Espinel-Correal, C.; Léry, X.; Villamizar, L.; Gómez, J.; Zeddam, J.L.; Cotes, A.M.; López-Ferber M. (2010). *Genetic and biological analysis of Colombian Phthorimaea operculella granulovirus isolated from Tecia solanivora (Lepidoptera: Gelechiidae)*. Appl Environ Microbiol, 2010 Nov, vol 76, pp. 7617-1625. ISSN: 0099-2240.

Faktor, O. and Raviv, D. (1996). *A polymerase chain reaction for the detection of nucleopolyhedroviruses in infected insects: the fate of the Spodoptera littoralis virus in Locusta migratoria*. J Virol Methods, 1996 Sept, vol 61, pp. 95-101. ISSN: 0166-0934.

Crooks, G.E.; Hon, G.; Chandonia, J.M.; Brenner, S.E. (2004). *WebLogo: a sequence logo generator*. Genome Res, 2004 Jun, vol. 14, pp.1188-1190. ISSN: 1088-9051.

Gilligan, C.A. (2008). *Sustainable agriculture and plant diseases: an epidemiological perspective*. Philos Trans R Soc Lond B Biol Sci. 2008 Feb 27, vol 363, pp. 741-759. ISSN: 1471-2970.

Inceoglu, A.B.; Kamita, S.G.; Hammock, B.D. (2006). *Genetically modified baculoviruses: a historical overview and future outlook*. Adv Virus Res. 2006; vol. 68, pp. 323-360. ISSN: 0065-3527.

Jehle, J.A.; Blissard, G.W.; Bonning, B.C. (2006a). *On the classification and nomenclature of baculoviruses: a proposal for revision*, Archives of Virology, 2006 Jul, vol. 151, pp. 1257–1266. ISSN 0304-8608.

Jehle, J.A.; Lange, M.; Wang, H.; Hu, Z.; Wang, Y.; Hauschild, R. (2006b). *Molecular identification and phylogenetic analysis of baculoviruses from Lepidoptera*. Virology, 2006 Mar 1, vol 346, pp. 180-93. ISSN: 0042-6822.

Jinn, T.R.; Tu, W.C.; Lu, C.I.; Tzen, J.T. (2006). *Enhancing insecticidal efficacy of baculovirus by early expressing an insect neurotoxin, LqhIT2, in infected Trichoplusia ni larvae*. Appl Microbiol Biotechnol. 2006 Oct, vol. 72, pp. 1247-1253. ISSN: 0175-7598.

Jorio H.; Tran, R.; Meghrous, J.; Bourget, L.; Kamen, A. (2006). *Analysis of baculovirus aggregates using flow cytometry*. J Virol Methods. 2006 Jun, vol 134, pp. 8-14. ISSN: 0166-0934.

Kalendar, R.; Lee, D.; Schulman, A.H. (2009). *FastPCR Software for PCR Primer and Probe Design and Repeat Search*. Genes, Genomes and Genomics, vol 3, pp. 1-14. ISSN: 1749-0383. [http://primerdigital.com/fastpcr.html]

Lasa, R.; Williams, T.; Caballero, P. (2008). *Insecticidal properties and microbial contaminants in a Spodoptera exigua multiple nucleopolyhedrovirus (Baculoviridae) formulation stored at different temperatures*. J Econ Entomol, 2008 Feb, vol 101, pp. 42-49. ISSN: 0022-0493

Lee, H.Y.; Krell, P.J. (1992). Generation and analysis of defective genomes of Autographa californica nuclear polyhedrosis virus. J Virol. 1992 July, vol. 66, pp. 4339-4347. ISSN: 0022-538X.

Li, H.; Bonning, B.C. (2007). *Evaluation of the insecticidal efficacy of wild-type and recombinant baculoviruses*. Methods Mol Biol, vol 388, pp. 379-404. ISSN: 1064-3745.

Lin, T.; Yu, M.; Wu, W.; Yu, Q.; Weng, Q.; Yang, K.; Yuan, M.; Pang Y. (2010). *Functional analysis of Spodoptera litura nucleopolyhedrovirus p49 gene during Autographa californica nucleopolyhedrovirus infection of SpLi-221 cells*. Virus Genes. 2010 Dec, vol 41, pp. 441-449. ISSN: 0920-8569.

Manzán, M.A.; Aljinovic, E.M.; Biedma, M.E.; Sciocco-Cap, A.; Ghiringhelli, P.D. and Romanowski, V. (2008). *Multiplex PCR and quality control of EpapGV production*. Virus Genes, 2008 Oct, vol 37, pp. 203-211. ISSN: 0920-8569.

Mengual Gómez, D.L.; Belaich, M.N.; Rodríguez, V.A.; Ghiringhelli, P.D. (2010). *Effects of fetal bovine serum deprivation in cell cultures on the production of Anticarsia gemmatalis multinucleopolyhedrovirus*. BMC Biotechnol. 2010 Sep, vol. 15; 10:68. ISSN 1472-6750.

Micheloud, G.A.; Gioria, V.V.; Pérez, G.; Claus J.D. (2009). *Production of occlusion bodies of Anticarsia gemmatalis multiple nucleopolyhedrovirus in serum-free suspension cultures of the saUFL-AG-286 cell line: influence of infection conditions and statistical optimization.* J Virol Methods. 2009 Dec, vol 162, pp. 258-266. ISSN: 0166-0934.

Miele, S.A.B.; Garavaglia, M.J.; Belaich, M.N.; Ghiringhelli, P.D. (2011). *Baculovirus: Molecular Insights on Their Diversity and Conservation.* International Journal of Evolutionary Biology. Vol. 2011, Article ID 379424, 15 pages, 2011 Feb. doi:10.4061/2011/379424. ISSN: 2090052X.

Moscardi, F. (1999). *Assessment of the application of baculoviruses for control of Lepidoptera.* Annu Rev Entomol. Vol 44, pp. 257-489. ISSN: 0066-4170.

Murillo, R.; Muñoz, D.; Williams, T.; Mugeta, N.; Caballero, P. (2006). *Application of the PCR-RFLP method for the rapid differentiation of Spodoptera exigua nucleopolyhedrovirus genotypes.* J Virol Methods, 2006 Jul, vol 135, pp. 1-8. ISSN: 0166-0934.

Notredame, C.; Higgins, D.G.; Heringa, J. (2000). *T-Coffee: A novel method for fast and accurate multiple sequence alignment.* J Mol Biol, 2000 Sep 8, vol 302, pp. 205-17. ISSN: 0022-2836

Parola, A.D.; Sciocco-Cap, A., Glikmann, G., Romanowski, V. (2003). *An immunochemical method for quantitation of Epinotia aporema granulovirus (EpapGV).* J Virol Methods. 2003 Sep, vol 112, pp. 13-21. ISSN: 0166-0934.

Peng, K.; van Oers, M.M.; Hu, Z.; van Lent, J.W.; Vlak, J.M. (2010). *Baculovirus per os infectivity factors form a complex on the surface of occlusion-derived virus.* J Virol, 2010 Sep, vol 84, pp. 9497-504. ISSN: 0022-538X.

Poirot, O.; O'Toole, E.; Notredame, C. (2003). *Tcoffee@igs: A web server for computing, evaluating and combining multiple sequence alignments.* Nucleic Acids Res, 2003 Jul 1, vol 31, pp. 3503-3506. ISSN 0305-1048.

Rodríguez, V.A.; Belaich, M.N.; Gómez, D.L.; Sciocco-Cap, A; Ghiringhelli, P.D. *Identification of nucleopolyhedrovirus that infect Nymphalid butterflies Agraulis vanillae and Dione juno.* J Invertebr Pathol, 2011 Feb, vol 106, pp. 255-262. ISSN: 0022-2011.

Rohrman, G.F. (2008). *Baculovirus Molecular Biology,* National Library of Medicine (US), NCBI, Bethesda, Md, USA, 2008.

Rosinski, M.; Reid, S.; Nielsen, L.K. (2002). *Kinetics of baculovirus replication and release using real-time quantitative polymerase chain reaction.* Biotechnol Bioeng, 2002 Feb, vol 77, pp. 476-480. ISSN: 0572-6565.

Rowley, D.L.; Farrar, R.R. Jr; Blackburn, M.B.; Harrison, R.L. (2010). *Genetic and biological variation among nucleopolyhedrovirus isolates from the fall armyworm, Spodoptera frugiperda (Lepidoptera: Noctuidae).* Virus Genes, 2010 Jun, vol 40, pp. 458-468. ISSN: 0920-8569.

Shen, C.F.; Meghrous, J.; Kamen, A. (2002). *Quantitation of baculovirus particles by flow cytometry.* J Virol Methods. 2002 Sep, vol 105, pp. 321-30. ISSN: 0166-0934.

Simón, O.; Williams, T.; López-Ferber, M.; Caballero, P. (2004). *Genetic structure of a Spodoptera frugiperda nucleopolyhedrovirus population: high prevalence of deletion genotypes.* Appl Environ Microbiol, 2004 Sep, vol 70, pp. 5579-5588. ISSN: 0099-2240.

Song, J.; Wang, R.; Deng, F.; Wang, H.; Hu, Z. (2008). *Functional studies of per os infectivity factors of Helicoverpa armigera single nucleocapsid nucleopolyhedrovirus.* J Gen Virol, 2008 Sep, vol 89, pp. 2331-2338. ISSN: 0022-1317.

Szewczyk, B.; Hoyos-Carvajal, L.; Paluszek, M.; Skrzecz, I., Lobo de Souza, M. (2006). *Baculoviruses, re-emerging biopesticides.* Biotechnol Adv. 2006 Mar-Apr, vol. 24, pp. 143-60. ISSN: 0734-9750.

Theilmann, D.A.; Blissard, G.W.; Bonning, B.C.; Jehle, J.A.; O'Reilly, D.R.; Rohrmann, G.F.; Thiem, S.M.; and Vlak, J.M. Baculoviridae, In C. M. Fauquet, M. A. Mayo, J. Maniloff, U. Desselberger, and L. A. Ball. (2005). Virus Taxonomy: (eds.), *Taxonomy of Viruses: VIII Report of the International Committee on Virus Taxonomy*. Elsevier, London, pp. 177-185.

van Beek, N.; Davis, D.C. (2007). *Baculovirus insecticide production in insect larvae*. Methods Mol Biol. 2007, vol 388, pp. 367-378. ISSN: 1064-3745.

van Oers, M. M. and Vlak, J. M. (2007). *Baculovirus genomics*. Current Drug Targets. 2007 Oct vol. 8, pp. 1051–1068. ISSN 1389-4501.

Wang, C.H.; Yang, H.N.; Liu, H.C.; Kou, G.H.; Lo, C.F. (2000). *Nested polymerase chain reaction and in situ hybridization for detection of nucleopolyhedrosis*. J Virol Methods, 2000 Jan, vol 84, pp. 65-75. ISSN: 0166-0934.

Wang, H.; Deng, F.; Pijlman, G.P.; Chen, X.; Sun, X.; Vlak, J.M.; Hu, Z. (2003). *Cloning of biologically active genomes from a Helicoverpa armigera single-nucleocapsid nucleopolyhedrovirus isolate by using a bacterial artificial chromosome*. Virus Res, 2003 Nov, vol 97, pp. 57-63. ISSN: 0168-1702.

Wu, W.; Liang, H.; Kan, J.; Liu, C.; Yuan, M.; Liang, C.; Yang, K.; Pang, Y. (2008). *Autographa californica multiple nucleopolyhedrovirus 38K is a novel nucleocapsid protein that interacts with VP1054, VP39, VP80, and itself.* J Virol, 2008 Dec, vol 82, pp. 12356-12364. ISSN: 0022-538X.

Part 6

Quality Control in Engineering

23

Quality Control and Characterization of Scintillating Crystals for High Energy Physics and Medical Applications

Daniele Rinaldi[1], Michel Lebeau[2],
Nicola Paone[3], Lorenzo Scalise[3] and Paolo Pietroni (formerly with[3])
[1]*Università Politecnica delle Marche/*
Dipartimento di Scienze e Ingegneria della Materia, dell'Ambiente e Urbanistica
[2]*European Organization for Nuclear Research (CERN)/Department PH*
[3]*Università Politecnica delle Marche/*
Dipartimento di Ingegneria Industriale e Scienze Matematiche
[2]*Switzerland*
[1,3]*Italy*

1. Introduction

To the discovery and first use of scintillators are linked the names of W. Crookes, A. Becquerel and E. Rutherford. Since then phosphors have been used to materialise information generated by various scientific, medical and industrial apparata.

Phosphorescence, luminescence and scintillation are basically the same phenomenon, differing by the internal mechanisms involved and by their decreasing time scales. Energy carried by radiative phenomena is converted into light in the phosphor when excited electrons turn back to their equilibrium state by the release of photons in the visible (or near-visible) range. In the first applications, observation was only visual and result qualitative. By their growing demand in space and time resolution, applications themselves prompted the increased performance of the phosphors. The early X-ray radioscopic devices, with their slowly glowing zinc sulphide screens, gave way to faster and safer means of observation.

Thanks to the development of adapted technologies (invention of the photomultiplier tube by Curran and Baker in 1944), light could be converted into electric analog signal and ultimately into manageable data. At that stage, the major interest of the phenomenon i.e. the proportionality of the light output to the incident energy, could be fully exploited. Experimentation in nuclear and particle physics began using plastic and liquid scintillators of increasing light yield and fast response. The discovery of NaI(Tl) in 1949 (Hofstadter, 1949) owed Hofstadter a Nobel Prize. Despite its hygroscopy and radioactive dopant, NaI(Tl) reached mass production scale, and is still in wide use today. With the discovery of BGO ($Bi_4Ge_3O_{12}$) scintillation in 1973, M.Weber and R.Monchamp (Weber, M. & Monchamp, R., 1973) opened the era of fast, dense synthetic mineral crystalline scintillators at the industrial scale: by 1989 1.4m^3 (11 400 pieces) of BGO had been produced for the L3 experiment at CERN. This achievement initiated the steady supply for a growing medical imaging market. Thanks to focused progress in solid state physics, deeper understanding of the physical phenomenon led to light production levels competing with the best organic scintillators without their weak sides. The role of R&D collaborations and dedicated conference cycles was crucial in this progress (SCINT conferences since 1991, Crystal Clear Collaboration since 1992). Attempts were made to produce amorphous (glass) and even polycrystalline (ceramics) scintillators, to try and gain from available profitable mass production methods. Finally the many advantages of monocrystalline structure has turned to be the mainstream in scintillator development. After NaI(Tl) and CsI(Tl) followed a sequel of new, better performing crystals. BGO, CeF_3, BaF_2, $PbWO_4$, LuAP, LSO, LYSO paved the way of a continued progress in performance but also in quality and quantity, the early formulae mostly thanks to the growing scale of high energy physics instruments, the more recent sustained by the growing demand and important economic prospects of medical imaging equipment.

This new generation of materials is characterised by an excellent time resolution, with a steep rise and short persistence (no afterglow), a chemical structure that guarantees reproducible properties and reliable performance over time and a resistance to working conditions (no aging, radiation resistance). By choosing high Z chemicals, high density crystals can be synthesised, ensuring the tight containment of deposited energy -thus reducing the instrument dimensions. High purity raw materials, sophisticated production processes and adapted quality control methods ensure production of high grade crystals. Among several quality criteria the optical transparency in the scintillation wavelengths is a severe limiting factor as light attenuation and non-uniformity may deteriorate the crystal performance.

Not only intrinsic scintillators have been produced. Passive crystalline lattices have been designed to host specific chemical species –dopants- responsible for the light production, either by themselves or by their specific bonds with the lattice. This is a strong economic incentive as the host lattice may be made of less expensive materials, and budget better spent on costly dopants (e.g. rare earths). The dopant fraction is usually of a few percent. Dopants are selected to match the lattice properties to the best (crystal symmetry, lattice parameters). Two obstacles have to be overcome: segregation that makes light production uneven over the crystal volume, and lattice distortion that may induce mechanical stress and be detrimental to the production yield. Depending on the application and the selected scintillator, crystal sizes may be very different. From bulky 150cm^3 prisms used in high energy physics electromagnetic calorimeters, to mm^3 scale in medical imaging, production problems are quite different and may stress on different features for the optimisation of the material. The latter application has recently seen the development of crystalline fibres of the order of 1mm that may be the present industrial optimum at that scale.

In order to present methods for quality control of scintillating crystals, it is necessary to understand the production process and to identify the characteristic features which determine crystal performance and therefore will be subject to specifications that need to be verified. The following paragraph 2 describes crystal production process, paragraph 3 outlines applications of scintillating crystals, then paragraph 4 treats the methods for quality control based on photoelastic analysis and their applicability to process and product control of crystals. A final paragraph will resume the content of the chapter.

2. Production process of scintillating crystals

Scintillating crystal production has evolved from the chemical laboratory to the industrial plant scale. Crystals are produced by growth methods specific (optimal) of every chemical compound, part size and quantity (Lecoq, P. et al., 2006).

2.1 Raw materials

Upstream the preparation of raw materials is a prerequisite of the ultimate crystal quality. Quality control and traceability have to secure a supply of tightly specified ingredients. Purity is not an absolute criterion but rather an economic compromise of innocuous and poisonous impurities, affecting scintillation (afterglow, light yield) transparency (colour centres) radiation resistance and built-in stress level (cell distortion). Stoichiometric proportions may not be the optimum as some components may be lost during the growth process, either by evaporation in the furnace atmosphere, or by combination with the crucible material.
Raw materials are ground to specific granularity distributions, thoroughly mixed to required proportions. Preparation is completed by melting the components and producing a polycrystalline compound that shall be used to fill the crucible for the growth operation.

2.2 Growth methods

A variety of techniques is used to grow scintillating inorganic crystals. They are all derived from two main methods that shall be briefly described.

2.2.1 Czochralsky method

In Czochralsky method (fig. 1), raw materials are molten in a metallic crucible and kept slightly above fusion point.

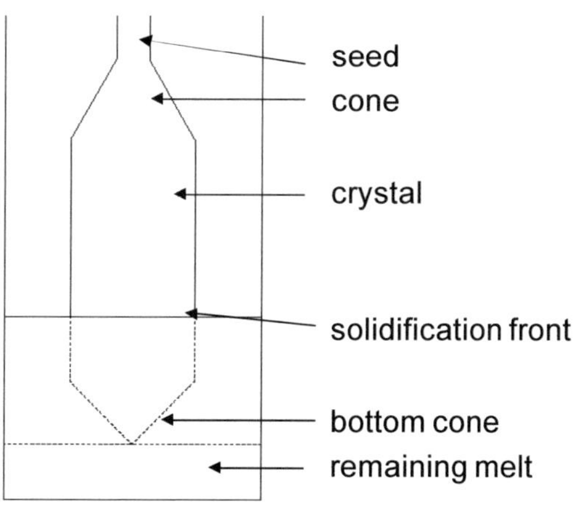

Fig. 1. Czochralsky growth method (pull-from-melt)

A small monocrystal of the same material (seed) is put into contact with the molten bath and pulled up to lift a small meniscus of liquid by capillarity. Solidification occurs at a position and a rate fixed by several parameters. The thermal gradient is regulated by an induction loop heating the melt with help of the crucible mass. The melt temperature is homogenised by crucible regulation. The growth ("pulling") rate is monitored by a weighting system connected to the seed support. The rising solidified bulk takes the shape of the desired ingot (or boule). The pulling rate gives the ingot its conical top and bottom ends and its cylindrical overall shape. Optimal conditions result in turning the full melt into the ingot volume. The growth of a 1kg PWO ingot took of the order of 1 day. In spite of the circular symmetry of the whole system, the ingot shape may be facetted or oval (fig. 2), depending on the crystal lattice and on the seed orientation. The latter is a critical parameter.

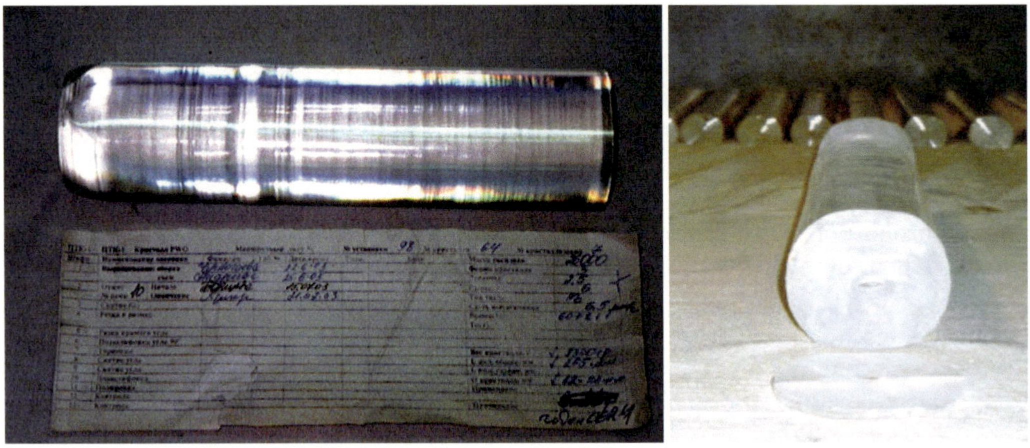

Fig. 2. Czochralsky crystals. The ingot section results from growth parameters and lattice dimensions. Important material scrap is due to difference between ingot and piece, but "peel-like" scraps are recyclable

The crucible is metallic to allow heating by induction. The metal is selected for a high melting point, and for a low chemical affinity with the melt. Platinum is the panacea but poses an economic problem because of its high cost and stock price volatility. The crucible raw material is therefore usually leased from a specialised bank. Even after a large production, the crucible material loss remains small and amounts for a minor fraction of the production cost. Some less expensive metals like palladium or osmium may be used depending on the crystal. The melt free surface is subject to chemical exchanges with the furnace atmosphere, with the risk of losing the design proportions. This is therefore a critical feature of the method and controlled atmosphere is the rule. Depending on the crystal material, vacuum, inert, reducing or oxidizing atmospheres are chosen. For instance, in the case of PWO growth, lack of oxygen induces mechanical stress, whereas its excess is detrimental to crystal transparency.

2.2.2 Bridgman method

This method is an evolution of the fusion zone refining method (fig. 3).

Work piece shape is given by grinding off, powder scrap is not recyclable.

The polycrystalline material is shaped to the final part proportions with extra thickness for the mechanical processing to follow. The shape is usually an elongated prism. A monocrystal seed is placed at one end. This assembly is contained in a thin (a few tenths of a mm) metallic envelope, usually made of platinum. One or several of these assemblies are positioned in the furnace enclosure. The fusion zone (thermal gradient) moves from the seed interface along the ingot length by combining several induction loops. The gradient displacement velocity is comparable to the Czochralsky growth rate, but the operation has to be repeated several times (at the week scale) to reach monocrystallinity and purity at the required level. The Bridgman process competes with the Czochralsky method only by the use of multiple ingot furnaces. The last passes are considered as an annealing phase. After cooling down the metallic envelope is peeled off and recycled.

Fig. 3. Bridgman crystals. The ingot shape is given by an envelope-like crucible

2.2.3 Annealing

The solidification process results in thermally-induced stresses. They are supposed to have the conventional parabolic profile, with moderate compression in the core, and higher tension at the periphery. The latter is a matter of concern as crystals are especially prone to breaking in tension (crack opening). Annealing is necessary to reduce these tensions. The annealing temperature is very close to the melting point to ease dislocation movements and resolution. The thermal effect is sometimes combined with a chemical one with help of a controlled atmosphere, as a chemical correction is expected to restore the crystal lattice equilibrium. In that case annealing time is governed by diffusion and may take several days. The annealing furnace is therefore a complex device and annealing a costly operation in time and money.

2.3 Mechanical processing

Ingots are the base material for expensive parts tightly specified in shape and dimensions. Mechanical processing is designed to achieve these two goals. For calorimeters, ingots are usually dimensioned for the yield of one piece. Attempts were made to get two and even four pieces out of an ingot. The tiny pixels of medical imaging are either obtained by the many from a single ingot, or cut from fibres.

2.3.1 Cutting (shaping)

The cutting operation provides the part its general shape. Precision is required to limit the amount of material to be removed later by lapping and polishing. It is therefore an economic target to obtain the best geometry (planarity, correct angles) at cutting. Processing parameters must be optimised to save on time, but not to the price of too thick a damaged sub-surface, and increased risk of edge chipping. The surface finish obtained at cutting ranges from 0,2 to 0,05 µm Ra. Finer finish is not worth doing at cutting as lapping easily produces values below 0,02 µm Ra. A very delicate feature of the cutting operation is the balancing of released thermally-induced tensions as the ingot periphery is removed, face by face. Operations have to be sequenced in such a way that this transitory unbalance is not aggravated (fig. 4).

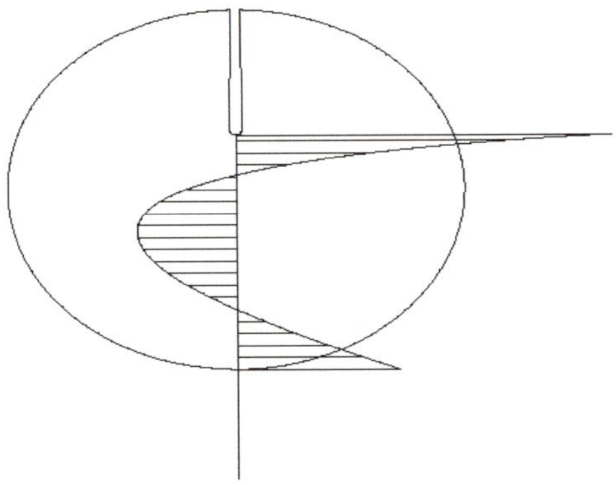

Fig. 4. Effect of cutting on tensional state balance; Cutting a tensioned crystal "frees" mechanical tensions: New boundary conditions transform mechanical tensions into material displacements and deformations, and new mechanical tension distribution. Complex problem with risk of even higher concentrated stresses and breaking when cutting does not follow crystal lattice symmetry

Fig. 5. Crystal shaping by grinding off

Thanks to their transparency, scintillating crystals may be subject to photoelastic stress observation. This method will be discussed in paragraph 4, and it was specifically applied to BGO (Rinaldi et al., 1997) and later to PWO (Cocozzella et al. 2001, Lebeau et al., 2005). The resulting information, as well as lattice orientation and mechanical properties are critical inputs for determining the processing conditions (Lebeau, 1985, Ishii & Kobayashi, 1996, Pietroni et al., 2005). Grinding-off is the oldest method, that consists in removing excess material by multiple passes of a grinding wheel (fig. 5), on a conventional planar grinder.

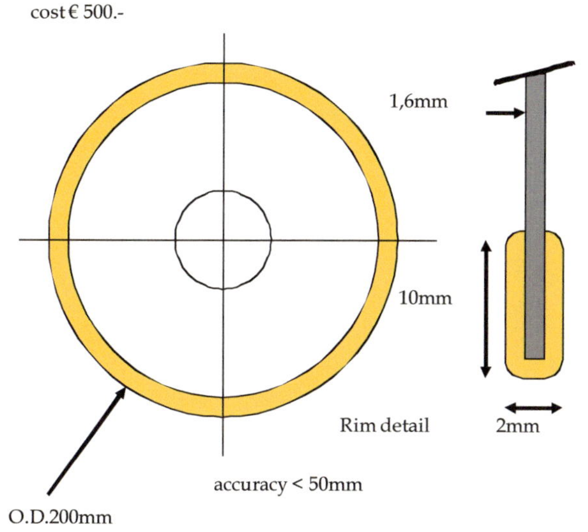

(a)

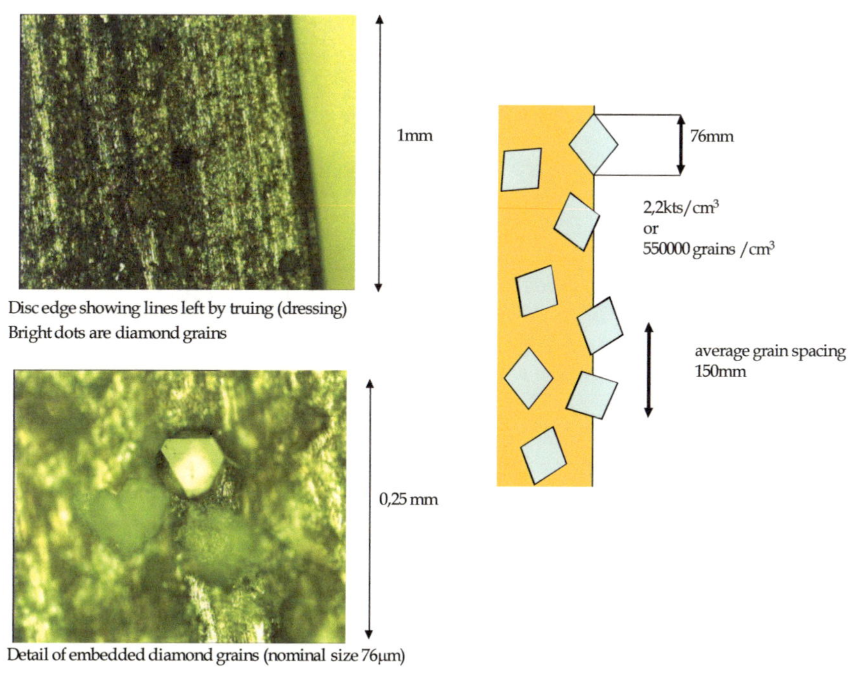

(b)

Fig. 6. Cutting disc (a) and disc details (b)

This is a safe, very accurate but lengthy operation that has been replaced by more economic solutions. Thin disk saws with abrasive-loaded rim are now the most commonplace tool for crystal cutting (fig. 6).

The abrasive grain selection is capital. A good formula is 50 to 100 μm diamond grain sintered in high density in a bronze matrix. The disc diameter is selected to achieve a complete cut in one pass, e.g. 200mm for a 30mm cut depth (fig. 7).

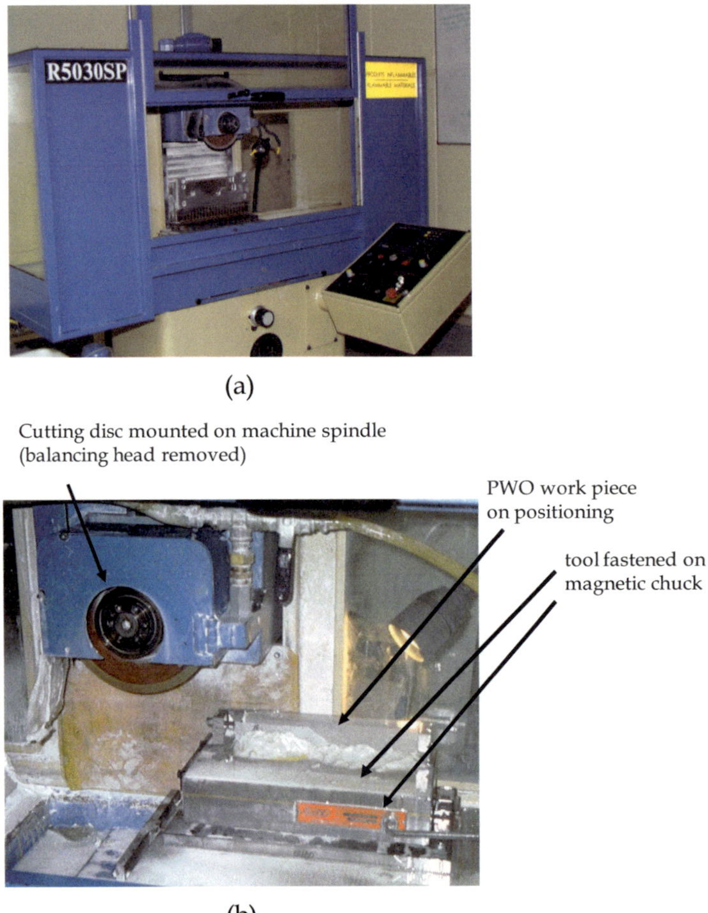

Fig. 7. Crystal cutting machine (a) adapted grinder, CERN, 2000, detail of cutting machine (b) Details of the cutting machine

The best planarity achieved with this method is of the order of 20 μm, and depending on the tooling and machine quality, a dimensional precision of the order of 50 μm is expected. The disc thickness is a compromise between the cutting precision and the ingot material loss (as fall-off is recycled, but not cutting swarf). The disc thickness also contributes to sub-surface damage depth. A thickness of 2mm is the optimum for a 200mm disc. To achieve the above-mentioned precision the best discs on the market have to be purchased. The machine tool is derived from a conventional grinder, with improved casing for intensive lubrication (fig. 8, a).

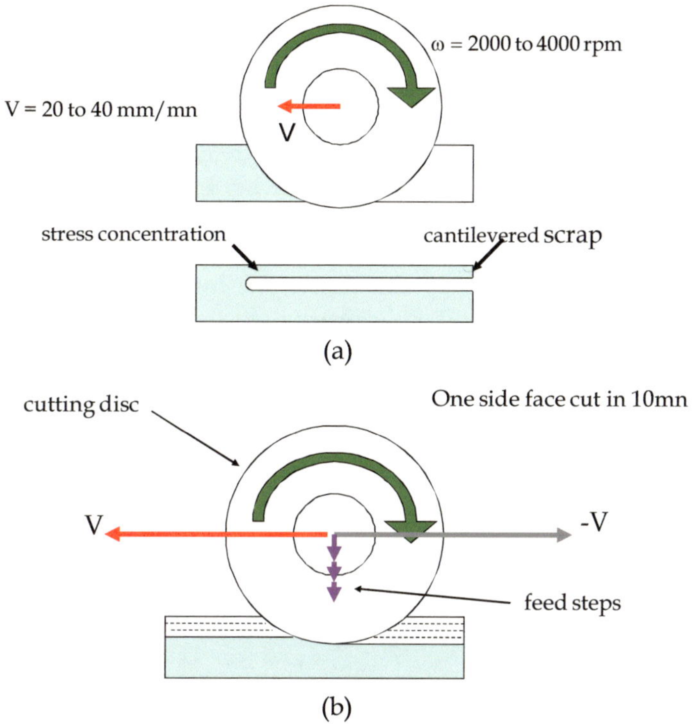

Fig. 8. Full-depth cutting (a); sweeping mode (b) keeps some stress balance

The disc velocity is about 3000rpm, the feed speed about 40mm/min. Sweeping mode: instead of cutting one crystal face in one stroke of the cutting disc, several passes at higher feed (meter per second) with a small vertical take (0,1mm) at every pass are combined for a similar processing time (fig. 8, b). The main advantage is a smaller pressure in the cutting zone, and a balanced release of tensions as the work piece remains symmetrical during the operation. The internal disc saw method was widely used for wafer cutting before the generalisation of plain wire cutting. A very thin (0,3mm) metallic circular membrane tensioned on its periphery is tensioned to achieve a stable planar shape. The edge of the central circular opening is bordered with sintered abrasive grains (fig. 9).

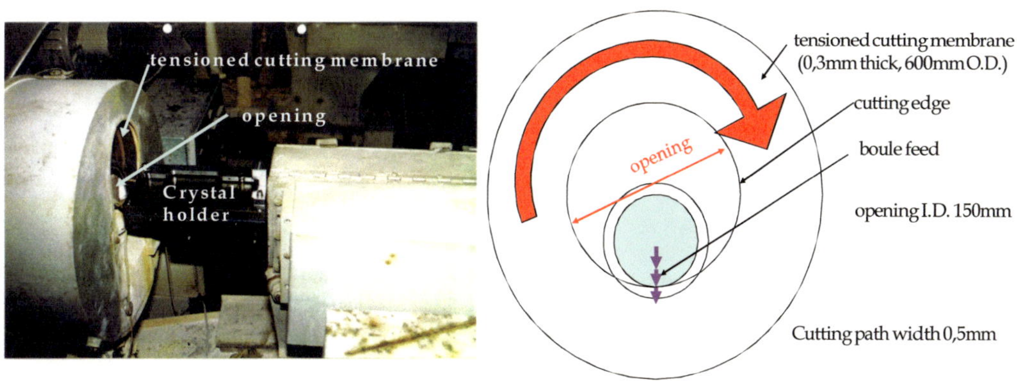

Fig. 9. Inner cutting machine

Only bars of small section compared to the disc opening can be cut with this method. The advantage is an economy in material thanks to the narrow cutting path (disc thickness plus 0,1mm max.). Processing of prismatic pieces is only possible for end cuts. Because of the limited free space within the central opening supporting tooling design is critical.
The plain wire saw method, as said above, is now widely used for mass production of silicon wafers, for electronics as for solar cells. The km long wire runs back and forth and follows a complex path to achieve multiple cutting planes on several ingots (today up to seven with diameters exceeding 320mm). The abrasive slurry (usually cheap corundum) is poured on the wire where it holds by capillarity. The cutting action depends on the grain adherence to the wire, with the result of decreasing efficiency with cut depth. Wire diameter and cutting path are comparable to internal disc saw. This method therefore requires a correction of the planarity afterwards. The specific arrangement of this equipment is only fit for slicing and has no interest for the shaping of prismatic scintillators. Wires with sintered abrasive have been successfully developed to correct the weak sides of the plain wire. Typical diameter is 0,25mm with an 80 µm diamond grain coating. The 2km wire is expensive (1 €/m order) and fragile: processing parameters and lubrication have to be carefully adapted to dedicated machine tools (fig. 10).

Wire length 2km @ 2€/m
Wire O.D. 0,25mm
Diamond grain 80 Hm
Wire path 0,3mm
Wire speed 5 to 10m/s
Feed 25 to 50Hm/s

Fig. 10. Wire saw (abrasive wire)

Feeds of 50 µm per minute can be achieved with an excellent planarity and a very low sub-surface damage. Parameter optimisation also aims at reducing the wire wear. By combining the feed with the crystal rotation, a symmetric end-cut is possible (cropping), with a balanced stress relief (fig. 11). The machine open configuration allows cutting long side faces (up to 300mm).

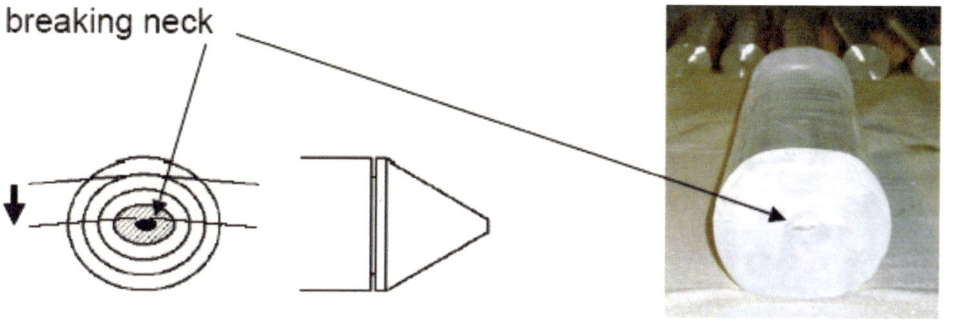

Fig. 11. Rotary wire end-cut (solves boule-ends tensions release); Large ingots have to be put to length before annealing because of annealing furnace dimensions. Cutting un-cured ingots is very delicate and a rotary method is used to keep some symmetry. The cutting wire is slowly fed down while ingot rotates until the end breaks at the thin remaining neck

Crystal cutting is an abrasive process at the microscopic scale. Every abrasive grain works as a gross tool with a negative cutting angle that locally induces high compressive stress. To prevent high crack density and possible propagation, reduce tangential forces, keep work piece temperature low and ease chip removal, the appropriate lubricant must be applied in abundant flow. pH, chemical polarity and affinity may be adapted to the crystal material in a profitable way. Filtering, sedimentation and recycling are environmental constraints.

Lapping is free abrasive action between the crystal face and the surface of a rotary table, the lap (fig. 12).

Combined rotations of the lap and the crystal result in an even distribution of the abrasive action and a regular material removal. Working parameters are the lap and crystal rotation velocities (a few m/s), the pressure exerted on the crystal face (a few N/cm^2), the abrasive material and granularity (usually about 15 μm corundum or diamond), the lubricant mixed with the abrasive (slurry) , and finally the lap material. A typical stock removal for PWO was 50 μm/min. With a 0,02 μm Ra finish reached after 3 min, the damaged sub-surface layer from cutting was easily removed. This finish Ra is a good value to start polishing. To prevent edge chipping and resulting deep scratches on the surface, chamfers are necessary on every sharp edge of the crystal before lapping (and polishing): 0,2-0,3mm bevels are usually sufficient. Polishing produces optically transparent faces, that are necessary for scintillating light collection (Auffray et al. 2002). The polish quality can be specified according to a maximum number of visible scratches per view field at a given magnification. The value is far less demanding than for conventional lens polishing. Scintillator polishing operates in similar configuration as lapping. The main differences are the abrasive grain size (from 3 down to fraction of a μm), and the lap cover. Because of the abrasive fine grain, stock removal is slow (less than 1 μm/min) and polishing takes 10 to 20 min per face. This is the critical path in a crystal processing line (Auffra et al. 2002). In mechanical polishing, the material removal results from grain abrasion as for lapping, but at a smaller scale (fig.13). Diamond is the best abrasive in that case. Cooling and lubrication are critical to avoid sub-surface damage.

This method was developed for electronic chips and finds interesting developments for scintillators. The abrasive action is enhanced by specific chemical conditions. For instance, a suspension of very fine grains of quartz (20 nm) in pH9 colloidal silica produces an efficient polishing free of sub-surface damage (Mengucci et al., 2005). Soda and potash were also

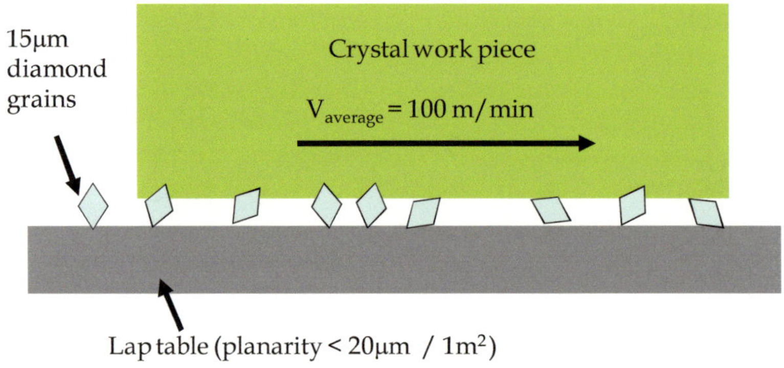

(a)

(b)

Fig. 12. Lapping principle (a): Abrasive grains are bumped and tilted between lap and work piece and present fresh cutting edges to work. Lapping tooling (PWO, CERN, 2000) (b): Three crystal shapes are cut out in the lapping mask (or holder). A satellite ring keeps the mask (and crystals inside) in radial position on the lap. Crystal length 230mm, ring I.D. 320mm

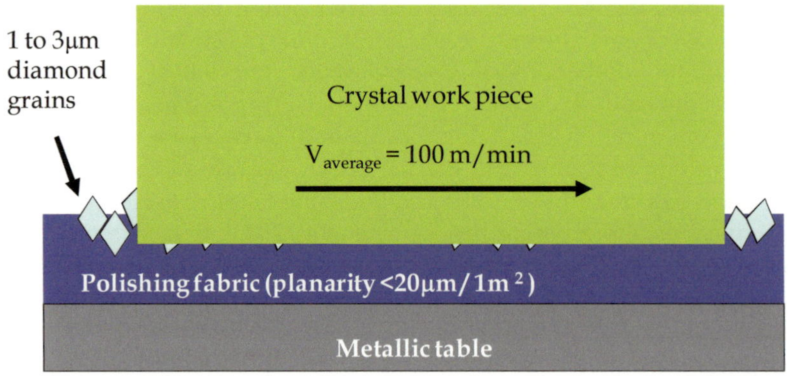

Fig. 13. Polishing schematics. Diamond grains are taken in stable cutting orbits by the fabric. Material removal operates at microscopic scale and some ductile effect results, with limited subsurface damage

tested but surface etching sometimes happened when mechanical action was not properly balanced. The use of aggressive chemicals (bases) poses difficult safety and environment conditions that prevent the spread of these methods. Cerium oxide is known for its combined abrasive action and chemical reaction with conventional lens glass. It is less delicate in use and has been successfully tested with PWO and LSO, but its chemical action remains unclear to date.

3. Scintillating crystals: Applications fields

In recent years, scintillating crystals have found numerous applications in different fields; hereafter we will briefly recall the main areas: Nuclear and high energy physics, medicine (imaging of biological tissues), geology and security.

3.1 Nuclear and high energy physics

This domain is where scintillators were discovered and where most of their development took place. The early 20th century saw their use in fundamental research.
The atomic era opened by the 2nd World War multiplied the use of scintillating counters, also necessary in nuclear energy production. The quick development of fundamental research in high energy and particle physics after the war was a stimulating motivation for increased performance, quantity and economy. The most striking example is in electromagnetic calorimeters (fig. 14), with new projects involving tons of the most recent scintillators (LYSO). Medical imaging benefits of the spin-off of this striving discipline.

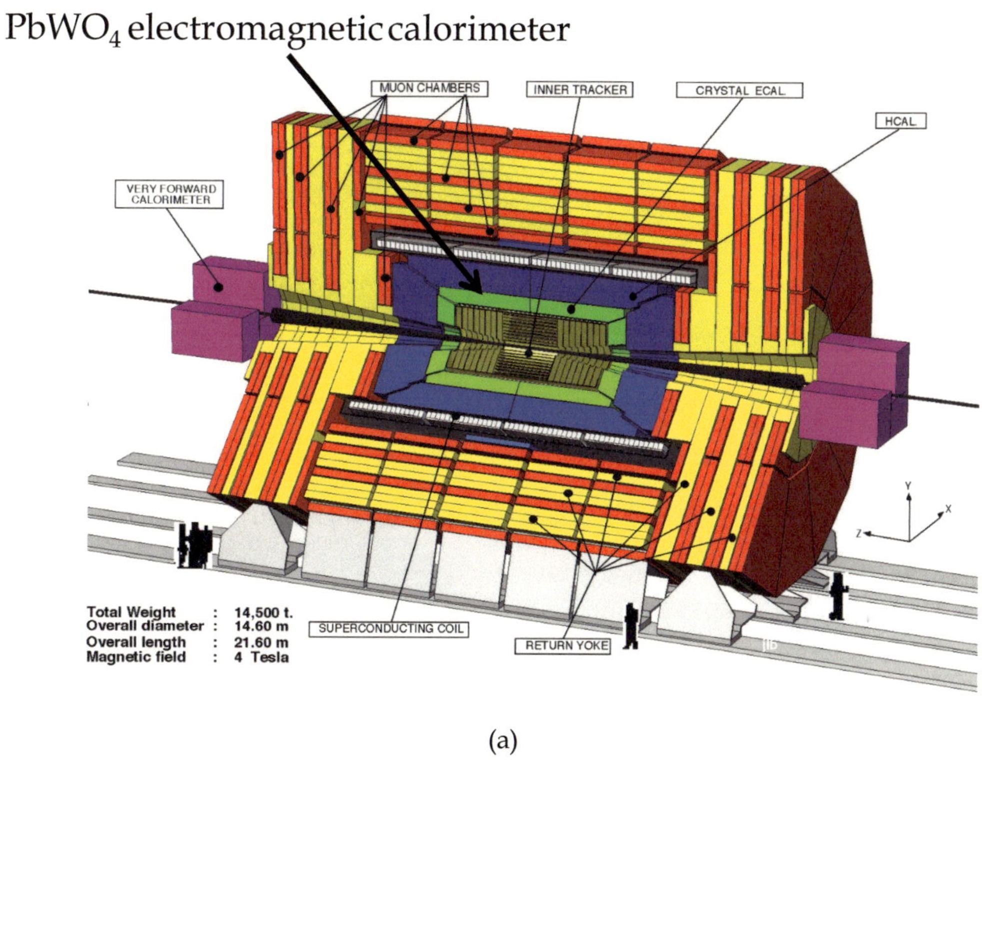

(a)

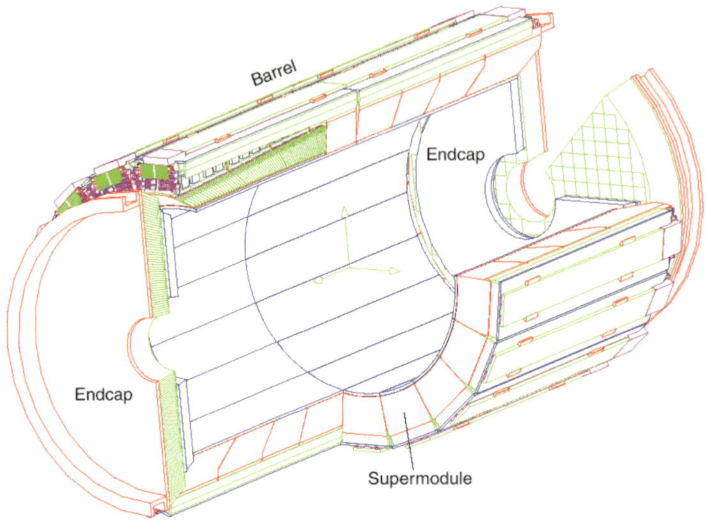

(b)

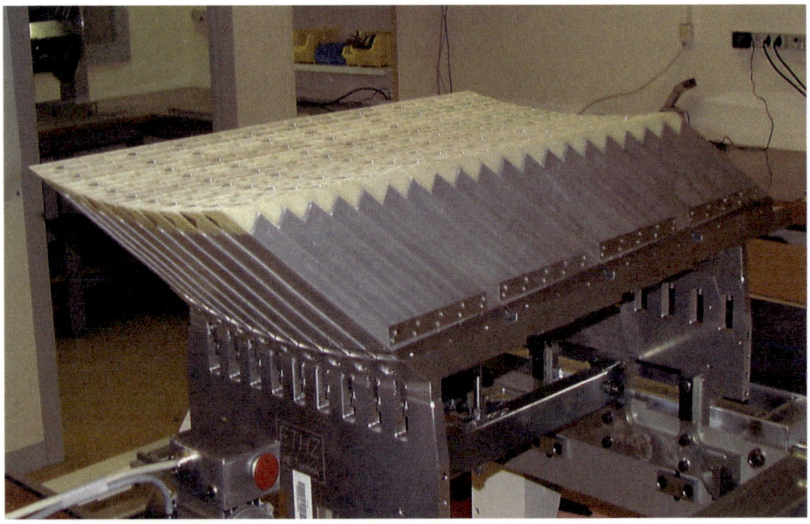

(c)

Fig. 14. The CMS experiment at CERN LHC with $PbWO_4$ electromagnetic calorimeter (a), the CMS $PbWO_4$ electromagnetic calorimeter Modularity crystal sub-module of 10 crystals module of 400 (500) sub-modules super-module of 4 modules barrel of 36 super-modules (b) and (c) a module of the CMS $PbWO_4$ electromagnetic calorimeter (10 x 20 crystals arranged in pointing geometry)

3.2 Medical imaging

Modern radiography is characterised by lower radiation doses (and/or shorter exposures), 3-D information (tomography), real-time observation, tissue or function identification, with the help of large arrays of fine scintillators (pixels) surrounding the patient or covering the organ (mammography) and powerful reconstruction software. The two main types are projective imaging (e.g. X-ray CT) and PET scanners. In X-ray CT radiation scans the patient from outside and projective information is reconstructed. In PET the patient ingests some radio-element that releases positrons (beta decay). The recombination of the positron with an ambient electron produces two opposite gamma rays of well defined energy detected in opposite scintillator pixels. The radio-element is combined in a chemical (tracer) specific of of tiny 2 x 2 x 10 mm^3 prisms. Processing was dominated by material loss and sub-surface damage because of the crystal scale compared to processing tools. Scintillating fibres are an interesting solution. New fast scintillators with high light output like LuAP, LSO and LYSO are very promising for the medical domain, where economic prospects are high.

3.3 Geologic research

Mining, gas and oil logging are very active economic domains, because of the increasing demand in raw materials and fuel, opposed to the shortage forecast and resulting crisis. Research is performed by prospective drilling. The drill hole may be scanned with two types of detectors. The simpler one is a radiation detector used for the research of radioactive minerals. B.Pontecorvo already proposed such a device with a ionizing chamber counter in 1941. The other type contains a powerful neutron source (Cf, Am, Be, Cs) that irradiates the underground vicinity of the hole. Stimulated gamma emission reaching the detector is typical of the chemical bonds in the mineral (hydrocarbon). Obviously the detector is shielded from the source. The energy spectrum typical of the concerned mineral, and the signal intensity may give a quantitative information. The detector transmits its signal to the recording station on the surface. The source is usually left in the drill hole bottom for safety reasons. Today detectors are of the scintillator type (NaI(Tl), BGO, GSO, LuYAP). Literature is scarce because of patent protection.

3.4 Security

Quick, non-invasive inspection of transport loads, containers, but also luggage and passengers is familiar to everybody in today's life. In the latter case, soft X-ray scanners are used to reveal hidden weapons or hazardous objects, thanks to their density or specific form. Nuclear explosives are also detected by portable radiation detectors. Scintillators are used in these applications. Dual (or multiple) energy systems combined with colour coding display help material identification (different Z) against apparent density. Visual training for qualitative identification are crucial to the efficiency of these detectors. Scientific literature lacks because of patent protection

4. Photoelastic methods for the quality control on scintillating

Paragraph 2 has presented the various phases of crystal production process. It emerges that internal stress distribution is influenced by the mechanical and thermal processes that the crystal undergoes. This brings to the issue of quality control, which implies the assessment of internal stress state in order to set up a proper production process and during its operation.

Residual stress is indeed a major hazard in crystal processing. Crystals are brittle materials, therefore residual tensile stress may easily lead to fracture and breakage during processing or, even worse, during the following assembly of many crystals into complex geometry detectors. Scintillating materials are transparent and usually are optically anisotropic. Internal stress causes lattice strain and deformations, which manifest as stress induced birefringence; this means that the piezo-optic properties of the material can be observed to verify its internal strain (or stress) state. Photoelasticity is a classical measurement technique suited to observe stress induced birefringence in transparent materials (Wood E., 1964, Dally J. & Riley W. 1987). Therefore photoelasticity is a natural candidate method for quality control of scintillating crystals; of course this measurement method provides only information on quality related to mechanical strain and stress, and requires an accurate knowledge of piezo-optic properties of the material, which is not always available. Furthermore, it is a volumetric technique which provides information on the spatial integral of the stress distribution along the light path through the crystal and not local values.

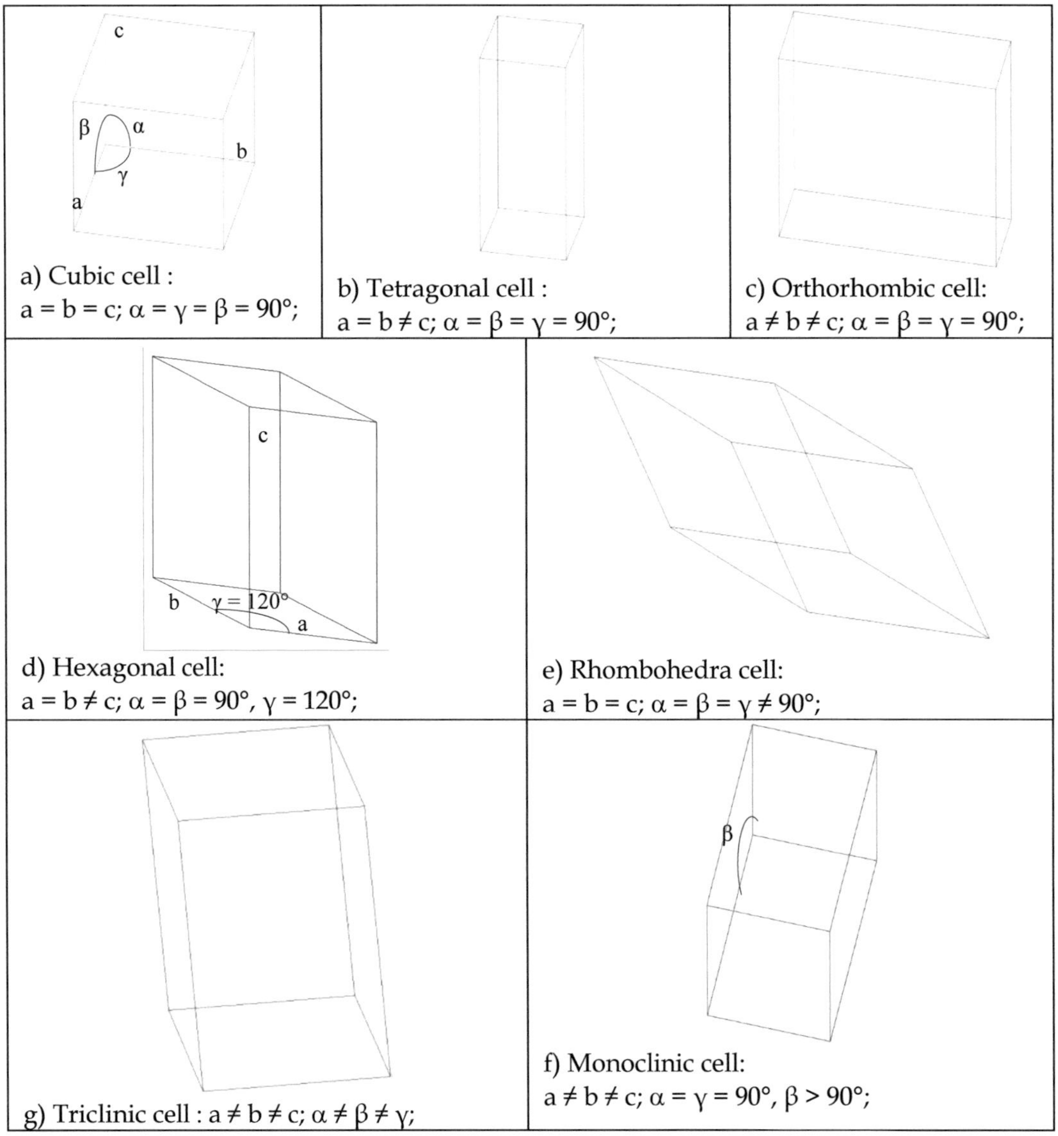

a) Cubic cell : $a = b = c$; $\alpha = \gamma = \beta = 90°$;

b) Tetragonal cell : $a = b \neq c$; $\alpha = \beta = \gamma = 90°$;

c) Orthorhombic cell: $a \neq b \neq c$; $\alpha = \beta = \gamma = 90°$;

d) Hexagonal cell: $a = b \neq c$; $\alpha = \beta = 90°$, $\gamma = 120°$;

e) Rhombohedra cell: $a = b = c$; $\alpha = \beta = \gamma \neq 90°$;

g) Triclinic cell : $a \neq b \neq c$; $\alpha \neq \beta \neq \gamma$;

f) Monoclinic cell: $a \neq b \neq c$; $\alpha = \gamma = 90°$, $\beta > 90°$;

Fig. 15. Geometrical shapes of crystals

Hereafter it is discussed crystal optics, optical anisotropy, piezo-optic behaviour and then photoelasticity is presented for crystal quality control.

4.1 Geometric, mechanical and optical proprieties of crystals

4.1.1 Crystal lattice and symmetry

A crystal is a solid material constituted by a 3D ordered structure which has the name of *crystal lattice*. Each crystal lattice is formed by the repetition of a fundamental element, *the primitive unit cell*: thanks to its replication, it produces the crystal structure (Wood E., 1964, Hodgkinson W., 1997, Wooster W., 1938.). From a geometrical point of view, it is possible to build up the crystal lattice simply translating the unit cell in parallel way with respect to its faces. Indeed the cell geometry should have peculiar characteristics: in particular, the opposite faces should be parallel and, for this reason, it should be a parallelepiped. Possible geometrical shapes are hereafter reported.
The crystal physical and optical properties depend on the typology of unit cell and on the atomic bondages strength. Indeed those properties have the same symmetries of the crystal structure.

4.1.2 Elastic properties of crystals

Crystals undergoing a mechanical stress will deform, so they will exhibit an internal strain distribution. If the mechanical stress is below a limit, named *elastic limit*, crystal deformation is reversible. The strain is proportional with the applied stress for low level stresses. If the crystal undergoes an arbitrary uniform stress $[\sigma_{\kappa\lambda}]$ the generated strain components ε_{ij} is linearly correlated with the stress tensor (Wood E., 1964). This means that:

$$\varepsilon_{ij} = s_{ijkl}\,\sigma_{kl} \quad (i, j, k, l = 1, 2, 3) \tag{1}$$

Equation 1 is the generalized Hook law. Here, s_{ijkl} factors are crystal elastic compliances. The total number of the elastic compliances s_{ijkl} is 81. The Hook law can be written in the following way:

$$\sigma_{ij} = c_{ijkl}\,\varepsilon_{kl} \quad (i, j, h, l = 1, 2, 3) \tag{2}$$

Where c_{ijkl} are crystal elastic stiffness coefficients. The coefficients c_{ijkl} and s_{ijkl} form a forth order tensor. This means that in a coordinate system transformation from a coordinate system X_1, X_2, X_3 to X'_1, X'_2, X'_3 the coefficients s_{ijkl} (c_{ijkl}) are transformed into s'_{mnop} (c'_{mnop}) throughout the law:

$$s'_{mnop} = C_{mi}\, C_{nj}\, C_{ok}\, C_{pl}\, s_{ijkl} \tag{3}$$

where C_{mi}, C_{nj}, C_{ok}, C_{pl} are direction cosine which define the X_1,X_2,X_3 axes orientation with respect to X'_1,X'_2,X'_3. axes. Each s_{ijkl} (c_{ijkl}) coefficient has a precise amplitude and correlation with respect to a specific coordinate system, linked to the crystal. If this coordinate system is coincident with the crystallographic one, the coefficients are the *basic* ones. Since the strain and stress tensors are symmetrical, the tensor coefficients c_{ijkl} and s_{ijkl} are symmetrically coupled according to the subscript *i* and *j*, *k* and *l*, so:

$$s_{ijkl} = s_{jikl}, \quad c_{ijkl} = c_{jikl}, \tag{4}$$

$$s_{ijkl} = s_{jilk}, \quad c_{ijlk} = c_{ijkl}, \tag{5}$$

The equations (4) and (5) reduce the number of independent components of c_{ijkl} and s_{ijkl} to 36. Since c_{ijkl} and s_{ijkl} are symmetrical with respect to the first two subscripts and the second ones, the equations (4) and (5) can be written in more compact way:

$$\left.\begin{aligned} s_{ij} \ (i,j=1,\,2,\,3,\,4,\,5,\,6) \\ c_{ij} \ (i,j=1,\,2,\,3,\,4,\,5,\,6) \end{aligned}\right\} \tag{6}$$

This notation reduces the number of terms of (1) and (2)

$$\left.\begin{aligned} e_i = s_{ij}s_j \ (i,j=1,\,2,\,3,\,4,\,5,\,6) \\ s_i = c_{ij}e_j \ (i,j=1,\,2,\,3,\,4,\,5,\,6) \end{aligned}\right\} \tag{7}$$

but the following rules should be respected:

$$\left.\begin{aligned} s_{ijkl} = s_{mn} \text{ when m and n are equal to } 1,\,2, \text{ or } 3 \\ 2s_{ijkl} = s_{mn} \text{ when m or n are equal to } 4,\,5, \text{ or } 6 \\ 4s_{ijkl} = s_{mn} \text{ when m and n are equal to} 4,\,5, \text{ or } 6 \end{aligned}\right\} \tag{8}$$

It is necessary to underline that the symmetry further reduces the number of independent coefficients c_{ij} and s_{ij} . The following formula relates the elastic compliances s_{ij} to the elastic stiffness c_{ij}:

$$s_{ij} = \frac{(-1)^{i+j}\Delta c_{ij}}{\Delta^c} \tag{9}$$

Where Δ^c is a determinant composed of elastic stiffness:

$$\begin{vmatrix} c_{11} & c_{12} & c_{13} & c_{14} & c_{15} & c_{16} \\ c_{12} & c_{22} & c_{23} & c_{24} & c_{25} & c_{26} \\ c_{13} & c_{23} & c_{33} & c_{34} & c_{35} & c_{36} \\ c_{14} & c_{24} & c_{34} & c_{44} & c_{45} & c_{46} \\ c_{15} & c_{25} & c_{35} & c_{45} & c_{55} & c_{56} \\ c_{16} & c_{26} & c_{36} & c_{46} & c_{56} & c_{66} \end{vmatrix}$$

and Δc_{ij} is the minor obtained from this determinant by crossing out the i-th row and j-th column. Likewise:

$$c_{ij} = \frac{(-1)^{i+j}\Delta s_{ij}}{\Delta^s} \tag{10}$$

The following constants are often used for a description of elastic properties of both isotropic and anisotropic media. *Young's modulus E*, characterizing elastic properties of a medium in a specific direction, is defined as the ratio of the mechanical stress in this direction to the strain it produces in the same direction. The *Poisson ratio* ν is defined as the ratio of the transverse compression strain to the longitudinal tensile strain caused by a mechanical stress. The *Shear modulus* μ is defined as the ratio of shear stress and shear strain, it produces in a material. In isotropic bodies only two of the above-mentioned constants are independent. For this reason, elastic properties of isotropic bodies are often described using the constants λ and μ, called the *Lame constants.* The constant λ and μ are related to the stiffness matrix components as follows:

$$\mu = \frac{c_{11} - c_{12}}{2} \tag{11}$$

$$\lambda = c_{12} \tag{12}$$

Considering the matrix (s_{ij}) then it is possible to write the following formulas E and μ in isotropic case:

$$s_{11}=1/E, \quad s_{12}= -\nu/E \tag{13}$$

$$2(s_{11}- s_{12}) =1/\mu \tag{14}$$

In anisotropic medium Young's modulus in a arbitrary direction X'_3 is:

$$E = 1/s'_{3333} \tag{15}$$

where $s'_{3333} = C_{3i}\, C_{3j}\, C_{3k}\, C_{3l}\, s_{ijkl}$ and C_{3i} ,C_{3j} ,C_{3k} ,C_{3l} are the direction cosine of the axis X'_3 with respect to the crystallographic coordinate system and s_{ijkl} are the basic compliances referred to crystallo-physical coordinate system. Young's modulus is a function of direction for all crystallographic classes, including cubic class.
In anisotropic media the Poisson ratio is equal to

$$\nu^{hk} = \frac{s_{hk}}{s_{kk}} \tag{16}$$

and it represents an estimation of lateral compression parallel to X_h with respect to accompanied elongation parallel to X_h .

4.1.3 Piezo-optical properties of crystals

The piezoptical effect consists of changes in the optical properties of crystals throughout static and alternating external mechanical stresses and it is described in terms of the index ellipsoid. The general equation of the index ellipsoid in an arbitrary coordinate system X1, X2, X3, whose origin coincides with that of the main (crystallophysical) coordinate system, can be written in the following form:

$$B_{11}x_1^2 + B_{22}x_2^2 + B_{33}x_3^2 + 2B_{23}x_2x_3 + 2B_{13}x_1x_3 + 2B_{12}x_1x_2 = 1 \tag{17}$$

Where B_{ij} are the dielectric impermeabilities or polarization constants. Equation 17 is related to anisotropic crystal without any applied mechanical stress.

An applied mechanical stress produces variation ΔB_{ij} in the dielectric impermeabilities:

$$\Delta B_{ij} = B_{ij} - B_{ij}^{0} \tag{18}$$

Considering a first-order approximation, the increments in the dielectric impermeability tensor components are proportional to mechanical stresses:

$$\Delta B_{ij} = \pi_{ijkl}\sigma_{kl} \tag{19}$$

On the other hand, the same increments can be expressed in terms of strain:

$$\Delta B_{ij} = p_{ijkl}\varepsilon_{kl} \tag{20}$$

Such a change in the optical index ellipsoid of the crystal due to the straining is called the elasto-optical effect. The coefficients π_{ijkl} and p_{ijkl} form a rank four tensor and they are called the piezo-optical and elasto-optical constants, respectively. In the matrix notation eqs. (19) and (20) can be rewritten in the following form

$$\Delta B_{m} = \pi_{mn}\sigma_{n} \tag{21}$$

$$\Delta B_{m} = P_{mn}\varepsilon_{n} \tag{22}$$

where if n = 1, 2, or 3:

$$\pi_{mn} = \pi_{ijkl}$$

while if n= 4, 5, or 6:

$$\pi_{mn} = 2 \cdot \pi_{ijkl}$$

and P_{mn} are the elasto-optical coefficients, $P_{mn} = p_{ijkl}$ for all m and n. In the general case:

$$\pi_{mn} \neq \pi_{nm} \quad P_{mn} \neq P_{nm}$$

The piezo-optic and elasto-optic coefficients are related by the following formulas:

$$P_{mn} = \pi_{mr} c_{rn}, \quad \pi_{mn} = P_{mr} s_{rn}$$

Where c_{rn} are the elastic stiffness and s_{rn} are the elastic compliances.

4.1.4 Isotropy and anisotropy in crystal optical properties

As it is described in previous paragraphs, the symmetry on the crystal lattice influences the symmetry on the optical properties. The Isotropy and anisotropy affect changes on refraction index and the consequent variations of the light velocity with respect to the direction inside the crystal material. Crystals, having a cubic cell, can be considered isotropic for their optical properties. All the rest of crystals cells has an anisotropic behaviour in terms of optical properties (Wood, E. 1964, Hodgkinson I., 1997, Wooster W., 1938). The optical anisotropy allows to classify crystals in two categories: uni-axial anisotropic crystals and biaxial ones

according to the index ellipsoid which provides the value of the refraction index along a specified direction in the crystal (Wood, E. 1964, Hodgkinson I., 1997).
A fundamental representation of *Uniaxial anisotropic crystals* is the *optical indicatrix* or *ellipsoid of the refraction indices*. As far as the uniaxial crystal, there are two principal indices: *ordinary index of refraction* n_o *and extraordinary index of refraction* n_e. Indeed the optical indicatrix is a rotation ellipsoid, where the both the axes are proportional to n_o *and* n_e. It is possible to state that an indicatrix is positive, when $n_e > n_o$, and negative, when $n_e < n_o$.

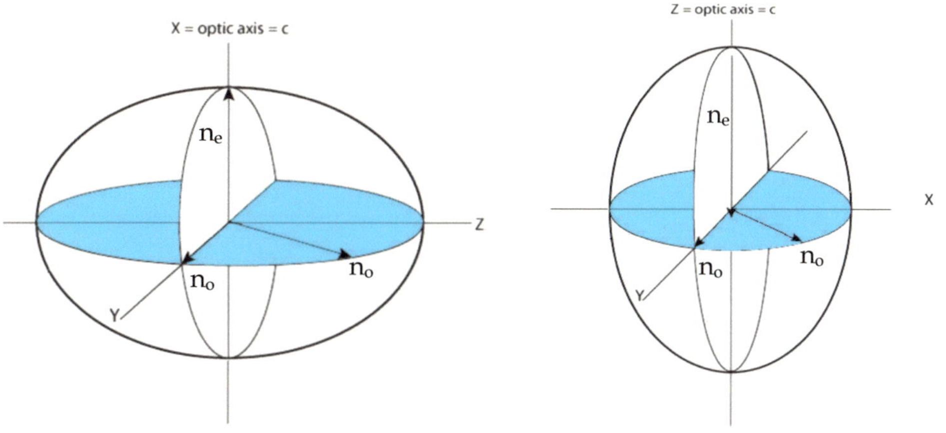

Fig. 16. Optical indicatrix for uniaxial positive and negative crystal

Hereafter three examples (fig. 17-19) are reported in order to explain the concept of uni-axial anisotropy and consequently of birefringence, assuming to study a positive crystal. In first case (fig. 17), the crystal lattice is oriented so that the optic axis is along the light travelling direction. In this case, considering the indicatrix its section perpendicular to the wave vector is circular with radius is n_o. Since in all the vibration direction of the electromagnetic field the refraction index is constant the birefringence is zero.

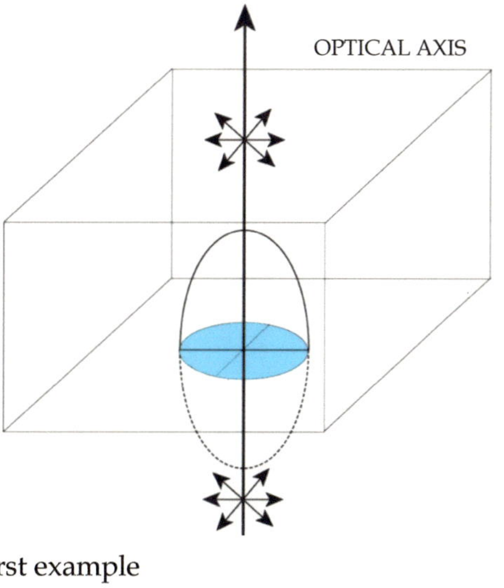

Fig. 17. Uniaxial crystal first example

In the second example (fig. 18), the crystal lattice is oriented in a random orientation so that the light path is at angle θ to the optic axis. The section through the indicatrix parallel to the incoming light wave is an ellipse whose axes are n_o and n_e. The extraordinary ray electromagnetic field vibrates parallel to the trace of the optic axis as seen from Fig. 18, while the ordinary ray one vibrates at right angles.

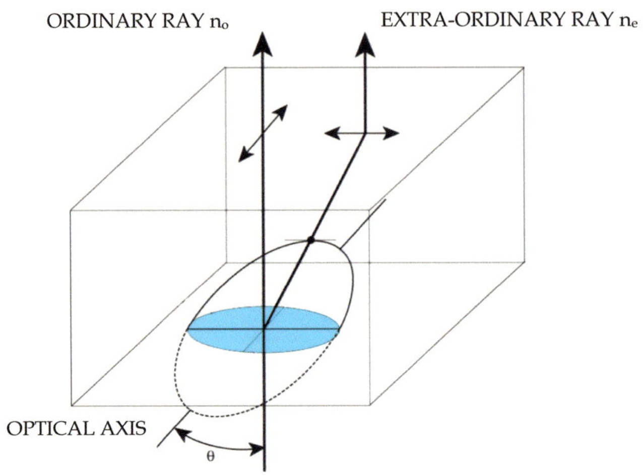

Fig. 18. Uniaxial crystal second example

In a third case (fig. 19), the crystal lattice is oriented so that its optic axis is parallel to the light wavefront. Because the optic axis has this orientation, this section is a principal elliptic section whose axes are n_o and n_e. The ordinary ray therefore has index of refraction n_o and the extraordinary ray n_e, which is its maximum because the crystal is optically positive. The extraordinary ray vibrates parallel to the trace of the optic axis (c axis) and the ordinary ray vibrates at right angles.

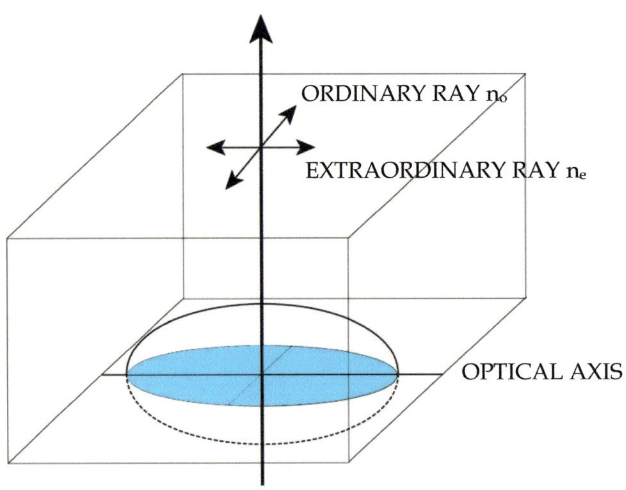

Fig. 19. Uniaxial crystal third example

These uniaxial crystals have tetragonal, rhombohedra and hexagonal cell.

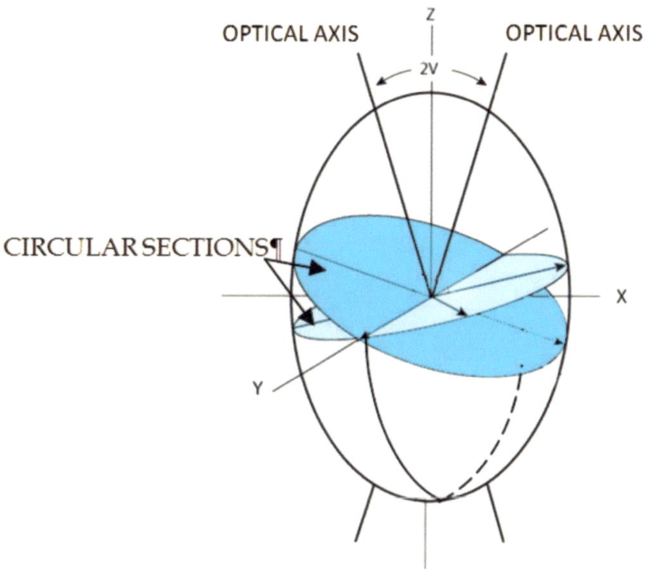

Fig. 20. Optical indicatrix for biaxial crystals

Those crystals having orthorhombic, monoclinic e triclinic cells are biaxial crystals. They have three different principal indices of refraction n_x, n_y and n_z, so that the indicatrix becomes a triaxial ellipsoid. Assuming that $n_x < n_y < n_z$, ZX plane is the optical plane and the Y axis is the optical normal (see fig. 20).
Between the two optical axes an acute angle, 2V, is the optical angle. The bisector of such angle is the acute bisector Bxa (see fig. 21): in the positive crystal that is Z axis, while in negative ones that is the X axis. The bisector of the other obtuse angle between the optical axes is the obtuse bisector Bxo.

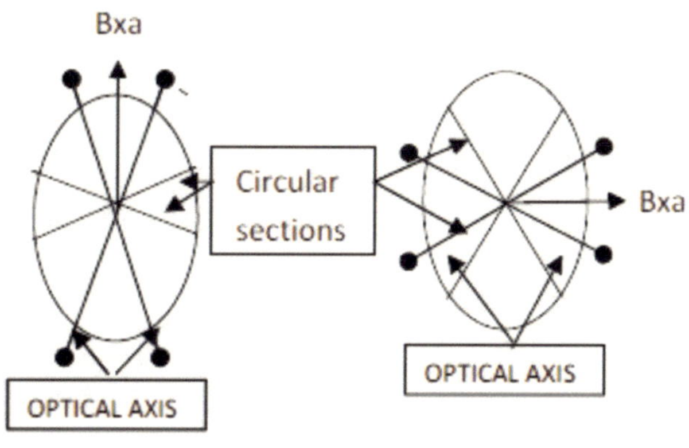

Fig. 21. Optical indicatrix for biaxial positive and negative crystals

The study of birefringence in this type of crystal can be conducted as in the case of uniaxial crystals. The indicatrix has the property that the axial sections normal to the optical axes are circular with a radius equal to n_y: then, a wave that propagates along the optical axis will behave as if they were moving in an isotropic medium. Any other section is elliptical wave moving along a direction different from the optical axis, therefore, will split into two beams, with vibration directions parallel to the major and minor semi-axis of the ellipse. The optical angle can be determined experimentally, but there are approximated formulas for its calculation according to the value of the indices of refraction:

$$tg^2 V = \frac{\frac{1}{n_x^2} - \frac{1}{n_y^2}}{\frac{1}{n_y^2} - \frac{1}{n_z^2}} \tag{23}$$

The eq. (23) is valid for Z axis bisector: if V > 45° the crystal is negative, while if V < 45° the crystal is positive.

4.2 Photoelasticity

Photoelasticity is a classical technique that allows to visualize internal stress/strain states in transparent materials; it exploits the changes in refractive indices induced by strain within transparent materials

4.2.1 General scheme of polariscope

The polariscope is an optical instrument which utilizes polarized light in inspecting a specimen subject to strain; usually it is used to explore a two-dimensional planar stress state, with stress components orthogonal to the optical axis *z*. Light travels across the material of the specimen and its polarization state is affected by the spatial distribution of refraction index, which depends on strain. According to the kind of polarization, it is possible to consider a *plane polariscope* or a *circular polariscope.*

In a plane polariscope, devices known as *plane or linear polarizer* are utilized: they are optical elements which divide an incident electromagnetic wave in two components which are mutually perpendicular (fig. 22). The component, which is parallel to the polarization axis is transmitted, while the perpendicular one is absorbed or totally reflected internally.

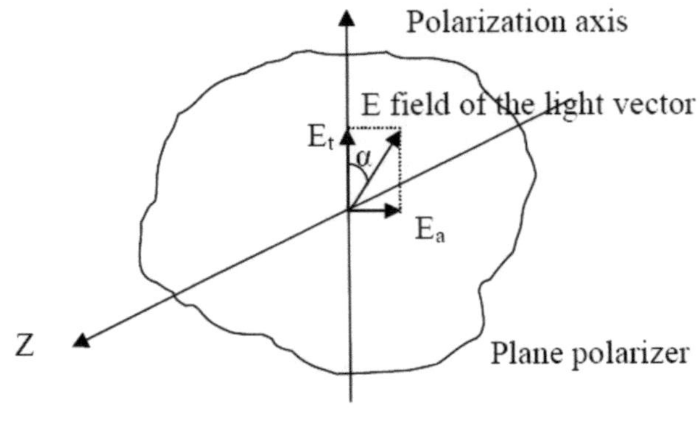

Fig. 22. Plane polarizer

It is possible to make the assumption that the polarizer is placed at the z_0 coordinate along the z axis, the equation of the light vector can be written:

$$E = a\cos\frac{2\pi}{\lambda}(z_0 - ct) \tag{24}$$

Since the initial phase is not important for this treatment, it is possible to rewrite it in the following way (it is assumed that $f = c/\lambda$ = wave frequency)

$$E = a\cos 2\pi ft = a\cos\omega t \tag{25}$$

where $\omega = 2\pi f$ is the wave angular frequency. The absorbed and transmitted components of light vector are:

$$E_a = a\cos\omega t\sin\alpha \quad E_t = a\cos\omega t\cos\alpha \tag{26}$$

where α is the angle between the light vector and the polarization axis.
In the plane polariscope two linear polarizers are used. Between those ones, the crystal under inspection is placed: the linear polarizer which is close to the light source is called the "*polarizer*", while that one placed on the opposite side with respect to the crystal is called the "*analyser*".

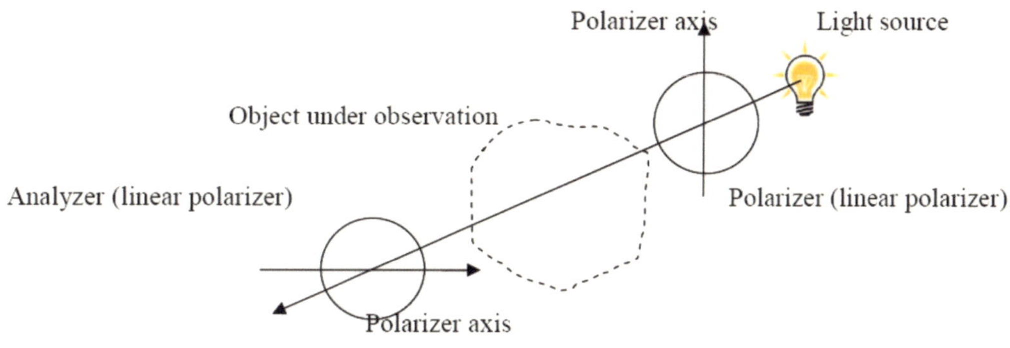

Fig. 23. Plane polariscope scheme

The usual configuration is the one where the two axes of polarization of analyser and polarizer are orthogonal to each other. The specimen to be analysed is put between the two, so that light goes through it. If the specimen is optically anisotropic, then light polarization is affected (see fig. 23). In our case the sample will be a crystal cut with plane surfaces. The advantage of such configuration is that what is observed is totally due to the crystal lattice effect: in fact, without crystal, light reaches the analyser could not be transmitted due to its perpendicular polarization with respect to the analyser polarization axis. Indeed this condition is also named "dark field". On the other hand, when a crystal is introduced, the crystal birefringence produces a light vector rotation of each light wave so that part of the light can pass the analyser.
In the circular polarizer (and in general in the elliptical one), a *wave plate* is used: it divides the light vector in two orthogonal components at different velocities. Such plate is produced with birefringence materials (Dally & Riley, 1987; Wood, 1964). The wave plate has two

principal axes, identified with number 1 and 2 (fig. 24): the transmission of the polarized light along the axis 1 occurs with the velocity c_1, while that one along the axis 2 occurs at c_2. In general $c_1 > c_2$, for this reason the axis 1 is the *fast axis*, while the axis 2 is the *slow axis*.

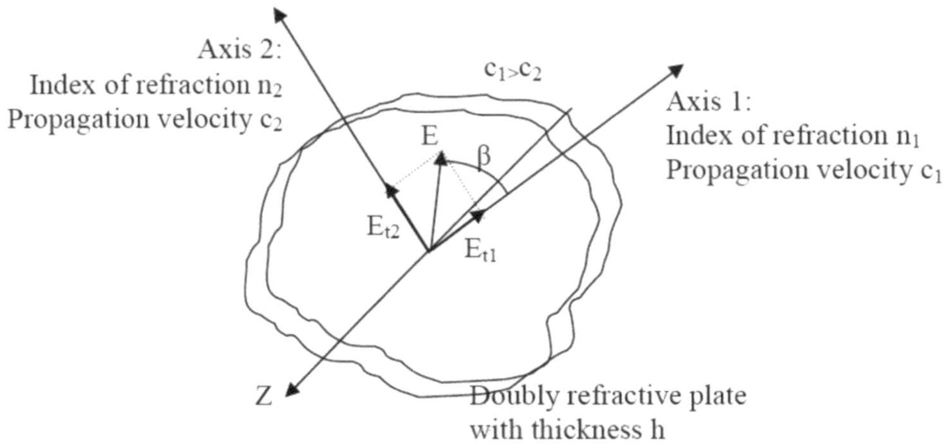

Fig. 24. Optical scheme of a wave plate

If a wave plate is placed after a polarizer, it is necessary to consider that the transmitted wave vector E_t forms an angle β with the fast axis 1. After that it has passed through the plate, E_t is divided in two components E_{t1} and E_{t2}, which are parallel respectively to 1 and 2 axis. The amplitudes of each resulting vector are:

$$\begin{aligned} E_{t1} &= E_t \cos\beta = a\cos\alpha \cos\omega t \cos\beta = k \cos\omega t \cos\beta \\ E_{t2} &= E_t \sin\beta = a\cos\alpha \cos\omega t \sin\beta = k \cos\omega t \sin\beta \end{aligned} \tag{27}$$

Where k = a cosα. Since the two components travel with different velocities (c_1 and c_2), they cross the plate at different times, implying a relative phase offset. Considering h the plate thickness, the relative delay, that both the wave, travelling across the plate, have with respect to a wave travelling in air (n is considered the air index of refraction), is respectively:

$$\delta_1 = h\,(n_1 - n) \qquad \delta_2 = h\,(n_2 - n)$$

And then the phase difference is

$$\delta = \delta_2 - \delta_1 = h\,(n_2 - n_1)$$

The angular phase difference Δ results:

$$\Delta = \frac{2\pi}{\lambda}\delta = \frac{2\pi h}{\lambda}(n_2 - n_1) \tag{28}$$

When $\Delta = \pi/2$ the wave plate is called a quarter wave plate (*λ/4*). Once the two waves have abandoned the plate, they can be described by the following equations:

$$\begin{aligned} E_{t1}' &= k \cos\beta \cos\omega t \\ E_{t2}' &= k \sin\beta \cos(\omega t - \Delta) \end{aligned} \tag{29}$$

Recombining the two waves, the amplitude of the resulting wave vector considering these two components is expressed as following:

$$E_t^{'} = \sqrt{(E_{t1}^{'})^2 + (E_{t2}^{'})^2} = k\sqrt{\cos^2\beta\cos^2\omega t + \sin^2\beta\cos^2(\omega t - \Delta)} \tag{30}$$

The angle with respect to the axis 1 of the plate is:

$$\tan\gamma = \frac{E_{t2}^{'}}{E_{t1}^{'}} = \frac{\cos(\omega t - \Delta)}{\cos\omega t}\tan\beta \tag{31}$$

In order to obtain a circular polarization the $\lambda/4$ plates are used, with β equal to $\pi/4$. In this configuration it is possible to write:

$$E_t^{'} = \frac{\sqrt{2}}{2}k\sqrt{\cos^2\omega t + \sin^2\omega t} = \frac{\sqrt{2}}{2}k \tag{32}$$

$$\gamma = \omega t$$

It is possible to observe that the amplitude of the light vector is constant, while its direction (which is indicated by the angle γ with axis 1 of the plate) varies linearly with time: therefore, the tip of the vector forms a circle. In particular, if $\beta = \pi/4$ the rotation is counter-clockwise, while if $\beta = 3\pi/4$ the rotation is clockwise. In order to obtain an elliptic polarization a $\lambda/4$ plate is used oriented in such way that $\beta \neq n\pi/4$ (with n integer). It is possible to have:

$$E_t^{'} = k\sqrt{\cos^2\beta\cos^2\varpi t + \sin^2\beta\sin^2\varpi t} \tag{33}$$

$$\tan\gamma = \tan\beta\tan\varpi t$$

therefore, the tip of the light vector forms an ellipse. In general, $\lambda/4$ plate is used in order to obtain the circular polariscope see Fig. 25.

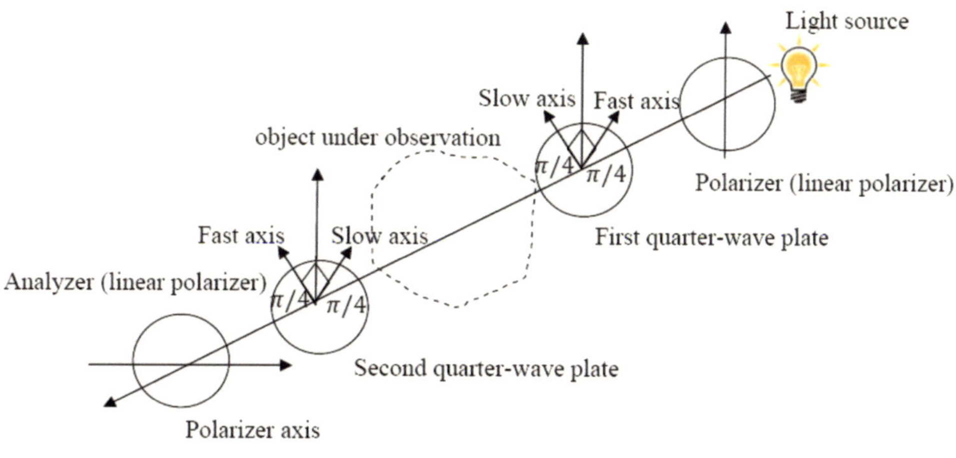

Fig. 25. Circular polariscope scheme

The first element is the polarizer which converts light in linearly polarized light with vertical direction. Then there is the $\lambda/4$ plate which is placed with an angle $\beta = \pi/4$ with respect to the polarization axis of the polarizer. In this way light undergoes circular polarization. Another $\lambda/4$ plate is place with the fast axis parallel to the slow axis of the previous one: the task is to convert the circular polarization in linear one with vertical direction. As last element, there is the analyzer with horizontal polarization axis which produces the dark field. The presence of the crystal between the two $\lambda/4$ plates let light pass through the analyser. In this way it is possible to observe interference fringes.
The interference figures belong to two families: isochromatics and isogyres. Intercepting the light coming from the analyser of the polariscope with a screen or plane of the observer, the isochromatic curves represent the loci where all rays with the same difference in optical path strike on such plane, while interference figure where the light vibration directions through the specimen are parallel to the polarization directions of polariscope are the isogyres.

4.3 Photoelasticity for quality control of crystal samples

The use of photoelastic techniques for quality control involves a knowledge of the piezo-optical properties of the crystal. As a matter of fact, the number of parameters concerning the piezo-optical effect, as far as refraction index variation cannot be directly calculated without the piezo-optic matrix Π, that relates stress and refraction indices. The components of the piezo-optic matrix Π depend on the symmetry group for each crystal (Nye, 1985, Sirotin et al., 1982). Due to the complexity of the three-dimensional problem of piezo-optic response of scintillating crystals, the values of the single components of Π, at present, are unknown for most crystals (to our best knowledge); therefore the procedure presented hereafter is essentially a semi-emipirical approach, which provides qualitative information and an integral indicator of the internal stress state which we call a quality index. It is not an accurate measurement of internal stress distribution, but nevertheless provides useful information for assessing if residual stress state developed in the crystal has reached critical values.
The methodology for quality control of the internal stress in scintillating crystals has been developed and demonstrated on the uniaxial $PbWO_4$ (PWO) crystal, but it can be extended to the whole class of uniaxial crystals. (PWO) is an optically uniaxial birefringent crystal with ordinary and extraordinary refraction indices $n_o = 2.234$ and $n_e = 2.163$ respectively, for $\lambda = 632.8$ nm (Baccaro et al., 1997). The development has been carried out on long prismatic samples, cut from an ingot and polished. They can be represented in a (x,y,z) Cartesian coordinate reference system with a solid body having rectangular cross-section (in the x-z plane) and length L (along y axis). The crystallographic c axis coincides with the optical axis (Born & Wolf, 1975; Walhstrom E., 1960), in a stress-free condition, that in the (x,y,z) reference coincides with the z axis. This is also the observation direction.
When the crystal sample is subjected to a uniform monoaxial compressive stress σ_y, this compressive stress induces the crystal to became biaxial, and, following the classical interference theory concerning anisotropic crystals, as stated by Born and Wolf (Born & Wolf, 1975), applied to bi-axial crystals, it can be found a fourth-order polynomial expression (34). Eq. 34 (Rinaldi et al., 2009) represents a model for the loci of the interference surfaces called the Bertin surfaces (Walhstrom, 1960):

$$(N\lambda)^2 = (x^2+y^2+z^2)\cdot(n_x-n_z)^2\left(1-\left[\frac{z\cdot\cos\beta + x\cdot\sin\beta}{\sqrt{x^2+y^2+z^2}}\right]^2\right)\cdot\left(1-\left[\frac{z\cdot\cos\beta - x\cdot\sin\beta}{\sqrt{x^2+y^2+z^2}}\right]^2\right) \quad (34)$$

where n_x, n_z are the refraction indices along the x and z axes, N is the fringe order, λ is the light wavelength of the light source for the observations, β is the semi-angle between the two optical axes when the crystal becomes biaxial under stress. β is represented by the following function of the three refraction indices n_x, n_y, n_z (Walhstrom E., 1960). Equation (35) holds for negative crystals (i.e. $n_e < n_o$) like PWO is (Walhstrom, 1960).

$$\tan^2\beta = \frac{\frac{1}{n_y^2}-\frac{1}{n_x^2}}{\frac{1}{n_z^2}-\frac{1}{n_y^2}} \quad (35)$$

For a sample of fixed thickness, equation (34) represents interference images given by "Cassini-like" 4th-order curves in the x-y plane. From equation (34) it can be obtained a family of fringes (isochromatic) that are parameterized by the fringe order N. Attention can be focussed on the first-order fringe (N = 1), as visible in a dark-field configuration of the plane polariscope (i.e. analyser perpendicular to laser polarisation). From a phenomenological point of view, supported by experiment evidence, it can be observed a linear dependence of the refraction index n_y on the applied stress σ, along the direction of application, at least for low stress levels. As a matter of fact, an applied stress along y affects all three refraction indices and the refraction index variations depends on the tensor ΔB (variation of dielectric impermeability) as expressed the matrix equation:

$$\begin{Bmatrix}\Delta B_{xx}\\ \Delta B_{yy}\\ \Delta B_{zz}\\ \Delta B_{xz}\\ \Delta B_{yz}\\ \Delta B_{xy}\end{Bmatrix} = \begin{bmatrix}\pi_{xxxx} & \pi_{xxyy} & \pi_{xxzz} & 0 & 0 & \pi_{xxxy}\\ \pi_{xxyy} & \pi_{xxxx} & \pi_{xxzz} & 0 & 0 & -\pi_{xxxy}\\ \pi_{zzxx} & \pi_{zzxx} & \pi_{zzzz} & 0 & 0 & 0\\ 0 & 0 & 0 & \pi_{xzxz} & \pi_{xzyz} & 0\\ 0 & 0 & 0 & -\pi_{xzyz} & \pi_{xzxz} & 0\\ \pi_{xyxx} & -\pi_{xyxx} & 0 & 0 & 0 & \pi_{xyxy}\end{bmatrix}\begin{Bmatrix}\sigma_{xx}\\ \sigma_{yy}\\ \sigma_{zz}\\ \sigma_{xz}\\ \sigma_{yz}\\ \sigma_{xy}\end{Bmatrix} \quad (36)$$

where the Voigt notation is used, and the Π components depend on 4/m point group symmetry concerning the PWO (Nye, 1985, Sirotin et al., 1982).
The dielectric tensor $[\varepsilon]$ is obtained by the relation:

$$[B] = [B_0] + [\Delta B] = [\varepsilon]^{-1} \quad (37)$$

In a principal reference system, the refractive indices n can then be derived by:

$$n_i = \sqrt{\varepsilon_i}$$

Numerical simulations, based only on the variation of $\tan\beta$ (eq.35) in equation (eq.34), produce results in agreement with the experimental observation in calculating the isochromatic interference fringes. Therefore it appears a possibility to relate internal stress state to fringe geometry; the quality control methods developed are based on this experimental evidence, supported by theory. The interference images in the case of stress free uniaxial samples are families of circles. An applied load on the crystal sample induces a distortion that in the simple uniaxial stress case is a Cassini-like curve (Rinaldi et al., 2009), as the crystal becomes biaxial owing to the applied stress. For low stress level, these curves resembles ellipses.

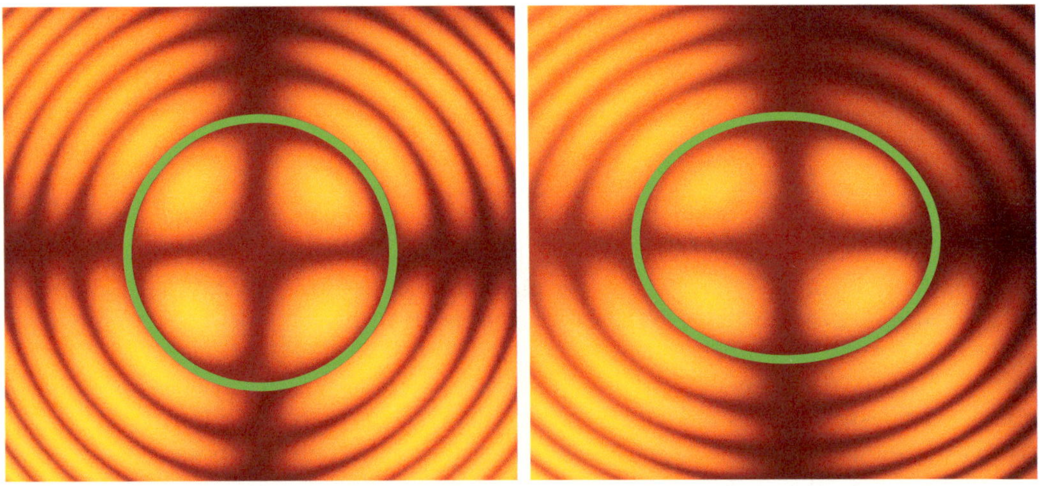

Fig. 26. PWO interference images. The highlighted first order fringe is fitted by the model from eq. 34. On the left, the crystal is stress free, on the right, it is subjected to a uniform, uniaxial compressive stress

Since the evaluation of the refraction index variations by means of (eq. 36) is a hard task, owing to the lack in the knowledge of the Π matrix, an alternative option is to evaluate the fringe distortion by means of an experimental index correlated to fringe distortion; to this purpose it was defined an elliptical ratio C_{ell} (Cocozzella N. et al., 2001) as:

$$C_{ell} = \frac{a}{b} - 1 \tag{38}$$

where a and b are the major and minor axes (along x and y respectively) of the first order isochromatic fringe obtained by observing the crystal in a plane polariscope in dark field (fig. 27). Therefore, in an empirical way, it is established a link between internal stress and fringe distortions by defining the photoelastic constant f_σ (Cocozzella N. et al., 2001) as:

$$\sigma_y \cdot f_\sigma = C_{ell} \tag{39}$$

In a series of works (Cocozzella N. et al., 2001; Lebau M. et al., 2005), it was experimentally verified in PWO samples that f_σ is a constant for a sample with thickness z = d, as C_{ell} depends linearly on σ_y. So it is necessary to systematically evaluate f_σ for PWO samples with different thicknesses d to relate (eq. 39) with the parameter d. Once this photoelastic parameter is known by calibration, which means experimental loading of a crystal sample with known loads, then the same parameter can be used on unloaded samples to assess if an internal stress state is present; the amount of distorsion of the isochromatic fringe provides an empirical assessment on the existence of internal stress.

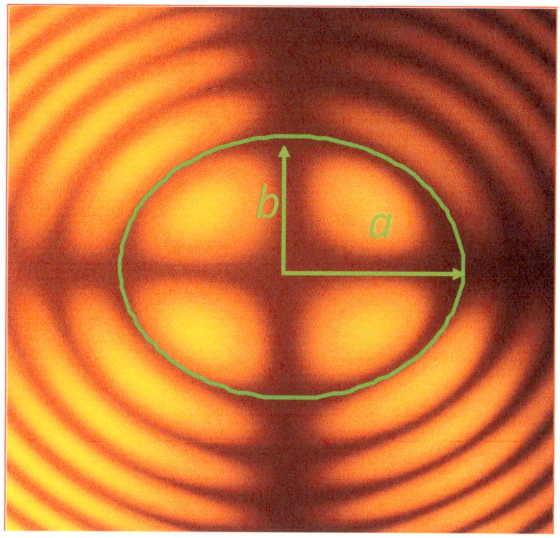

Fig. 27. Parameters used to define the elliptical ratio

In order to know f_σ as a function of d it is necessary to have a set of good quality PWO samples to which a known uniaxial load has to applied. This procedure was described in (Davì & Tiero, 1994); in that case samples have been chosen respecting the "De Saint Venant" conditions with thickness ranges from 5 to 15 mm and a dedicated compression loading machine was used. A dedicated polariscope employing a He-Ne laser source (λ = 632.8 nm) to perform the quality control tests can be designed (fig. 28) according to the classical polariscope theory (Born M., 1975).

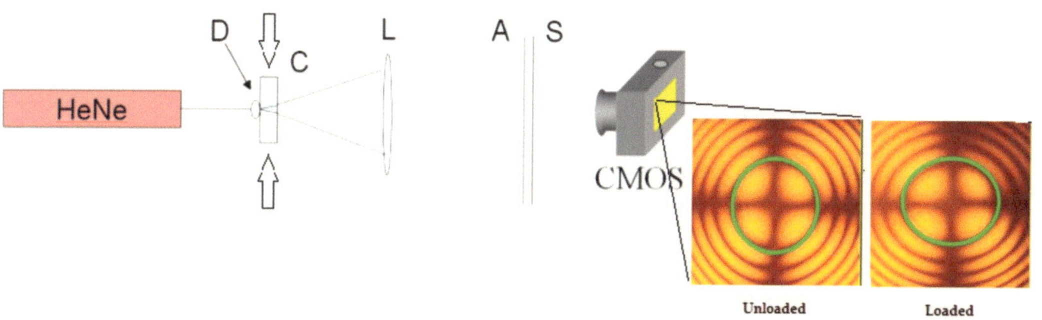

Fig. 28. Laser-light plane polariscope. In the dark field configuration, the analyser is set perpendicular to the light polarisation. D = glass diffuser. C = sample. L = convergent lens. A = analyser. S = ground glass screen

Some changes must be introduced when using laser source instead of a non-coherent diffused light source with respect of classical polariscope (Lebeau et al., 2005; Lebeau et al.,

2006; Frocht, 1941). As the laser light is already linearly polarized, the polarizer is not required. Moreover, interference fringes are obtained in convergent light so a small ground-glass diffuser was positioned just before the sample. Finally, all the parallel rays emerging from the crystal are focused on a ground glass screen, creating a bijective correspondence between the propagation direction inside the crystal and a point on the screen (Born & Wolf., 1975). C_{ell} is systematically measured versus f_σ applying uniaxial stresses along the y axis crystals, for different thicknesses of the PWO samples. All measured C_{ell} values exhibit a linear trend with the stress σ in the loading range 0-4 MPa. Four cases are reported in figure 29.

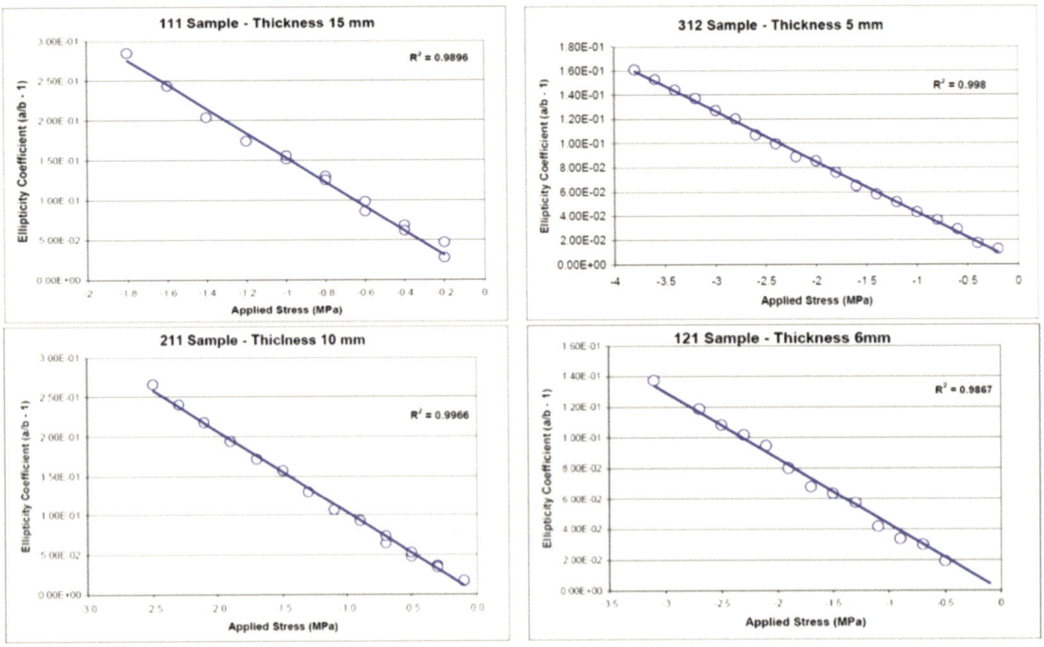

Fig. 29. The elliptical ratio C_{ell} as a function of applied stress for the four samples of different thickness

Numerical simulations, based on equation (34), and experimental data confirm that, C_{ell} linearly scales with the thickness. The resulting f_σ must also follow a linear dependence on crystal thickness z. The experimental data obtained using a laser with wavelength $\lambda = 632.8$ nm, lead to the evaluation of f_σ through a linear regression (correlation coefficient $R = 0.997$):

$$f_\sigma = 0.0172(\pm 0.0049) - 0.0114(\pm 0.0005)\, z \qquad (40)$$

Numerical simulations from equations (eqs. 34 - 40), using realistic values of n_x, n_y, n_z, confirm the linear behaviour of f_σ versus z at least for z value not too close to z = 0 (for z=0 the analysis loose physical meaning) and for z lower than 20 mm. For larger samples (up to 30 mm), only simulations have been performed , in this case a second-order law rules the variation of f_σ versus z. For z ranging from 5 to 15 mm the linear law can be applied (Ciriaco et al., 2007).

4.3.1 Mapping residual stress distribution in a crystal boule

The knowledge of the photoelastic constant f_σ allows the evaluation of the internal stresses for PWO samples by means of the determination of the elliptical ratio, observing the crystals by means of plane polariscope using the same light length. The residual stresses developed during crystal growth tend to increase proportionally to the boule diameter, due to the thermal gradients resulting from growth conditions. The presence of residual stress can hardly be solved by process control. The knowledge of the stress distribution inside the sample during or after growth, can be used as a quality control technique and provide feed-back for growth process optimization; furthermore, it can address useful information for planning the mechanical processing.

In order to prove this concept, PWO samples have been studied by the photoelastic method explained above. For each sample, it is therefore determined the ellipticity coefficient C_{ell} of the first isochromatic fringe and through the photoelastic constant it is derived a stress estimate by:

$$\sigma = C_{ell} / f_\sigma \tag{41}$$

The boule was grown using the Czochralski method, with optical axis orthogonal to the sample axis (fig. 30); 8 samples were cut from the boule, with reference to the X-Y-Z cartesian frame of the figure. The stress distribution has been mapped in the samples. at different locations (x,y,z).

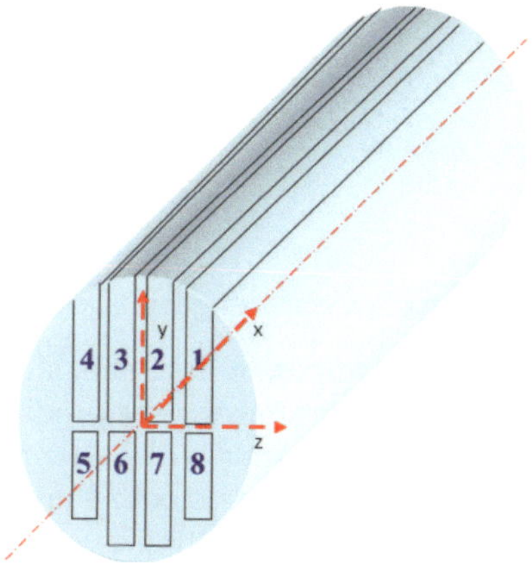

(a)

(b)

Fig. 30. Samples position respect to the boule and reference axes (a); a photo of the 8 samples (b)

The mapping gives the possibility of the 3D recostruction of the stress distribution inside the boule; of course, one should taking into account that what observed is a stress state in the samples after cutting, therefore it represents the stress state after relaxation. If an appropriate model would be available, it could be possible to reconstruct the effective stress inside the boule after the growth before the cutting. In fig 31 the data for sample 2 are reported. From the overall data we can deduce that the stress decreases from seed to boule end. As expected, owing to delicate initial growth phase, the classical constant gradient distribution is due to high peripheral tension compared to low internal compression. It should be noted that values are taken after the cutting off of crystal volume in a larger boule; this changes the internal stress state. The original tensional state is unknown and certainly can be higher because of stress relief from cutting. However, the method gives information about the growth of crystals that are fundamental to understanding the growth phenomenon and for the design of the production cycle. Therefore it provides a method for quality control and tuning of crystal production through the analysis of samples taken from the process.

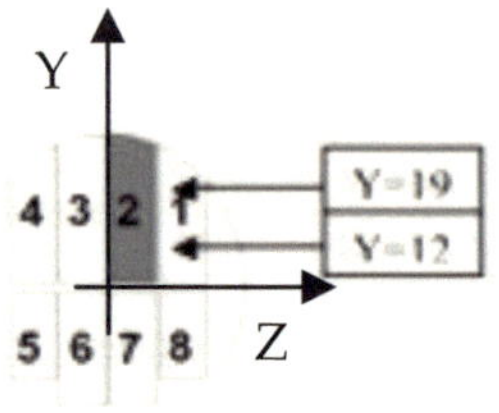

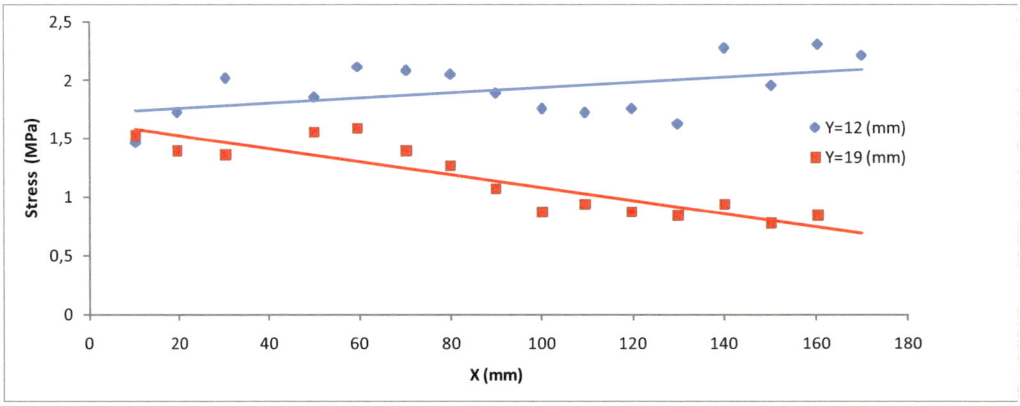

Fig. 31. Stress map inside the sample 2 as a function of sample length x, starting from the seed

A theoretical model of stress is needed in order to extrapolate the initial stress in the boule. Given this limit, this method allows for modifying and optimising the production process. It may lead to re-shaping the boule cone profile to obtain crystals with lower residual stresses. Moreover the obtained data pave the way to balance the process before and after annealing (stress-relief pre-cuts) and for appropriately sequencing the cutting operations, in order to reduce the losses by breaking.

4.3.2 Quality index

Given the characteristics of the quality control method presented above, it emerges that if a synthetic quality index would be available, the procedure could be helpful in tuning a production process.

It is worth to remark again that photoelasticity provides information about stress components in the plane normal to the polariscope axis, aligned in the direction of the optical axis. It is necessary to stress that what is observed, results from the effect of the principal stress difference absolute value, e.g. $|\sigma_{11}-\sigma_{22}|$, in the plane normal to the observation direction. So, we cannot measure, as a matter of fact, the value of each stress component. A single stress component is measured in the case of clearly monoaxial stress field, as in the classical case of an imposed stress by a loading device (Rinaldi et al., 2009, Lebeau et al., 2005) In what follows, for convenience the expression stress-level has been used instead of the absolute principal stress difference. The reported analysis is applied to PWO, but can be extended to the overall family of similar uniaxial crystals. The family of circular isochromatic fringes are in a plane normal to the optical axis, in the case of unstressed crystals. Stress components contained in this plane induce a biaxial state and we detect a family of Cassini-like fringes as explained before, and following the protocol explained in this chapter , the internal stress level was computed from the measurement of the elliptical ratio.

The production of a large number of high quality crystals is a hard goal to perform owing to the complex production route involving different parameters to be controlled. For this, a quality control plan is needed. The need to produce while minimizing costs and production time, leads to the implementation of a fast and easy feed-back on growth parameters, such as temperature distribution and solidification-front velocity. In the following it is proposed a quality feedback for process optimization, obtained by a fast characterization of sample crystals taken from the pre-serial production using the photoelastic methods outlined above. For this purpose, it is possible to use also a classical plane polariscope (in alternative to the previous explained laser light polariscope). In the proposed analysis a green monochromatic light (λ = 530 nm) has been used.

Quality control dedicated to the selection of the better production process is made by taking crystals randomly extracted from every delivered batch.

In this example the ingots were produced by a "modified" Bridgman method: the platinum crucible has a 35x35 mm^2 section and is about 300 mm long. The fusion front temperature is controlled at 1250 (nominal) ± 0.5 °C. A slow and steady shift of the solidification front is produced by the movement of a thermal gradient estimated 30°C/cm at the solid-liquid interface. Using the Bridgman method, the optical axis is along the longitudinal axis, for this, each ingot was cut in 10 mm-thick slices, numbered starting from the seed as shown in fig. 32.

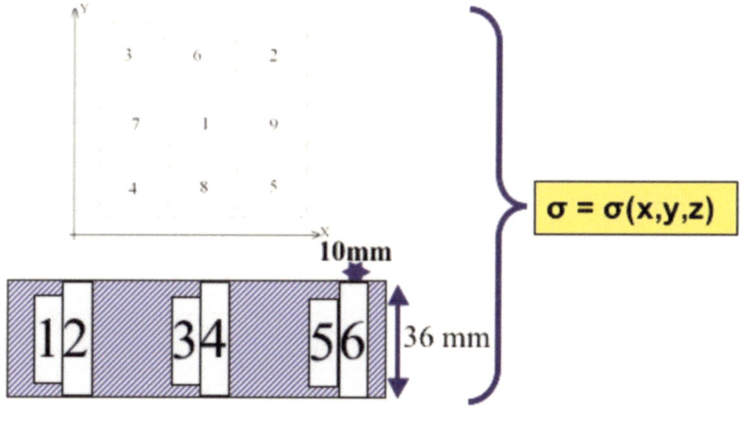

Fig. 32. Typical positions observed in each slice. Position 1 is called centre, positions 2-5 corner and position 6-9 "edge middle" (upper figure). Position of the slices cut normal to the crystallographic C-axis (optical axis) (lower figure)

The stress can be measured in typical locations in each slice, as is shown in Fig. 32. The cut of the slices orthogonal to the growth axis (Z), induces a stress relaxation in the axial components σ_z and in the shear stresses τ_{xz} and τ_{yz}.

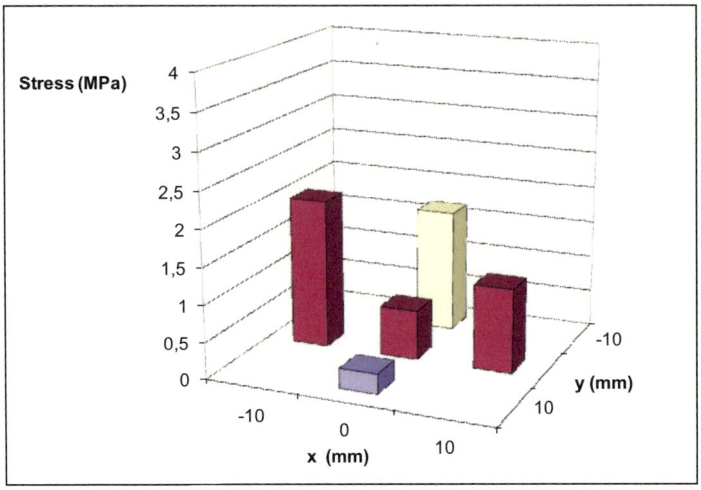

Fig. 33. Principal stress along x and y directions on slice 5

A 3-D reconstruction of stress distribution is possible staring from the measurements for each slice in the typical positions. The reported example leads to the following data (fig. 33, 34 and 35).

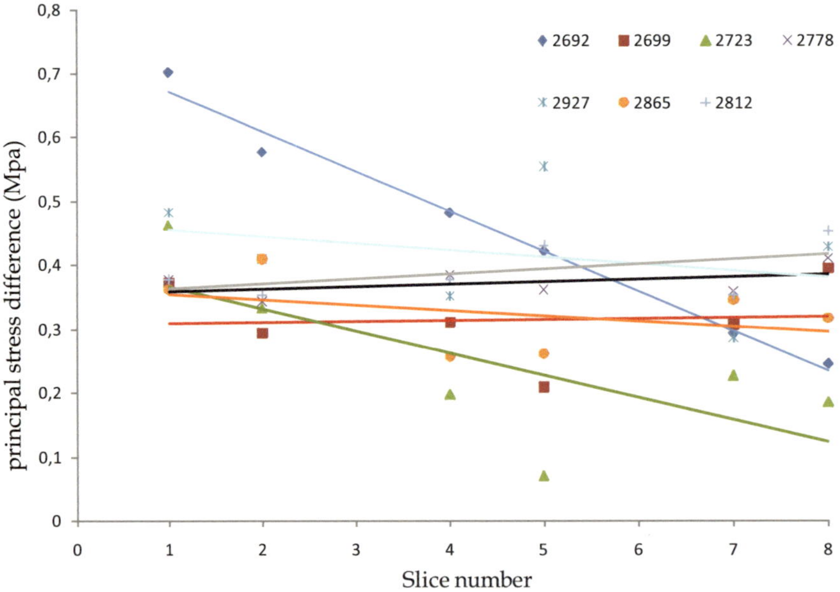

Fig. 34. Longitudinal average stress at slice corners for the 7 samples labeled in the figure. Generally decreasing trend from the seed to the end of crystal is confirmed. Linear regression have been shown for each crystal

The large stress difference at the corners for Sample 2692 in the first slice can suggest the presence of thermal problems after the solidification in the initial part or a crucible defect. The single crystal stress homogeneity seems to be restored at the crystal end. A not optimal thermal field displacement can also be supposed. The speed of temperature change could have been too high in the early growth stages of 2692 sample.

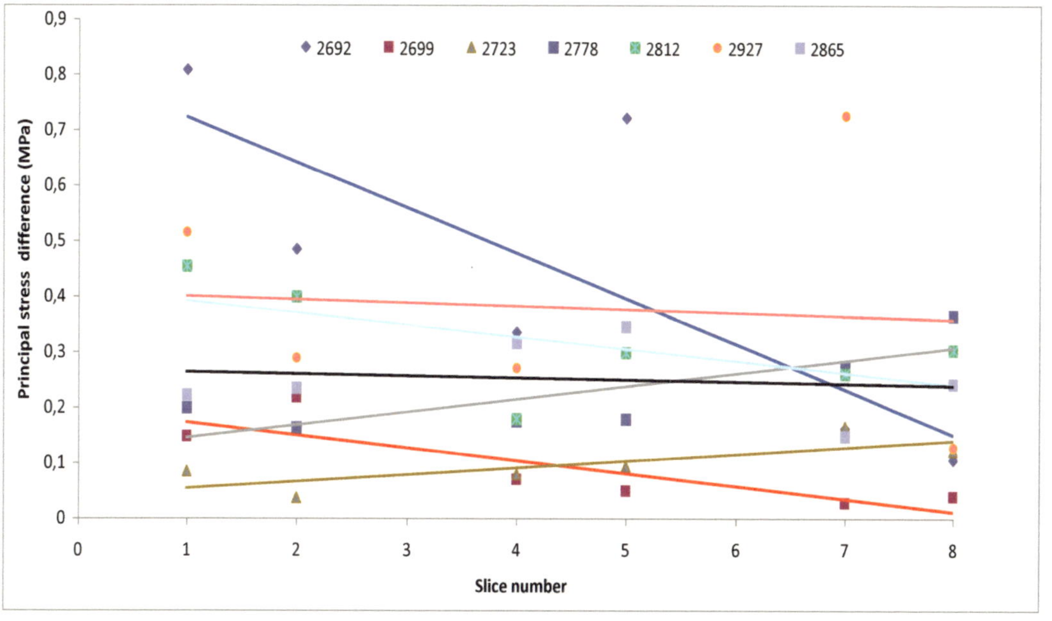

Fig. 35. Longitudinal trend at centre of the slices for the 7 samples labeled in the figure

The data collected can provide an analytical overview of the problem by assessing the evolution of stress in the mapping of the crystals. This analysis puts emphasis on individual problems that emerged during growth. The need of a synthetic index of quality derives from the necessity to understand what production parameters must be changed in the direction of better overall quality. The problem lies in handling a large quantity of data, as in the example shown. Extracting the average measures of stress from 6 samples we can write the following table:

SAMPLE	St. Dev. [MPa]	Mean [MPa]
2692	0.38	0.62
2699	0.38	0.36
2723	0.37	0.29
2778	0.33	0.54
2812	0.23	0.52
2865	0.33	0.39

Table 1. Mean stress and standard deviation of 6 samples

Quality control dedicated to the selection of the better production process is made by taking crystals randomly selected from every delivered batch. A large set of data can be collected from ingots produced in different conditions.
The residual stress inside the crystals can vary heavily depending on the type of process used but also as a function of the position inside the crystal.
In order to identify the best method of production and to guide the development of production methods, it's necessary to find a summary measure of the quality of the samples. To evaluate in synthetic manner the overall quality of each crystal, the data can be put in a Cartesian plane, with X axis the average stress level (σ_{av}) of all the measured points from each ingot, and Y axis the corresponding standard deviation (S). It is possible to define an "index of quality", R (Rinaldi et al., 2010):

$$R = \sqrt{(kS)^2 + \sigma_{av}^2} \tag{42}$$

In brittle materials, as in particular single crystals, the stress level variation may cause breaking risk, due to the gradient, as dangerous as the than an high average value σ_{av}. For this, larger values of R correspond to the lower crystal quality. The standard deviation S is weighted with a coefficient k (≥ 1) to be set from the producer in order to achieve the results identified on the basis of the required quality standards. k amplifies the standard deviation with respect to the average stress value highlighting the existence of gradients and inhomogeneity.
The general purpose of the (eq. 42) need some explanations linking the mathematical properties of (eq. 36) to the physical meaning of the parameters.
In the plane S-σ_{av}, the R=const. is a boundary delimiting an area related to the ingots quality. Each crystal is identified by its σ_{av} and S. The accepted crystals are in the area below the curve R=const. From the producer point of view, R will be the maximum σ_{av} acceptable (that is σ_M). Choosing k following the experimental experience and related to the quality, in terms of homogeneity required, it is easy to calculate the maximum S, that is $S_M = R/k = \sigma_M/k$. It is evident that the maximum standard deviation, S_M , must be lower than, or at least equal to the average stress level in the limit case. From the quality point of view the need that the stress variation must be controlled leads to a new constrain:

$$S = c\sigma_{av} \tag{43}$$

with c the slope of a line in the plane S-σ_{av}, c must be chosen for $c \leq 1$, as explained above. The acceptance area is below the curves. These constrains depend both from the average stress and the weighted stress standard deviation. The condition that the standard deviation cannot exceed the average stress value is thought fundamental in brittle materials, in fact, the stress variation, due to the crystal heterogeneity, might increase the breaking risk. In fig. 36 an example of this technique is shown.
The boundary curves delimiting the acceptability area are obtained with k = 1.5 and the limit quality value is R = 0,7 MPa. $c = 1$ was chosen as a limit case. These parameters can be set by the manufacturer according to statistical data of accidental cracking. Only two samples (highlighted by the blue oval) lie in the acceptability area, that is below the curves

42 and 43. The fact that three samples (2865, 2723 and 2699) are close to the line where standard deviation is equal to the average stress value clearly highlights the existence of high stress gradient. In particular the 2723 and 2699 samples are below the curve (eq. 42) but above the line (eq. 43), therefore are not accepted due to the high stress gradient. From this analysis it is possible to conclude that the process 2865 and 2812 have the best production parameter and indicates the development direction to improve the crystals quality. As final analysis it is possible to perform a comparison between the best and worst samples to put in evidence the specific critical points, as shown in the figure 37 (Rinaldi et al., 2010).

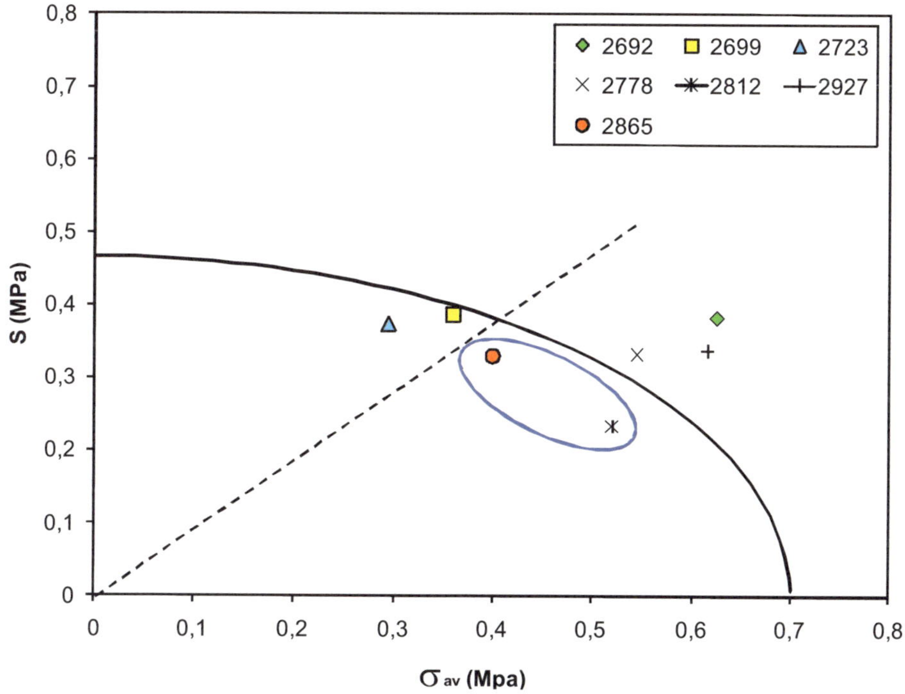

Fig. 36. Different samples in the plane S-σ_{av}. Only samples 2865 and 2812 are accepted, meeting the quality requirements

Fig. 37 clearly put in evidence that the low quality sample 2692 exhibits high absolute stress values but also high stress variation. On the other hand, the higher level quality sample 2865 has lower absolute stress values and it appears more homogenous. It indicates that the specific production process is well tuned and that probably the production parameter and gradients are well controlled yielding an homogeneous sample.

4. Conclusions

Scintillating crystals are widely used in radiographic systems, in computerized axial tomography devices and in calorimeter used in high energy physics. Scintillating crystals are cut to their final shape from an ingot, which is grown by classical crystal growth techniques. From a mechanical point of view, the quality of a crystal is closely related to its geometry, to the surface finish and moreover to its internal state of residual stresses. In particular an excessive residual stress is a major cause of crystal breakage, which often may occur during crystal cut, during surface finishing or, even worse, only when the crystal is assembled into the detector units.

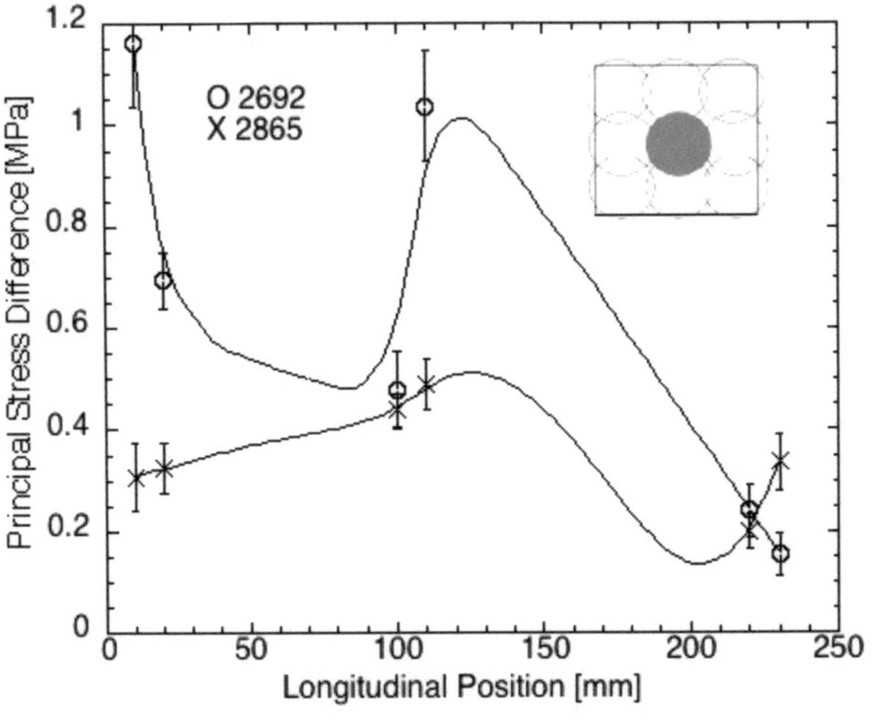

Fig. 37. Comparison between better/worst samples at centre position of the slices (as shown by the inset) as a function of the longitudinal position from the seed

The need to produce high-quality crystals is therefore fundamental both to avoid damage during assembly and finishing of crystals. Crystal performance in terms of production of light strongly depends on surface finish, therefore crystal tool machining is a crucial process to achieve the high performance needed in the case of scintillating crystals for high energy physics and medical applications.

For optimal crystals performance, attention has therefore to be paid to the mechanical aspects of the production process; from the mechanical point of view this can be guaranteed by adequate quality control methods. If adequate quality inspection of crystals is achieved, this has the potential to prevent breaking during the assembly in an array. The authors have reported the experience which was made within the collaboration with CERN to the development of the electromagnetic calorimeter of the Compact Muon Solenoid (CMS) presently working at CERN. From an industrial point of view, the trend is to use smaller and smaller crystals for biomedical instrumentation; in such crystals the surface plays an even more relevant role in the production of light. For this reason, the final mechanical processing is important for producing high quality crystals. Therefore the experience made for the large crystals of CMS is in general valuable to guide the development of suitable quality control methods for scintillating crystals and in particular for biomedical industry.

An increasing attention to limit production costs requires an assessment of crystal quality by a fast and possibly non-destructive methodology, finalized to tune and keep under control the crystal growth and finishing processes, and to eliminate from the production process the crystals which are produced out of tolerance, thus reducing downtime and waste.

Internal residual stress is not only the most important causes of breaking, but may be interpreted as an overall quality indicator.

Residual stresses, induced by temperature spatial and temporal distribution during the growth and by complex interaction of the melt material and the growing ingot with the crucible, play an important role in production yield in terms of cracking risk during mechanical processing and heterogeneity in finished crystal properties. A regular production of good crystals requires a quality control plan leading to a fast and easy feed-back on growth parameters, such as temperature distribution and solidification-front velocity. The developed methodology for quality control consists in providing the producer a quality feedback for process control and optimization, obtained by experimental characterization of sample crystals taken from the pre-serial production by photoelasticity. Photoelasticity is a measurement technique aiming to study and evaluate the stress state inside a transparent medium. In traditional photoelasticity a plane stress state distribution is studied, by means of a plane polariscope. Usually it is applied to optically isotropic media, Perspex, glass or optically isotropic crystals, which become birefringent under stress.

Referring to naturally anisotropic media, such as uniaxial and biaxial scintillating crystals, the observation of unstressed crystals, by means of a plane polariscope, shows a symmetrical interference pattern due to the symmetry of the lattice. An internal stress state induces a lattice symmetry distortion. The modelling of the interference image obtained from an anisotropic uniaxial crystal when a stress state is present, and the measurement of characteristic parameters of the interference fringe pattern offers a mean for quality control able to provide spatially integrated information on the internal stress.

Although a mathematical modelling of the piezo-optical effects is possible, the knowledge of the coefficients of the model is not complete and accurate; therefore a semi-empirical approach is proposed. This leads to the definition of a parameter correlated to the deformation of the fringe pattern of a crystal under stress. The ellipticity, introduced into the fringe pattern is due to the stress state. Linear regression of experimental data of ellipticity vs. stress, collected with crystals undergoing known stress states, allows to build an experimental relationship which can then be used for quality assessment of unknown crystal samples. If the internal stresses are residual stresses, this allows to develop a quality control method to detect the presence of residual stresses non invasively. The method could be applicable on samples taken from the production, for process optimization and control, or it can be applied on the finished crystal as a pass-fail filter for removing from the batch all samples which exceed prescribed limits.

The statistical analysis of many data from samples randomly taken from a pre-serial production allows to build a quality index depending on mean stress value and on its standard deviation, which are quantities related to residual stress intensity and gradient. This index can be used as a global indicator of process capacity to produce crystals with acceptable residual stress state.

This method suggests therefore a quality indicator to synthetically evaluate the production by means of a criterion of acceptability, useful in general crystal production.
The procedure and the quality index have been validated on $PbWO_4$ (PWO) uniaxial scintillating crystals; they have been intensively studied owing to the necessity of large amount of them (about 82000 large crystals) for the CMS. In fact the production effort needed a fast and reliable quality control.
In that case study, the attention was focused on the measurement of residual stresses over the whole crystal volume, particularly in sections cut perpendicularly to the optical axis. The collected data enabled the construction of a 3-dimensional stress map for each crystal from a pre-serial production. The detection of internal stress and defects, can be related to the corresponding production parameters and may suggest improvements in the production process or highlight criticalities to be solved before a serial production is started.
What presented is demonstrated for uniaxial crystals, but the same approach can be extended to all types of crystals, particularly those of a new generation (LYSO, LuYAP) as a function of their applications in high energy physics and for medical diagnostics.
As conclusive remarks, we have to consider that other techniques should be taken into account to analyse crystals quality. In particular researchers are paying attention to experimental methods for the assessment of the surface damage, which are not treated in this chapter: X-ray diffraction (XRD), grazing incidence X-ray diffraction (GID) and RX reflectometry (XRR), (Mengucci et al., 2005).

5. Acknowledgment

This work has seen the contribution of many colleagues, amongst which we thank prof. Giuseppe Majni and Prof. Fabrizio Davì, who contributed through many fruitful discussions. A relevant part of the work has been developed with the direct contribution of students amongst which we warmly thank Nicola Cocozzella, who was the first to deal with this topic, and PhD candidates, in particular dr. Andrea Ciriaco, whose PhD thesis constitutes a milestone in our work.

6. References

Auffray E., Cavallari F., Lebeau M., Lecoq P., Schneegans M., Sempere-Roldan P. (2002). *Crystal conditioning for high-energy physics detectors*, Nuclear Instruments and Methods in Physics Research Section A (NIM A) 486, pp. 22-34.

Baccaro S., Barone L. M., Borgia B., Castelli F., Cavallari F., Dafinei I., de Notaristefani F., Diemoz M., Festinesi A., Leonardi E., Longo E., Montecchi M., Organtini G. (1997). *Ordinary and extraordinary complex refractive index of the lead tungstate ($PbWO_4$) crystal*. Nuclear Instruments and Methods in Physics Research Section A, 385, pp. 209-214.

Born M., Wolf E., (1975). *Principles Of Optics*, 6th ed., Pergamon press, New York, USA.

Cocozzella N., Lebeau M., Majni G., Paone N., Rinaldi D. (2001). *Quality inspection of anisotropic scintillating lead tungstate ($PbWO_4$) crystals through measurement of interferometric fringe pattern parameters*. Nuclear Instruments and Methods in Physics Research Section A (NIM A) 469 3 pp.331-339.

Ciriaco A., Davì F., Lebeau M., Majni G., Paone N., Pietroni P., Rinaldi D. (2007). *PWO photo-elastic parameter calibration by laser-based polariscope.* Nuclear Instruments and Methods in Physics Research A 570, 55–60

Dally J. W., Riley W. F., (1987). *Experimental Stress Analysis,* 2nd ed., McGraw-Hill Book Company, Singapore.

Davì F. and Tiero A., (1994). *The Saint-Venant's problem with Voigt's hypotheses for anisotropic solids.* J. Elasticity 36, pp. 183-199.

Frocht, M.M. *Photoelasticity,* Wiley, New York, 1941.

Hodgkinson I. J., Wu Q. H., (1997). *Birefringent Thin Films and Polarizing Elements,* World Scientific, New Jersey, USA.

Hofstadter, R (1949). *The detection of gamma-rays with thallium-activated sodium iodide crystals.* Phys.Rev. 75, pp. 796-810.

Ishii M., Kobayashi M. (1996). *Mechanical properties of PWO.* Nuclear Instruments and Methods in Physics Research Section A (NIM A) 376, pp. 203-207

Lebeau M. (1985). *Monocrystalline bismuth germanate $Bi_4Ge_3O_{12}$ (BGO) recent results on mechanical properties.* J.Mat.Sci.letters 4, 779-782.

Lebeau M., Ciriaco A., Gobbi L., Majni G., Paone N., Pietroni P., Rinaldi D. *Quality monitoring in PWO scintillating crystal production during R&D phase* Proceedings of the 8th International Conference on Inorganic Scintillators and their Use in Scientific and Industrial Applications, Publisher: National Academy of Sciences of Ukraine, Kharkov (2006) 334-337.

Lebeau M. (2003). Crystal Growth Technology. In *Methods and Tools for Mechanical Processing of Anisotropic Scintillating Crystals,* pp.561-586. Wiley and Sons, London.

Lebeau M., Pietroni P., Gobbi L., Majni G., Paone N., Rinaldi D. (2005). Mapping residual stresses in $PbWO_4$ crystals using photoelastic analysis., *Proceedings of Scint'03 7th International Conference on Inorganic Scintillators,* September 8-12, 2003, Valencia, Spain. NIM A537 154-158

Lecoq P. et al. (2006). *Inorganic Scintillators for Detector Systems.* ISBN-10 3-540-27766-8 Springer Berlin Heidelberg New York.

Mengucci P., Di Cristoforo A., Lebeau M., Majni G., Paone N., Pietroni P., Rinaldi D. (2005) *Surface quality inspection of $PbWO_4$ crystals by grazing incidence X-ray diffraction.* Nuclear Instruments and Methods in Physics Research Section A (NIM A) 537, 207-210.

Perelomova N. V. and Tagieva N. M., (1983). *Problems in Crystal Physics with solutions,* Mir Publishers, Moscow, Russia.

Pietroni P., Lebeau M., Majni G., Paone N., Rinaldi D. (2005). *Development of Young's modulus non-destructive measurement techniques in non-oriented CeF_3 crystals.* Nuclear Instruments and Methods in Physics Research Section A (NIM A) 537, 203-206.

Rinaldi D., Lebeau M., Majni G., Paone N. (1997). *Photoelasticity for the investigation of internal stress in BGO scintillating crystals.* Nuclear Instruments and Methods in Physics Research Section A (NIM A) 317-322.

Rinaldi D., P. Pietroni, F. Davì (2009). *Isochromate fringes simulation by Cassini-like curves for photoelastic analysis of birefringent crystals.* Nuclear Inst. and Methods in Physics Research, A 603, 294–300

Rinaldi D., Ciriaco A., Lebeau M., Paone N. (2010). *Quality control on pre-serial Bridgman production of $PbWO_4$ scintillating crystals by means of photoelasticity* Nuclear Inst. and Methods in Physics Research, A 615, 254–258

Walhstrom E.E., (1960). *Optical Crystallography*, Wiley, New York, (USA).

Weber M., Monchamp R. (1973). *Luminescence of $Bi_4Ge_3O_{12}$*- Journal of Applied Physics 44: 5495-5499.

Wood E. A. (1964). *Crystal And Light*, Van Nostrand Company, New Jersey.

Wooster, W. A. (1938). *A test-book on Crystal Physics*, Cambridge University Press.

24

Effect of Last Generation Additives on the Concrete Durability

Ana M. Carvajal, M. Soledad Gómez, Pablo Maturana and Raul Molina
Pontificia Universidad Católica de Chile
Chile

1. Introduction

The influence of carbonation on corrosion of reinforcement depends on the degree of ease of diffusion of CO_2 through the concrete from its surface, also on environmental conditions, on the pore structure of concrete (cement, aggregates, and water (without additives)) and on the W / C, where a high ratio generates porous and permeable mortar and concrete (Duran C., 2003; Troconis O. et al., 2006).

Permeability is not necessarily related to porosity, but depends on the geometry of the pores and the distribution of pore sizes: two porous bodies can have similar porosities but different permeability, so it is important to consider the penetration of CO_2 into the concrete. If the concrete is not permeable, the attack will be relatively superficial and limited to the surface. The attack in concrete is governed by molecular diffusion, which is much slower than convection processes. The use of concrete with low permeability is the primary means to prevent or minimize the effects of external attack (Morin et al., 2001; Papadakis et al., 1992).

A well-proportioned mix of aggregate, which follows a continuous grading curve will produce concrete of good workability, high cohesion and a reduced tendency to segregation. At the same time it will be slightly porous and therefore possess a prolonged durability. Superplasticizer additives added to the mix, filling the interstitial space between large particles, which can cause a high density, high strength and resilient material, with a smaller amount of mixing water (Erdogdu S., 2000; Morin et al., 2001).

The main mechanism for CO_2 transport in concrete is difusion, and with moisture, carbonation leads, a phenomenon to be considered from the viewpoint of durability of reinforced concrete (Carvajal et al., 2006).

There are expressions that relate the diffusion coefficient of concrete with compressive strength, where increase of resistance, decrease of diffusion coefficient. Because the phenomenon of diffusion of gases is of long-term, resistance in ancient age must to be taken into account and not the resistance usually specified at 28 days.

1.1 Carbon dioxide

The CO_2 could form carbonic acid with water. The entry of CO_2 inside the concrete is produced through the pores and capillaries of the cement paste. As a result, the pH of carbonated concrete decreases and once the carbonation front reaches the armor begins to dissolve the passive film that protects steel from corrosion.

1.2 Carbonation of concrete

The importance of considering the carbonation in reinforced concrete structures, increases in holding that causes a chemical imbalance and a decrease in pH of water in the pores of the concrete from 12.6 to 13.5 to values around 9, causing depassivation strengthening reinforcements adverse reactions of chlorides and sulfides, and exposing them to corrosion.
Without the passive layer, the steel is corroded as if it were exposed to the environment without any protection, and, the carbonation depends on many factors, but those with a higher incidence are: type of cement, concrete permeability, W/C ratio, concrete curing, relative humidity and CO_2 concentration in the environment (Barrera et al., 2003; Carvajal et al., 2003; da Silva et al., 2002).
Carbonation is the process by which atmospheric CO_2 is combined with calcium hydroxide [$Ca(OH)_2$] to form calcium carbonate, losing its alkalinity by decreased pH.

$$Ca(OH)_2 + CO_2 \rightarrow \underset{\text{Insoluble carbonate}}{\underset{\downarrow}{CaCO_3}} + H_2O$$

In a mass of plain concrete, the carbonation can be beneficial, improving some of its properties, such as breaking loads between 22% to 78% higher, to obtain a denser concrete generated by an open porosity that is closed (5 to 12%).
On the other hand, the attack produced by carbonic acid, which reacts with calcium hydroxide released from the hydration process of concrete which promotes its alkalinity, will form acid carbonates or bicarbonates (more soluble than carbonates) that has lower pH. Due to this decrease in alkalinity of the concrete, it loses the passivity of the reinforcement, leaving them prone to corrosion.
The CO_2 present in polluted environments produces carbonic acid that diffuses into the concrete mixing with pore water (Knopf et al., 1999).

$$CO_2 + H_2O \rightarrow H_2CO_3$$

$$2H_2CO_3 + Ca(OH)_2 \rightarrow \underset{\text{Soluble bicarbonate}}{\underset{\downarrow}{Ca(HCO_3)_2}} + 2H_2O$$

The water in the atmosphere, rain or fog, contains a slight amount of carbonic acid by absorption of atmospheric CO_2 and are exceptions the industrial areas and cities, where the fumes, especially heating, mixed with steam and the fog for a longer period of time, it depositing on all surfaces.
A depth that CO_2 has penetrated and reactions have occurred that has changed the pH, usually it`s called "carbonation front" (Thiery et al., 2007).
The alkalinity of concrete is mainly due to calcium hydroxide ($Ca(OH)_2$, pH 13 approx.) formed during hydration of cement silicates and alkalis that may be part of the cement. These substances place the pH of the aqueous phase contained in the pores between 12 and 14, most alkaline of pH range.
Corrosion will occur in concrete that has a permeability such that allow the carbonation to reach the concrete in contact with steel or soluble chlorides can penetrate to the steel. If the concrete is in a dry atmosphere (below 40% RH) or submerged in water (without air intake), the risk of corrosion to the reinforcement decreases. An optimum for the corrosion process is 50 to 70% RH (Troconis O. & Duracon Collaboration, 2006).

Considering the effect of carbonation, the pH decreases to values close to 9, which causes the passive iron oxide layer is destroyed (Duran C., 2003).

1.3 Accelerated carbonation chamber

As the carbonation is a long-term process, it was implemented a test system of accelerated carbonation, to attack the concrete more quickly and effectively, obtaining experimental results with more speed than the real time. The accelerated carbonation chamber was designed in many countries for this purpose and in general is to expose the concrete samples and continuous ideal environment for the development of carbonation, where four variables can be controlled: temperature, CO_2 concentration, relative humidity and pressure (Carvajal et al., 2003, 2006; Duran C., 2003). The conditions of T ° and RH ranges are 20 and 25°C and 50-70% respectively, due to these are the environmental conditions of higher penetration rate of CO_2.

To generate a constant environment in the system, the CO_2 pressure has not changes. Respect to the concentration of CO_2, the atmosphere of the chamber is saturated with 100% CO_2 (Carvajal et al., 2003, 2006).

The carbonation chamber, is in acrylic, 6 mm thickness and dimensions 1.00 x 0.50 x 0.50 m. The addition of pure CO_2 through pipes made of PVC previously adapted.

1.4 Rate of carbonation

A simple model to predict the rate of carbonation of concrete is that which relates the depth of carbonation with the square root of exposure time.

$$XCO_2 = KCO_2 \sqrt{t}$$

Where:

XCO_2 = depth of carbonation, mm

KCO_2 = carbonation constant : mm * year $^{-0.5}$

t = time: years

The information obtained can provide the time that is associated with a certain depth of carbonation. Likewise, it can gets the time associated to generate a greaterdamage, that is, reaching the reinforcement of the structure (CYTED, 1998; Carvajal et al., 2006).

1.5 Additives

The additives are chemicals added to concrete. Additives are defined as "a material other than water, aggregates and hydraulic cement used as a component of concrete or mortar and added to the mixture immediately before or during mixing" (American Concrete Institute, 1991).

1.5.1 Additives used in the study

The additives tested are classified as water-reducing admixtures of high rank. According to ASTM C494 classification are type A and F.

Higher Reducing Water- admixtures (HRWR) reduce the water content of concrete between 12 and 25%, which is why they are used to increase strength and reduce permeability of concrete by reducing water content in the mixture, or to greatly increase the settlement and produce a fluid concrete without adding water. Its use is essential for high-strength concrete with high contents of cementitious materials and silica fume mixtures.

1.5.1.1 Polycarboxylate-based additive

The polycarboxylate based additive is an additive high water reducing capacity, based on synthetic polymers allows maximum flow, high cohesion and maintain the workability of the mixture for long periods.

1.5.1.2 Nanosilica based additive

Nanosilica is a nano additive in liquid silica-based nano-sized particles. It is recommended as much water reducer, high activity. Belongs to a last generation additives, where chemical reactions in the concrete make nanoparticles of silica nanoparticles cement.

1.6 Capillary absorption

Capillary absorption is a reaction that has a concrete (porous solid) from having contact with a liquid, which penetrates and goes into their pores as well as the relationship between their section and the surface tension permits.

According CYTED (1998), the Manual Inspection Evaluation and Diagnosis of Corrosion in Reinforced Concrete Structures, is defined as follows: "capillary absorption is the mass of water per unit area that can be absorbed into the capillaries when the concrete is in contact with liquid water. Represents the effective porosity or accessible to water and therefore to an aggressive environment.

To measure the absorption of concrete, tests performed on samples previously conditioned or witnesses to this effect, to measure the mass absorved for differents times, since it comes in contact with the liquid

This test is simple to implement and to determine the absorption coefficient of the material according to the amount of water absorbed per unit area at a given time (root of time).

2. Experimental procedure

2.1 Materials

Pozzolanic cement, potable water and crushed gravels were used for the manufacture of concrete with and without additives.

The gravels with size range of 6-40 mm were used. The fine aggregate was river sand with a maximum size of 4 mm.

Additives: nanosilica and polycarboxylate.

The chemical composition of the Pozzolanic cement is shown in Table 1.

Com	SiO_2	Al_2O_3	Fe_2O_3	CaO	MgO	Na_2O	K_2O	SO_3
%	29.7	4.6	3.3	56.6	1.5	0.2	0.4	2.2

Table 1. Composition of Pozzolanic Cement

2.2 Specimens preparation

2.2.1 Specimens cure

Specimens were demounted 2 days after casting, and then they were cured in humid chamber for 28-days, with a 95 + 3 % R H and 20 + 2°C temperature range.

2.2.2 Grade of concretes

The concrete without additives was H25 with w/c 0.60 and a slump cone of 19 cm.

The concrete with polycarboxylate was H25 with w/c 0.48 and a slump cone of 19 cm.

The concrete with nanosilica was H25 with W/C 0.55 and a slump cone of 10 cm.
The different W/C ratios are the result of the search for a particular settlement for each concrete, due to the properties of the additives used.

2.3 Experimental method

2.3.1 Compressive strength

Compressive strength test of the concrete was made according to NCh 1037 (ASTM C-39)

2.3.2 Accelerated carbonation test

The specimens that enter to accelerated carbonation chamber, must be conditioned to favour the entrance of CO_2 to their inside through drying in oven at 60°C, temperature that does not damage the capillarity of the concrete, for 24 hours and/ or to invariable weight. The process of carbonation was accelerated using a controlled environment in a special apparatus: temperature (25+ 2 °C), Relative Humidity (50 - 70 %) and CO_2 concentration (100%), to expose the specimens for 5, 7, 9 and 11 days.
The method used to determine the carbonation depth was the application of a phenolphthalein solution in alcohol/water (50/50). For the measure of the carbonation depth, the methodology recommended by RILEM (1988) was used.

2.3.3 Capillary absorption

The test was made according to the standard DIN 18550-Part 1, drying four quarters of the specimens for a period of 48 hours at 50°C ± 10°C, until obtaining a constant weight. The dry specimens were isolated with a plastic film to avoid the humidity absorption from the environment. The time of the test was 48 hours.
The test was applied to the internal faces as well as the external ones with the purpose of discuss the possible differences in capillary absorption between both faces.
It was determined the coefficient of water adsorption (C_i), from the curve of water absorption accumulated in function of the root of time.

3. Results

For the specimens with nanosilica and polycarboxylate, without accelerated carbonation, it has a minimum evolution of strength between the ages of 28 and 58 days which are considered negligible. When they were carbonated presented an increment in the strength. For the specimens without additives and no carbonated, they get strength to late ages (26,5% of difference) while the carbonated specimens presented a decrease of the strength (Table 2).
The specimens manufactured with polycarboxylate additive show lesser carbonation depth and consequently lesser carbonation coefficient than the others. The results are in Table 3.
Abbreviations: P: polycarboxylate
N1: nanosilica
N2: without additives
Numbers: 5, 7, 9 and 11 are days of carbonation
The concrete with nanosilica presents an intermediate carbonation; higher than the concrete with polycarboxylate and lesser than the concrete without additive thus it shows coefficients of carbonation.

Type of concrete	28	56	58	Age of concretes (days)
Carbonated Policarboxilate	58.3	61.9	61.8	**Compressive Strength** **(MPa)**
No Carbonated Policarboxilate	58.3	58.3	58.3	
Carbonated Nanosilica	37.8	44.1	44.9	
No Carbonated Nanosilica	37.8	37.8	37.9	
Carbonated, without additives.	30.3	30.9	30.5	
No Carbonated without additives	25.3	32.0	32.0	

Table 2. Compressive strength with age of carbonated and no carbonated concretes

Type of concrete	Carbonation coefficient
5P	0.85
7P	3.30
9P	4.43
11P	4.92
5N1	7.19
7N1	9.69
9N1	9.55
11N1	10.32
5N2	12.74
7N2	11.79
9N2	11.12
11N2	12.72

Table 3. Accelerated carbonation coefficient for concretes with different days of carbonation

Days of carbonation	Polycarboxylate (P)	Nanosilica (N1)	Without additive (N2)
0	0,32	0,91	0,80
5	0,29	0,95	0,75
7	0,36	0,88	0,97
9	0,61	0,79	1,19
11	0,50	0,53	1,60

Table 4. Average capillary absorption coefficient for different type of concrete and time of carbonation

The concrete without additive presents the highest carbonation, thus it shows the highest carbonation coefficients.
The specimens with polycarboxylate additive had better response in capillary absorption, in all the times of carbonation. The results are in Table 4.
The concrete with nanosilica is the one that shows a higher coefficient in the initial days of the carbonation, decreasing as time passes, that indicates a higher absorption of water at the beginning than any of the other two.
The concrete without additive shows a higher absorption coefficient in the last days of the carbonation.
It was demonstrated that the specimens with high density show less carbonation. (Figure 1)
The highest strength shows the least capillary absorption and carbonation depth, and the highest densities, as it is possible to deduce with the results obtained summarized in the figures 2 to 6.

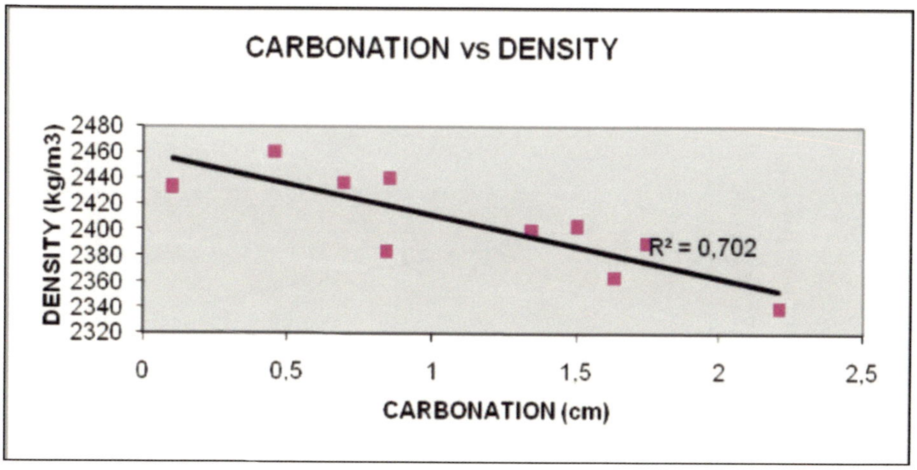

Fig. 1. To higher density, lesser carbonation is produced

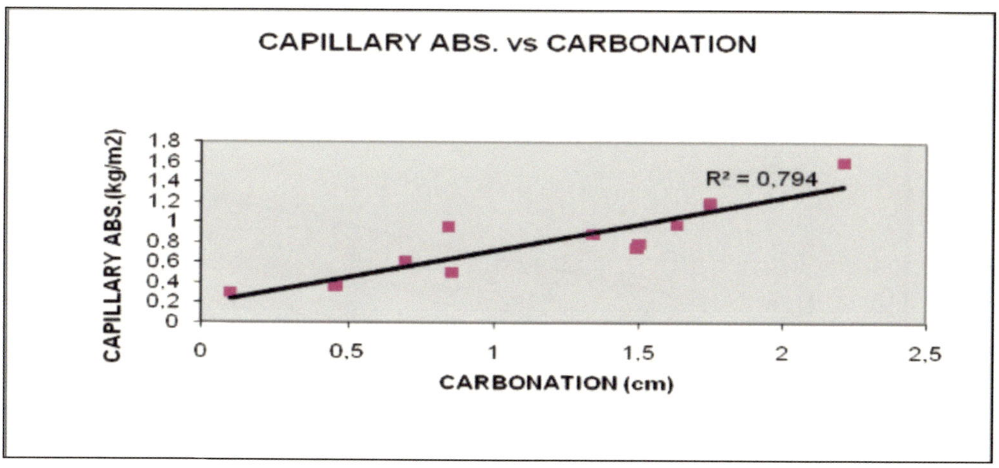

Fig. 2. Relation between capillary absorption and carbonation depth

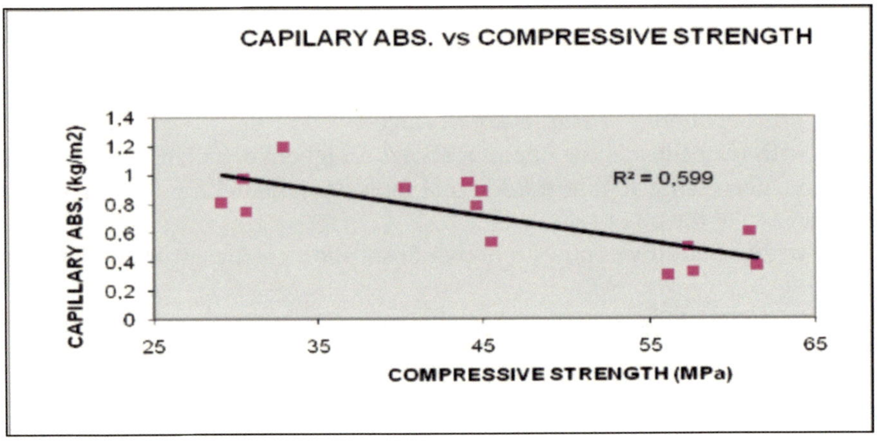

Fig. 3. Relation between capillary absorption and compressive strength

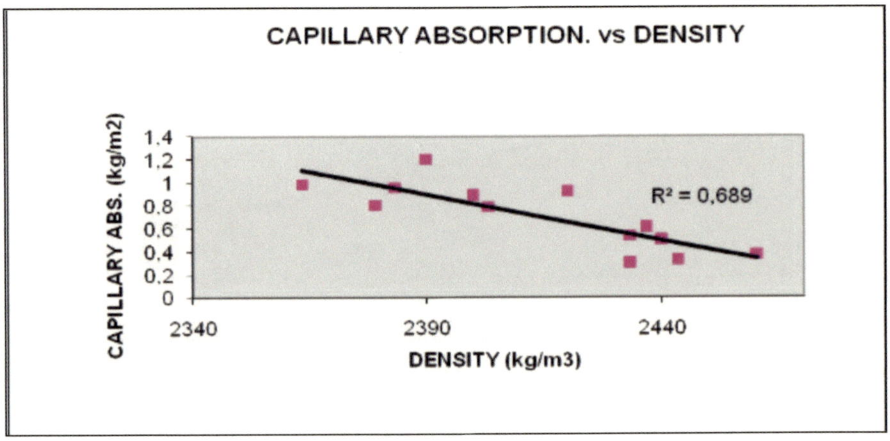

Fig. 4. Relation between capillary absorption and density

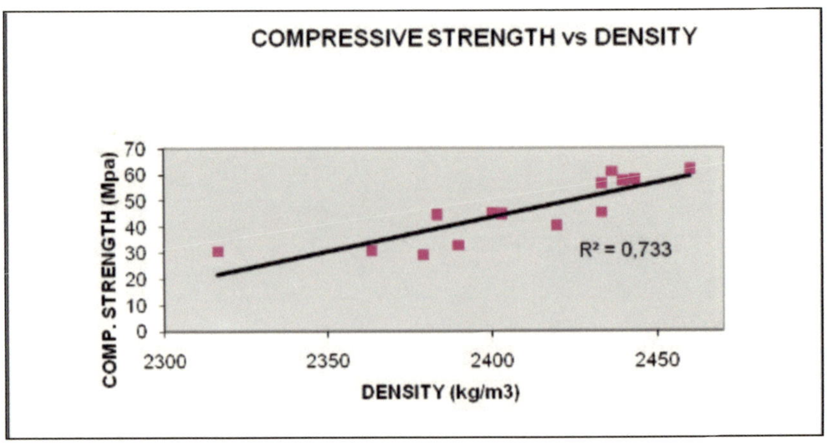

Fig. 5. Relation between compressive strength and density

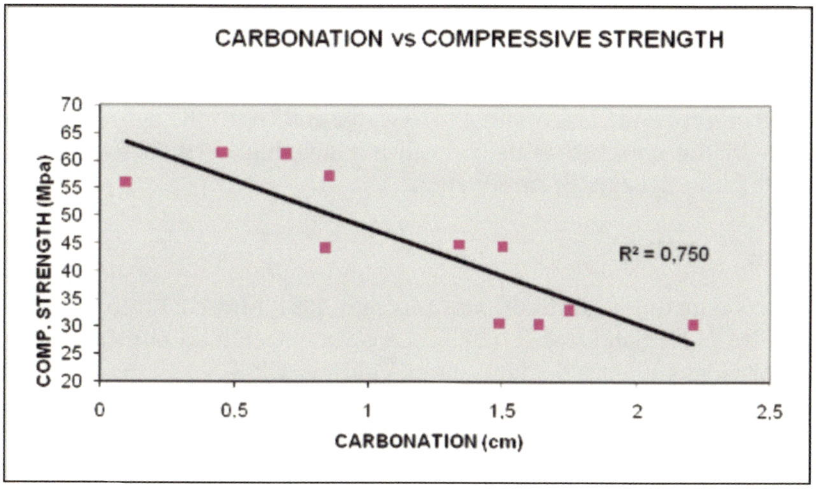

Fig. 6. Relation between carbonation and compressive strength

4. Conclusions

From the results obtained it was possible to conclude that the specimens of higher density correspond to the ones of higher strength which show lesser depth of carbonation and lesser capillary adsorption.

It is worth to mention the importance of the composition of concrete where include the parameter of density to know its behaviour before the results of compressive strength, it could be possible in the future.

In the specimens without additives it is seen that accelerated carbonation tends to produce higher water absorption. This is explained chemically, because the acid carbonates produced by excess of CO_2 have higher water solubility than the carbonates formed in not contaminated environments, where they are insoluble and help to seal the pores of the concrete.

The specimens which have additives present a carbonation coefficient lesser than the ones composed by concrete without additives, may suggest a higher durability in the long term for the specimens with additives.

The specimens with additives, although they were carbonated, show a better behaviour; the penetration tends to stop in time, giving protection to the concrete mass and indirectly to the reinforcing steel.

Regarding the compressive strength, the concrete with polycarboxylate got the highest values of compressive strength and resisted in a better way the accelerated carbonation process. As to the concrete with nanosilica in spite of not having a lesser settlement, it shows lesser strength. At the same time the concrete without aggregate, for the same age shows the least strength.

The concrete with nanosilica presents a higher absorption in spite of having a settlement much lesser than the concrete with polycarboxylate.

However the decrease of the absorption coefficient for higher time of carbonation that matches with a lesser speed of the advance facing the carbonation depth can be explained if it is accepted that the capillaries can have lesser diameter in the concrete mass, and therefore the capacity of forming carbonates to the inside may be seen as decreased although to be

able to explain convincingly this behaviour more exhausting studies in future projects should be carried out.
Regarding the capillary absorption, the specimens with additive absorb lesser quantity of water resulting in a concrete less attainable to water and/or to the aggressive agents, being the absorption in the concrete with nanosilica much higher than in the concrete with polycarboxylate for initial ages of carbonation.

5. References

American Concrete Institute. Aditivos para concreto. 1991. México: Editorial Limusa.

Barrera H., Pérez, H. & Sandoval R. (2003). La carbonatación en edificios de Santiago. In: XIV Jornadas Chilenas del Hormigón. Valdivia, Chile.

Carvajal A., Benavides F., Silva C., Valiente J. & Venegas A. (2007) Efectos de La carbonatación acelerada en distintos tipos de cemento y hormigones. Revista de la Construcción. 6(1) 88-97.

Carvajal A., Maturana P, Pino C. & Poblete J. (2009) Analysis of the relation between compressive strength, carbonation and porosity of concrete, in the search of a new control method by durability. Revista de la Construcción. 8 (2) 129-135

CYTED. Manual de Inspección, evaluación y diagnóstico de corrosión en estructuras de hormigón armado; DURAR, 1998. vol 1.

Da Silva R.; Pedrosa R.; Soares F. & Luiz, W. (2002). Cambios Microestructurales Relacionados con la Carbonatación en Concreto de Larga Durabilidad. Revista Ingeniería de Construcción. 17 (3), 144-150.

Duran, C. (2003). Accelerated carbonation and testing of concrete made with fly ash. Construction and Building Materials. 17 (147-153).

Erdogdu, S. (2000) Compatibility of super plasticizers with cements different in composition, Cement and Concrete Research. 30, 767-773.

Knopf F., Roy A., Samrow H. A. & Dooley K. M. (1999). Materials and Interfaces. High-Pressure Molding and Carbonation of Cementitius Materials. Industrial & Engineering Chemistry Research. 38 (7). p 2641-2649

Morin V., Cohen F., Feylessouifi A. & Richard P. (2001). Super plasticizer effects on setting and structuring mechanisms of ultrahigh-performance concrete, Cement and Concrete Research; 31 (63-71).

Papadakis V.G., Fardis, M.N. y Vayenas C.G. (1992). Effect of composition, environmental factors and cement-lime mortar coating on concrete carbonation. Materials and Structures. 25; 293-304.

RILEM (1988). TC 14, CPC-18 Measurement of hardened concrete carbonation depth. Materials and Structures, Vol 21, N°126, pp 453-455.

Thiery M., Villain G., Dangla, P. and Platret G. (2007). Investigation of the carbonation front shape on cementitious materials: Effects of the chemical kinetics. Cement and Concrete Research. 37,7. 1047-1058

Trocónis O. & Duracon Collaboration. (2006). Durability of concrete structures: DURACON, an Iiberoamerican Project. Preliminary results. Building and Environment. 41 925-962.

25

A Convenient and Inexpensive Quality Control Method for Examining the Accuracy of Conjugate Cam Profiles

Wen-Tung Chang[1] and Long-Iong Wu[2]
[1]*Opto-Mechatronics Technology Center, National Taiwan University of Science and Technology, Taipei 10607*
[2]*Department of Power Mechanical Engineering, National Tsing Hua University, Hsinchu 30013*
Taiwan

1. Introduction

The cam mechanism, basically consisting of a frame, a cam and a translating or oscillating follower with a roller in contact with the cam, is a simple and reliable device for motion control in machinery. Being a high-value-added product, a conjugate cam mechanism consists of a pair of disk cams that their profiles must be mutually conjugate to contact their respective follower. The conjugate cam mechanism is therefore a positive-drive mechanism (Wu, 2003; Rothbart, 2004; Norton, 2009) that can eliminate the use of return springs. As a benefit of positive-drive, the conjugate cam mechanism can ensure the contact between the cam and the follower roller with lower contact stresses between them. Such a situation can further contribute to the reduction of excessive noise, wear and vibrations occurred in the mechanism. In other words, reasonably designed conjugate cam mechanisms are especially suited to high-speed applications. However, since a conjugate cam mechanism is a so-called kinematically overconstrained arrangement (Wu, 2003), to ensure its movability condition and its ability to run without backlash (Rothbart, 2004; Norton, 2009), its cam profiles must be accurately designed and machined. The machined cams must then be carefully examined to check whether their profile errors fall within a specified tolerance range in order to achieve high quality and performance of the mechanism.
Up to the present time, using a highly sensitive and accurate coordinate measuring machine (CMM) to examine the accuracy of machined cam profiles is an industry-recognized technique, although it is still costly and time-consuming. For the quality control of machined cams, the cam profile must be directly measured by using a CMM, while the path planning and/or the coordinate measuring data are dealt with by some mathematical approaches to evaluate the profile errors (Lin & Hsieh, 2000; Qiu et al., 2000a; Qiu et al., 2000b; Qiu et al., 2000c; Hsieh & Lin, 2007; Chang et al., 2008). As an alternative quality control method, a special conjugation measuring fixture, which is improved from the one proposed by Koloc and Václavík (1993) and further investigated by Chang and Wu (2008), is developed by Chang et al. (2009) for indirectly evaluating the profile errors of conjugate disk cams. The conjugation measuring fixtures are based on the means of measuring the conjugate variation

of the assembled conjugate cam mechanism. According to the concept proposed by Chang et al. (2009), for a conjugate cam mechanism with an oscillating roller follower as shown in Fig. 1, if the constant center distance between the cam and follower pivots, f, is intentionally changed to be a variable parameter, f^*, by enabling the follower (link 3) being pivoted on a slider (link 4), as shown in Fig. 2, the mechanism will no longer be overconstrained. In other words, the follower subassembly (links 3 and 4) can serve as a conjugation measuring fixture. For the assembled conjugate cams with profile errors, the magnitude of distance f^* will vary with respect to the cam rotation angle θ, and the variation of the center distance between the cam and follower pivots, Δf ($= f^* - f$), can be detected by directly measuring the positional variation of the slider with the use of an inexpensive linear displacement meter, such as a dial (or digimatic) indicator or a linear scale, and the meter reading can indicate the variation of cam profile errors. Such a measurement method should be more convenient and inexpensive than the use of a CMM. By applying this concept, Chang et al. (2009) have presented a rapid indirect method for examining profile deviations of conjugate disk cams. In their work, an analytical approach called conjugate variation analysis (or conjugate condition analysis), based on the mechanical error analysis of disk cam mechanisms (Wu and Chang, 2005; Chang and Wu, 2006), has been developed for relating the center distance variation with the profile deviations of a pair of conjugate disk cams. Then, conservative criteria for qualify control of assembled conjugate cams with the measurement of the center distance variation have been proposed and an experimental verification had also been conducted. However, the rapid indirect method itself is mainly applied for evaluating whether the conjugate variation induced by a pair of machined conjugate disk cams is acceptable, but not able to examine the profile errors of each individual machined cam.

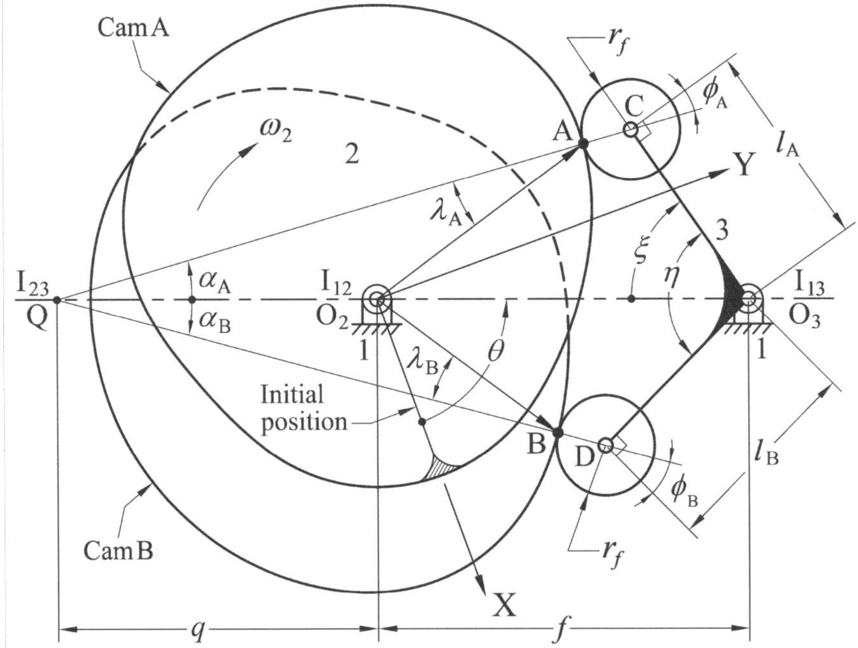

Fig. 1. Conjugate disk cams with an oscillating roller follower

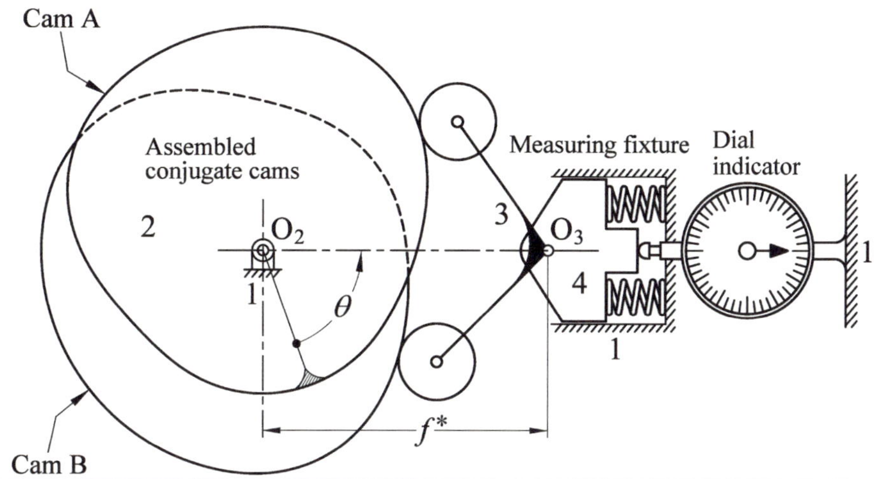

Fig. 2. Assembled conjugate cams with measuring fixture

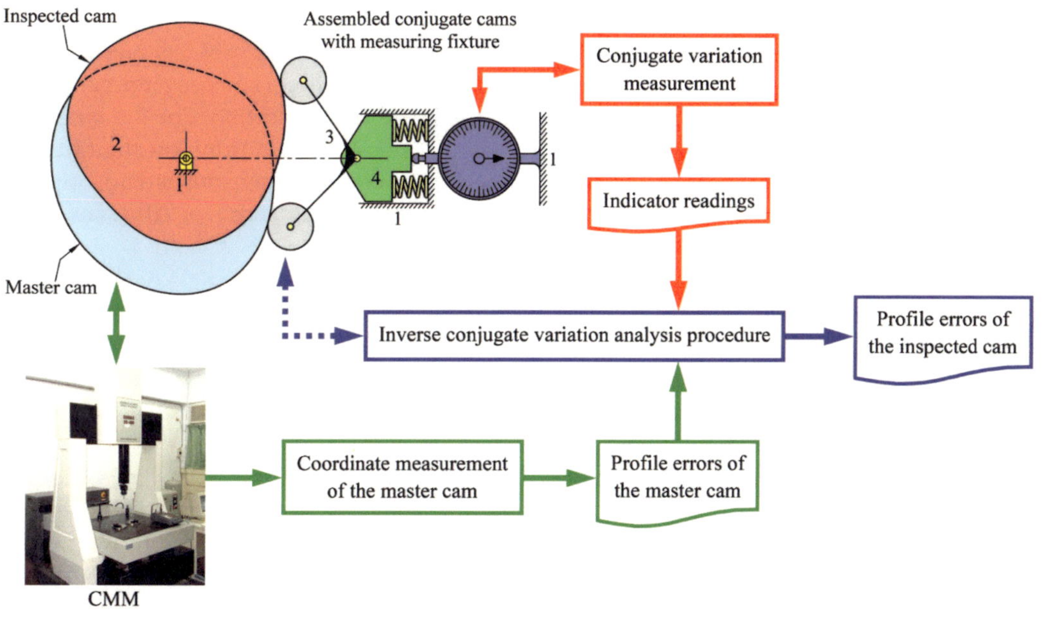

Fig. 3. Procedure for profile error estimation of the inspected cam

From the practical perspective of cam design and manufacture, a pair of conjugate disk cams can be machined in one piece or each cam be machined individually and then assembled together. The latter is usually a relatively easy and inexpensive manner, especially for mass production of conjugate cams. When the design of assembled conjugate cams is adopted, based on the concept of the rapid indirect method (Chang et al., 2009), an improved manner for examining the profile errors of each individual machined cam can be further developed. That is, if a pair of master conjugate cams with known profile errors is additionally available, through the measured center distance variations induced by a pair of

assembled conjugate cams that consists of one master cam and the other being the inspected cam, then the profile errors of each inspected cam can be estimated and examined. Such a concept is abstractly shown in Fig. 3; in which, for a pair of assembled conjugate cams consisting of one master cam, whose profile errors have been measured by using a CMM, and the other being the inspected cam, through the measurement of the center distance variation and the "inverse conjugate variation analysis procedure" of the assembled conjugate cam mechanism, the profile errors of the inspected cam can be estimated and then examined by an analytical manner. For the quality control in mass production of assembled conjugate disk cams, simply a pair of master conjugate cams with known profile errors and a set of conjugation measuring fixture must be prepared.
The objective of this study is to demonstrate how to examine the profile accuracy of assembled conjugate disk cams by applying the conjugate variation measurement and the inverse conjugate variation analysis. In order to verify the feasibility of the presented concept, an experiment meant to examine profile errors of a pair of machined conjugate cams was conducted. The profile errors of the machined cams estimated by using the presented method were compared with the measuring results obtained by using a CMM.

2. Parametric expressions for the conjugate cam profiles

In order to evaluate the dimensional variations of the machined cam profiles, the analytical expressions for the theoretical cam profiles must be derived first. For easy reference, the analytical expressions derived by Wu (2003) are provided in this section. For the conjugate cam mechanism shown in Fig. 1, its two cams A and B are fixed on a common shaft. Its two follower rollers C and D, mounted to a common follower, are each pushed in opposite directions by the conjugate cams. In the figure, f represents the distance from the cam center O_2 to the follower pivot point O_3, r_f represents the radii of rollers C and D, l_A and l_B represent the arm lengths of the follower, and η is the fixed subtending angle of the follower arms. By setting up a Cartesian coordinate system X-Y fixed on the cam and with its origin at the fixed pivot O_2, the cam profile coordinates may be expressed in terms of θ, which is measured against the direction of cam rotation from the reference radial on cam to the line of centers (line O_2O_3). In order to let θ have a counterclockwise angle, the cam is to rotate clockwise with a constant angular velocity of $\omega 2$.
As referred to in Fig. 1, the two normal lines through the points of contact and line of centers must always intersect at the instant center I_{23} (Wu, 2003), where "I" denotes the instant center and subscripts indicate the related links. For simplicity, in the following, the frame will be consistently numbered as 1, the cam as 2 and the follower as 3. By labeling instant center I_{23} as Q and $O_2Q = q$, the parametric vector equations of the cam profile coordinates are (Wu, 2003)

$$\mathbf{O_2A} = \begin{Bmatrix} X_A(\theta) \\ Y_A(\theta) \end{Bmatrix} = \begin{Bmatrix} (QC - r_f)\cos(\theta + \alpha_A) - q\cos\theta \\ (QC - r_f)\sin(\theta + \alpha_A) - q\sin\theta \end{Bmatrix} \tag{1}$$

$$\mathbf{O_2B} = \begin{Bmatrix} X_B(\theta) \\ Y_B(\theta) \end{Bmatrix} = \begin{Bmatrix} (QD - r_f)\cos(\theta - \alpha_B) - q\cos\theta \\ (QD - r_f)\sin(\theta - \alpha_B) - q\sin\theta \end{Bmatrix} \tag{2}$$

where

$$q = \frac{f\dfrac{d\xi(\theta)}{d\theta}}{1-\dfrac{d\xi(\theta)}{d\theta}} \tag{3}$$

$$\mathrm{QC} = \sqrt{l_{\mathrm{A}}^2 + (f+q)^2 - 2l_{\mathrm{A}}(f+q)\cos\xi(\theta)} \tag{4}$$

$$\mathrm{QD} = \sqrt{l_{\mathrm{B}}^2 + (f+q)^2 - 2l_{\mathrm{B}}(f+q)\cos[\eta-\xi(\theta)]} \tag{5}$$

$$\alpha_{\mathrm{A}} = \sin^{-1}\left[\frac{l_{\mathrm{A}}\sin\xi(\theta)}{\mathrm{QC}}\right] \tag{6}$$

$$\alpha_{\mathrm{B}} = \sin^{-1}\left\{\frac{l_{\mathrm{B}}\sin[\eta-\xi(\theta)]}{\mathrm{QD}}\right\} \tag{7}$$

in which, $\xi(\theta)$ is the angular displacement function of the follower:

$$\xi(\theta) = \cos^{-1}\left[\frac{l_{\mathrm{A}}^2 + f^2 - (r_b + r_f)^2}{2l_{\mathrm{A}}f}\right] + S(\theta) \tag{8}$$

where r_b is the radius of the base circle of cam A, and $S(\theta)$ is the follower angular motion program. Thus, in Eq. (3),

$$\frac{d\xi(\theta)}{d\theta} = \frac{dS(\theta)}{d\theta} = V(\theta) \tag{9}$$

in which, $V(\theta)$ is the follower angular velocity program. Also, the pressure angles ϕ_{A} and ϕ_{B} of the conjugate cam mechanism can be expressed as (Wu, 2003)

$$\phi_{\mathrm{A}} = 90^\circ - \alpha_{\mathrm{A}} - \xi(\theta) \tag{10}$$

$$\phi_{\mathrm{B}} = 90^\circ - \alpha_{\mathrm{B}} - [\eta - \xi(\theta)] \tag{11}$$

In addition, the shift angles λ_{A} and λ_{B} of the cam profiles, that is, the subtending angles between the radial and normal lines through the points of contact, can be expressed as (Chang et al., 2008; Chang & Wu, 2008; Chang et al., 2009)

$$\lambda_{\mathrm{A}} = \angle \mathrm{O_2AQ} = \sin^{-1}\left(\frac{q\sin\alpha_{\mathrm{A}}}{\|\mathbf{O_2A}\|}\right) = \sin^{-1}\left\{\frac{fV(\theta)\sin\alpha_{\mathrm{A}}}{[1-V(\theta)]\|\mathbf{O_2A}\|}\right\} \tag{12}$$

$$\lambda_{\mathrm{B}} = \angle \mathrm{O_2BQ} = \sin^{-1}\left(\frac{q\sin\alpha_{\mathrm{B}}}{\|\mathbf{O_2B}\|}\right) = \sin^{-1}\left\{\frac{fV(\theta)\sin\alpha_{\mathrm{B}}}{[1-V(\theta)]\|\mathbf{O_2B}\|}\right\} \tag{13}$$

These two angles are derived geometric parameters for correlating radial-dimension errors and normal-direction errors of disk cam profiles (Chang et al., 2008; Chang & Wu, 2008; Chang et al., 2009).

3. Conjugate variation measurement and the examination of profile accuracy

The measurement of the conjugate variation of the assembled conjugate cam mechanism can indirectly reveal the cam profile errors. By applying the analytical approach of the conjugate variation analysis (Chang et al., 2009), a convenient and inexpensive means for examining the profile accuracy of each individual machined cam can be developed.

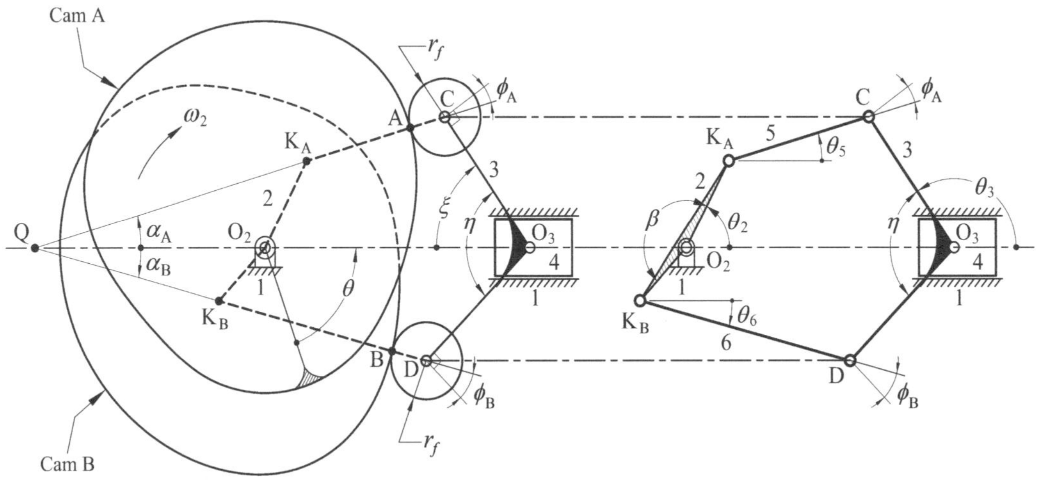

Fig. 4. An assembled conjugate cam mechanism and its equivalent six-bar linkage

3.1 Basic concepts

As referred to in Figs. 1 and 2, the center distance between the cam and follower pivots in the conjugation measuring fixture is designed to be variable. The difference between the variable center distance f^* (that is between the cam and follower pivots) and its ideally constant distance f may be induced by the radial-dimension errors of cams A and B, Δr_A and Δr_B, the roller-radius errors of rollers C and D, Δr_{fC} and Δr_{fD}, the errors of the arm lengths, Δl_A and Δl_B, and the subtending angle error of the follower arms, $\Delta\eta$. As a special case of the mechanical error analysis of disk cam mechanisms (Wu and Chang, 2005; Chang and Wu, 2006), by employing the concept of equivalent six-bar linkage of this assembled conjugate cam mechanism, as shown in Fig. 4, the analytical expressions of the center distance variations, Δf_r caused by Δr_A and Δr_B, Δf_{rf} caused by Δr_{fC} and Δr_{fD}, Δf_l caused by Δl_A and Δl_B, and Δf_η caused by $\Delta\eta$, respectively, have been derived as (Chang et al., 2009)

$$\Delta f_r \approx \frac{\Delta r_A(l_B\cos\phi_B\cos\lambda_A)+\Delta r_B(l_A\cos\phi_A\cos\lambda_B)}{l_A\cos\phi_A\cos\alpha_B+l_B\cos\phi_B\cos\alpha_A} \tag{14}$$

$$\Delta f_{rf} = \frac{\Delta r_{fC}(l_B\cos\phi_B)+\Delta r_{fD}(l_A\cos\phi_A)}{l_A\cos\phi_A\cos\alpha_B+l_B\cos\phi_B\cos\alpha_A} \tag{15}$$

$$\Delta f_l = \frac{\Delta l_A(l_B\cos\phi_B\sin\phi_A)+\Delta l_B(l_A\cos\phi_A\sin\phi_B)}{l_A\cos\phi_A\cos\alpha_B+l_B\cos\phi_B\cos\alpha_A} \tag{16}$$

$$\Delta f_{\eta} = -\frac{\Delta\eta(l_A l_B \cos\phi_A \cos\phi_B)}{l_A \cos\phi_A \cos\alpha_B + l_B \cos\phi_B \cos\alpha_A} \tag{17}$$

in which, the correlations of $\theta_5 = \alpha_A$ and $\theta_6 = \alpha_B$ exist as shown in Fig. 4. Also, parameters θ_2 and β depending on the locations of points K_A and K_B, which are the centers of curvatures of cams A and B respectively, are not involved in the derived results of Eqs. (14)-(17). Note that in practice, depending on the value of cam rotation angle θ, the magnitudes of the cam profile errors Δr_A and Δr_B may vary, while Δr_{fC}, Δr_{fD}, Δl_A, Δl_B and $\Delta\eta$ remain constant. In other words, $\Delta r_A = \Delta r_A(\theta)$ and $\Delta r_B = \Delta r_B(\theta)$. Assuming the small manufacturing or assembly errors $\Delta r_A(\theta)$, $\Delta r_B(\theta)$, Δr_{fC}, Δr_{fD}, Δl_A, Δl_B and $\Delta\eta$ in the assembled conjugate cam mechanism have been precisely measured, the overall center distance variation can be approximated by the sum of the derived center distance variations:

$$\Delta f_{est} = \Delta f_r + \Delta f_{rf} + \Delta f_l + \Delta f_{\eta} \tag{18}$$

Ideally, the estimated variation Δf_{est} will be equal to the measured value Δf_{mea} that can be obtained by means of a dial indicator as shown in Fig. 2. In the following context, the subscript "est" indicates estimated or calculated terms, while the subscript "mea" indicates actually measured ones.

The measurement of the center distance variation can be inversely applied to develop a convenient and inexpensive means for examining the conjugate cam profile errors. From Eq. (18) and considering the correlation of $\Delta f_{mea} \approx \Delta f_{est}$, it follows that

$$\Delta f_r \approx \Delta f_{mea} - (\Delta f_{rf} + \Delta f_l + \Delta f_{\eta}) \tag{19}$$

If the error terms Δr_{fC}, Δr_{fD}, Δl_A, Δl_B, $\Delta\eta$ and Δf_{mea} have been precisely measured and then Δf_{rf}, Δf_l and Δf_{η} have been evaluated by using Eqs. (15)-(17), respectively, Eq. (19) itself can accurately predict the center distance variation Δf_r without knowing the actual cam profile errors Δr_A and Δr_B. In order to calculate the unknown cam profile error Δr_A, however, the radial profile error of cam B must be measured in advance. From Eqs. (14) and (19), the estimated (calculated) radial profile error of cam A will be

$$\Delta r_{A,est} \approx \frac{1}{l_B \cos\phi_B \cos\lambda_A}\{(l_A \cos\phi_A \cos\alpha_B + l_B \cos\phi_B \cos\alpha_A)[\Delta f_{mea} - (\Delta f_{rf} + \Delta f_l + \Delta f_{\eta})] - \Delta r_{B,mea}(l_A \cos\phi_A \cos\lambda_B)\} \tag{20}$$

where $\Delta r_{B,mea}$ is the measured radial profile error of cam B. Likewise, if the radial profile error of cam A has been measured, the unknown cam profile error Δr_B can be estimated (calculated) by

$$\Delta r_{B,est} \approx \frac{1}{l_A \cos\phi_A \cos\lambda_B}\{(l_A \cos\phi_A \cos\alpha_B + l_B \cos\phi_B \cos\alpha_A)[\Delta f_{mea} - (\Delta f_{rf} + \Delta f_l + \Delta f_{\eta})] - \Delta r_{A,mea}(l_B \cos\phi_B \cos\lambda_A)\} \tag{21}$$

where $\Delta r_{A,mea}$ is the measured radial profile error of cam A. In order to proceed with such a cam profile error estimation, it is necessary to have two master cams $A_{(m)}$ and $B_{(m)}$ whose profiles are precisely measured and thus the magnitudes of $\Delta r_{A,mea}$ and $\Delta r_{B,mea}$ in the above two equations, respectively, can be known. Then, for a conjugate cam mechanism, the profile errors of each cam can be estimated subsequently by means of the conjugate variation measurement. The process presented above can be regarded as the "inverse conjugate variation analysis procedure" of the assembled conjugate cam mechanism.

As referred to in Fig. 3, for good cam profile control in mass production of conjugate cams, one must prepare a pair of master cams $A_{(m)}$ and $B_{(m)}$ whose profiles are accurately machined and also precisely measured by using a CMM to obtain each of their small cam profile errors. Then, if the finished products of cam A are to be examined, the inspected cam A and the master cam $B_{(m)}$ are mounted together as a pair to be measured. Once the center distance variations induced by this pair of cams have been measured, the actual profile of the inspected cam A can be estimated by means of the above presented inverse conjugate variation analysis procedure. On the other hand, if the finished products of cam B are to be examined, the inspected cam B and the master cam $A_{(m)}$ must be mounted together as a pair to be measured.

Based on the presented concept, criteria for determining whether the machined cam profiles are qualified can be established as follows. For the examination of cam A, after its upper and lower bounds of the radial-dimension errors, $\Delta r_{A(u)}$ and $\Delta r_{A(l)}$, are specified, the upper and lower acceptable extreme deviations of the center distance will be

$$\Delta f_{A(u),est} = \Delta f_{r,A(u)} + \Delta f_{rf} + \Delta f_l + \Delta f_\eta \tag{22}$$

and

$$\Delta f_{A(l),est} = \Delta f_{r,A(l)} + \Delta f_{rf} + \Delta f_l + \Delta f_\eta \tag{23}$$

in which,

$$\Delta f_{r,A(u)} \approx \frac{\Delta r_{A(u)}(l_B \cos\phi_B \cos\lambda_A) + \Delta r_{B(m),mea}(l_A \cos\phi_A \cos\lambda_B)}{l_A \cos\phi_A \cos\alpha_B + l_B \cos\phi_B \cos\alpha_A} \tag{24}$$

and

$$\Delta f_{r,A(l)} \approx \frac{\Delta r_{A(l)}(l_B \cos\phi_B \cos\lambda_A) + \Delta r_{B(m),mea}(l_A \cos\phi_A \cos\lambda_B)}{l_A \cos\phi_A \cos\alpha_B + l_B \cos\phi_B \cos\alpha_A} \tag{25}$$

where $\Delta r_{B(m),mea}$ is the known radial-dimension error of the master cam $B_{(m)}$. Then, the necessary condition of a qualified cam A is

$$\Delta f_{A(l),est} \le \Delta f_{mea} \le \Delta f_{A(u),est} \tag{26}$$

That is, if the profile deviation of an inspected cam A falls within its specified tolerance range, the measured value of the center distance variation, Δf_{mea}, will also fall within the range of $\Delta f_{A(l),est} \sim \Delta f_{A(u),est}$. Likewise, for the examination of cam B, after its upper and lower bounds of the radial-dimension errors, $\Delta r_{B(u)}$ and $\Delta r_{B(l)}$, are specified, the upper and lower acceptable extreme deviations of the center distance will be

$$\Delta f_{B(u),est} = \Delta f_{r,B(u)} + \Delta f_{rf} + \Delta f_l + \Delta f_\eta \tag{27}$$

and

$$\Delta f_{\mathrm{B}(l),\mathrm{est}} = \Delta f_{r,\mathrm{B}(l)} + \Delta f_{rf} + \Delta f_{l} + \Delta f_{\eta} \tag{28}$$

in which,

$$\Delta f_{r,\mathrm{B}(u)} \approx \frac{\Delta r_{\mathrm{A}(m),\mathrm{mea}}(l_{\mathrm{B}} \cos\phi_{\mathrm{B}} \cos\lambda_{\mathrm{A}}) + \Delta r_{\mathrm{B}(u)}(l_{\mathrm{A}} \cos\phi_{\mathrm{A}} \cos\lambda_{\mathrm{B}})}{l_{\mathrm{A}} \cos\phi_{\mathrm{A}} \cos\alpha_{\mathrm{B}} + l_{\mathrm{B}} \cos\phi_{\mathrm{B}} \cos\alpha_{\mathrm{A}}} \tag{29}$$

and

$$\Delta f_{r,\mathrm{B}(l)} \approx \frac{\Delta r_{\mathrm{A}(m),\mathrm{mea}}(l_{\mathrm{B}} \cos\phi_{\mathrm{B}} \cos\lambda_{\mathrm{A}}) + \Delta r_{\mathrm{B}(l)}(l_{\mathrm{A}} \cos\phi_{\mathrm{A}} \cos\lambda_{\mathrm{B}})}{l_{\mathrm{A}} \cos\phi_{\mathrm{A}} \cos\alpha_{\mathrm{B}} + l_{\mathrm{B}} \cos\phi_{\mathrm{B}} \cos\alpha_{\mathrm{A}}} \tag{30}$$

where $\Delta r_{\mathrm{A}(m),\mathrm{mea}}$ is the known radial-dimension error of the master cam $\mathrm{A}_{(m)}$. Then, the necessary condition of a qualified cam B is

$$\Delta f_{\mathrm{B}(l),\mathrm{est}} \leq \Delta f_{\mathrm{mea}} \leq \Delta f_{\mathrm{B}(u),\mathrm{est}} \tag{31}$$

When the profile deviation of an inspected cam B falls within its specified tolerance range, the measured value of the center distance variation, Δf_{mea}, will also fall within the range of $\Delta f_{\mathrm{B}(l),\mathrm{est}} \sim \Delta f_{\mathrm{B}(u),\mathrm{est}}$. Because $\Delta f_{\mathrm{A}(u),\mathrm{est}}$, $\Delta f_{\mathrm{A}(l),\mathrm{est}}$, $\Delta f_{\mathrm{B}(u),\mathrm{est}}$ and $\Delta f_{\mathrm{B}(l),\mathrm{est}}$ will vary with respect to the cam rotation angle θ, their corresponding values should be calculated for the cam profile examination.

3.2 Simulated example

The presented method will be illustrated by the following simulated example.

A conjugate cam system requires the oscillating roller follower to oscillate 30° clockwise with cycloidal motion (Rothbart, 2004; Norton, 2009) while the cam rotates clockwise from 0° to 120°, dwell for the next 40°, return with cycloidal motion for 120° cam rotation and dwell for the remaining 80°. The distance between pivots, f, is 120 mm. The lengths of the follower arms, l_{A} and l_{B}, are both equal to 66 mm, and both follower rollers have the same radius of 16 mm. The base circle radius, r_b, is 60 mm and the theoretical subtending angle of the follower arms, η, is 100°.

The profiles of cams A and B, with respective maximum radial dimensions of 93.793 mm and 93.859 mm, are shown in Fig. 1. For a tolerance grade of IT6, the cam profiles may have tolerance amounts of $\pm\Delta r_{\mathrm{A}} = \pm\Delta r_{\mathrm{B}} = \pm 22$ μm (i.e., $\Delta r_{\mathrm{A}(u)} = \Delta r_{\mathrm{B}(u)} = 22$ μm and $\Delta r_{\mathrm{A}(l)} = \Delta r_{\mathrm{B}(l)} = -22$ μm), the follower arm lengths may have tolerance amounts of $\pm\Delta l_{\mathrm{A}} = \pm\Delta l_{\mathrm{B}} = \pm 19$ μm, the radius errors of the follower rollers, $\Delta r_{f\mathrm{C}}$ and $\Delta r_{f\mathrm{D}}$, may have tolerance amounts of $\pm\Delta r_{f\mathrm{C}} = \pm\Delta r_{f\mathrm{D}} = \pm 11$ μm, and the subtending angle of the follower arms may have a tolerance amount of $\pm\Delta\eta = \pm 0.022°$. Note that this work is to estimate (calculate) the cam profile deviations Δr_{A} and Δr_{B} of being inspected ones. Therefore, for a pair of master conjugate cams and a conjugation measuring fixture constructed according to the presented method, all constant design parameters as well as the master cam profiles should have been precisely measured. Accordingly, the profile errors of the master cams, $\Delta r_{\mathrm{A}(m),\mathrm{mea}}(\theta)$ and $\Delta r_{\mathrm{B}(m),\mathrm{mea}}(\theta)$, and the five constant deviations Δl_{A}, Δl_{B}, $\Delta r_{f\mathrm{C}}$, $\Delta r_{f\mathrm{D}}$ and $\Delta\eta$ may be assumed to be known, and then the magnitudes of center distance deviations Δf_{rf}, Δf_{l} and Δf_{η} can be evaluated by using Eqs. (15)-(17), respectively, before the examination of inspected cams.

In this example, $\Delta l_{\mathrm{A}} = \Delta l_{\mathrm{B}} = 19$ μm, $\Delta r_{f\mathrm{C}} = \Delta r_{f\mathrm{D}} = 11$ μm, and $\Delta\eta = 0.022°$ are assumed. The master cams $\mathrm{A}_{(m)}$ and $\mathrm{B}_{(m)}$ are assumed to have variable profile deviations with the following forms: $\Delta r_{A(m),mea}(\theta)$ = (18.5 + 3.5sinθ) μm and $\Delta r_{B(m),mea}(\theta)$ = (17.5 + 4.5cos2θ) μm. Then, the measured center distance variation $\Delta f_{mea}(\theta)$ caused by a pair of assembled conjugate cams consisting of a master cam and a being inspected one (either a pair of cams A and $B_{(m)}$ or the other pair of cams $A_{(m)}$ and B) is considered to have an invariant value of 22 μm, which is the corresponding value of tolerance grade IT6 of the theoretical center distance f, when the cams rotate a complete revolution. Figure 5 shows some evaluated results of this example,

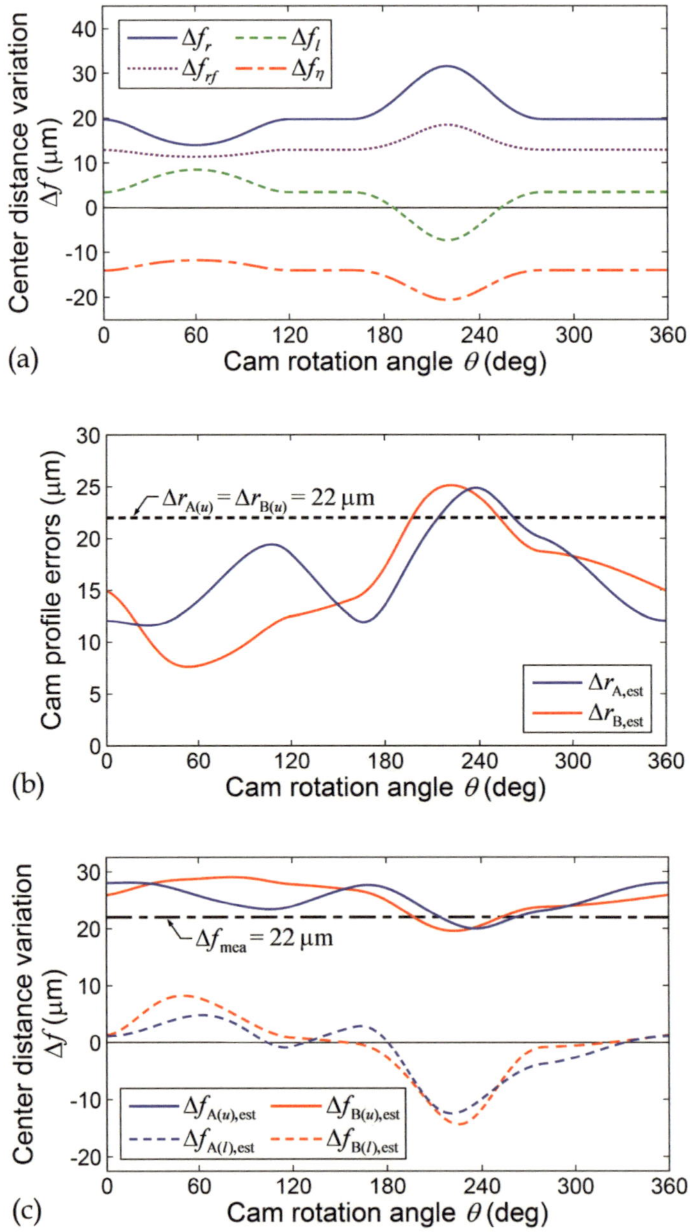

Fig. 5. Evaluated results of a simulated example

while their extreme values are also listed in Table 1. The calculated center distance variations with respect to the cam rotation angle θ are shown in Fig. 5(a), in which, Δf_{rf}, Δf_l and Δf_η are calculated by using Eqs. (15)-(17), respectively, while Δf_r is calculated by using Eq. (19). In this case, the extreme values of Δf_l and Δf_{rf} are slighter than those of Δf_r and Δf_η, while Δf_r, Δf_{rf} and Δf_η have similar variation trends if their signs are ignored. Figure 5(b) shows the estimated cam profile errors, $\Delta r_{A,est}$ and $\Delta r_{B,est}$, with respect to the cam rotation angle θ, in which, $\Delta r_{A,est}$ is calculated by using Eq. (20) with the information of $\Delta r_{B(m),mea}(\theta)$ = (17.5 + 4.5cos2θ) μm and $\Delta f_{mea}(\theta)$ = 22 μm, while $\Delta r_{B,est}$ is calculated by using Eq. (21) with the information of $\Delta r_{A(m),mea}(\theta)$ = (18.5 + 3.5sinθ) μm and $\Delta f_{mea}(\theta)$ = 22 μm. It can be seen that $\Delta r_{A,est}$ ranges between 11.62 and 24.9 μm, and $\Delta r_{B,est}$ ranges between 7.62 and 25.15 μm. Apparently, when θ = 213.97° ~ 261.78°, $\Delta r_{A,est}$ exceeds its specified upper bound $\Delta r_{A(u)}$ (= 22 μm), and when θ = 197.36° ~ 252.63°, $\Delta r_{B,est}$ also exceeds its specified upper bound $\Delta r_{B(u)}$ (= 22 μm). Such situations can also be judged through the results shown in Fig. 5(c), in which, the magnitude of Δf_{mea} (= 22 μm) is out of the range between $\Delta f_{A(l),est}$ ~ $\Delta f_{A(u),est}$ when θ = 213.97° ~ 261.78° and also out of the range between $\Delta f_{B(l),est}$ ~ $\Delta f_{B(u),est}$ when θ = 197.36° ~ 252.63°. As a result, both cams A and B in this example are partially unqualified and whose profile errors can be estimated and examined by means of the inverse conjugate variation analysis procedure.

Cam angle	Extreme value	Cam angle	Extreme value
θ = 15.6°	$(\Delta f_{A(u),est})_{max}$ = 28.14 μm	θ = 219.99°	$(\Delta f_r)_{max}$ = 31.6 μm
θ = 26.52°	$(\Delta r_{A,est})_{min}$ = 11.62 μm	θ = 219.99°	$(\Delta f_\eta)_{min}$ = −20.67 μm
θ = 49.62°	$(\Delta f_{B(l),est})_{max}$ = 8.25 μm	θ = 220°	$(\Delta f_l)_{min}$ = −7.37 μm
θ = 53.33°	$(\Delta r_{B,est})_{min}$ = 7.62 μm	θ = 220°	$(\Delta f_{rf})_{max}$ = 18.44 μm
θ = 59.92°	$(\Delta f_\eta)_{max}$ = −11.79 μm	θ = 220.75°	$(\Delta f_{A(l),est})_{min}$ = −12.43 μm
θ = 59.97°	$(\Delta f_r)_{min}$ = 13.95 μm	θ = 222.03°	$(\Delta r_{B,est})_{max}$ = 25.15 μm
θ = 59.98°	$(\Delta f_l)_{max}$ = 8.49 μm	θ = 222.78°	$(\Delta f_{B(u),est})_{min}$ = 19.58 μm
θ = 59.99°	$(\Delta f_{rf})_{min}$ = 11.35 μm	θ = 225.28°	$(\Delta f_{B(l),est})_{min}$ = −14.35 μm
θ = 62.09°	$(\Delta f_{A(l),est})_{max}$ = 4.84 μm	θ = 236.41°	$(\Delta f_{A(u),est})_{min}$ = 19.99 μm
θ = 81.65°	$(\Delta f_{B(u),est})_{max}$ = 29.08 μm	θ = 237.72°	$(\Delta r_{A,est})_{max}$ = 24.9 μm

Table 1. Extreme values of a simulated example

4. Experimental details

In order to test the feasibility and effectiveness of the presented method, an experiment meant to examine profile errors of a pair of machined conjugate cams was conducted.

4.1 Experimental apparatus

An assembled conjugate cam mechanism, whose center distance between the cam and follower pivots is variable, had been designed and built for experimental work. The built mechanism is shown in Fig. 6, which was the identical one used for the experiment of measuring the center distance variation to verify the theoretical derivation results of the conjugate variation analysis (Chang et al., 2009). The specified design parameters of this built mechanism are identical to those of the cam system illustrated in Sub-section 3.2. The conjugate cams, identical to those used for experiments conducted in previous studies

(Chang et al., 2008; Chang and Wu, 2008; Chang et al., 2009), were made of stainless steel JIS SUS304/AISI 304. Both cams had the same thickness of 12 mm and whose profiles were manufactured by a computer numerical control (CNC) electro-discharge wire-cutting (EDWC) machine. In order to make the center distance variation large enough to be easily sensed and read in the experiment, both cams had been specified to have a radial-dimension tolerance of ±220 μm (i.e., $\Delta r_{A(u)} = \Delta r_{B(u)} = 220$ μm and $\Delta r_{A(l)} = \Delta r_{B(l)} = -220$ μm in this case), a considerably large tolerance grade of IT11.

Fig. 6. Built assembled conjugate cams with measuring fixture

The experimental apparatus and instrumentation are schematically shown in Fig. 7. To drive the built conjugate cam mechanism, an Animatics SM2315D 0.13 kW DC servomotor coupled with an Apex Dynamics AB060-S1-P1 gear reducer with a reduction ratio of 9:1 were used. The servomotor was powered by a DC power supply. A personal computer was prepared to control the servomotor through a communication cable (Animatics CBLSM1) connecting the servomotor and one RS-232 port of the computer. A Mitutoyo ID-C112M 543-251 digimatic indicator, whose resolution and accuracy are 1 μm and ±3 μm, respectively, was employed to measure the center distance variation between the cam and follower pivots. The digital measuring data read from the digimatic indicator were inputted to the same computer by using a Mitutoyo MUX-10F Multiplexer data transfer device connecting the digimatic indicator and another RS-232 port of the computer. A Keyence FU-25/FS-V31 fiber optic sensor module, powered by the same DC power supply, was applied to identify the initial angular position for the cam rotation and also to ensure that the conjugate cams can actually return to the initial angular position in each revolution. The fiber optic sensor module beamed one end face of cam A for sensing and calibrating the initial angular position of the conjugate cams.

4.2 Experimental procedure

Before the experiment of examining the profile accuracy of assembled conjugate cams was conducted, the cam profiles had been measured by using a Giddings & Lewis Sheffield Measurement Cordax RS-25 CMM with a Renishaw touch-trigger probe (PH9 probe head and TP200 probe with a stylus for its ruby ball diameter of 2 mm) (Chang et al., 2008; Chang and Wu, 2008), as shown in Fig. 8. The measuring time of each cam with 3600 points on the cam contour being measured had taken about 3 hours. The radial-dimension errors of the cams had then been evaluated from the coordinate measurement data by using the analytical approach proposed by Chang et al. (2008). Before the built conjugate cam mechanism had been assembled, the dimensions of the follower arms and the rollers had also been measured by using the CMM. Thus, the measured radial dimension errors of cams A and B, $\Delta r_{A,mea}(\theta)$ and $\Delta r_{B,mea}(\theta)$, the roller-radius errors of rollers C and D, Δr_{fC} and Δr_{fD}, and the errors of the arm lengths, Δl_A and Δl_B, had been obtained.

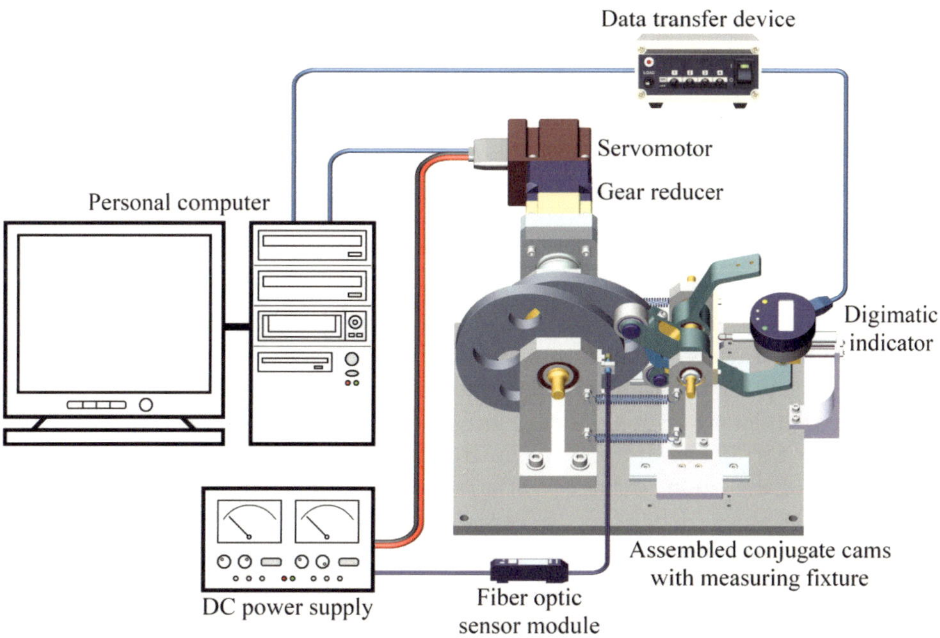

Fig. 7. Schematic of the experimental apparatus and instrumentation

Fig. 8. Measuring the conjugate cam profiles by using a CMM

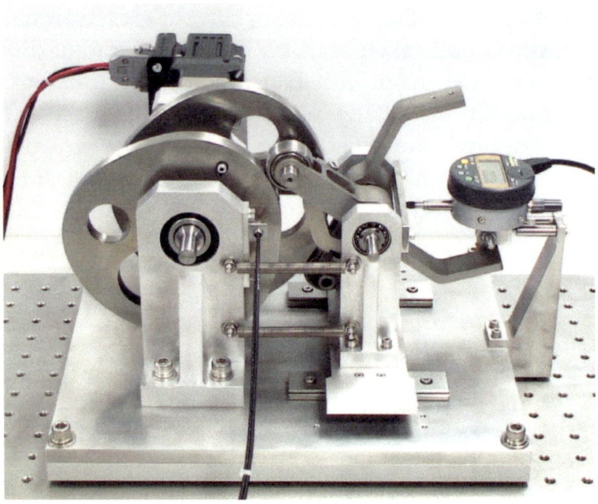

Fig. 9. Measuring the center distance variation by using a digimatic indicator

After the built conjugate cam mechanism was assembled and set for the measurement of the center distance variation in this study, as shown in Fig. 9, the fixed subtending angle between the follower arms was also measured by using the CMM to obtain the subtending angle error, $\Delta\eta$. During the measurement of the center distance variation, the conjugate cams rotated continuously with a constant angular velocity of 4 rev/hour (≈ 0.0667 rev/min ≈ 0.007 rad/sec) for 10 revolutions, while the data sampling rate was set to 5 Hz; the measuring time of the center distance variations for each revolution took 15 minutes. Ten data sets of 4500 values of the motion variations of the digimatic indicator for each cam revolution were recorded. For each revolution, 3600 interpolated values of the indicator readings corresponding to the cam angles with an equal interval of 0.1° based on the original 4500 measured values were calculated by using linear interpolation. Then, the 10 sets of the interpolated indicator reading data were obtained as the interpolated center distance variations. The averages of the interpolated center distance variations with respect to each corresponding cam rotation angle were calculated and considered as representatives of the experimental data function of angle θ, $\Delta f_{\text{mea}}(\theta)$.

The experimental data of $\Delta f_{\text{mea}}(\theta)$, $\Delta r_{\text{A,mea}}(\theta)$ and $\Delta r_{\text{B,mea}}(\theta)$ were then adopted for examining the cam profile error with the use of the presented method. For the profile error examination of cam A, data of $\Delta f_{\text{mea}}(\theta)$ and $\Delta r_{\text{B,mea}}(\theta)$ were adopted to calculate $\Delta r_{\text{A,est}}(\theta)$ by using Eq. (20). Likewise, for the profile error examination of cam B, data of $\Delta f_{\text{mea}}(\theta)$ and $\Delta r_{\text{A,mea}}(\theta)$ were adopted to calculate $\Delta r_{\text{B,est}}(\theta)$ by using Eq. (21).

5. Results and discussion

The actual dimensions of the constant parameters (i.e., l_A, l_B, r_{fC}, r_{fD} and η) and their corresponding errors in the built mechanism are listed in Table 2. [Note that the subtending angle error $\Delta\eta$ (= 0.275°) in the experiment was about 95.8 percent of that in previous study (Chang et al., 2009) because of the reassembling of the follower subassembly of the built mechanism; while the other four errors remained identical to their previous ones.] The measured cam profile errors by using a CMM (Chang et al., 2008; Chang and Wu, 2008) are shown in Fig. 10. By using Eqs. (14)-(17) with the error terms in Table 2 and Fig. 10 being involved, the evaluated center distance variations for the experiment, Δf_r caused by $\Delta r_{A,mea}$ and $\Delta r_{B,mea}$, Δf_{rf} caused by Δr_{fC} and Δr_{fC}, Δf_l caused by Δl_A and Δl_B, and Δf_η caused by $\Delta\eta$, are shown in Fig. 11. Extreme values of related functions shown in Figs. 10 and 11 are also listed in Table 3. In Fig. 10, it can be seen that the magnitude of $\Delta r_{A,mea}$ exceeded its upper bound of $\Delta r_{A(u)}$ = 220 μm at about θ = 80° ~ 110°; while the magnitude of $\Delta r_{B,mea}$ fell within the range of its specified tolerance. Figure 11 shows that the magnitudes and variation ranges of Δf_r and Δf_η were much greater than those of Δf_{rf} and Δf_l. Thus, the cam profile errors, $\Delta r_{A,mea}$ and $\Delta r_{B,mea}$, and the subtending angle error, Δf_η, mainly dominated the trend of the overall center distance variation, Δf_{est} (= $\Delta f_r + \Delta f_{rf} + \Delta f_l + \Delta f_\eta$), calculated by using Eq. (18).

Parameter	Nominal value	Actual value (in average)	Error (in average)
l_A	66 mm	65.984 mm	16 μm
l_B	66 mm	65.932 mm	−68 μm
r_{fC}	16 mm	15.989 mm	−11 μm
r_{fD}	16 mm	15.990 mm	−10 μm
η	100°	100.275°	0.275° (≈ 0.00479 rad)

Table 2. Nominal and actual values of the constant parameters

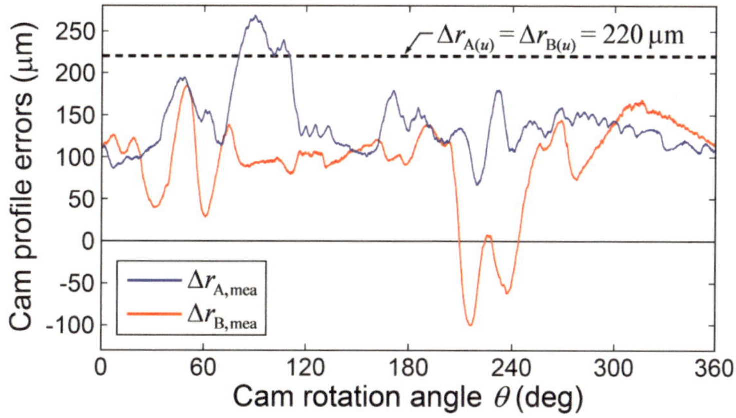

Fig. 10. Measured cam profile errors by using a CMM (Chang et al., 2008; Chang and Wu, 2008)

Figure 12 shows the measured and estimated results of the experiment. The measured and estimated center distance variations, Δf_{mea} and Δf_{est}, and their difference ($\Delta f_{est} - \Delta f_{mea}$) are shown in Fig. 12(a). The estimated and measured profile errors of cam A, $\Delta r_{A,est}$ and $\Delta r_{A,mea}$, and their difference ($\Delta r_{A,est} - \Delta r_{A,mea}$) are shown in Fig. 12(b), while the estimated and measured profile errors of cam B, $\Delta r_{B,est}$ and $\Delta r_{B,mea}$, and their difference ($\Delta r_{B,est} - \Delta r_{B,mea}$) are shown in Fig. 12(c). Extreme values and root-mean-square values of related functions shown in the figure are also listed in Tables 3 and 4, respectively. As shown in Fig. 12(a), Δf_{mea} and Δf_{est} were very close to each other, while their difference ($\Delta f_{est} - \Delta f_{mea}$), once again shown in Fig. 13(a) for clarity of illustration, ranged between −7.70 and 6.91 μm and had a root-mean-square value of 2.75 μm. Considering that Δf_{mea} ranged between −264.59 and 5.41 μm and had a root-mean-square value of 80.85 μm, the statistically relative deviation between Δf_{mea} and Δf_{est} was evaluated as 3.4% [= (2.75/80.85) ×100%]. Such results implied well agreement between the measured results and the estimated ones.

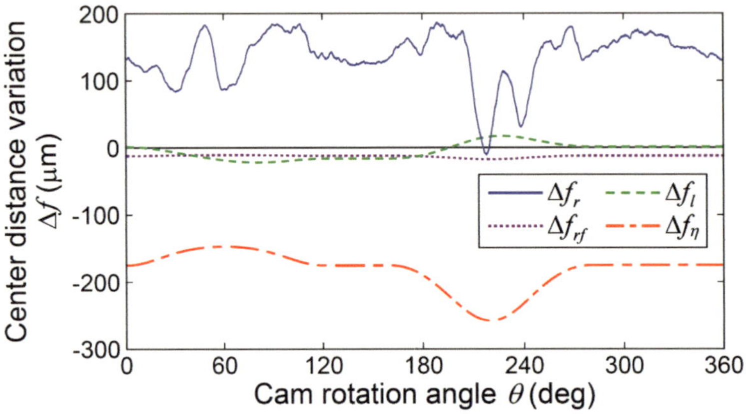

Fig. 11. Evaluated center distance variations for the experiment

Cam angle	Extreme value	Cam angle	Extreme value
$\theta = 45.6°$	$(\Delta f_{B(l),est})_{max} = -175.18$ μm	$\theta = 216.4°$	$(\Delta f_{A(u),est})_{min} = -165.5$ μm
$\theta = 48.4°$	$(\Delta f_{est})_{max} = 9.31$ μm	$\theta = 216.4°$	$(\Delta f_{A(l),est})_{min} = -504.35$ μm
$\theta = 48.7°$	$(\Delta f_{mea})_{max} = 5.41$ μm	$\theta = 217.8°$	$(\Delta f_r)_{min} = -11.25$ μm
$\theta = 49.4°$	$(\Delta r_{B,est})_{max} = 180.51$ μm	$\theta = 217.8°$	$(\Delta f_{est})_{min} = -270.79$ μm
$\theta = 49.7°$	$(\Delta r_{B,mea})_{max} = 185.12$ μm	$\theta = 218.2°$	$(\Delta f_{mea})_{min} = -264.59$ μm
$\theta = 49.7°$	$(\Delta f_{A(l),est})_{max} = -199.43$ μm	$\theta = 219.3°$	$(\Delta r_{A,mea})_{min} = 66.95$ μm
$\theta = 59.9°$	$(\Delta f_\eta)_{max} = -147.14$ μm	$\theta = 219.3°$	$(\Delta f_{B(u),est})_{min} = -40.56$ μm
$\theta = 61.7°$	$(\Delta r_{A,est} - \Delta r_{A,mea})_{max} = 13.95$ μm	$\theta = 219.3°$	$(\Delta f_{B(l),est})_{min} = -375.73$ μm
$\theta = 61.7°$	$(\Delta r_{B,est} - \Delta r_{B,mea})_{max} = 13.75$ μm	$\theta = 219.9°$	$(\Delta f_{rf})_{min} = -17.6$ μm
$\theta = 62.4°$	$(\Delta f_{rf})_{max} = -10.84$ μm	$\theta = 220°$	$(\Delta f_\eta)_{min} = -257.89$ μm
$\theta = 80.7°$	$(\Delta f_l)_{min} = -22$ μm	$\theta = 220.2°$	$(\Delta r_{A,est})_{min} = 69.47$ μm
$\theta = 89.2°$	$(\Delta r_{A,est})_{max} = 260.94$ μm	$\theta = 223.7°$	$(u_f)_{max} = 3.7$ μm
$\theta = 89.6°$	$(\Delta r_{A,mea})_{max} = 268.89$ μm	$\theta = 226.6°$	$(\Delta f_l)_{max} = 17.28$ μm
$\theta = 90°$	$(\Delta f_{B(u),est})_{max} = 66.86$ μm	$\theta = 279.6°$	$(u_f)_{min} = 0.43$ μm
$\theta = 179.9°$	$(\Delta f_{est} - \Delta f_{mea})_{min} = -7.7$ μm	$\theta = 308.6°$	$(\Delta f_{est} - \Delta f_{mea})_{max} = 6.91$ μm
$\theta = 188.6°$	$(\Delta f_r)_{max} = 187.17$ μm	$\theta = 308.6°$	$(\Delta r_{A,est} - \Delta r_{A,mea})_{min} = -11.29$ μm
$\theta = 215.5°$	$(\Delta r_{B,mea})_{min} = -100.46$ μm	$\theta = 308.6°$	$(\Delta r_{B,est} - \Delta r_{B,mea})_{min} = -12.36$ μm
$\theta = 215.5°$	$(\Delta r_{B,est})_{min} = -103.14$ μm	$\theta = 315.9°$	$(\Delta f_{A(u),est})_{max} = 42.38$ μm

Table 3. Extreme values of the experiment

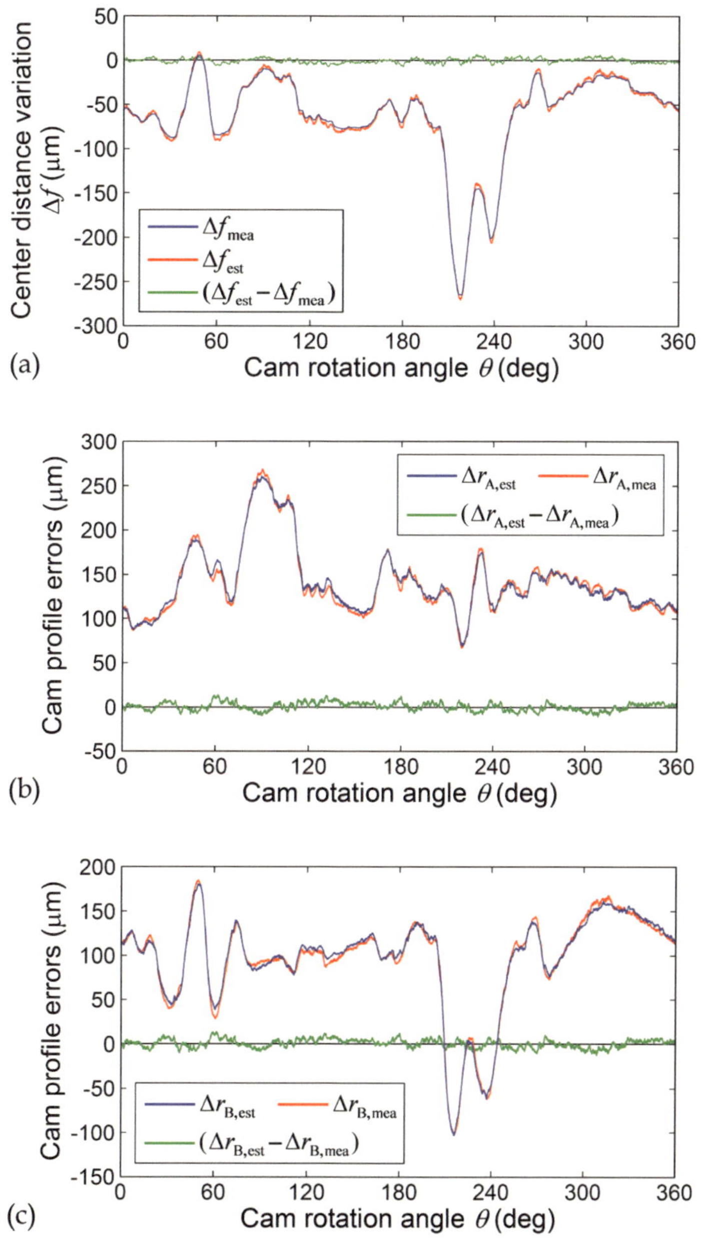

Fig. 12. Measured and estimated results of the experiment

As shown in Figs. 12(b) and 12(c), the trends and magnitudes of the estimated profile errors were well consistent with those of the measured ones. The differences between the estimated and measured profile errors are once again shown in Figs. 13(b) and 13(c) for clarity of illustration. The difference ($\Delta r_{A,est}$ − $\Delta r_{A,mea}$) ranged between −11.29 and 13.95 μm and had a root-mean-square value of 4.68 μm. Considering that $\Delta r_{A,mea}$ ranged between 66.95 and 268.89 μm and had a root-mean-square value of 146.13 μm, the statistically relative deviation between $\Delta r_{A,est}$ and $\Delta r_{A,mea}$ was evaluated as 3.2% [= (4.68/146.13) ×100%]. Also, the difference ($\Delta r_{B,est}$ − $\Delta r_{B,mea}$) ranged between −12.36 and 13.75 μm and had a root-mean-square value of 4.69 μm. Considering that $\Delta r_{B,mea}$ ranged between −100.46 and 185.12 μm

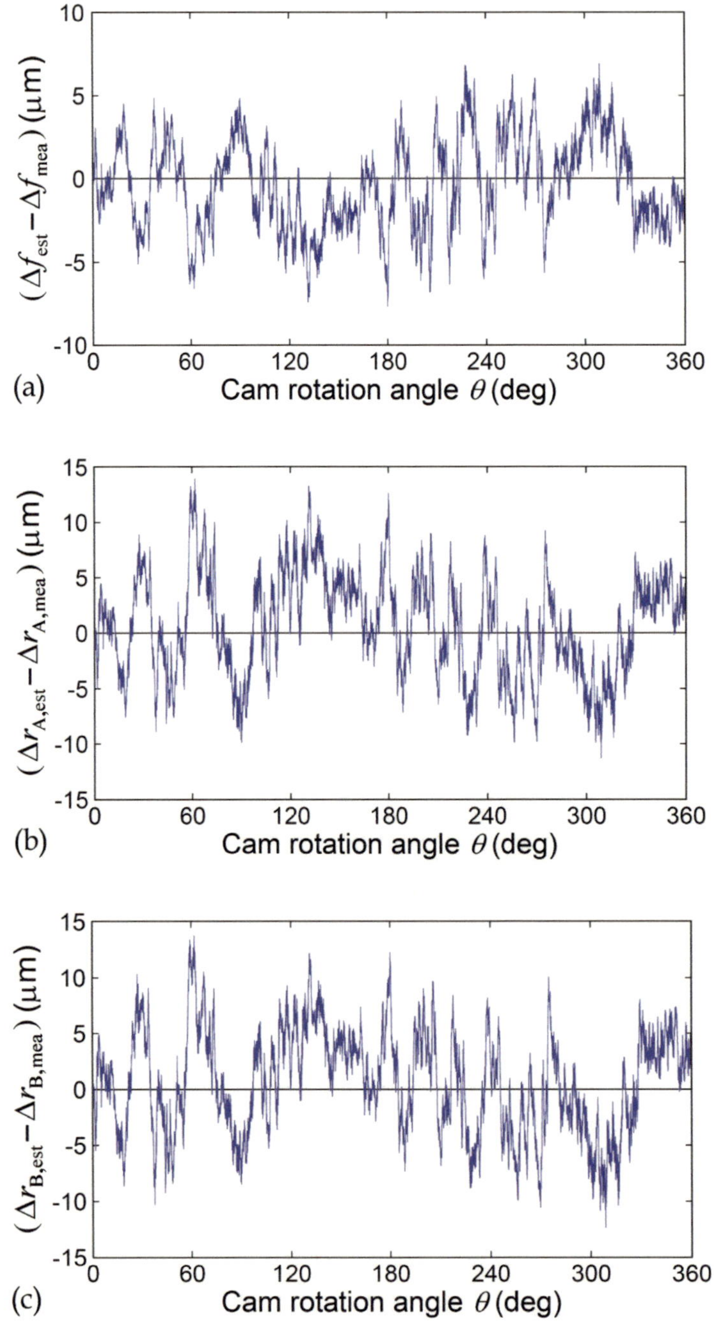

Fig. 13. Differences between the measured and estimated results

and had a root-mean-square value of 109.5 μm, the statistically relative deviation between $\Delta r_{B,est}$ and $\Delta r_{B,mea}$ was evaluated as 4.28% [= (4.69/109.5) ×100%]. Thus, from a statistical viewpoint, the differences and relative deviations in root-mean-square forms between the estimated and measured profile errors were less than 5 μm or 4.3%. Such results showed the effectiveness of the presented method for the profile error examination.

Center distance variations	Profile errors of cam A	Profile errors of cam B
$(\Delta f_{mea})_{rms}$ = 80.85 μm	$(\Delta r_{A,mea})_{rms}$ = 146.13 μm	$(\Delta r_{B,mea})_{rms}$ = 109.5 μm
$(\Delta f_{est})_{rms}$ = 81.48 μm	$(\Delta r_{A,est})_{rms}$ = 146.42 μm	$(\Delta r_{B,est})_{rms}$ = 109.8 μm
$(\Delta f_{est} - \Delta f_{mea})_{rms}$ = 2.75 μm	$(\Delta r_{A,est} - \Delta r_{A,mea})_{rms}$ = 4.68 μm	$(\Delta r_{B,est} - \Delta r_{B,mea})_{rms}$ = 4.69 μm

Table 4. Root-mean-square values of the experiment

In Fig. 13, it is found that without considering the scale, the wave of difference ($\Delta f_{est} - \Delta f_{mea}$) was upside down to the waves of their corresponding differences ($\Delta r_{A,est} - \Delta r_{A,mea}$) and ($\Delta r_{B,est} - \Delta r_{B,mea}$), respectively. In other words, the deviations between the measured and estimated center distance variations should proportionally influence the accuracy of the estimated profile errors. Figure 14 shows the uncertainty of the measured center distance variations, u_f, which is evaluated from the 10 data sets of the interpolated center distance variations through using the three-standard-deviation-band approach (Beckwith et al., 2004) with respect to each corresponding cam rotation angle. The evaluated uncertainty u_f ranged between 0.43 and 3.7 μm and had a root-mean-square value of 1.97 μm. The statistical representatives of the measured center distance variations, $\Delta f_{mea,SR}$, can be expressed as

$$\Delta f_{mea,SR} = \Delta f_{mea} \pm u_f \tag{32}$$

Thus, the upper and lower bounds of $\Delta f_{mea,SR}(\theta)$, $\Delta f_{mea,SR(u)}(\theta)$ and $\Delta f_{mea,SR(l)}(\theta)$, are defined as terms $[\Delta f_{mea}(\theta) + u_f(\theta)]$ and $[\Delta f_{mea}(\theta) - u_f(\theta)]$, respectively. Considering one of the worst cases, when data of $\Delta f_{mea,SR(u)}(\theta)$, $\Delta r_{A,mea}(\theta)$ and $\Delta r_{B,mea}(\theta)$ were adopted to calculate $\Delta r_{A,est}(\theta)$ and $\Delta r_{B,est}(\theta)$ by using Eqs. (20) and (21), respectively, the evaluated difference ($\Delta r_{A,est} - \Delta r_{A,mea}$) as shown in Fig. 15(a) ranged between −6.8 and 17.57 μm and had a root-mean-square value

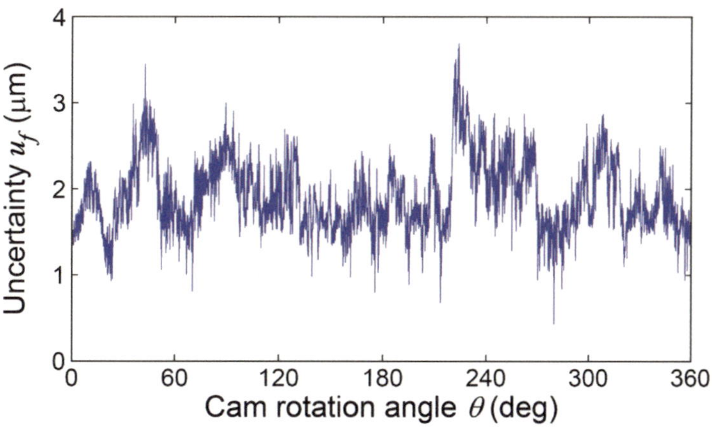

Fig. 14. Uncertainty of the measured center distance variations

of 5.97 μm, and the evaluated difference ($\Delta r_{B,est} - \Delta r_{B,mea}$) as shown in Fig. 15(b) ranged between −7.44 and 17.32 μm and had a root-mean-square value of 5.93 μm. The statistically relative deviation between $\Delta r_{A,est}$ and $\Delta r_{A,mea}$ was evaluated as 4.09% [= (5.97/146.13) ×100%], and that between $\Delta r_{B,est}$ and $\Delta r_{B,mea}$ was evaluated as 5.42% [= (5.93/109.5) ×100%]. Likewise, considering the other of the worst cases, when data of $\Delta f_{mea,SR(l)}(\theta)$, $\Delta r_{A,mea}(\theta)$ and $\Delta r_{B,mea}(\theta)$ were adopted to calculate $\Delta r_{A,est}(\theta)$ and $\Delta r_{B,est}(\theta)$ by using Eqs. (20) and (21), respectively, the evaluated difference ($\Delta r_{A,est} - \Delta r_{A,mea}$) as shown in Fig. 16(a) ranged between −15.78 and 10.32 μm and had a root-mean-square value of 5.55 μm, and the evaluated difference ($\Delta r_{B,est} - \Delta r_{B,mea}$) as shown in Fig. 16(b) ranged between −17.27 and 10.18 μm and had a root-mean-square value of 5.64 μm. The statistically relative deviation between $\Delta r_{A,est}$ and $\Delta r_{A,mea}$ was evaluated as 3.8% [= (5.55/146.13) ×100%], and that between $\Delta r_{B,est}$ and $\Delta r_{B,mea}$ was evaluated as 5.15% [= (5.64/109.5) ×100%]. In other words, when considering the worst cases, the differences and relative deviations in root-mean-square forms between the estimated and measured profile errors were still less than 6 μm or 5.5%. Therefore, the uncertainty of the measured center distance variations in this experiment should have merely slight effect on influencing the accuracy of the estimated profile errors.

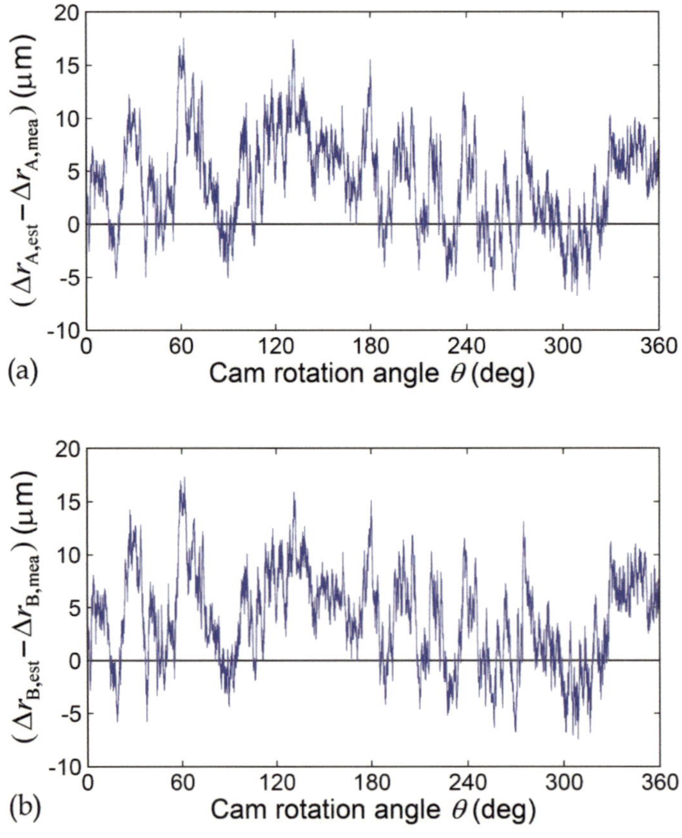

Fig. 15. Differences between the measured and estimated results evaluated by considering the upper bounds of the statistical representatives of the measured center distance variations

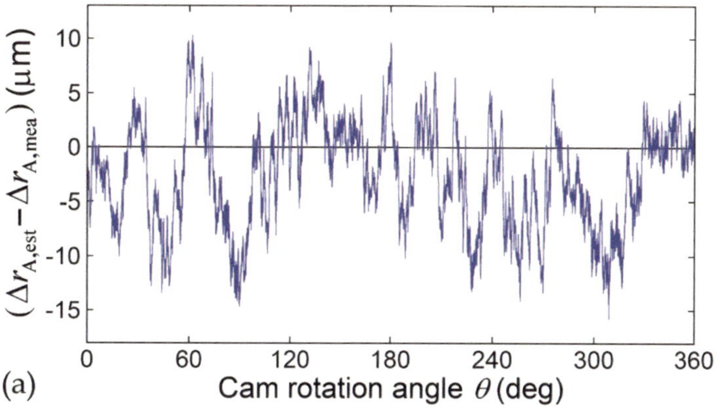

(a)

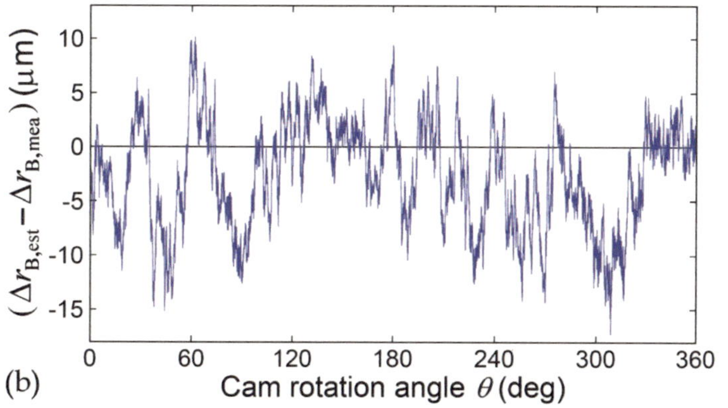

(b)

Fig. 16. Differences between the measured and estimated results evaluated by considering the lower bounds of the statistical representatives of the measured center distance variations

In addition, by applying the criteria established in Sub-section 3.1, the allowable upper and lower limits of the measured center distance variations are shown in Fig. 17, and whose extreme values are also listed in Table 3. As shown in the figure, the measured values of Δf_{mea} exceeded their allowable upper bound, $\Delta f_{A(u),est}$, when $\theta = 80° \sim 110°$ but totally fell within the range of $\Delta f_{B(l),est} \sim \Delta f_{B(u),est}$. Recall from Fig. 10 that the magnitude of $\Delta r_{A,mea}$ exceeded the specified tolerance of ±220 μm at about $\theta = 80° \sim 110°$, while the magnitude of $\Delta r_{B,mea}$ fell within the range of its specified tolerance. Obviously, the profile error evaluating results by using the established criteria agreed with the measuring results by using a CMM. As a result, the method presented in this study has been verified a feasible means for examining profile errors of assembled conjugate disk cams.

As compared with the use of a CMM to examine profile errors of conjugate disk cams that had taken 3 hours for measuring each cam, the presented method that took 15 minutes for examining each cam through the rotation of the assembled conjugate cams for 1 revolution could provide acceptable results with efficiency. Although the presented method cannot completely replace the use of CMMs, but in certain aspects it should be a more convenient and inexpensive means for the quality control in mass production of assembled conjugate disk cams.

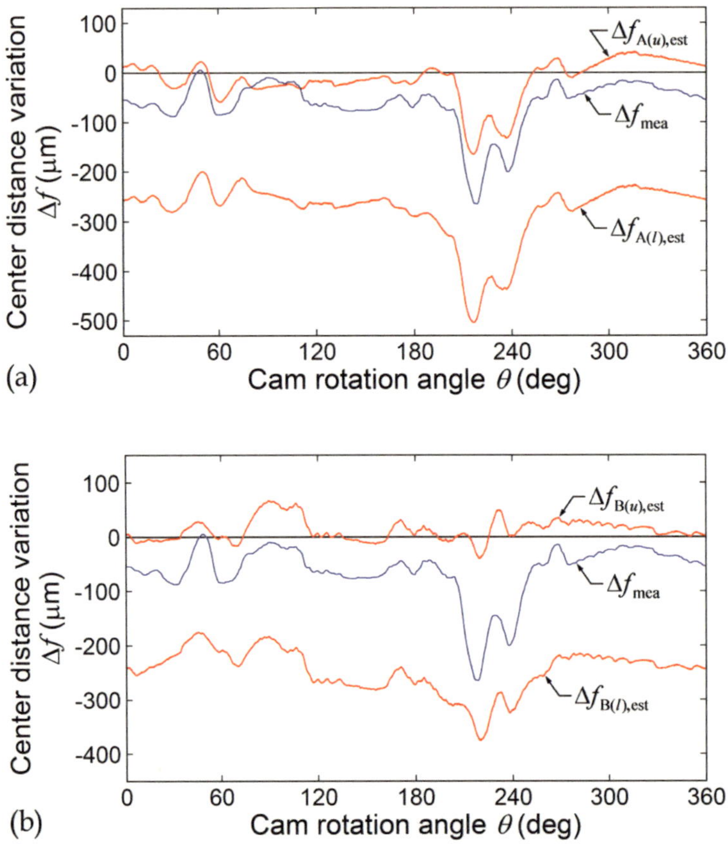

Fig. 17. Allowable upper and lower limits of the measured center distance variations

6. Conclusion

Based on combining the concepts of conjugate variation measurement and inverse conjugate variation analysis, the profile accuracy of assembled conjugate disk cams can be examined by a convenient and inexpensive manner. If a pair of master conjugate cams with known profile errors and a set of conjugation measuring fixture are available, by means of the measured center distance variations between the cam and follower pivots induced by a pair of assembled conjugate cams consisting of one master cam and the other being the inspected cam, then the profile errors of the inspected cam can be estimated with the use of the analytical equations derived in this study. Then, the accuracy of the inspected cam can be examined through the information of the measured center distance variations with the use of the criteria established in this study. An experiment meant to examine the profile errors of a pair of machined conjugate cams had been conducted. The machined conjugate cams had been examined by the presented method to compare with the measuring results obtained by using a CMM. The experimental results showed that the estimated profile errors were well consistent with those of the measured ones by using a CMM. From a statistical viewpoint, the differences and relative deviations in root-mean-square forms between the estimated and measured results of the cam profile errors were less than 6 μm and 5.5%, respectively, even though the machined cams had been intentionally specified to have a large tolerance grade of IT11. In conclusion, the method presented in this study has been verified a feasible and efficient alternative means for examining profile errors of assembled conjugate disk cams. Therefore, the presented method could be useful for the quality control in mass production of assembled conjugate disk cams and may replace the use of expensive CMMs in certain aspects. Integrating the presented method with machine system design to develop a specialized quality control system could be possible future work.

7. Acknowledgment

The authors are grateful to the National Science Council of Taiwan for supporting this research under Grant No. NSC-95-2221-E-007-012-MY2 and Grant No. NSC-98-2221-E-007-015-MY2.

8. References

Beckwith, T.G.; Marangoni, R.D. & Lienhard V, J.H. (2004). *Mechanical Measurements* (5th edition), pp. 45-125, Pearson Education Taiwan, ISBN 986-154-022-9, Taipei, Taiwan

Chang, W.T. & Wu, L.I. (2006). Mechanical Error Analysis of Disk Cam Mechanisms with a Flat-Faced Follower. *Journal of Mechanical Science and Technology*, Vol.20, No.3, (March 2006), pp. 345-357, ISSN 1738-494X

Chang, W.T.; Wu, L.I.; Fuh, K.H. & Lin, C.C. (2008). Inspecting Profile Errors of Conjugate Disk Cams with Coordinate Measurement. *Transactions of the ASME, Journal of Manufacturing Science and Engineering*, Vol.130, No.1, (February 2008), 011009, ISSN 1087-1357

Chang, W.T. & Wu, L.I. (2008). A Simplified Method for Examining Profile Deviations of Conjugate Disk Cams. *Transactions of the ASME, Journal of Mechanical Design*, Vol.130, No.5, (May 2008), 052601, ISSN 1050-0472

Chang, W.T.; Wu, L.I. & Liu, C.H. (2009). Inspecting Profile Deviations of Conjugate Disk Cams by a Rapid Indirect Method. *Mechanism and Machine Theory*, Vol.44, No.8, (August 2009), pp. 1580-1594, ISSN 0094-114X

Hsieh, J.F. & Lin, P.D. (2007). Application of Homogenous Transformation Matrix to Measurement of Cam Profiles on Coordinate Measuring Machines. *International Journal of Machine Tools and Manufacture*, Vol.47, No.10, (August 2007), pp. 1593-1606, ISSN 0890-6955

Koloc, Z. & Václavík, M. (1993). *Cam Mechanisms*, pp. 411-413, Elsevier, ISBN 0-444-98664-2, New York, USA

Lin, P.D. & Hsieh, J.F. (2000). Dimension Inspection of Spatial Cams by CNC Coordinate Measuring Machines. *Transactions of the ASME, Journal of Manufacturing Science and Engineering*, Vol.122, No.1, (February 2000), pp. 149-157, ISSN 1087-1357

Norton, R.L. (2009). *Cam Design and Manufacturing Handbook* (2nd edition), pp. 27-30, pp. 433-440, Industrial Press, ISBN 978-0-8311-3367-2, New York, USA

Qiu, H.; Li, Y.; Cheng, K. & Li, Y. (2000). A Practical Evaluation Approach towards Form Deviation for Two-Dimensional Contours Based on Coordinate Measurement Data. *International Journal of Machine Tools and Manufacture*, Vol.40, No.2, (January 2000), pp. 259-275, ISSN 0890-6955

Qiu, H.; Li, Y.B.; Cheng, K.; Li, Y. & Wang, J. (2000). A Study on an Evaluation Method for Form Deviations of 2D Contours from Coordinate Measurement. *The International Journal of Advanced Manufacturing Technology*, Vol.16, No.6, (May 2000), pp. 413-423, ISSN 0268-3768

Qiu, H.; Cheng, K.; Li, Y.; Li, Y. & Wang, J. (2000). An Approach to Form Deviation Evaluation for CMM Measurement of 2D Curve Contours. *Journal of Materials Processing Technology*, Vol.107, No.1-3, (November 2000), pp. 119-126, ISSN 0924-0136

Rothbart, H.A. (Ed.) (2004). *Cam Design Handbook*, pp. 8-9, pp. 44-46, McGraw-Hill, ISBN 0-07-137757-3, New York, USA

Wu, L.I. (2003). Calculating Conjugate Cam Profiles by Vector Equations. *Proceedings of the Institution of Mechanical Engineers, Part C: Journal of Mechanical Engineering Science*, Vol.217, No.10, (October 2003), pp. 1117-1123, ISSN 0954-4062

Wu, L.I. & Chang, W.T. (2005). Analysis of Mechanical Errors in Disc Cam Mechanisms. *Proceedings of the Institution of Mechanical Engineers, Part C: Journal of Mechanical Engineering Science*, Vol.219, No.2, (February 2005), pp. 209-224, ISSN 0954-4062

26

Material Characterization and Failure Analysis for Microelectronics Assembly Processes

Chien-Yi Huang[1,2], Ming-Shu Li[1],
Shan-Yu Huang[1], Cheng-I Chang[1] and Min-Hui Huang[1]
[1]*Process Technology Enabling & Materials Characterization Div. Operations, Wistron Corporation, Hsinchu 300, Taiwan,*
[2]*Department of Industrial Engineering and Management, National Taipei University of Technology, Taipei 106, Taiwan, R.O.C.*

1. Introduction

In recent decades, the electronic industry has shown a clear trend towards miniaturization with increasing functionality. In the context of essential competition within the market, the reliability of long term operations has become a popular issue. This study examines the properties of printed circuit board (PCB) and its failure phenomena. PCB reliability is characterized through verifications taken from various process conditions. Notably, results can be used as selection criteria for PCB materials, helping to reduce PCB delamination during the assembly process. In addition, surface finish is a key factor seen to affect a product's durability, as the microstructure between solder and the metallized layer varies between surface finish types and has been shown to affect overall solder joint strength. Notably, the black pad phenomenon will reduce the strength of solder joint significantly and affect product's durability.

Characterization of failures and materials related to chemical and soldering processes used in microelectronics assembly are also discussed in this study. Analytical techniques used for chemical structures, compositions, and soldering properties including Fourier transform infrared spectrometer (FTIR), scanning electron microscopy/energy-dispersive x-ray spectroscopy (SEM/EDX), and dye staining are conducted. The jumper pillow speaker connector (JPSPK) connector pins show an obvious difference in color between the clean and contaminated areas. The contaminants on the connector pin were identified as the flux used in the assembly processes following a comparison of the FTIR spectra database. Additionally, the incoming plastic housings showed different bright and dark surfaces whose chemical structures were shown to be polycarbonate (PC) and acrylic ester, respectively. It indicates that varied surface treatments for the incoming housings.

To determine whether any cracks in the solder joints occurred in the CPU BGAs, a dye staining analysis was carried out. The crack size percentage is classified according to the crack's (dyed) area. The establishment of the infrared spectra database for fluxes and process materials helps determine the root cause of the contaminants in order to reduce the chance of a re-occurrence of similar problems thereby enhancing the manufacturing capability. The infrared spectrophotometry technique can be used by professional design manufacturers and/or electronics manufacturing service (ODM and/or EMS) providers to investigate board/component defects during product pilot run stages as well as during full-volume production.

2. PCB evaluation of lead free soldering process

The performance and heat resist-ability of PCB are investigated. This study focuses on the inner layers of the boards, looking at critical properties such as the glass transition (Tg) temperature and curing agents. Two factors, each with two levels, are used in the experimental design and are shown in Table 1. Other factors which remained constant are also shown in Table 1. Tg refers to the temperature at which material changes from glass-like to rubber-like, and where the coefficient of expansion (CTE) increases dramatically. Excessive time durations for processing temperatures above the Tg may lead to cracking at the plated-through-hole (PTH). This study considers PCBs with Tg's in the range of 110~150°C (normal Tg) and above 170°C (high Tg). The curing agent is critical to the polymerization of epoxy resins, and includes two categories: Dicy and Phenolic. Based on the experimental matrix, we select PCB types that are available to the current market. Four types of PCBs are used: ND (normal Tg / Dicy), NP (normal Tg / Phenolic), HD (high Tg / Dicy), and HP (high Tg / Phenolic). The material properties are shown in Table 2. All types of PCBs meet the IPC-4101B industry standard requirements.

Level / Factor	Level 1	Level 2
Tg	Normal (110 ~ 150℃)	High (>170℃)
Curing agent	Dicy	Phenolic
Constant		
Decomposition temperature	320 ~ 340℃	
Expansion rate	3.0 ~ 4.0% (25 ~ 260℃)	
Moisture absorption rate	0.13 ~ 0.18 %	
Process condition	Reflow and wave soldering	

Table 1. Experimental design

PCB types	ND	NP	HD	HP
Tg	140℃ (normal Tg)	150℃ (normal Tg)	170℃ (high Tg)	180℃ (high Tg)
Curing agent	Dicy	Phenolic	Dicy	Phenolic
Decomposition temperature	323℃	340℃	329℃	340℃
Expansion rate	3.98%	3.55%	3.37%	3.0%
Moisture absorption rate	0.14	0.13	0.15	0.18

Table 2. Material properties

2.1 Test vehicle

The surface finish of the PCB is organic solderability preservatives (OSP). Sample boards are composed of eight layers with a thickness of 1.6 mm and length and width of 224 mm x 114 mm, respectively. There are eight samples on each test board, and each sample has a total of 1922 PTHs. The PTH is 10 mils in diameter, with 40 mil spacing; the width of the annular ring is 6 mils (Figure 1). The large dimensionality combined with fine spacing simulates the worst-case scenario. The thickness of copper in the PTH is in the range of 0.8 mil and 1.0 mil to ensure that no damage occurs during the PCB fabrication process.

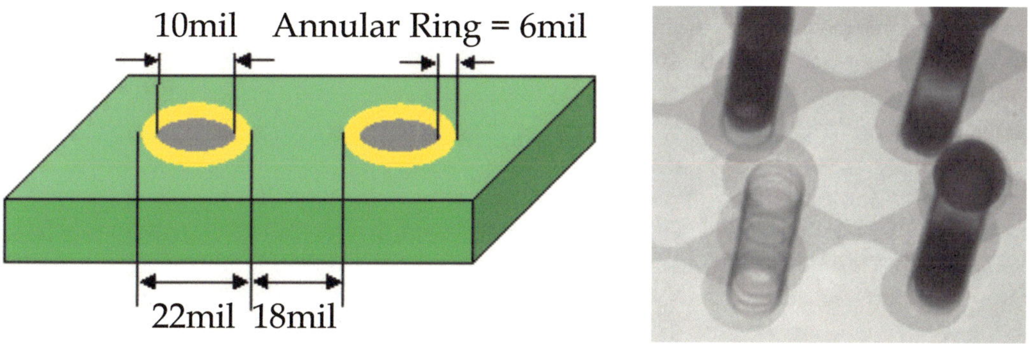

Fig. 1. PTH on the test board

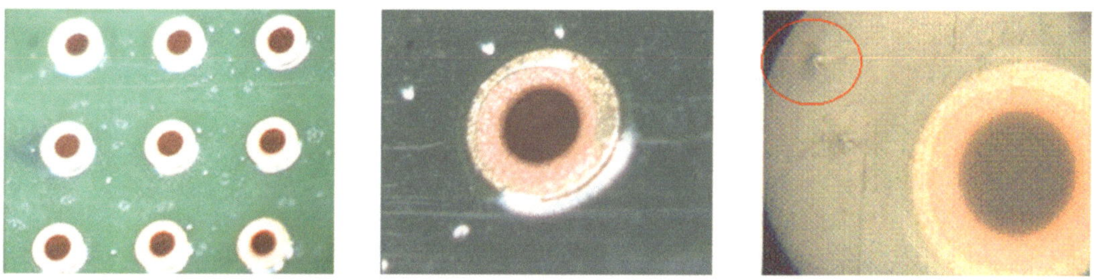
Fig. 2. Surface of the HD samples

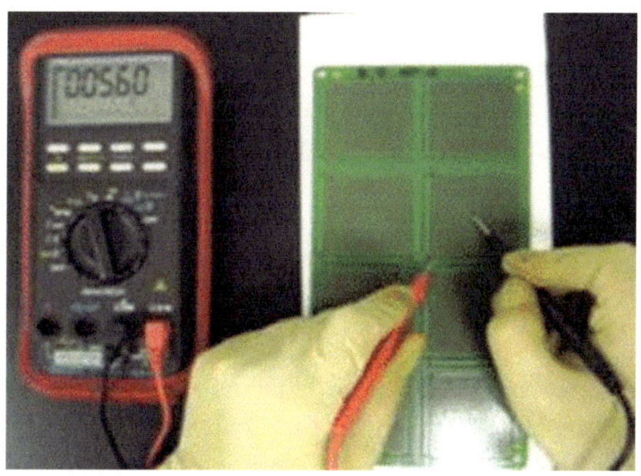
Fig. 3. Resistance measurements for the test vehicles

The test boards are subjected to the Standard Inspection Process (SIP) which includes the incoming quality control (IQC) for inspection of trace, solder mask, labelling, board thickness, warpage, electrical resistance, and PTH dimension. The HD (high Tg / Dicy, HD) sample shows blisters on the board surface (Fig. 2).
The meter used for measuring the resistance is correct within 0.02% (Fig. 3). The average resistances are: ND= 5.77 Ω, NP= 5.66 Ω, HD= 5.55 Ω, HP= 5.50 Ω. The normal Tg (ND, NP) material shows a higher resistance compared to the high Tg material (HD, HP). Failure is defined as: the value of resistance increased by 15%. Failed samples are then cross-sectioned for failure analysis. Samples are then subject to two tests. In the first test, heat resist-abilities of different PCB materials are evaluated. The second test involves the simulation of potential environments through thermal shock testing followed by a failure analysis.

2.2 Assembly process verification

The melting temperatures of lead-free solder alloys are usually higher than that of traditional tin-lead solder. Therefore, the reflow and wave soldering temperatures for PCB assembly processes are also higher. This may have a negative impact on the solder joint, the electronic components, and the board. In this study, we verify the performances and effectiveness of various types of PCBs used in lead-free applications. The heat resistibility and corresponding failures are investigated and are followed by an analysis of failure modes. Sample size of 48 arranged in six panels is prepared. The initial resistance is measured when the PCB is taken from the dry package. Sample boards are then processed through two reflow cycles with one wave soldering in a random sequence. Resistances are again measured. Samples are cross-sectioned if the resistance increases by 15% or more.

2.2.1 Assembly process

The assembly process includes two reflow cycles (for two sided PCBs) and one wave soldering. The reflow oven has nine temperature zones. The conveyor velocity is 65 cm/min. TAL (time above liquid) is 90 s while process spec is between 40 s and 90 s (Fig. 4). The oxygen level is 350 ppm. During the wave soldering process, the conveyor velocity is 50 cm/min. The dwell time is 10 s while process spec is between 4 s and 6 s. The soldering temperature is 265°C. The preheat temperatures are 170°C, 190°C and 210°C (Fig. 5).
A profile board is made to ensure that the desired temperature profile is achieved (Fig. 6). Thermal couples are attached at locations near the conveyor edge, the center of the oven, the PCB top side, and the PCB bottom side. The temperature variation across the board is within 10°C.

2.2.2 Results

The crack of solder mask is observed in samples of the above mentioned four types of PCB materials. A black attached substance is also observed at the edge of the samples. The analysis through scanning electronic microscope (SEM) equipped with the energy dispersive spectrometer (EDS) indicates that the substance is carbide (Fig. 7). The results of resistance measurement show that the variation in resistance increases as the assembly process progresses, but all within the 15% criterion.

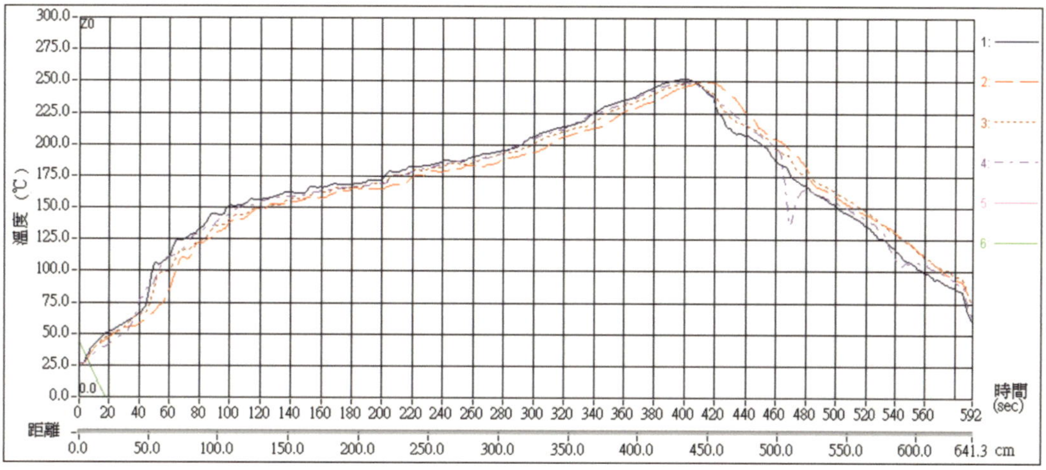

Fig. 4. Reflow temperature profile

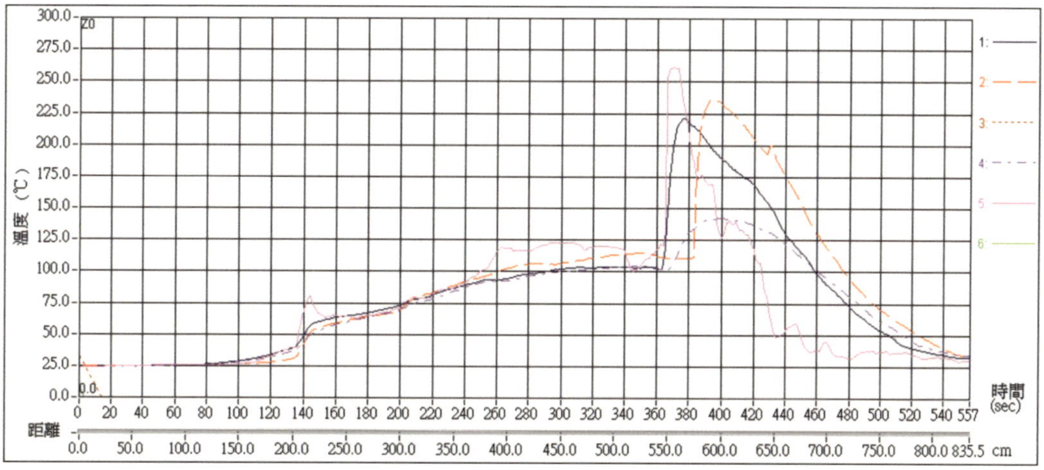

Fig. 5. Wave soldering profile

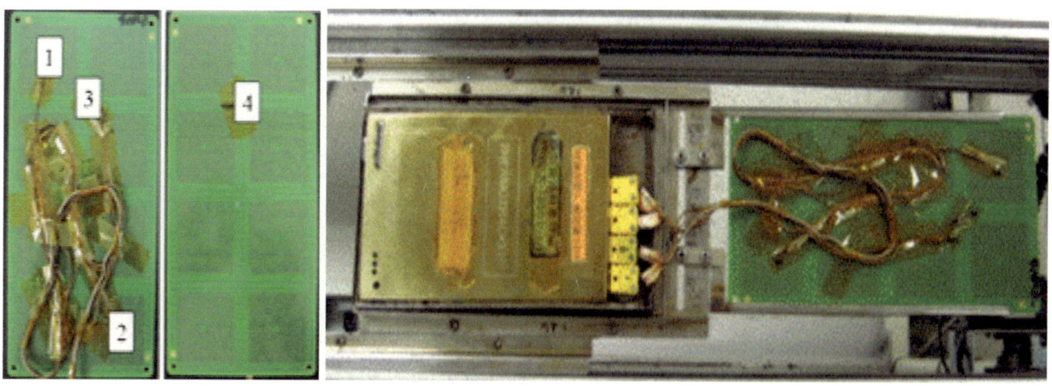

Fig. 6. Profile board and temperature recorder

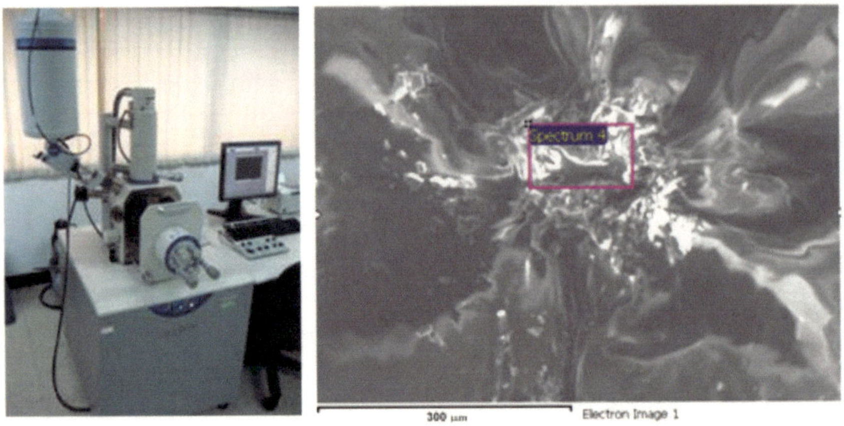

Fig. 7. SEM analysis

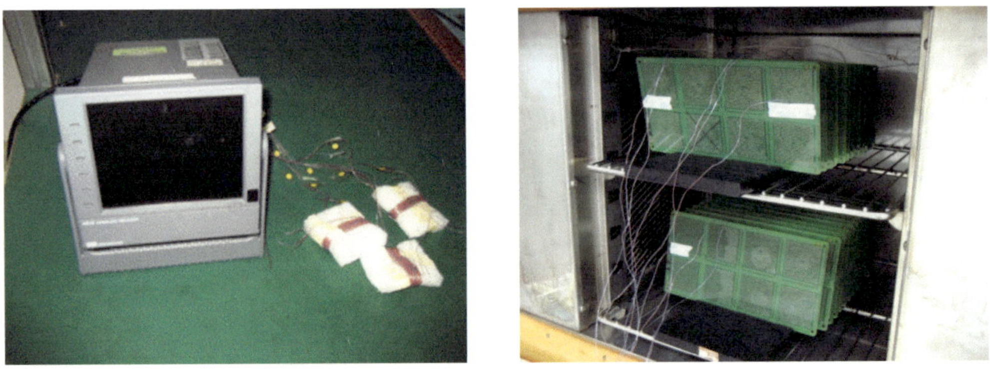

Fig. 8. Real time monitoring system

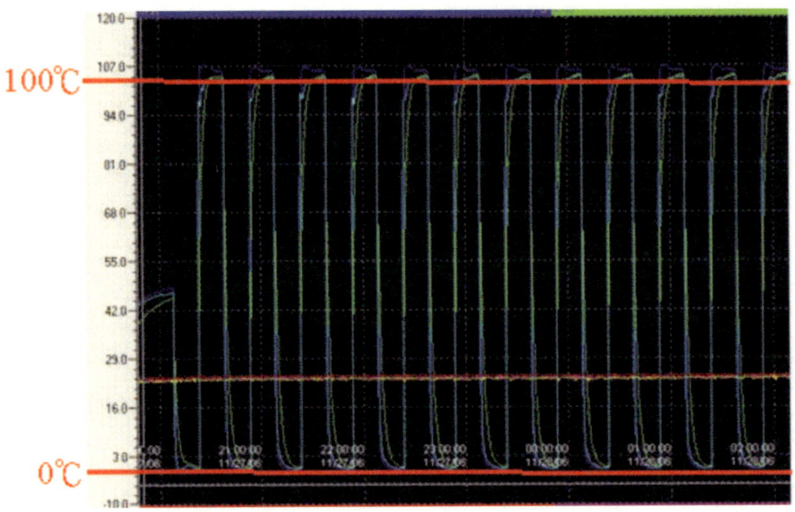

Fig. 9. Temperature profile during thermal shock test

2.3 Thermal shock test

The thermal shock test simulates the change of environmental temperatures that the electronics products may encounter during field service (Bevilacqua, 2004; Jayatilleka, 2003; Zhao, 2004). Each material is represented by a total of 48 samples (6 test boards). Samples are subject to the two reflow and one wave soldering profiles prior to the thermal shock test. The temperature ranges from 0°C to +100°C (temperature change within 3 min) and the time at both extreme temperatures is 12 min. A total of 900 cycles with real time temperature monitoring systems are used (Fig. 8). Samples are removed from the chamber after every 12 temperature cycles for resistance measurement (Fig. 9).

2.3.1 Results

Solder mask cracking is the only symptom observed during the appearance check within the 900 temperature cycles. The measured resistance shows that failure only obtained in the normal Tg PCB (ND = 6.25%, NP = 20.83%). High Tg PCB shows better performance during the thermal shock test (Fig. 10). Cross-sectional analysis of the normal Tg PCB also shows that delamination and cracking occurred at the ND inner layers (Fig. 11). These defective scenarios did not occur to either Dicy or Phenolic of the high Tg PCBs.

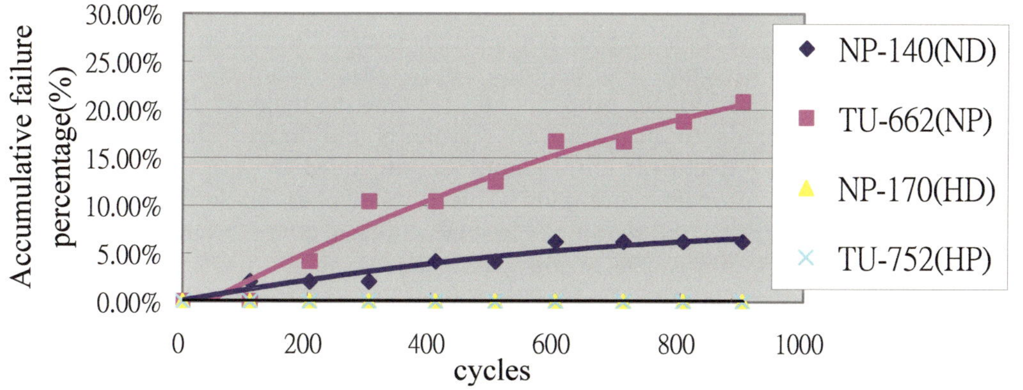

Fig. 10. Accumulative failure percentages after 900 thermal shock cycles

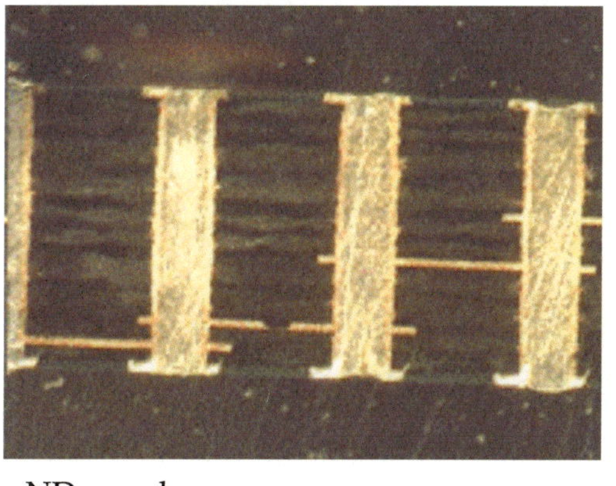

Fig. 11. Inner layers of the ND samples

2.3.2 Life time estimation

There are three and ten failed samples for the ND and NP PCBs, respectively. The accumulative failure data are shown in Figs. 12 and 13. The Weibull probability plots are used to estimate the parameters of the failure distributions (Figs. 14 and 15).

The shape parameters β and characteristic life θ of the ND PCBs are found to be 0.5858 and 84570 cycles, respectively. For the NP PCBs, β and θ of the ND PCBs are 1.1955 and 2770 cycles, respectively. Through the use of Gamma function, the expected life for ND and NP PCBs are 129,027 cycles and 2,603 cycles of testing. Finally, the modified Coffin Mason formula is used to calculate the accelerated factor (AF). Assuming that the normal-use environment is within the temperature range of 25°C~40°C, and the frequency of temperature change is five, the AF is found to be 40. As a result, the expected life for ND and NP PCBs are 2,557,142 h and 51,580 h during their normal service environment.

3. Black pad

Surface finished types that include organic solderability preservatives (OSP), electroless nickel immersion gold (ENIG), immersion tin, and immersion silver were adapted for certain applications. The finished surface provides another function which is to protect the soldering pad from oxidation during preservation and assembly processes. If oxidation takes place at the soldering pad, the system will suffer a reduction of solderability during the soldering process and even trigger non-wetting issues. Electroless nickel immersion gold, ENIG, is a popular surface treatment used in the print circuit board (PCB) industry. Specifically, it is used in handheld products and for antenna material due to its excellent performance with regards to its corrosion resistance and longer storage capabilities. However, ENIG surface treatment is frequently accompanied by a specific defect if the plating process is not properly controlled. This is known as black pad. Solder joints suffering from this symptom will see significant reductions in strength which will affect the product's durability (IPC-7095A, 2004; Zeng et al., 2004).

3.1 Black pad symptom

Figure 16 presents the black pad symptom occurring on the ENIG surface finished from raw materials. Through a cross-sectional analysis using scanning electron microscopy (SEM) equipment, several spikes can be observed on the Ni layer and which have penetrated into the Ni layer around 25% of the total thickness. Based on author's experience, ENIG soldering pads with this symptom can be wetted by solder, but their reliability performance may have been decreased to one-half or one-third of its original. As it stands, there are no criteria with which to judge this symptom. A careful company usually screens out the potential risk factors through reliability tests before mass production.

3.2 Practical failure analysis case

Figure 17(a) presents the Ni surface after removal of the Au layer using a chemical etching method. Several dark lines can be seen on the Ni surface. A more detailed magnification is shown in Fig. 17(b). Compared with the cross-sectional view of raw material shown in Fig. 17(c), these dark lines were groove structured and penetrated into the Ni layer. Several studies have indicated that this symptom was caused by corrosion during the Au plating process (Bulwith et al., 2002; Xu et al., 2008). PCB vendors claim that a Ni layer without this spike symptom is not achievable. Based on the literature as well as our practical experience,

Ni layers with fewer corrosion symptoms could obtain better performance in reliability tests (Zeng et al., 2004).

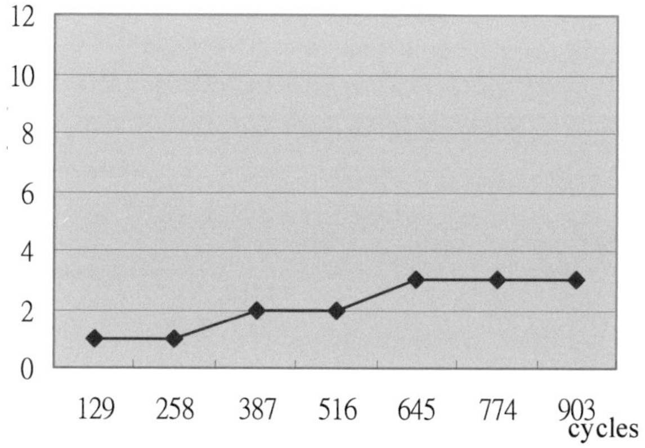

Fig. 12. Accumulative failure data for the ND PCBs

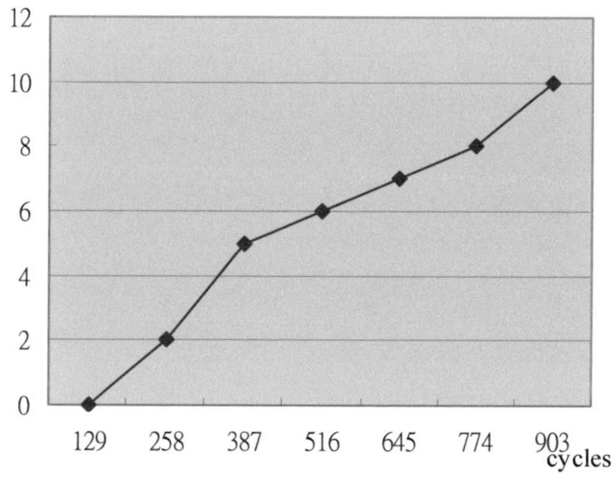

Fig. 13. Accumulative failure data for the NP PCBs

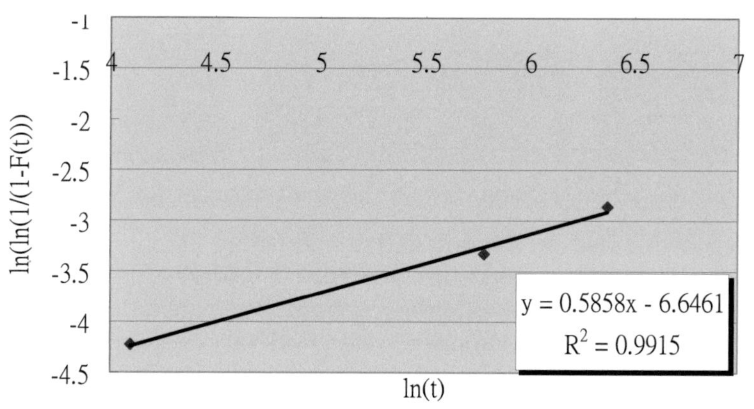

Fig. 14. Weibull probability plots of ND PCBs

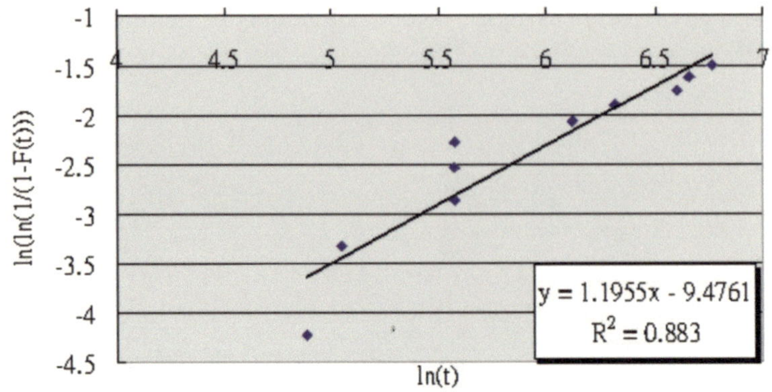

Fig. 15. Weibull probability plots of the NP PCBs

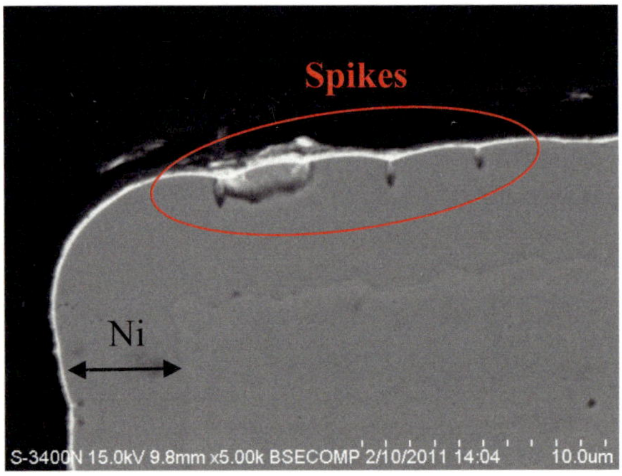

Fig. 16. Black pad symptom, cross-section view

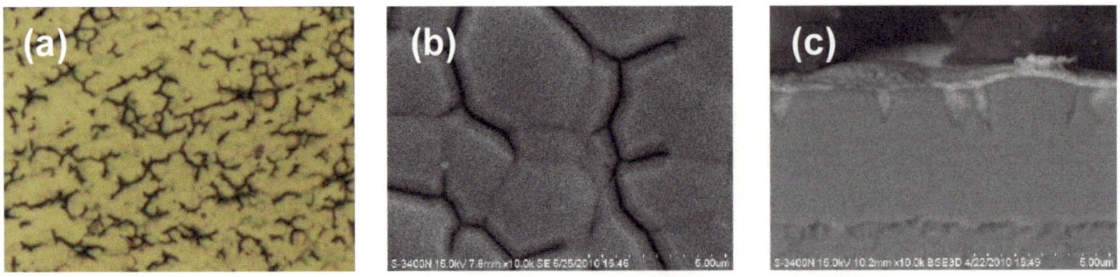

Fig. 17. (a) Plane view, Ni layer (b) plane view, Ni layer (c) cross-section view

A typical cross-sectional view of the solder joint structure is shown in Fig. 18. The microstructures from top to bottom and sides were the Cu soldering pad, the Ni layer, and the IMC layer, respectively. These spikes observed from raw material remained on the Ni layer after the soldering process. Additionally, a thin dark layer between the Ni layer and inter-metallic compound (IMC) was observed; this layer was a P-rich layer. A P-rich layer was actually developed during the Au plating process due to Ni atoms nearby. The Ni layer surface was replaced by Au atoms, but P atoms within the Ni layer did not act in the Au plating bath which resulted in an increased P concentration. After the post-assembly process, the P-rich layer will increase in thickness again due to Ni atoms diffused in the solder. Normally, a P-rich layer is composed of P concentrations with 15~25 w.t.% and the Ni element. Figure 19 presents the component fall-off issue during the drop test. This view presents the soldering pad from the component side as it was supposed to be jointed with the PCB after the assembly process. The solder joint between this component and PCB were unable to endure external stressing during the drop test which resulted in component fall-off issues.

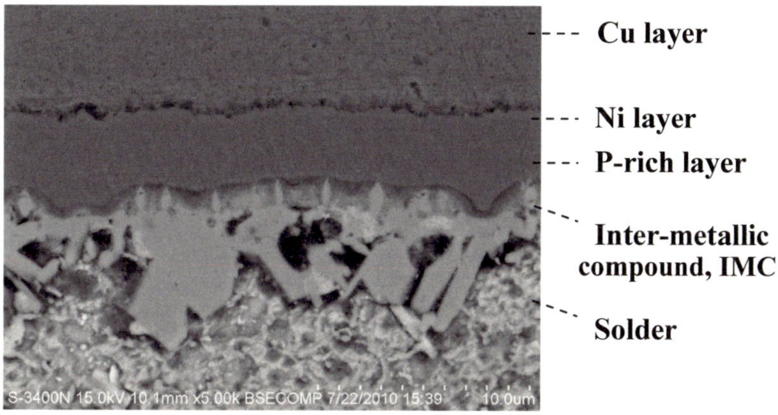

Fig. 18. Solder joint structure, cross-section view

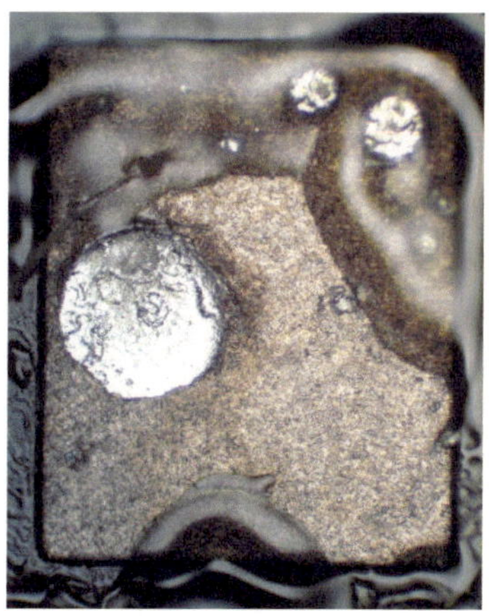

Fig. 19. Fracture appearance

Through cross-sectional analysis (not shown in this article), we found that the fracture was located between the P-rich layer and the IMC layer. Several studies have reported that the ENIG with black pad was fractured at this layer, and that its failure mechanism was not clearly defined (IPC-7095A, 2004; Zeng et al., 2004). PCB manufactures have eliminated the threat to the product's durability and solderability issue by developing several new technologies to avoid P-rich layer formation, for instance, electroless Ni/electroless Pd/immersion Au, ENEPIG.

Figure 20 shows a more detailed view than Figure 19. We can observe several moiré on the soldering pad. This phenomenon may be attributed to deeper spikes forming on the Ni layer during the Au plating process.

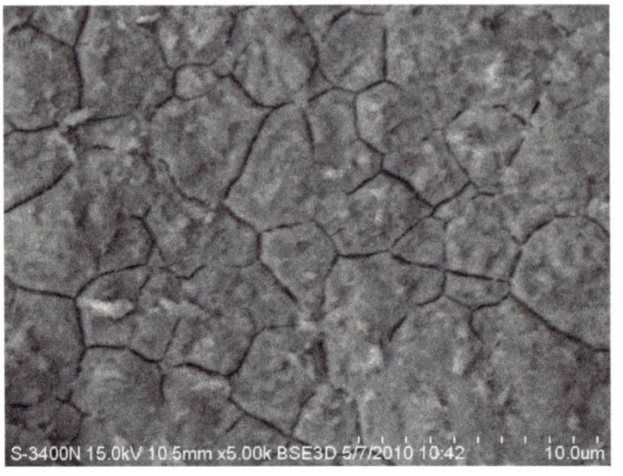

Fig. 20. Fracture morphology, SEM view (5000 x)

3.3 Enhanced solder joint strength by underfill

Figure 21 illustrates a method for enhancing solder joint strength. The free space between the component and the PCB were filled with an underfill material which becomes solid after the baking process. This kind of reinforced method is commonly used in PCB assembly for enhanced bonding strength in bigger components. So far, there has been no published academic research related to using underfill enhancement for improved solder joint strength when assembly plants encounter black pad symptoms. Hence, we have performed the test to understand whether underfill enhancement is suited to mitigate the influence of the black pad phenomenon. The testing component for the drop-test was prior diagnosed as the black pad symptom, and the defect rate was almost 100%. Before drop-testing, we enhanced the bonding strength for this component by pouring underfill material into the free space between the component and the PCB.

The drop-test resulted in failure; the suspected component was still attached to the PCB during drop-testing. Through cross-section analysis, we could found that the crack penetrated the entire solder joint at the component side, Fig. 22. This implied that the underfill enhancement was unable to eliminate the black pad threat.

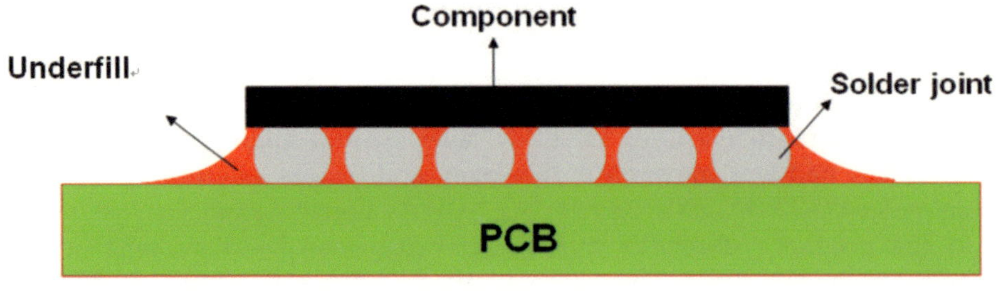

Fig. 21. Illustration of underfill enhancement

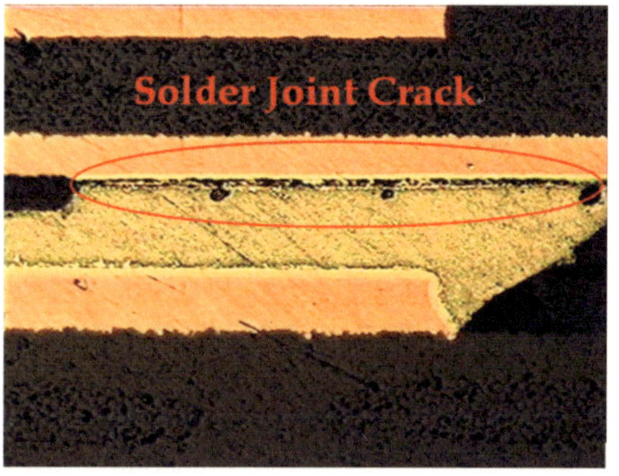

Fig. 22. Solder joint crack after drop-test

4. Chemical characterization of failures and process materials

Along with the increasing development and advancement within the electronics industry, the basic properties of printed circuit boards (PCBs) and related materials become more and more critical. The characterization of these material compositions and various contamination sources are extremely important (Dunaway, 1989; Smith, 2007a; Welch, 1995; Shangguan and Gao, 1997; Huang et al., 2009). Even minor deterioration and/or variation of materials could impact the quality of the final product.

In microelectronics assembly, organic materials (and their compounds) are usually employed and sometimes mixed with inorganic solders. The chemical analysis of these materials, including: polymers (epoxies, PVC, ABS), organic metals (OSP), and flux (liquid and paste fluxes), are necessary for obtaining desired soldering processes. Infrared absorption spectrometry (or infrared spectrophotometry) is one of the most common and powerful tools to identify organic and inorganic materials (Smith, 2007b). Surface organic contaminant identification testing through the use of infrared analytical methodology was also specified in IPC standards (IPC-TM-650 2.3.39, 1988).

The infrared spectrophotometry technique can be used by professional original design manufacturing (ODM) and/or electronics manufacturing service (EMS) providers to investigate board/component defects during product pilot run stages and volume production. Identification of the root cause of failures helps reduce re-occurrence of similar problems and enhance manufacturing capabilities. The analytical work can be used for quality control and/or in house process verification.

In this study, the failures resulted from the residues and contaminants left on the connector pins. The database comprising Fourier transform infrared spectrometer (FTIR) spectra of fluxes and plastic housings are established to facilitate the mapping in order to identify the root cause(s) of failure.

4.1 FTIR spectroscopy and SEM/EDX

Fourier transform infrared spectrometer (FTIR) is applied to investigate the absorption characteristics of molecules to IR radiation and provides a qualitative and quantitative analysis method. FTIR spectrophotometry offers a quick, accurate, non-destructive approach and sophisticated sample preparation is not necessary. Therefore, less chemical pollution results from sample pre-treatment. FTIR can be used by PCB manufacturers and systemic assembly houses for the analysis of solder mask curing, flux residue, and various possible contaminations on the board (Dunaway, 1989; Shangguan and Gao, 1997).

With micro FTIR, the scope and type of test samples can be expanded for various applications. The regular FTIR has a beam size in the range of 6-8 mm and may not be applicable for sample dimensions less than 100 μm. The IR beam needs to be condensed in order to cope with a magnified image for micro size samples. The sample dimension limitations of micro FTIR is around 10 μm squared, depending on the IR natures and sample conditions. This is critical, especially for high density applications. Notably, the small beam size helps ensure that the spectrum collected is on the target material and is not subjected to influence from adjacent materials. Therefore, the micro FTIR is suitable for the analysis of surface contamination from particles, fibers, coatings, and solvents in the microelectronic industry.

Following restrictions of the hazardous substances (RoHS) compliance test, FTIR with accuracy of percentage (%) order is recommended for screening tests before gas chromatograph (GC)/mass spectrometer (MS)/inductively coupled plasma (ICP) tests with accuracy of part per million (ppm) is ordered (Bromine Science and Environment Forum http://www.bsef.com; RoHS and WEEE information pages http://uk.farnell.com/static/rohs/). FTIR provides a quick examination of polymer type and brominated flame retardant (BFR) rough contents. While sample preparations for GC/MS/ICP are time consuming, FTIR offers a cost/time effective screening solution.

It needs to be pointed out that employing multiple analytical techniques is essential to reach a more accurate conclusion. FTIR is frequently associated with other analytic techniques such as energy-dispersive X-ray spectroscopy (EDX), X-ray fluorescence spectroscopy (XRF), and X-ray photoelectron spectroscopy (XPS) in order to investigate chemical structures, compositions, and binding states (Hirt and Artaki, 1991).

The appearance of the failure samples was observed at all stages of component disassembly. This is followed by a conformance inspection with the aid of an optical microscope (Leitz Ergolux AMC) rated between 20 and 2000X magnification before the chemical analyses.

FTIR is suggested to be performed prior to SEM/EDX analysis since the existence and/or properties of micro contaminant could vary according to electron beam energy during the EDX analysis. This is due to the extremely short wave length of the electron beam with a relatively strong energy. While 1 keV and 30 keV acceleration voltages are applied, the wavelength of the electron beam is 0.3876 Å and 0.0698 Å, respectively (Lee, 1993).

4.1.1 FTIR

FTIR usually has three operation modes including transmission, reflectance, and attenuated total reflectance (ATR). While transmission mode is used for the transparent sample (thickness within 2~5 micron), reflectance mode is used for the opaque sample. As for bulk materials, ATR mode collecting spectra from the surface (depth of 2~6 micron) should be used.

In this study, Jasco type MFT-2000 and Perkin Elmer Spectrum 100 FT-IR, both with microscope capability, were used to characterize the chemical structure of the failure samples and materials. Spectrum 100 FT-IR using a Ge-coated KBr beam splitter is used for detecting spectra at the frequency range 8,300-350 cm^{-1}. This equipment is capable of achieving a signal-to-noise ratio of 10,200: 1 peak to peak within a five-second measurement. The measurements were made using a deuterated tryglycine sulphate (DTGS) detector at 4 cm-1 spectral resolution. The spectra of single reflection ATR with ZnSe crystal were collected on a Spectrum 100 FT-IR spectrometer equipped with an Auto-image microscope system. The microscope is with liquid nitrogen cooled mercury cadmium telluride (MCT) detector.

4.1.2 SEM/EDX

The morphologies of the residues and contaminants were observed with a Hitachi S-3400N scanning electron microscopy (SEM). The electron source is pre-centered cartridge type tungsten hairpin filament. The magnification can be adjusted within the range of 5 to 300,000X. Magnifications are dependent upon accelerating voltage, working distance, and scan speed. The secondary electron image resolution is 3 and 10 nm for 30 and 3 kV acceleration voltages (in high vacuum mode), respectively. The reflection electron image resolution is 4 nm (30 kV acceleration voltages, in low vacuum mode).
The chemical compositions of the failures were performed using an EDX analyzer with a Horiba EX-250 instrument. The detectable range of elements is between B5 and U92. The detector used was an INCA X-sight Series Si (Li).

4.2 Failure scenario & analysis results

4.2.1 Case I - Residues on the connector pins

A failure symptom of open contact was found to the JPSPK connector of the audio station bed connector (ASBC) main board. This had resulted in in-circuit test (ICT)/automatic test equipment (ATE) testing failure after the test fixture was attached (or during the basic electric conduction and functional tests). The failure rate for the connector test is about 15%. The connector pin unit can be dislocated from the plastic holding and metal housing (outer layer), as shown in Figs. 23(a) and (b). The appearance observation by OM and CCD shows severe residues on the connector pins. This also indicates the obvious difference in color between the clean and the contaminated areas in Fig. 23(c).
The chemical analysis was carried out to examine the residual substances that may impact the electrical conductivity of the contact. The contaminated areas of the connector pin unit for micro FTIR analysis are shown in Fig. 23(c).
The micro FTIR spectra at the contaminated area from two different connector pins (named P1 and P2) (random sampling) are shown in Fig. 24. IR analysis was carried out at three adjacent locations for each pin (marked dash -1 to -3) to confirm the uniformity of the chemical structures (functional groups). Organic compositions and properties appeared on the contaminated areas. The characteristic absorption peaks of C-H bonds at around 2950 and 2870 cm^{-1} may assign to the symmetric and asymmetric CH_3 and CH_2 stretching vibration bands. The C-H bending vibrations for CH_3 and CH_2 are seen at around 1380 and 1460 cm^{-1}. The carbonyl band (C=O) appeared at 1710 cm^{-1}. The various contaminated areas from different pins showed a very similar chemical structure.
Two different solder pastes (a sticky mixture of powdered solder and flux) were used in reflow soldering during the assembly process, namely paste P1 and P2, respectively. Contrary to metal solders, organic fluxes have the absorption characteristics of molecules with respect to IR radiation. After comparing with the FTIR spectra database, the spectra of the contaminated areas seem to be consistent with that of the paste P1 in Figs. 25(a) and (b)

(sorted by the similarity of chemical structures). Their characteristic functional groups appeared at the same absorption peaks. The paste P2 showed similar structures, but showed few differences compared with the paste P1 (Fig. 25(c)). This suggests that the residues on the connector pins are attributed to the flux from the solder paste used in the assembly processes. Therefore, the micro FTIR was found to be an effective tool to identify the root cause of the micro residue left on the connector pins (Huang et al., 2009).

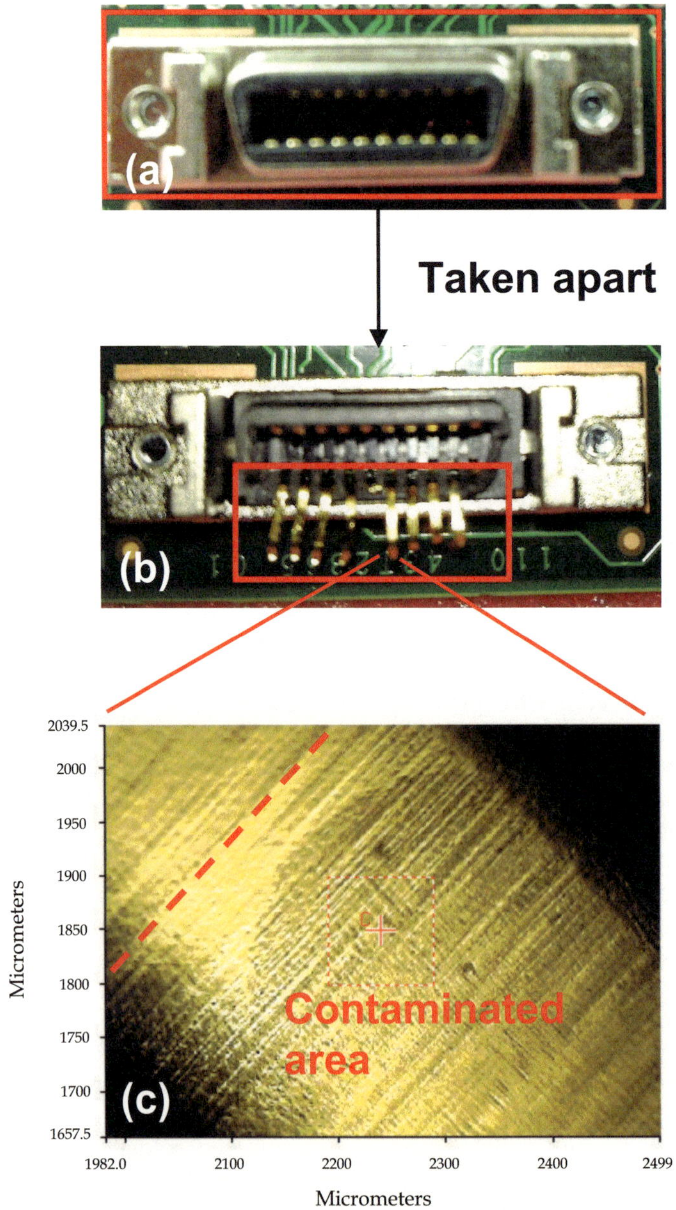

Fig. 23. JPSPK connector (a) plastic holding with metal housing, (b) pins unit, (c) contaminated area on pin for micro FTIR analysis

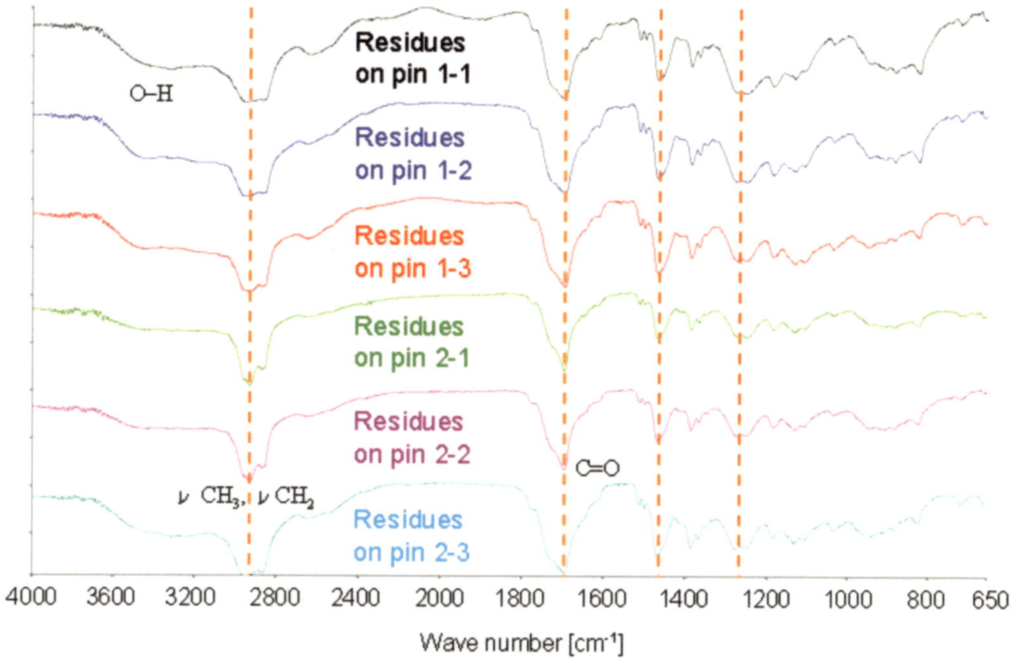

Fig. 24. Micro FTIR spectra of the contaminated connector pin 1 and 2 from different locations (marked dash -1 to -3)

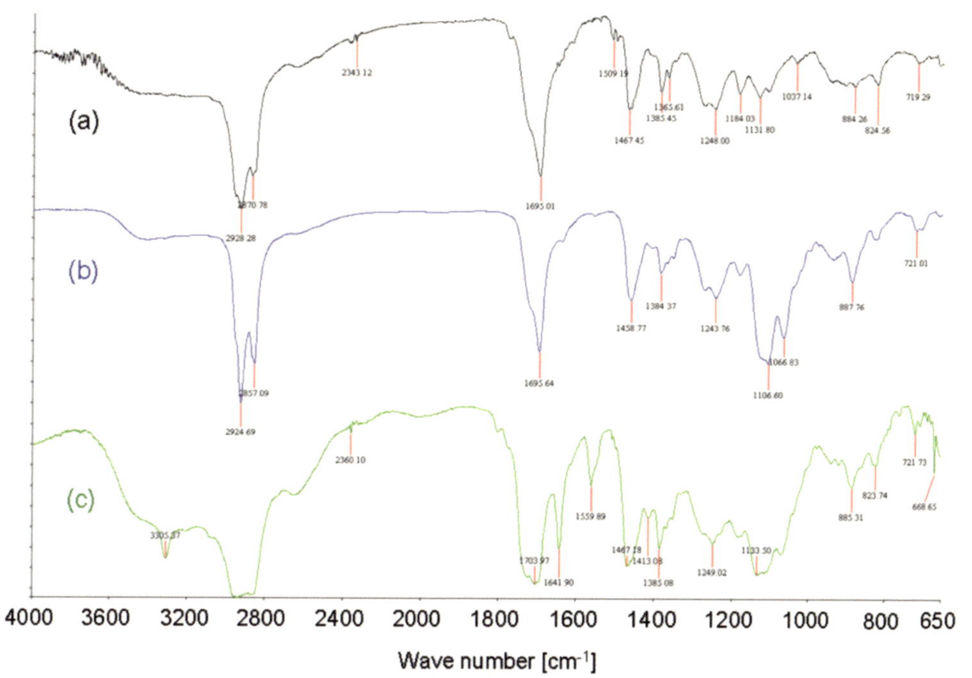

Fig. 25. Micro FTIR spectra of the (a) contaminated connector pin, (b) paste P1, and (c) paste P2

4.2.2 Case II - Plastic housing qualification

Incoming quality control (IQC) and supplier quality management (SQM) found that the suspected spray-painted plastic housings had mixed in, with the raw materials, primary colors. The defect rate is about 6% for this discoloration issue after sorting. The suspected housings presented darker surfaces than the normal incoming conditions. It is difficult to recognize the difference in color between the two kinds of housings through visual inspection. Therefore, an instrumentation analysis was applied to help confirm the properties and compositions that caused the cosmetic variation. An FTIR analysis was performed to investigate the chemical structures and compositions from the housing surfaces. It is important to find out whether improper or unnecessary processes were done for the incoming materials. Clients usually attach great importance to the surface textures and dispositions of their products. The appearance of the plastic housing is shown in Fig. 26.

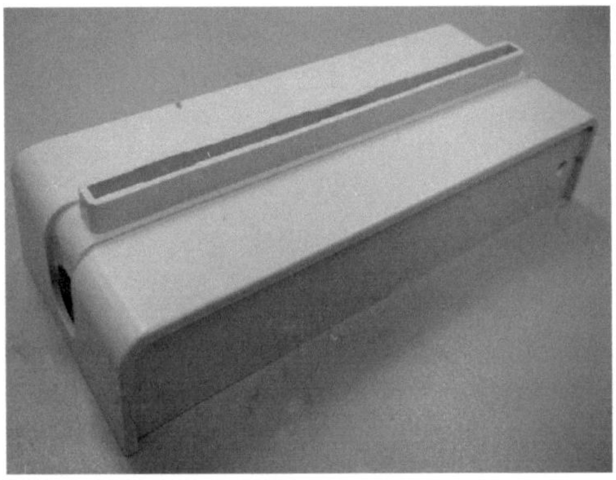

Fig. 26. The appearance of the plastic housing

The FTIR spectra under analysis with ATR mode from two different plastic housing surfaces are shown in Fig. 27. IR analysis was carried out at three adjacent locations for each surface to confirm the uniformity of the chemical structures. This showed very similar results from the various locations. The spectra were then compared to the FTIR built-in standard database (Fig. 27). The raw plastic housings with primary colors showed a bright surface and consistent spectrum (chemical structures) with PC (polycarbonate). The spray-painted plastic housings showed a little dark surface and consistent spectrum with acrylic ester.

5. The dye staining test

A variety of BGA packages are used in the electronic industry, for example, molded array process BGAs (MAPBGAs), plastic BGAs (PBGAs), and ceramic BGAs (CBGAs). The feature of BGA packages is such that solder balls are made to connect the packages to the printed circuit board (PCB). Once BGA packages are attached to a PCB, the inspection of the solder joints becomes necessary and is essential for failure analysis (Castello et al., 2006).

Identifying a failed solder joint under the package is of some significance to the industry. However, one can only inspect the BGA outer ball row (perimeter) for failed solder joints using a microscope. X-ray inspection can provide transmission images for inner rows of balls farther under the package, but shows less information related to solder joint fractures and open interconnects. Cross-sectioning of a BGA mounted to a PCB is also considered to be a time consuming process.

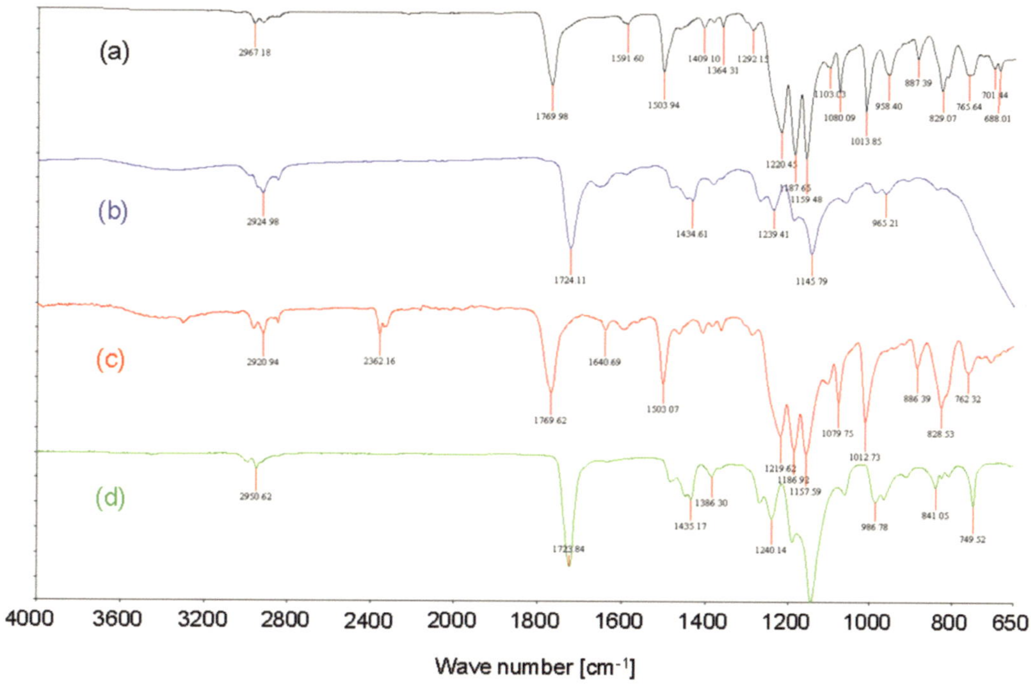

Fig. 27. FTIR spectra of the (a) raw plastic housing surface, (b) suspected spray-painted plastic housing surface, (c) PC (polycarbonate), and (d) acrylic ester

The dye staining test is commonly applied to verifying whether solder joint cracks occur under BGA packages (Liu et al., 2011). The dye staining technology, which is a destructive test method used to reveal the defects on the solder balls, relies on a liquid dye that penetrates into existing micro cracks or under open solder balls. After the dyeing process, a heating process is used to dry the dye followed by prying off the BGA. The presence of the dye on solder balls shows the location of the failure within the solder joint and the degree of the crack propagation. It is a commonplace method applied to grid array soldering problems.

5.1 Experimental

For analyzing solder joints, the dye staining test, also known as the dye and pry test, has become a general method due to its many advantages. These include an uncomplicated operation, ease of results interpretation, and simple implementation. Additionally, the test provides rich evidences of the failed solder joints such as solder joint cracks, cold welding, micro voids, and solder bridges. Sometimes, the root cause for failure can be seen from the test results directly (Lau et al., 2004).

The operation procedures of dye staining for a BGA sample in this work are as follows.

1. Detach the cell from the PCB
2. Remove flux residue using flux remover spray around the BGA solder bumps and then blow-dry with air gun

3. Clean the sample around the BGA solder bumps using isopropyl alcohol and then blow-dry with air gun
4. Repeat step 2 and 3 twice
5. Inject Dykem steel red layout fluid around the BGA component with pipe
6. Bake sample for 2 h at 125±5°C
7. Remove sample from oven and allowed to cool at room temperature
8. Bend the PCB to remove the sample mechanically

Optical microscopy (Keyence VH-5500) was used to identify dyed areas from the separated package and PCB to determine the failed solder joints. For a dyed solder joint, the evidence of dye penetration shows the failure mode and the magnitude of the dye penetration.

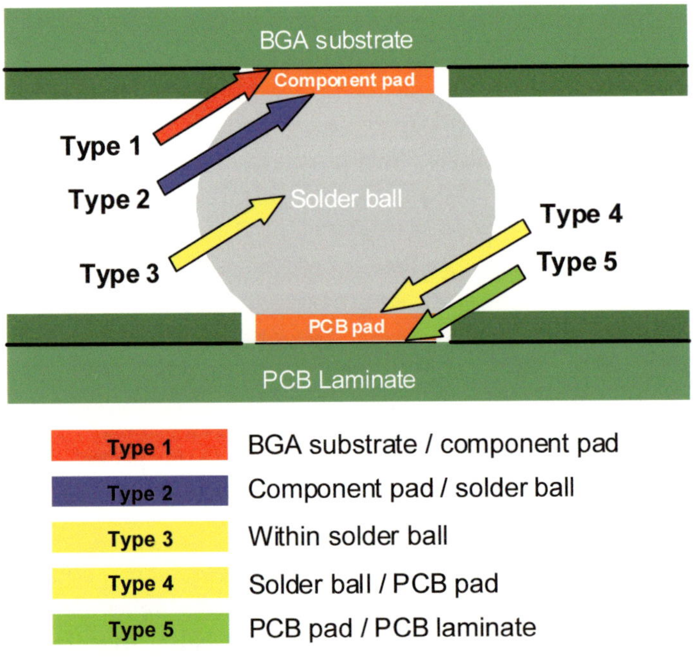

Fig. 28. Failure mode classification for a solder joint

The failure mode of a solder joint can be classified into five types, as illustrated in Fig. 28. Type 1 indicates that the fracture is between the BGA substrate and the component pad. Type 2 shows that the fracture is between the component pad and the solder ball. Types 3 and 4 both show that the fracture is within the solder ball and the solder ball/PCB pad, respectively. A fracture between the PCB pad and the PCB laminate indicates Type 5.

The magnitude of the dye penetration for each solder ball shows the degree of the crack propagation and can be classified into four types. Type A is such that the magnitude of dye penetration for a solder ball is below 25%. Type B falls between 26% and 50%. 51% to 75% indicates Type C. Type D has over 76% dye penetration.

5.2 Results and discussion

In-circuit test (ICT) was performed on the mass-produced PCBA boards. Probes were placed back into ICT fixture locations under the CPU socket and then the press was actuated; this

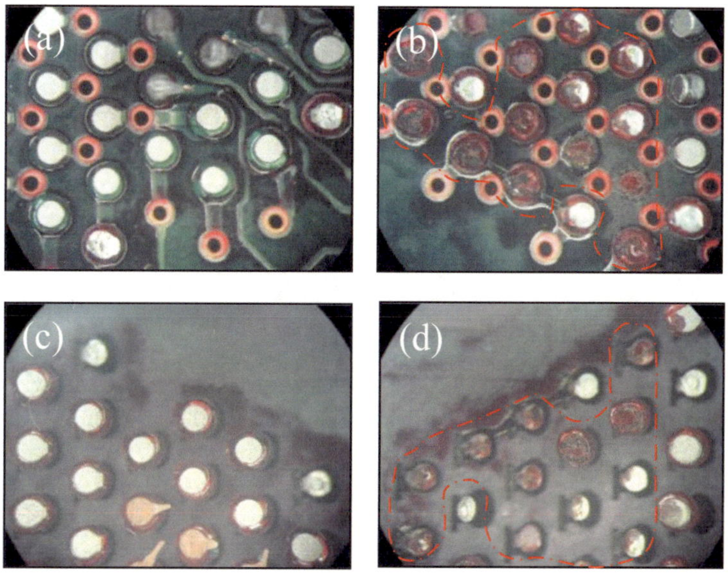

Fig. 29. Optical picture of (a) PCB pad side of 1st corner (b) PCB pad side of 2nd corner (c) component BGA side of 1st corner (d) component BGA side of 2nd corner after dye staining process

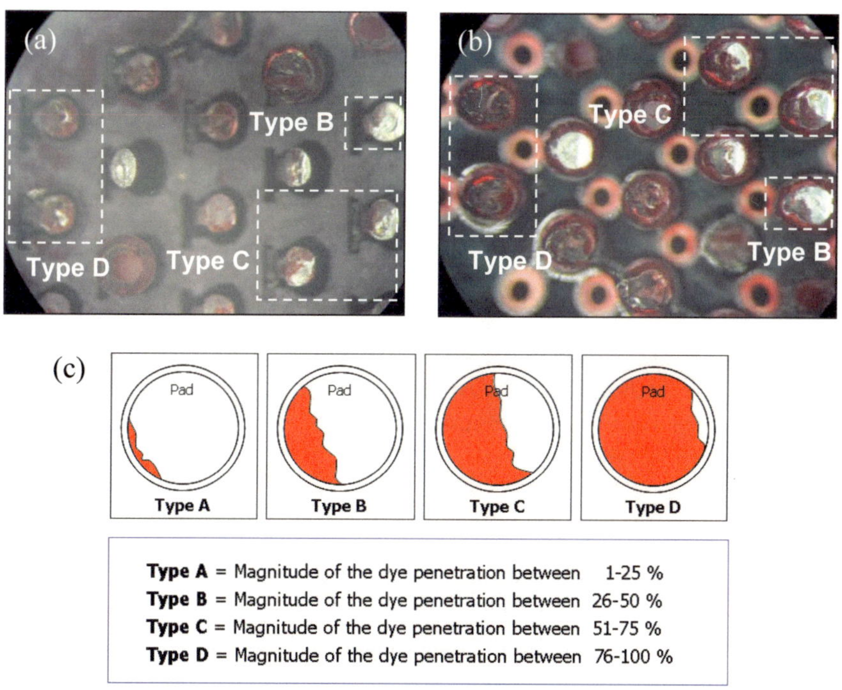

Fig. 30. (a) Enlarged picture of PCB pad side of 2nd corner (b) enlarged picture of component BGA side of 2nd corner (c) the classification of magnitude regarding dye penetration for each solder balls (crack size)

was repeated 20 times for every board. The dye staining analysis was carried out to confirm if any solder cracks occurred in CPU BGAs on the PCB. Optical microscopy was used to inspect the dyed areas and determine the failure mode classification. The test results showed that the solder balls at one outermost corner had been dyed, indicating that many solder joint cracks were found in the corner of the component.

Figure 29 is an optical picture taken after the dye staining process. In order to accurately interpret the dye staining results, two photos from the PCB pad side and component BGA side are compared together to find out the right failure mode of the dyed areas. For example, Fig. 29(a) corresponds with Fig. 29(c). Dye propagation and magnitude from both sides provide the main judgment criteria regarding whether cracks occur. Following the dyeing process, the failed solder joints are identified and the failure mode classification can be defined.

Figs. 30(a) and (b) are enlarged photos showing several failed solder joints. The failure mode of these solder joint cracks is between the component pad and the solder ball, which is Type 2, based on the classification of failure mode. Figure 30(c) shows the classification of the magnitude of the dye penetration. With respect to the magnitude of the dye penetration, Types B, C, and D of high percentage crack sizes can be seen in the photos indicating severe solder joint cracks occurring after a slight variation to the ICT fixture.

6. Conclusions

This study investigates PCB material performance and failure phenomenon during harsh assembly processes such as thermal shock and moisture exposure. Materials with a combination of Tg levels and Dicy / Phenolic curing agent were considered. All materials passed the assembly process verification and no PCB failure was observed. The high Tg material with a Phenolic curing agent is suggested for use in lead-free processes.

Black pad is a notable failure symptom within the PCB industry and not only causes an assembly quality issue but also significantly affects product durability. For overall quality control in PCB assembly, performing reliability testing during pilot runs is essential in ensuring product quality.

In this study, FTIR, SEM/EDX, and dye staining tests have been successfully used to characterize the failure samples and process materials associated with microelectronics assembly. The IR spectroscopic technique is capable of analyzing miniature samples as small as 100 μm. The characterization of process materials helps to determine the handling and/or process parameters. The sources of contaminants, such as flux, can be identified and then containment actions can be taken.

7. Acknowledgement

The authors are thankful to Jimmy Yang, Chen-Liang Ku, Hao-Chun Hsieh and all the other members of the Process Technology Team of Global Operations, Wistron Corp. for their valuable comments and involvement throughout this research. The authors are also thankful to Dr. Harvey Chang and Kenny Wang, vice-presidents of Wistron Corporation, for their great foresights to establish the analytic capabilities/expertise and financial support for the material Lab. The authors would also like to thank Dr. Kevin Chang, FT-IR specialist of PerkinElmer Taiwan Corporation, for his technical support.

8. References

Arimoto, H. (1964). α-ϒ Transition of nylon 6. *Journal of Polymer Science Part A: General Papers,* Vol. 2, Iss. 5, pp. 2283-2295

Bevilacqua, M. & Ciarapica, F.E. (2004). Experiment Investigation on Electric Fans for Domestic Appliance: Product Development and Reliability Tests. *10th ISSAT International Reliability and Quality in Design Conference,* Les Vegas, NEVADA, U.S.A., pp.45-49

Biehl, E. (2006). How to select base material for lead free applications, *EIPC Winter Conference,* Budapest

Bulwith, R.A., Trosky, M., Picchione L.M. & Hug, D. (2002). The black pad failure mechanism - from beginning to end. *Global SMT &packaging,* pp. 9-13

Castello, T., Rooney, D., & Shangguan, D. (2006). Failure analysis techniques for lead-free solder joints. *Soldering & Surface Mount Technology,* Vol. 18, No. 4, pp. 21-27

Dunaway, J.B. (1989). Residue on printed wiring assemblies: An overview and a case history. *Circuit World,* Vol. 15, No. 4, pp. 28-30

Hirt, A. & Artaki, I. (1991). Non-ionic water soluble flux residue detection by XPS. *Circuit World,* Vol. 17, No. 2, pp. 4-8

Huang, Y.T. & Lin, J.S. (2005). A Chi-square Test for Randomly Truncated Data. *Journal of Statistics and Computing,* p67-80

Huang, C.Y., Li, M.S., Ku, C.L., Hsieh, H.C., & Li, K.C. (2009), "Chemical characterization of failures and process materials for microelectronics assembly", *Microelectronics International,* Vol. 26 No. 3, pp. 41-48

IPC-4101B (2006). Specification for base materials for rigid and multilayer printed boards.

IPC-7095A (2004). Design and Assembly Process Implementation for BGAs. pp. 32-33

IPC-A-600G (2004). Acceptability of Printed Boards

IPC-TM-650 2.3.39 (1988). Surface organic contaminant identification test (infrared analytical method). *The Institute for Interconnecting and Packaging Electronic Circuits,* 2215 Sanders Road, Northbrook, IL, U.S.A., pp. 60062-6135

Jayatilleka, S. & Okogbaa, G. (2003). Use of Accelerated Life Test on Transmission Belts for Prediction Product Life, Identifying Better Designs, Materials and Suppliers. *Annual Reliability and Maintainability Symposium,* Ottawa, Ontario, Canada, pp. 101-105

Lau, J., Shangguan, D., Castello, T., Horsley, R., Smetana, J., Hoo, N., Dauksher, W., Love, D., Menis, I., & Sullivan, B. (2004). Failure analysis of lead-free solder joints for high-density packages. *Soldering & Surface Mount Technology,* Vol. 16, No. 2, pp. 69-76

Lee, R.E. (1993). *Scanning electron microscopy and X-ray microanalysis,* PTR Prentice Hall, New Jersey

Lin, Y. & Shih, T.Y. (2007). Morphological and Microstructural Evolution of Phosphorous-Rich Layer in SnAgCu/Ni-P UBM Solder Joint. *Journal of Electronic Materials,* Vol. 36, No. 11, pp. 1469-1475

Liu, F., Meng, G., & Zhao, M. (2010). Experimental investigation on the failure of lead-free solder joints under drop impact. *Soldering & Surface Mount Technology,* Vol. 22, No. 3, pp. 36.41

Shangguan, D. & Gao, G. (1997). Lead-free and no-clean soldering for automotive electronics. *Soldering & Surface Mount Technology,* Vol. 9, No. 2, pp. 5-8

Smith, C.A. (2007a). Failure analysis of electronic components and interconnection systems. *Circuit World*, Vol. 33, No. 1, pp. 15-21

Smith, C.A. (2007b). Chemical characterisation of materials in electronic systems using infrared spectroscopy. *Circuit World*, Vol. 33, No. 3, pp. 38-47

Welch, C.S. (1995). Feasibility study of OSEE inspection for flux residue on electronics assemblies. *Soldering & Surface Mount Technology*, Vol. 7, Iss. 1, pp. 8-11

Xu, L., Pang, J.H.L., & Che, F. (2008). Impact of Thermal Cycling on Sn-Ag-Cu Solder Joints and Board-Level Drop Reliability. *Journal of Electronic Materials*, Vol. 37, No. 6, pp. 880-886

Zeng, K., Stierman, R., Abbott, D., & Murtuza M. (2006). The Root Cause of Black Pad Failure of Solder Joints with Electroless Ni/Immersion Gold Plating. *JOM Journal of the Minerals, Metals and Materials Society*, Volume 58, No. 6, pp. 75-79

Zhao, W., Mettas, A., Zhao, X., Vassiliou, P. & Elsayed, E.A. (2004). Generalized step stress accelerated life model. *International Conference on Business of Electronic Product Reliability and Liability*, Tucson, AZ, U.S.A., pp. 19-25